A Visual Analogy Guide to Human Anatomy & Physiology

Paul A. Krieger

Grand Rapids Community College

Morton Publishing Company

925 W. Kenyon Ave., Unit 12

Englewood, CO 80110

800-348-3777

http://www.morton-pub.com

Book Team

Publisher Douglas N. Morton
Biology Editor David Ferguson
Cover & Design Bob Schram, Bookends, Inc.
Illustration Paul Krieger
Copyeditor Carolyn Acheson
Composition Ash Street Typecrafters, Inc.

Printed in the United States of America

10 9 8 7 6 5 4 3 2 1

ISBN: 0-89582-801-4

ISBN-13: 978-089582-801-9

To those teachers, students, and friends

who have inspired me

Acknowledgments

So many people contributed to this book that I would like to acknowledge them. My editor, David Ferguson, for his support and thoughtfulness in giving me the time and freedom to complete the book I originally envisioned. To Jeff McKelvey, my content editor for physiology, for his outstanding scholarship and attention to detail. His numerous insightful comments greatly improved the final manuscript. To Dan Matusiak for his diligence in faithfully creating the glossary. I would also like to offer a special thank you to reviewers Faith Vruggink, Steve Bassett, Cynthia Herbrandson, Chris Sullivan, and David Canoy for their valuable feedback during the writing process. My wife, Lily Krieger, kindly accepted all my long working hours away from the family. Without her help and support this book would not have been written. Thanks to Mike Timmons for helping me crystallize my initial ideas about the visual analogy guides and turn them into a tangible reality. To my friend, Kevin Patton, for graciously responding to all my authoring questions along the way. Finally, for my students, friends in the Human Anatomy & Physiology Society (HAPS), and everyone else who offered suggestions, support, and encouragement. Thanks to all of you. I am truly grateful.

Contents

Nervous System 185

Endocrine System 237

A Visual Analogy Guide to Human Anatomy & Physiology

How To Use this Book

Purpose

This book was written primarily for students of human anatomy and physiology; however, it will be useful for teachers or anyone else with an interest in this topic. It was designed to be used in conjunction with any of the major anatomy or anatomy and physiology textbooks. What makes it unique, creative, and fun is the **visual analogy** learning system. This will be explained later. The modular format allows you to focus on one key concept at a time. Each module has a text page on the left with corresponding illustrations on the facing page. Most illustrations are unlabeled so that you can quiz yourself on the structures. A handy key to the illustration is provided on the text page. While this book covers most all major organ systems, the topics are weighted more toward areas that typically give students difficulty. It uses a variety of learning activities such as labeling, coloring, and mnemonics to help instruct. In addition, it offers special study tips for mastering difficult topics.

What Are Visual Analogies?

A visual analogy is a helpful way to learn new material based on what you already know from everyday life. It compares an anatomical structure to something familiar such as an animal or a common object. For example, the vertebral column has three different types of vertebrae. One type looks like a giraffe. Comparing the vertebra to a giraffe allows you to mentally correlate the *unknown* (*vertebra*) with the *known* (*giraffe*). Doing this accomplishes several things.

1. It reduces your anxiety about learning the material and helps you focus on the task at hand.
2. It forces you to observe anatomical structures more carefully. After all, being a good observer is the first step to becoming a good anatomist—or any type of scientist!
3. It makes the learning more fun, relevant, and meaningful so you can better retain the information.

Whenever a visual analogy is used in this book, a small picture of it appears in the upper righthand corner of the illustration page for easy reference. This allows you to quickly reference a page visually, simply by flipping through the pages.

Icons Used

The following icons are used throughout this book:

Microscope icon—indicates any illustration that is microscopic.

Crayon icon—indicates illustrations that were specially made for coloring. Even though students may color any of the illustrations to enhance their learning, they may benefit more by referring to this icon. In some cases, written instructions appear next to this icon with directions about exactly what to color or how to color it.

Scissors icon—indicates that something is either cut or broken. For example, it may be used to show that a chemical bond between two molecules is broken.

3-D Three-dimensional icon—Indicates a three-dimensional view of an anatomical structure.

2-D Two-dimensional icon—Indicates a two-dimensional view of an anatomical structure.

Dashed lines usually are used to indicate a structure that is behind another structure.

Abbreviations and Symbols

The following abbreviations are commonly used throughout this book:

l.	=	left	n.	=	nerve
r.	=	right	ex.	=	example
a.	=	artery	sing.	=	singular
v.	=	vein	pl.	=	plural
m.	=	muscle			

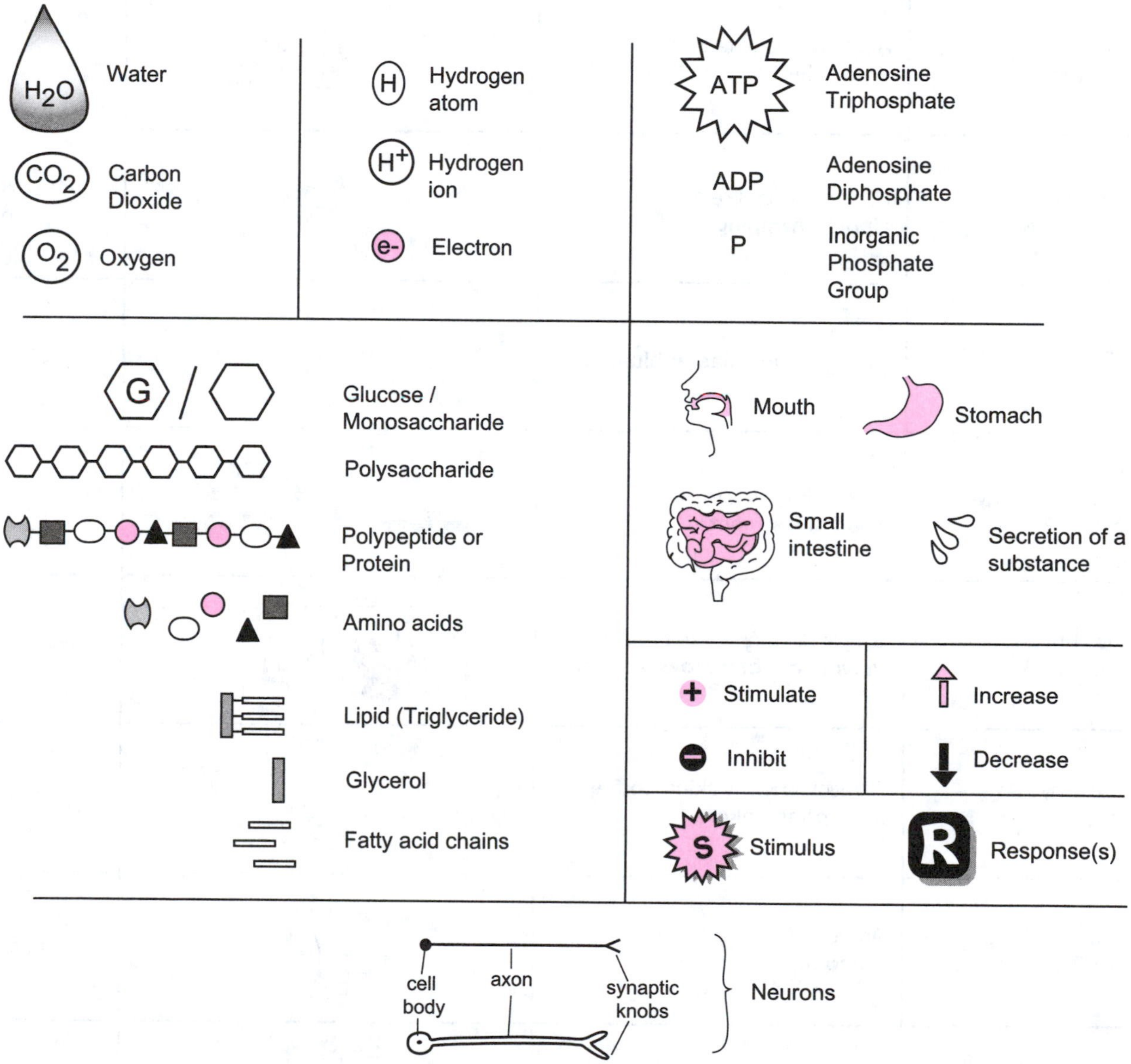

I sincerely hope that this book will be a fun, helpful tool for anyone who is interested in learning human physiology. Many of the visual analogies used in the book have been tested with students in both lab and lecture settings to ensure that they are useful.

Enjoy learning with visual analogies!

VISUAL ANALOGY INDEX

TOPIC	ANALOGY	ICON(S)	PAGE NO.
1. Language of Anatomy—Body Cavities and Membranes	A serous membrane is like a **fist in a balloon**		p. 21
2. Physiology Concepts—Homeostasis	The variable is like a **teeter-totter**		p. 29
3. Cells—Plasma Membrane	A plasma membrane is like a **waterbed**		p. 35
4. Cellular Physiology—DNA Replication	DNA opens up like a **zipper unzipping**		p. 39
5. Cellular Physiology—Protein Synthesis	DNA is like the **master blueprint**		p. 41
6. Cellular Physiology—Role of ATP	ATP hydrolysis is like a **good investment**		p. 43
7. Cellular Physiology—Simple Diffusion	Simple diffusion is like **people moving out of a crowded room**		p. 47
8. Cellular Physiology—Filtration	Filtration is like **making coffee in a coffeemaker**		p. 51
9. Cellular Physiology—Active Transport	Active transport is like a **sump pump**		p. 53
10. Cellular Physiology—Membrane Potentials	A membrane potential is like a **battery**		p. 57
11. Cellular Physiology—Sodium-Potassium Pump	Sodium-potassium pump is like a **revolving door**		p. 59

VISUAL ANALOGY INDEX

TOPIC	ANALOGY	ICON (S)	PAGE NO.
12. Cellular Physiology—Epithelial Cells	Tight junctions are like **rivets**	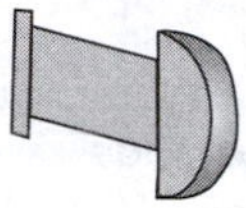	p. 63
13. Tissues—Simple Squamous Epithelium	Each cell looks like a **fried egg**	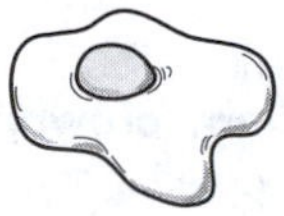	p. 65
14. Tissues—Simple Cuboidal Epithelium	Each cell looks like an **ice cube**		p. 67
15. Tissues—Simple Columnar Epithelium	Each cell looks like a **column**	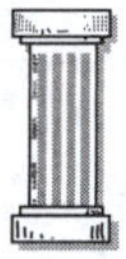	p. 69
16. Tissues—Variations in Connective Tissues	Collagen is like a **steel cable**; elastin is like a **rubberband**	 	p. 77, 85
17. Tissues—Reticular Connective Tissues	Reticular fibers are like **cobwebs**		p. 83
18. Tissues—Cartilage	Cartilage is like a block of **Swiss cheese**		p. 89, 91, 93
19. Tissues—Bone	Osteon looks like a **tree stump**		p. 95, 113
20. Tissues—Skeletal Muscle	Each skeletal muscle cell looks like a **birch tree log**	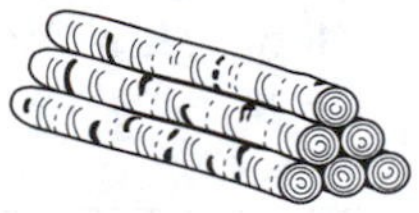	p. 97
21. Tissues—Cardiac Muscle	The intercalated disc is like **two pieces of a jigsaw puzzle**		p. 99
22. Tissues—Smooth Muscle	Smooth muscle stack in a sheet like **bricks in a wall**	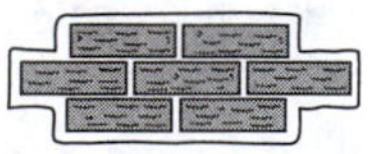	p. 101

VISUAL ANALOGY INDEX

TOPIC	ANALOGY	ICON(S)	PAGE NO.
23. Tissues— Nervous Tissue	Multipolar neuron looks like an **octopus**		p. 103
24. Skeletal System— Skull	The coronal suture is like a **tiara (crown)**; The sutures on the posterior aspect of the skull looks like a **modified peace sign**	 	p. 116
25. Skeletal System— Skull	The sella turcica of the sphenoid bone looks like a **horse's saddle**	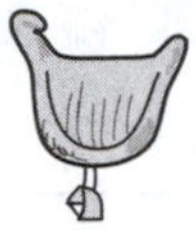	p. 117
26. Skeletal System— Temporal Bone	Temporal bone resembles a **rooster's head**		p. 121
27. Skeletal System— Ethmoid Bone	Superior view looks like a **door hinge**; crista galli looks like a **shark fin**	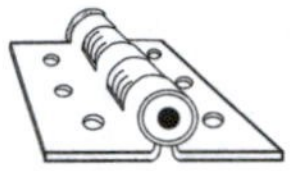 	p. 123
28. Skeletal System— Sphenoid Bone	Sphenoid resembles a **bat**		p. 125
29. Skeletal System— Sphenoidal Foramina	Remember the sphenoidal foramina with **Ros the cowboy**		p. 127
30. Skeletal System— Palatine Bones	Palatine bones are like **2 letter "L"s**—one the mirror image of the other		p. 129
31. Skeletal System— Numbers of Vertebrae	Total number of each type of vertebrae correspond to **meal times**		p. 131
32. Skeletal System— Atlas and Axis	Atlas the **turtle head**; Axis the **football player**	 	p. 133

VISUAL ANALOGY INDEX

TOPIC	ANALOGY	ICON (S)	PAGE NO.
33. Skeletal System—Lumbar Versus Thoracic	Thoracic and lumbar vertebrae are like a goose with **wings in different positions**		p. 135
34. Skeletal System—Lumbar Versus Thoracic	**"Thoracic giraffe; Lumbering moose"**		p. 137
35. Skeletal System—Humerus	Distal end of the humerus looks like the **hand of a hitchhiker**		p. 143
36. Skeletal System—Radius and Ulna	Head of radius—**hockey puck** Ulna—**crescent wrench**		p. 145
37. Skeletal System—Pubis Bones	Pubis bones—**mask**; Coccyx—**rattlesnake tail**		p. 131, 149
38. Muscular System—Actin Filament	Each actin filament is like a **double-stranded chain of pearls**		p. 163
39. Nerve and Muscle—The Contraction Cycle	Myosin head is like a **double-headed cotton swab**		p. 167
40. Muscular System—Muscles of Neck, Shoulder, Thorax, and Abdomen	Pectoralis major—**fan**; Serratus anterior—**serrated knife**; Abdominal muscles—**sandwich**		p. 175
41. Muscular System—Muscles of Thigh	Sartorius is like a **sash**		p. 179
42. Muscular System—Muscles that Move Ankle, Foot, and Toes	Soleus is like a **sole flatfish**		p. 181
43. Nervous System—Multipolar Neuron	Axon of a neuron is like an **electrical cord**		p. 191

VISUAL ANALOGY INDEX

TOPIC	ANALOGY	ICON(S)	PAGE NO.
44. Nerve and Muscle—Nerve Impulse Conduction	Action potential is like a **wave of electric current**	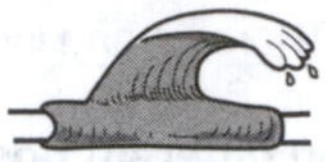	p. 195
45. Nerve and Muscle—Continuous Conduction	Continuous conduction is like a **domino effect**	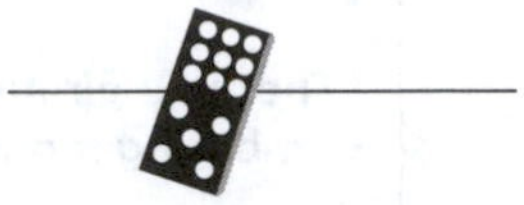	p. 197
46. Nerve and Muscle—Continuous Conduction	Saltatory conduction is like **skipping a stone**		p. 199
47. Nervous System—Peripheral Nerve	A peripheral nerve is like **tubes within tubes**	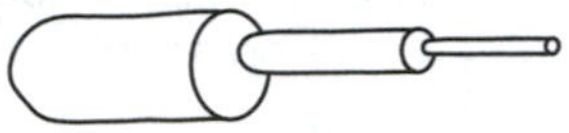	p. 163, 203
48. Nervous System—Spinal Cord	The gray matter in the spinal cord is shaped like a **butterfly**		p. 207
49. Nervous System—Reflexes	A reflex arc involves input, processing, and output, like **typing a letter on your laptop**		p. 209
50. Nervous System—Brain Ventricles	The ventricular system can be compared to the **neck, head, and horns of a ram**		p. 215
51. Nervous System—Sympathetic Division of ANS	Sympathetic division is like **stepping on the gas**		p. 231
52. Nervous System—Parasympathetic Division of ANS	Parasympathetic division is like **stepping on the brake**	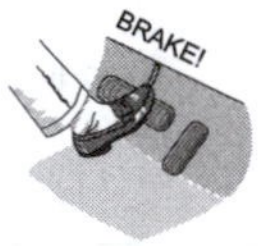	p. 233
53. Special Senses—Eye—Internal	Macula lutea is like a **target**; Fovea centralis—**bullseye**		p. 265
54. Special Senses—Ear—General Structure	Malleus—**hammer**; Incus—**anvil**; Stapes—**stirrup**; Cochlea—**snail shell**		p. 267

VISUAL ANALOGY INDEX

TOPIC	ANALOGY	ICON(S)	PAGE NO.
55. Special Senses—Tongue	Fungiform papilla—**mushroom cap**; Filiform papilla—**flame**		p. 269
56. Cardiovascular System—Heart	A-V valves—**parachute**; Valve flap—**kangaroo pouch** Semilunar valves—**modified peace sign**		p. 285
57. Cardiovascular System—Intrinsic Conduction System	The SA node is like the **spark plug** that sets the heartbeat		p. 267, 361
58. Cardiovascular System—Cardiac Cycle	Ventricular contraction is like **wringing out a wet rag**	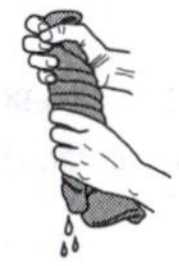	p. 291
59. Cardiovascular System—Blood Flow	Normal blood flow moves like **water through a garden hose**	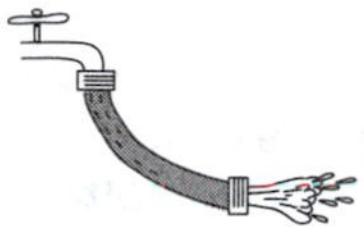	p. 299
60. Cardiovascular System—Arterial Pressure	Blood pressure is like the **force of water against the wall of a garden hose**	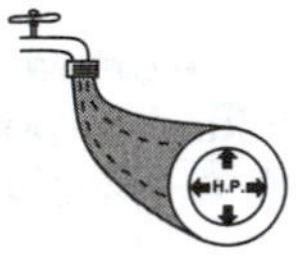	p. 301
61. Cardiovascular System—Capillary Function—Filtration	The filtration process is like **water gushing out of the holes in the garden hose**	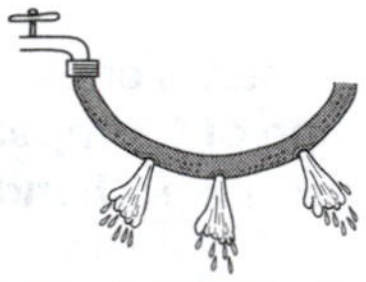	p. 305
62. Cardiovascular System—Return of Venous Blood	Venous pumps are like **squeezing a rubber tube with your hands**	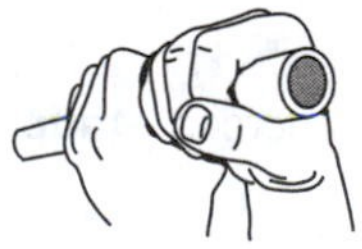	p. 307
63. Lymphatic System—Creation of Lymph	Lymph nodes filter debris like an **oil filter** filters oil in a car engine		p. 335
64. Immune System—Overview	Immune response is like an **army attacking a medieval castle**	*(no icon—written explanation only)*	p. 336
65. Respiratory System—Overview	Alveoli—**bubblewrap**; Larynx—**head of snapping turtle**		p. 345

VISUAL ANALOGY INDEX

TOPIC	ANALOGY	ICON(S)	PAGE NO.
66. Respiratory System—Mechanics of Breathing	Ribs swing upward like the **swinging handle on a pail**; recoil effect is like tension released from a **rubberband**	 	p. 349
67. Respiratory System—Alveolus	Alveolar macrophages are like the **housekeepers** of the alveoli		p. 351
68. Respiratory System—Hemoglobiin	Hemoglobin functions like a **taxicab**		p. 355
69. Respiratory System—Neural Control	Pneumotaxic center is like the **traffic cop**		p. 361
70. Digestive System—Physiology of Stomach	Stomach mixes up the chyme-like batter like a **hand mixer mixes batter**		p. 379
71. Digestive System—Small Intestine	Plicae circularis—**folded carpet sample**; Villus—**single carpet fiber**	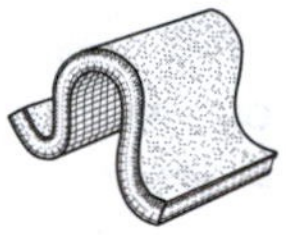 	p. 381
72. Digestive System—Movement Through the Small Intestine	Peristalsis propels like **squeezing** a **tube of toothpaste**; segmentation mixes like a **hand mixer**	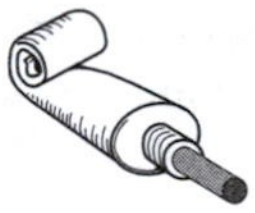 	p. 383
73. Digestive System—Pancreas	Pancreas—**tadpole**	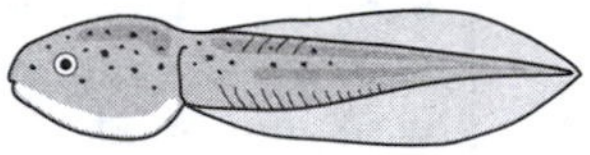	p. 387
74. Digestive System—Liver	Hepatic lobule is like a **Ferris wheel**	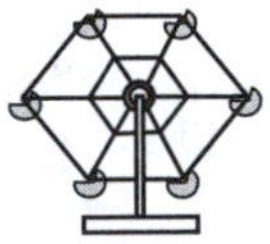	p. 391
75. Digestive System—Appendix	Appendix is like a **worm**		p. 395
76. Metabolic Physiology—Carbohydrate Metabolism	Reduced coenzymes are like a car with a **trailer carrying an electron as its cargo**		p. 403

VISUAL ANALOGY INDEX

TOPIC	ANALOGY	ICON (S)	PAGE NO.
77. Metabolic Physiology ETS	Electron transport chain is like **passing a hot potato**; hydrogen ion gradient is like **water behind a dam**	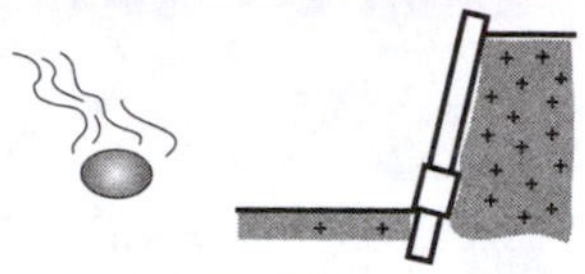	p. 409
78. Urinary System—Kidney	The calyces are like a **plumbing system**	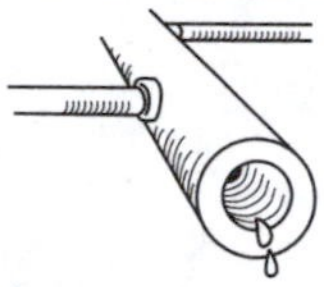	p. 421
79. Kidney—Nephron Function—Filtration	GHP is like a **sumo wrestler**		p. 424
80. Kidney—Regulation of the Glomerular Filtration Rate	The glomerulus is like a **loop of garden hose**	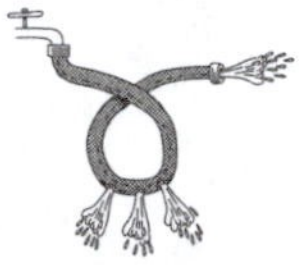	p. 427
81. Kidney—Nephron Function—Reabsorption and Secretion	Reabsorption and secretion is like **workers sorting products on a conveyor belt**		p. 429
82. Kidney—Countercurrent Multiplier	Osmolarity is like the "**saltiness**" of the solution		p. 439, 441
83. Male Reproductive System—Penis	The penis in cross section looks like a **monkey face**		p. 451
84. Female Reproductive System—Ovary	Ovulation is like a **water balloon bursting**		p. 465

Notes

Language of Anatomy

Description

An essential skill in anatomy is being able to visualize a sliced section of a tissue, organ, or region of the human body. This requires you to mentally jump from the three-dimensional to the two-dimensional. The prerequisite to developing this skill is being able to visualize the different ways that an object can be sliced. The three basic planes that can pass through an object to section it are:

- **Sagittal** *(median)* **plane**—This plane slices an object down the middle, making a left half and a right half.
- **Frontal plane**—This is also called the coronal plane. The term coronal means "crown." This plane splits an object into a front half and a back half.
- **Transverse plane**—This plane divides the body into an upper half and a lower half.

Key to Illustration

1. Frontal plane
2. Sagittal (*median*) plane
3. Transverse plane

Planes of the Body
(through a 2-year old boy)

1. ____________________

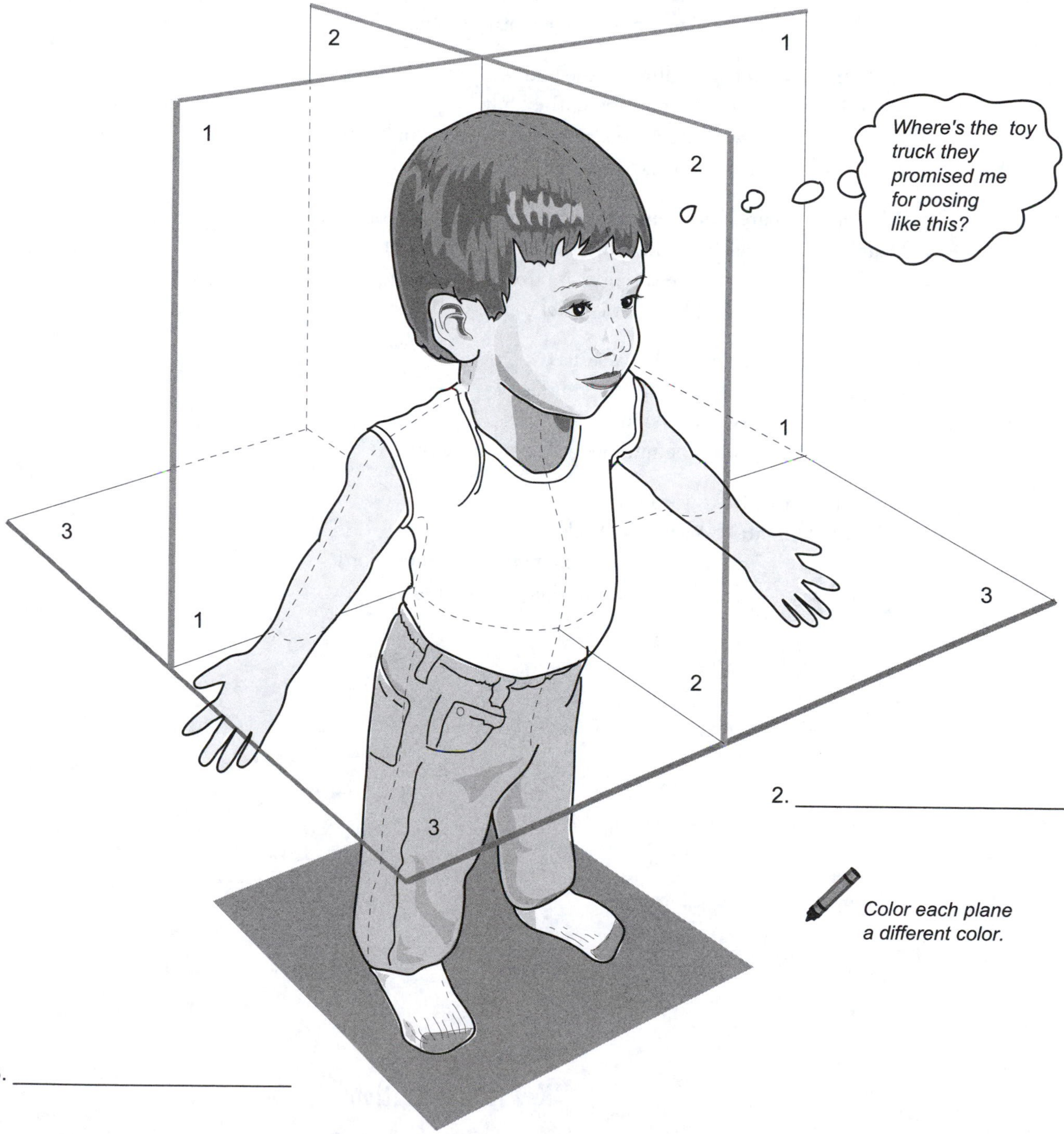
2
1
1
2
Where's the toy truck they promised me for posing like this?
1
3
1
3
2
3

2. ____________________

Color each plane a different color.

3. ____________________

Description

Directional terms are part of the working language of anatomy. They are commonly used to describe the position of an anatomical structure or the position of one body part in relation to another. Some of the most common terms for body orientation/direction are given below, along with their meanings and examples of how to use them correctly in a sentence.

Superior: above
Inferior: below
ex: The brain is *superior* to the lungs.
The lungs are *inferior* to the brain.

Medial: toward the midline of the body
Lateral: away from the midline of the body
ex: The nose is *medial* to the external ear.
The external ear is *lateral* to the nose.

Proximal: nearer the trunk of the body or point of attachment
Distal: farther from the trunk of the body or point of attachment
ex: The shoulder is *proximal* to the wrist.
The wrist is *distal* to the shoulder.

Superficial: toward the body surface
Deep: away from the body surface
ex: The skin is *superficial* to the muscle.
The muscles are *deep* to the skin.

Anterior/ventral: toward the front
Posterior/dorsal: toward the back
ex: The heart is *anterior* to the spinal cord.
The spinal cord is *posterior* to the heart.

Key to Illustration

1. Superior
2. Inferior
3. Anterior (ventral)
4. Posterior (dorsal)
5. Proximal
6. Distal
7. Medial
8. Lateral

Directional Terms

1. ______________________
2. ______________________
3. ______________________
4. ______________________

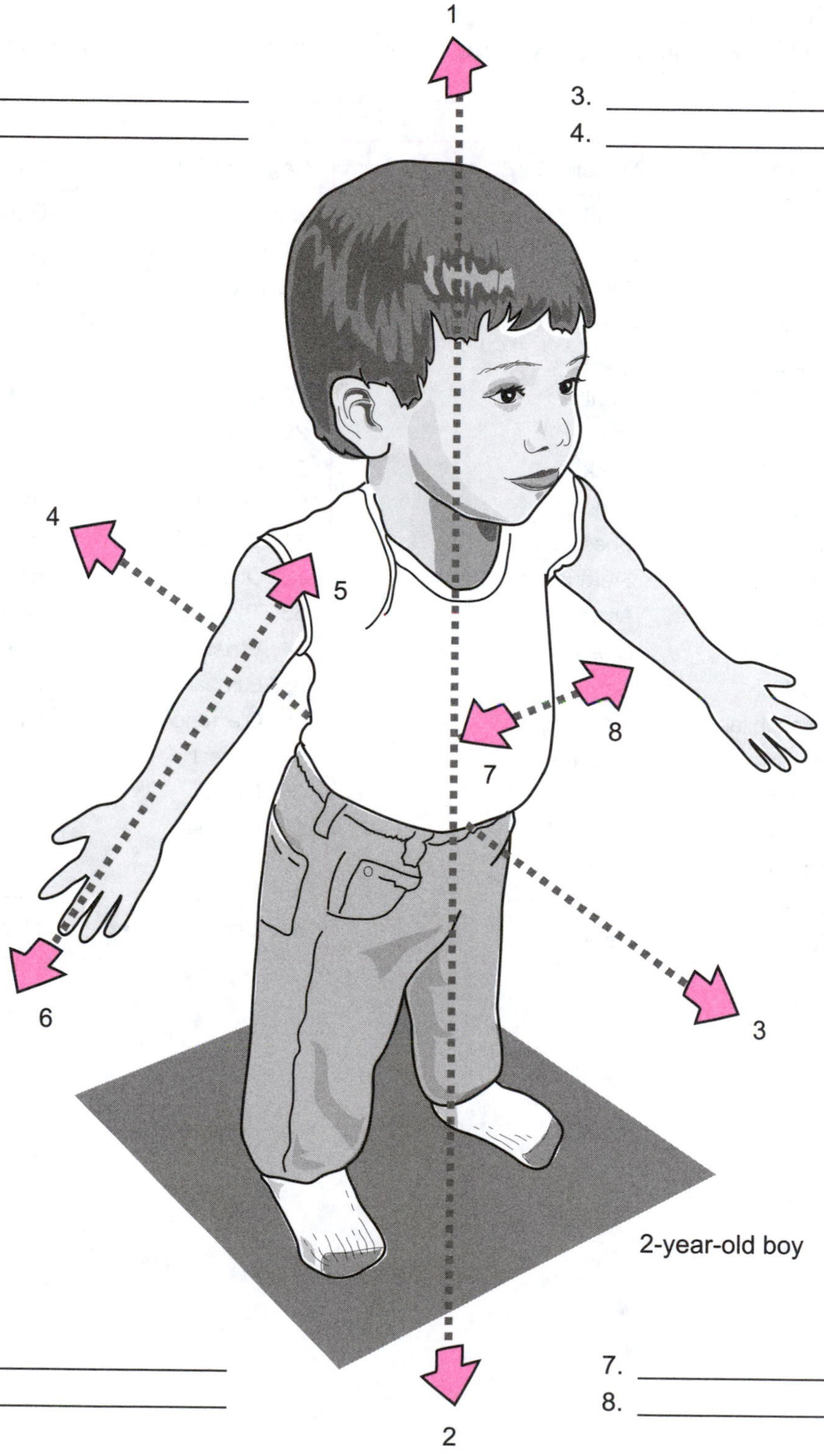

5. ______________________
6. ______________________
7. ______________________
8. ______________________

Description

Regional terms are part of the working language of anatomy. Many anatomical structures are named after the region of the body in which they are found. For example, there is the *brachial* artery, the *femoral* nerve, and the *deltoid* muscle. Memorizing these terms now will serve you well throughout your study of anatomy.

The table below gives the key to the numbers on the illustration as well as the definition of each regional term.

Regional Term	Description	Regional Term	Description
1. Cephalic	Head	22. Inguinal	Groin
2. Frontal	Forehead	23. Coxal	Hip
3. Orbital	Eye	24. Pubic	Anterior region of the pelvis
4. Nasal	Nose	25. Femoral	Thigh
5. Buccal	Cheek	26. Patellar	Knee
6. Oral	Mouth	27. Crural	Leg
7. Mental	Chin	28. Tarsal	Ankle
8. Cervical	Neck	29. Pes	Foot
9. Deltoid	Shoulder	30. Cranial	Skull
10. Pectoral	Chest	31. Occipital	Back of head
11. Sternal	Sternum	32. Otic	Ear
12. Axillary	Armpit	33. Thoracic	Chest (*thorax*)
13. Mammary	Breast	34. Vertebral	Spinal column
14. Brachial	Arm	35. Lumbar	Lower back
15. Antecubital	Front of arm	36. Olecranon	Elbow
16. Abdominal	Abdomen	37. Gluteal	Buttock
17. Antebrachial	Forearm	38. Manus	Hand
18. Carpal	Wrist	39. Perineal	Region between the anus and the genitals
19. Palmar	Palm	40. Popliteal	Back of knee
20. Digital	Finger	41. Calcaneal	Heel
21. Pelvic	Pelvis		

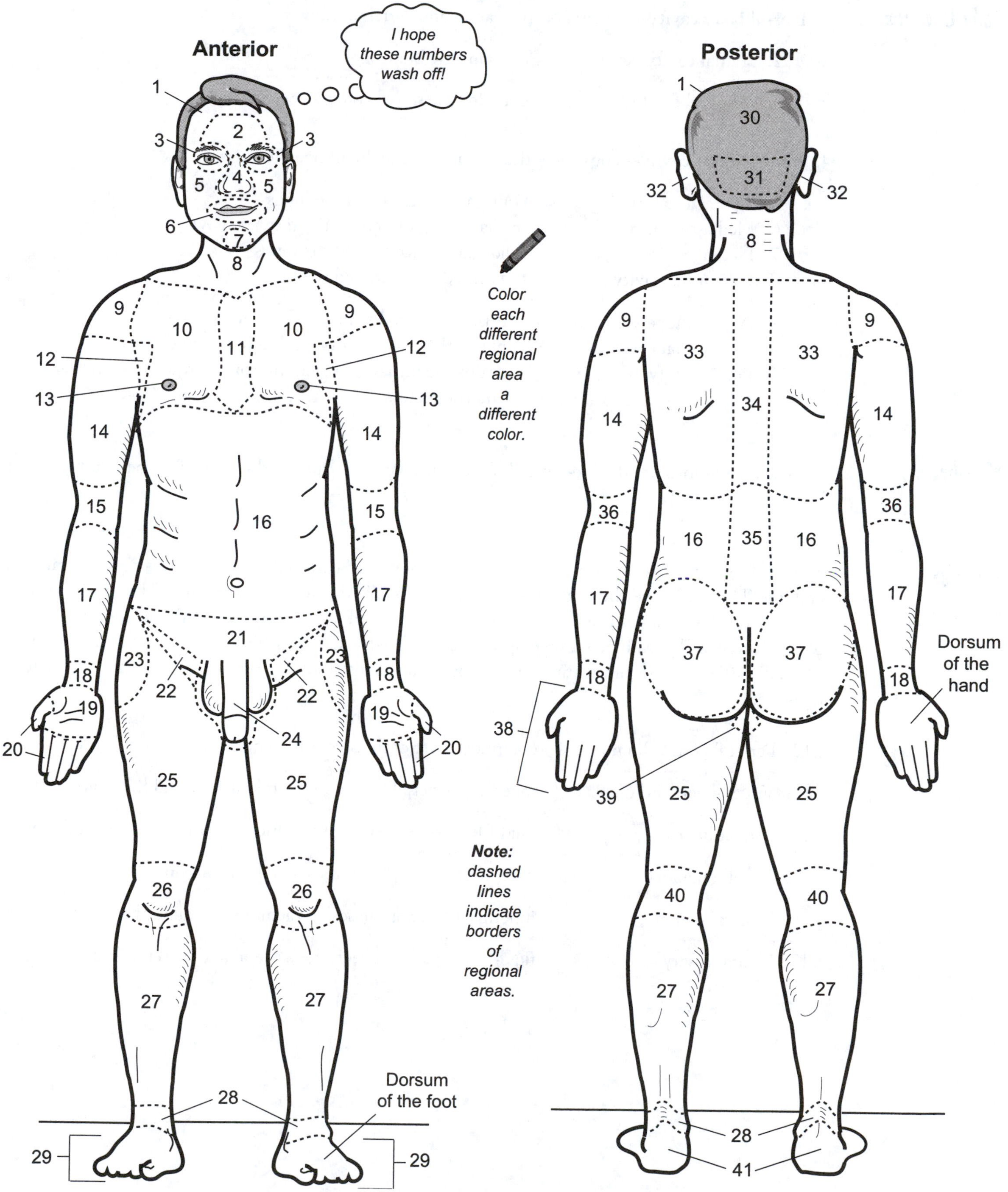
Anterior
I hope these numbers wash off!
Posterior
Color each different regional area a different color.
Dorsum of the hand
Note: dashed lines indicate borders of regional areas.
Dorsum of the foot

Body Cavities

1. **Dorsal body cavity**—consists of the cranial and vertebral cavities

2. Cranial cavity	contains the brain
3. Vertebral cavity	contains the spinal cord

4. **Ventral body cavity**—consists of the thoracic and abdominopelvic cavities

5. Thoracic cavity	contains the heart and the lungs
6. Mediastinum	median compartment of the thoracic cavity
7. Pleural cavity	fluid-filled space around the lungs
8. Pericardial cavity	fluid-filled space around the heart
9. Abdominopelvic cavity	consists of the abdominal and pelvic cavities
10. Abdominal cavity	contains the digestive organs, kidneys, and ureters
11. Pelvic cavity	contains the urinary bladder, internal reproductive organs, and the rectum

Membranes

Serous membranes are double-layered, fluid-filled sacs that surround organs like the heart and the lungs.

Analogy

To visualize how a serous membrane surrounds an organ, imagine a fist pushed into a partially inflated balloon. The fist is like the **organ** and the balloon is like the **serous membrane**. The **inner layer** of the balloon that touches the fist is like the **visceral layer**. The **outer wall** of the balloon is like the **parietal layer**. The balloon's **inner space** that is filled with air is like the **serous cavity** which is normally filled with a lubricant called serous fluid. This fluid is made by cells within the serous membrane.

12. Parietal pericardium	outermost layer of the serous membrane around the heart
13. Visceral pericardium	innermost layer of the serous membrane around the heart
14. Pericardial cavity	fluid-filled space between the parietal and visceral pericardium
15. Parietal pleura	outermost layer of the serous membrane around the lungs
16. Visceral pleura	innermost layer of the serous membrane around the lungs
17. Pleural cavity	fluid-filled space between the parietal and visceral pleura

BODY CAVITIES

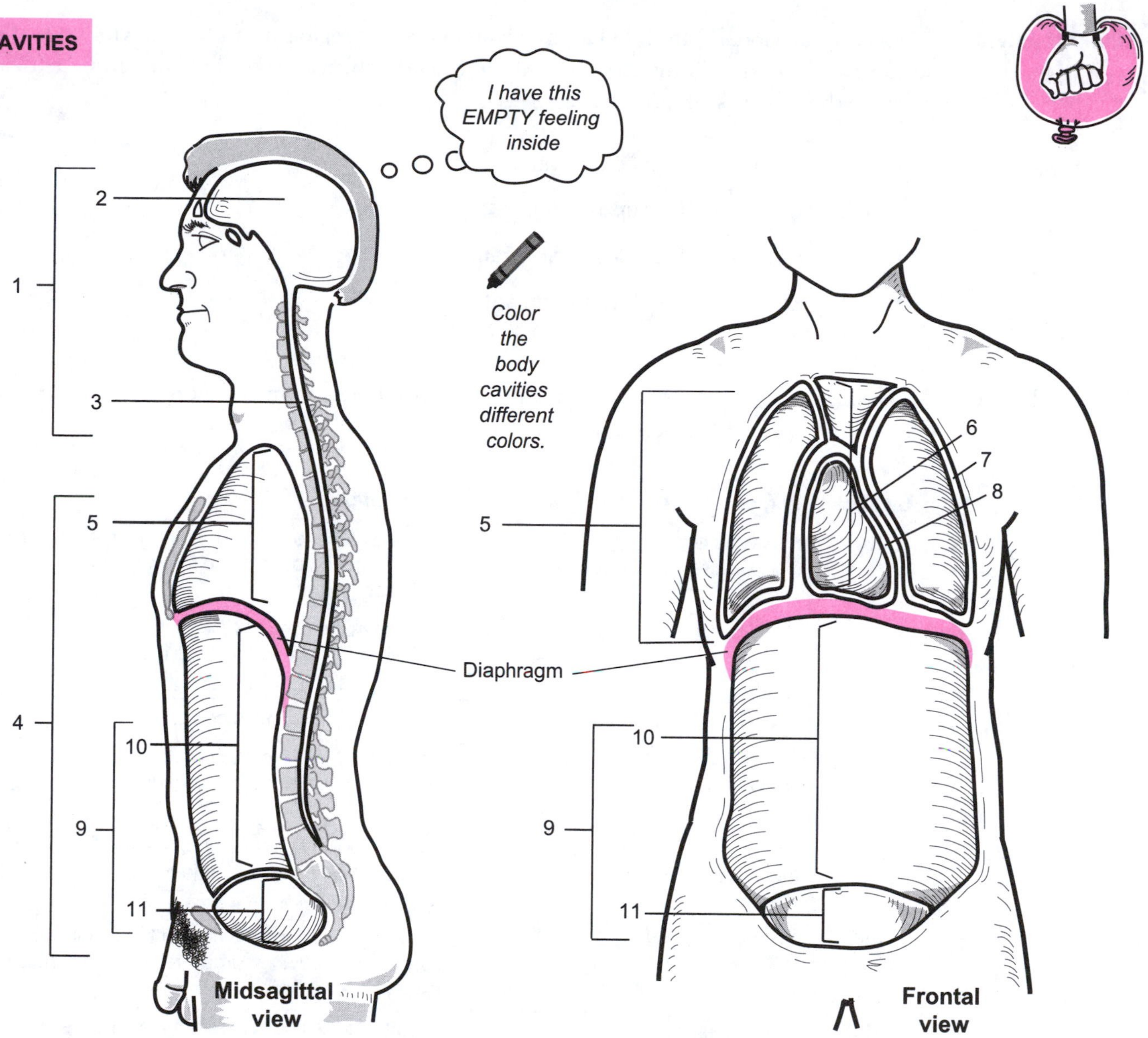

MEMBRANES

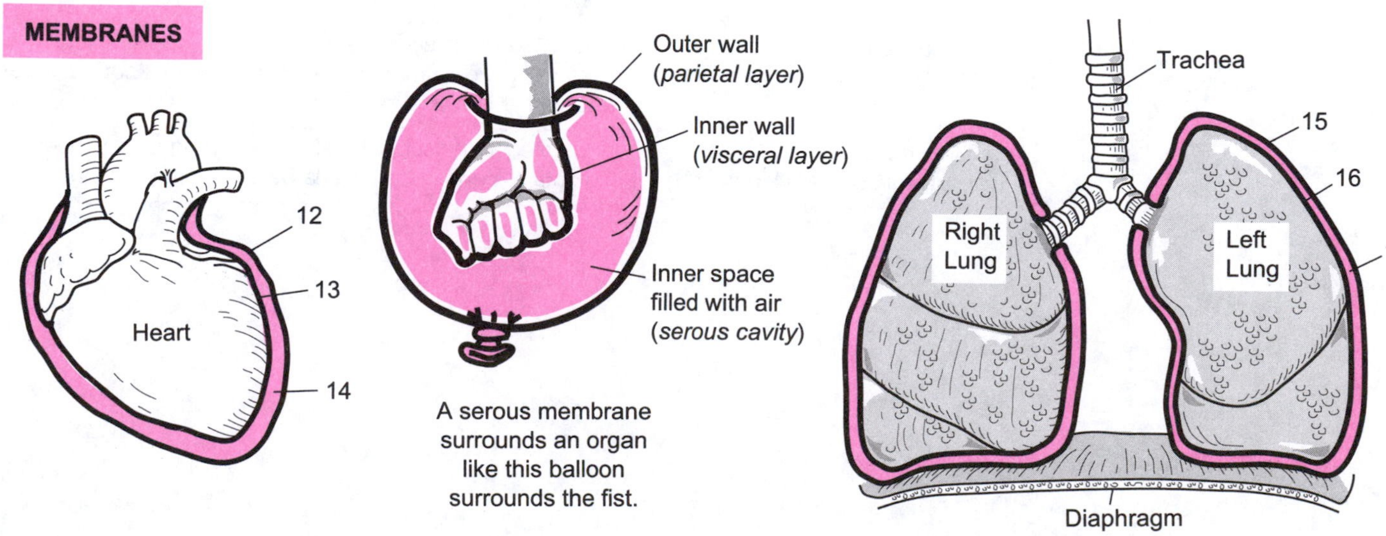

A serous membrane surrounds an organ like this balloon surrounds the fist.

Abdominopelvic quadrants

Medical professionals often divide the **abdominopelvic region** into the following four areas to help locate pain sites due to injuries and other medical problems. Note that the terms "left" and "right" refers to the subject's left and right.

RUQ	=	Right upper quadrant
LUQ	=	Left upper quadrant
RLQ	=	Right lower quadrant
LLQ	=	Left lower quadrant

Abdominopelvic regions

Anatomists often prefer to divide the abdominopelvic region into a more detailed nine square grid. The table below explains each region.

Symbol	Regional Name	Description
E	Epigastric region (*epi* = above; *gastri* = belly)	Located at the top of the middle column; contains the duodenum and parts of the following: liver, stomach and pancreas.
U	Umbilical region (navel)	Located in the center of the grid; contains parts of both the transverse colon and the small intestine.
H	Hypogastric region (*hypo* = below; *gastri* = belly)	Located at the bottom of the middle column; contains the urinary bladder, sigmoid colon, and part of the small intestine.
RH	Right hypochondriac region (*hypo* = below; *chondro* = cartilage)	Located to the subject's right of the epigastric region; contains the gallbladder and parts of both the right kidney and the liver.
LH	Left hypochondriac region (*hypo* = below; *chondro* = cartilage)	Located to the subject's left of the epigastric region; contains the spleen and parts of the following: stomach, left kidney, and large intestine.
RL	Right lumbar region (*lumbus* = loin)	Located to the subject's right of the umbilical region; contains parts of the following: large intestine, small intestine, and right kidney.
LL	Left lumbar region (*lumbus* = loin)	Located to the subject's left of the umbilical region; contains parts of the following: large intestine, small intestine, and left kidney.
RI	Right iliac region (*iliac* = largest part of hip bone)	Located to the subject's right of the hypogastric region; contains the bottom of the cecum, the appendix, and part of the small intestine
LI	Left iliac region (*iliac* = largest part of hip bone)	Located to the subject's left of the hypogastric region; contains parts of both the large and small intestine

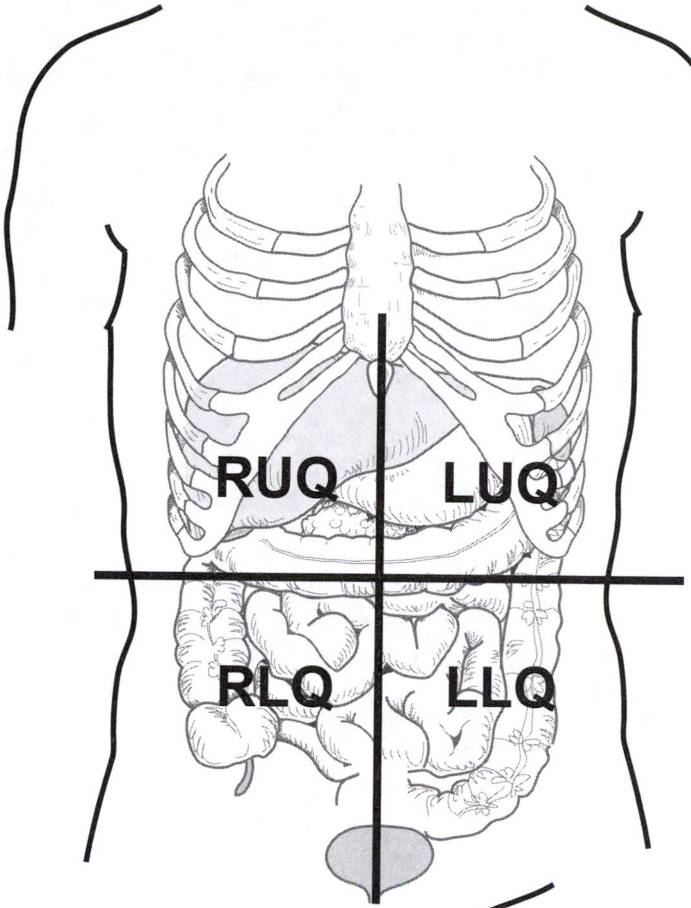

Color the quadrants different colors.

The 4 abdominopelvic quadrants

Color the regions different colors.

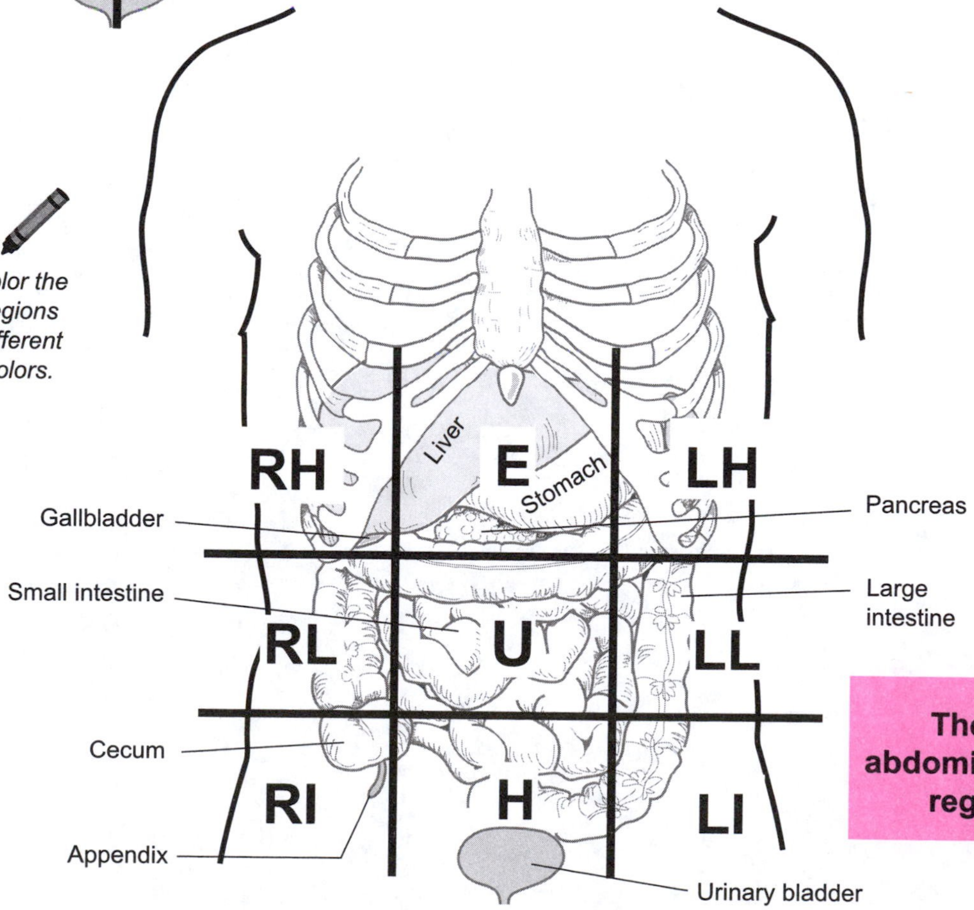

The 9 abdominopelvic regions

Notes

Physiology Overview

Description

The human body has different levels of structural organization. **Atoms** (*ex: oxygen*) combine to form **molecules** *(ex: water)*. Very large molecules called **macromolecules** *(ex: carbohydrates, lipids, proteins, nucleic acids)* are the building blocks used to form any **cell** (ex: *cardiac muscle cell*). A group of similar cells working together is called a **tissue** (ex: *cardiac muscle tissue*). A variety of different tissues working together forms an **organ** *(ex: the heart)*, and a group of organs working together forms a complex **organ system** *(ex: cardiovascular system = heart and the blood vessels)*. These complex organ systems all working together form an **organism** *(ex: human organism)*. For convenience, teachers and textbook authors sometimes discuss each organ system independent of the others. But it should be emphasized that no organ system exists as an island unto itself because they are all interdependent. Substances such as water, nutrients, and waste products have to move between organ systems to maintain **homeostasis** for the entire organism. Keep in mind when studying physiology that it is common to jump back and forth between these different levels of structure when learning about physiological mechanisms.

The following ten organ systems are shown in the illustration on the facing page: **integumentary system, musculoskeletal system, nervous and endocrine systems, cardiovascular system, lymphatic/ immune system, respiratory system, digestive system, urinary system, and reproductive systems**. This schematic illustration attempts to show the interrelationships between organ systems. The white rectangular band around the periphery represents the **integumentary system**. This includes the skin and all its components such as hair follicles, oil glands and a host of other structures. Our skin functions as a thin mantle to protect us from our external environment. The **musculoskeletal system** provides an internal framework for support and allows for body movements. Consider that movement would be impossible without muscles acting on bones as a kind of lever system.

Two organ systems work together to act as the regulators of body function—the **nervous** and **endocrine** systems. Because they both serve a similar purpose, they were placed together in the same box. Without the nervous system we would have no way to detect stimuli from either our external or internal environments, no way to process this information, and no way to respond to it. The endocrine system is composed of glands (*ex: pancreas*) that release chemical messengers called **hormones** *(ex. insulin)*, which travel through the bloodstream to target a particular structure such as an organ to induce a response. Working together, these two systems help maintain the vital activities that keep us alive.

The **cardiovascular system** consists of the heart and all the blood vessels *(ex. arteries, veins, arterioles, venules, and capillaries)*. The heart functions as a pump to constantly circulate blood throughout the body. The blood vessels must connect to all the other organ systems to deliver nutrients (*ex. oxygen*) to all the body cells via the bloodstream.

The following four (4) organ systems contain organs/structures that are hollow: **respiratory system**, **digestive system**, **urinary system**, and **reproductive systems**. Structurally speaking, all of these systems have an internal chamber called a **lumen** that extends to the external environment. Any substance within this lumen is still considered part of the external environment until it crosses the wall of that structure and enters body tissues. The major function of the **respiratory system** is to exchange the respiratory gases, oxygen and carbon dioxide, between the body and the external environment. Oxygen is brought in from the external environment, transferred to the blood, and delivered to body cells. Carbon dioxide is a normal waste product made by body cells. It is transferred to the blood, then sent back to the respiratory system, where some of it is released to the external environment.

The **digestive system** functions to take in nutrients, break them down to their simplest components, and absorb them into the blood. Then they are delivered to cells. In addition, the digestive system eliminates waste products from the body. The **urinary system** constantly filters and processes the blood to eventually form urine, which contains nitrogen wastes.

The **lymphatic** and **immune systems** typically are grouped together *(even though they are not illustrated that way)*. One major function of the **lymphatic system** is to take the **interstitial fluid** found between cells and to cleanse it of debris and possible pathogens, then to deliver it back to the bloodstream. The **immune system** functions to protect the body from foreign pathogens such as bacteria and viruses. For example, when a break in the skin occurs, pathogens may enter the body and the immune system must launch a response akin to soldiers going to war. It accomplishes this by making antibodies and other substances that help fight off these invaders.

Trace the pathway oxygen follows until it enters a body cell. How many barriers must it cross?
O2
(Oxygen)
CO2 (Carbon dioxide)
Integumentary system
Food
Respiratory system
Musculoskeletal system
Digestive system
(Gray arrows indicate direction of blood flow)
CO2
O2
Heart
Lymphatic system
nutrients
Cardiovascular system
Body cell
Color the different organ systems different colors.
Lymph node
CO2
Nervous and Endocrine systems
Interstitial fluid
O2
Sperm (male); Egg, (female)
Urinary system
Reproductive system
Immune system
Urine containing nitrogen wastes
Feces, undigested materials

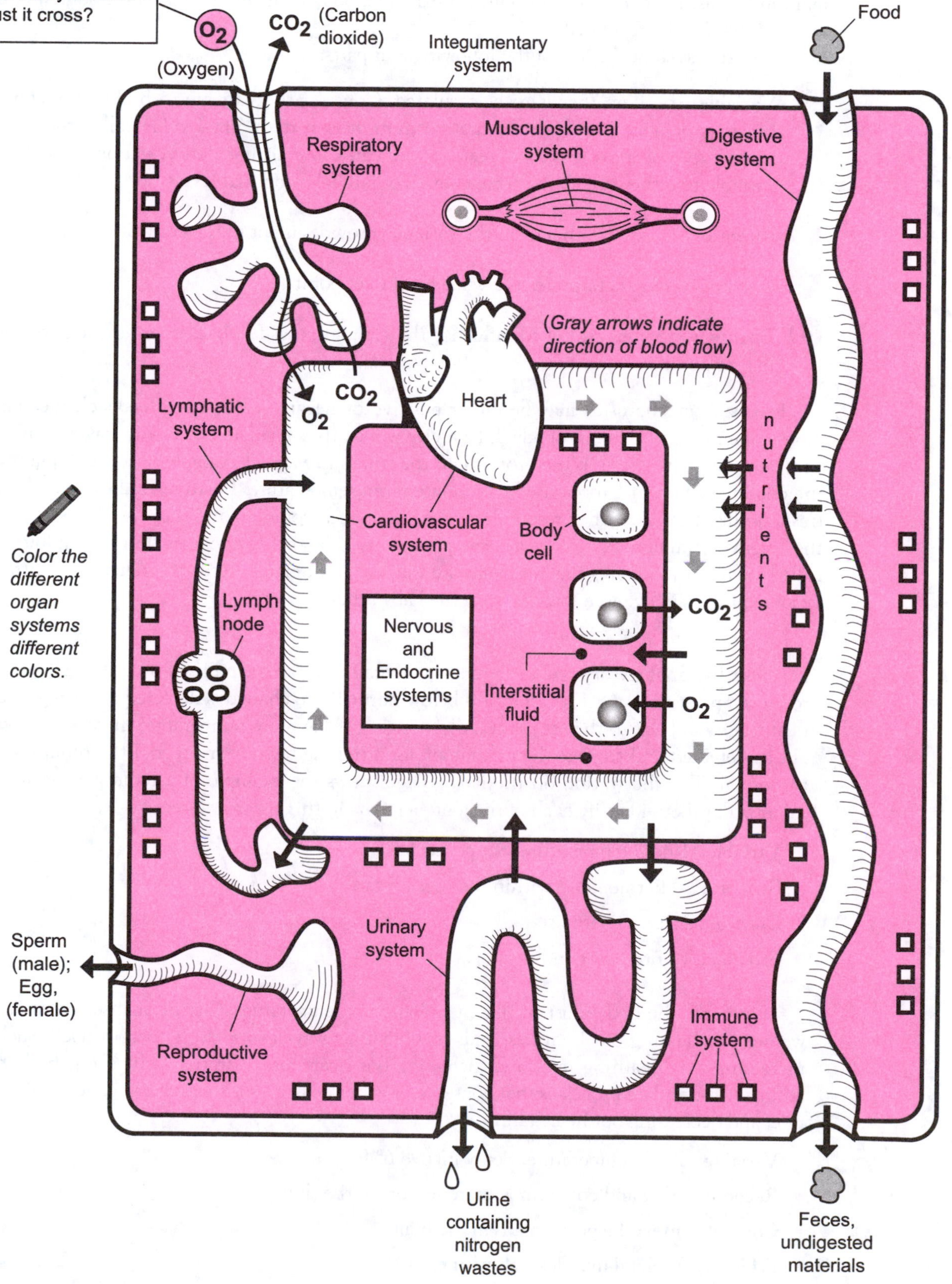

Description

Homeostasis is defined as maintaining a stable, constant internal environment within an organism. As changes constantly occur in the external environment, organisms must have control systems so they can detect these changes and respond to them. Anything that must be maintained in the body within a normal range must have a control system. For example, body temperature, blood pressure, and blood glucose levels all have to be regulated.

A **control system** consists of the following four parts:

1. **Variable:** the item that is to be regulated. Consider the variable to be like a teeter-totter. If it is perfectly balanced in the horizontal position, it is at the normal value called the **set point**. Just as a single person resting on the teeter-totter can cause an imbalance, various stimuli can cause the variable to either rise above normal levels or fall below normal levels.

2. **Receptor:** senses the change in the variable; provides input to the control center.

3. **Control center:** determines the value for the set point.

4. **Effector:** provides output to influence the stimulus; uses feedback to return the system to the original set point.

Feedback mechanisms may be either positive or negative. **Negative feedback** mechanisms are the more common in the human body. Like a city works, they regulate the day-to-day activities of the body. The output in this system is used to turn off the stimulus. A good example of this is explained below, with the cooling system in a room used to regulate room temperature. **Positive feedback** mechanisms, in contrast, are like the emergency responders. They are not as common and they use the responses to intensify the original stimulus. A biological example of this is labor contractions during childbirth. A hormone called oxytocin stimulates the muscle in the wall of the woman's uterus to strengthen contractions so they become progressively more forceful until the baby is born.

Analogy

Let's use the analogy of regulating room temperature as an example of a control system. If we set the room temperature at 68°F, then it falls below normal, the change is detected and triggers the furnace to heat the room back to the set point. Alternatively, if the room temperature increases above this level, it also is detected and triggers the central air conditioning to turn on to cool the room temperature back to normal. The illustration on the facing page shows the example of a room getting warmer on a hot summer day. Let's identify each of the four elements in this control system.

- **Variable:** Room temperature. Set point? 68°F.
- **Receptor:** Thermometer within the thermostat.
- **Control center:** Thermostat.
- **Effector:** Central air conditioning.

Similarly, the body must regulate its own body temperature. Our normal set point is 98.6°F. When we are overheating our body responds in numerous ways in an effort to cool itself, such as sweating, increasing our breathing rate, and increasing our heart rate. Alternatively, if our body temperature is falling, we stimulate processes that conserve or generate heat, such as shivering, decreasing our breathing rate, and decreasing our heart rate.

- **Variable:** Body temperature. Set point? 98.6° F.
- **Receptor:** Hot and cold temperature sensors in the skin.
- **Control center:** Hypothalamus in the brain.
- **Effector:** Skeletal muscles—shivering.

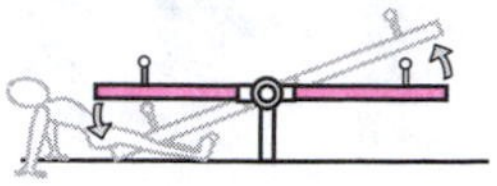

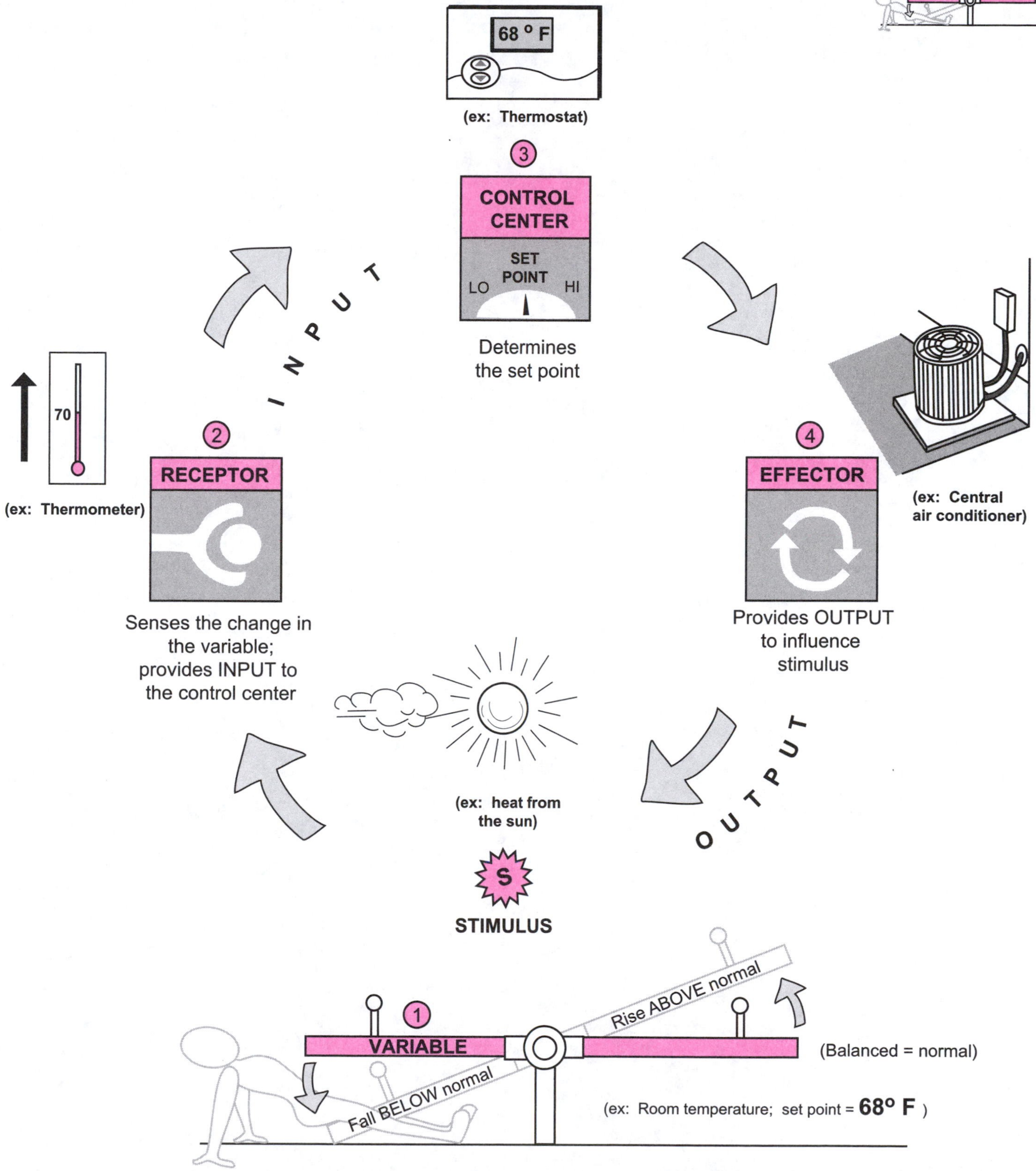

Control system: This illustration shows all the features of a control system. As an analogy, regulating room temperature on a hot day is shown.

Notes

Cells—
Structure and Function

Description

The cell is the basic unit of life. The human organism begins as a single cell called a **zygote** (*fertilized egg*). Then it divides over and over again to produce more than a trillion cells in the adult. Though there are many different types of cells with a variety of sizes and shapes, all share basic structural features in common. Each cell is surrounded by a **plasma** (*cell*) **membrane** and contains many different types of organelles (*small organ*), each with its own special function(s).

NONMEMBRANOUS ORGANELLES

Organelle	Function
Centrosome (composed of two centrioles)	Forms poles in cell for movement of chromosomes in cell division.
Cilia	**Plasma membrane** extensions containing microtubules that move materials over the cell.
Microfilament	Connects **cytoskeleton** to **plasma membrane**; contraction allows for movement of part of a cell or a change in cell shape.
Microtubules	Hollow tubes of protein which can act as tracks along which organelles move.
Microvilli	Plasma membrane extensions that assist in absorption of nutrients and other substances.
Ribosomes	Composed of two subunits, functions in protein synthesis.

MEMBRANOUS ORGANELLES

Organelle	Function
Endoplasmic reticulum (ER)—(two types) Smooth (has no ribosomes attached) Rough (has ribosomes attached)	Smooth: lipid and carbohydrate synthesis Rough: modification and packaging of newly made proteins
Golgi complex (Golgi body; Golgi apparatus)	Finishes, stores, distributes chemical products of the cell (*e.g.,* proteins)
Lysosome	Vesicles containing digestive enzymes to remove pathogens and broken organelles
Mitochondrion	Site of aerobic cellular respiration; produces ATP for the cell
Nucleus • nucleolus (region within nucleus containing DNA and RNA)	Large structure that contains DNA—the genetic material for making proteins
Peroxisome	Vesicles containing enzymes to break down substances such as hydrogen peroxide, fatty acids, amino acids; detoxifies toxic substances

Key to Illustration

1. Microvilli
2. Plasma *(cell)* membrane
3. Smooth endoplasmic reticulum
4. Nuclear pores
5. Peroxisome
6. Nucleus
7. Nucleoplasm
8. Nucleolous
9. Ribosomes
10. Rough endoplasmic reticulum
11. Cytoskeleton
12. Mitochondrion
13. Lysosome
14. Vacuole
15. Microfilament
16. Golgi complex
17. Cytoplasm
18. Cilia
19. Microtubule
20. Centrioles

Composite Cell

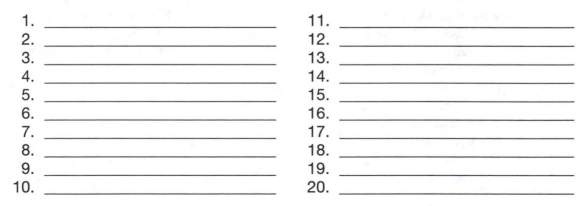

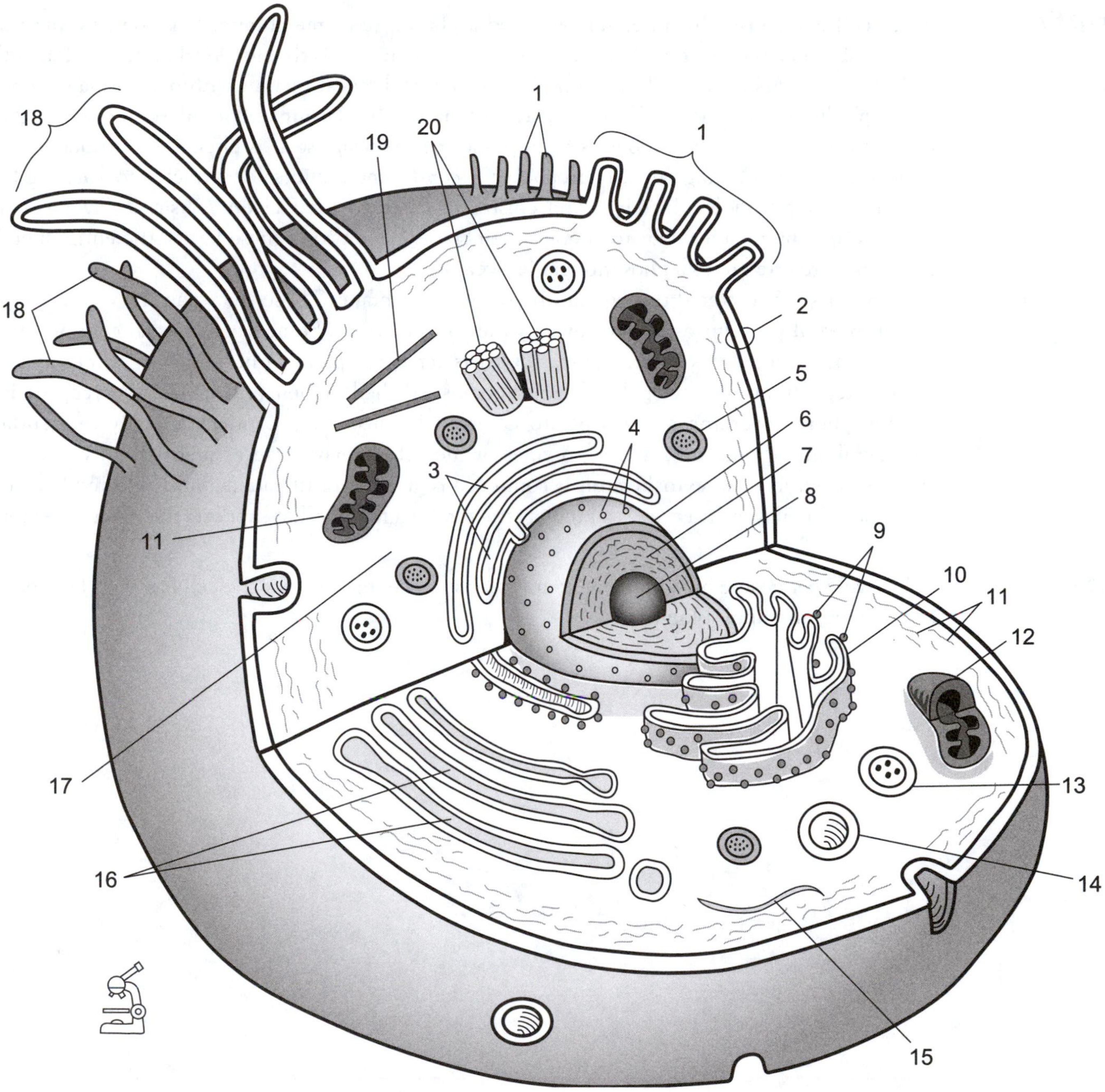

1. ____________________
2. ____________________
3. ____________________
4. ____________________
5. ____________________
6. ____________________
7. ____________________
8. ____________________
9. ____________________
10. ____________________
11. ____________________
12. ____________________
13. ____________________
14. ____________________
15. ____________________
16. ____________________
17. ____________________
18. ____________________
19. ____________________
20. ____________________

Description

Every cell is enclosed by an envelope called a **plasma** (*cell*) **membrane**. It is the gateway through which substances enter or exit any cell. Because this structure is the first level of interaction with any cell, it is very important in physiology. The fundamental repeating unit within a plasma membrane is a **phospholipid molecule** that has two parts: a spherical **headgroup** and a **tailgroup** consisting of two fatty acid chains. The headgroup is hydrophilic (*water-loving*) so it is chemically attracted to water molecules, while the tailgroup is hydrophobic (*water-fearing*), so it is not attracted to water molecules.

Phospholipid molecules align and cluster together to form layers. A single layer is called a **phospholipid monolayer**. Because plasma membranes have two of these layers, the entire membrane is generally referred to as a **phospholipid bilayer**.

Proteins are scattered throughout the plasma membrane. Those that span the entire bilayer are called **integral proteins** and they have a variety of functions. Some act as special channels through which only certain types of ions can pass. Others act as receptors for hormones or neurotransmitters. Some integral proteins have polysaccharide chains attached to them and are called **glycoproteins**.

Peripheral proteins are a separate category because they connect to only one surface of the membrane. Some may act as enzymes, while others may serve as a structural component of the cytoskeleton. The **cytoskeleton** is composed of a series of filaments and is located beneath the phospholipid bilayer. It serves as a kind of structural scaffolding that supports the plasma membrane.

Function

Selectively permeable membrane; it controls what can enter and exit the cell based on factors such as size, charge, and lipid solubility.

Key to Illustration

Plasma (Cell) Membrane

1. Polysaccharide chain
2. Glycoprotein
3. Phospholipid monolayer
4. Cholesterol molecule
5. Filaments of cytoskeleton
6. Peripheral protein
7. Integral protein
8. Glycolipid
9. Phospholipid bilayer

Phospholipid Molecule

10. Headgroup of phospholipid
11. Tailgroup of phospholipid

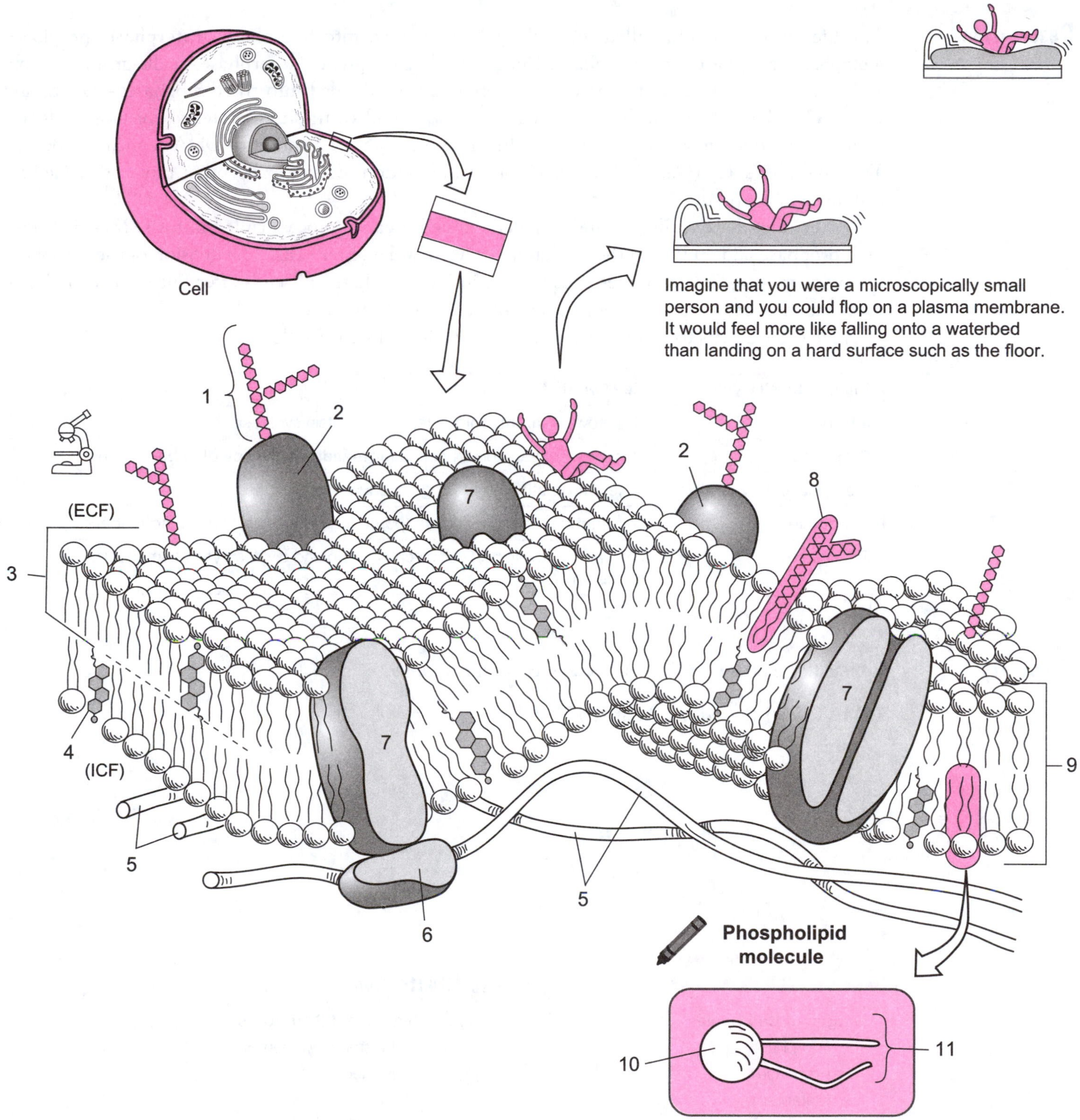
Cell
Imagine that you were a microscopically small person and you could flop on a plasma membrane. It would feel more like falling onto a waterbed than landing on a hard surface such as the floor.
1
2
7
2
8
(ECF)
3
4
(ICF)
7
7
9
5
5
6
Phospholipid molecule
10
11

KEY
(ICF) = intracellular fluid
(ECF) = extracellular fluid

Description

The life cycle of a cell, called the **cell cycle**, is divided into five phases: **interphase**, **prophase**, **metaphase**, **anaphase**, and **telophase**. During the longest phase—**interphase**—cells grow, develop, and make copies of their chromosomes. Interphase is subdivided into three subphases—$\mathbf{G_1}$, **S**, and $\mathbf{G_2}$. In $\mathbf{G_1}$ **(Gap 1)**—when the cell is actively growing and synthesizing proteins. For most cells this period lasts from a few minutes to hours. In the S phase (S for **S**ynthetic) DNA replication occurs. The final phase, $\mathbf{G_2}$ **(Gap 2)** is relatively short in duration. Enzymes and other proteins needed for cell division are produced.

After interphase, cells prepare to undergo a cell division process called **mitosis**. Mitosis begins with **prophase** and ends with the formation of two new **daughter cells**. The division of the cytoplasm during mitosis, called **cytokinesis**, can begin in late anaphase and ends after mitosis is completed. Each daughter cell is an exact cloned copy of the original parent cell.

A few key events to identify each stage of mitosis are listed as follows:

Phase of Cell Cycle	Key Event
Interphase	Cell growth; **chromosomes** replicate; protein synthesis
Prophase	Nuclear membrane disappears; **chromatids** visible; **spindle fibers** appear
Metaphase	Sister **chromatids** align along the equator of the cell
Anaphase	**Chromatids** are separated and pulled to opposite poles by **spindle fibers**
Telophase	Cleavage furrow forms; cell pinches in two and divides **cytoplasm**; nuclear membrane reappears

Study Tip

To recall the phases of the cell cycle, use this mnemonic:

IPMAT = "I Passed My Anatomy Test"

Key to Illustration

Stages of Cell Cycle

1. Interphase
2. Prophase
3. Metaphase
4. Anaphase
5. Telophase *(and cytokinesis)*

Final Products

6. Daughter cells

Significant Structures

a. Plasma *(cell)* membrane
b. Nucleolus
c. Nucleus
d. Centriole
e. Chromosome
f. Centromere
g. Spindle fibers
h. Sister chromatids
i. Cleavage furrow
j. Nuclear membrane *(still forming)*

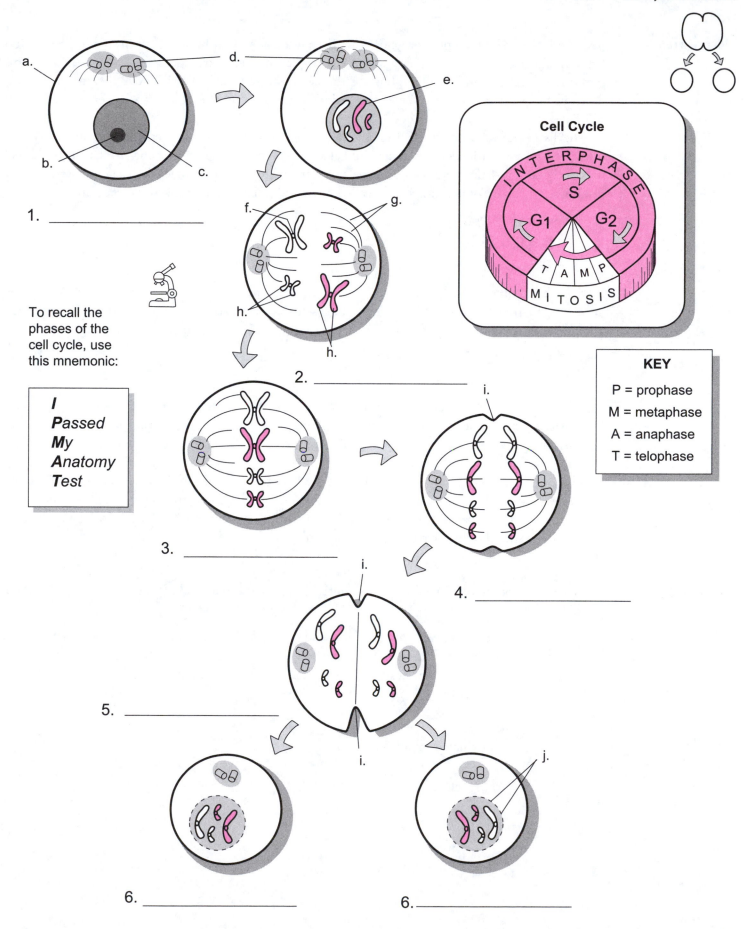

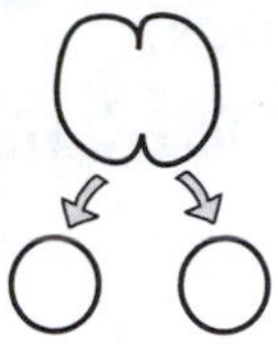

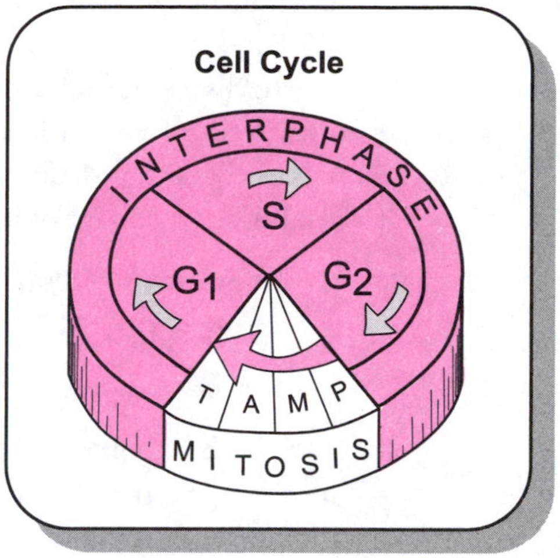

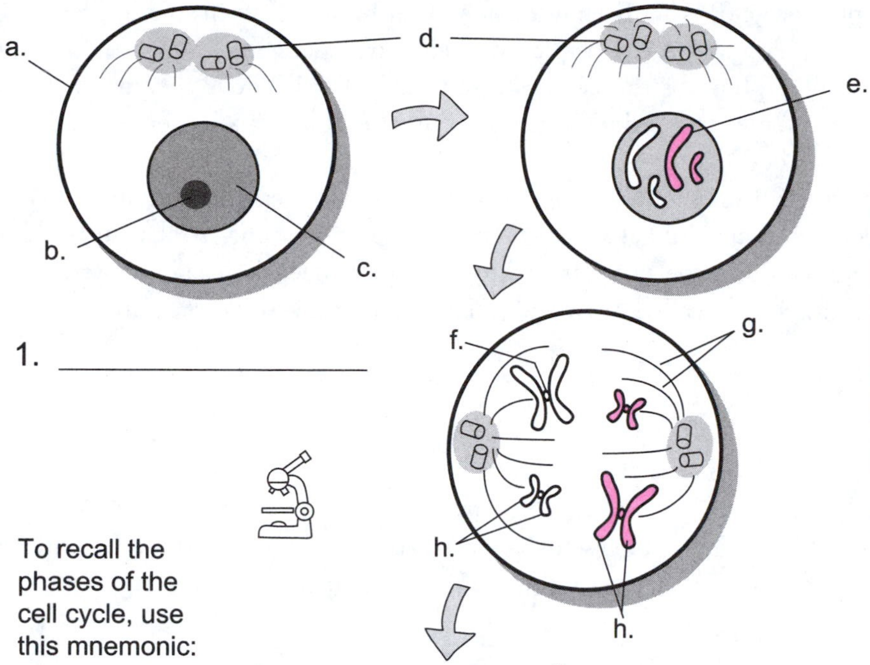

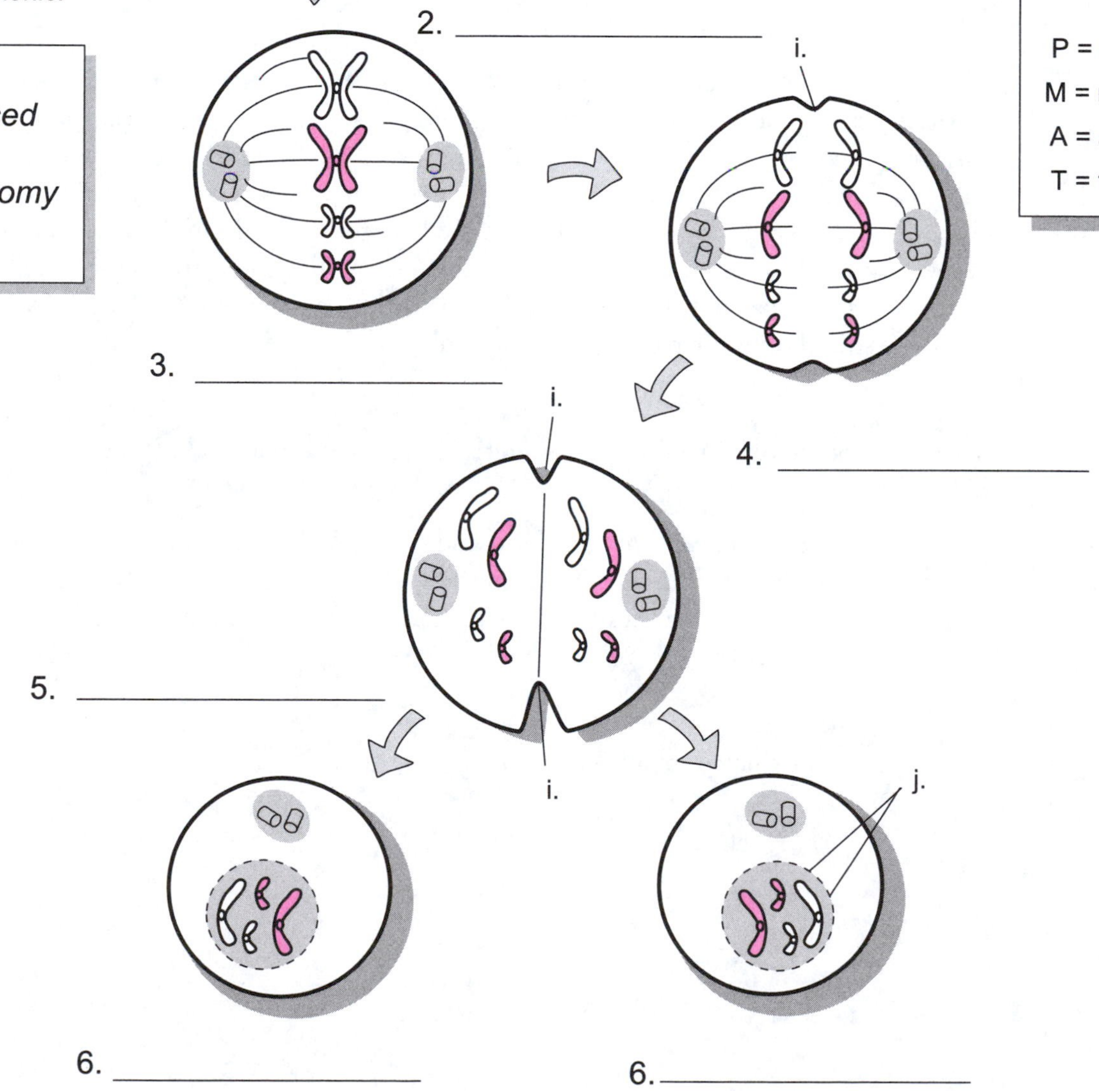

To recall the phases of the cell cycle, use this mnemonic:

I
***P**assed*
***M**y*
***A**natomy*
***T**est*

KEY

P = prophase

M = metaphase

A = anaphase

T = telophase

Description

Because DNA (deoxyribonucleic acid) is THE genetic material, it has to be replicated in preparation for normal cell division so that a copy of it can be placed in each new daughter cell (see p. 36). This replication occurs during the S phase of the cell cycle. Structurally, DNA is a complex macromolecule located in the cell nucleus in a double helix form resembling an extremely tall spiral staircase.

To discuss its structure, we will simplify it. Imagine that the spiral staircase is flexible so it can be untwisted into a tall ladder. Then cut the ladder down the middle of the rungs so it is in two equal halves. Each half-ladder is bonded to the other by hydrogen bonds. Each half-ladder is created by linking a long chain of building blocks called **nucleotides**. Each nucleotide has three parts:

1. **Sugar** (5-carbon)
2. **Phosphate group**
3. a **Base** (containing nitrogen).

The backbone of the ladder is made of alternating sugars and phosphate groups bonded to one another. The rungs on the ladder are made of base pairs. Each base can be one of four different types: adenine (A), thymine (T), guanine (G), or cytosine (C). The base pairing is as follows:

- **A** bonds to **T** (or **T** to **A**)
- **G** bonds to **C** (or **C** to **G**)

This base pairing is the basis for the genetic code. This message directs the creation of proteins by specifying the sequence of amino acids that are to be joined to make that protein. The type of proteins produced determines the type of organism, such as a goldfish, maple tree, or human. These proteins also account for individual differences between members of the same species.

Replication occurs with the help of specialized enzymes called **DNA polymerases**. They attach to one section of DNA, and it unzips. Free nucleotides within the cytoplasmic soup are paired with their proper bonding partners (A to T, G to C) to synthesize a new strand. Each new DNA molecule contains one strand from the parent DNA molecule.

Analogies

- The DNA **double helix** is like a spiral **staircase**. The **DNA molecule** (simplified version) is like a **ladder**.
- When DNA prepares to replicate itself it is said to **UNZIP**. This is an accurate description. When you **unzip** your coat, each **half of the zipper** is like one-half of the DNA molecule, called the **DNA template**, used to synthesize a new strand. As the new strand is synthesized, DNA **ZIPS** back together again.

Location

DNA is found in the nucleus of all body cells.

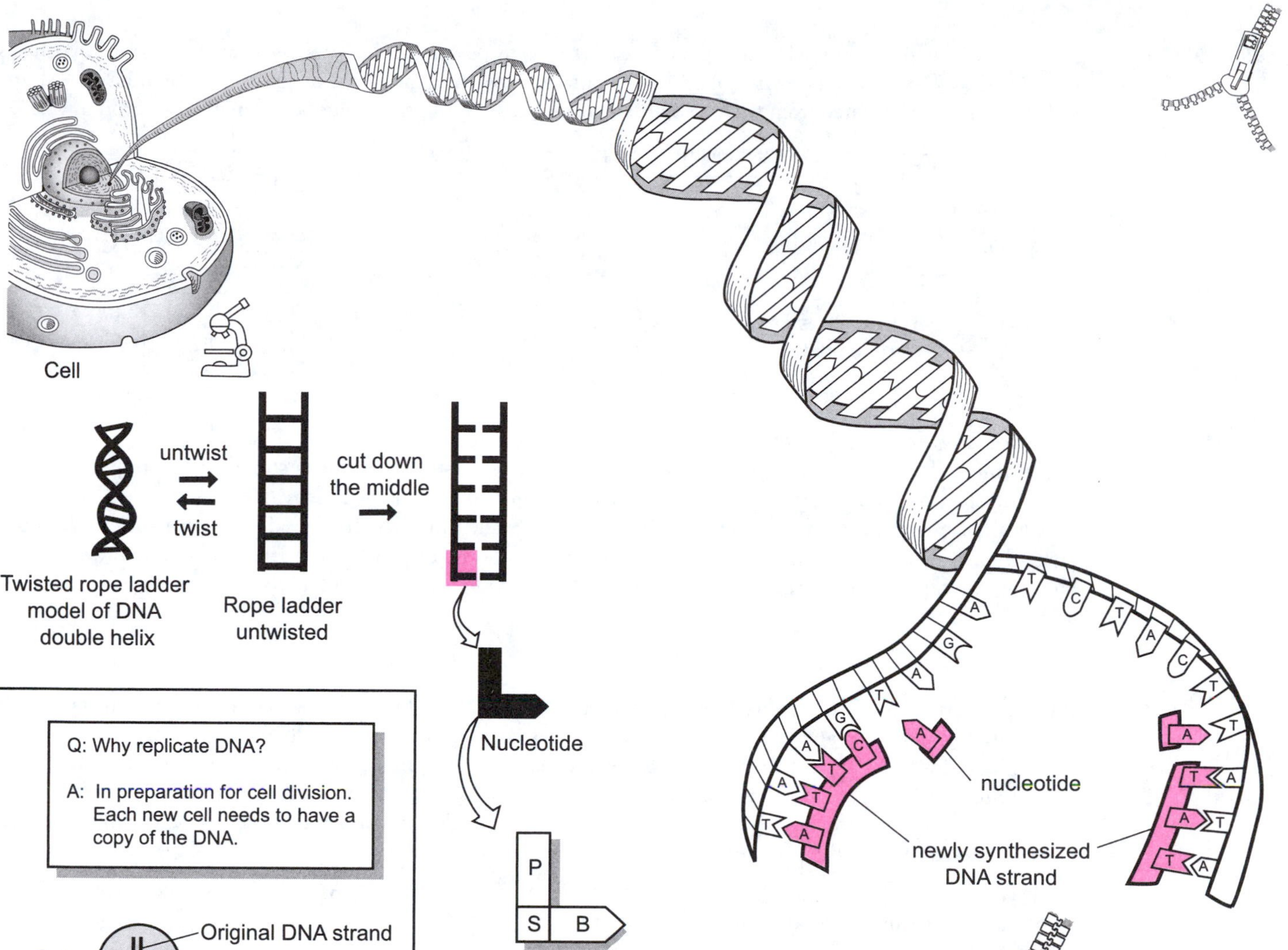

Nucleotide (detail)

Q: Why replicate DNA?

A: In preparation for cell division. Each new cell needs to have a copy of the DNA.

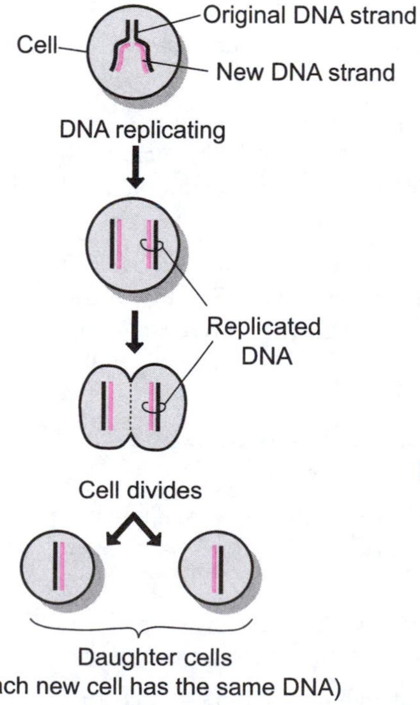

KEY

P = phosphate group
S = sugar (5 carbons)
B = nitrogen base
(can be one of four types: adenine (A), thymine (T), guanine (G), or cytosine (C))

Base pairing in DNA

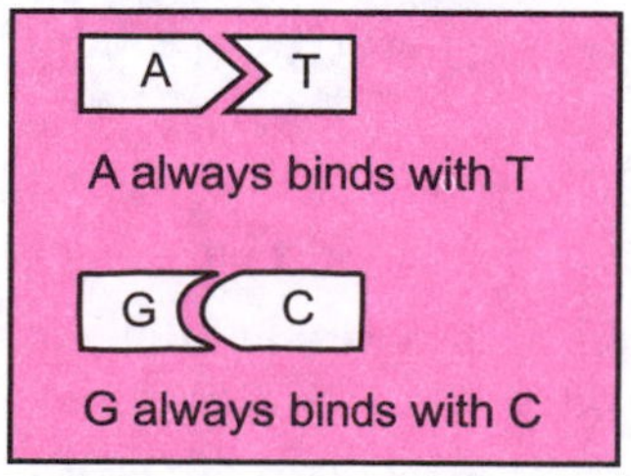

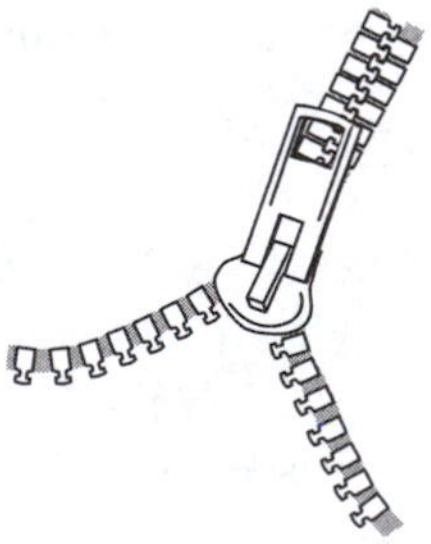

DNA "unzips" to replicate by breaking the hydrogen bonds between the base pairs. Each separated strand of DNA then serves as a template for making a new strand of DNA.

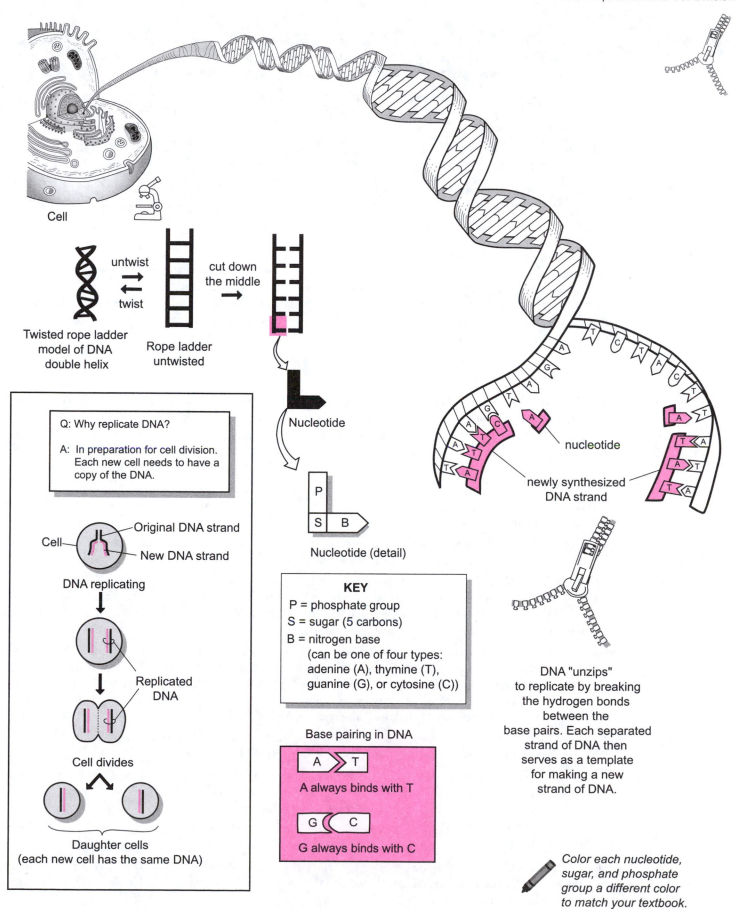

Color each nucleotide, sugar, and phosphate group a different color to match your textbook.

Description

Most cells in your body have one primary task—to make proteins. This process is called **protein synthesis**. It requires that **DNA** first be transcribed into a similar molecule called messenger RNA or **mRNA**. Then the mRNA must be translated from the base pair language of RNA into the amino acid language of proteins.

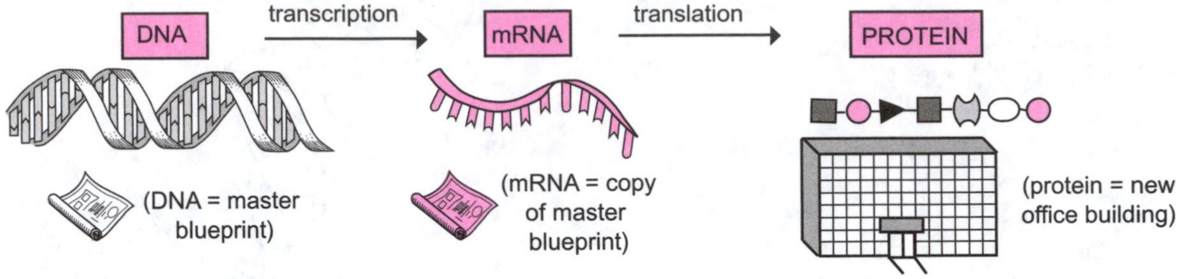

Analogy

The process of protein synthesis in a cell is like building a new office building. The construction site is like the **cytoplasm** and the master blueprint is like the DNA. The master blueprint is stored in the trailer on the construction site, just like DNA is stored in the **nucleus** because it is too large to leave. Just like copies can be made of the original blueprint, a smaller copy of DNA is made, called mRNA. It is small enough to leave the nucleus, enter the cytoplasm and bind to the construction workers of the cell, called the **ribosomes**. The raw materials for building the office building are steel, concrete, and glass, to name a few. To build a protein, the raw materials are **amino acids**.

Basic Steps in Protein Synthesis

1. Transcription: The enzyme **RNA polymerase** travels along DNA and separates the base pairs by "unzipping" one portion at a time for the purpose of making a sister molecule to DNA called mRNA. The building blocks for making mRNA are called **RNA nucleotides**. The enzyme uses one side of the DNA molecule as a template to match the base of the RNA nucleotide to the corresponding base in DNA. As the RNA nucleotides are lined up next to each other, they are covalently bonded together to eventually form a short, single-stranded mRNA molecule.
2. RNA processing: Like film being edited, the single-stranded mRNA molecule is spliced. Small segments called **introns** are removed from the molecule.
3. The single-stranded mRNA is small enough to leave the nucleus through the **nuclear pore** and move into the cytoplasm, where it binds to the small subunit of a ribosome.
4. A transfer RNA, tRNA, molecule has an **amino acid** binding site at one end and three bases at the other end, called the **anti-codon**. The purpose of the tRNA is to deliver the proper **amino acid** to the ribosome. With the help of an enzyme and free energy from ATP hydrolysis, one specific amino acid is bonded to a specific tRNA molecule based on its anti-codon sequence.
5. The RNA message is read in groups of 3 base pairs. Each group is called a **codon**. Note that in RNA, the base uracil (U) replaces thymine (T), which explains why A bonds to U. The tRNA with the amino acid called methionine always binds to the beginning of the message called the **start codon**. Next, the large ribosomal subunit binds to the small ribosomal subunit to activate the ribosome. The large subunit contains two binding sites for tRNAs called the **P site** and the **A site**.
6. The initial tRNA binds to the P site. One by one, amino acids are positioned next to each other, then bonded together. This results in a growing **polypeptide chain**. Here is how it works:
7. A tRNA with the anti-codon matching the codon arrives at the A site with the next amino acid to be added to the chain. The growing polypeptide chain at the P site is transferred to the tRNA at the A site as it binds to the next amino acid to be added to the chain.
8. As the mRNA slides to the left, it moves the tRNA at the A site over to the **P site**. Then the tRNA with the next amino acid to be added to the chain arrives at the A site and the process repeats itself.
9. The process of growing the polypeptide chain ends when a special codon called the **stop codon** is reached at the A site. The final polypeptide is released and the last tRNA is released as well. Finally, the large and small ribosomal subunits separate.

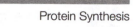

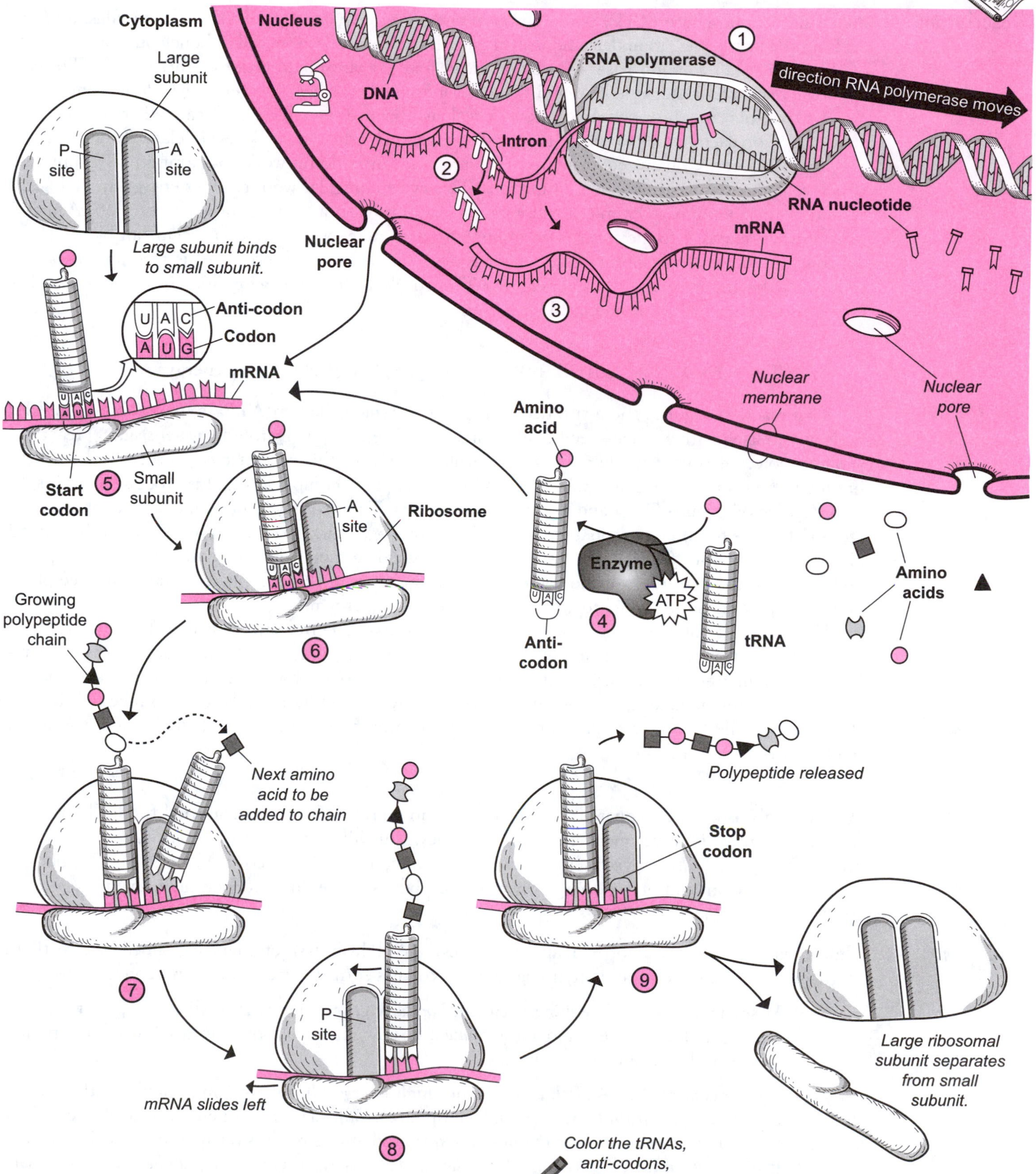
Cytoplasm
Nucleus
DNA
RNA polymerase
1
direction RNA polymerase moves
Intron
2
RNA nucleotide
mRNA
3
Nuclear pore
Nuclear membrane
Nuclear pore
Large subunit
P site
A site
Large subunit binds to small subunit.
Anti-codon
Codon
U A C
A U G
mRNA
Start codon
5
Small subunit
A site
Ribosome
6
Amino acid
Enzyme
ATP
4
Anti-codon
tRNA
Amino acids
Growing polypeptide chain
Next amino acid to be added to chain
7
P site
mRNA slides left
8
Polypeptide released
Stop codon
9
Large ribosomal subunit separates from small subunit.

Color the tRNAs, anti-codons, and ribosomes.

Description

Adenosine triphosphate (ATP) is the universal energy currency for all cells. Look at the illustration of ATP on the facing page to understand its structure. It is classified as a specialized nucleotide and consists of three parts: (1) **adenine base**, (2) **ribose sugar**, and (3) **phosphate groups**—three total. The term *adenosine* refers to the adenine base and ribose sugar bonded together. The term *triphosphate* comes from the fact that there are three phosphate groups covalently bonded to each other. Because there is a net negative charge on each of the phosphate groups, they repel each other and make ATP relatively unstable.

Cells are highly organized structures that are constantly doing work to maintain their physical structure and carry out their general functions. This cycle of regular work requires energy. In biological systems, it is common to couple a spontaneous reaction with a nonspontaneous reaction. A metal rusting when exposed to moist air is an example of a spontaneous reaction (it occurs all by itself). In contrast, muscle cell attempting to contract is an example of a nonspontaneous reaction (it occurs only with the input of additional energy). ATP hydrolysis occurs as a spontaneous reaction and is represented by the following chemical equation:

$$\text{ATP} + H_2O \xrightarrow{\textbf{ATPase}} \text{ADP} + P_i \textit{ (inorganic phosphate group)} + \textbf{free energy}$$

This reaction requires the assistance of a special enzyme called an **ATPase**, which binds an ATP molecule to itself with a shape-specific fit like a lock and key. In any hydrolysis (*water splitting*) reaction, a water molecule is used to cleave a single covalent bond. In this case, it cleaves the covalent bond linking the terminal phosphate group to the second phosphate group. During this process, water is split into a **hydroxyl group** (OH^-) and a hydrogen atom (H). The hydroxyl group binds to the phosphorus atom (P) in the terminal phosphate group while the hydrogen binds to the oxygen atom on the second phosphate group (*see illustration*). As a result, a more stable molecule called **adenosine diphosphate (ADP)** is formed as a result of the decreased repulsion between the phosphate groups. The free phosphate group often is transferred to another substrate or to an enzyme.

In addition, some free energy is released in the process. The free energy is used to drive processes that must occur in cells, such as muscle contraction, active transport of substances across plasma membranes, movement of a sperm cell's tail, manufacturing a hormone, or anything else a cell has to do.

ATP is manufactured inside the cells via the energy derived from foods that are ingested. During this process called cellular respiration, ATP is formed by bonding ADP and Pi through a series of **oxidation-reduction reactions**.

Analogy

ATP hydrolysis can be compared to an **investment in a profitable stock or mutual fund**. Though you must make an initial investment of your own money, you will reap a greater reward (dividend) in the end. Similarly, an initial investment of energy is required to break the covalent bond in ATP, but the result is a larger amount of free energy when the new bonds in the final products are formed.

Study Tips

Unfortunately, the function of ATP and the process of ATP hydrolysis often are explained incorrectly in many biology and anatomy and physiology textbooks. Let's correct some common misconceptions:

- **Misconception #1**: ***Breaking chemical bonds releases energy.*** Actually, the opposite is true. During a chemical reaction the *formation* of new chemical bonds in a more stable product results in the release of some energy.
- **Misconception #2**: ***ATP has a special "high-energy phosphate bond" between the second phosphate group and the terminal phosphate group.*** The bond here is actually a covalent bond. This statement gives the false impression that this bond is ready to fly apart like a jack-in-the-box. In fact, bonds are a force that hold atoms together so an *input of energy* is necessary to break a chemical bond. Instead, the three phosphate groups in ATP are all linked together by relatively strong bonds called *covalent* bonds, which represent shared pairs of electrons.

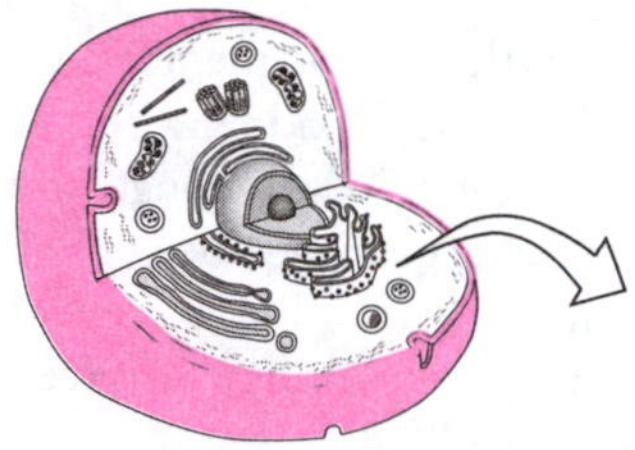

(ATP is found inside body cells)

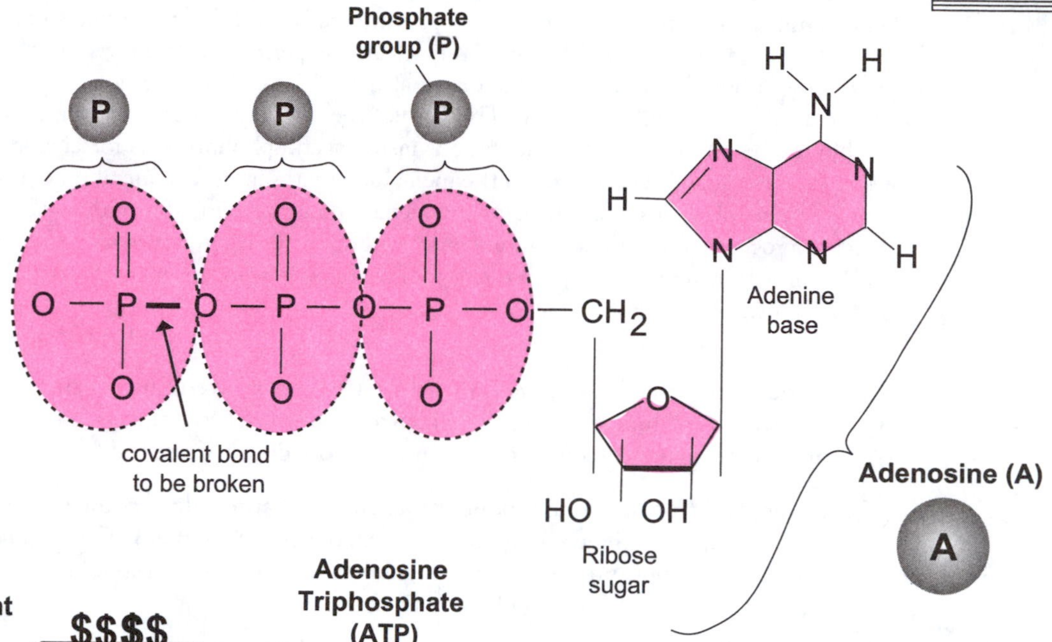

Adenosine Triphosphate (ATP)

Investment Analogy

ATP hydrolysis is like a good investment. Though it requires an initial investment, it will be worth it because a good stock will return more money in its dividend. Similarly, an initial investment of energy is required to break the covalent bond in ATP, but it's also worth the initial investment of energy because the amount of energy released upon formation of the new bonds in the products is greater than the inital investment. In summary, there is a net release of energy in ATP hydrolysis.

ATP hydrolysis equation:

$$\text{ATP} + \text{H}_2\text{O} \xrightarrow{\text{ATPase}} \text{ADP} + \text{Pi} + \text{free energy}$$

(Adenosine triphosphate) (plus) (water) (and ATPase enzyme forms) (Adenosine diphosphate) (plus) (inorganic phosphate ion) (plus) (free energy)

(OR)

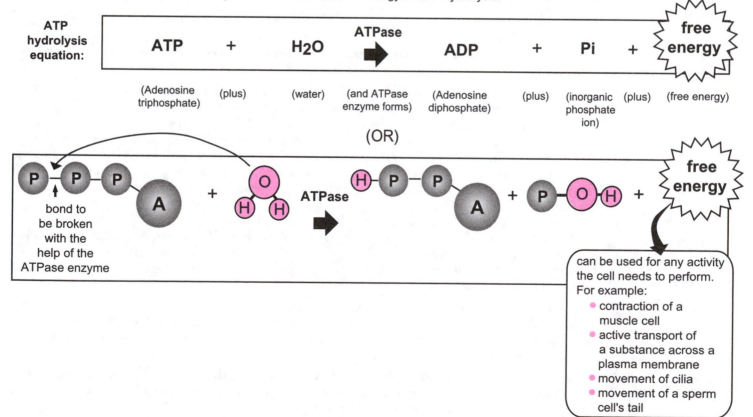

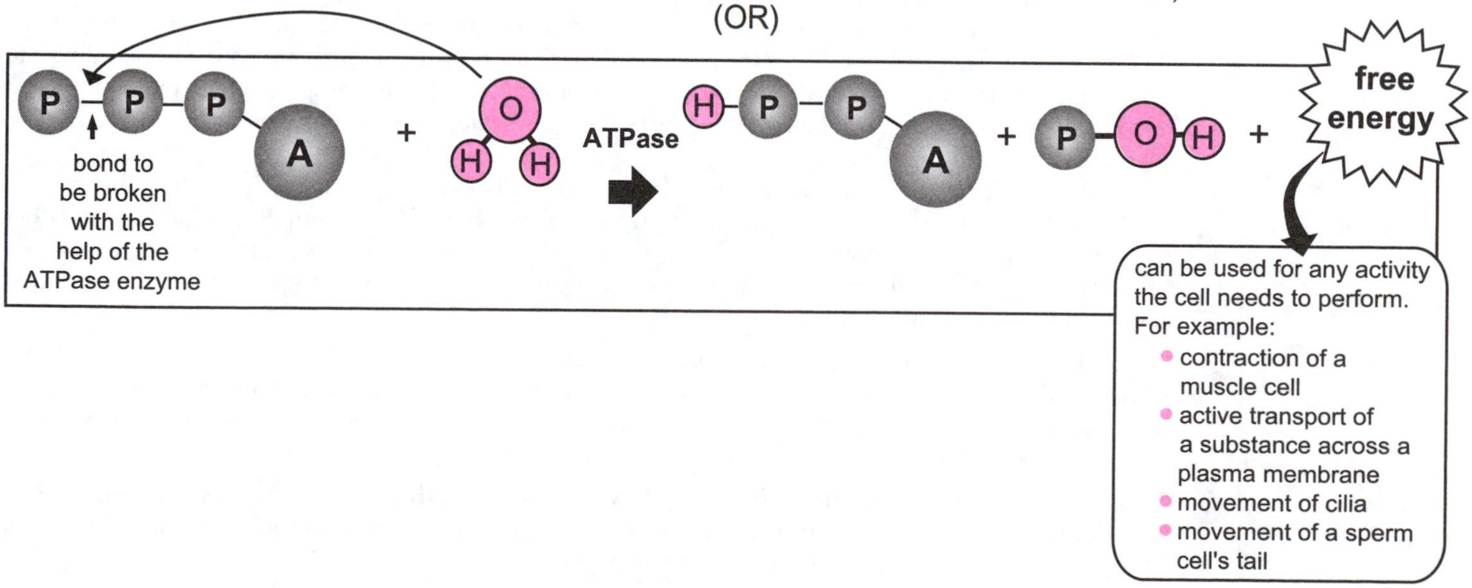

Description

Any substance that is able to either enter or leave the cell must do so by passing through the plasma membrane. In general, nutrients such as oxygen have to pass into cells, while waste products such as carbon dioxide have to pass out of them. Because the cell membrane is selectively permeable, it allows some substances to pass through while preventing others from doing so. Factors that determine whether or not a substance can pass include: (1) size, (2) lipid solubility, and (3) charge. The smaller the size, the easier it is for a molecule to cross. The more lipid-soluble the better, as the cell membrane is made of phospholipids. As for charge, uncharged substances have an easier time crossing directly through the membrane (without using a membrane protein).

Transport of a substance through the membrane is categorized as either passive or active. **Passive transport** refers to processes that occur spontaneously without any energy investment by the cell in terms of energy from ATP hydrolysis. **Active transport** refers to any transport process that occurs only when the cell invests energy from ATP hydrolysis to force transport to occur.

Passive Transport

The illustration on the facing page shows three types of passive transport through the plasma membrane. All of them deal with the diffusion process, in which a solute particle spontaneously moves from an area of higher solute concentration to an area of lower solute concentration. Each is described below:

1. In **simple diffusion** a small, nonpolar, uncharged particle diffuses directly through the phospholipid bilayer of the plasma membrane. Oxygen gas enters our cells by simple diffusion, and carbon dioxide gas leaves our cells by the same process. Other substances that diffuse across plasma membranes include fatty acids, steroids, and fat-soluble vitamins (A, D, E, and K).

2. **Simple diffusion** also can be a channel-mediated type **protein channel**. This is how small substances such as ions diffuse through a plasma membrane. For example, sodium ions diffuse through sodium channels to enter a cell. Similarly, potassium ions pass through potassium channels to leave the cell.

3. **Facilitated diffusion** differs from simple diffusion in the following ways: (1) involves the transport of larger solutes, (2) uses a protein carrier that has a shape-specific fit for a specific solute, (3) involves a shape change in the protein carrier. The solute must bind to the shape-specific binding site. This puts a limitation on how quickly diffusion can occur. The action of the binding of the solute to its binding site induces a shape change in the protein carrier. This allows the solute to be released on the opposite side of the plasma membrane. Glucose enters cells via facilitated diffusion.

Active Transport

The illustration on the facing page shows three types of active transport processes through the plasma membrane. Each is described below:

4. **Primary active transport** refers to the movement of a specific substance against its concentration gradient. In other words, a solute particle is moved from an area of low solute concentration to an area of high solute concentration. Like paddling against the current in a river, this requires energy. The protein used in this process is generally referred to as a pump, which appropriately implies the need for energy. Free energy from ATP hydrolysis is the fuel that keeps the pump working. This protein pump always contains a binding site for the substance to be transported. For example, the iodine pump actively transports iodine out of the blood and into the cells in our thyroid gland. The thyroid requires iodine as a raw material for manufacturing the hormones it normally produces.

   Unlike primary active transport, **secondary active transport** does not directly use free energy from ATP hydrolysis. Instead, it gets its energy indirectly from the established concentration gradients from either sodium ions (Na^+) or hydrogen ions (H^+). These ion gradients were established by primary active transport. Movement of these ions down their concentration gradients is then coupled to the simultaneous transport of another substance via either symport or antiport.

5. **Symport** refers to the simultaneous transport of two substances in the same direction across the membrane with the aid of a specific type of membrane protein. One example is the sodium-glucose symporter located in the epithelial cells in the intestinal mucosa. This transport process helps glucose get absorbed into the blood from the digestive tract.

6. **Antiport** refers to the simultaneous transport of two substances in the opposite direction across the membrane with the aid of a specific type of membrane protein. One example is the sodium-calcium antiporter, which keeps calcium levels low inside cells by pumping it out.

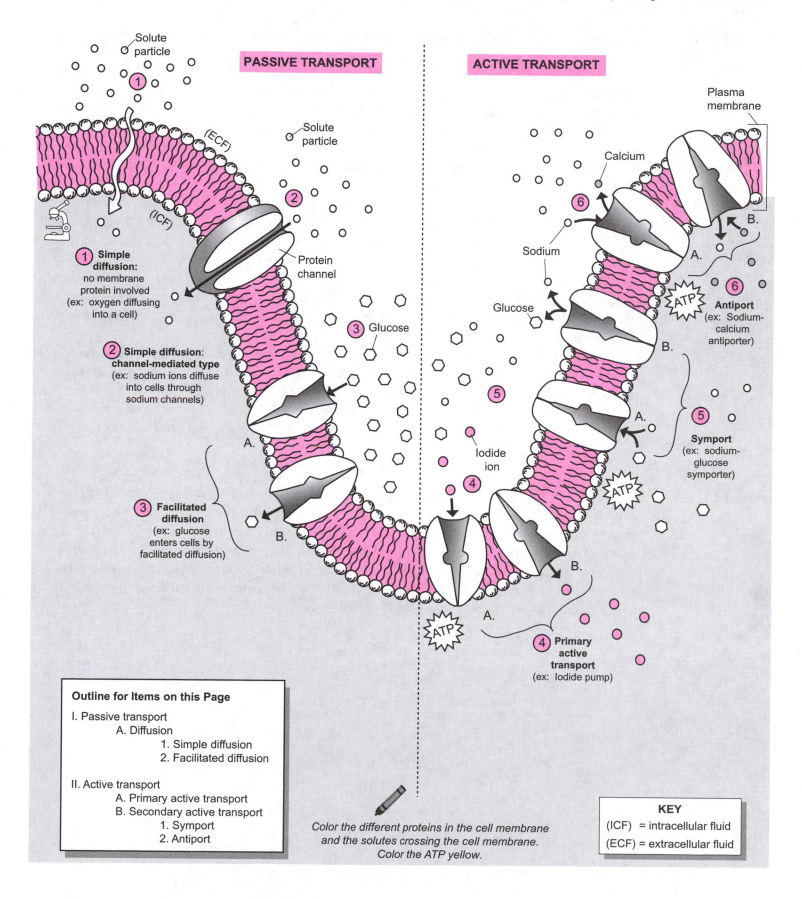

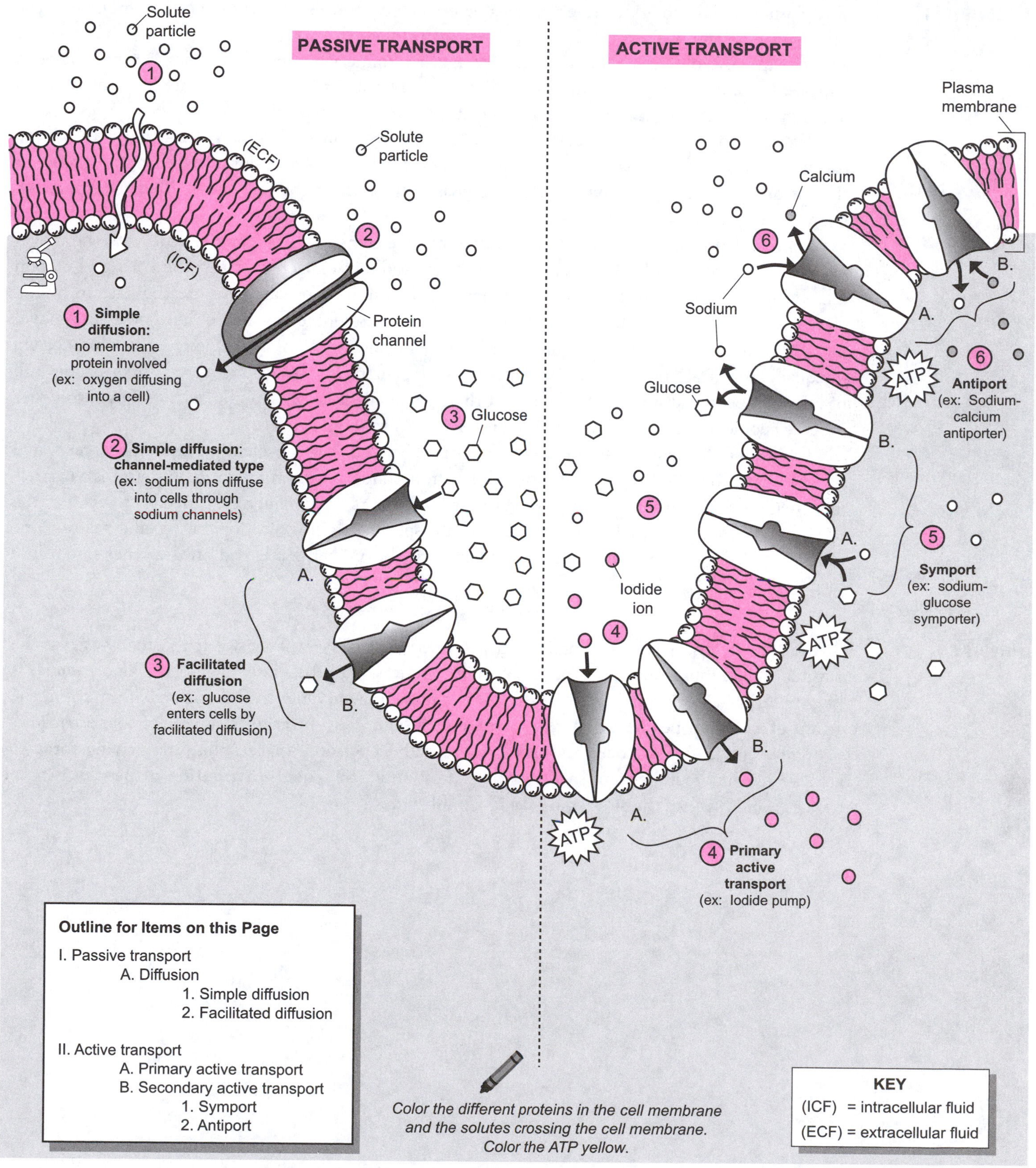

Outline for Items on this Page

- I. Passive transport
  - A. Diffusion
    - 1. Simple diffusion
    - 2. Facilitated diffusion
- II. Active transport
  - A. Primary active transport
  - B. Secondary active transport
    - 1. Symport
    - 2. Antiport

Color the different proteins in the cell membrane and the solutes crossing the cell membrane. Color the ATP yellow.

KEY
(ICF) = intracellular fluid
(ECF) = extracellular fluid

Description

The plasma membrane is selectively permeable because it controls what substances enter and exit the cell. Nutrients and waste products both must pass through this structure, but not all substances are able to cross the plasma membrane. Factors that determine whether a substance can pass through include: (1) size, (2) lipid solubility, and (3) charge. If a substance is smaller, more lipid-soluble, and does not have a charge, it is more likely to be able to cross the membrane.

Transport of a substance is classified as either **passive transport** or **active transport**. In passive transport, the cell does not have to expend energy for the process to occur. Active transport requires the use of cellular energy. Typically, this energy is in the form of the free energy liberated from ATP hydrolysis.

Simple diffusion falls under the category of **passive transport**. Typically, it deals with the movement of a solute particle. For example, this process allows oxygen to enter the cells and carbon dioxide to exit them. In diffusion, substances tend to move from areas where they are in higher concentration to areas of lower concentration. The driving force for simple diffusion is the movement toward a state of dynamic equilibrium, in which concentrations of the designated substance are in equal concentrations on both sides of the plasma membrane. This **driving force** is like a **room full of crowded people at a party who all need their personal space**. As the illustration shows, this *need* induces people to move from the crowded room A over to room B.

Facilitated diffusion is also a **passive transport** process, but it typically deals with larger solute particles than those in simple diffusion. It differs from simple diffusion primarily in that it is a **carrier-mediated** process. This means that a protein carrier in the plasma membrane is needed to assist in the diffusion process. The carrier protein has a shape-specific binding site for the specific solute to be transported. The solute binds to this site temporarily before being released on the other side of the membrane.

Analogy

In the illustration, **a person** represents **a solute particle.** The **wall** between the rooms is like the **plasma membrane. Room A** is like a **solution** on one side of the plasmamembrane, and **room B** is like the **other solution** on the other side of the plasma membrane. In the "Before" illustration, the **group of crowded people in room A** represents a **high concentration gradient** of solute particles relative to room A. The **need** for people to have their personal space is like the **driving force** to achieve dynamic equilibrium. In the "After" illustration, the **equal distribution of people** in both rooms A and B is like the **state of dynamic equilibrium**.

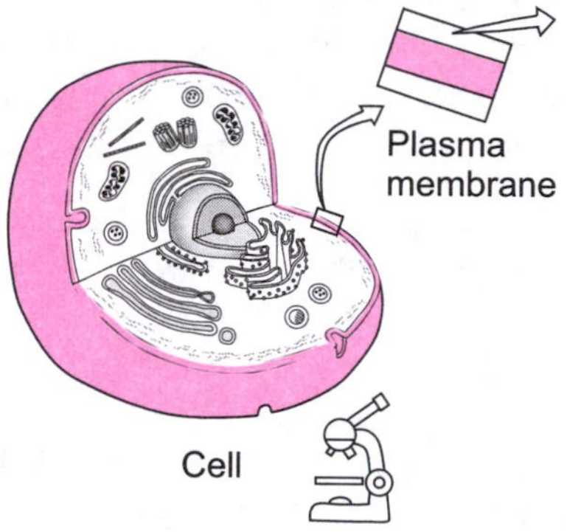

SIMPLE DIFFUSION:

(A) Non-pore-mediated

(B) Pore-mediated

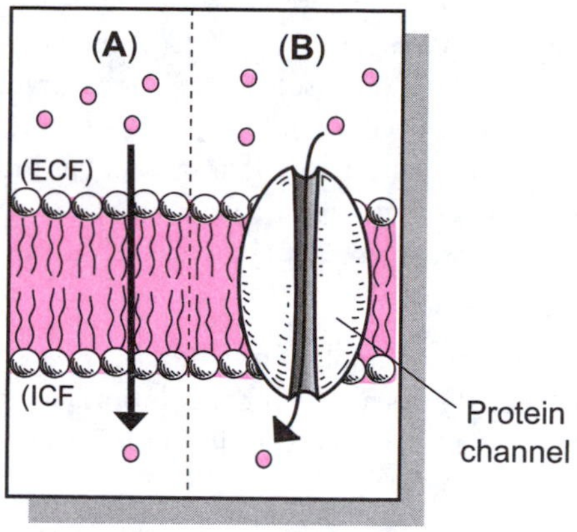

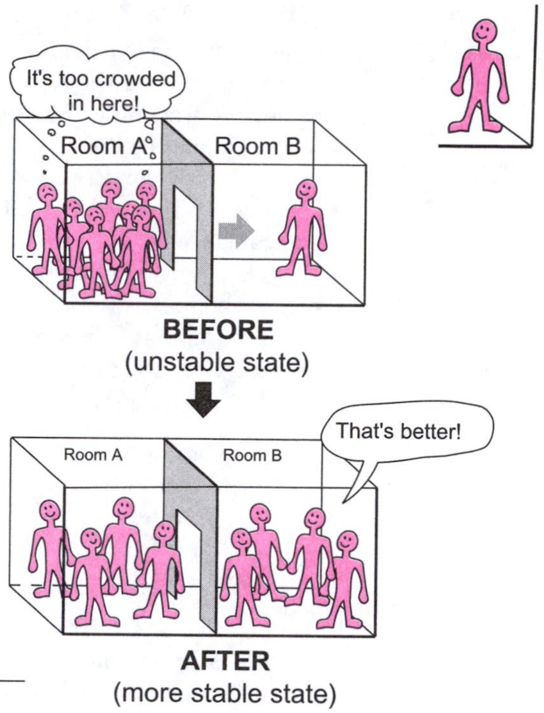

FACILITATED DIFFUSION

- Carrier-mediated
- Involves larger solute particles

The driving force that causes diffusion is movement toward dynamic equilibrium or an equal concentration of solute particles on both sides of the cell membrane. In this sense, the solute particles are like a bunch of crowded guests at a party who all want their personal space. As they move from Room A to Room B, they tend to keep moving until they have maximized their personal space. After this occurs, diffusion stops because the state of dynamic equilibrium has been achieved.

Color the protein channel and the protein carriers different colors.

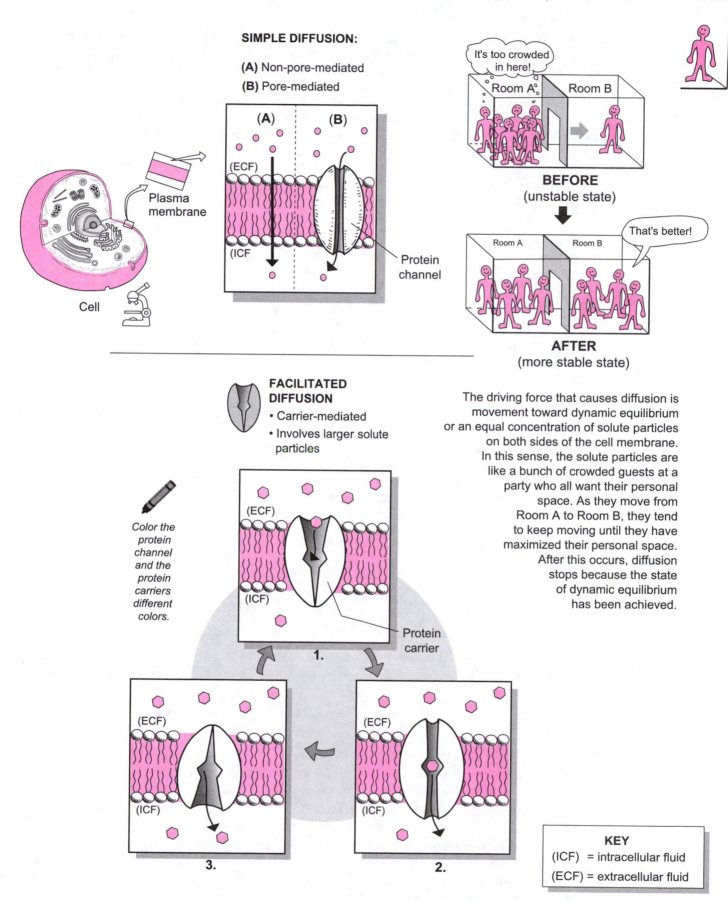

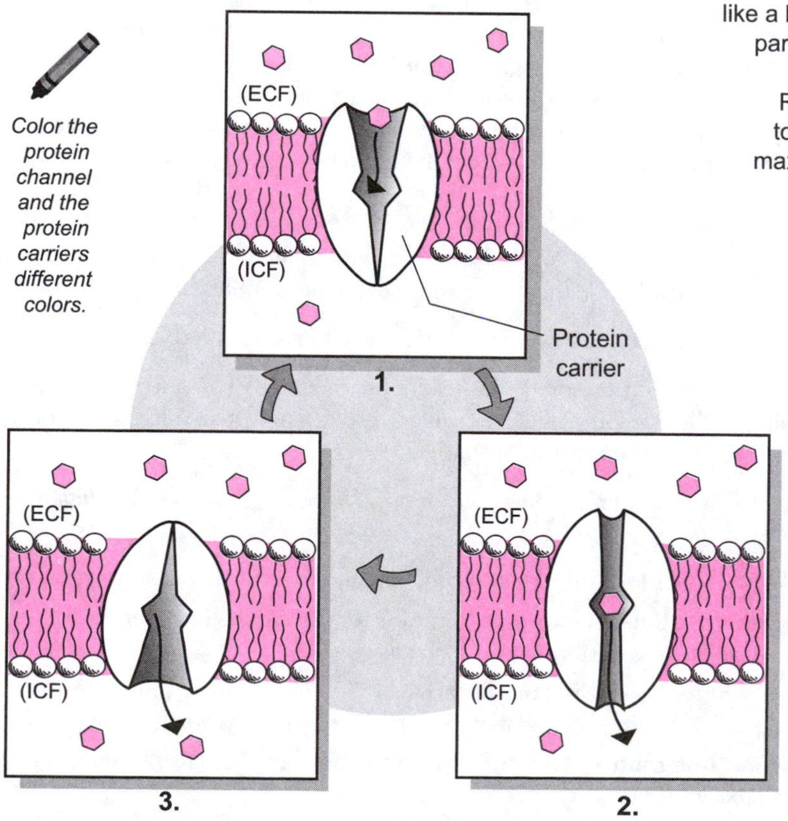

KEY

(ICF) = intracellular fluid

(ECF) = extracellular fluid

Description

Osmosis falls under the category of **passive transport**. It is the flow of a solvent (typically, **water**) across a plasma membrane toward the solution with the higher solute concentration. The driving force is the tendency toward achieving dynamic equilibrium in which there is an equal concentration of solute across the plasma membrane. When examining osmosis, the solute in question is impermeable to the plasma membrane. It is also essential to realize that the *intracellular* solution always is compared to the *extracellular* solution.

If you tell me I am tall, I might say, "Compared to what?" Compared to a toddler, I may be tall but not when compared to Mt. Everest. Similarly, if you tell me a solution is concentrated, I could also say, "Compared to what?"

The word "concentrated" makes sense only in the context of a relative comparison. While your body has regulatory mechanisms to keep the concentration of the intracellular and extracellular solutions relatively stable, the extracellular solution is more likely to change in its solute concentration because of external influences. Sometimes it is slightly more dilute and other times it is more concentrated.

Three different terms describe all the possibilities.

1. The body's extracellular solutions normally are **isotonic** (*iso* = equal) **solutions**, which have the same concentration of impermeable solutes as the other solution. This is a stable state for body cells. A cell in this solution already has achieved dynamic equilibrium, so the net movement of water in and out of the cell is zero.
2. A **hypertonic solution** (*hyper* = more, greater) has a *greater* concentration of impermeable solutes than the other solution. A cell placed in this solution will shrink (*crenate*) because of water leaving the cell.
3. A **hypotonic** (*hypo* = less) **solution** has a *lesser* concentration of impermeable solutes than the other solution. A cell placed in this solution will swell and possibly burst (*lyse*) because of water rushing into the cell.

By convention, biologists typically use these terms to describe the extracellular solution though they technically can be used to describe either intracellular or extracellular solutions as a relative comparison. Note that *when water moves*, it *always moves toward the solution with the higher solute concentration*. Think of this as water's attempt to *dilute the more concentrated solution* to try to get closer to dynamic equilibrium.

The *rate* at which water moves is determined by the **concentration gradient**—the degree of difference in solute concentration across the plasma membrane. This concentration gradient can be either **high** (large difference) or **low** (small difference). *The rule is that the higher the concentration gradient, the faster water will move across the membrane*. The converse also is true—namely, *the lower the concentration gradient, the slower water will move*. This explains why a cell placed in a hypotonic solution may either swell or burst. It is more likely to burst with a high concentration gradient. Researchers exploit this by *osmotically shocking* (*lysing*) cells when they want to extract their internal contents.

Study Tips

- The *solute* in question is always *impermeable* to the plasma membrane. If it were permeable, the solute could simply diffuse its concentration gradient to achieve dynamic equilibrium all on its own (see p. 46).
- Always compare the solute concentration in the intracellular solution to that of the extracellular solution.
- Convert a "word problem" into a simple illustration: If given a basic word problem about osmosis, it often helps to draw out a simple illustration to help you figure it out. Draw a simple picture of a cell (circle) in a beaker (box or square). Practice this on the facing page.
- Remember that the osmotic terminology (*hypertonic*, *hypotonic*, and *isotonic solutions*) is always in reference to the *solute*.
- For linear thinkers, use the following step-by-step method to solve any basic osmosis problem:
  (1) Compare the intracellular solute concentration to the extracellular solute concentration. Then, correctly label your extracellular solution as *hypertonic*, *hypotonic*, or *isotonic*.
  (2) Are the two solutions equal in concentration? If "yes", dynamic equilibrium has been achieved and there will be no significant shift in water movement. If "no", water is going to move (so go to #3).
  (3) Which way will water move? Use this rhyme to remember: "*Osmosis! Osmosis! It's really quite a hoot. Water always moves toward the greater solute!*"
  (4) What happens to the cell as a result of water movement? If water leaves the cell, it will shrink (*crenate*). If water enters the cell, it will become more turgid and may even burst (*lysis*).

water (H_2O) molecules
H
O
H_2O
semipermeable aritificial membrane
pore
protein (impermeable to membrane)
BEFORE
Osmosis! Osmosis! It's really quite a hoot. Water always moves toward the greater solute!
AFTER
Three possibilities:
Isotonic solution
I feel happy and normal!
H_2O
I feel dehydrated!
Hypertonic solution
H_2O OUT
I feel like I am ready to pop!
Hypotonic solution
H_2O IN
(1)
(2)
(3)
My drawings of the three (3) possibilities:
(1)
(2)
(3)

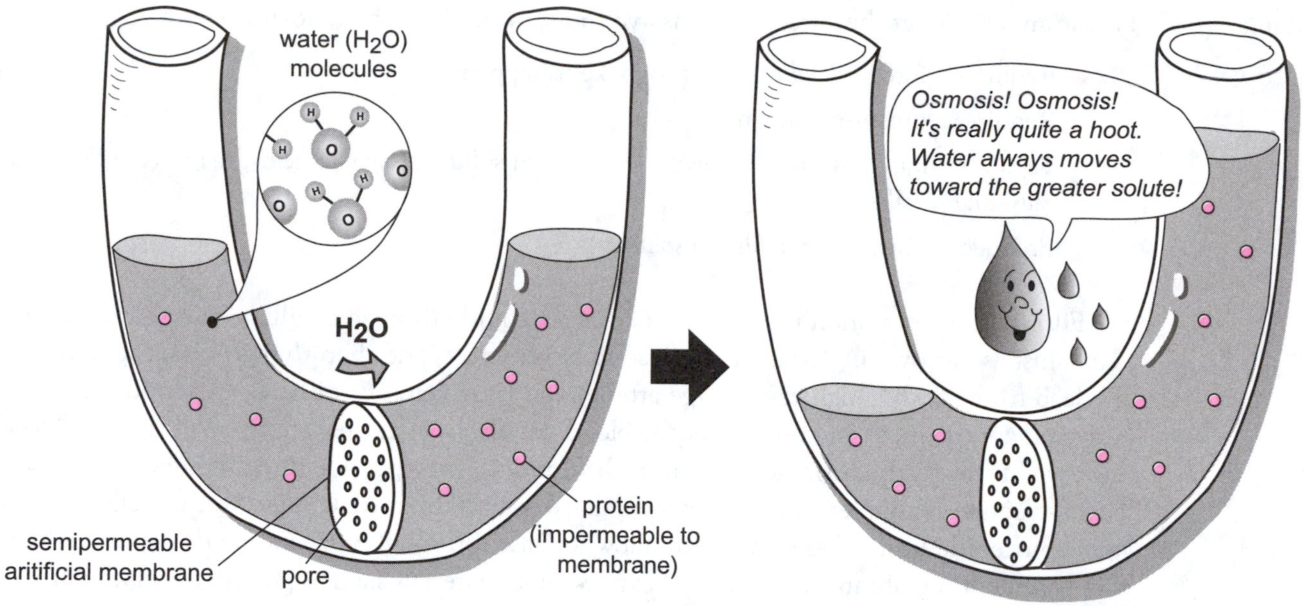

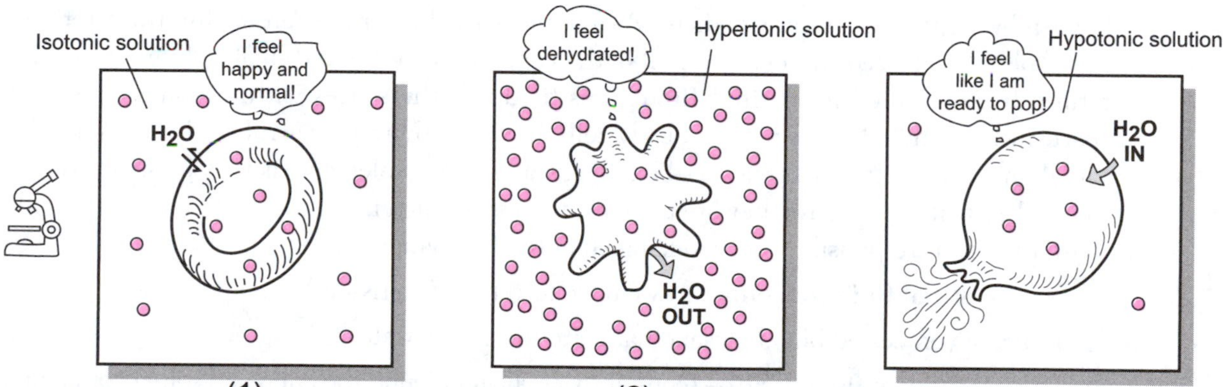

Description

Filtration falls under the category of **passive transport** and has the following features:

- Requires a force (*blood pressure*) to make it happen
- Requires a pressure gradient.
- Separates liquids (*water and small solutes*) from solids (*plasma proteins, red blood cells, white blood cells, and platelets*).
- Produces a filtrate (*final filtered solution*)

Filtration is a common process that occurs constantly through capillary beds that link arteries to veins. Just as narrow, single-lane roads lead to larger streets and then to highways, capillaries eventually lead to the body's highways—large arteries and large veins. Capillaries are found throughout the body and are the most microscopic of the blood vessels. In fact, they are so small that red blood cells must pass through them single-file. Through filtration, they provide direct contact with body cells. Their structure is simple: The wall of a capillary is made of a single, flat layer of cells called **simple squamous epithelium**. These thin cells allow for easy filtration. Surrounding this epithelial layer is a thin **basement membrane**, composed of extracellular materials such as protein fibers that are secreted by epithelial cells.

Almost all tissues have an ample supply of capillary beds—a branching network of capillaries with a higher pressure arteriole end and a lower pressure venule end. Like a one-way street, blood always moves from the arteriole end to the venule end. Because filtration is completely dependent on pressure, most of it occurs nearer the arteriole end. The blood pressure is the force that makes it all happen by creating a pressure gradient. As a result of the higher pressure within the arteriole end, the liquid plasma is forced through the wall of the capillary, so water and small solutes leave the blood to form a **filtrate** called **interstitial fluid**. This fluid fills the spaces within tissues called **interstitial spaces**, which include gaps between body cells as well as those between blood vessels and body cells. Simultaneously, large solids such as blood cells and large molecules such as plasma proteins remain inside the capillary because they are too large to pass through it.

Filtration is responsible for producing all of the following:

- **interstitial fluid** (the only way we have to get fluid to tissues)
- **urine** (processed blood plasma that contains mostly water)
- **cerebrospinal fluid (CSF)**—(constantly circulates around the brain and spinal cord and has many functions: provides a protective cushion, distributes nutrients, and removes waste products)

There are three different types of capillaries (see p. 302). Some are more permeable than others, which determines how easily filtration occurs. For example, specialized capillaries in the kidneys called **glomeruli** (sing. *glomerulus*) (see p. 422) are highly permeable to allow for easy filtration of the plasma to form urine, and capillaries in the brain are the least permeable in the body to protect the brain's precious neurons from being damaged by toxic substances that may have entered the bloodstream.

Analogy

The process of filtration is the same, generally, as a coffeemaker making coffee. Compared to filtration through a capillary, the **force** is **gravity** instead of **blood pressure.** The **coffee filter** is like the **simple squamous epithelium**. The **coffee grounds** are like the **solid** materials in the blood (**red blood cells, white blood cells, plasma proteins**, etc.). The **water** is like the **plasma**. The **coffee solution** is called the *filtrate*, which is like the **interstitial fluid**.

Capillary network
Arteriole end
Venule end
Artery
Vein
Black arrows indicate the direction of blood flow.
Coffee filter
Coffee grounds
Water
Coffemaker (cutaway view)
Filtrate
A coffeemaker uses the same filtration process that occurs in your body
Basement membrane
Color the simple squamous epithelium and the coffee filter the same color, as they are analogous structures.
Simple squamous epithelium
White arrows indicate filtration.
Interstitial fluid (filtrate)
Small solute particles in filtrate
Body cell
Red blood cell in lumen
Nucleus of body cell
Interstitial space
Interstitial space

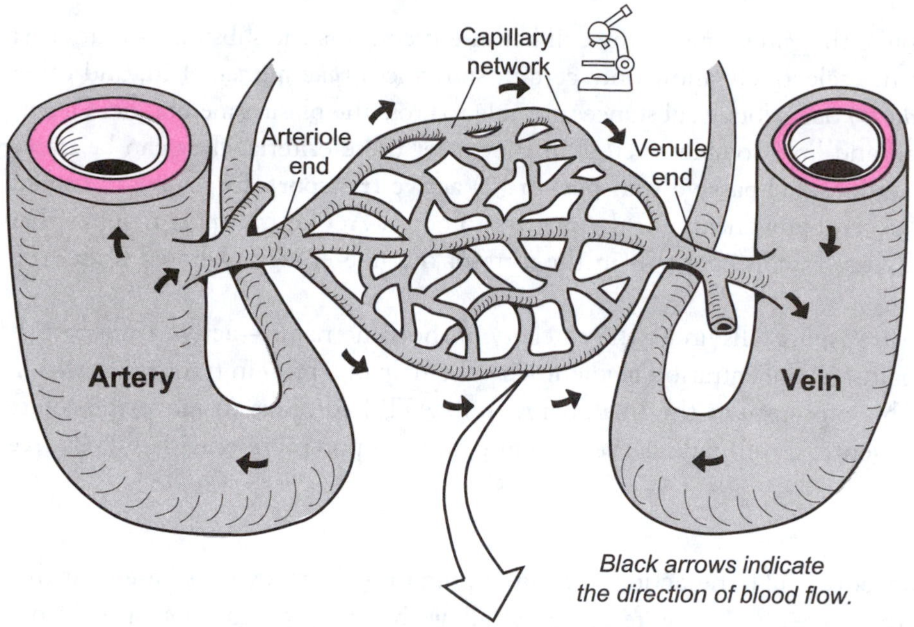

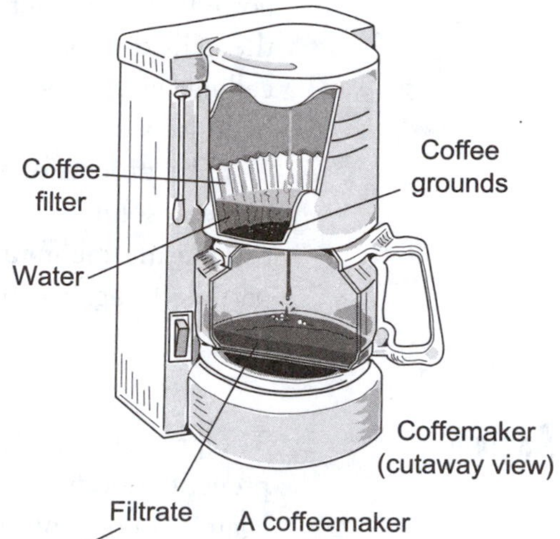

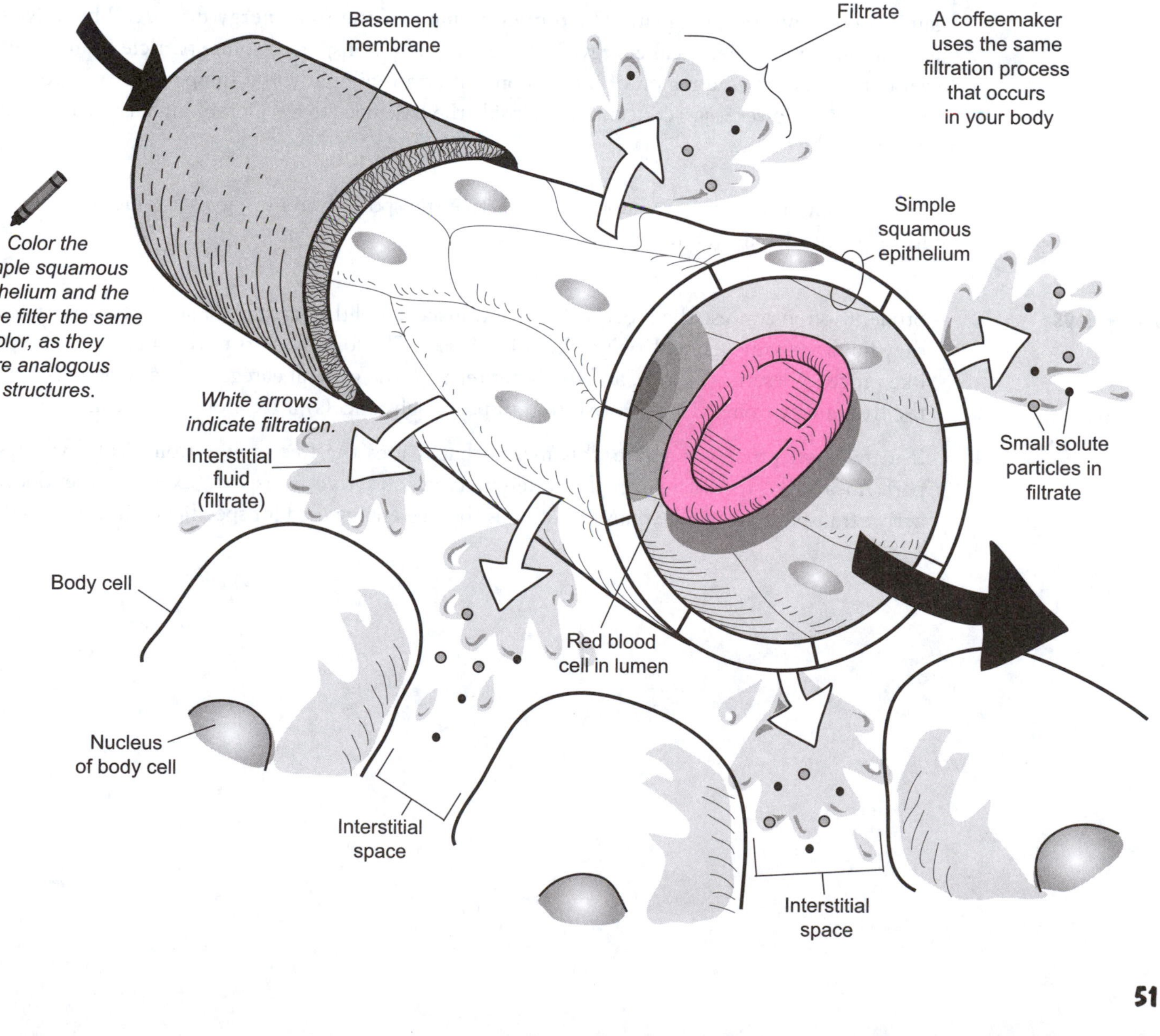

Description

The plasma membrane is like the gatekeeper of the cell because it controls the substances that enter and exit. Plasma membranes selectively allow only certain substances (*like nutrients*) in, and other substances (*like waste products*) out. Not all substances are able to cross the plasma membrane. Generally, they have to be small and lipid-soluble to cross. If they meet these criteria, they can be transported by two different methods: (1) **passive transport** or (2) **active transport**. In passive transport, the cells do not have to expend any energy for the process to occur. Active transport requires that cellular energy be used. Typically, this energy is in the form of the free energy liberated from ATP hydrolysis.

The process of **active transport** falls under the category of the same name—**active transport**. It moves a solute particle against a concentration gradient with the help of a protein transporter within the plasma membrane. This protein uses the free energy from ATP hydrolysis to move the solute particle(s) against the gradient. An example is the sodium-potassium pump present in all cells (see p. 58).

Analogy

The process of **active transport** is like the action of a **sump pump** in a homeowner's basement that prevents flooding inside the home. Both require energy to allow the pump to transport a substance against a concentration gradient. The **protein pump** uses the **free energy** from ATP hydrolysis and the **sump pump** uses **electrical energy**. The protein pump transports a solute particle from a region of lower concentration to a region of higher concentration. Similarly, the sump pump transports *water* from a region of lower *water* concentration inside the basement to a region of higher *water* concentration outside the home.

Note: the limitation of this analogy is that active transport always involves a solute particle rather than a solvent such as water.

Study Tips

- Students often confuse the process of **active transport** and the category of **active transport** because they have the same name. Don't make this mistake. The former refers to the specific "sump pump-like process" described above, and the latter refers to the general category for any process that uses the free energy from ATP hydrolysis to transport a substance(s) into or out of the cell.
- If a transport process is active, this means that it uses the free energy from ATP hydrolysis to perform a task. To distinguish the different types of active transport (exocytosis, endocytosis, active transport, etc.), ask yourself: "What is the free energy used for specifically in each case?"

KEY
(ICF) = intracellular fluid
(ECF) = extracellular fluid

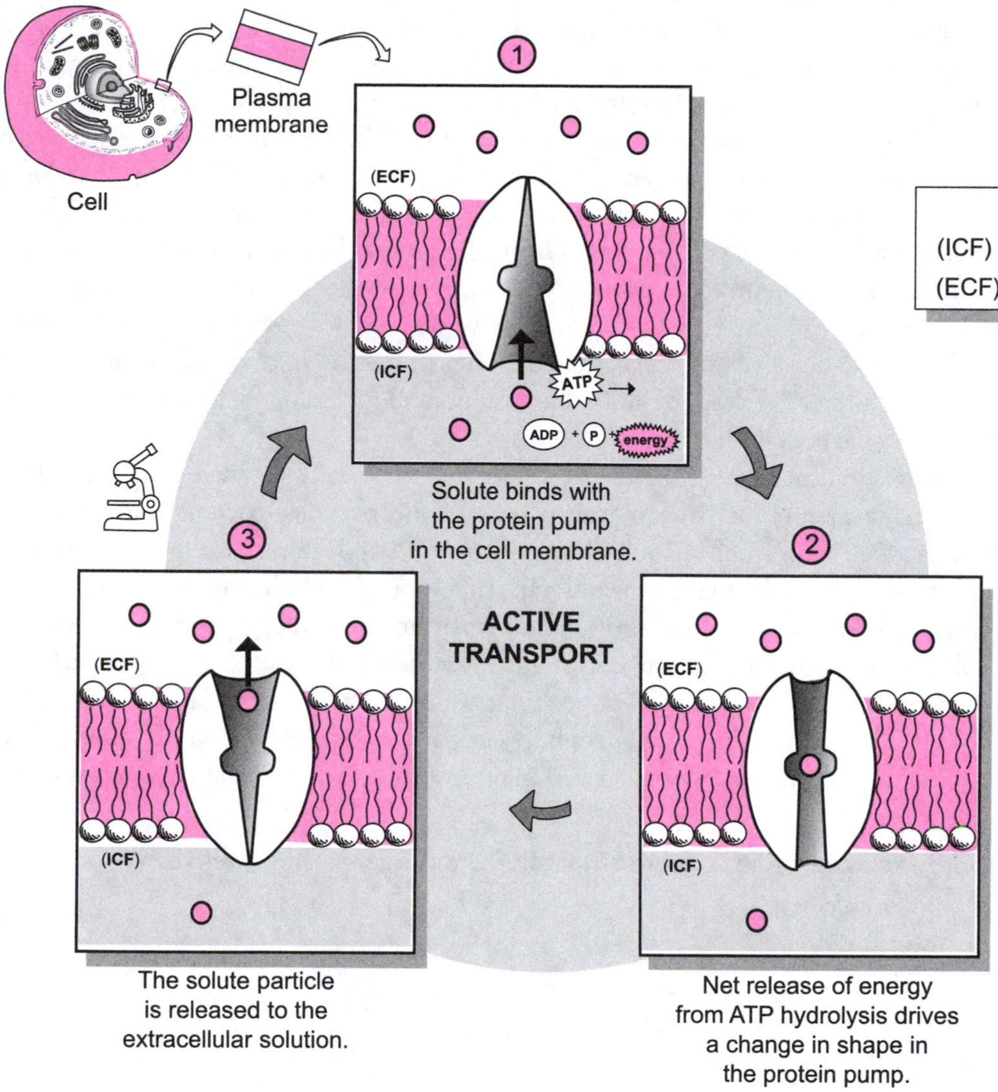

Color the protein pump.

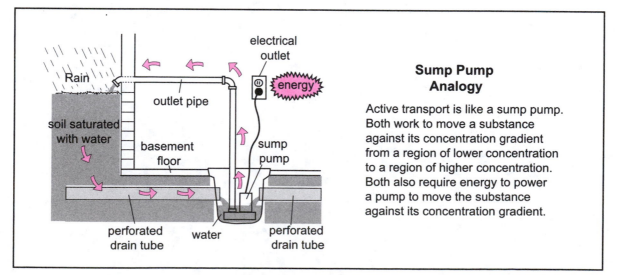

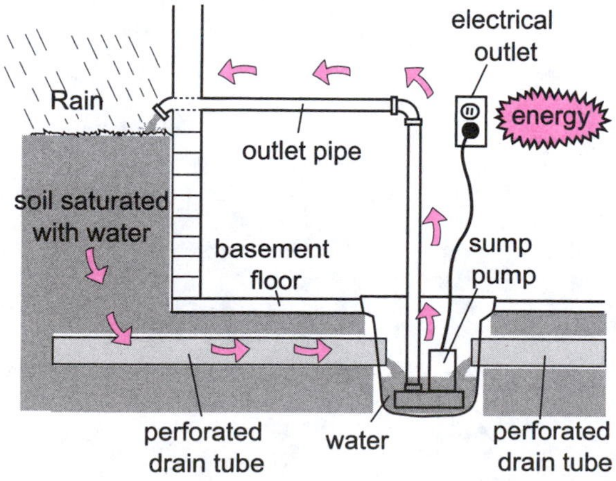

Sump Pump Analogy

Active transport is like a sump pump. Both work to move a substance against its concentration gradient from a region of lower concentration to a region of higher concentration. Both also require energy to power a pump to move the substance against its concentration gradient.

Description

The plasma membrane is like the gatekeeper of the cell because it controls which substances enter and exit. Plasma membranes selectively allow only certain substances (*like nutrients*) in and other substances (*like waste products*) out. Not all substances are able to cross the plasma membrane. Generally, they have to be small and lipid-soluble to cross. If they meet these criteria, they can be transported by two different methods: (1) **passive transport** or (2) **active transport**. In passive transport, the cell doesn't have to transport any energy for the process to occur. Active transport requires cellular energy be used. Typically, this energy is in the form of the free energy liberated from ATP hydrolysis.

Exocytosis is a type of active transport. This is the process by which substances made by the cell (e.g., a hormone) are concentrated within a vesicle. Special regulatory proteins associated with the plasma membrane use cellular energy to move the vesicle toward the plasma membrane so the two can merge. Then the contents of the vesicle are released to the extraceullar solution. This is how cells within your pancreas release the hormone insulin, and for example, how neurons release neurotransmitters such as acetylcholine.

Endocytosis is another type of active transport that is actually the reverse of exocytosis. In this process the plasma membrane invaginates to trap a solid particle (*bacteria, virus*) or some liquid (*extracellular solution*) in a small, pouch-like structure. This pouch uses cellular energy to gradually close, pinch itself off from the plasma membrane, and become a vesicle within the cell.

There are two types of endocytosis: (1) **pinocytosis** and (2) **phagocytosis**. If the material in the pouch is a liquid solution such as the extracellular solution, it is called **pinocytosis**. Cells use this process to obtain nutrients from the extracellular solution. If the material is a solid object like a bacterium, it is called **phagocytosis**. Macrophages are a type of white blood cell that attacks foreign pathogens. Because they normally are engaged in phagocytosis, they are called *phagocytic* cells.

Study Tip

To distinguish exocytosis from endocytosis, recall the meaning of these prefixes:

exo = *outer; outside*

endo = *inner; inside*

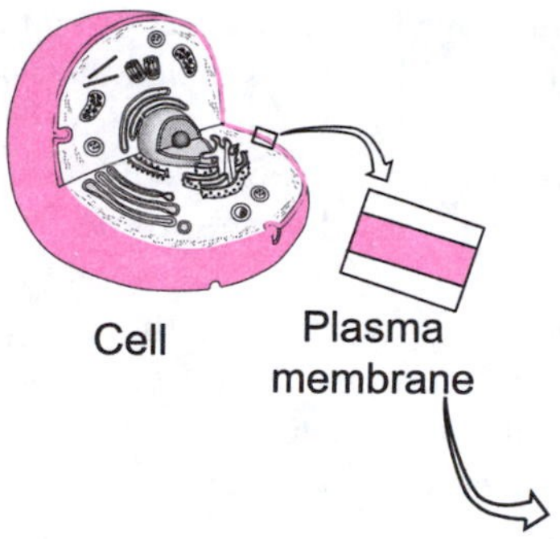
Cell
Plasma membrane

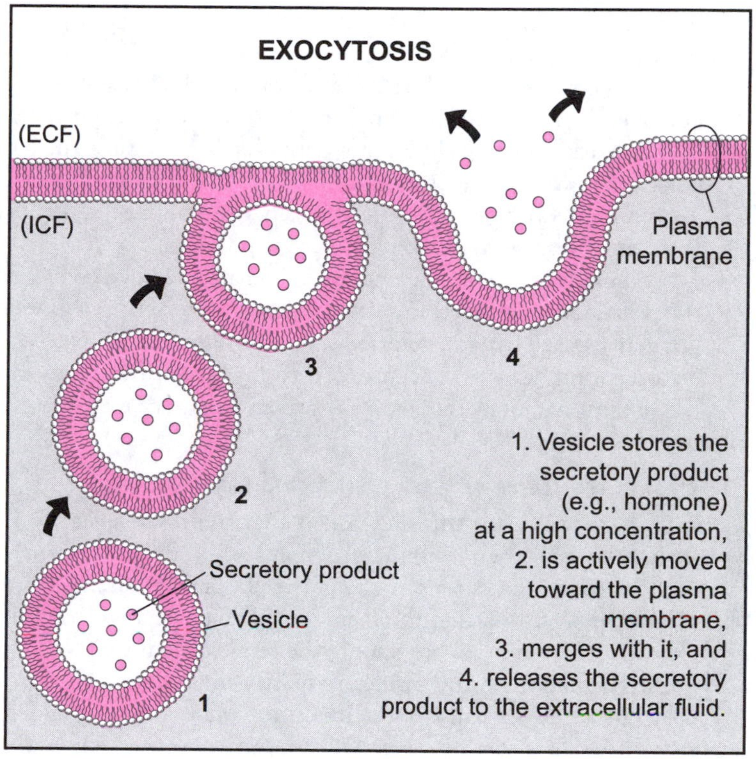
EXOCYTOSIS
(ECF)
(ICF)
Plasma membrane
Secretory product
Vesicle
1
2
3
4
1. Vesicle stores the secretory product (e.g., hormone) at a high concentration,
2. is actively moved toward the plasma membrane,
3. merges with it, and
4. releases the secretory product to the extracellular fluid.

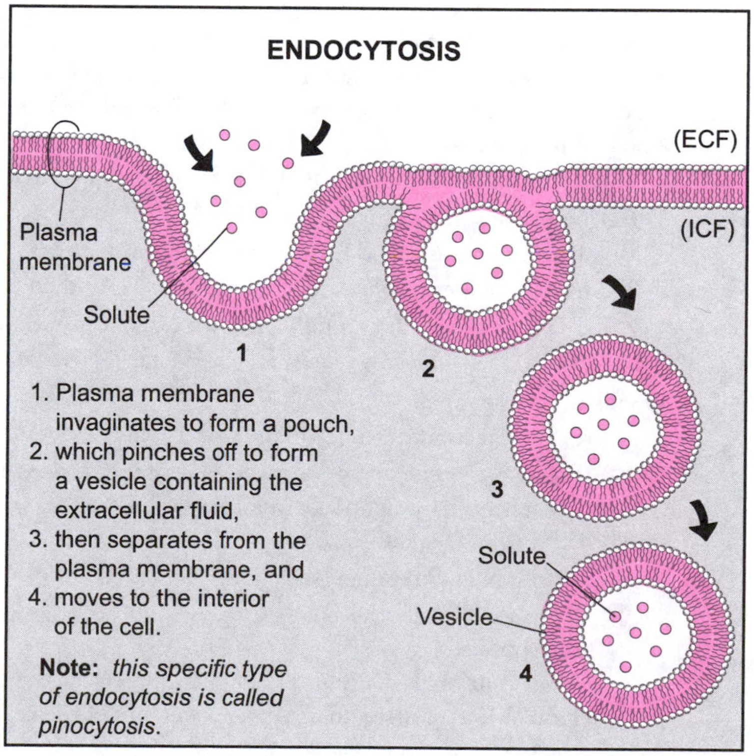
ENDOCYTOSIS
(ECF)
(ICF)
Plasma membrane
Solute
Solute
Vesicle
1
2
3
4
1. Plasma membrane invaginates to form a pouch,
2. which pinches off to form a vesicle containing the extracellular fluid,
3. then separates from the plasma membrane, and
4. moves to the interior of the cell.
Note: this specific type of endocytosis is called pinocytosis.

KEY

(ICF) = intracellular fluid

(ECF) = extracellular fluid

Description

A **membrane potential** (MP) is a separation of positive (+) and negative (−) charges across a **plasma membrane** that results in an *electrical potential* or a *voltage*. This voltage is measured in units called *millivolts* (mV)—much smaller than that used by a battery. Typically, the outside of the plasma membrane is positively charged and the inside is negatively charged. Most human body cells have an MP, but the exact value is different for different cell types and ranges between –5 mV and –100 mV. The negative sign means that the inside of the membrane is negative relative to the outside.

Let's compare the MP to a 1.5 V. battery, which also has an electrical potential because of its positive (+) pole at one end and negative (−) pole at the other. The battery has **potential energy** because of its *potential* to do work.

Potential energy is like water behind a dam. Because the *position* of the water level is higher on one side of the dam and lower on the other, it also has the *potential* to do work. When the floodgate opens, this potential energy gets converted into kinetic energy or energy in motion. As the water flows through the gate, it can turn a turbine that generates electricity—hydroelectric power.

Similarly, the potential energy in the battery can be converted into a flow of electricity to power a flashlight. For the **plasma membrane**, this potential energy is stored—not as water behind a dam but as a cation (+) concentration gradient. Based on the rule for simple diffusion, ions passively flow down their concentration gradients from regions of higher ion concentration to regions of lower ion concentration. This cation concentration gradient is the potential energy source for cells to do work. For example, in a neuron this gradient is used to generate a nerve impulse.

What is the source of these positive and negative charges?

The short answer to this question is that the *extracellular fluid* (ECF) and *intracellular fluid* (ICF) are electrically neutral because they contain equal numbers of *cations* (+ ions) and *anions* (− ions). The only place this isn't true is at the inner and outer surfaces of the **plasma membrane**. Here, there is an accumulation of positive (+) charges on the outside surface of the membrane and an accumulation of negative (−) charges on the inside. The positive charges on the outside are mainly the result of **sodium (Na^+)** ions—the most dominant cations in the **ECF**. The negative charges on the inside are mostly the result of negatively charged **proteins** that are too large to leave the cell. But it's also because of the loss of the most dominant cations in the ICF—**potassium (K^+)** ions.

The real situation is actually more complex (surprise, surprise!). Let's summarize the key factors that contribute to the MP:

- **Ion concentration gradients for Na^+ and K^+ across the plasma membrane**
  - **Na^+** has a greater concentration in the ECF than the ICF, so it could diffuse *into* the cell.
  - **K^+** has a greater concentration in the ICF than the ECF, so it could diffuse *out* of the cell.
  - Na^+ and K^+ can pass through gated-channels only in the membranes that are usually closed but can be stimulated to open.
  - These ion concentration gradients are a source of potential energy to do work.
- **Differing permeabilities for Na^+ and K^+**
  - Membrane proteins called *leakage channels* allow both Na^+ and K^+ to diffuse across the membrane.
  - Plasma membranes are *much* more permeable to K^+ than Na^+.

    (1) K^+ leaves the cell as it diffuses down its concentration gradient.
    The result? This loss of positive charge makes the interior of the cell more negative (see illustration).
- **Opposite charges attract**
  - As previously mentioned, as K^+ diffuses out of the cell, this loss of positive charge makes the interior more negative.

    (2) Because opposite charges attract, some of the K^+ ions are drawn back into the negatively charged interior.
    The result? This relatively small gain in positive charge makes the interior less negative than it might be otherwise. (see illustration)
- **Action of the Na^+-K^+ pump** (see p. 58)

    (3) With each cycle, the **Na^+–K^+ pump** actively transports 3 Na^+ ions out for every 2 K^+ ions it brings into the cell.
    The result? This maintains the MP by preserving the charge imbalance across the membrane.
    How? More positive charges are moved out of the cell (+3), and fewer positive charges are brought into the cell (+2) (*see illustration*).

Study Tip

To distinguish *cations* (+ ions) from *anions* (− ions) remember this phrase: "I ***positively*** hate **cats**."

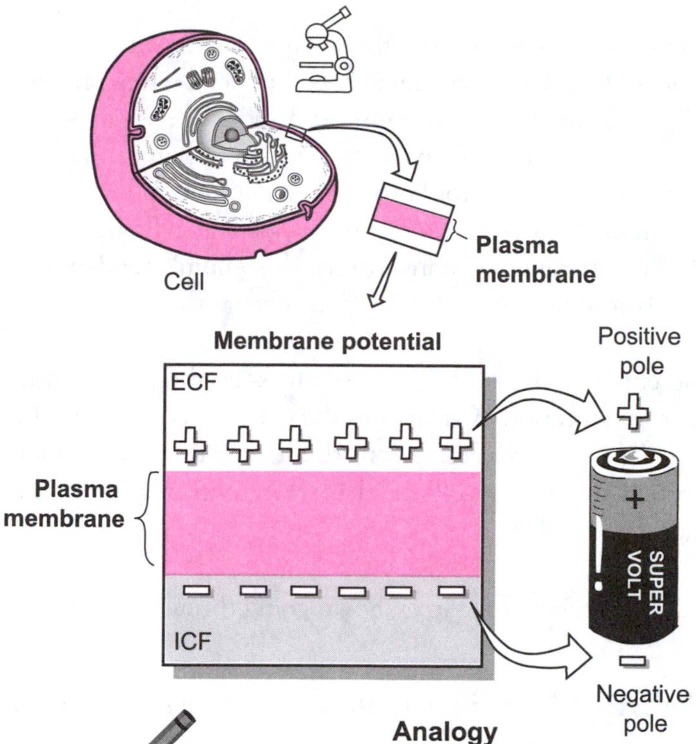

POTENTIAL ENERGY...

...is like water behind a dam. It is ready to do work.

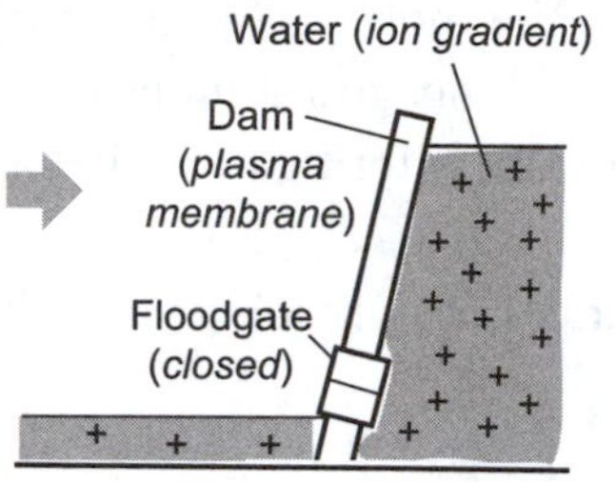

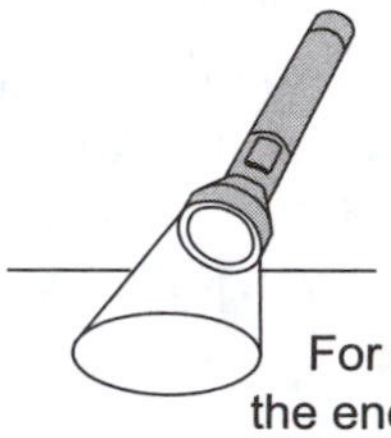

For a battery, the energy can be used to generate electricity for a flashlight

ENERGY CONVERSION

For a **plasma membrane**, the energy can be used to generate a **nerve impulse**

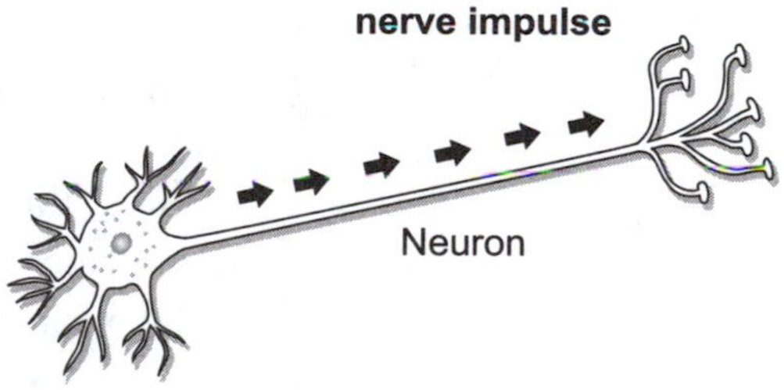

Color the positive (+) and negative (–) signs different colors.

Analogy

Both a plasma membrane and a battery have a positive (+) end and a negative (-) end. By separating positive and negative charges, both are able to store **potential energy**.

Where do the positive and negative charges come from?

Color the K^+ ions.

KEY	
Na^+	= Sodium
K^+	= Potassium

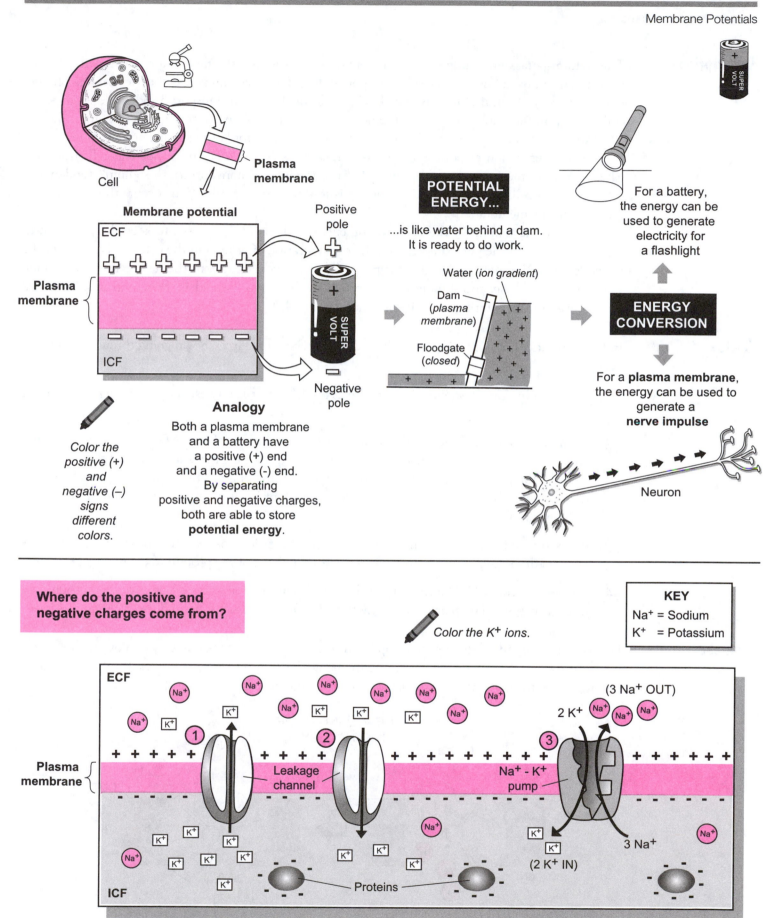

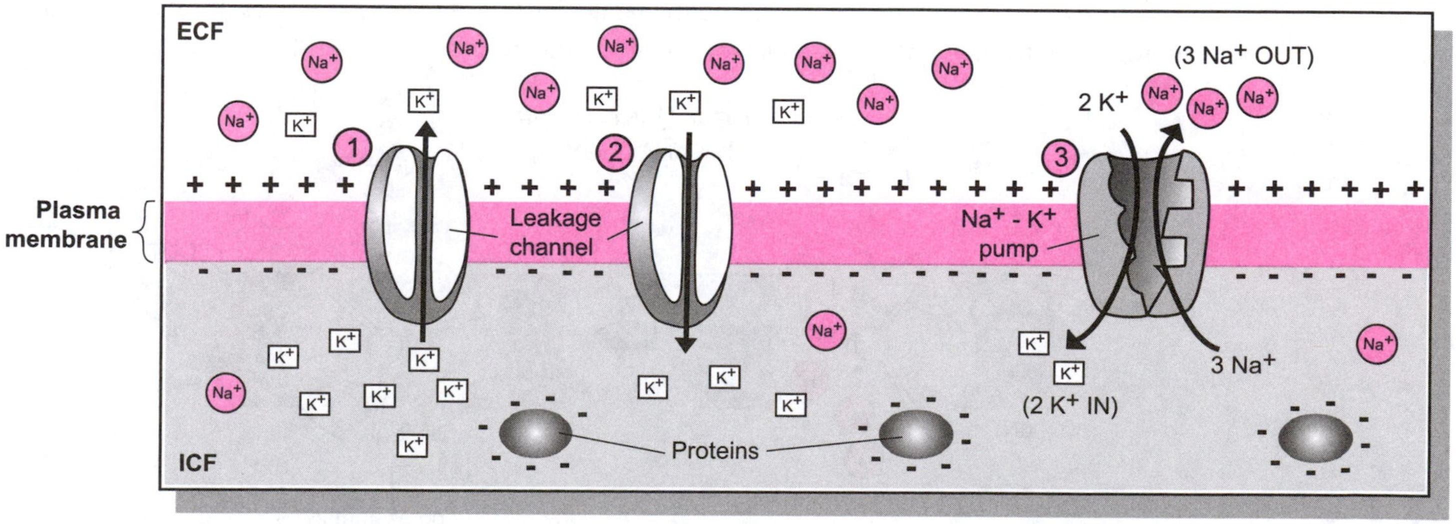

Description

The **sodium-potassium pump** is a protein found in the plasma membrane of all animal cells. It functions as an ion exchanger to actively transport sodium for potassium across the membrane. The net release of energy from the hydrolysis of 1 ATP molecule is used to transport 3 sodium ions *out* of the cell and 2 potassium ions *into* the cell. Since a single cell has many of these protein pumps, the net effect of all of them working together is to maintain a gradient for both sodium and potassium.

The concentration of potassium ions is normally 10–20 times greater inside the cells than out. The reverse is true for sodium. Because of this difference in concentration, sodium's tendency is to diffuse into cells and potassium's tendency is to diffuse out of cells. This occurs through leakage channels in the membrane.

A significant amount of the cell's resting energy—about 33%—is used to keep the pump working. The activity of the pump depends on the concentration of sodium in the cytosol. The greater the concentration, the more active the pump; the lesser the concentration, the less active. If the pump were to stop working, the sodium concentration in the cytosol would increase and nothing could prevent water from entering the cell via osmosis (see p. 48).

Cycle

In the illustration on the facing page, the sodium-potassium pump is shown going through a four-step cycle:

1. **Sodium binding:** 3 sodium ions in the cytosol bind at their respective binding sites on the pump.
2. **Shape change:** The binding of sodium causes the hydrolysis of one ATP molecule into ADP, a phosphate group and a net release of free energy. The free energy is used to bind the phosphate group to the pump, which induces a shape change in the pump. This shape change both enables the release of the 3 sodium ions and makes it easier for 2 potassium ions to bind from the extracellular solution.
3. **Potassium binding:** The binding of the 2 potassium ions causes the phosphate group to be released from the pump, which, in turn, causes the shape of the protein to change again.
4. **Potassium released:** The restoring of the pump to its original shape releases the 2 potassium ions into the cytosol. The pump is now ready to bind 3 new sodium ions, and the cycle repeats itself.

Analogy

Functionally, the sodium-potassium pump is like a revolving door. It takes energy to move the revolving door just like it takes ATP energy to drive the pump. Each cycle of the pump result in 3 sodium ions moving *out* of the cell and 2 potassium ions moving *into* the cell.

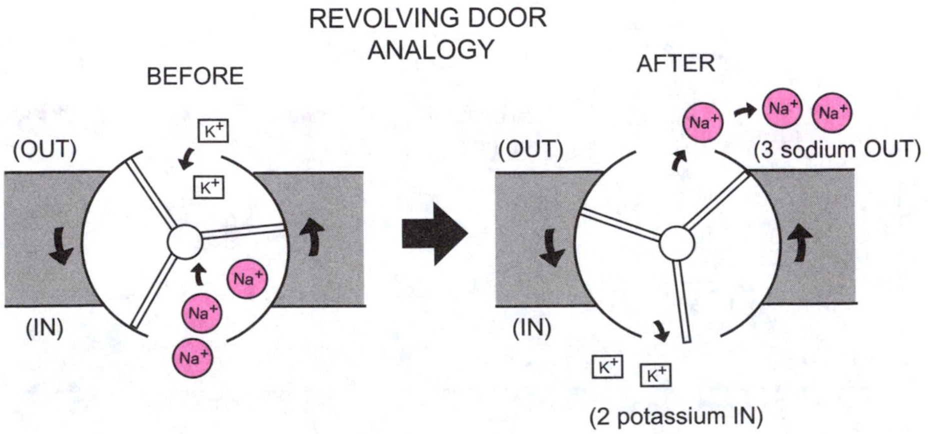

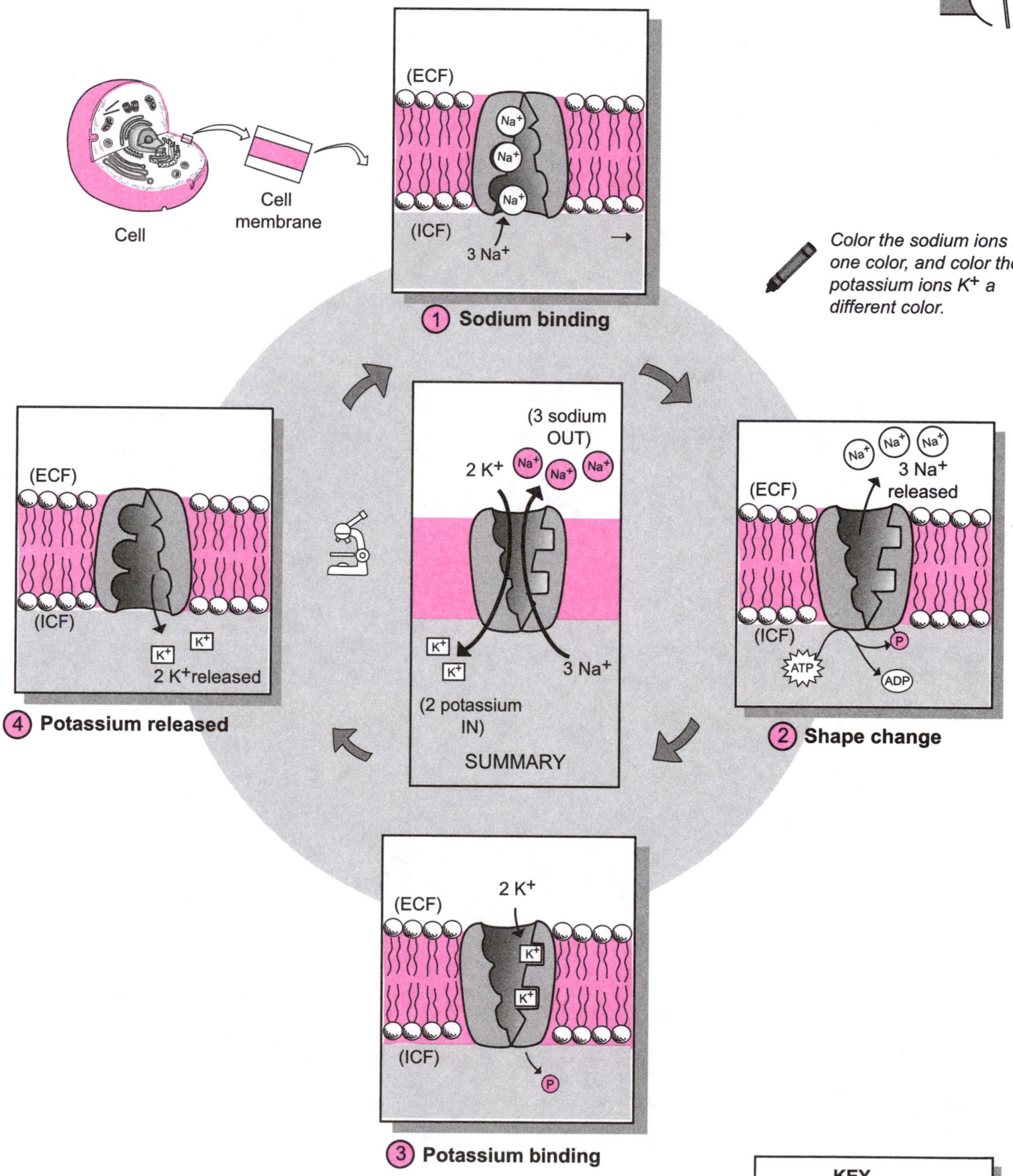

Color the sodium ions Na^+ one color, and color the potassium ions K^+ a different color.

KEY
(ICF) = intracellular fluid
(ECF) = extracellular fluid

Notes

Tissues

Description

Tissues are a group of similar cells working together for a common purpose. The four different types of tissues are: (1) **epithelial**, (2) **connective**, (3) **muscular**, and (4) **nervous**. Let's consider only the first tissue type—**epithelial tissues.** There are many different kinds of epithelial tissues, and they are often named after their cell shape. They are found lining internal body cavities and passageways and covering body surfaces. Epithelial tissues are composed mostly of cells that rest on a thin **basement membrane**. In physiology, epithelial cells are an important theme because substances have to pass through them before they can enter the blood. Every organ system contains epithelial tissues. For example, they line (*are inside of*) the following structures: blood vessels, digestive organs, the urinary bladder, and the microscopic air sacs in the lungs. Moreover, they cover the surface of the skin. In short, they are everywhere in the body.

Each cell in an epithelial tissue has three different surfaces that serve as reference points on the cell. Each surface represents a different side of the cell and is composed of the **plasma membrane**. The top of the cell, which faces the lumen or body surface, is called the **apical surface**. The sides of the cell are called the **lateral surfaces**. The bottom of the cell that rests on the basement membrane and is closer to the blood is called the **basal surface**. The typical pathway that a substance follows through an epithelial cell is as follows: **apical surface** to **cytosol** to **basal surface** to **basement membrane** to **capillary**. (**Note:** a good way to distinguish the apical surface from the basal surface is to remember the alliteration—*Basal*, *Basement*, *Blood*. These three structures are in close proximity to each other).

Intercellular Junctions

The lateral surfaces of adjacent epithelial cells are connected one to another by different kinds of junctions. The four major types of junctions are: (1) **tight junctions**, (2) **adhering junctions**, (3) **desmosomes**, and (4) **gap junctions**. Each is summarized below.

(1) Tight junctions loop around the whole cell and are located near the apical surface. They look like lines of rivets that form a seal between adjacent cells by stapling their plasma membranes together. Each rivet-like structure is actually a **transmembrane protein**. Tight junctions prevent substances from passing between epithelial cells. This forces substances to move through cells instead of squeezing between them. For example, tight junctions in the digestive tract keep digestive enzymes inside the intestine and prevent it from entering the blood.

(2) Adhering junctions function like seam welds. They are typically located below tight junctions so they are also near the apical surface. They look like a belt that wraps around the whole cell and contain band-like proteins called **plaques**. Running along the length of the plaques are thin, contractile proteins called **microfilaments**. Together, the microfilaments and plaques form a structure called an **adhesion belt**. Adhering junctions prevent lateral surfaces from separating while providing a space for substances to enter. For example, a substance that already passed through the apical surface can enter this space as it continues on to the basal surface.

(3) Desmosomes are like spot welds. They consist of disk-like proteins called **plaques** that are held together and spaced apart by **linker proteins**. Instead of surrounding the entire cell, desmosomes are found at discrete points on the cell. They function as structural reinforcements at specific stress points along the cell. **Intermediate filaments** are long, strong cable-like proteins that extend into the cytosol and connect the plaque of one desmosome to the plaque of another desmosome on the opposite side of the cell. This helps maintain the structural integrity of the cell and the tissue. Desmosomes are common in the cells within the epidermis of the skin.

Half-desmosomes or **hemidesmosomes** are located on the basal surface, where they serve to anchor this surface to the basement membrane.

(4) Gap junctions allow for small substances to be transported between adjacent cells. Each gap junction consists of a small group of tubular structures. Each tubular structure is composed of a cluster of 6 proteins called **connexins** and has a fluid-filled **channel** running down its center. This channel allows small substances such as ions and glucose to travel between adjacent cells. In ciliated epithelial cells, this may help to coordinate the movement of the cilia. Gap junctions are also found in other tissues such as cardiac and smooth muscle.

Nucleus
Basement membrane
Epithelial tissue
Black arrows show two different pathways a substance may follow through an epithelial cell.
Substance
Apical surface (lumen side)
Metal plates
Rivet
Cell membrane
Transmembrane protein
1 Tight junction
2 Adhering junction
Adhesion belt
Plaque
Microfilament
Tight junction
Desmosome
Lateral surface
Gap junction
Hemidesmosome
Basal surface (blood side)
Nucleus
Nucleus
Basal surface
Basement membrane
Color the capillary red.
Capillary
Linker protein
Plaque
Intermediate filament
Intercellular space
3 Desmosome
Channel between cells
Connexin
Cell membranes
4 Gap junction
Color the basement membrane, apical, lateral, and basal surfaces, using different colors for each.

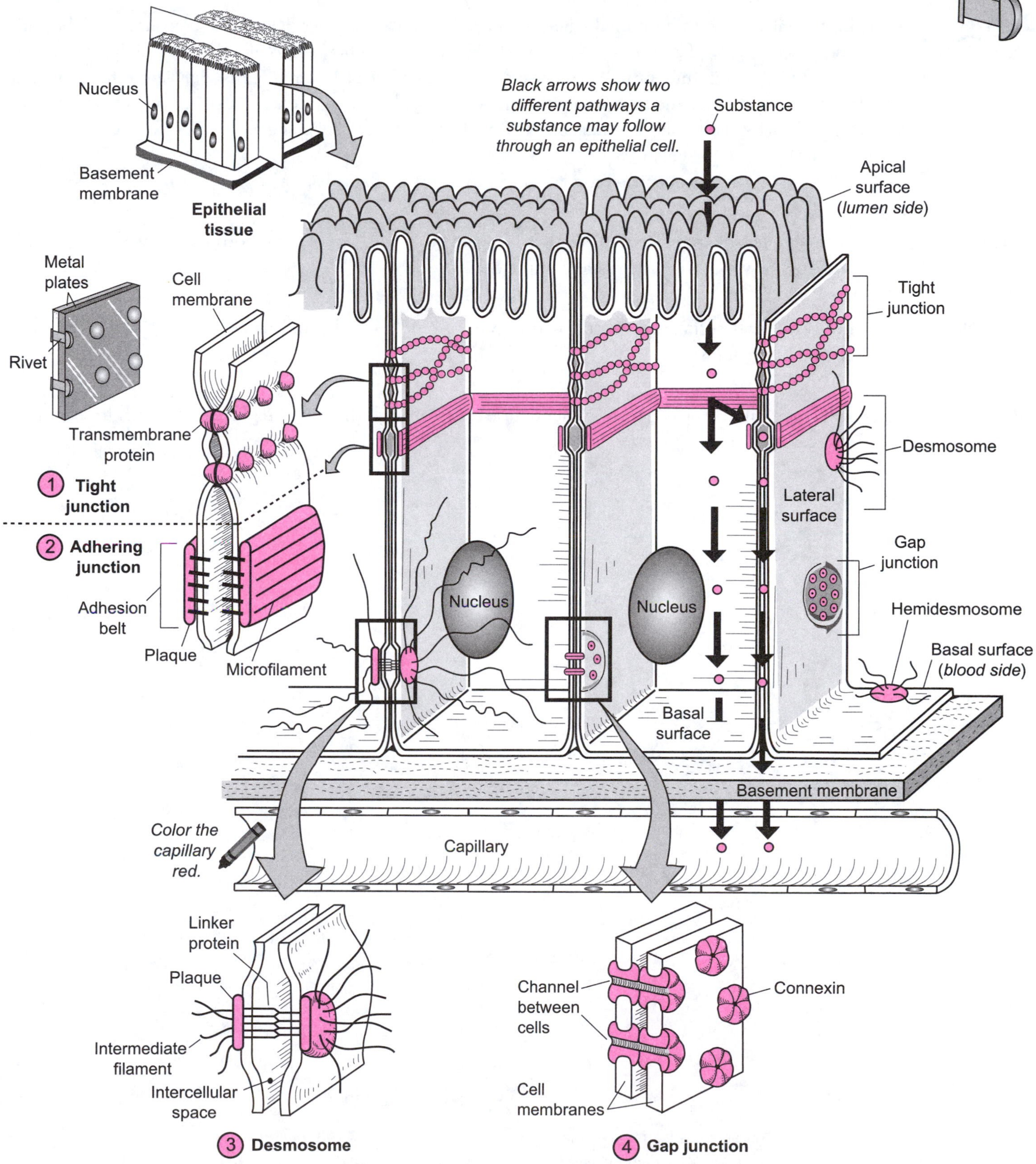

Description

Epithelial tissues line internal cavities and passageways and cover external body surfaces. They are composed mostly of cells that rest on a thin **basement membrane**. No blood vessels are present. One method of classifying epithelial tissues is by the number of layers of cells. *Simple* epithelia have a single layer of cells, and *stratified* epithelia have multiple layers of cells. Classification is also based on the following cell shapes: *squamous* (thin, flat), *cuboidal* (cube-shaped), and *columnar* (column-shaped).

Simple squamous epithelium is a single row of thin, flat cells.

Analogy

Each **simple squamous epithelial cell** is compared to a **fried egg** because both are flat with an irregular border. The yolk is like the nucleus of the cell.

Location

Lines internal surface of ventral body cavities, blood vessels, and heart; parts of kidney tubules; alveoli of the lungs.

Function

Flat shape allows substances to either diffuse easily through the cell or be filtered through it; secretion; reduces friction.

Key to Illustration

1. Individual simple squamous epithelial cell
2. Nucleus of simple squamous epithelial cell
3. Alveoli

Location
Lung
Alveolus
Fun Fact: Each lung contains about 150 million tiny, microscopic air sacs called alveoli.
3-D
Nucleus
Egg yolk
Color the egg yolks yellow.
One simple squamous epithelial cell (superior view)
Each cell looks like a fried egg.
The tissue is like a large group of fried eggs arranged side by side
Basement membrane
One simple squamous epithelial cell (side view)
2-D
Color the epithelial cells the same color.
1
2
3
(approx. 220x)
1. ___
2. ___
3. ___
My drawing of simple squamous epithelium

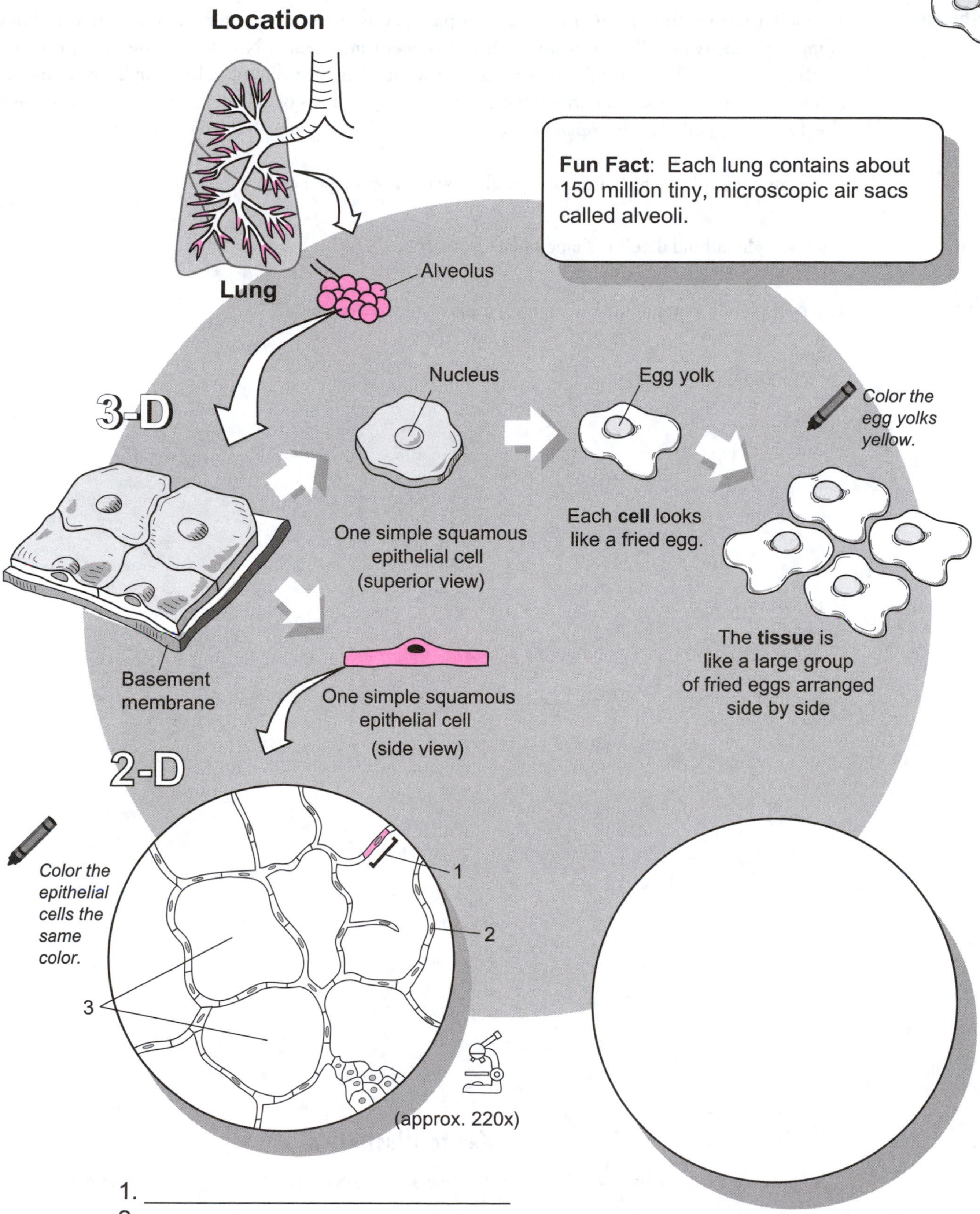

Description

Epithelial tissues line internal cavities and passageways and cover external body surfaces. They are composed mostly of cells that rest on a thin **basement membrane**. No blood vessels are present. One method of classifying epithelial tissues is by the number of layers of cells. *Simple* epithelia have a single layer of cells, and *stratified* epithelia have multiple layers of cells. Classification is also based on the following cell shapes: *squamous* (thin, flat), *cuboidal* (cube-shaped), and *columnar* (column-shaped).

Simple cuboidal epithelium is a single row of cube-shaped cells.

Analogy

Each **simple cuboidal cell** is shaped like an **ice cube**.

Location

Ducts of glands, parts of kidney tubules; follicles of thyroid gland

Function

Secretion; absorption

Key to Illustration

1. Simple cuboidal epithelial cell
2. Nucleus of simple cuboidal epithelial cell
3. Connective tissue

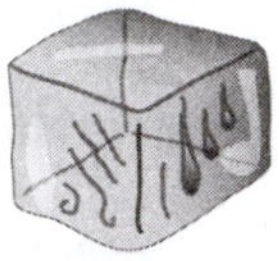

Location

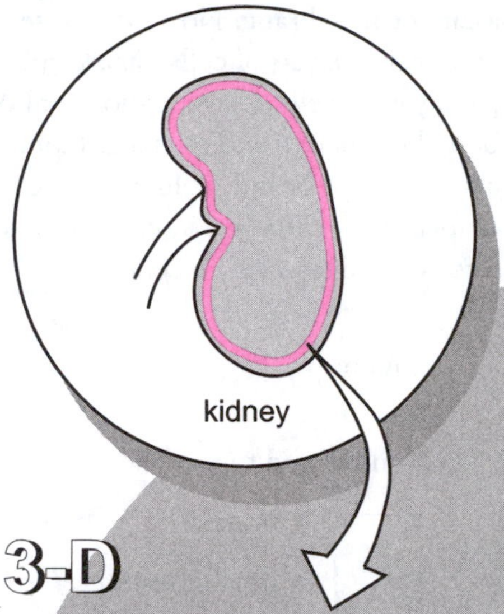

Fun Fact: On average, an adult's kidneys process about 165 liters of filtered blood plasma every day, but only 1% is excreted as urine.

3-D

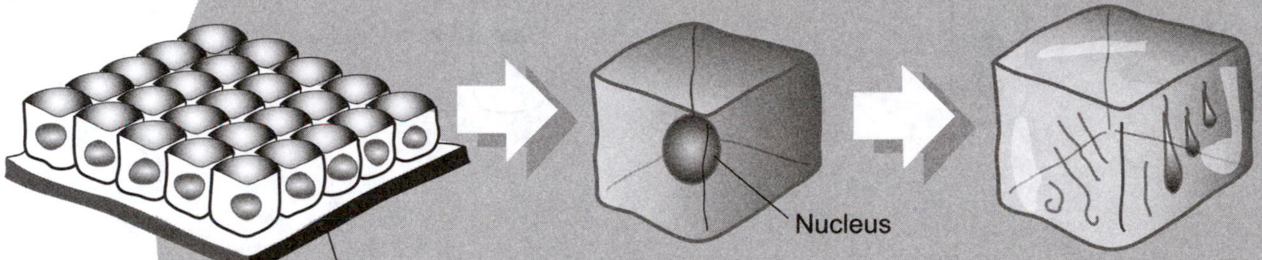

Basement membrane

One simple cuboidal epithelial cell

Each cell is shaped like an ice cube.

2-D

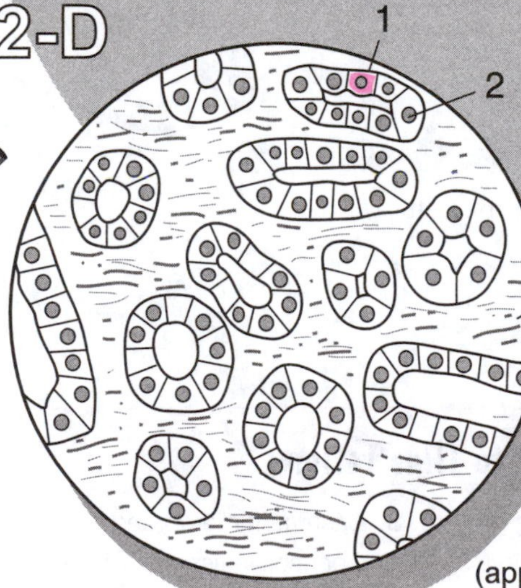

Color the epithelial cells the same color.

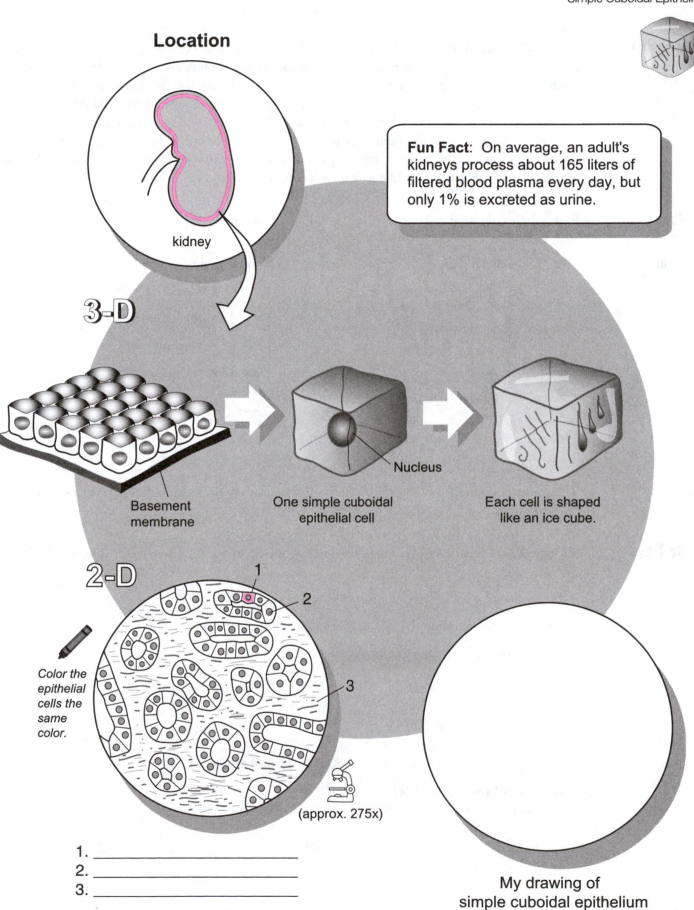

(approx. 275x)

1. ______________________
2. ______________________
3. ______________________

My drawing of simple cuboidal epithelium

Description

Epithelial tissues line internal cavities and passageways and cover external body surfaces. They are composed mostly of cells that rest on a thin **basement membrane**. No blood vessels are present. One method of classifying epithelial tissues is by the number of layers of cells. *Simple* epithelia have a single layer of cells, and *stratified* epithelia have multiple layers of cells. Classification is also based on the following cell shapes: *squamous* (thin, flat), *cuboidal* (cube-shaped), and *columnar* (column-shaped).

Simple columnar epithelium appears as a single layer of tall, column-shaped cells with oblong nuclei. They are of two types: *ciliated* and *non-ciliated*. **Cilia** are numerous folds in the **plasma membrane** that appear as hair-like structures located on the top of each cell.

Analogy

Each simple columnar cell in this tissue looks like a column.

Location

Ciliated type

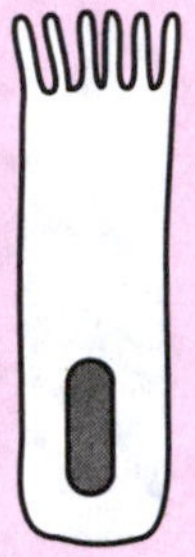

- uterine tubes

Non-ciliated type

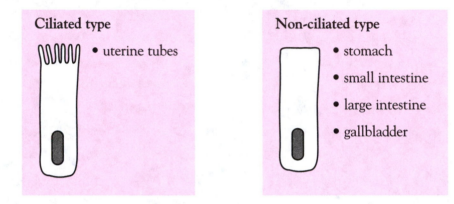

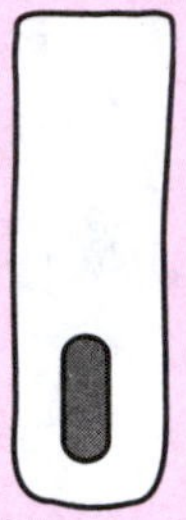

- stomach
- small intestine
- large intestine
- gallbladder

Function

The primary function is absorption; secretion of mucus, enzymes, and other substances; movement of mucus by cilia.

Study Tips

To identify this tissue either under the microscope or from a photograph or diagram, look for the following:

- tall rectangular cells
- oblong-shaped nucleus
- nucleus usually located in the lower half of the cell

Key to Illustration

1. Simple columnar epithelial cell
2. Nucleus
3. Basement membrane
4. Connective tissue

Location

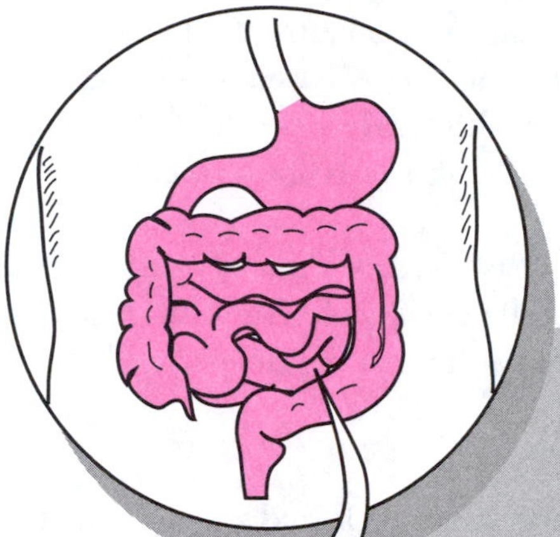

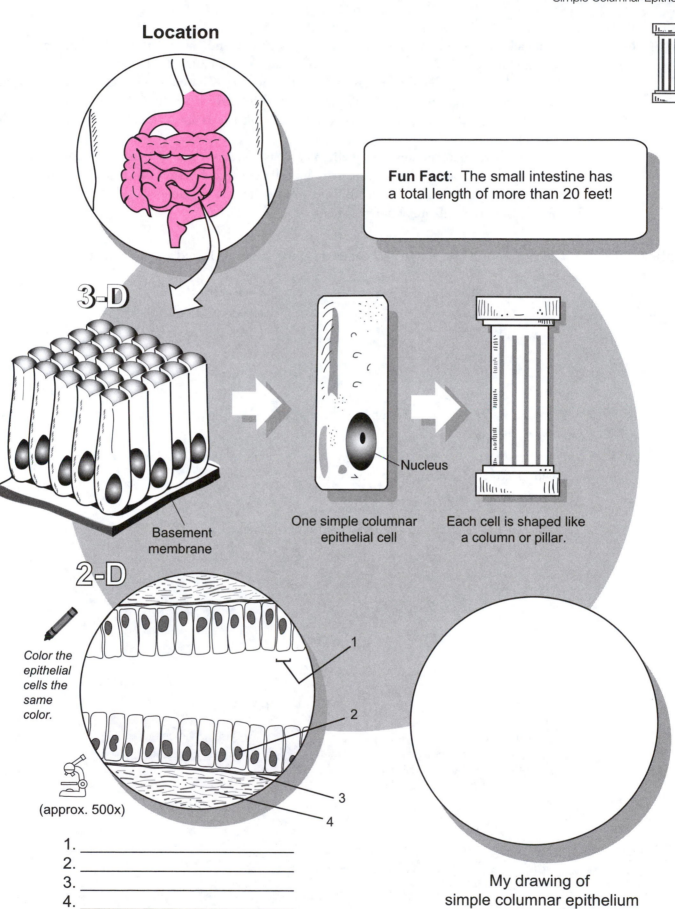

Fun Fact: The small intestine has a total length of more than 20 feet!

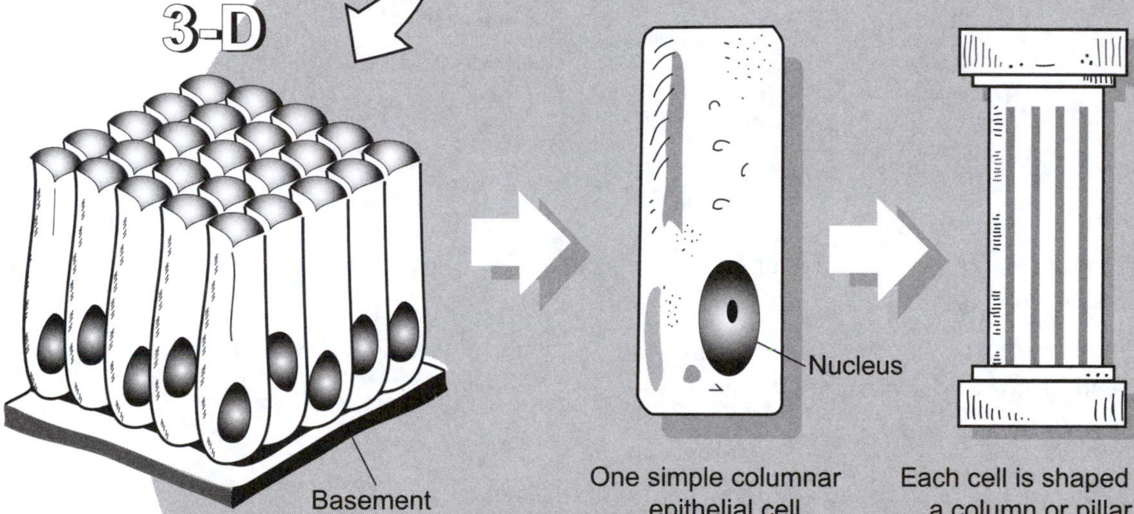

One simple columnar epithelial cell

Each cell is shaped like a column or pillar.

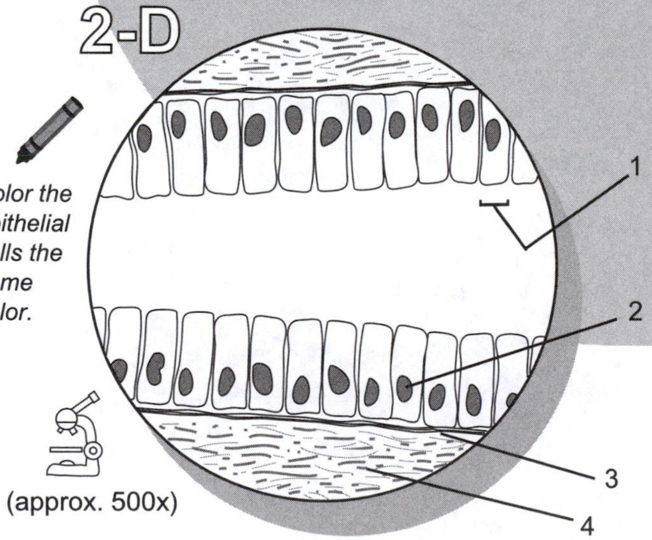

Color the epithelial cells the same color.

(approx. 500x)

1. ______________________
2. ______________________
3. ______________________
4. ______________________

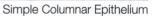

My drawing of simple columnar epithelium

Description

Epithelial tissues line internal cavities and passageways and cover external body surfaces. They are composed mostly of cells that rest on a thin **basement membrane**. No blood vessels are present. One method of classifying epithelial tissues is by the number of layers of cells. *Simple* epithelia have a single layer of cells, and *stratified* epithelia have multiple layers of cells. Classification is also based on the following cell shapes: *squamous* (thin, flat), *cuboidal* (cube-shaped), and *columnar* (column-shaped).

Pseudostratified columnar epithelium consists of a single row of cells. Most cells have a columnar shape, while other, shorter cells may look like cuboidal. The term "pseudostratified" literally means "falsely stratified." In other words, it looks as if it has multiple layers but actually has only one layer, because the cells are of differing heights.

Location

The two types of pseudostratified columnar epithelium are ciliated and non-ciliated.

Ciliated

- nasal cavity
- trachea
- bronchi

Non-ciliated

- ducts of male reproductive tract

Function

Protection; secretion; movement of mucus by **cilia**

Study Tip

Under the microscope at higher magnifications, you can use the following landmarks to distinguish pseudostratified columnar epithelium cells:

- Cells have differing heights.
- Nuclei are not in an organized row, but are more staggered.

Key to Illustration

1. Cilia
2. Nucleus of one pseudostratified ciliated columnar epithelial cell
3. Basement membrane
4. Connective tissue

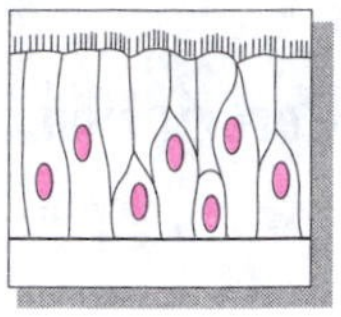

Location

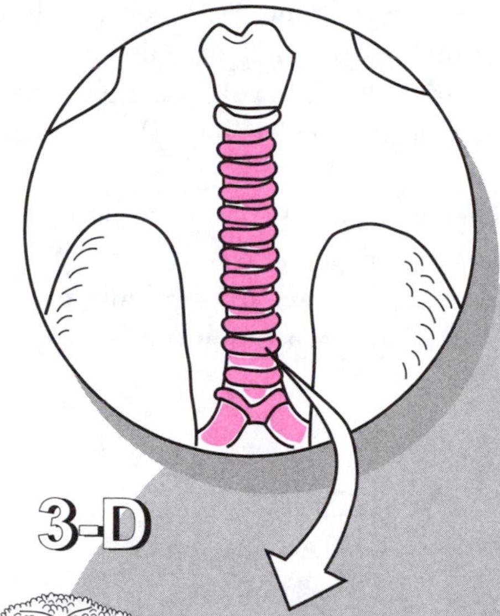

Fun Fact: In the lungs, the cilia on this tissue sweep a layer of mucus that traps dirt particles and moves it toward the throat to be coughed out or swallowed.

3-D

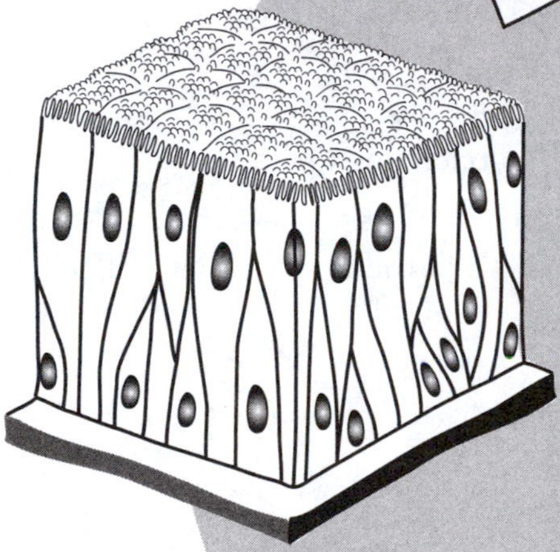

Look at the pattern—

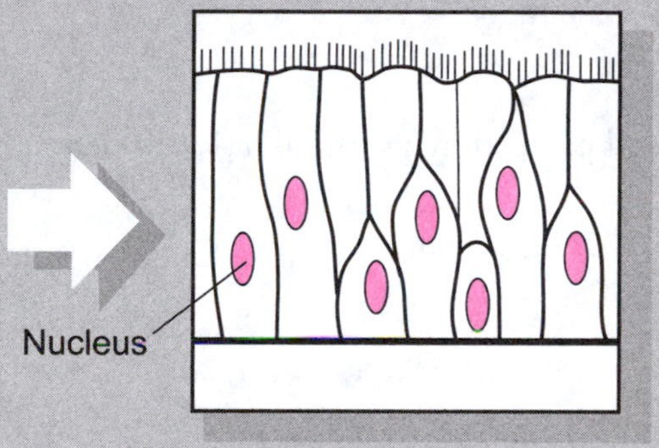

the nuclei are staggered because of the different heights of the cells.

All cells contact the basement membrane, but not all extend to the surface. Only those that reach the surface have cilia.

2-D

Color the epithelial cells the same color.

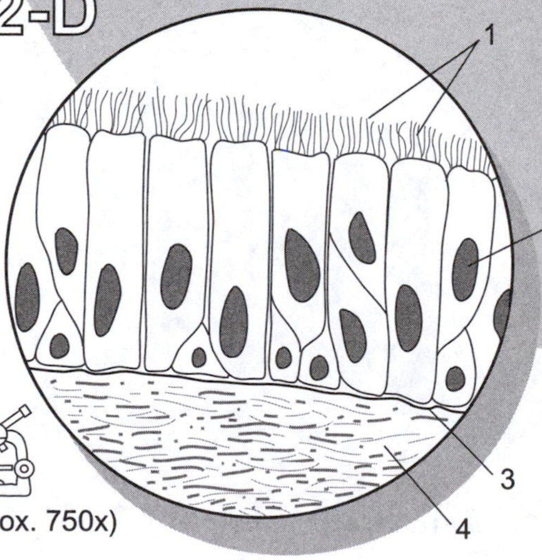

(approx. 750x)

1. ______________________________
2. ______________________________
3. ______________________________
4. ______________________________

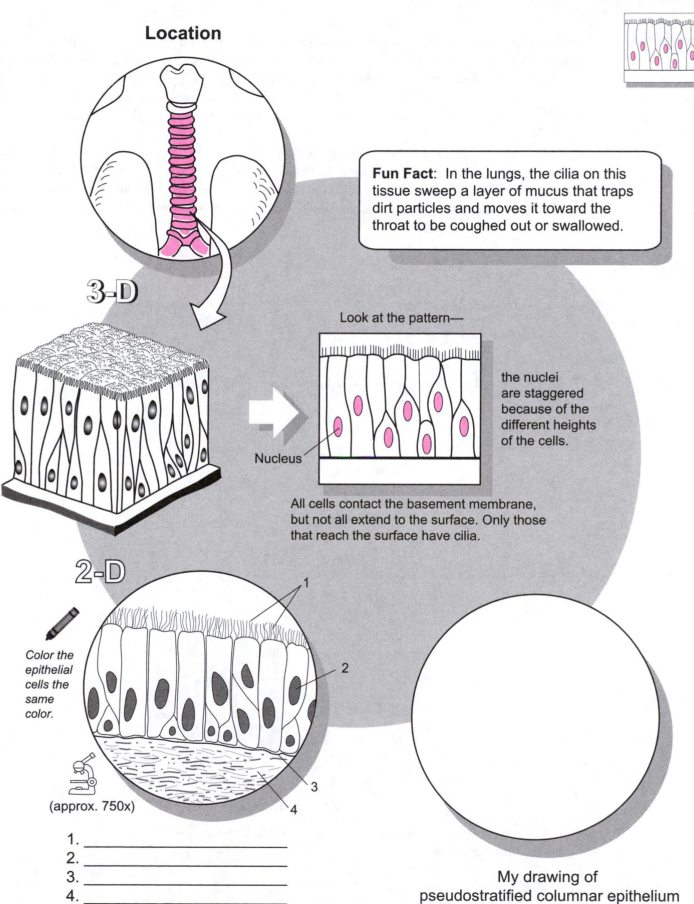

My drawing of
pseudostratified columnar epithelium

Description

Epithelial tissues line internal cavities and passageways and cover external body surfaces. They are composed mostly of cells that rest on a thin **basement membrane**. No blood vessels are present. One method of classifying epithelial tissues is by the number of layers of cells. *Simple* epithelia have a single layer of cells, and *stratified* epithelia have multiple layers of cells. Classification is also based on the following cell shapes: *squamous* (thin, flat), *cuboidal* (cube-shaped), and *columnar* (column-shaped).

Stratified squamous epithelium is of two different types—keratinized and non-keratinized. The bottom layer in either type is composed of cuboidal or columnar cells that are active in cell division. New cells are pushed upward toward the surface. In a process called **keratinization** the new cells in the keratinized type fill with a protein called **keratin**. The result is that the outer surface of this tissue is tough and water-resistant.

Location

Keratinized

- epidermis of skin

Non-keratinized

- lining of mouth, pharynx, esophagus, anus, and vagina

Function

Provides physical protection against abrasion and pathogens to underlying tissues.

Study Tip

Pattern: Transition in cell shape from cuboidal or columnar cells in the lower region to flat cells on the top.

Key to Illustration

1. Stratified squamous epithelium
2. Basement membrane
3. Connective tissue
4. Nuclei

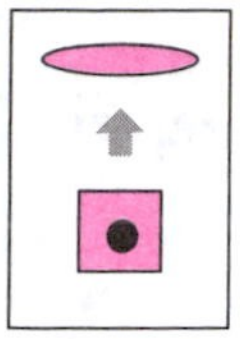

Location

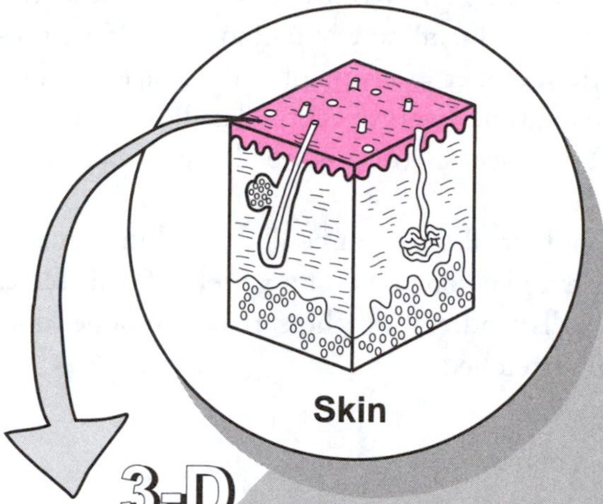

Fun Fact: On average, the epidermis of your skin regenerates about every 27 days.

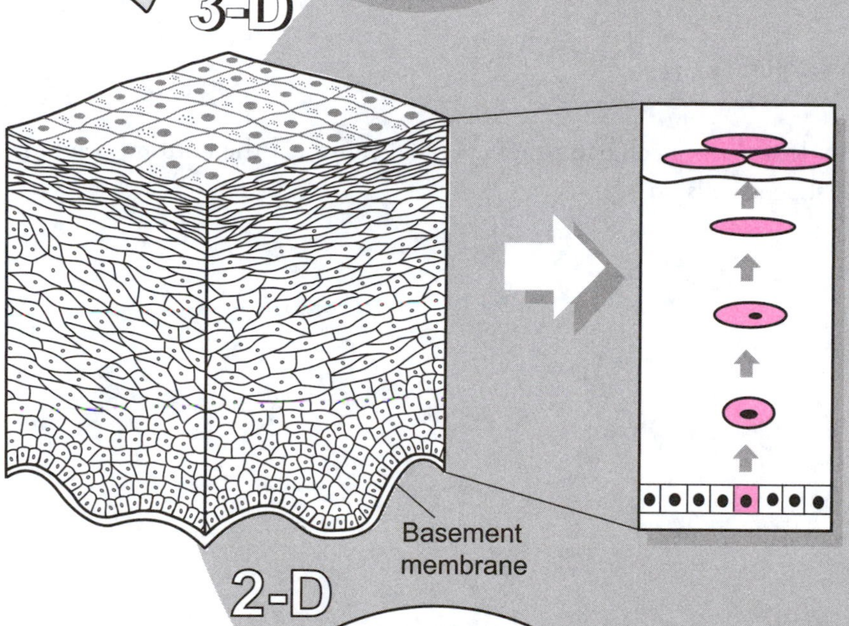

In the process of keratinization, new cells are pushed toward the surface as they are gradually transformed into flat, dead cells completely filled with a protein called keratin.

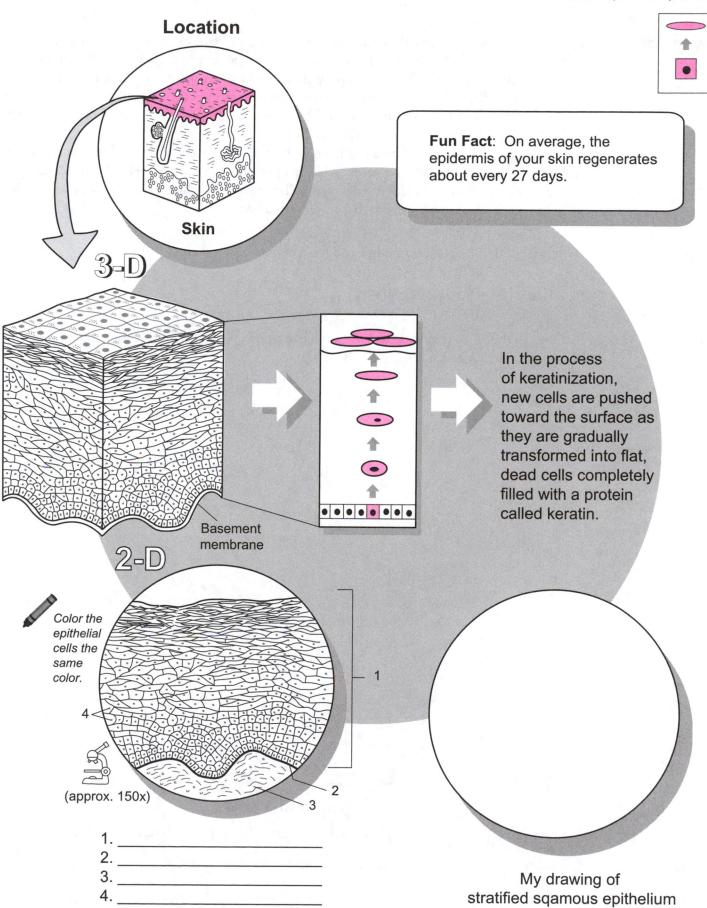

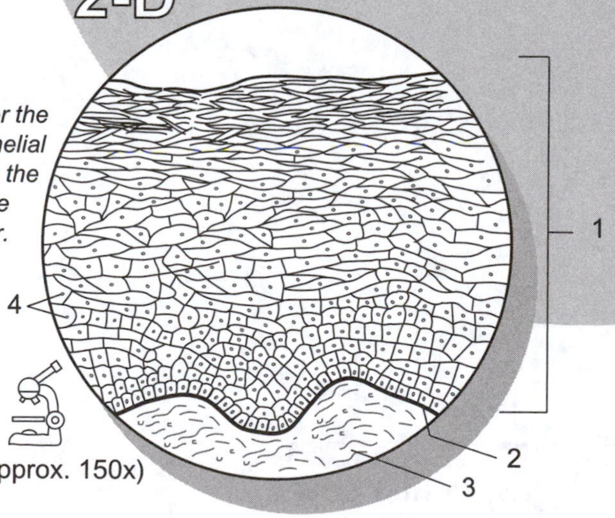

Color the epithelial cells the same color.

(approx. 150x)

1. ______________________
2. ______________________
3. ______________________
4. ______________________

My drawing of stratified sqamous epithelium

Description

Epithelial tissues line internal cavities and passageways and cover external body surfaces. They are composed mostly of cells that rest on a thin **basement membrane**. No blood vessels are present. One method of classifying epithelial tissues is by the number of layers of cells. *Simple* epithelia have a single layer of cells, and *stratified* epithelia have multiple layers of cells. Classification is also based on the following cell shapes: *squamous* (thin, flat), *cuboidal* (cube-shaped), and *columnar* (column-shaped).

Transitional epithelium is able to stretch and recoil so it can be illustrated in either a stretched or a relaxed state. In the relaxed state it appears to be composed of a variety of cell shapes. On the bottom it may contain cuboidal or columnar cells, and cells at the surface are large, dome-shaped cells that transform into a squamous shape when stretched.

Location

Lines ureters, urinary bladder, urethra, and renal pelvis.

Function

Easily allows stretching and recoiling.

Study Tip

Pattern: In the relaxed state, look for the cell pattern of cuboidal cells near the bottom, columnar in the middle, and large, dome-shaped cells on top.

Key to Illustration

1. Stratified transitional epithelium
2. Basement membrane
3. Connective tissue

Location

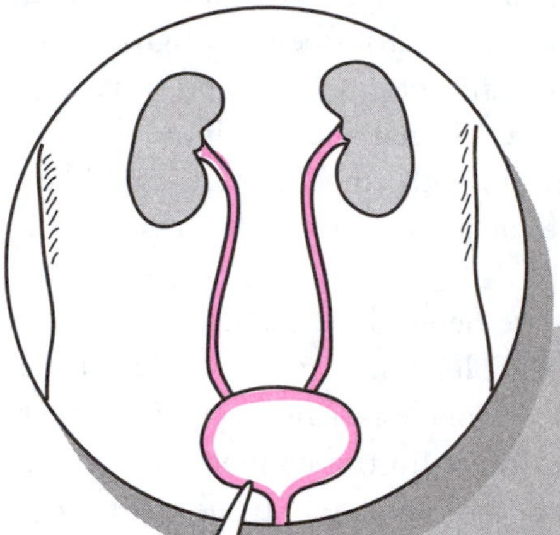

Fun Fact: The urinary bladder can expand to hold as much as 600–800 ml. of urine.

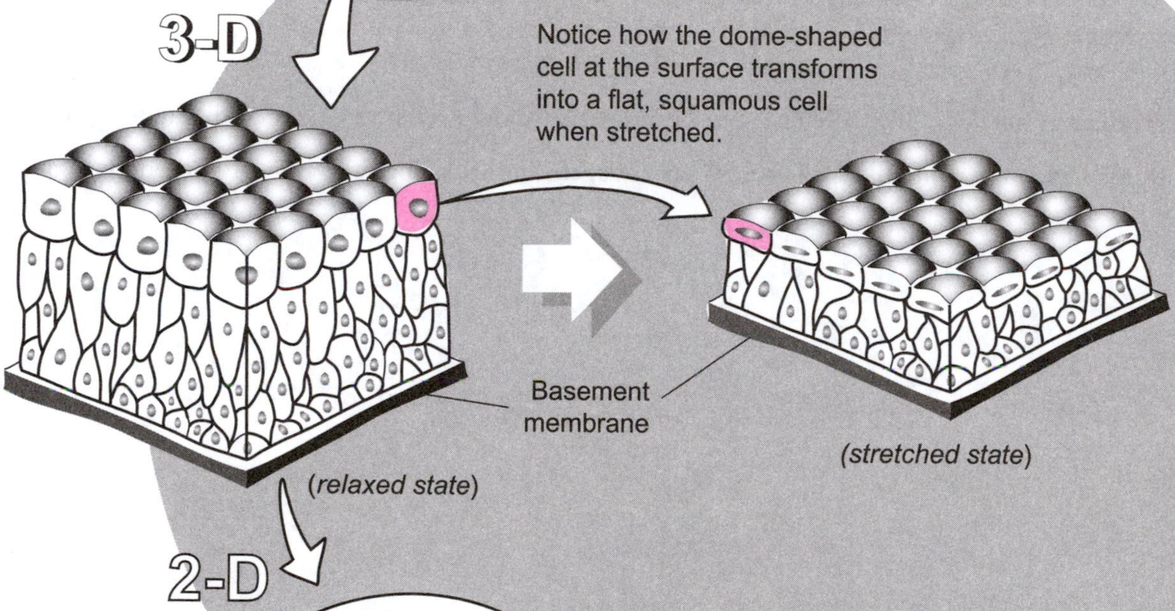
3-D
Notice how the dome-shaped cell at the surface transforms into a flat, squamous cell when stretched.
Basement membrane
(relaxed state)
(stretched state)
2-D

Color the epithelial cells the same color.

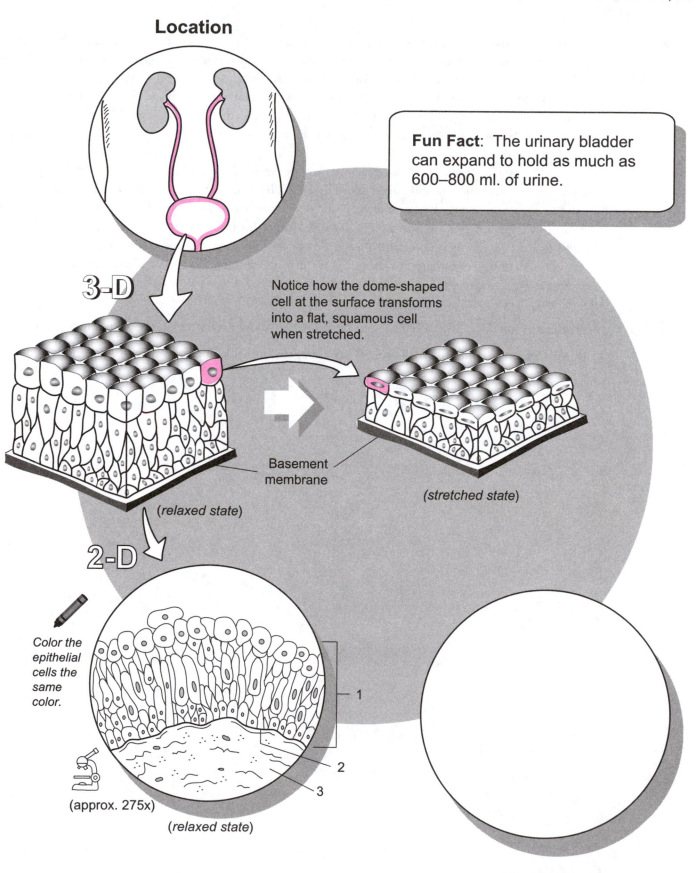

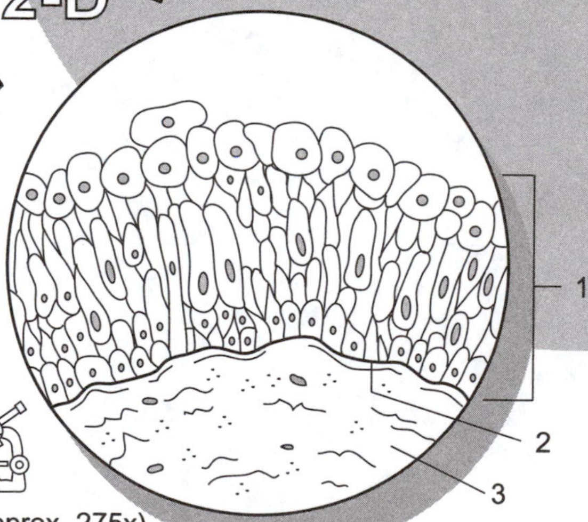
1
2
3

(approx. 275x)

(*relaxed state*)

My drawing of transitional epithelium

Description

Connective tissues primarily give structural support to other tissues and organs in the body. Though there are a wide variety of types, all are composed of cells, fibers, and ground substance. The most common cell type is called a **fibroblast**, which manufactures fibers and other extracellular material. The two most common types of protein fibers are **collagen** and **elastin**. Collagen is for strength, and elastin is for elasticity. The cells and the fibers are both embedded in a gel-like material called the **ground substance**. The ground substance varies in its consistency from gelatin-like to a much more rigid material.

The physical traits of a connective tissue are determined mainly by varying the proportion of cells, fibers, and the ground substance. For example, a strong connective tissue requires a greater proportion of collagen fibers and fewer cells. An example is dense regular connective tissue, which is found in tendons that anchor muscle to bone. In contrast, a connective tissue composed mostly of cells is not very strong. Such is the case with adipose connective tissue (*fat tissue*). The main purpose of adipose connective tissue is to store lipids (*fat*) in individual fat cells called **adipocytes**. This tissue contains numerous adipocytes and little else.

Analogy

- **Elastin fibers** are like **rubber bands** because they allow stretching and recoiling in a tissue.
- **Collagen fibers** are like **steel cables on a suspension bridge** because they give strength to a tissue.

All connective tissues contain the following basic components:

Cells	+	Fibers	+	Ground substance	=

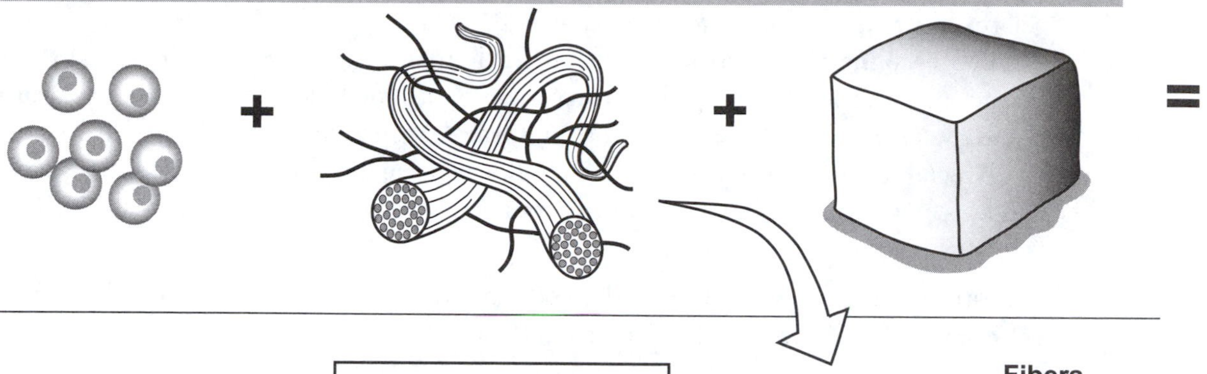

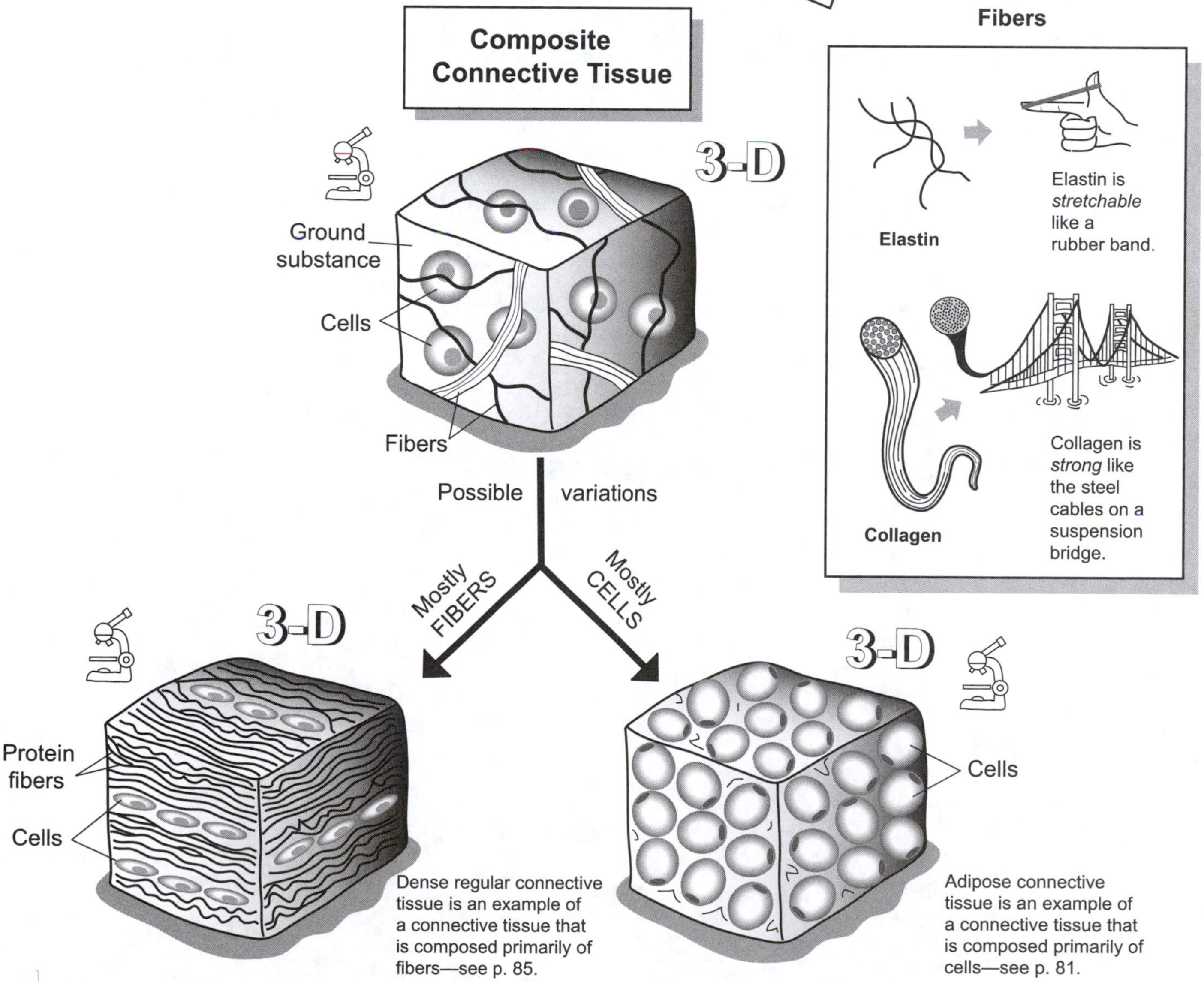

Dense regular connective tissue is an example of a connective tissue that is composed primarily of fibers—see p. 85.

Adipose connective tissue is an example of a connective tissue that is composed primarily of cells—see p. 81.

Description

Connective tissues primarily give structural support to other tissues and organs in the body. Though there are a wide variety of types, all are composed of cells, fibers, and ground substance. The most common cell type is called a **fibroblast**, which manufactures the fibers and other extracellular material. The two most common types of protein fibers produced are **collagen** and **elastin**. Collagen is for strength and elastin is for elasticity. The cells and the fibers are both embedded in a gel-like material called the **ground substance**. The ground substance varies in consistency from being gelatin-like to a much more rigid material.

Loose connective tissues have fewer fibers than other connective tissues and serve as a protective padding in the body. The three tissues classified as loose connective tissues are: *areolar connective tissue*, *adipose connective tissue*, and *reticular connective tissue*.

Areolar connective tissue has a random arrangement of cells, fibers, and ground substance. It contains all the basic components of any connective tissue without being specialized.

Location

Beneath epithelial tissues all over the body; between skin and skeletal muscles; surrounding blood vessels; within skin; around organs; around joints

Function

Cushions and protects organs; its phagocytes protect against pathogens; holds tissue fluid

Key to Illustration

1. Collagen fibers
2. Elastin fibers
3. Fibroblast nuclei

Location

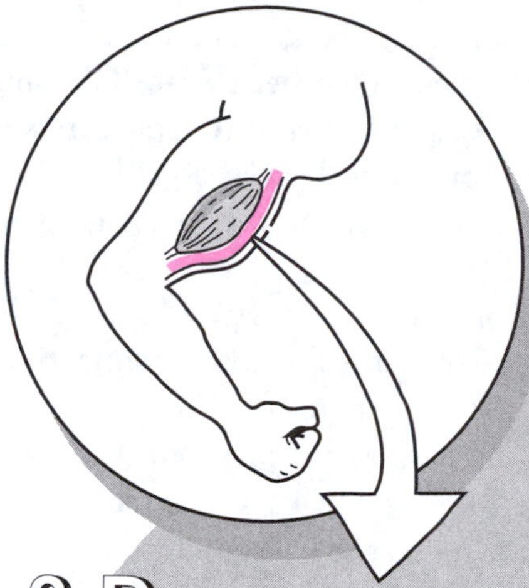

Fun Fact: When hunters skin an animal, the tissue they break to separate skin from muscle is areolar connective tissue.

3-D

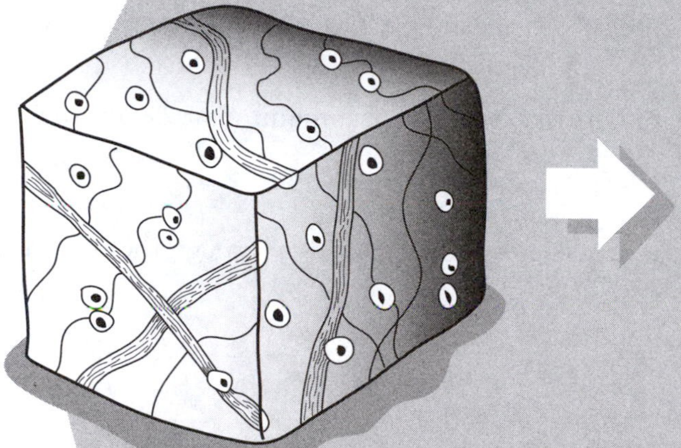

RANDOM

This tissue type is not specialized, so it does not look like anything in particular. It has a random arrangement of fibers and cells—nothing special!

2-D

Color the cells and fibers different colors.

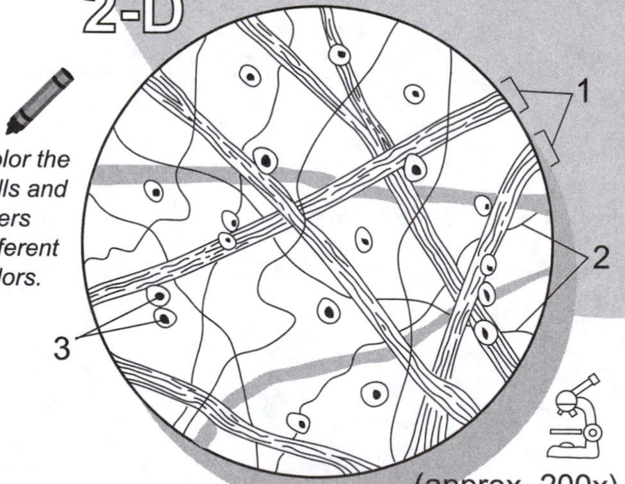

(approx. 200x)

1. ______________________________
2. ______________________________
3. ______________________________

My drawing of areolar connective tissue

Description

Connective tissues primarily give structural support to other tissues and organs in the body. Though there are a wide variety of types, all are composed of cells, fibers, and ground substance. The most common cell type, called a **fibroblast**, manufactures fibers and other extracellular material. The two most common types of protein fibers produced are **collagen** and **elastin**. Collagen is for strength and elastin is for elasticity. The cells and the fibers are both embedded in a gel-like material called the **ground substance**. The ground substance varies in its consistency from being gelatin-like to a much more rigid material.

Loose connective tissues have fewer fibers than other connective tissues and serve as a protective padding in the body. There are three tissues classified as loose connective tissues: *areolar connective tissue*, *adipose connective tissue*, and *reticular connective tissue*.

Adipose connective tissue is fat tissue. It is composed almost entirely of fat cells called **adipocytes** along with some blood vessels. These cells have a large vacuole to store lipids (*fats*). Though adipocytes are not able to divide, they do change in size by expanding or shrinking depending on the amount of lipid that is stored inside their vacuoles. For example, as a person loses weight, the amount of lipid in the adipocyte's vacuole decreases, causing the cell to shrink in size. Unfortunately, if a person regains that weight, the cells are able to expand back to their original size.

Location

Under all skin but especially in abdomen, buttocks, and breasts; around some organs such as eyeballs and kidneys.

Function

Protects certain organs and other structures; insulates against heat loss through the skin; stores energy as a reserve fuel.

Key to Illustration

1. Blood vessel
2. Nuclei of adipocytes *(fat cells)*
3. Vacuole for lipid storage
4. Plasma membrane of adipocyte *(fat cell)*

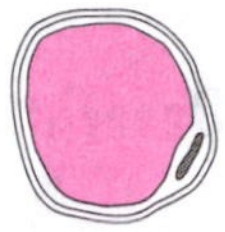

Location

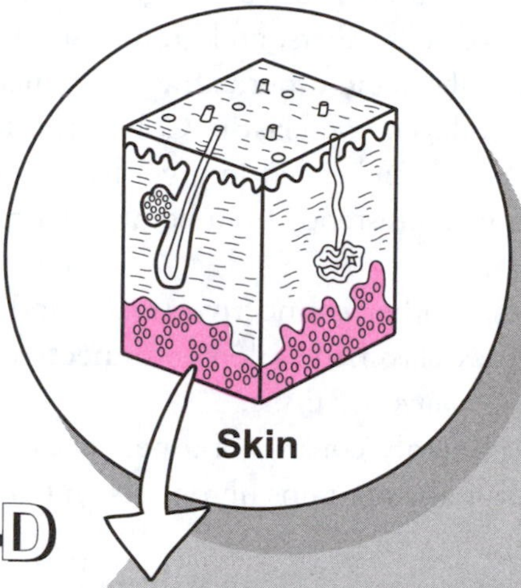

> **Fun Fact**: Excess adipose tissue can be removed by a surgical procedure called liposuction, but it does have some risks.

3-D

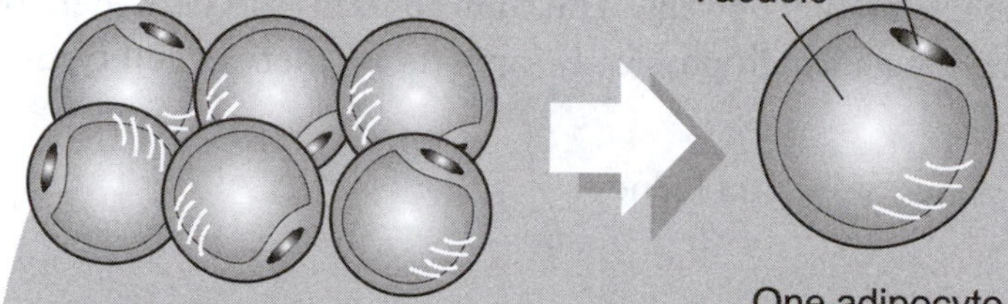

A cluster of six adipocytes *(fat cells)*

One adipocyte *(fat cell)*

Adipocyte

Why is an adipocyte's nucleus located at the outer edge of the cell? Because the vacuole for lipid storage expands as it fills, and displaces the nucleus and other organelles to the outer edge of the cell.

2-D

Color the adipocytes the same color.

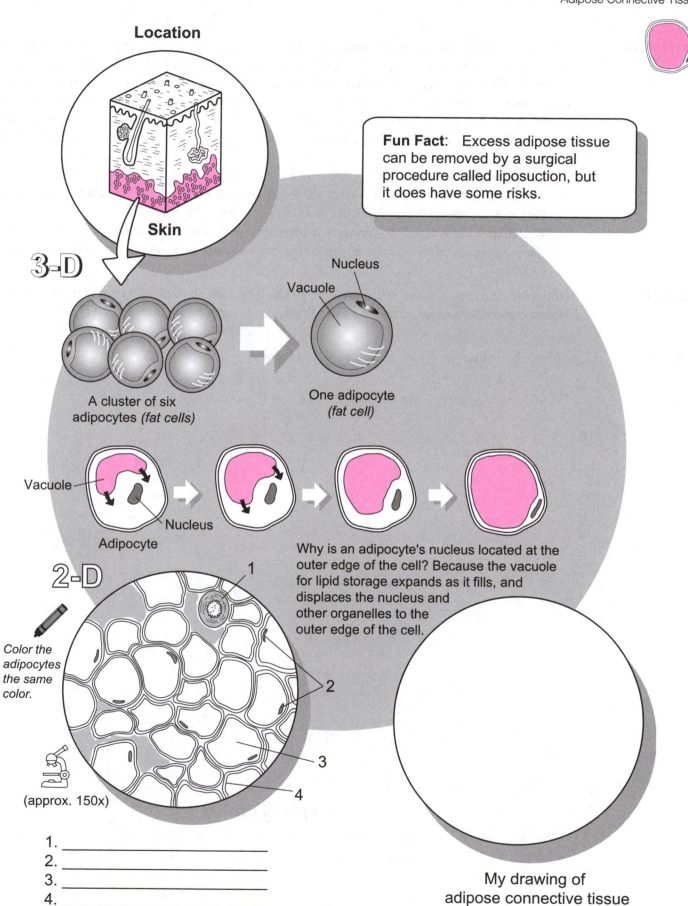

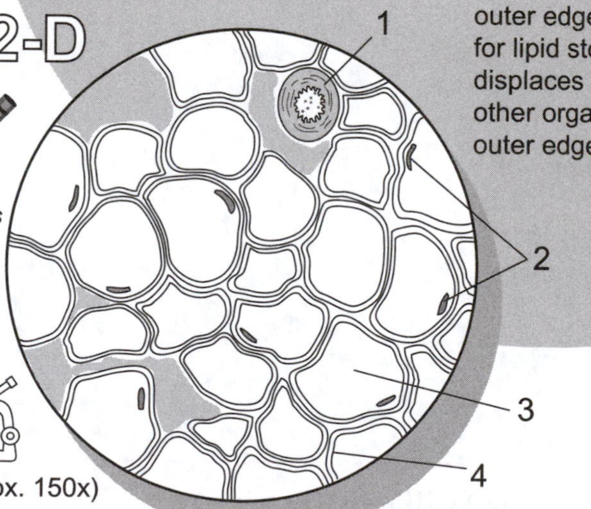

(approx. 150x)

1. ____________________
2. ____________________
3. ____________________
4. ____________________

My drawing of adipose connective tissue

Description

Connective tissues primarily give structural support to other tissues and organs in the body. Though there are a wide variety of types, all are composed of cells, fibers, and ground substance. The most common cell type, called a **fibroblast**, manufactures fibers and other extracellular material. The two most common types of protein fibers produced are **collagen** and **elastin**. Collagen is for strength and elastin is for elasticity. The cells and the fibers are both embedded in a gel-like material called the **ground substance**. The ground substance varies in its consistency from being gelatin-like to a much more rigid material.

Loose connective tissues have fewer fibers than other connective tissues and serve as a protective padding in the body. There are three tissues classified as loose connective tissues: *areolar connective tissue*, *adipose connective tissue*, and *reticular connective tissue*.

Reticular (*reticulata* = net) **connective tissue** primarily consists of a network of reticular fibers. The most common cell type is the reticular cell, but it also contains fibroblasts and macrophages.

Analogy

Reticular connective tissue is like **many cobwebs**. The **cobweb** itself is like the **network of reticular fibers** scattered throughout the tissue, which physically supports a variety of cell types.

Location

Spleen, bone marrow, lymph nodes, liver, and kidney

Function

Fibers form a supportive net-like structure for a variety of cell types.

Key to Illustration

1. Reticular fibers

Location

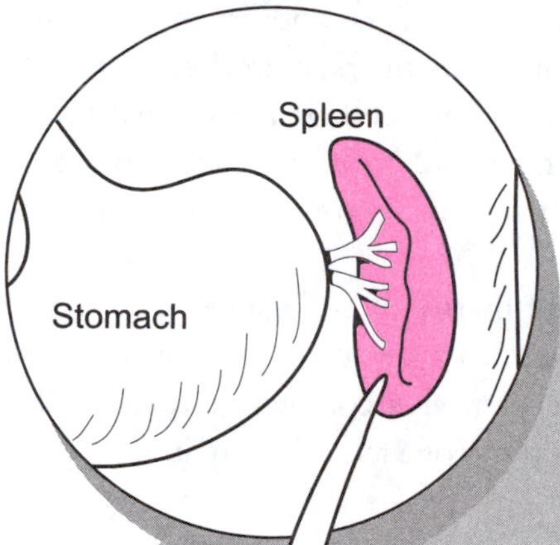

Fun Fact: If the spleen needs to be surgically removed, people can survive without it, but they may be more prone to getting various infections.

3-D

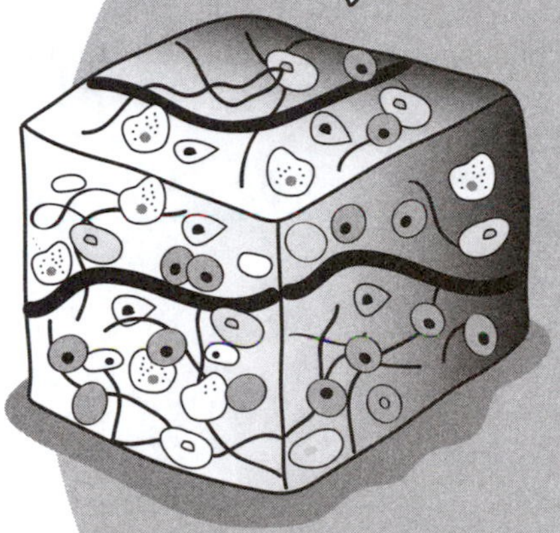

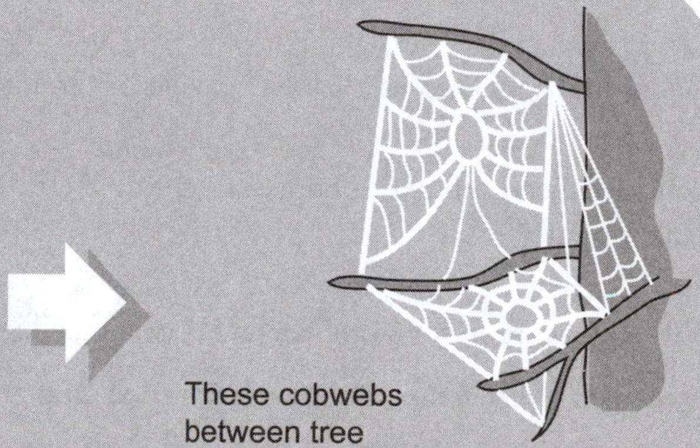

These cobwebs between tree branches are like the network of reticular fibers in reticular connective tissue. The reticular fibers provide a supportive framework for a variety of cell types.

2-D

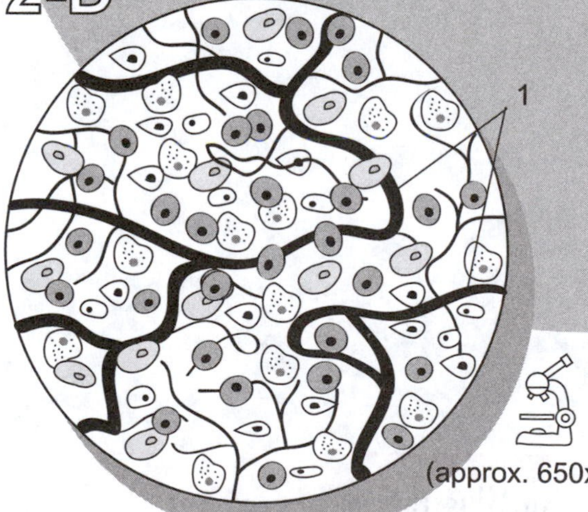

(approx. 650x)

1. ______________________________

My drawing of reticular connective tissue

Description

Connective tissues primarily give structural support to other tissues and organs in the body. Though there are a wide variety of types, all are composed of cells, fibers, and ground substance. The most common cell type is called a **fibroblast,** which manufactures fibers and other extracellular material. The two most common types of protein fibers produced are **collagen** and **elastin**. Collagen is for strength, and elastin is for elasticity. The cells and fibers are both embedded in a gel-like material called the **ground substance**. The ground substance varies in its consistency from being gelatin-like to a much more rigid material.

Dense regular connective tissue is composed primarily of collagen fibers, so it is also called *fibrous connective* or *collagenous tissue*. The body has two types of dense connective tissue: dense *regular* connective and dense *irregular* connective. Dense *regular* connective tissue is characterized by a large proportion of collagen fibers that are stacked on top of each other in an orderly arrangement.

Analogy

Layers of **collagen fibers** are strong like the **steel cables on a suspension bridge**.

Location

Tendons and aponeuroses; ligaments; covering around skeletal muscles.

Function

Anchors skeletal muscle to bone; attaches bone to bone; packages skeletal muscles; stabilizes bones within a joint.

Study Tips

- Fibroblasts are in rows sandwiched between collagen fibers.
- Collagen fibers are layered in an organized arrangement.

Key to Illustration

1. Collagen fibers
2. Nuclei of fibroblasts

Dense *(fibrous)* Regular Connective Tissue

Location

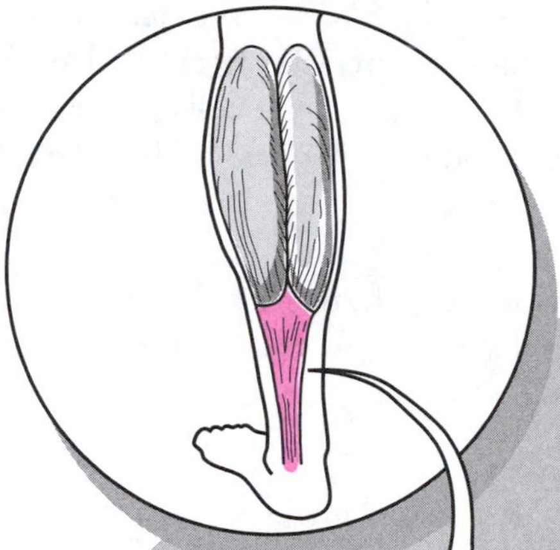

Fun Fact: The calcaneal, or Achilles' tendon is the strongest tendon in the body.

3-D

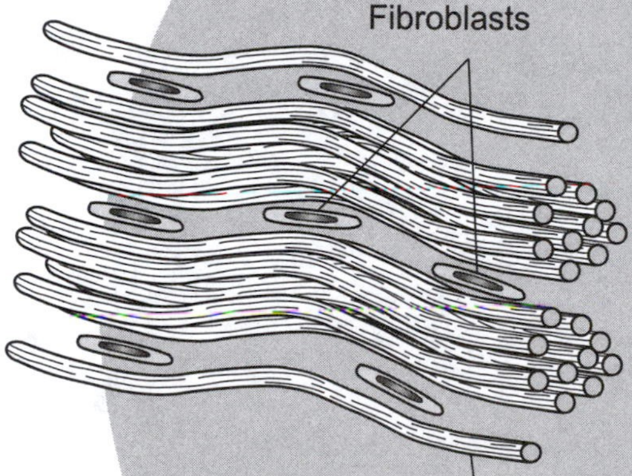

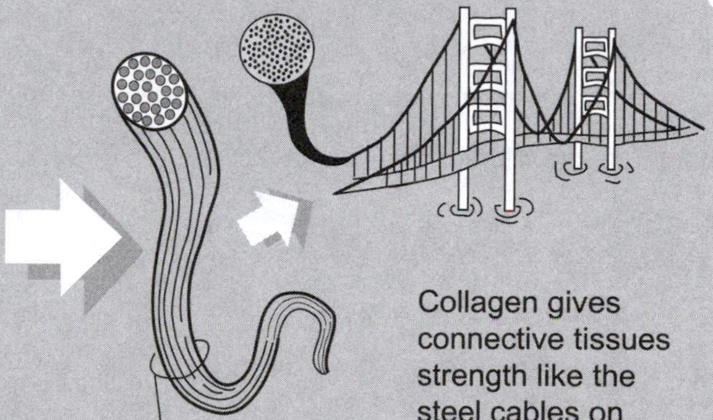

Collagen gives connective tissues strength like the steel cables on a suspension bridge.

2-D

Color the collagen fibers the same color.

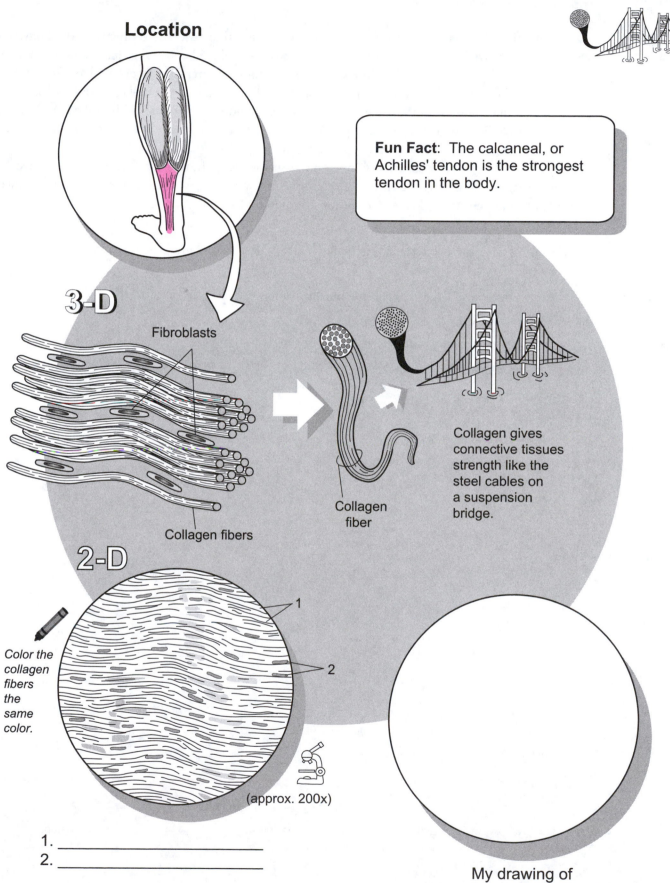

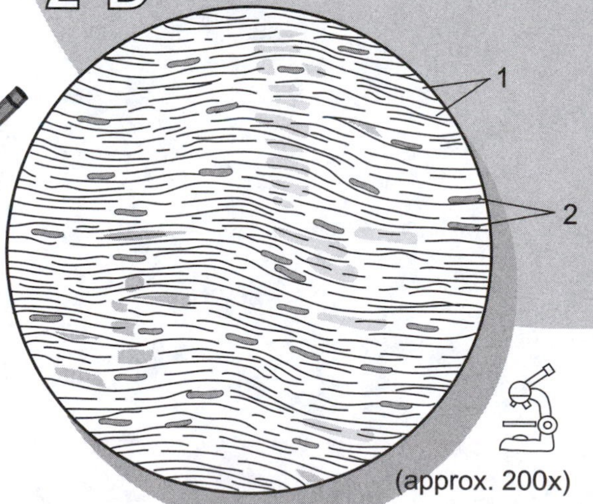

(approx. 200x)

1. ______________________
2. ______________________

My drawing of dense regular connective tissue

Description

Connective tissues primarily give structural support to other tissues and organs in the body. Though they are of a wide variety of types, all are composed of cells, fibers, and ground substance. The most common cell type is called a **fibroblast** which manufactures the fibers and other extracellular material. The two most common types of protein fibers produced are **collagen** and **elastin**. Collagen is for strength and elastin is for elasticity. The cells and the fibers are both embedded in a gel-like material called the **ground substance**. The ground substance varies in its consistency from being almost like gelatin to a much more rigid material.

There are two types of dense connective tissue in the body, namely, dense *regular* connective and dense *irregular* connective. Dense *irregular* connective tissue is characterized by a random arrangement of collagen fibers and a greater proportion of ground substance.

Location

Dermis of the skin; periosteum; visceral organ capsules; around muscles.

Function

Resists stresses applied in many different directions

Study Tips

- Fibroblasts are more scattered throughout the tissue
- Collagen fibers are **not** stacked on top of each other, randomly arranged

Key to Illustration

1. Nucleus of fibroblast
2. Collagen fibers
3. Ground substance

Location

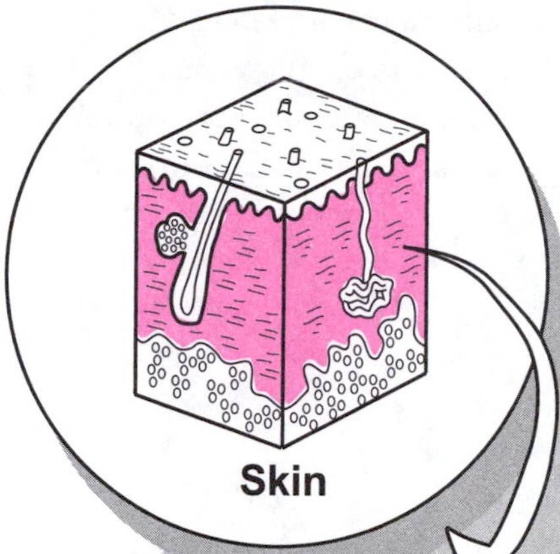

FUN FACT: Collagen in the dermis of the skin breaks down as we age. Collagen injections are a cosmetic procedure used to smooth facial lines.

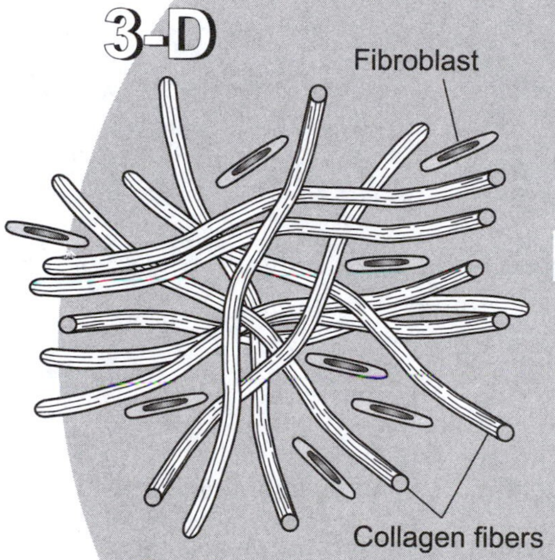

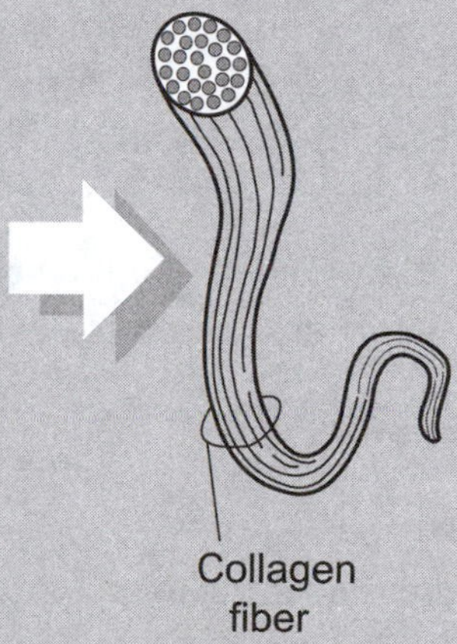

Irregular =

RANDOM

Compared to dense regular connective tissue, this tissue has its collagen fibers arranged more randomly

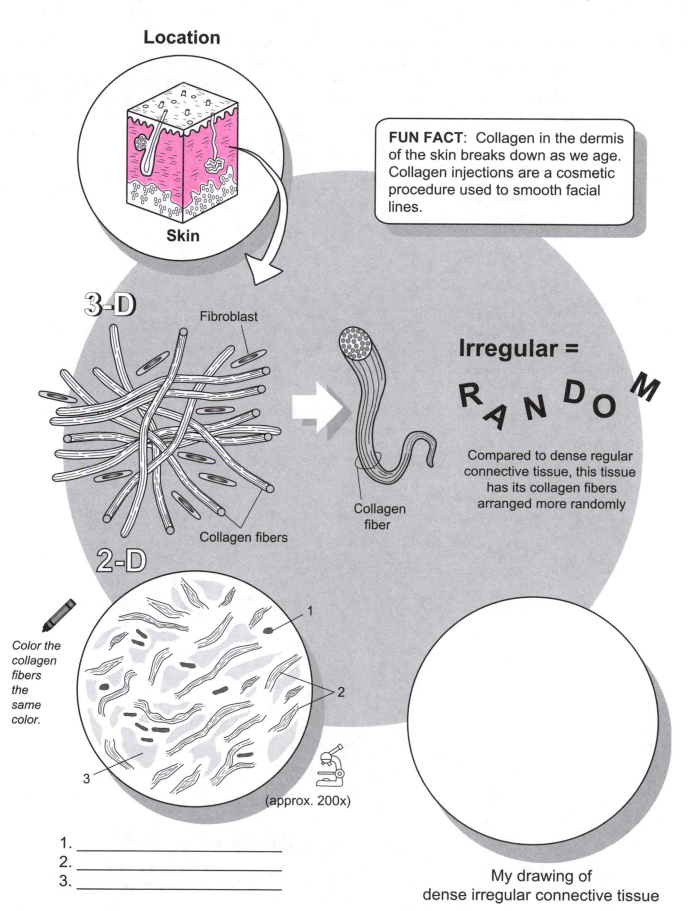

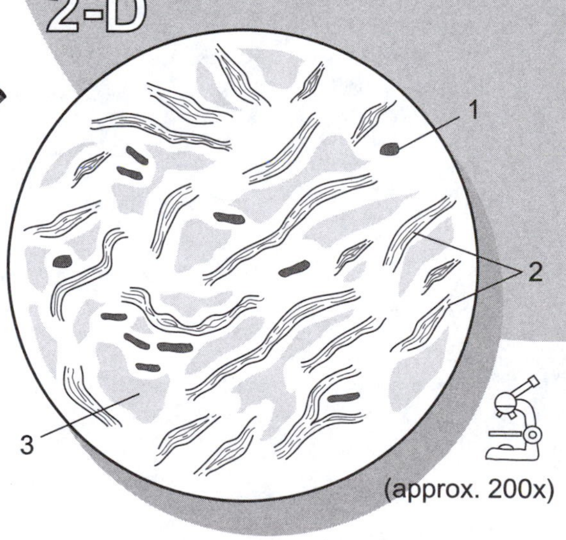

Color the collagen fibers the same color.

1. ______________________
2. ______________________
3. ______________________

My drawing of dense irregular connective tissue

Description

Connective tissues primarily give structural support to other tissues and organs in the body. Though there are a wide variety of types, all are composed of cells, fibers, and matrix.

Cartilage is a specialized type of connective tissue. It is characterized by three traits: **lacunae, chondrocytes,** and a rigid **matrix**. The **matrix** is a firm gel material that contains protein fibers and other substances. Within the matrix are small cavities called **lacunae**. Within the lacunae are living cartilage cells called **chondrocytes**. Because cartilage lacks blood vessels, chondrocytes rely on the diffusion of nutrients into the matrix to survive.

The three basic types of cartilage in the body are:

- **Hyaline cartilage**
- **Elastic cartilage**
- **Fibrocartilage**

Hyaline cartilage is the most common type of cartilage.

Analogy

Three dimensionally, a piece of **any type of cartilage** is similar to a **block of Swiss cheese** in its structure and general consistency. Though cartilage is much stronger, both are solid and flexible. The **cheese** itself is the **matrix** and the **holes** are the **lacunae**.

Location

Covers ends of long bones in synovial joints; between ribs and sternum; cartilages of nose, trachea, larynx, and bronchi; most portions of embryonic skeleton.

Function

Structural reinforcement, slightly flexible support; reduces friction within joints.

Study Tip

To identify this tissue either under the microscope or from a photograph, look for the following:

- This is the **only cartilage type with no apparent fibers** (*they are present but do not stain well*).
- Chondrocytes are evenly scattered within matrix.

Key to Illustration

1. Matrix
2. Lacunae
3. Chondrocytes *(cartilage cells)*
4. Nucleus of a chondrocyte

Location

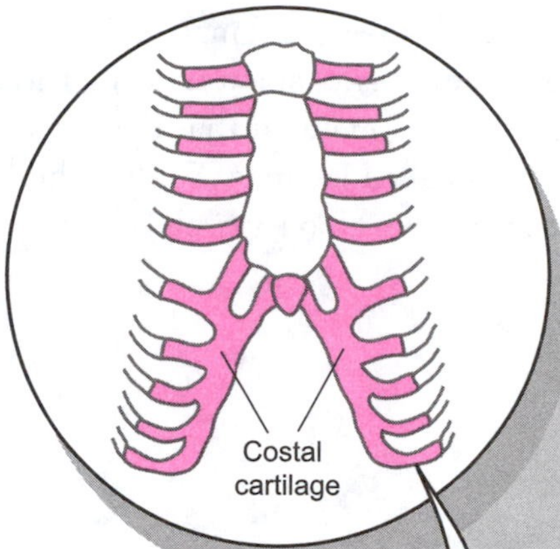

Fun Fact: Most of the bones in the fetus are made of hyaline cartilage and later ossify into bone.

3-D

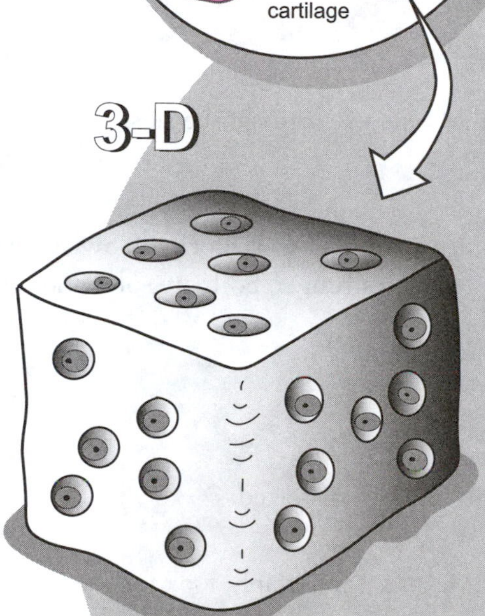

A block of cartilage is like a block of Swiss cheese.

2-D

Color the chondrocytes the same color.

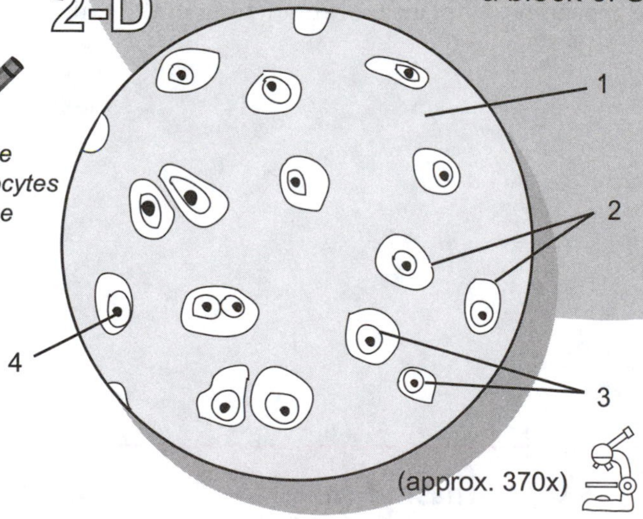

(approx. 370x)

1. ______________________________
2. ______________________________
3. ______________________________
4. ______________________________

My drawing of hyaline cartilage

Description

Connective tissues primarily give structural support to other tissues and organs in the body. Though there are a wide variety of types, all are composed of cells, fibers, and matrix.

Cartilage is a specialized type of connective tissue. It is characterized by three traits: **lacunae, chondrocytes,** and a rigid **matrix**. The **matrix** is a firm gel material that contains protein fibers and other substances. Within the matrix are small cavities called **lacunae**. Within the lacunae are living cartilage cells called **chondrocytes**. Because cartilage lacks blood vessels, chondrocytes rely on the diffusion of nutrients into the matrix to survive.

The three basic types of cartilage in the body are:

- **Hyaline cartilage**
- **Elastic cartilage**
- **Fibrocartilage**

Elastic cartilage is the most durable and flexible type of cartilage, because of the presence of many elastic fibers.

Analogy

Three dimensionally, a piece of **any type of cartilage** is similar to a **block of Swiss cheese** in its structure and general consistency. Though cartilage is much stronger, both are solid and flexible. The **cheese** itself is the **matrix** and the **holes** are the **lacunae**.

Location

External ear; epiglottis; auditory canal

Function

Provides support while easily returning to original shape when distorted.

Study Tip

To identify this tissue either under the microscope or from a photograph, look for the following:

- Chondrocytes appear larger than other cartilages.
- Numerous elastic fibers have appearance of plant roots branching in the soil.

Key to Illustration

1. Matrix
2. Lacunae
3. Chondrocytes *(cartilage cells)*
4. Elastin fiber
5. Nucleus of chondrocyte

Location

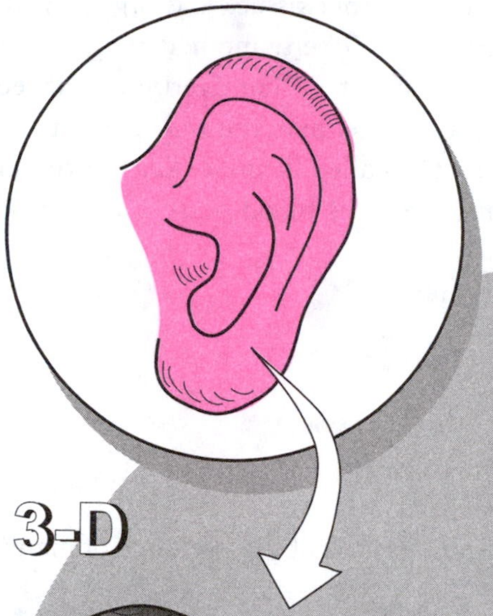

Fun Fact: The thyroid cartilage on the front of the larynx contains a protruding structure commonly called the "Adam's apple."

3-D

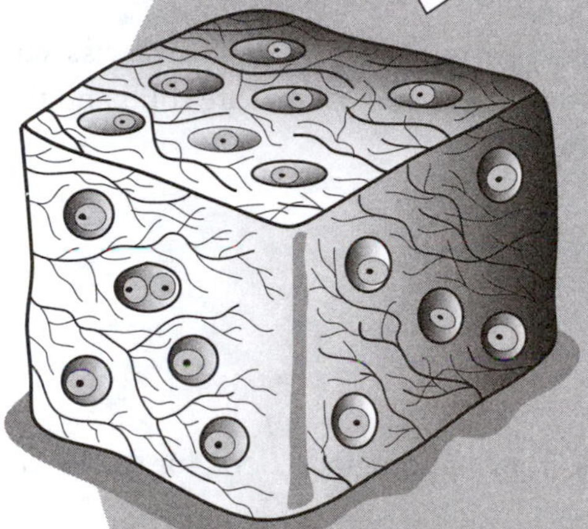

A block of cartilage is like a block of Swiss cheese.

2-D

Color the chondrocytes the same color.

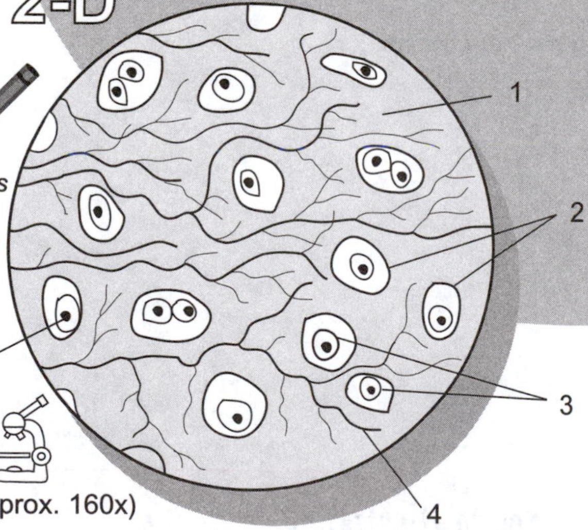

(approx. 160x)

1. ______________________
2. ______________________
3. ______________________
4. ______________________
5. ______________________

My drawing of elastic cartilage

Description

Connective tissues primarily give structural support to other tissues and organs in the body. Though there are a wide variety of types, all are composed of cells, fibers, and matrix.

Cartilage is a specialized type of connective tissue. It is characterized by three traits: **lacunae, chondrocytes,** and a rigid **matrix**. The **matrix** is a firm gel material that contains protein fibers and other substances. Within the matrix are small cavities called **lacunae**. Within the lacunae are living cartilage cells called **chondrocytes**. Because cartilage lacks blood vessels, chondrocytes rely on the diffusion of nutrients into the matrix to survive.

The three basic types of cartilage in the body are:

- **Hyaline cartilage**
- **Elastic cartilage**
- **Fibrocartilage**

Fibrocartilage is the strongest of the three types because of the presence of many **collagen** fibers.

Analogy

Three dimensionally, a piece of **any type of cartilage** is similar to a **block of Swiss cheese** in its structure and general consistency. Though cartilage is much stronger, both are solid and flexible. The **cheese** itself is the **matrix** and the **holes** are the **lacunae**.

Location

Intervertebral discs; pubic symphysis; pads within knee joint

Function

Shock absorber in a joint; resists compression

Study Tips

To identify this tissue either under the microscope or from a photograph, look for the following:

- Has the most collagen fibers of any cartilage.
- Collagen fibers often appear in a wavy pattern.
- Chondrocytes are often seen in rows and/or small clusters.

Key to Illustration

1. Lacuna
2. Chondrocyte
3. Nucleus of a chondrocyte
4. Matrix

Location

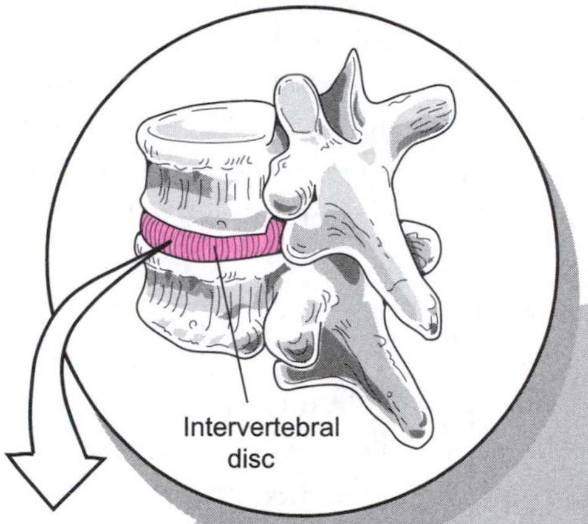

Fun Fact: Cartilage in joints such as the knee undergo the trauma of getting compressed every day.

3-D

A block of cartilage is like a block of Swiss cheese.

2-D

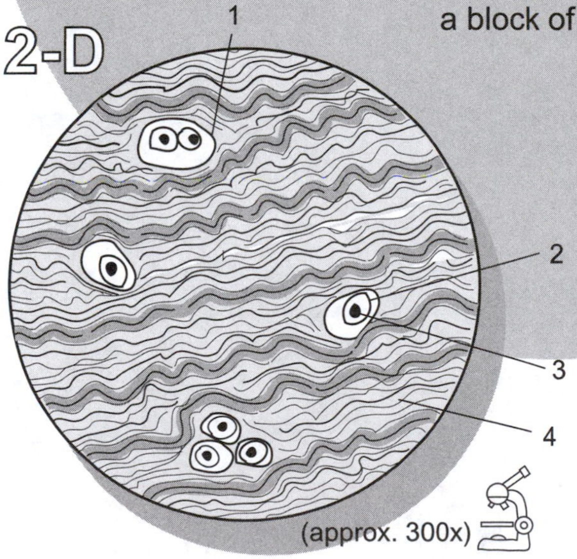

Color the chondrocytes the same color.

1. ____________________
2. ____________________
3. ____________________
4. ____________________

My drawing of fibrocartilage

Description

Connective tissues primarily give structural support to other tissues and organs in the body. Though there are a wide variety of types, all are composed of cells, fibers, and matrix.

Bone is a specialized type of connective tissue that has calcified into a hard substance. It is composed of organic and inorganic substances. The inorganic portion that constitutes about two-thirds of bone mass is made of modified calcium phosphate compounds called **hydroxyapatite,** while the organic portion is composed of **collagen** fibers. The two general types of bone are: spongy and compact. Spongy bone is less organized and is found in the ends of long bones and other places. Compact bone is more complex and orderly in structure and is found in the shaft of long bones and other locations.

Let's examine compact bone in more detail. The individual units in compact bone are tall, cylindrical towers called **osteons** (*Haversian systems*). In the middle of each osteon is a **central canal** that serves as a passageway for blood vessels. Around this canal are concentric rings of bony tissue called **lamellae.** Along each of these rings at regular intervals are small spaces called **lacunae** that contain a mature bone cell or **osteocyte.** Branching between individual lacunae are smaller passageways called **canaliculi,** which allow fluid with dissolved nutrients to travel to osteocytes.

Analogy

Each **surface of an osteon** looks like a **tree stump.** Both structures are made of hard, dense materials. Like the **growth rings** in a tree, the osteon has concentric rings called **lamellae.**

Location

Bones

Function

Bone supports body and protects vital organs; provides attachments for muscle to form a lever system for movement; stores calcium compounds and fat. Marrow contains stem cells that produce all blood cell types.

Key to Illustration

1. Osteon
2. Central canal
3. Osteocytes inside lacunae

Location

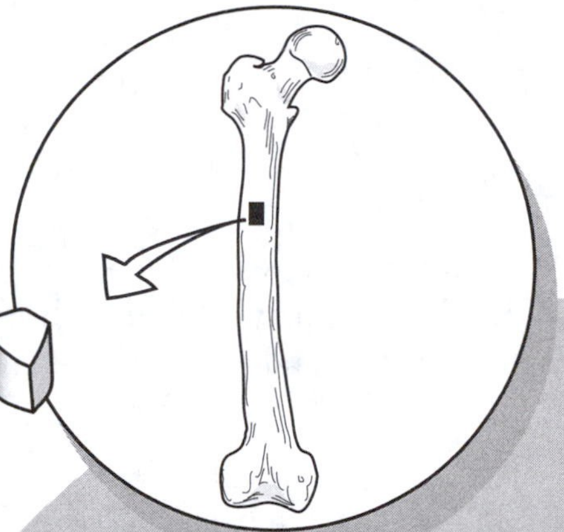

Fun Fact: Bone is stronger than concrete and nearly as strong as steel.

3-D

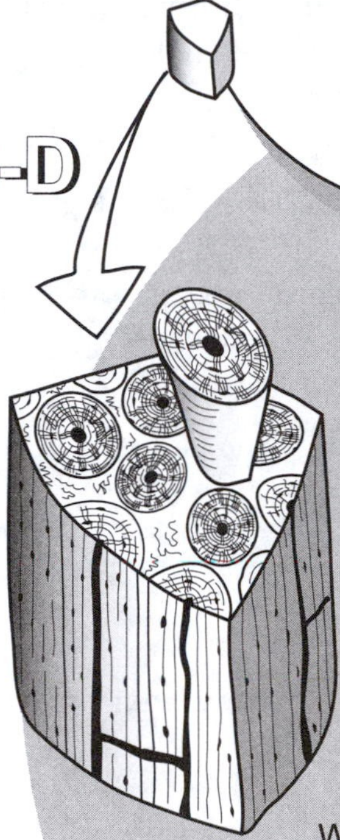

Wedge of compact bone

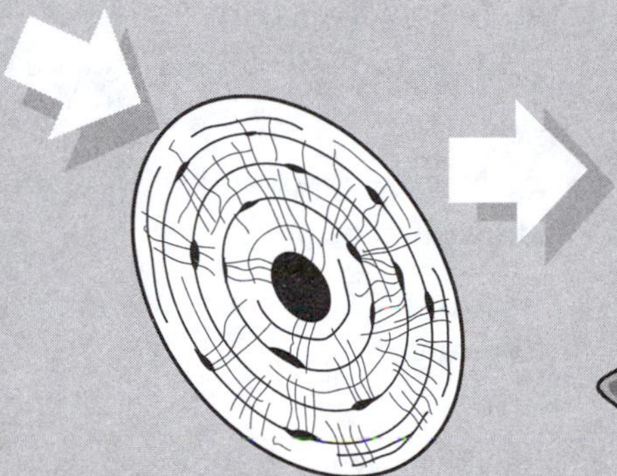

The surface of one osteon

From a superior view, each osteon looks like a tree stump.

2-D

Color the osteons the same color.

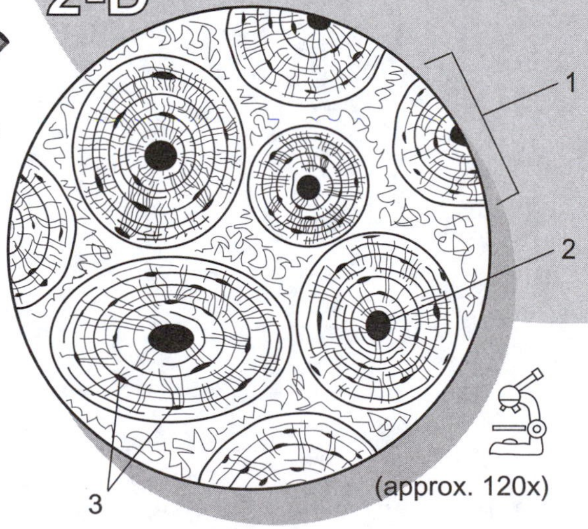

(approx. 120x)

1. ______________________________
2. ______________________________
3. ______________________________

My drawing of compact bone

Description

There are three different types of muscle tissue:

- **Skeletal muscle**
- **Cardiac muscle**
- **Smooth muscle**

Skeletal muscle is under conscious control, so it is also referred to as *voluntary* muscle. Each skeletal muscle cell is a long cylinder with a banding pattern, and each band is called a **striation**. Most body cells have only one nucleus per cell, but skeletal muscle has multiple nuclei in each cell—a unique feature.

Analogy

Under the microscope, **skeletal muscle** appears as a bunch of **stacked logs. Each log** is equivalent to one **skeletal muscle cell**. Consider them to be birch logs that have a striped pattern on them. These **stripes** are the **striations**. Note that the log doesn't show us the entire tree, just as the image under the microscope doesn't show us the entire cell. This is because the cells are very long.

Location

All the major muscles of the body are composed of skeletal muscle. Examples of skeletal muscle are the biceps brachii, gluteus maximus, and pectoralis major.

Function

Contraction of muscles (*conscious control*)

Study Tips

Under the microscope at higher magnifications, you can use the following landmarks to distinguish skeletal muscle tissue:

- Striations
- Multiple nuclei per cell
- Long cells. (Each cell is so long that you cannot see the ends of it under high magnification.)

Key to Illustration

1. Nuclei within one skeletal muscle cell
2. One skeletal muscle cell
3. Striation

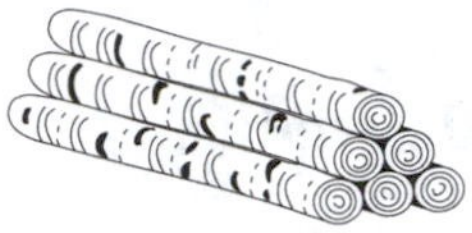

Location

Fun Fact: The human body is composed of more than 600 muscles. Together, they constitute about 40% of body mass.

3-D

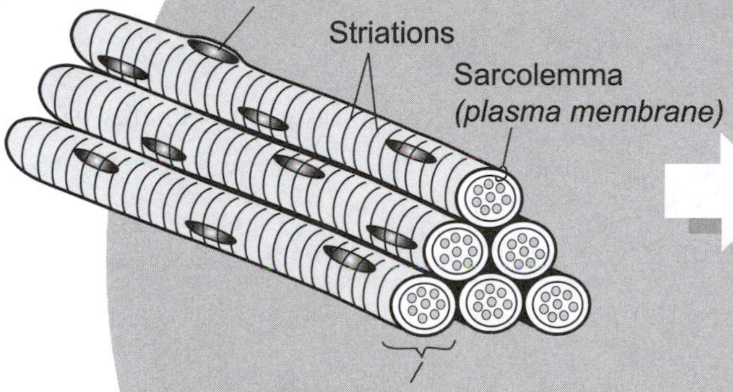

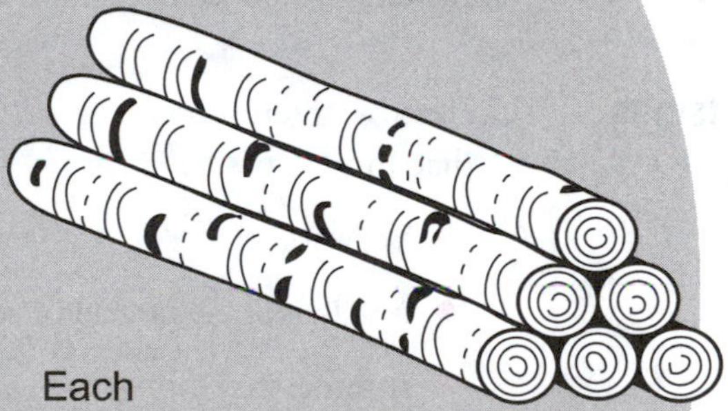

Each skeletal muscle cell (or fiber) is like a birch tree log because it is a long cylinder with a striped pattern.

2-D

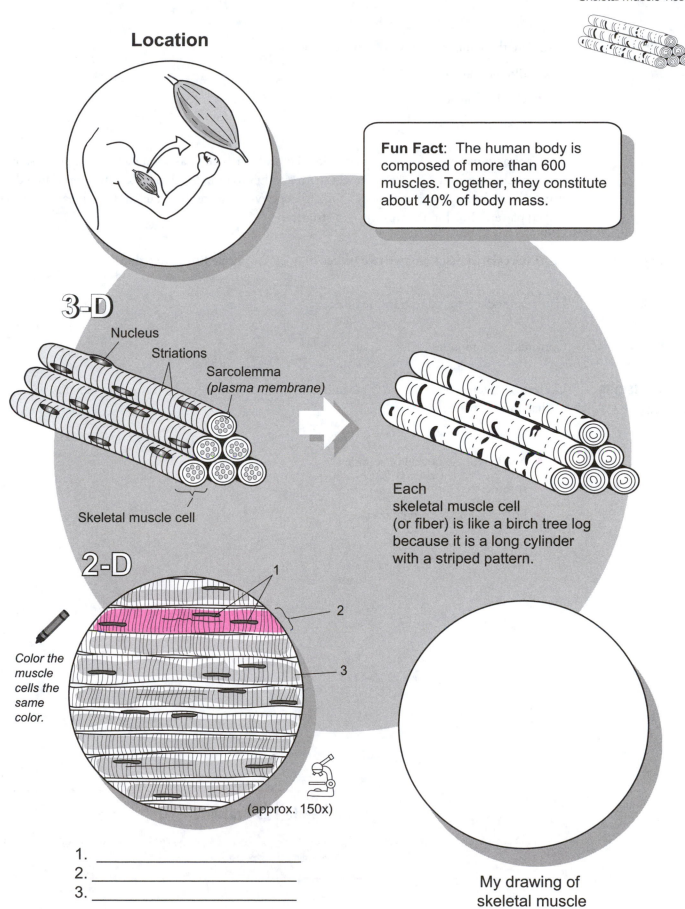

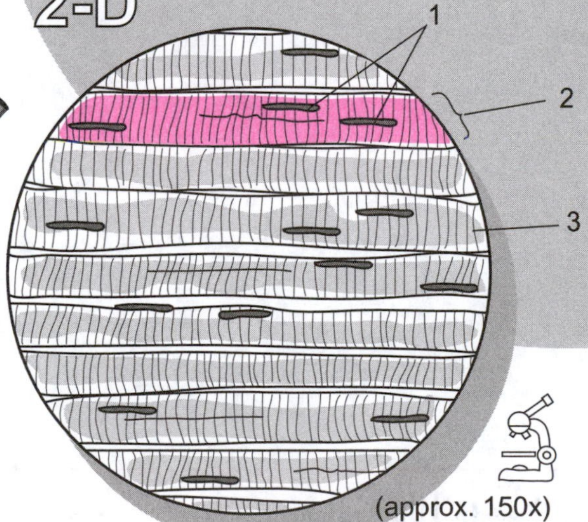

Color the muscle cells the same color.

1. ______________________
2. ______________________
3. ______________________

My drawing of skeletal muscle

Description

There are three different types of muscle tissue:

- **Skeletal muscle**
- **Cardiac muscle**
- **Smooth muscle**

Cardiac muscle is under our unconscious control. Each cell has a somewhat cylindrical shape and a single nucleus per cell. One cell connects with another to form a union called an **intercalated disc**. This structure can be seen with a compound microscope. Vertical bands run up and down each cell to form a striped pattern, in which each stripe is a **striation**.

Analogy

The **intercalated discs** are like **two pieces of a jigsaw puzzle** fitting together.

Location

Cardiac muscle is found *only* in the heart.

Function

Contraction of muscles (unconscious control)

Distinguishing Features

Under the microscope at higher magnifications, you can use the following landmarks to distinguish cardiac muscle tissue:

- Striations
- Nucleus appears oval or rounded
- Intercalated discs
- Forking or branching pattern

Key to Illustration

1. Nucleus of one cardiac muscle cell
2. Striation
3. Intercalated disc
4. Individual cardiac muscle cell

Location

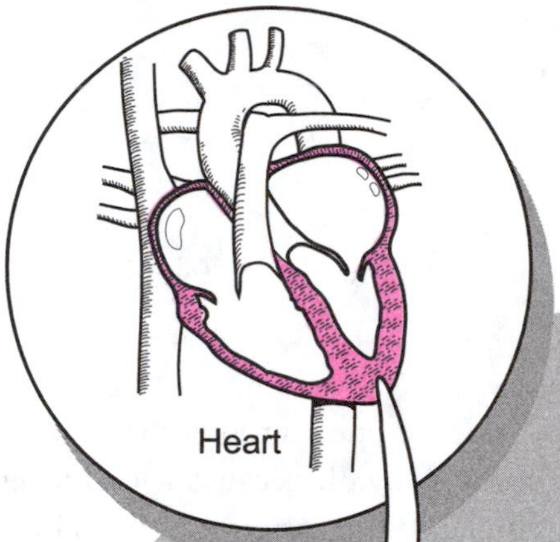

3-D

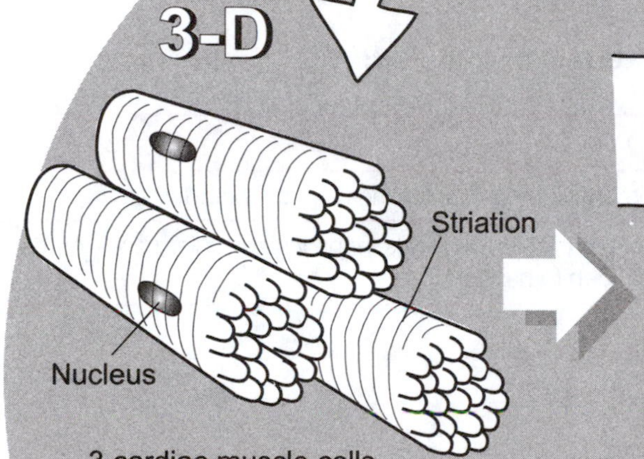

3 cardiac muscle cells

Fun Fact: Some cardiac muscle cells can initiate and maintain their own rhythmic beat.

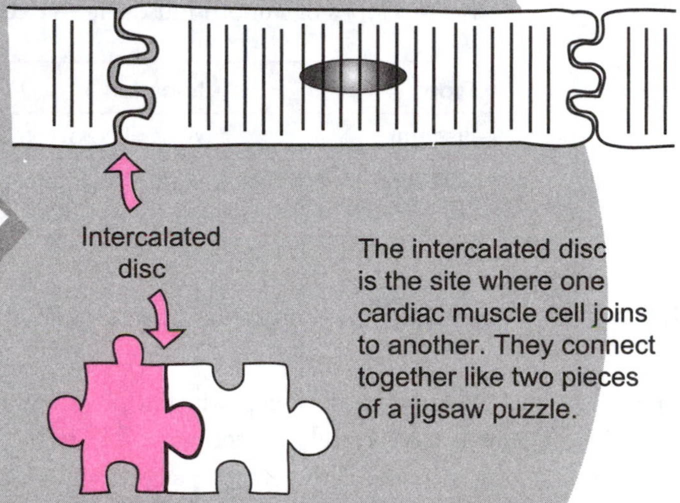

The intercalated disc is the site where one cardiac muscle cell joins to another. They connect together like two pieces of a jigsaw puzzle.

2-D

Color the muscle cells the same color.

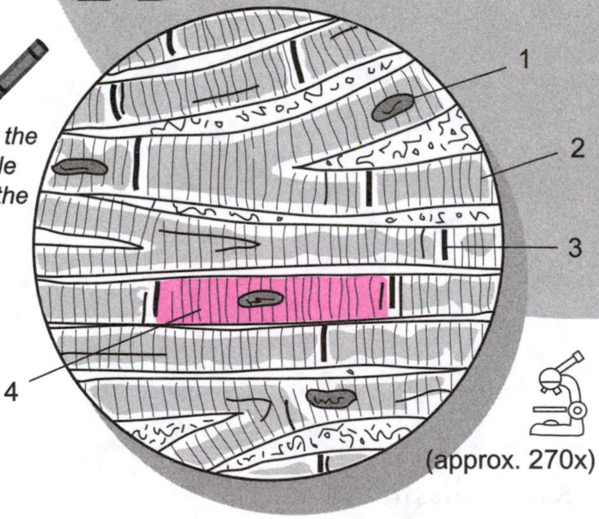

(approx. 270x)

1. ______________________
2. ______________________
3. ______________________
4. ______________________

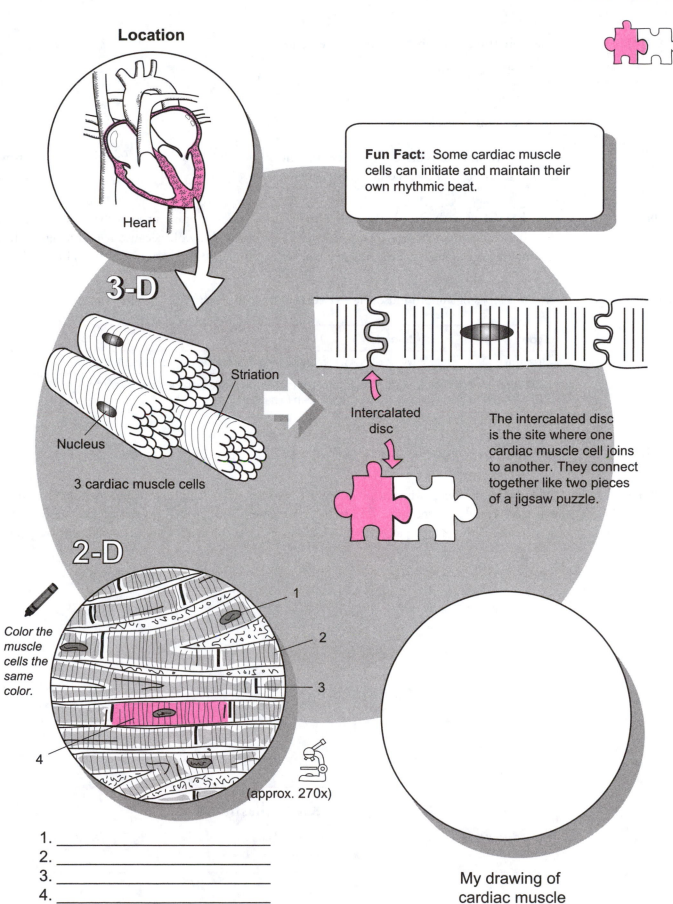

My drawing of cardiac muscle

Description

There are three different types of muscle tissue:

- **Skeletal muscle**
- **Cardiac muscle**
- **Smooth muscle**

Smooth muscle is under unconscious control. It lacks the striations found in the other two types of muscle tissue, and each cell has only one nucleus.

Analogy

In a **sheet of smooth muscle** the **individual cells** are stacked one on top of the other and staggered in their appearance. Each short cell is shaped like elongated ravioli because it is thicker in the middle and tapered on each end. This staggered pattern is similar to the pattern of **bricks in a wall**.

Location

The two types of smooth muscle are **visceral** and **multi-unit:**

Type	Location
Visceral	Walls of hollow organs (stomach, intestines, urinary bladder, etc.)
Multi-unit	Walls of large arteries, trachea, muscles in the iris and ciliary body of the eye, arrector pili muscles that attach to hair follicles

Function

Contraction of muscles (unconscious control)

Study Tip

Under the microscope, it will be difficult to see the cell membrane of individual cells. Instead, you will have to rely on the overall pattern of staggered cells stacked on top of one another to identify this tissue. The general pattern will show up in how the nuclei are arranged with respect to each other.

Key to Illustration

1. Individual smooth muscle cell
2. Nucleus of one smooth muscle cell
3. Plasma membrane

Location

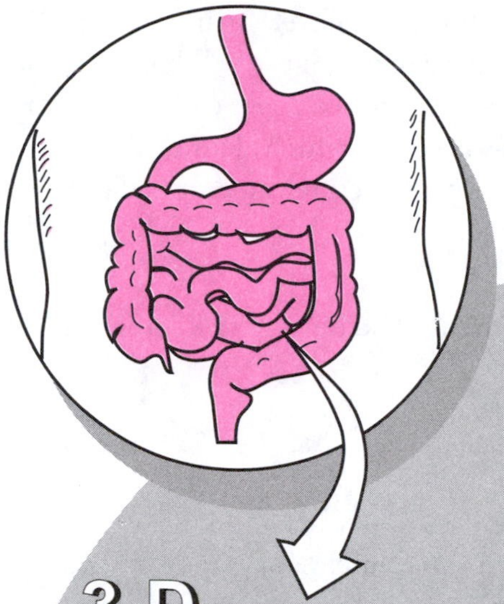

Fun Fact: The drug pitocin is a hormone used to induce labor. It causes the smooth muscle in the uterus to contract.

3-D

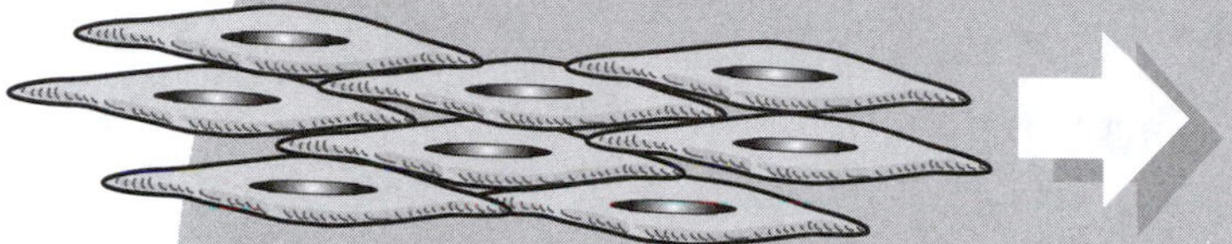

Smooth muscle is often arranged in sheets of cells stacked on top of each other.

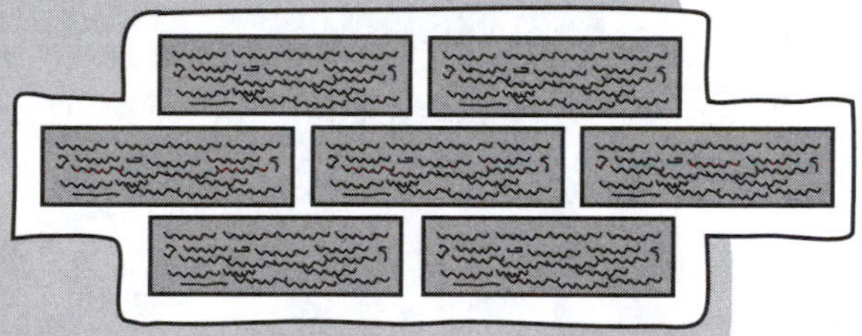

The staggered pattern of the cell arrangement looks like bricks in a wall.

2-D

Color the muscle cells the same color.

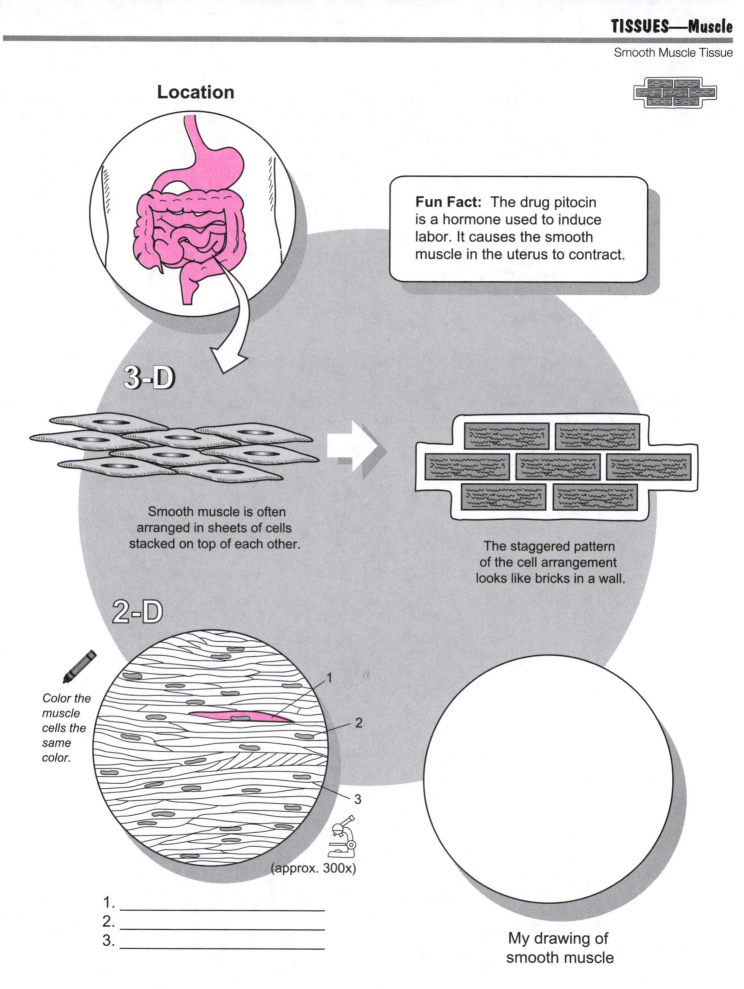

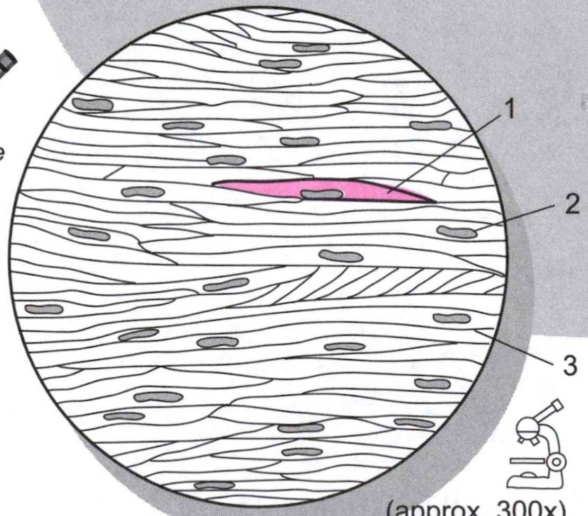

(approx. 300x)

1. ______________________
2. ______________________
3. ______________________

My drawing of smooth muscle

Description

Neurons, or nerve cells, are one of the fundamental cells in nervous tissue. Of the variety of types of neurons, all share certain features. Surrounding the nucleus of every neuron is a region called the **cell body**. Most of the organelles are found here. Branching out from the cell body are one of two types of processes—**dendrites** or an **axon**. Each neuron has only one axon per cell but may have one or more dendrites. At the end of every axon is the **synaptic knob**, which defines the end of the cell. Surrounding the neurons are various smaller cells that offer structural support and protection that constitute the **neuroglia**.

Analogy

Under the microscope, a **neuron** resembles an **octopus**. The **head** of the octopus is the **cell body** and the **tentacles** are the **cellular processes** (dendrites or axon).

Location

Brain, spinal cord, and peripheral nerves

Function

Conduct nervous (electrical) impulses to other neurons, muscles, or glands to regulate their function

Distinguishing Features

Under the microscope at higher magnifications, you can use the following landmarks to distinguish nervous tissue:

- Cellular processes (unique to nervous tissue)
- Cell body
- Rounded nucleus
- Nucleolus (may or may not be visible depending on slide quality and magnification)

Key to Illustration

1. Dendrites
2. Cell body
3. Nucleus
4. Nucleolus
5. Axon
6. Neuroglia

Location

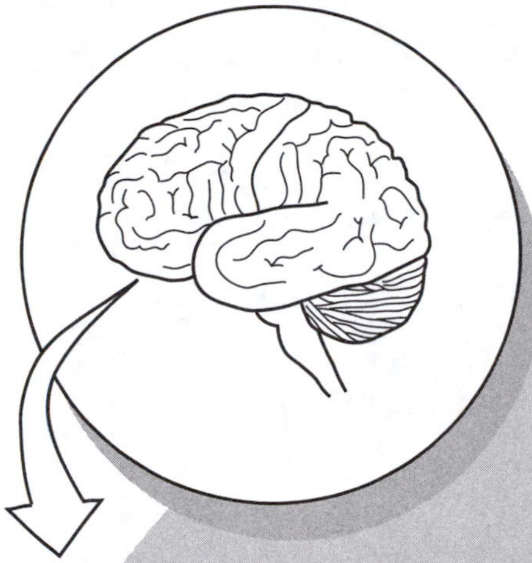

Fun Fact: Neurons are the longest cells in the body. Some are more than 3 feet in length.

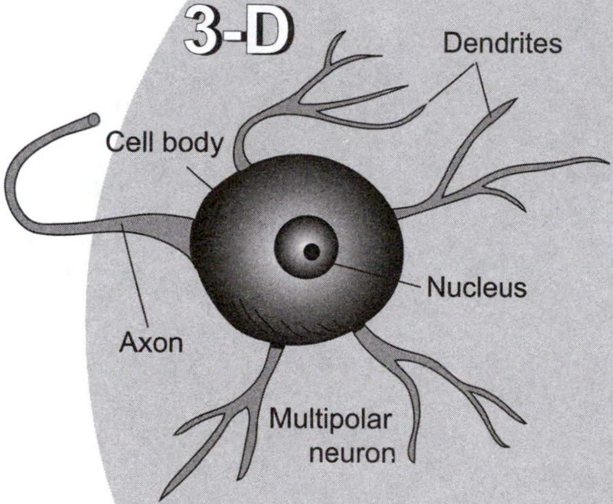

An individual multipolar neuron looks like an octopus.

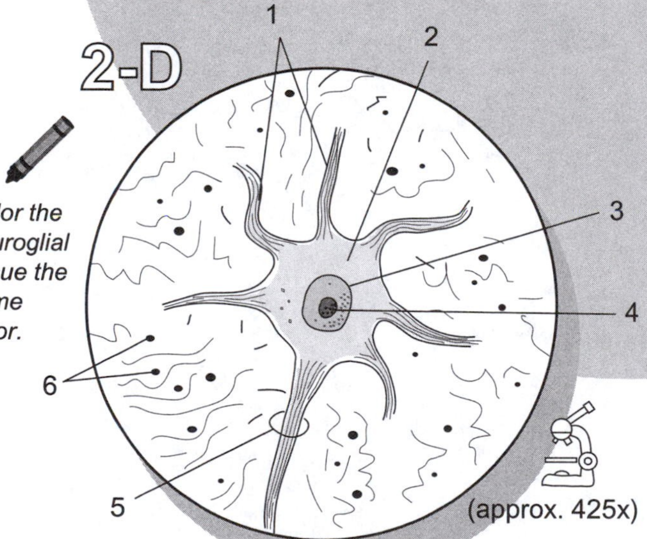

Color the neuroglial tissue the same color.

1. ______________________
2. ______________________
3. ______________________
4. ______________________
5. ______________________
6. ______________________

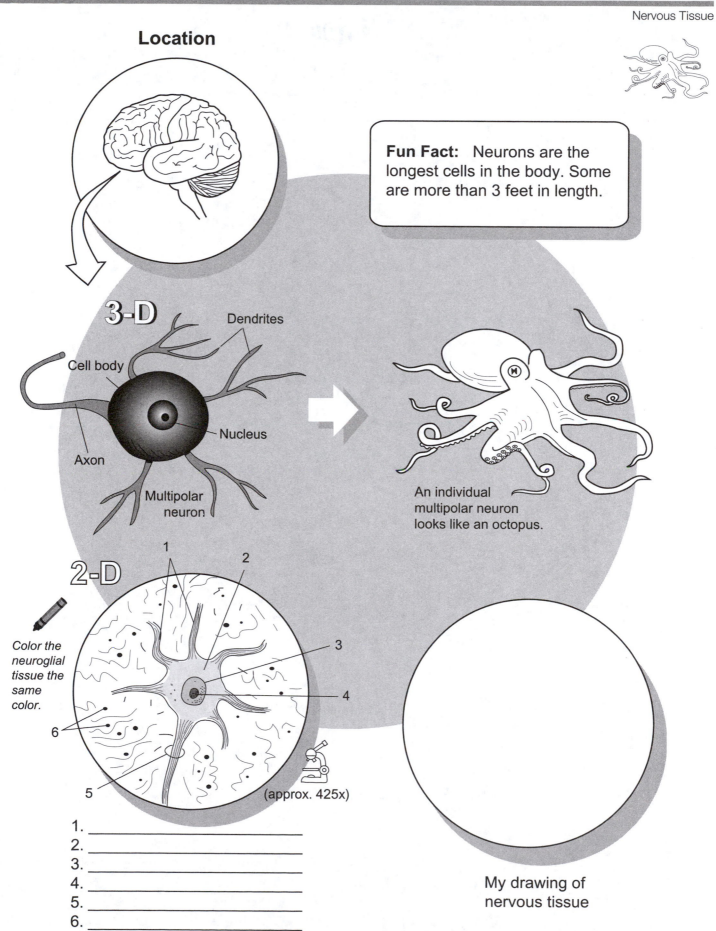

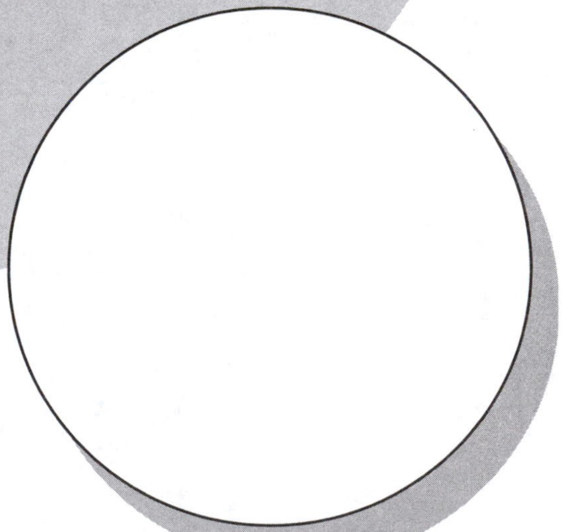

My drawing of nervous tissue

Notes

Integumentary System

Description

The skin, or **integumentary system**, is the largest organ in the body by total surface area. In the average adult it measures about 21 sq. ft. It is divided into three regional areas: **epidermis, dermis,** and **hypodermis** (*subcutaneous region*). The epidermis is the outermost layer and is composed of layers of cells. The dermis lies directly beneath and is subdivided into the upper **papillary region** and the lower **reticular region**. The papillary region consists of loose connective tissue, and the reticular region consists of dense irregular connective tissue. The dermis houses many of the glands and hair follicles within the skin. The hypodermis is below the dermis and is composed of both areolar and adipose connective tissues. Though not a true part of the skin, it is the foundation on which the skin rests. The upper region contains arteries and veins. When a *subcutaneous injection* is given with a *hypodermic* needle, it penetrates into these vessels so the drug can be delivered directly into the bloodstream.

Accessory structures include hair follicles, hair shafts, glands, and sensory receptors. **Hair follicles** produce **hair shafts**. An **arrector pili** muscle is a small cluster of smooth muscle cells connected to hair follicles. When they contract, they pull the hair follicle up, causing "goose bumps." Typical glands are **sebaceous glands** and **sweat glands**. Sebaceous (oil) glands are often connected to hair follicles. They produce an oily substance that lubricates hair shafts.

There are two types of sweat glands: **merocrine** and **apocrine**. Merocrine glands are more common and widely distributed throughout the skin. They secrete a watery secretion called sweat into a duct and release it directly onto the skin surface. This watery film absorbs heat, then evaporates, resulting in cooling the body. Apocrine glands are less common and are connected to hair follicles. They are located in the armpits, around nipples, and in the groin. Their secretion contains both lipids and proteins. When bacteria act on this secretion, the result is body odor.

Sensory receptors include **tactile corpuscles**, **free nerve endings**, and **lamellated corpuscles**. Tactile corpuscles detect mostly light touch, free nerve endings detect mostly pain and temperature changes, and lamellated corpuscles respond to changes in deep pressure.

Study Tips

Layers of Epidermis

- To remember the layers of the epidermis from the outermost to innermost layer, use the following mnemonic: "*Can Lucy Give Some Blood?*"

  (**Note**: This works only with thick skin samples such as in the sole of the foot because the *stratum lucidum* is not present in thin skin.)

- You can remember that the stratum **b**asale is at the **b**ottom of the epidermis because ***b**asale* and ***b**asement* both begin with the letter "**b**."

Function

Physical protection from environment, regulation of body temperature, synthesis of vitamin D, excretion of waste products, prevention of water loss, prevention of invasion by pathogens

Key to Illustration

Layers of Epidermis *(Thick Skin)*
1. Stratum corneum
2. Stratum lucidum
3. Stratum granulosum
4. Stratum spinosum
5. Stratum basale

Regional Areas of the Skin
6. Epidermis
7. Dermis
8. Papillary region
9. Reticular region
10. Hypodermis *(subcutaneous region)*

Accessory Structures
11. Dermal papillae
12. Tactile corpuscle
13. Free nerve ending
14. Sebaceous *(oil)* gland
15. Hair shaft
16. Hair follicle
17. Lamellated corpuscle
18. Arrector pili muscle
19. Sweat gland duct
20. Merocrine sweat gland
21. Apocrine sweat gland

Layers of Epidermis

1. ____________________
2. ____________________
3. ____________________
4. ____________________
5. ____________________

Regional Areas of the Skin

6. ____________________
7. ____________________
8. ____________________
9. ____________________
10. ____________________

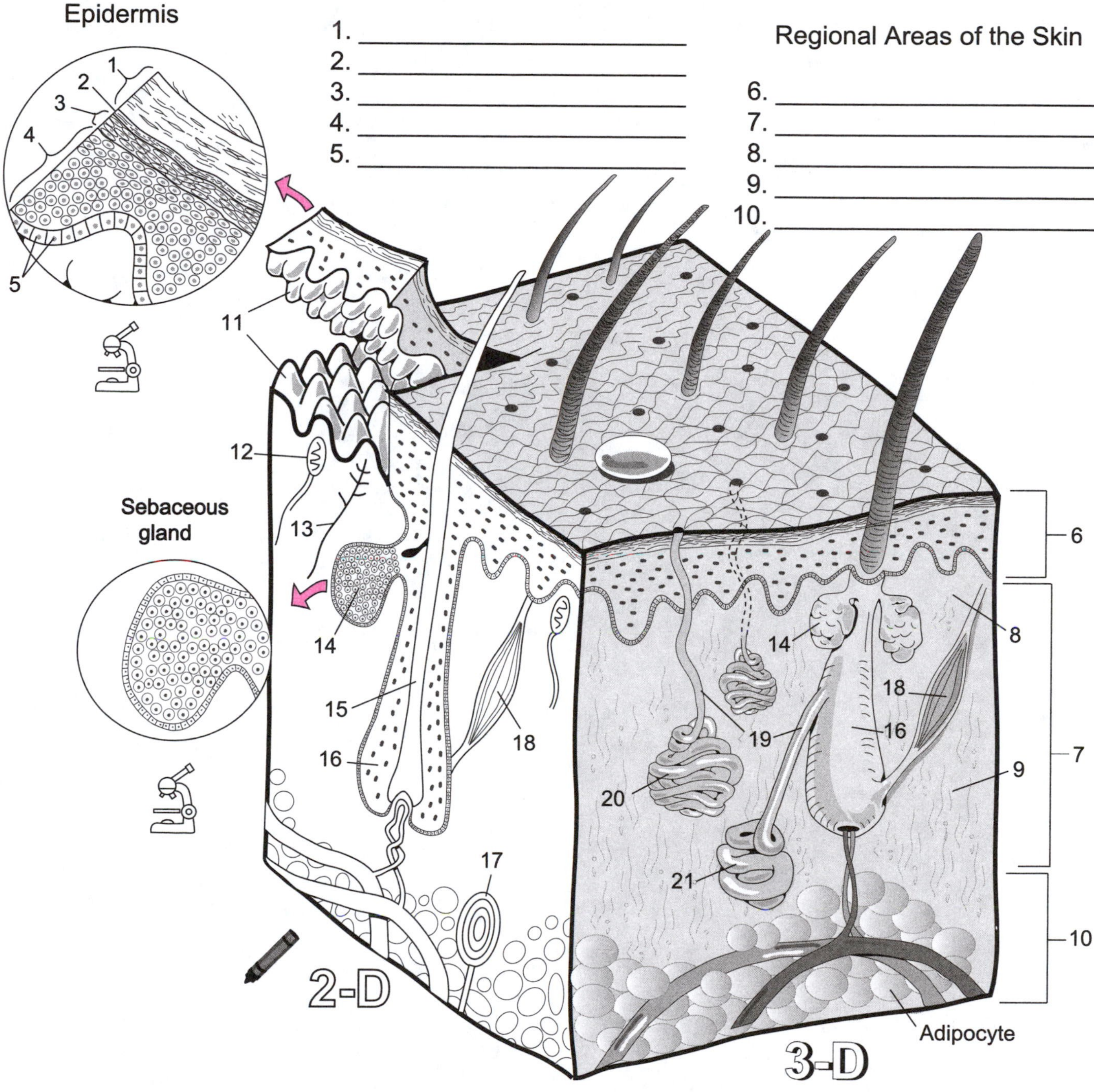

Accessory Structures

11. ____________________
12. ____________________
13. ____________________
14. ____________________
15. ____________________
16. ____________________
17. ____________________
18. ____________________
19. ____________________
20. ____________________
21. ____________________

Notes

Skeletal System

Description

The shaft of the long bone is called the **diaphysis**, while each end is called an **epiphysis**. Covering the diaphysis is a sheath of fibrous connective tissue called **periosteum**, which aids in the attachment of muscles to bone. Covering each epiphysis is a smooth layer of hyaline cartilage that is more generally referred to as **articular cartilage** because it is used to form joints. This smooth surface helps to reduce friction within the joint. Inside the diaphysis is a hollow chamber called the **medullary cavity**. It contains **yellow marrow**, which consists mostly of fatty tissue that acts as a reserve fuel supply for the body. Lining the inside of the medullary cavity is a thin cellular layer called the **endosteum**, which contains both **osteoblasts** and **osteoclasts**.

There are two types of bone within the body—**spongy** and **compact**. Spongy bone can be found within the epiphyses and lining the medullary cavity. Like a sponge, it is more porous, less organized, and contains many open spaces within it. By contrast, compact bone is much more organized and dense. It is much stronger than spongy bone and it constitutes the wall of the diaphysis. In adults, a thin layer of compact bone is also found at the **epiphyseal line**. This marks where the epiphyseal growth plate used to be located before it ossified.

Study Tip

Epiphysis is the end of a long bone—**"E**piphysis" and "**E**nd" both begin with the letter "**e**."

Key to Illustration

1. Proximal epiphysis
2. Diaphysis
3. Distal epiphysis
4. Articular cartilage
5. Spongy bone
6. Epiphyseal line
7. Medullary cavity
8. Periosteum
9. Endosteum
10. Yellow marrow
11. Compact bone
12. Wedge of compact bone

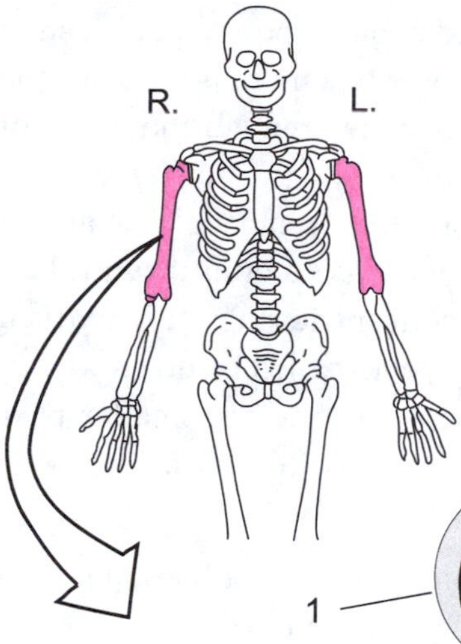

The humerus is an example of a long bone.

Color the different structures different colors.

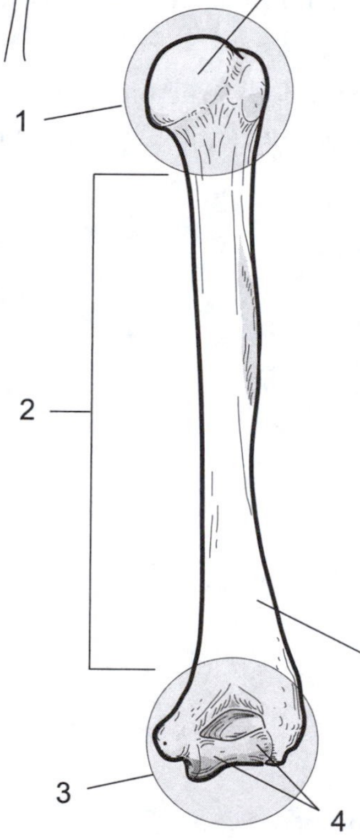

Right humerus
posterior view

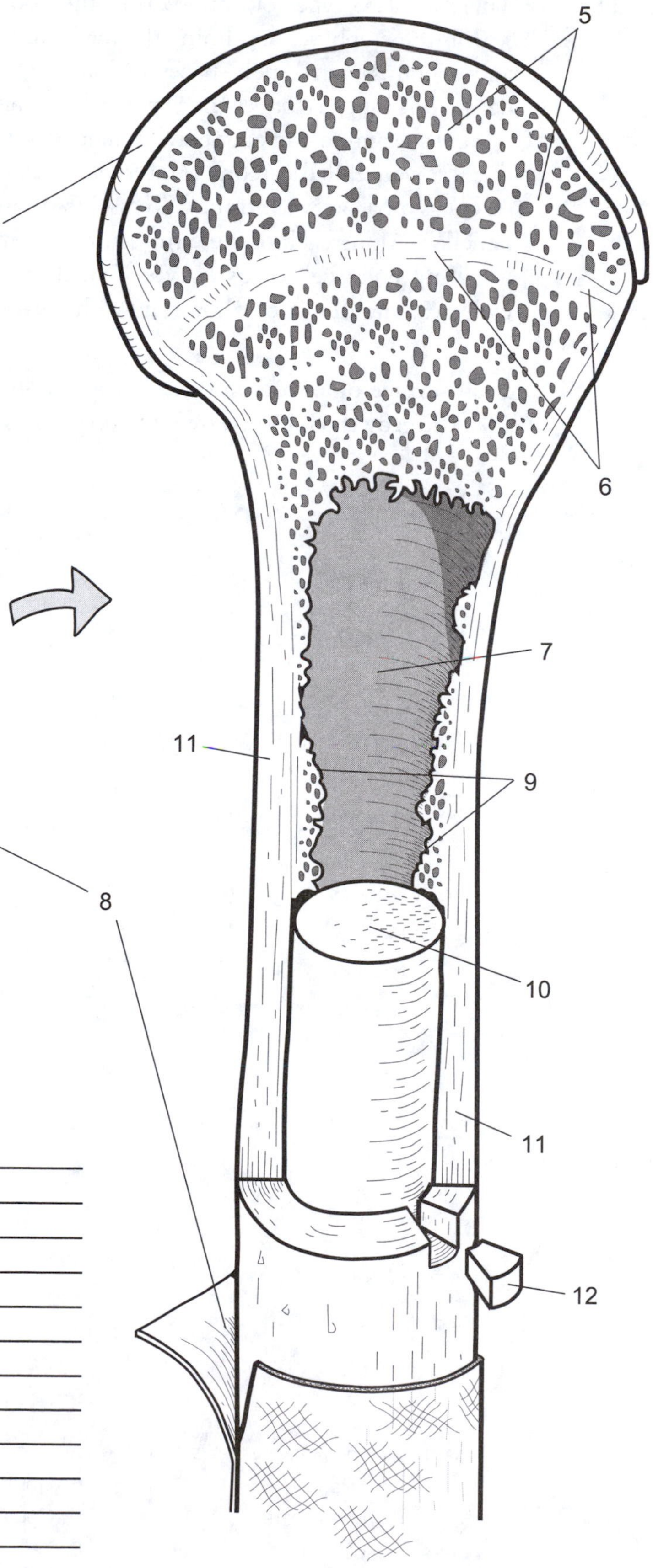

1. ______________________
2. ______________________
3. ______________________
4. ______________________
5. ______________________
6. ______________________
7. ______________________
8. ______________________
9. ______________________
10. ______________________
11. ______________________
12. ______________________

Description

There are two types of bone within the body—**spongy** and **compact**. Spongy bone can be found within the epiphyses and lining the medullary cavity. Like a sponge, it is more porous, less organized, and contains many open spaces within it. By contrast, compact bone is much more organized and dense. It is much stronger than spongy bone, and it constitutes the wall of the diaphysis.

Let's examine compact bone in more detail. The individual units in compact bone are tall, cylindrical towers called **osteons** (*Haversian systems*). In the middle of each osteon is a **central canal** that serves as a passageway for blood vessels. Around this canal are concentric rings of bony tissue called **lamellae**. Along each of these rings at regular intervals are small spaces called **lacunae**, which contain a mature bone cell or **osteocyte**. Branching between individual lacunae are smaller passageways called **canaliculi**, which allow fluid with dissolved nutrients to travel to osteocytes.

Analogy

Each **surface of an osteon** looks like a **tree stump**. Both structures are made of hard, dense materials. Like the **growth rings** in a tree, the osteon has concentric rings called **lamellae**.

Key to Illustration

1. Osteon *(Haversian system)*
2. Lamellae
3. Collagen fibers
4. Periosteum
5. Spongy bone
6. Communicating canal
7. Central canal
8. Lacuna
9. Osteocyte
10. Canaliculi

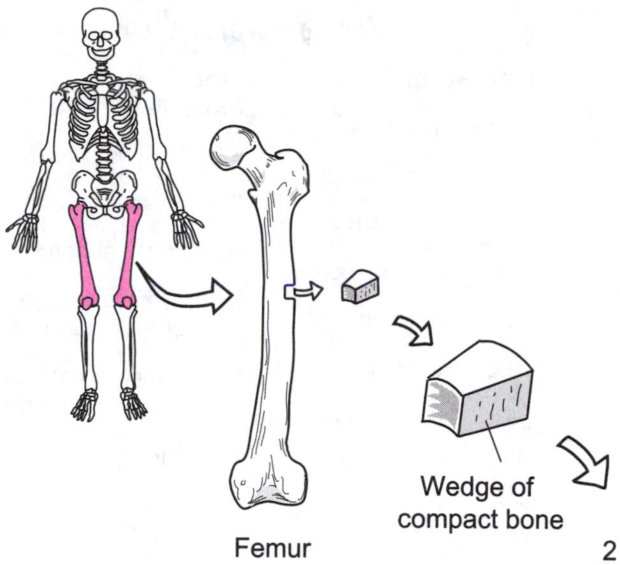

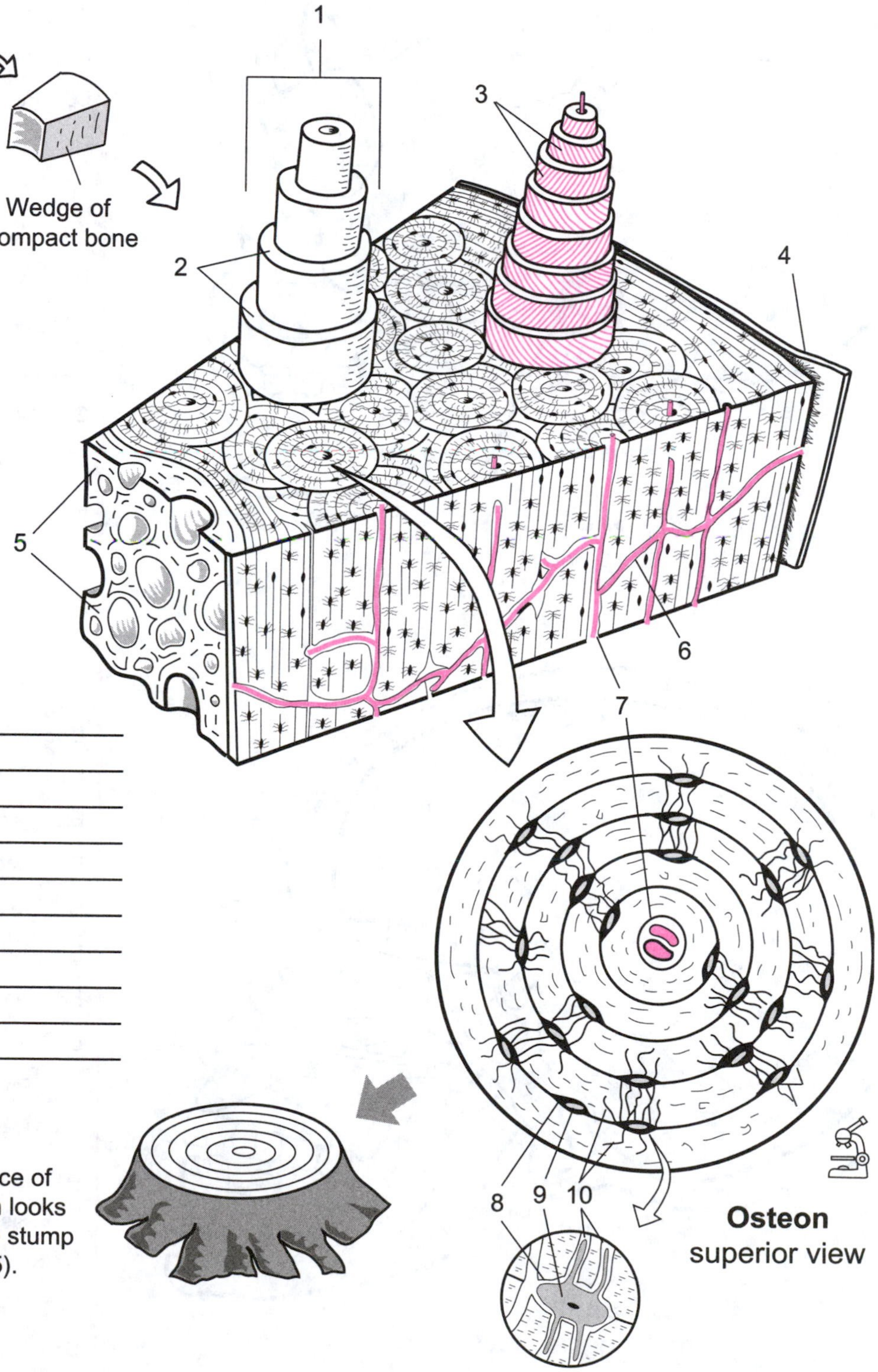

1. ______________________
2. ______________________
3. ______________________
4. ______________________
5. ______________________
6. ______________________
7. ______________________
8. ______________________
9. ______________________
10. ______________________

The surface of an osteon looks like a tree stump (see p. 95).

Skull—Lateral and Midsagittal Views

Color the different bones to match your textbook.

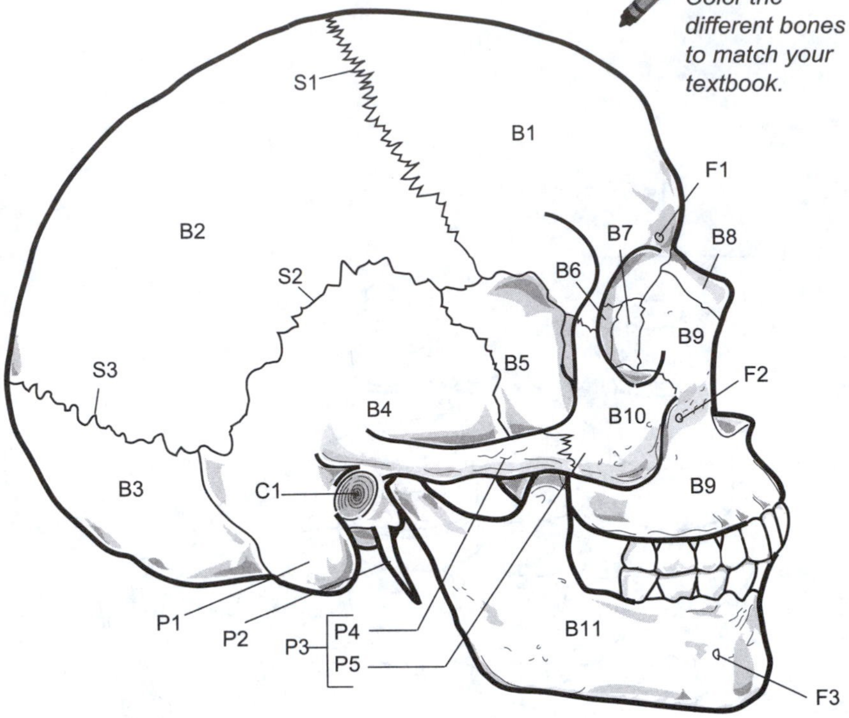

Skull
lateral view

Key to Lateral View

Bones (B)

- B1 Frontal bone
- B2 Parietal bone
- B3 Occipital bone
- B4 Temporal bone
- B5 Sphenoid
- B6 Ethmoid
- B7 Lacrimal bone
- B8 Nasal bone
- B9 Maxilla
- B10 Zygomatic bone
- B11 Mandible

Sutures (S)

- S1 Coronal suture
- S2 Squamous suture
- S3 Lambdoid suture

Processes, Projections (P)

- P1 Mastoid process
- P2 Styloid process
- P3 Zygomatic arch
- P4 Zygomatic process of temporal bone
- P5 Temporal process of zygomatic bone

Foramina (F)

- F1 Supraorbital foramen
- F2 Infraorbital foramen
- F3 Mental foramen

Canal (C)

- C1 External acoustic meatus (*canal*)

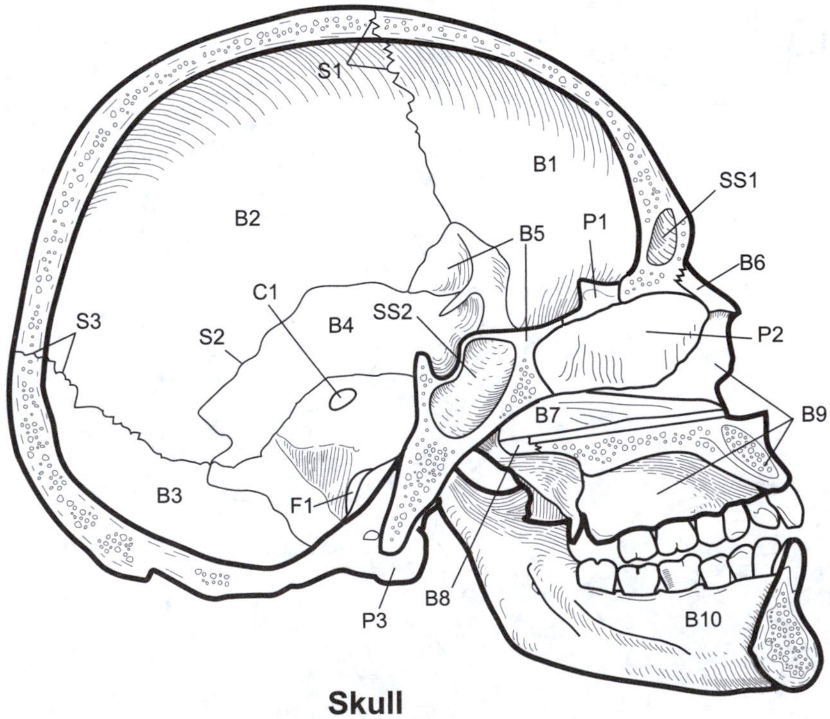

Skull
midsaggital view

Key to Midsaggital View

Bones (B)

- B1 Frontal bone
- B2 Parietal bone
- B3 Occipital bone
- B4 Temporal bone
- B5 Sphenoid
- B6 Nasal bone
- B7 Vomer
- B8 Palatine bone
- B9 Maxilla
- B10 Mandible

Sutures (S)

- S1 Coronal suture
- S2 Squamous suture
- S3 Lambdoid suture

Processes, Projections (P)

- P1 Crista galli (*of ethmoid*)
- P2 Perpendicular plate (*of ethmoid*)
- P3 Occipital condyle

Foramina (F)

- F1 Jugular foramen

Sinuses (SS)

- SS1 Frontal sinus
- SS2 Sphenoidal sinus

Canal (C)

- C1 Internal acoustic meatus (*canal*)

Color the different bones to match your textbook.

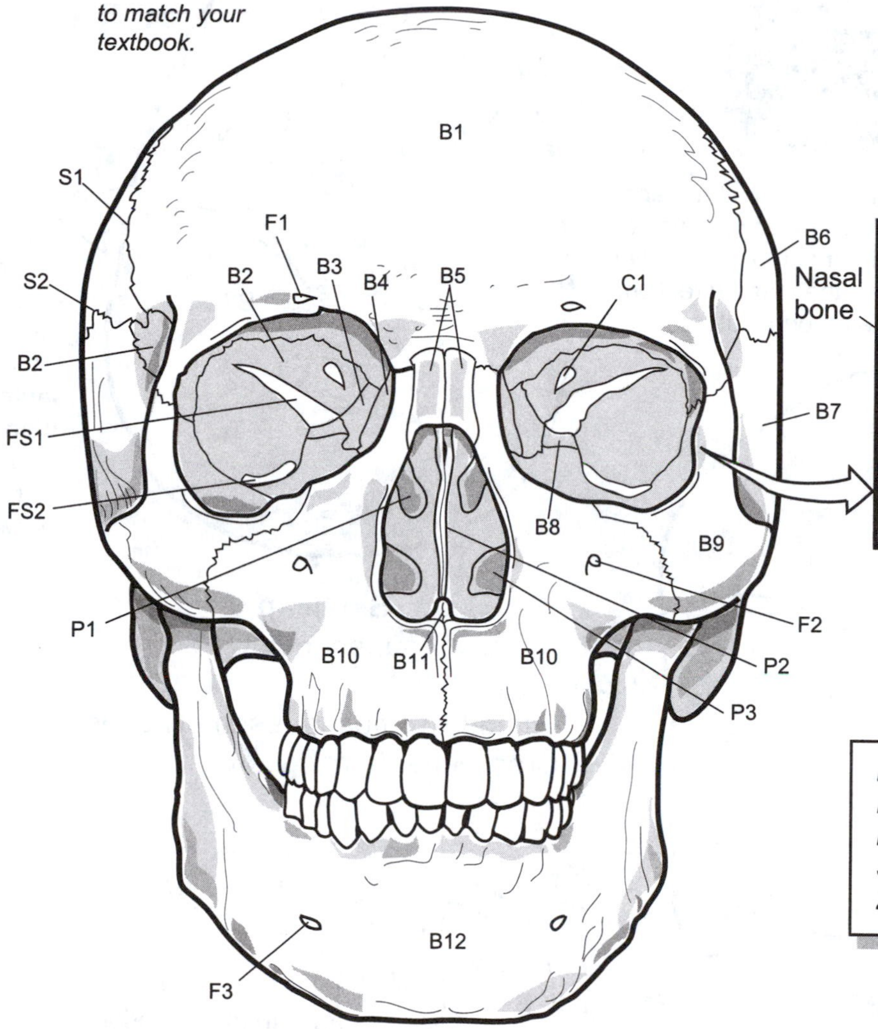

Orbital Complex

Seven bones (b.) form the bony orbit that encases the eye:

- Frontal b.
- Maxilla
- Lacrimal b.
- Ethmoid
- Palatine b.
- Sphenoid
- Zygomatic b.

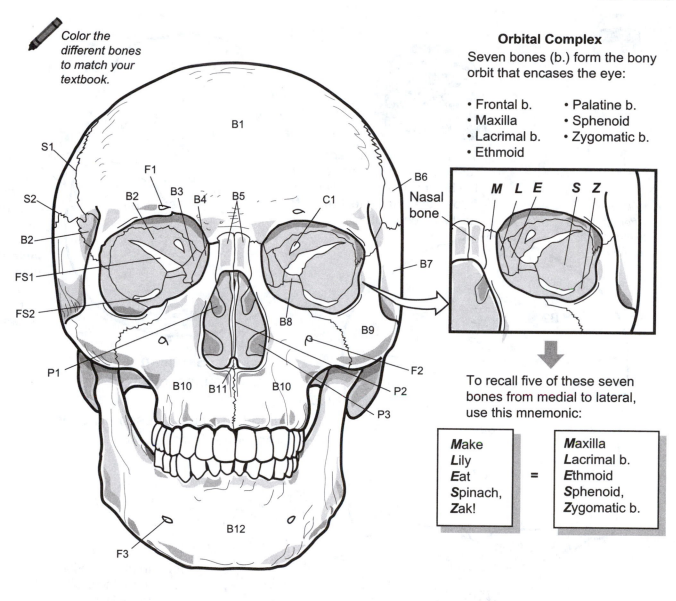

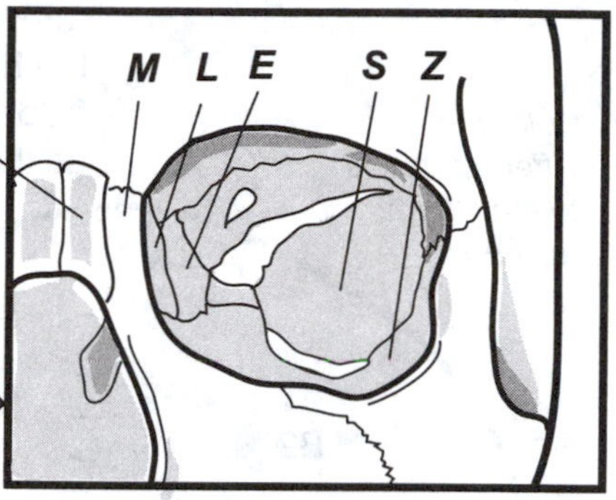

To recall five of these seven bones from medial to lateral, use this mnemonic:

***M**ake **L**ily **E**at **S**pinach, **Z**ak!*	=	***M**axilla **L**acrimal b. **E**thmoid **S**phenoid, **Z**ygomatic b.*

Key to Illustration

Bones (B)

B1 Frontal bone
B2 Sphenoid
B3 Ethmoid
B4 Lacrimal bone
B5 Nasal bone
B6 Parietal bone
B7 Temporal bone
B8 Palatine bone
B9 Zygomatic bone
B10 Maxilla
B11 Vomer
B12 Mandible

Sutures (S)

S1 Coronal suture
S2 Squamous suture

Foramina (F)

F1 Supraorbital foramen
F2 Infraorbital foramen
F3 Mental foramen

Canal (C)

C1 Optic canal

Processes, Projections (P)

P1 Middle nasal concha
P2 Perpendicular process of ethmoid
P3 Inferior nasal concha

Fissures (FS)

FS1 Superior orbital fissure
FS2 Inferior orbital fissure

Skull—Posterior View and Skullcap

The term "coronal" means "crown." A tiara (*crown*) on the head of a beauty queen follows the same direction as the coronal suture.

Color the different bones to match your textbook.

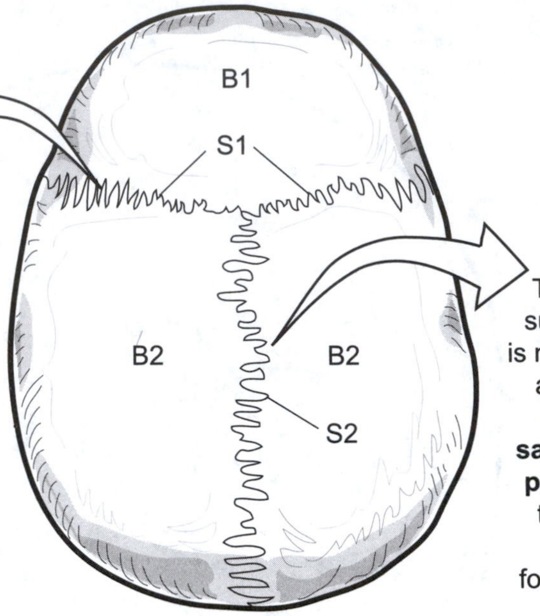

Skullcap
superior view

This suture is named after the **sagittal plane** that it follows.

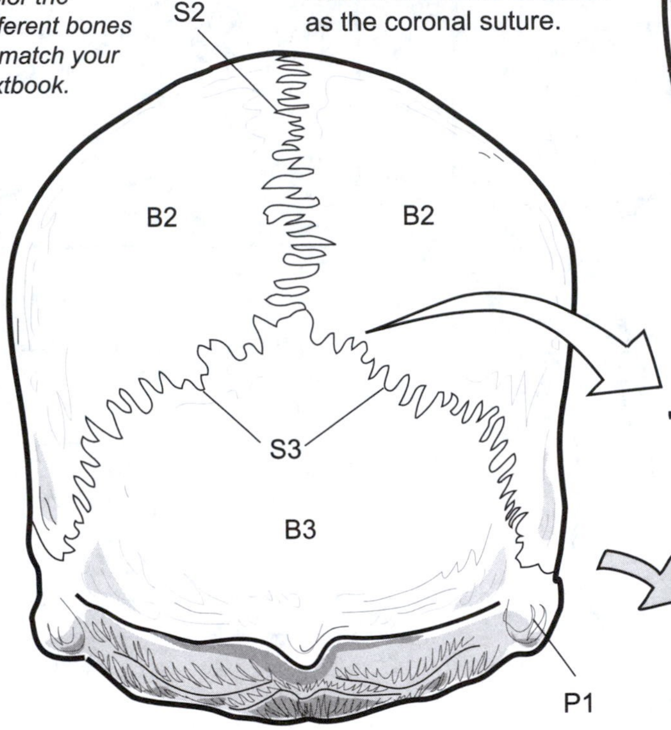

Skull
posterior view

"Lambdoid" means resembling the Greek letter lambda Λ

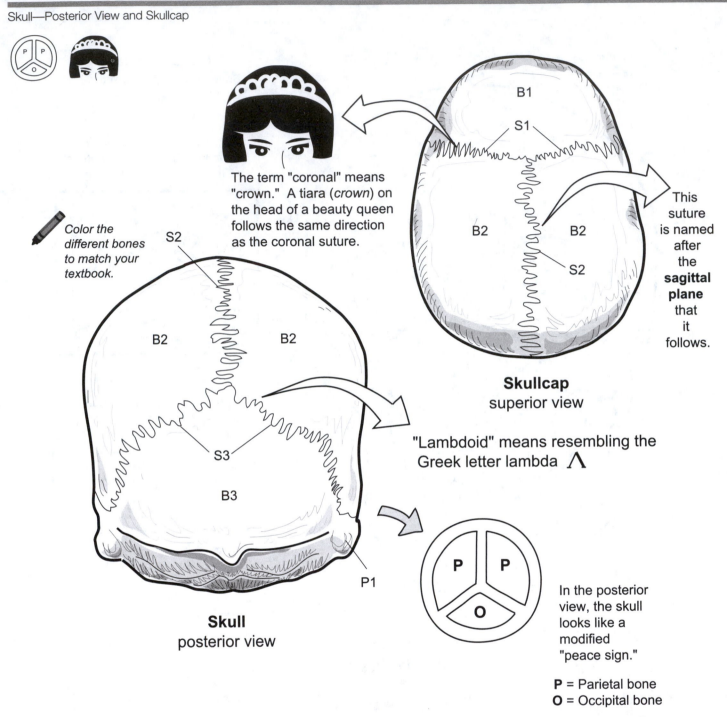

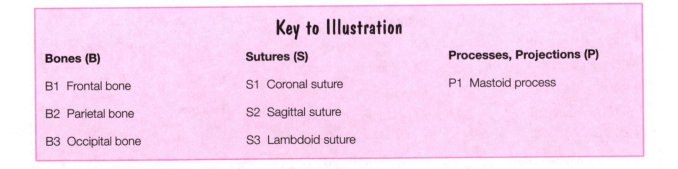

In the posterior view, the skull looks like a modified "peace sign."

P = Parietal bone
O = Occipital bone

Key to Illustration

Bones (B)	Sutures (S)	Processes, Projections (P)
B1 Frontal bone	S1 Coronal suture	P1 Mastoid process
B2 Parietal bone	S2 Sagittal suture	
B3 Occipital bone	S3 Lambdoid suture	

BONES of the SKELETON

The adult skeleton typically contains a total of **206** individual bones. Slightly more than half of these are found in the **HANDS** and **FEET**

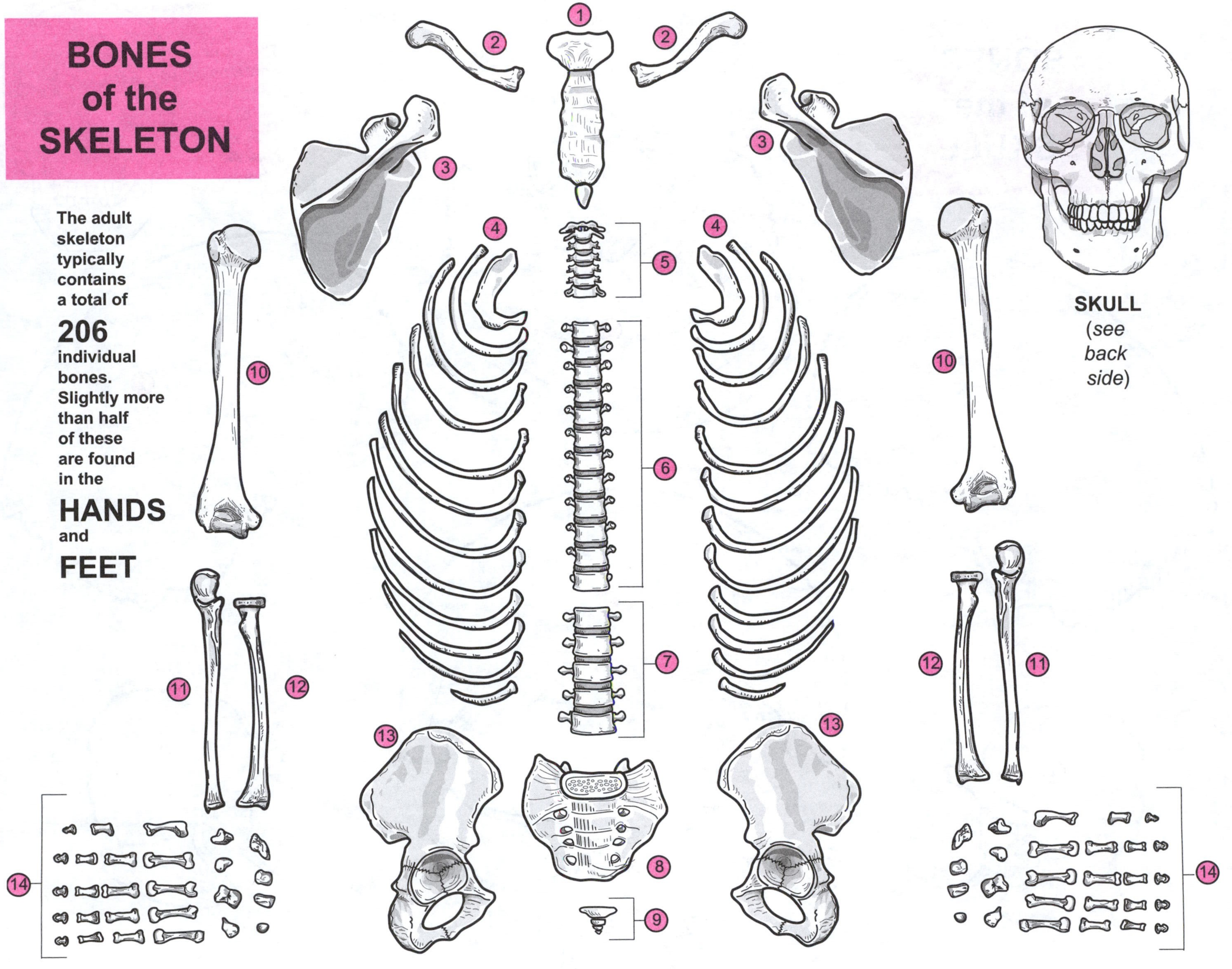

BONES of the SKULL

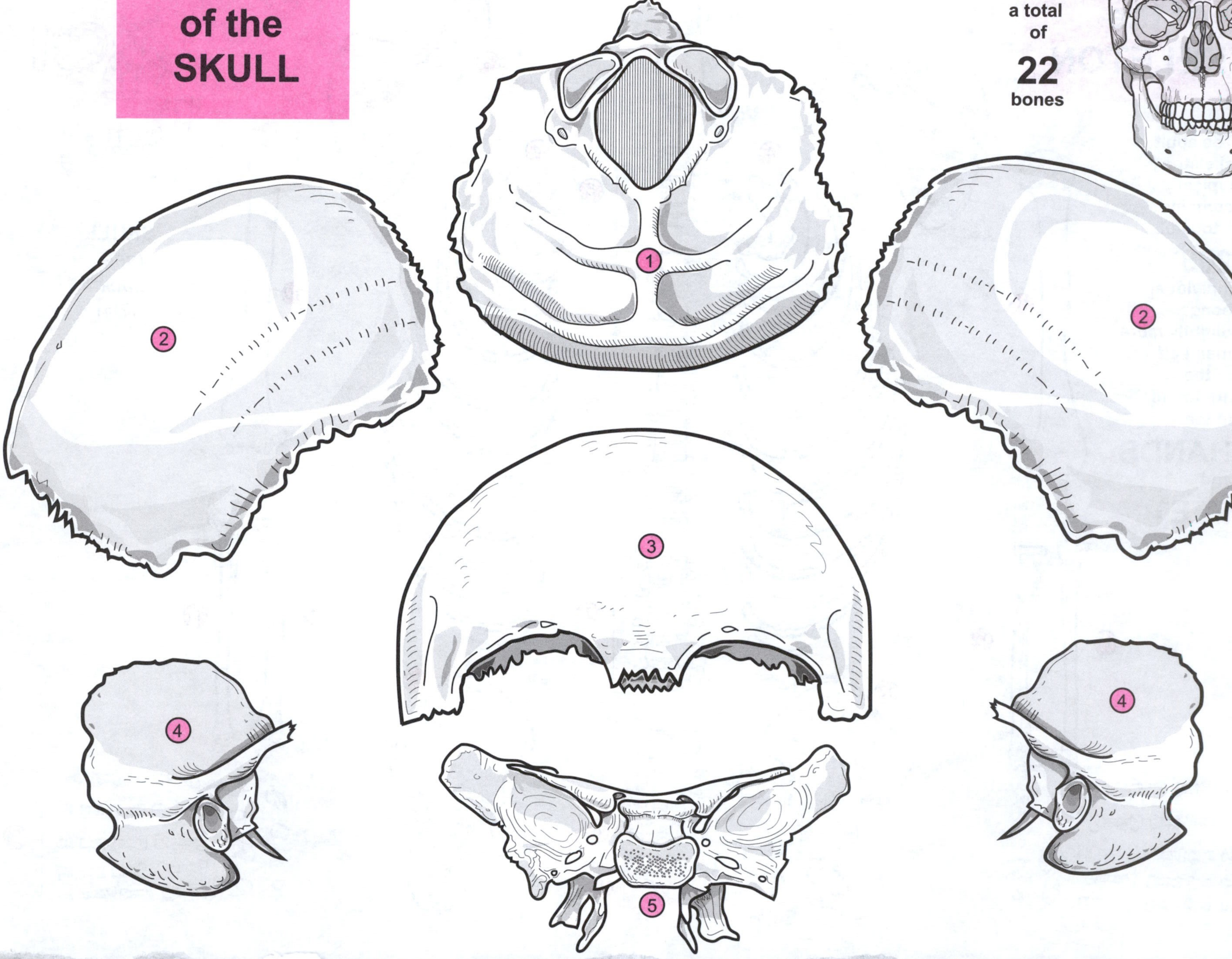

The skull contains a total of

22

bones

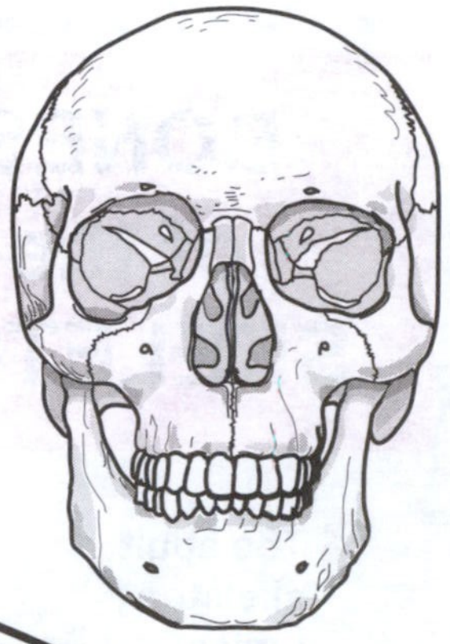

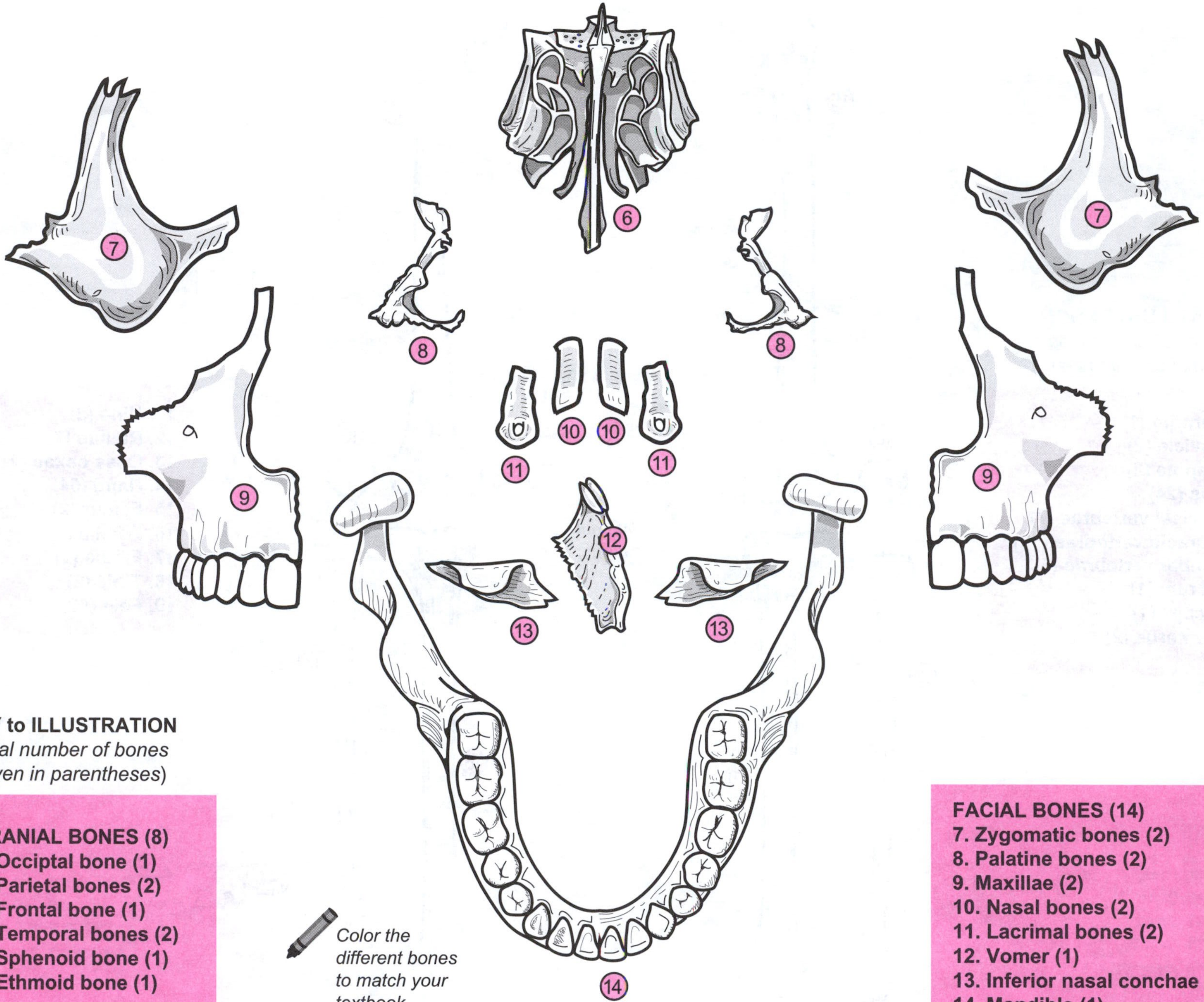

KEY to ILLUSTRATION
(total number of bones given in parentheses)

CRANIAL BONES (8)
1. Occiptal bone (1)
2. Parietal bones (2)
3. Frontal bone (1)
4. Temporal bones (2)
5. Sphenoid bone (1)
6. Ethmoid bone (1)

Color the different bones to match your textbook.

FACIAL BONES (14)
7. Zygomatic bones (2)
8. Palatine bones (2)
9. Maxillae (2)
10. Nasal bones (2)
11. Lacrimal bones (2)
12. Vomer (1)
13. Inferior nasal conchae (2)
14. Mandible (1)

KEY to ILLUSTRATION
(*total number of bones given in parentheses*)

1. Sternum (1)
2. Clavicle (2)
3. Scapula (2)
4. Ribs (24)
5. Cervical vertebrae (7)
6. Thoracic vertebrae (12)
7. Lumbar vertebrae (5)
8. Sacrum (1)
9. Coccyx (1)
10. Humerus (2)

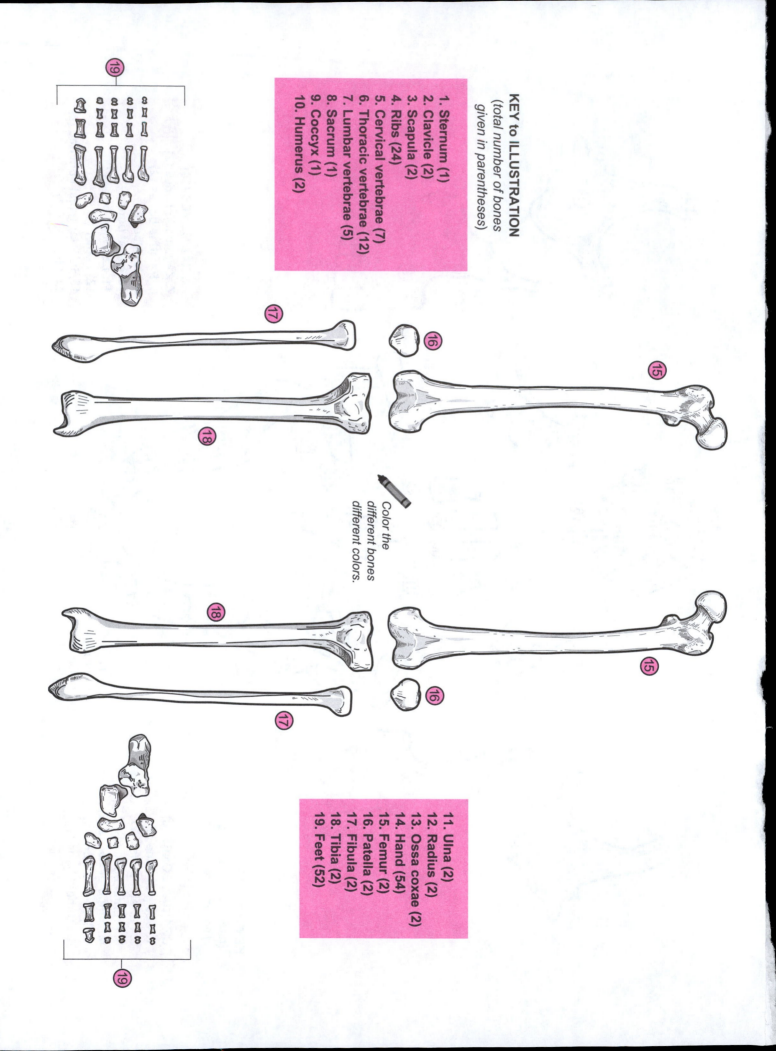

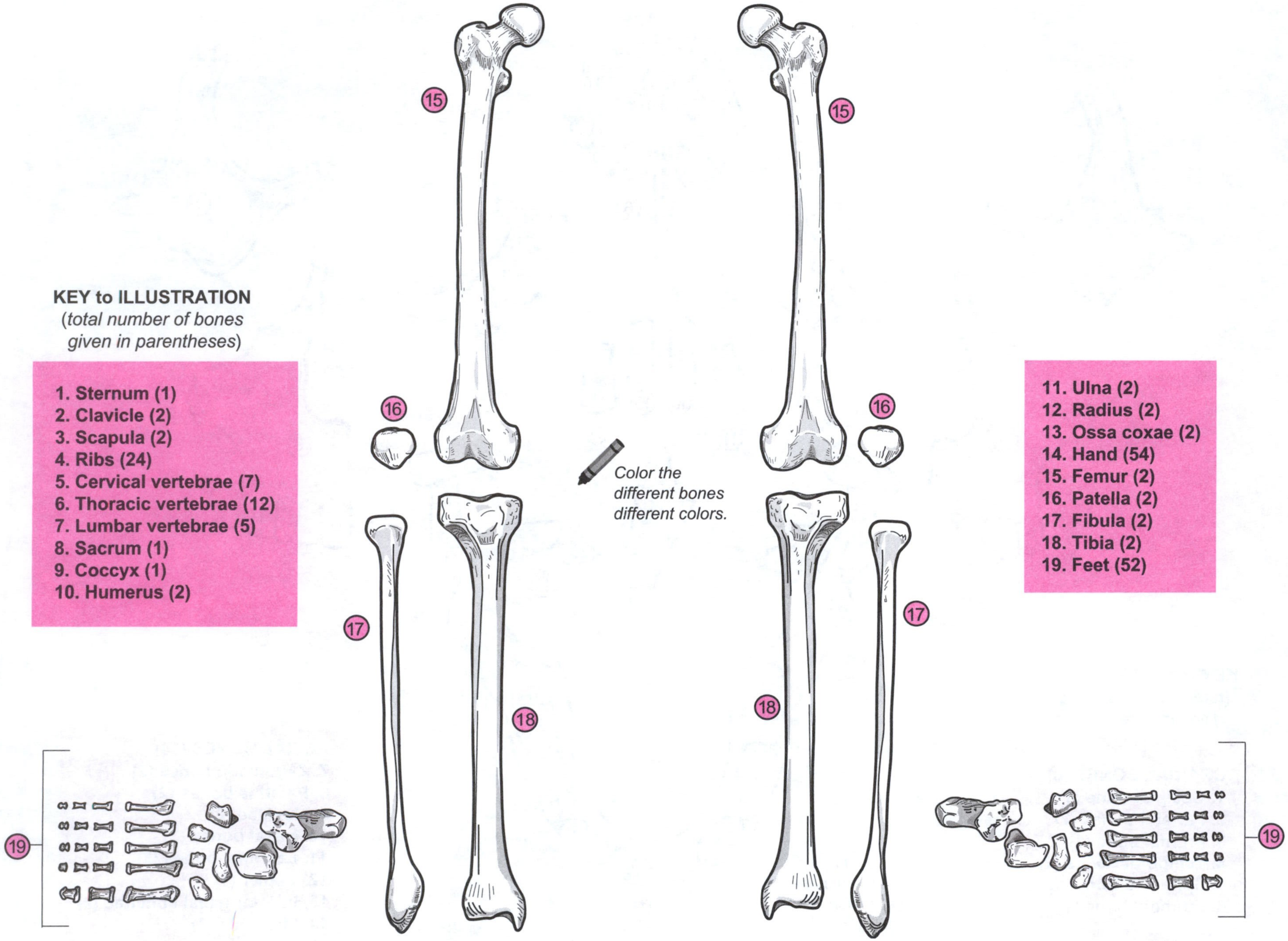

Color the different bones different colors.

11. Ulna (2)
12. Radius (2)
13. Ossa coxae (2)
14. Hand (54)
15. Femur (2)
16. Patella (2)
17. Fibula (2)
18. Tibia (2)
19. Feet (52)

Skull—Superior View without Skullcap

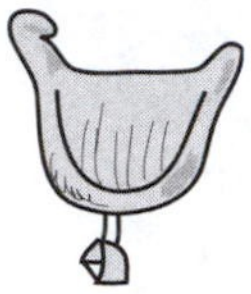

Simplify what you see, use the **sella turcica** and foramina as landmarks.

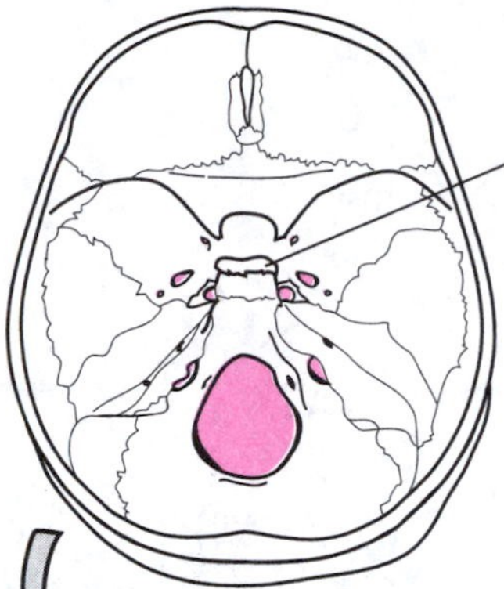

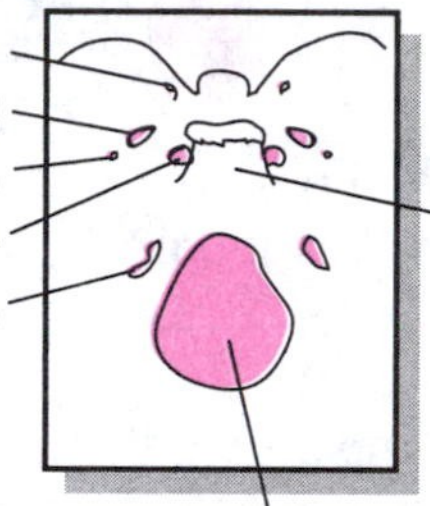

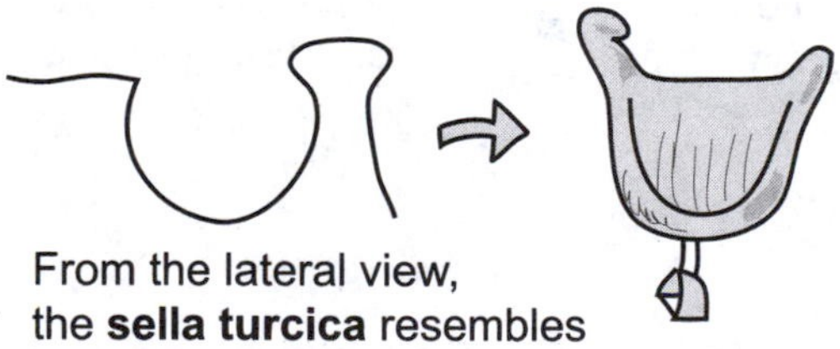

From the lateral view, the **sella turcica** resembles a horse's saddle.

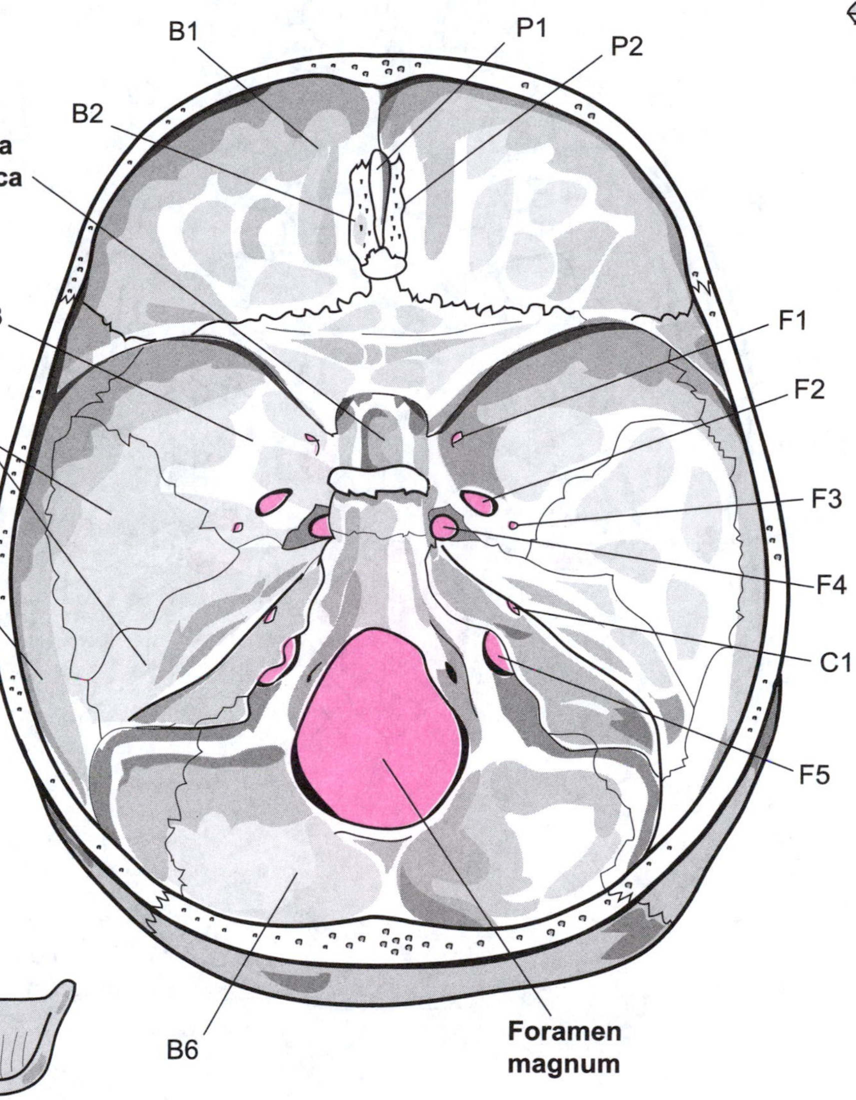

Key to Illustration

Bones (B)

B1 Frontal bone

B2 Ethmoid

B3 Sphenoid

B4 Temporal bone

B5 Parietal bone

B6 Occipital bone

Processes, Projections (P)

P1 Crista galli

P2 Cribriform plate

Canal (C)

C1 Internal acoustic meatus

Foramina (F)

F1 Foramen rotundum

F2 Foramen ovale

F3 Foramen spinosum

F4 Foramen lacerum

F5 Jugular foramen

Skull—Inferior View with Mandible Removed

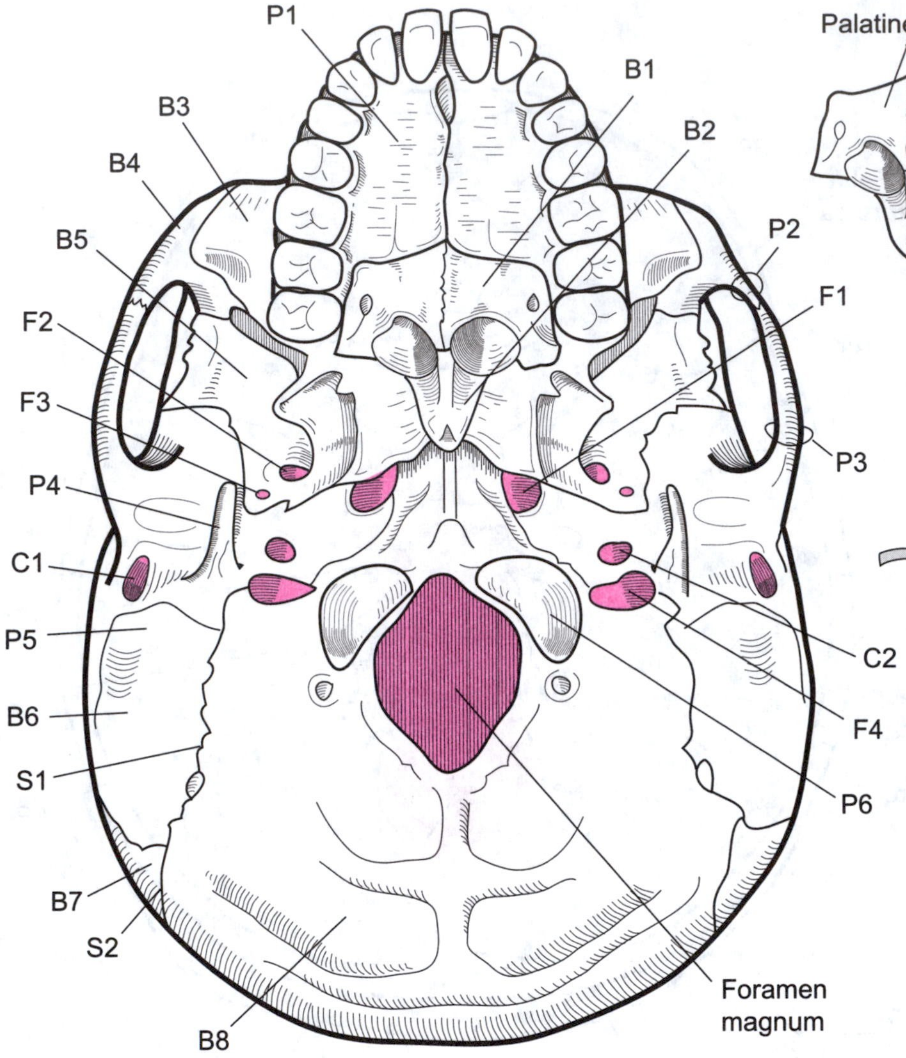

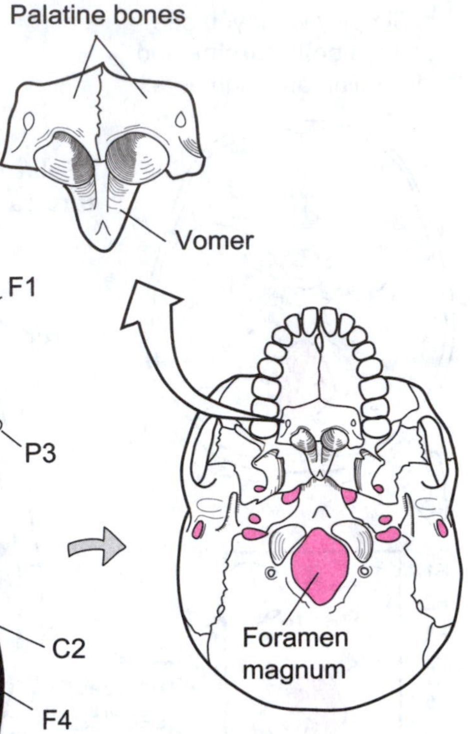

Simplify what you see and use major foramina as landmarks.

Color the different bones to match your textbook.

Key to Illustration

Bones (B)

B1 Palatine bone

B2 Vomer

B3 Maxilla

B4 Zygomatic bone

B5 Sphenoid

B6 Temporal bone

B7 Parietal bone

B8 Occipital bone

Sutures (S)

S1 Occipitomastoid suture

S2 Lambdoid suture

Foramina (F)

F1 Foramen lacerum

F2 Foramen ovale

F3 Foramen spinosum

F4 Jugular foramen

Processes, Projections (P)

P1 Palatine process of maxilla

P2 Temporal process of zygomatic bone

P3 Zygomatic process of temporal bone

P4 Styloid process (*temporal bone*)

P5 Mastoid process (*temporal bone*)

P6 Occipital condyle

Canal (C)

C1 External acoustic meatus (*canal*)

C2 Carotid canal

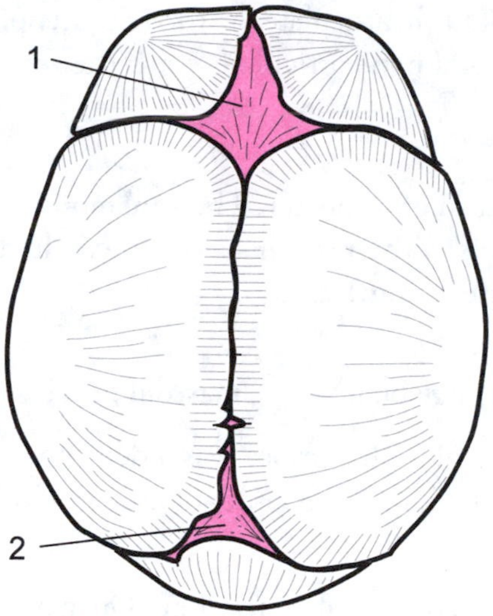

Fetal skull
superior view

Color the different bones to match your textbook.

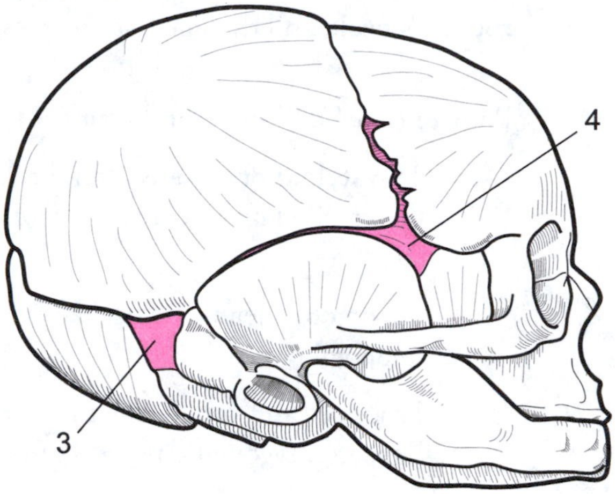

Fetal skull
lateral view

Fontanels are flat areas of fibrous connective tissue between bones of the skull that have not completed the ossification process. There are 4 major fontanels in the fetal skull. What is referred to as the "soft spot" on an infant's skull is actually the **anterior fontanel**, which typically closes by 2 years of age.

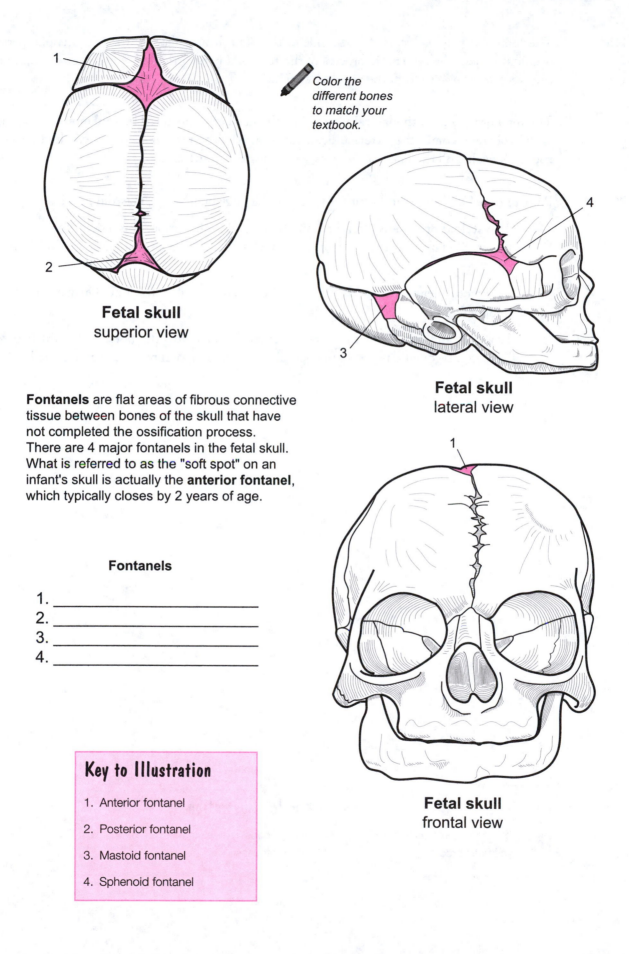

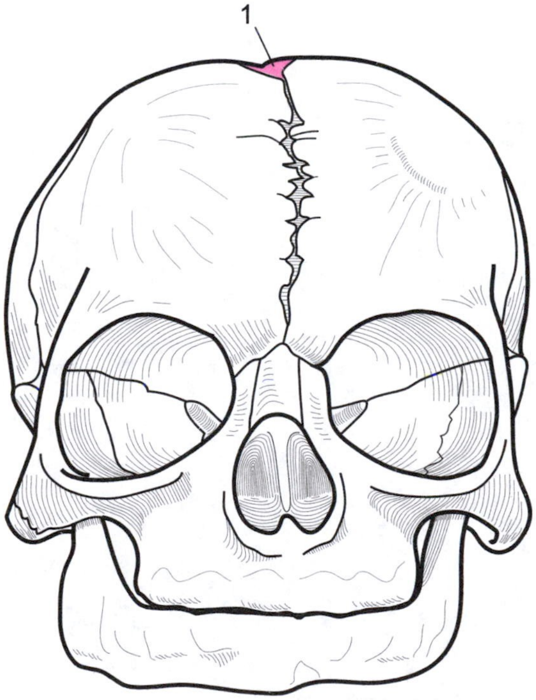

Fetal skull
frontal view

Fontanels

1. ______________________
2. ______________________
3. ______________________
4. ______________________

Key to Illustration

1. Anterior fontanel
2. Posterior fontanel
3. Mastoid fontanel
4. Sphenoid fontanel

Description

The temporal bone is located on the side of the skull. It articulates with the sphenoid, parietal, and occipital bones. The zygomatic process of the temporal bone articulates with the temporal process of the zygomatic bone to form the zygomatic arch.

Analogy

The **temporal bone in the lateral view** is like the **head of a rooster.** The **squamous portion** (*squama*) is the **rooster's comb**, the **external acoustic meatus** is the **rooster's eye**, the **styloid process** is the **rooster's beak**, and the **mastoid process** is the **rooster's wattles**.

Study Tips

Palpate (*touch*): Feel behind your ear to locate the large bump—the **mastoid process.**

- The **styloid process** is often broken off of a real skull because it is a delicate structure. Do not be surprised if you cannot locate it on a *real* skull, but a good-quality plastic skull will have this structure.
- Squamous means "scale-like," which accurately describes the flat, *squamous* portion of the temporal bone.
- Processes and other structures are sometimes named after the bones with which they articulate: *e.g.*, zygomatic process of temporal bone and temporal process of zygomatic bone.

Key to Illustration

1. Squamous part *(squama)*
2. Zygomatic process
3. External acoustic meatus
4. Mastoid process
5. Styloid process

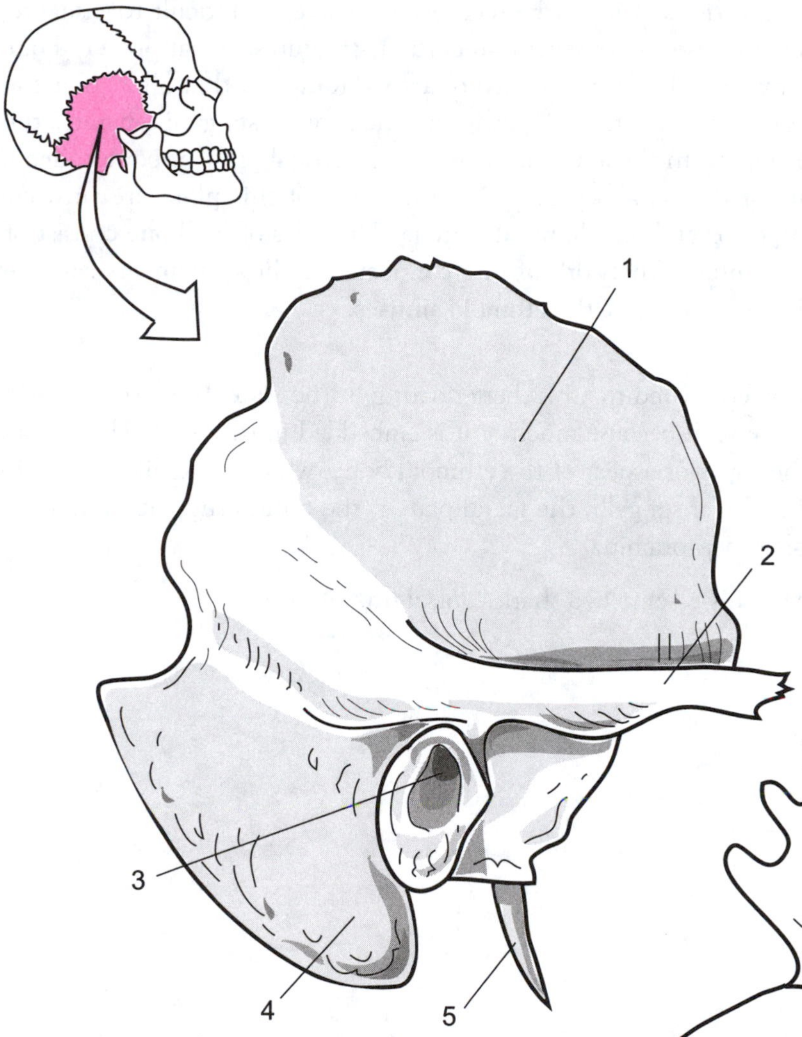

Temporal bone
lateral view

In the lateral view, a temporal bone looks like a rooster's head.

1. ______________________
2. ______________________
3. ______________________
4. ______________________
5. ______________________

Description

The **ethmoid** is embedded in the skull and located behind the bridge of the nose. The entire bone is not visible. Only specific portions of it can be seen, which makes it difficult to visualize how it fits into the skull. On its superior aspect are several important structures. A flat plate of bone called the **cribriform plate** has many small holes in it called **olfactory foramina** that allow olfactory nerves to pass from the olfactory organ to the brain. A partition called the **crista galli** separates the cribriform plate into left and right halves. In the anterior view, a long vertical plate of bone called the **perpendicular plate** forms part of the nasal septum. On either side of this plate are tiny, curved, bony projections called the **superior** and **middle nasal concha**. The sides of the bone consist of the lateral masses. These structures contain a network of interconnected, hollow chambers called **ethmoid air cells**. Together, these air cells constitute the **ethmoid sinuses**.

Analogy

- The entire bone can be compared to an iceberg floating in the water. Like the iceberg, the entire ethmoid bone cannot be seen because much of it is embedded in the skull. The tip of the iceberg that can be seen is the superior aspect of the ethmoid bone, which looks like a door hinge folded flat. The hinge itself is the crista galli, the metal plate is the cribriform plate, and the screw holes in the plate are the olfactory foramina.
- The crista galli more closely resembles a shark's dorsal fin.

Location

Skull

Key to Illustration

1. Crista galli
2. Cribriform plate
3. Olfactory foramina
4. Superior nasal concha
5. Ethmoid air cells
6. Middle nasal concha
7. Perpendicular plate

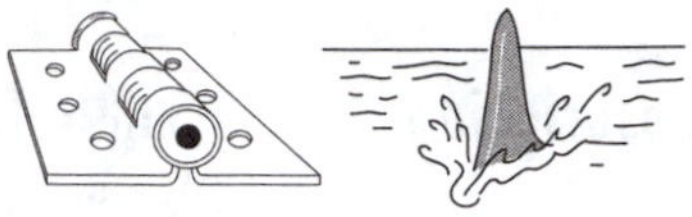

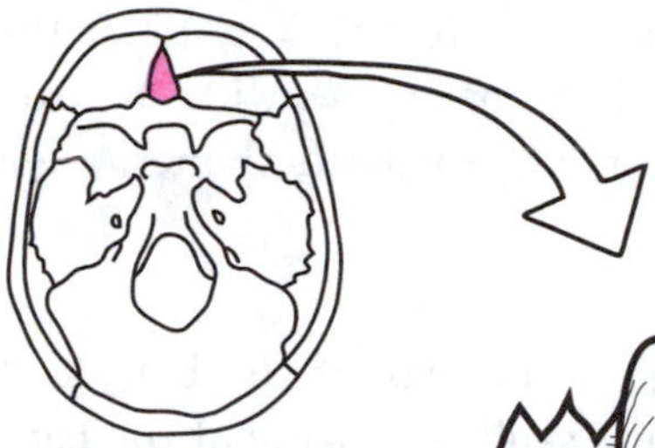

Skull
superior view
with skullcap removed

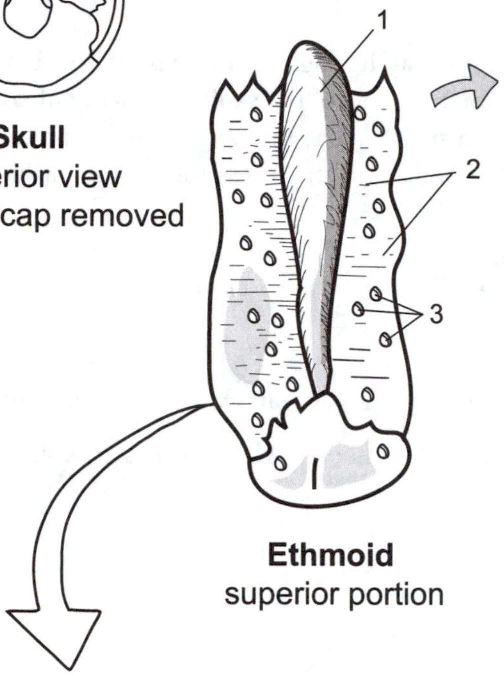

Ethmoid
superior portion

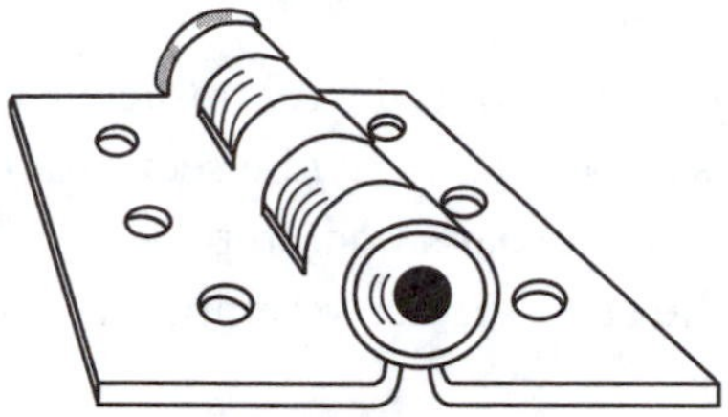

The superior view of the ethmoid looks like a door hinge.

1. ______________________________
2. ______________________________
3. ______________________________
4. ______________________________
5. ______________________________
6. ______________________________
7. ______________________________

In the superior view of the ethmoid, you are seeing only one small portion of the whole bone, like the tip of an iceberg.

The rest of the bone is embedded in the skull.

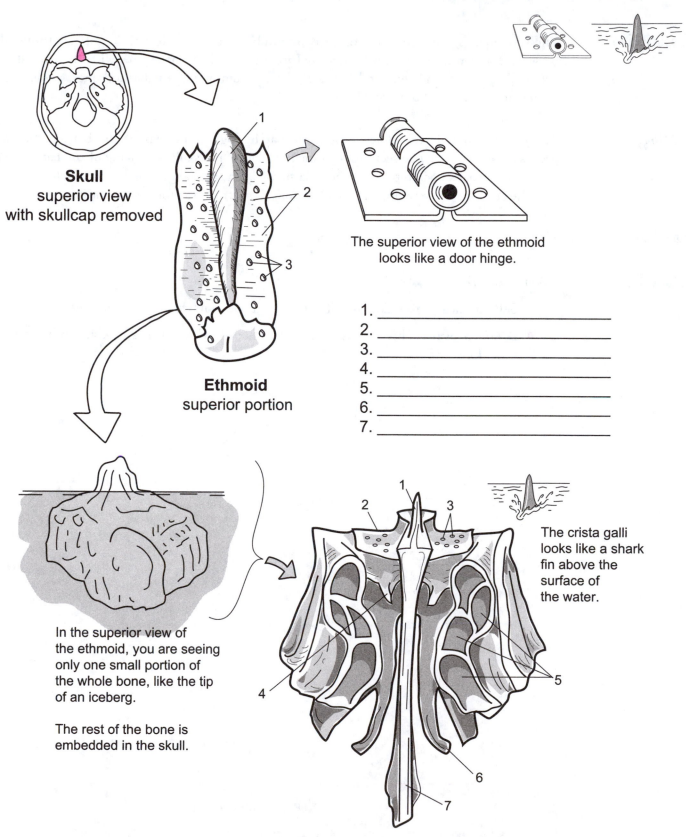

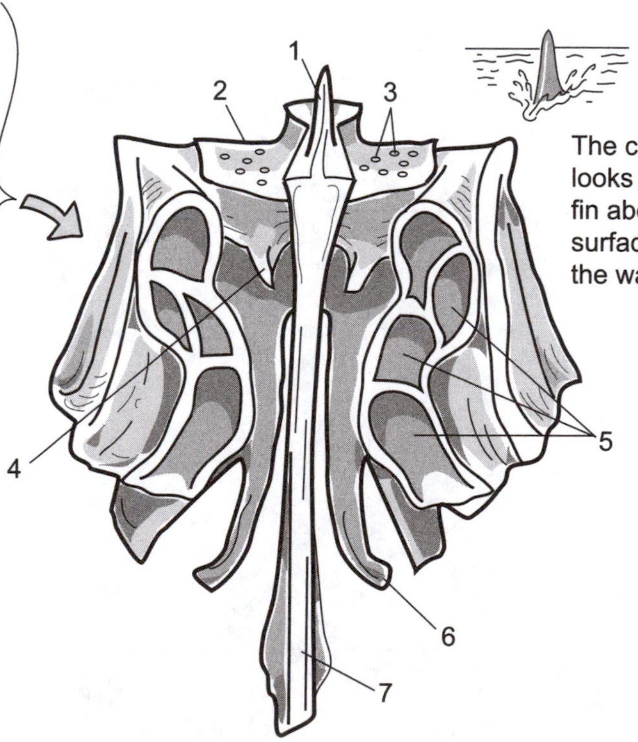

The crista galli looks like a shark fin above the surface of the water.

Ethmoid
anterior view

Description

The sphenoid (*sphenoid bone*) is embedded within the skull, so it can be difficult to visualize. To be seen in its entirety, it must be removed from the skull. It articulates with the frontal, parietal, occipital, ethmoid and temporal bones of the cranium and the palatine bones, zygomatic bones, vomer, and maxillae of the facial bones.

Analogy

The **entire sphenoid** resembles a **big-eared bat in flight**. The **legs of the bat** are the **pterygoid processes** and can only be seen from an inferior view of the skull. The **wing of the bat** is the **greater wing** and the **bat's body** represents the **sella turcica**. The big **ears of the bat** are the **lesser wings**.

The **sella turcica** resembles a **horse's saddle**. This structure protects the pituitary gland.

Location

Skull

Study Tips

Good landmarks:

- **Sella turcica.** This saddle-like structure is unique in its shape.
- **Foramen ovale.** This hole is usually oval-shaped, which makes it easier to distinguish from other foramina.

Key to Illustration

1. Greater wing
2. Lesser wing
3. Superior orbital fissure
4. Optic canal (foramen)
5. Sella turcica
6. Foramen rotundum
7. Foramen ovale
8. Foramen spinosum
9. Pterygoid process

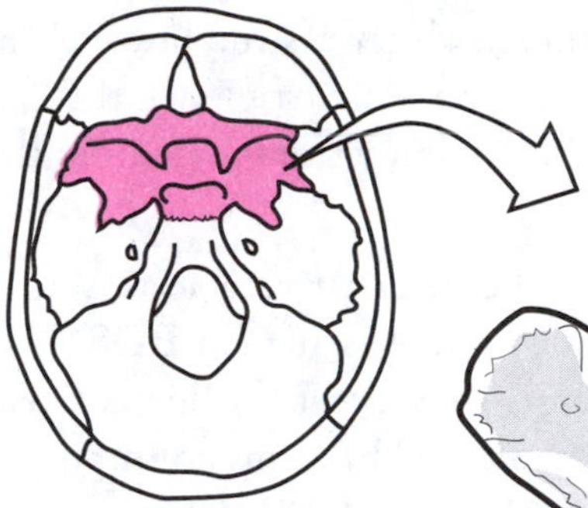

Skull
superior view
with skullcap removed

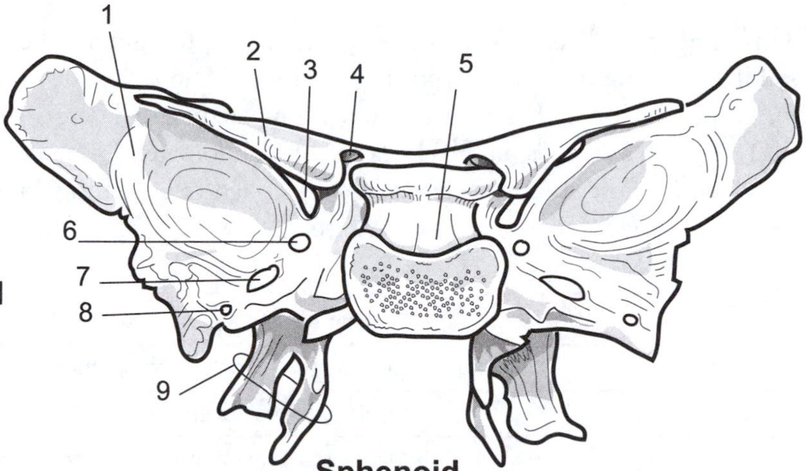

Sphenoid
posterior view

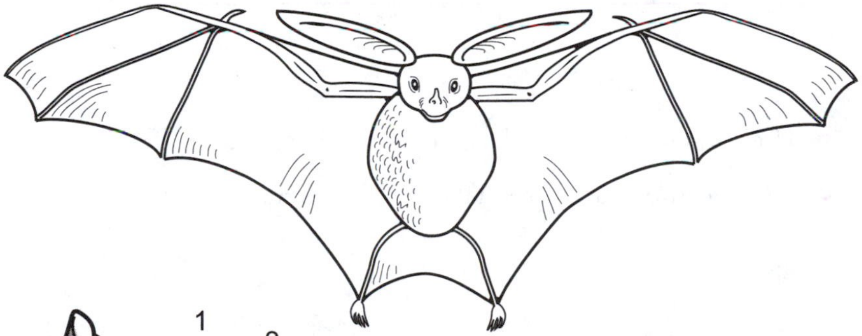

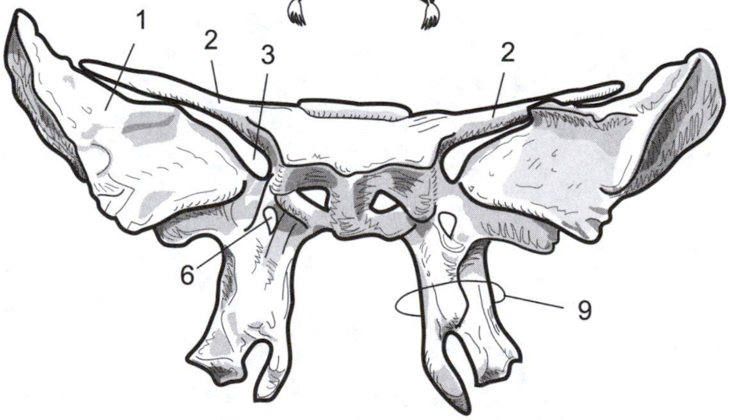

Sphenoid
anterior view

1. ______________________
2. ______________________
3. ______________________
4. ______________________
5. ______________________
6. ______________________
7. ______________________
8. ______________________
9. ______________________

Sphenoidal Foramina—Linking Them Together with Ros the Cowboy

Description

A **foramen** (sing., *foramina* = plural) is a hole in bone through which a structure such as a blood vessel or nerve passes. The **sphenoid** bone contains many foramina. At first glance they appear like many small, indistinguishable holes. The challenge is to be able to differentiate one from the other.

Analogy

This analogy will link five different sphenoidal foramina. The **sella turcica** looks like a **horse's saddle**. Imagine a miniature cowboy sitting on the sella turcica. He is named **ROS** for the three foramina (F.) on either side of him. From medial to lateral, they are as follows: F. **R**otundum, F. **O**vale, and F. **S**pinosum. His **Legs** go through foramen **Lacerum** and his arms go through the **Optic** foramen. Because the optic nerve passes through this opening, imagine ROS holding onto the optic nerves as if they were the reins of a horse.

Function

Foramen (F.)	Blood Vessel / Nerve Passing Through
Optic F.	Cranial nerve II (optic) and ophthalmic artery
F. rotundum	Maxillary branch of cranial nerve V (*trigeminal*)
F. ovale	Mandibular branch of cranial nerve V (*trigeminal*)
F. spinosum	Blood vessels to membranes around central nervous system
F. lacerum	Branch of ascending pharyngeal artery

Study Tips

- The **foramen ovale** tends to be shaped like an oval.
- The **F. spinosum** is the smallest of the foramina.

Key to Illustration

1. Optic foramen
2. Foramen rotundum
3. Foramen ovale
4. Foramen spinosum
5. Foramen lacerum
6. Sella turcica

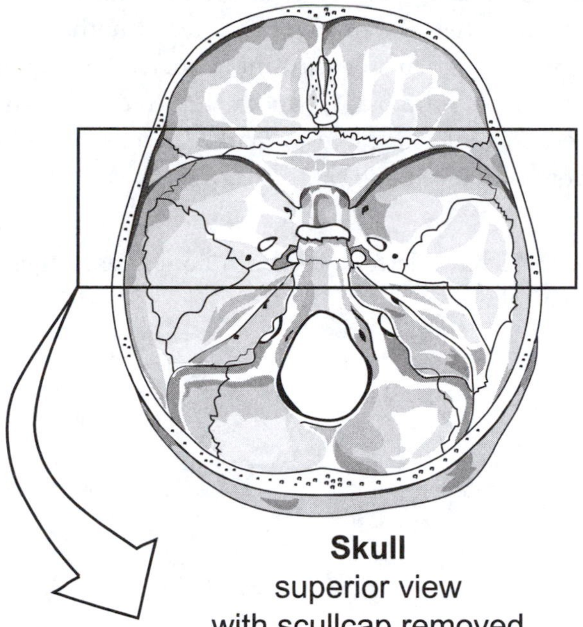

Skull
superior view
with scullcap removed

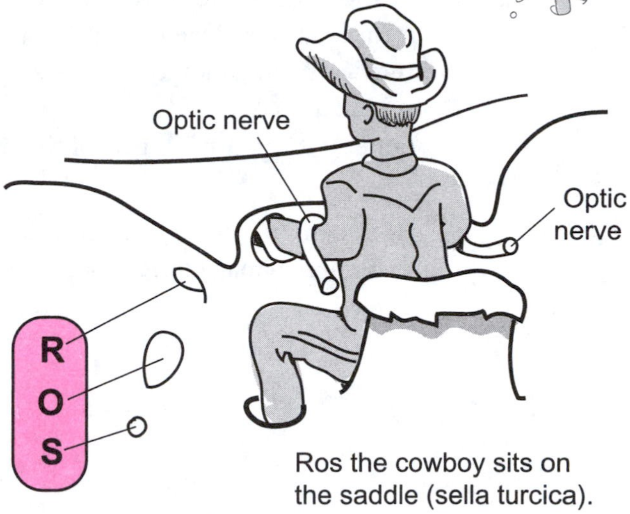

Ros the cowboy sits on the saddle (sella turcica).

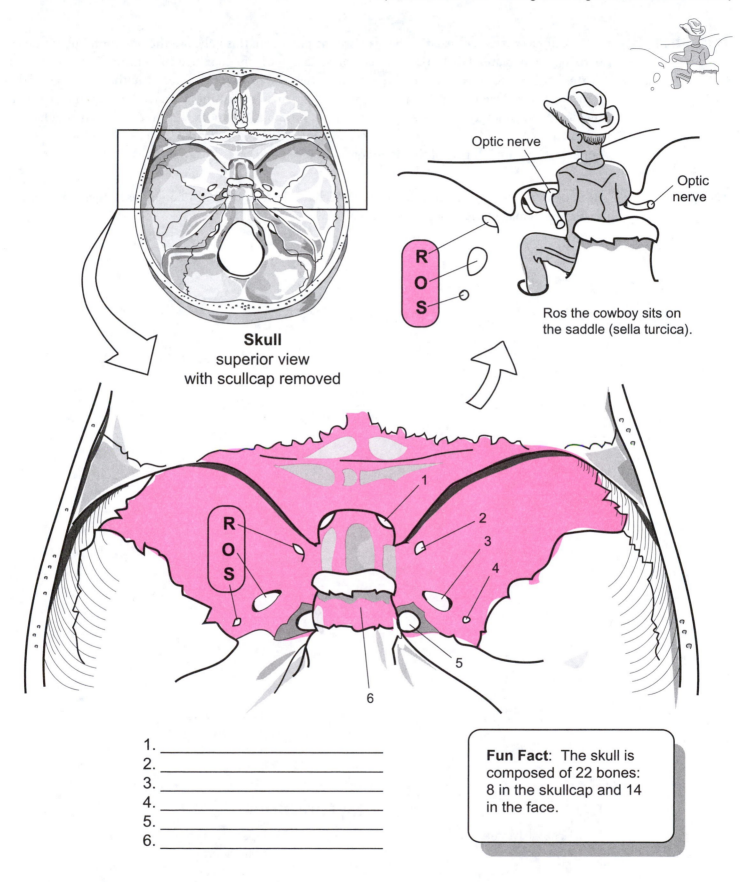

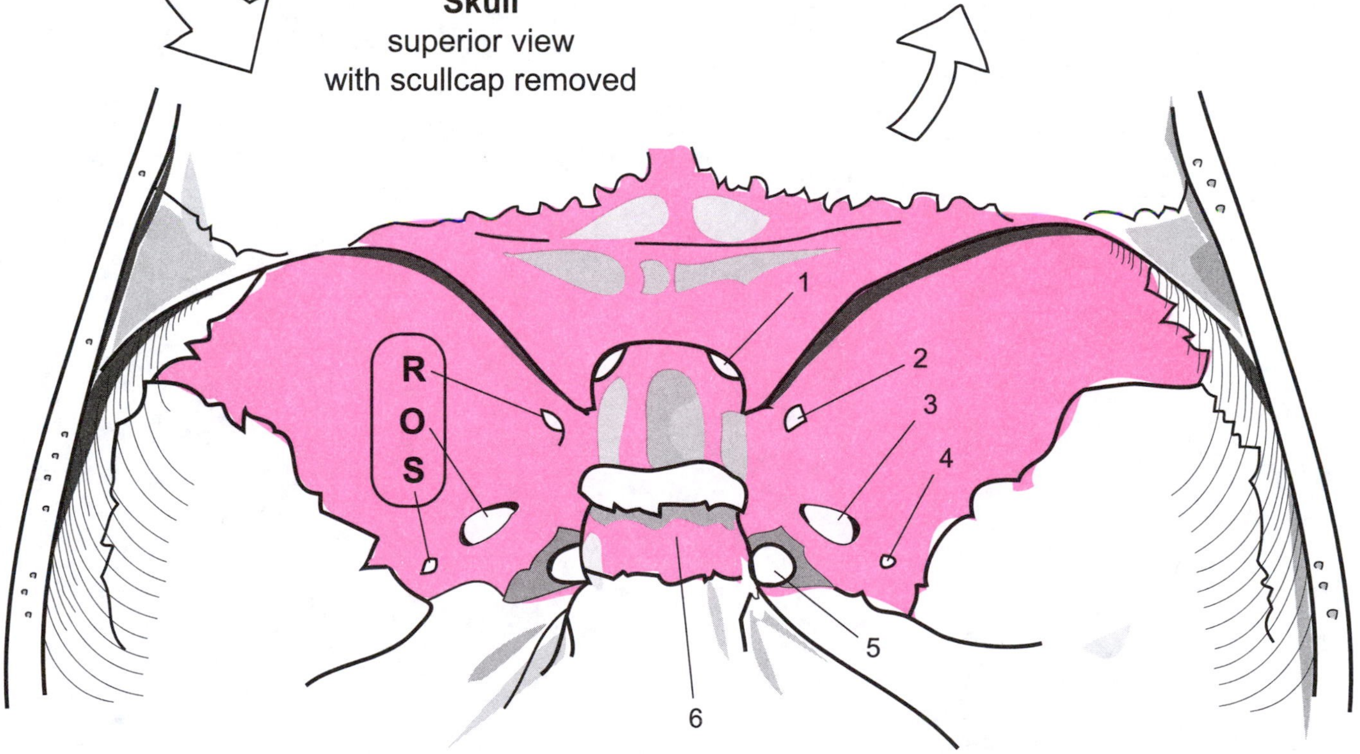

1. ______________________
2. ______________________
3. ______________________
4. ______________________
5. ______________________
6. ______________________

Fun Fact: The skull is composed of 22 bones: 8 in the skullcap and 14 in the face.

Description

The skull has two, small **palatine bones**. Two major parts of this bone are the **horizontal plate** and the **perpendicular plate**. The horizontal plate articulates with the maxillae to form the posterior portion of the hard palate in the roof of the mouth. When the horizontal plates of both bones touch, they form a narrow ridge called the **nasal crest**, that articulates with the vomer. The perpendicular plate runs vertically and articulates with the maxillae, ethmoid, sphenoid, and inferior nasal concha. The **orbital process** is located on top of the perpendicular plate to mark the most superior part of the bone. It forms a small part of the posterior portion of the orbit.

Analogy

In the anterior view, each **palatine bone** looks like a **letter "L." Both bones fused together** look like **two mirror-image letter L's touching each other**.

Location

Skull

Key to Illustration

1. Horizontal plate
2. Nasal crest
3. Perpendicular process
4. Orbital process

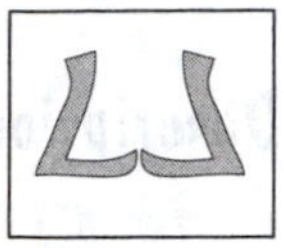

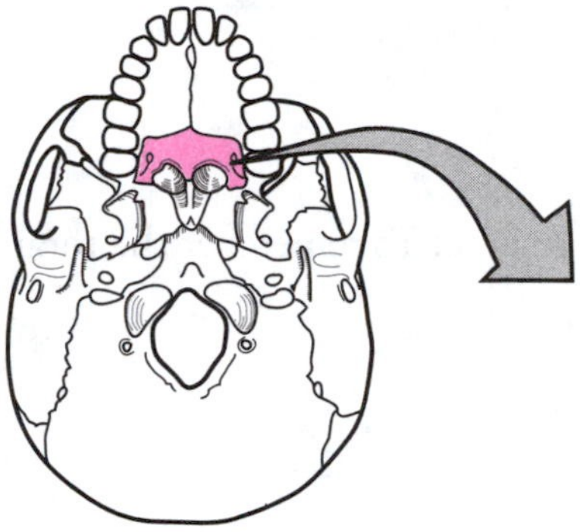

Skull
inferior view with
mandible removed

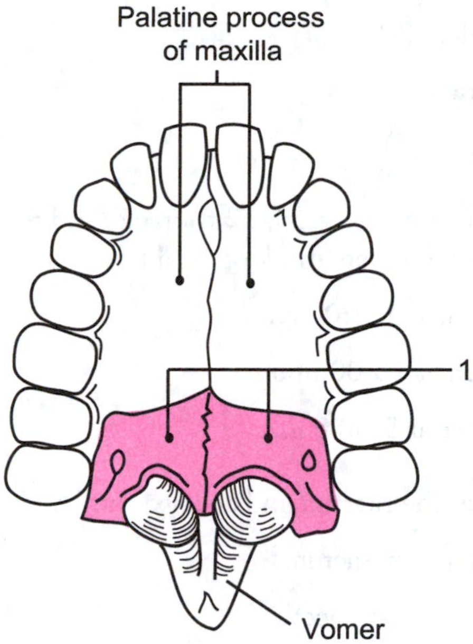

Palatine bones
inferior view

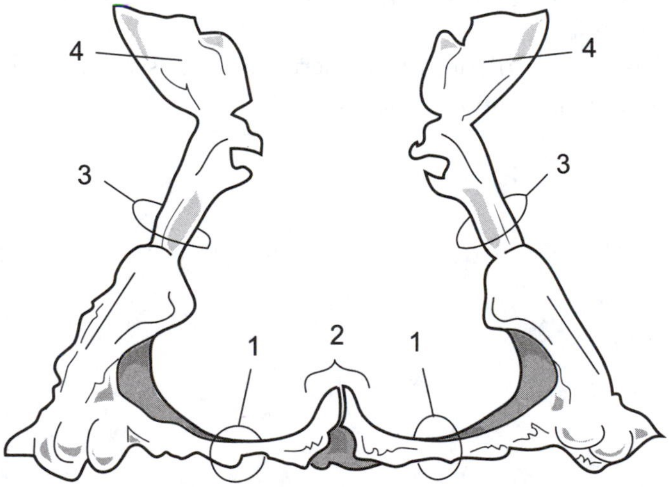

Palatine bones
anterior view

In the anterior view, each palatine bone resembles a letter "L." The two bones fused together are like two mirror-image L's touching each other.

1. ______________________________
2. ______________________________
3. ______________________________
4. ______________________________

Description

The vertebral column contains a total of **24** vertebrae of **three** different types:

- **Cervical** (7)
- **Thoracic** (12)
- **Lumbar** (5)

In total there are **7** cervical, **12** thoracic, and **5** lumbar vertebrae. To remember the total number of each type of vertebrae, think of meal times:

- **Breakfast** at **7:00 a.m.**
- **Lunch** at **12:00 noon**
- **Dinner** at **5:00 p.m.**

To remember the **sacrum** and **coccyx**, think of having 2 snacks in the evening:

- 1st snack = sacrum
- 2nd snack = coccyx

The **sacrum** results from the fusion of five vertebrae. On the top, in the anterior view, is a ridge of bone called the **sacral promontory**. This is an important landmark for a female's pelvic exam. The series of holes running through the bone are called the **sacral foramina**—nerves of the sacral plexus pass through them. On the posterior surface, there is an opening called the **sacral hiatus** that leads into a long passageway called the **sacral canal**. Nerves from the spinal cord run through this canal.

The **coccyx** (*tailbone*) is located inferior to the sacrum and consists of 3–5 bones. It serves as an anchor point for muscles, tendons, and ligaments.

Total Number of Each Type of Vertebrae

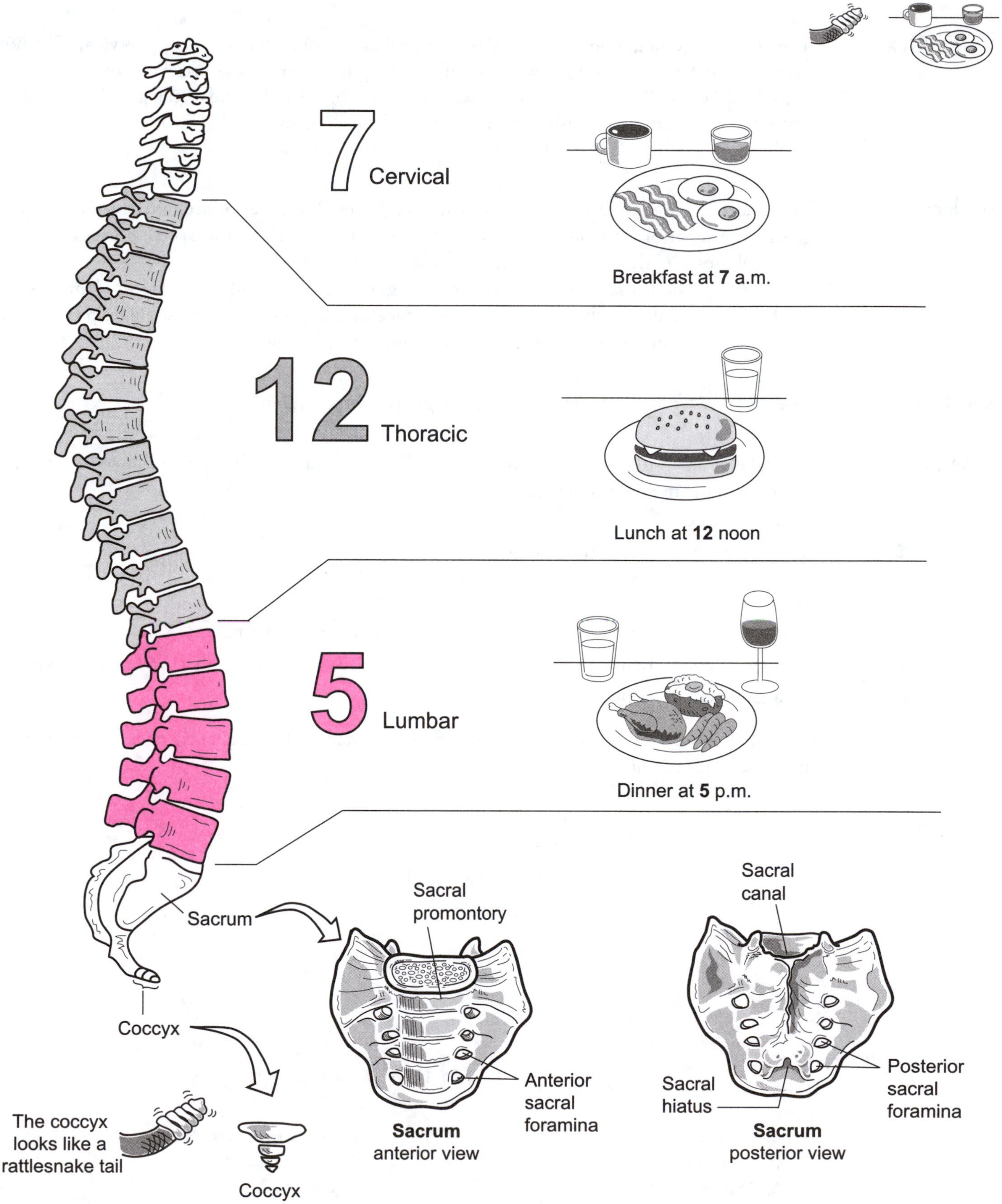

Description

The vertebral column contains a total of 24 vertebrae of three different types: cervical (7), thoracic (12), and lumbar (5). The first two cervical vertebrae at the top of the vertebral column are referred to as the **atlas** (*cervical 1 or C1*) and the **axis** (*cervical 2 or C2*). Like the Greek god Atlas held up the earth, the atlas vertebra is positioned at the base of the globe-like skull. The atlas is designed to pivot on the axis, which permits you to turn your head from side to side.

Analogy

The **atlas** (C1) resembles a **turtle's head with eyeglasses**. The **anterior arch** is the handle of the eyeglasses, the **superior articular facets** are the lenses of the glasses, the **transverse process** is the arm of the eyeglasses, and the **posterior arch** is the smile on the turtle's face.

The **axis** (C2) resembles a **football player grasping a football**. The **odontoid process** is the football player's helmet, the **superior articular facet** is the football player's shoulder pad, the **lamina** is the **forearm**, and the **spinous process** is the **hands grasping the football**.

Location

The first two cervical vertebrae at the top of the vertebral column

Function

The atlas and axis together form a pivot joint. When you turn your head from side to side, the atlas is rotating on the more stationary axis.

Study Tips

The following are good landmarks for these bones:

Atlas

- transverse foramen
- **large** superior articular facets
- **large** superior articular facets

Axis

- transverse foramen
- odontoid process (*dens*) (unique to axis only)

Note that *only* cervical vertebrae have a **transverse foramen**. This makes them easy to distinguish from thoracic or lumbar vertebrae.

Key to Illustration

Atlas

1. Anterior arch
2. Superior articular facet
3. Transverse process
4. Posterior arch
5. Transverse foramen

Axis

6. Odontoid process *(dens)*
7. Superior articular facet
8. Lamina
9. Bifid spinous process
10. Transverse foramen

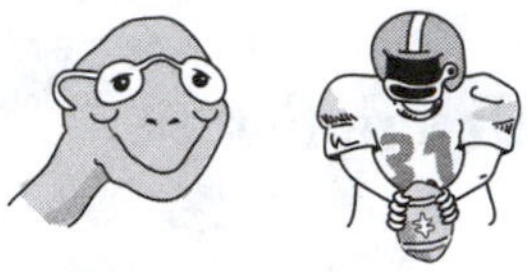

Atlas
superior view

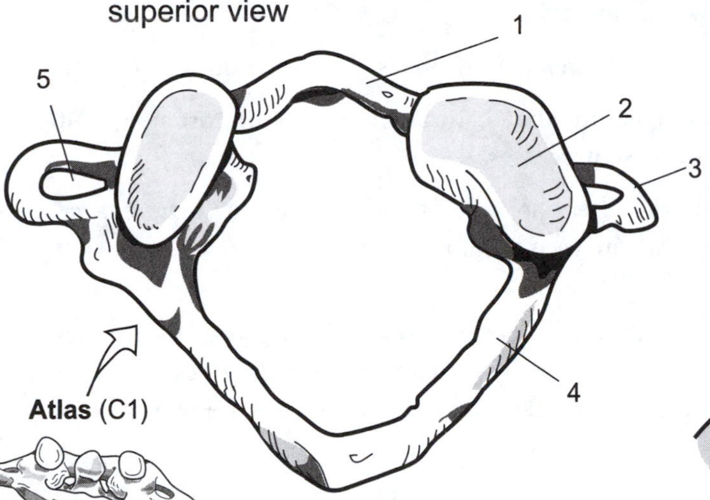

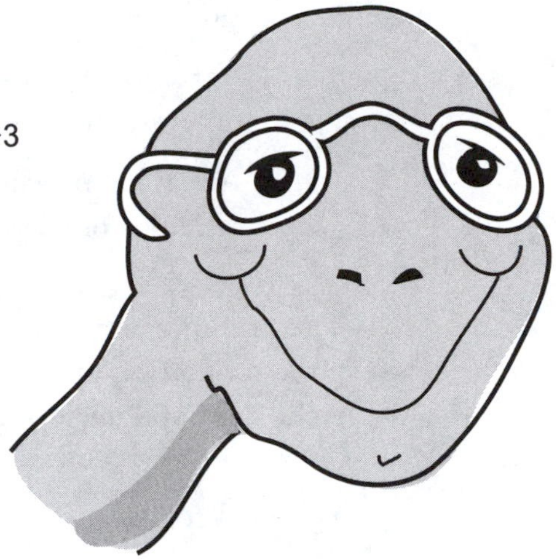

Atlas the turtle

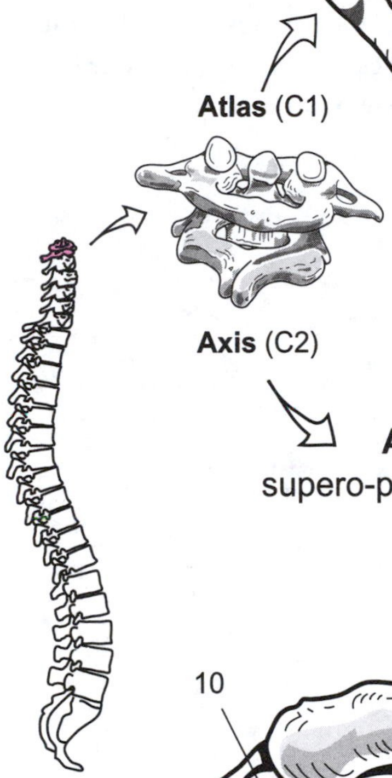

Axis
supero-posterior view

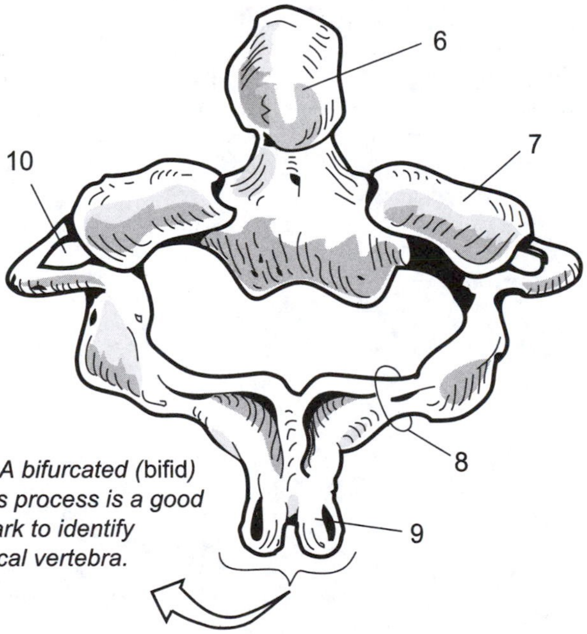

Note: *A bifurcated (bifid) spinous process is a good landmark to identify a cervical vertebra.*

Bubba Axis,
the football player

1. ______________________
2. ______________________
3. ______________________
4. ______________________
5. ______________________

6. ______________________
7. ______________________
8. ______________________
9. ______________________
10. ______________________

Description

The vertebral column contains three different types of vertebrae: cervical (7), thoracic (12), and lumbar (5). Each type has its own unique features to distinguish one from another, yet all of the types have three basic features in common:

1. **Body**—bears weight and increases in size as one moves down the vertebral column.
2. **Vertebral arch**—structure that contains a **vertebral foramen, pedicles, lamina, spinous process**, and **transverse processes.**
3. **Articular processes**—there are two of these—**superior articular process** and the **inferior articular process** that are used to join one vertebra to another.

Analogy

The posterior portion of a **thoracic vertebra** looks like a **goose with wings *arched forward***. The posterior portion of a **lumbar vertebra** looks like a goose with wings ***horizontal***. The **spinous process** is the **head and neck of a goose** and the **transverse process** is the **wing of a goose**.

Location

Vertebral column

Distinguishing Features

Feature	Thoracic	Lumbar
Location	Chest	Lower back
Body of vertebra	Medium-sized, heart-shaped; facets for ribs	Largest diameter, thicker, oval-shaped
Vertebral foramen	Medium-sized	Smaller-sized
Spinous process	Long, slender; points inferiorly	Broad; flat; blunt
Transverse process	10 of 12 have facets for rib articulations	Short; narrower, no articular facets or transverse foramina

Key to Illustration

1. Spinous process
2. Lamina
3. Vertebral foramen
4. Transverse process
5. Superior articular facet
6. Superior articular process
7. Body
8. Pedicle
9. Facet for rib articulation

Fun Fact: A human neck has the same number of vertebrae as a giraffe.

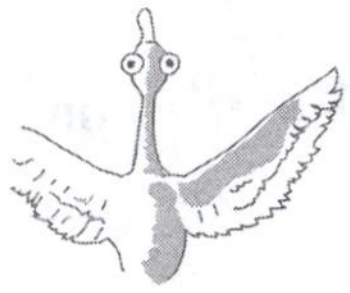

Thoracic vertebra
superior view

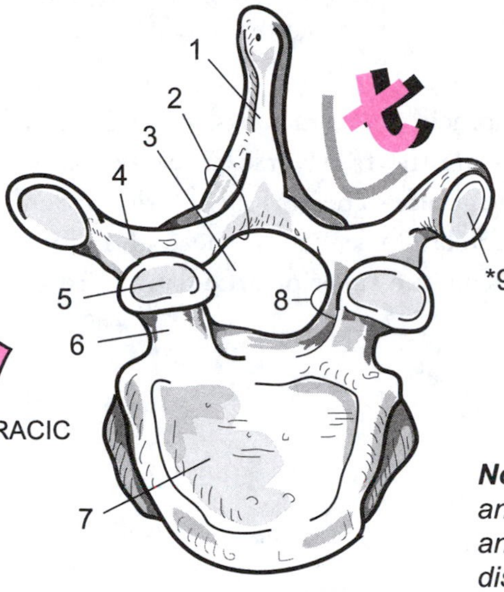

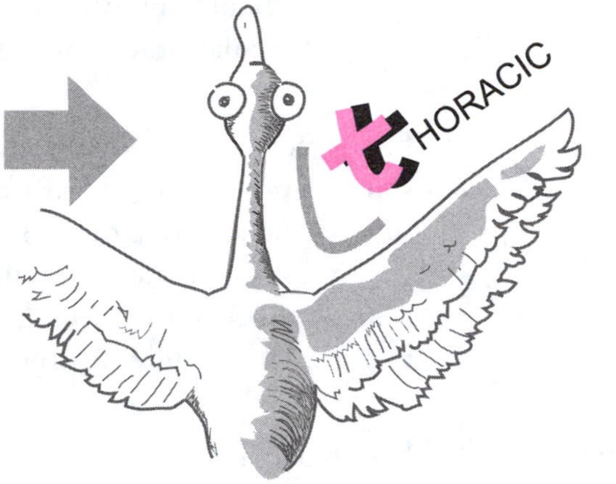

Location

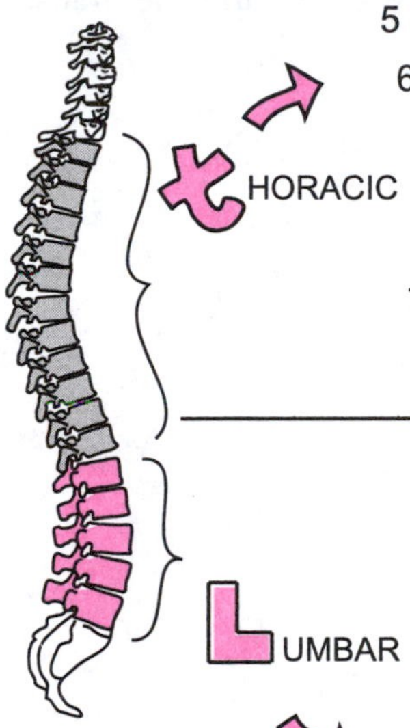

Note: *Most—but not all—thoracic vertebrae have an angle that is <90° between the spinous process and the transverse process. Be sure to use all the distinguishing features in the table on p. 134.*

Lumbar vertebra
superior view

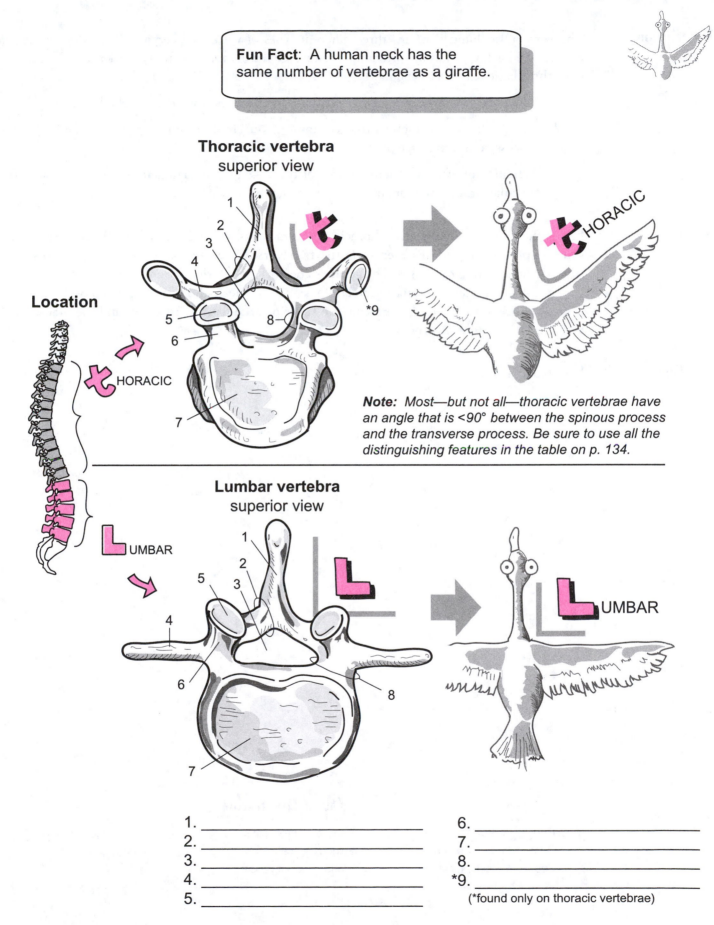

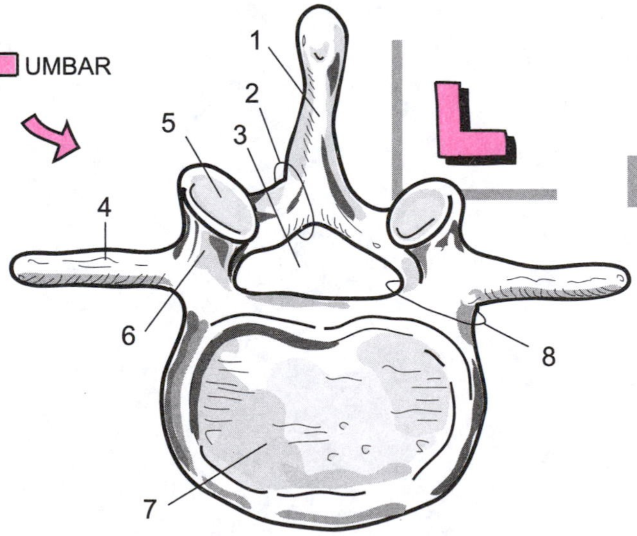

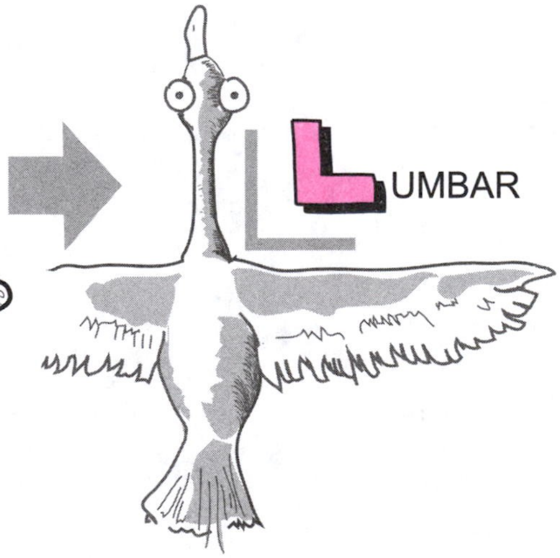

1. ______________________
2. ______________________
3. ______________________
4. ______________________
5. ______________________

6. ______________________
7. ______________________
8. ______________________

*9. ______________________
(*found only on thoracic vertebrae)

Description

The vertebral column contains three different types of vertebrae: cervical (7), thoracic (12), and lumbar (5). Each type has its own unique features to distinguish one from another, yet all of the types have three basic features in common:

1. **Body**—bears weight and increases in size as one moves down the vertebral column.
2. **Vertebral arch**—structure that contains a **vertebral foramen, pedicles, lamina, spinous process**, and **transverse processes.**
3. **Articular processes**—there are two of these—**superior articular process** and the **inferior articular process** that are used to join one vertebra to another.

Analogy

In the postero-lateral view, the **thoracic vertebra** looks like a **giraffe**. The **giraffe's snout** is the **spinous process**, the **giraffe's ears** are the **transverse processes**, and the **giraffe's horns** are the **superior articulating processes.** The **giraffe's cheek** is the **inferior articulating process.**

In the lateral view, a **lumbar vertebra** looks like the **head of a moose**. The **moose's snout** is the **spinous process**, the **moose's horns** are the **superior articulating processes**, and the **moose's bell** is the **inferior articulating process.**

Location

Vertebral column

Key to Illustration

1. Body
2. Superior articular process
3. Superior articular facet
4. Transverse process
5. Spinous process
6. Inferior articular process
7. Inferior articular facet

Thoracic Versus Lumbar Vertebrae—Distinguishing the Difference II

"Thoracic giraffe"
In this postero-lateral view, the posterior portion of a lumbar vertebra looks like the head of a giraffe.

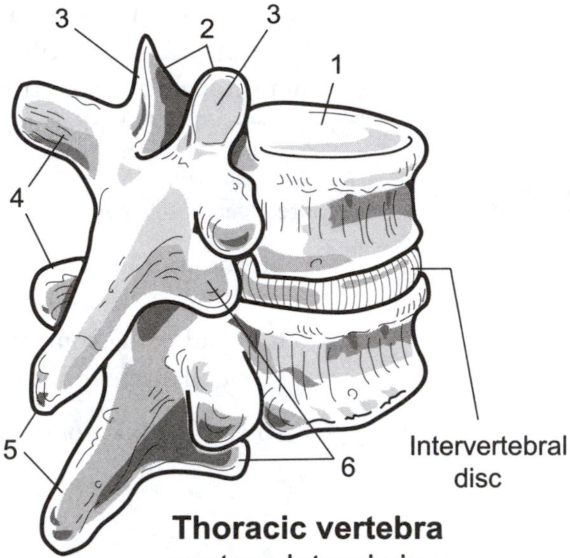

Thoracic vertebra
postero-lateral view

1. ______________________________
2. ______________________________
3. ______________________________
4. ______________________________
5. ______________________________
6. ______________________________
7. ______________________________

Location

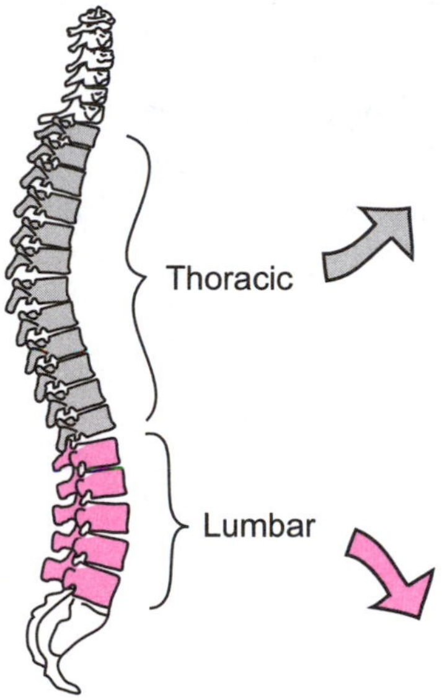

"Lumbering moose"
In the lateral view, the posterior portion of a lumbar vertebra looks like the head of a moose.

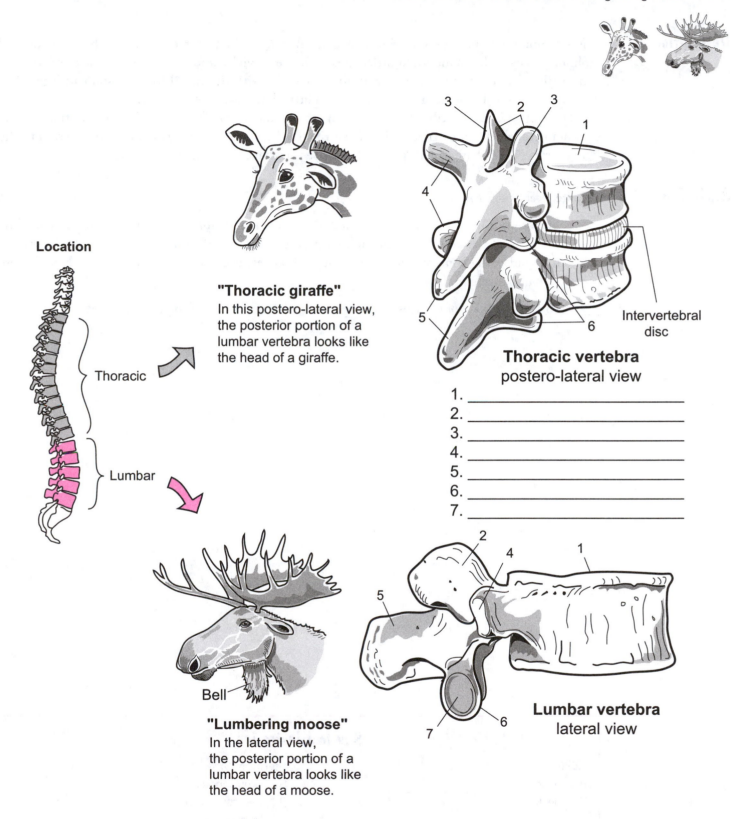

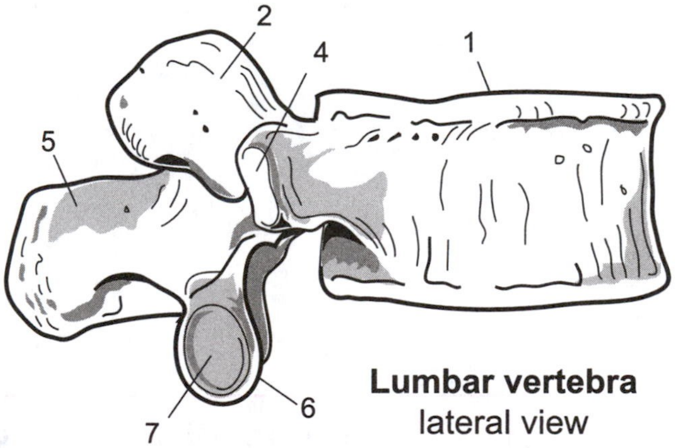

Lumbar vertebra
lateral view

Description

The **sternum** is commonly called the breastbone. It is divided into three parts: manubrium, body, and xiphoid process. The manubrium articulates with the clavicles and the costal cartilages of the first pair of ribs. The body is the largest part and articulates with the ribs through costal cartilages. The xiphoid process is the smallest part and does not articulate with another bone.

There are 12 pairs of **ribs**. The first seven are called **true ribs** because they directly attach to the sternum through costal cartilages. The **false ribs** (8–12) do not directly attach to the sternum. The last two ribs (11–12) are called "floating" because they have no connection to the sternum.

Study Tips

Palpate (*feel by touch*):

- **Sternum:** You can easily feel the middle of the **manubrium** and **body** of your sternum. The small tip at the end marks the **xiphoid process**. During CPR training, students are instructed to avoid doing compressions on the xiphoid process because it can break off easily and cause serious damage to the liver.
- **Ribs:** The body of the ribs can be felt on the lateral surface of the thoracic cage.

Key to Illustration

Rib (R)

R1. Head
R2. Neck
R3. Tubercle
R4. Body
R5. Sternal end *(attaches to costal cartilage)*

Sternum (S)

S1. Jugular notch
S2. Clavicular notch
S3. Manubrium
S4. Body
S5. Xiphoid process

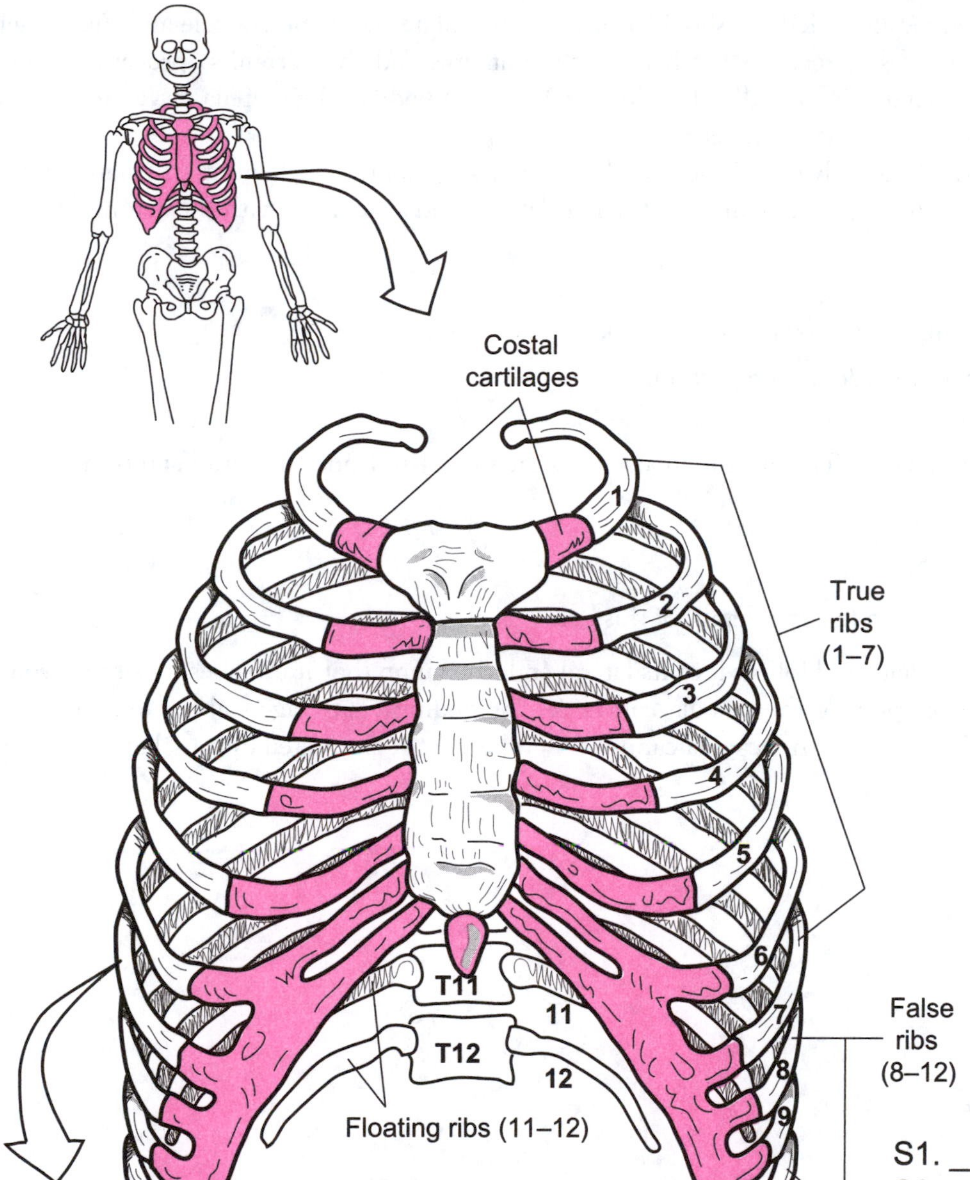

Sternum and ribs
anterior view

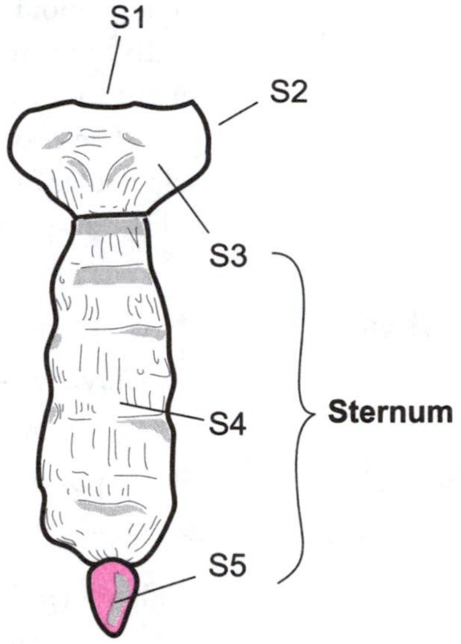

Sternum
anterior view

Sternum (S)

S1. ______________________
S2. ______________________
S3. ______________________
S4. ______________________
S5. ______________________

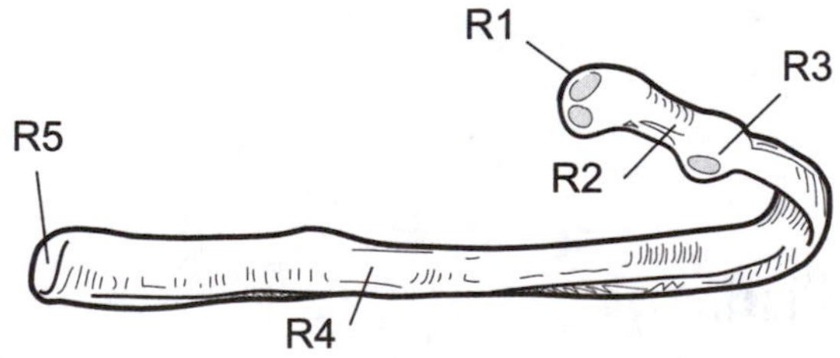

Rib
posterior view

Rib (R)

R1. ______________________
R2. ______________________
R3. ______________________
R4. ______________________
R5. ______________________

Description

The **scapula** is commonly called the shoulder blade. It articulates with the clavicle and the humerus. The **glenoid cavity** (*fossa*) receives the head of the humerus and the **acromion process** articulates with the acromial (lateral) end of the clavicle. The large, flat **body** of the scapula serves as an attachment for muscles, tendons, and ligaments.

The **clavicle** is commonly called the collarbone. At its sternal (*medial*) end it articulates with the clavicular notch of the sternum, while its acromial (*lateral*) end articulates with the acromion process of the scapula.

Location

- **Scapula**—acromial region (shoulder) and back
- **Clavicle**—superior border of the pectoral region

Study Tip

- To position the clavicle: The more rounded end is the sternal end, and the flatter end is the acromial end.

Palpate (*feel by touch*):

- **Clavicle:** You can feel the details of this bone.
- Feel your clavicle, and follow it to its lateral end, the bump that it attaches to is the **acromion process.** The **spine** of the scapula can be felt by gliding your fingers along the back of the shoulder. The body of the scapula cannot be felt because it is covered by muscle.

Key to Illustration

Scapula (S)
1. Acromion process
2. Coracoid process
3. Spine of scapula
4. Glenoid cavity *(fossa)*
5. Lateral border
6. Inferior angle
7. Infraspinous fossa
8. Medial border
9. Supraspinous fossa
10. Superior border
11. Suprascapular notch

Clavicle (C)

C1. Acromial (*lateral*) end

C2. Sternal (*medial*) end

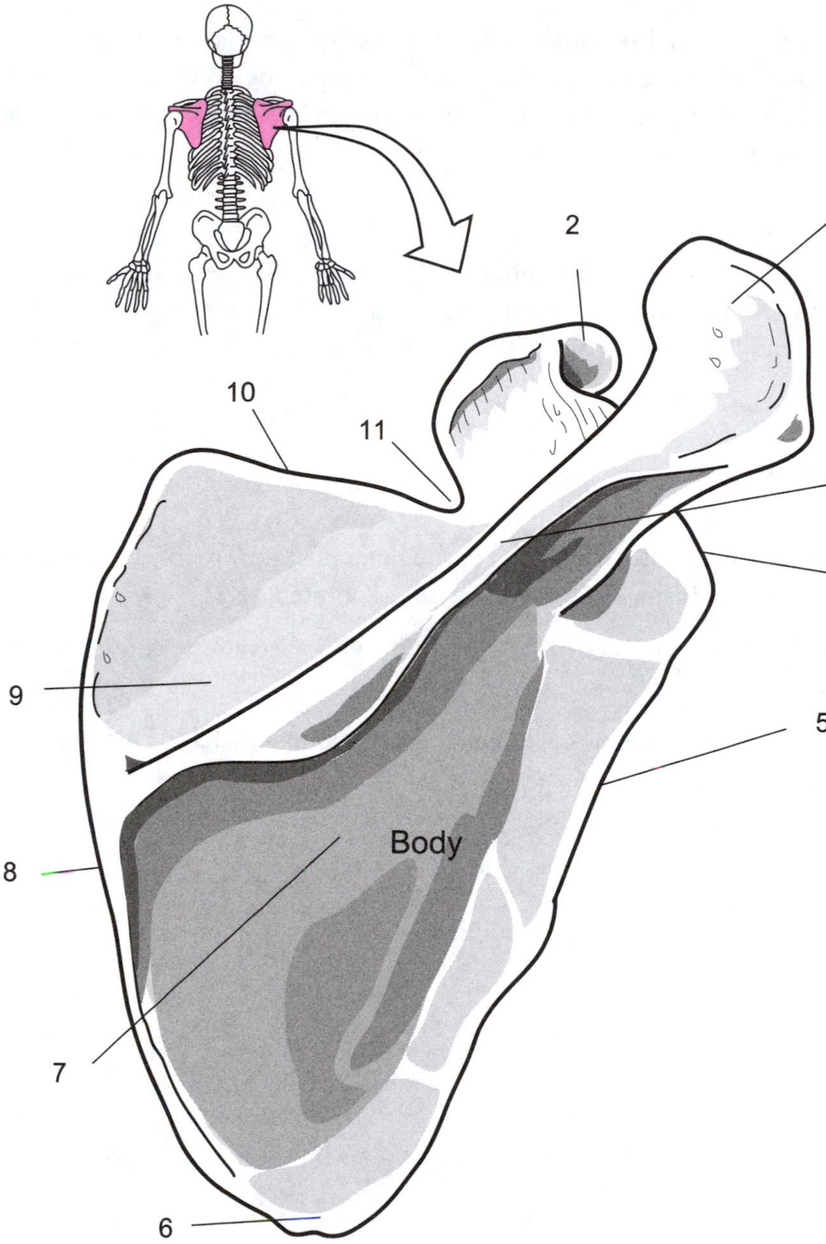

Right scapula
posterior surface

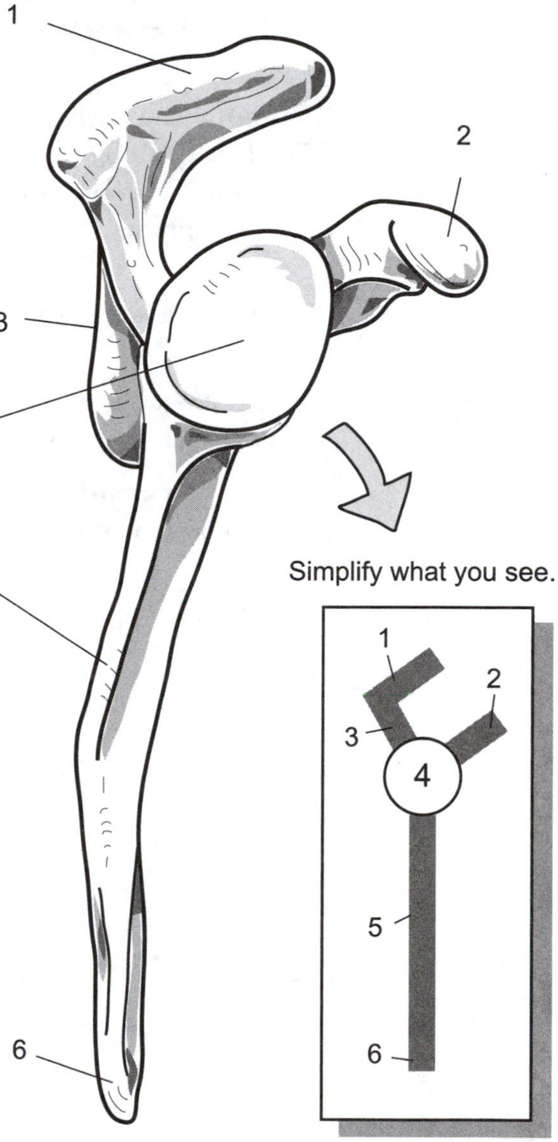

Right scapula
lateral view

Simplify what you see...

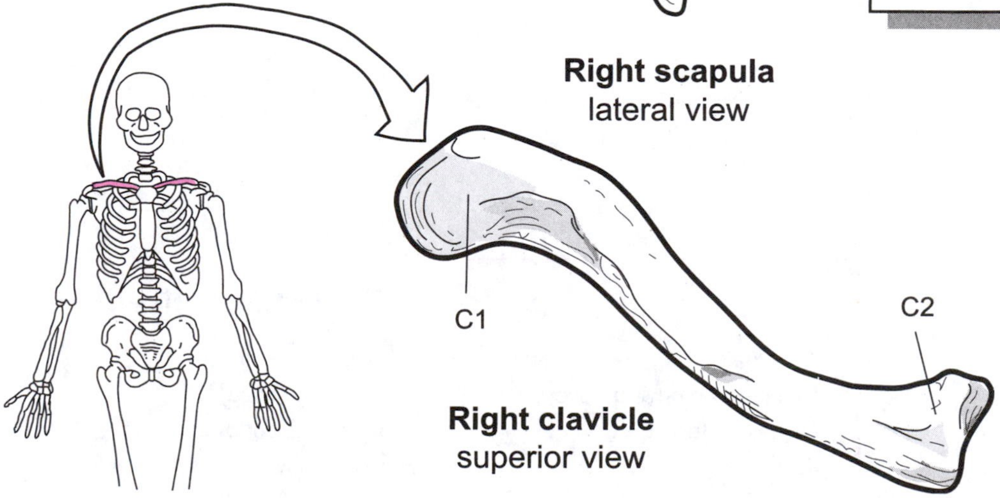

Right clavicle
superior view

Description

The **humerus** is the only bone in the brachial region. The **head** of the humerus articulates with the glenoid cavity in the scapula to form the shoulder joint. At the distal end the **trochlea** articulates with the ulna, while the **capitulum** articulates with the head of the radius. The **olecranon fossa** articulates with the olecranon process of the ulna, and the **coronoid fossa** articulates with the coronoid process of the ulna.

Analogy

The **distal end of the humerus** resembles a **hitchhiker's hand**. The thumb is the medial epicondyle. The first two fingers adjacent to the thumb are the **trochlea**, while the last two fingers are the **capitulum**. The depression in the middle of the palm is the **coronoid fossa**. Note that the thumb always points medially.

Location

Brachial region (between shoulder and elbow)

Study Tips

- To distinguish between the two condyles at the distal end of the humerus: the **T**rochlea has a more **T**riangular shape, while the **C**apitulum is simply **C**urved and not as pointed.
- The **olecranon fossa** is a good landmark to identify the posterior view because it is the deepest depression on the bone.
- Note that the **head** of the humerus and **medial epicondyle** always point *medially*. This helps to distinguish a *left* humerus from a *right* humerus.

Key to Illustration

1. Greater tubercle
2. Intertubercular groove
3. Lesser tubercle
4. Head
5. Deltoid tuberosity
6. Coronoid fossa
7. Lateral epicondyle
8. Capitulum
9. Trochlea
10. Medial epicondyle
11. Olecranon fossa

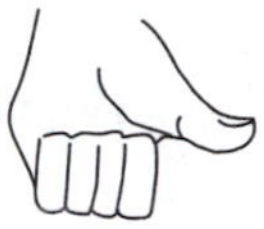

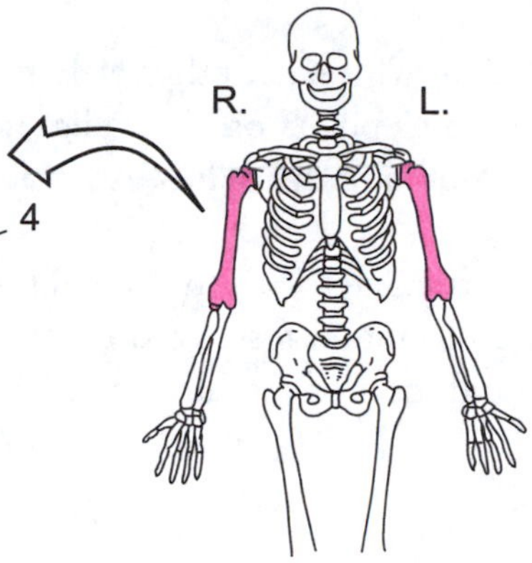

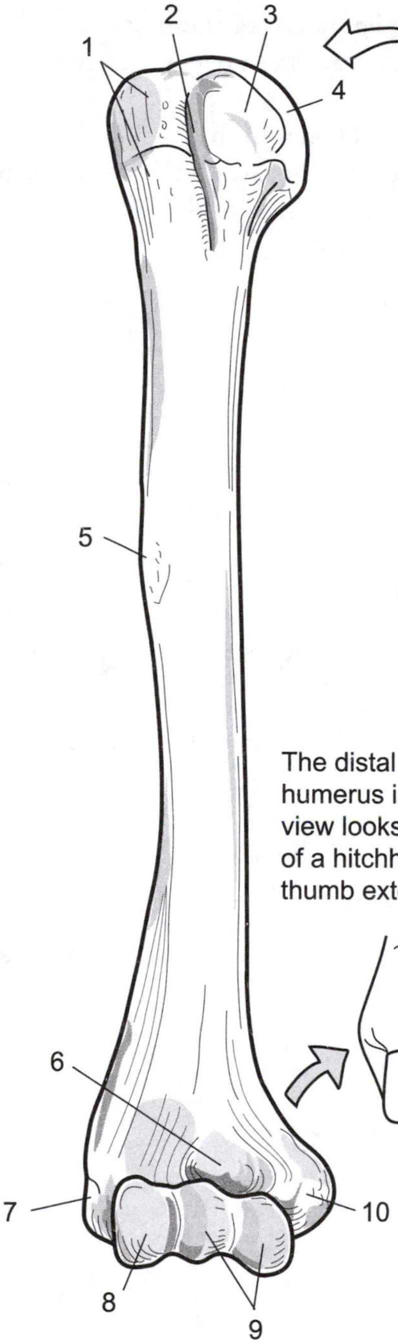

The distal end of the humerus in the anterior view looks like the hand of a hitchhiker with his thumb extended.

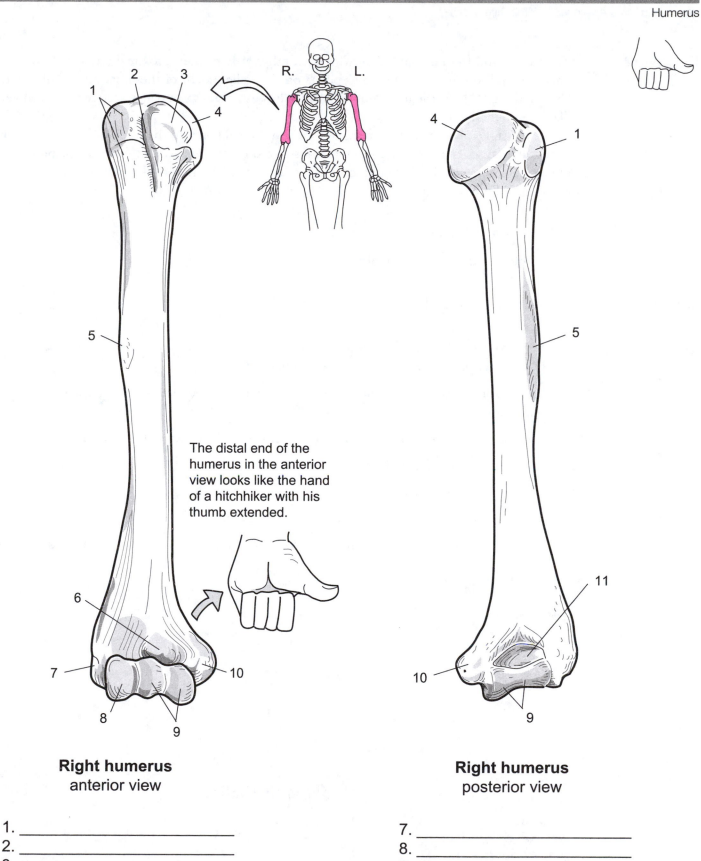

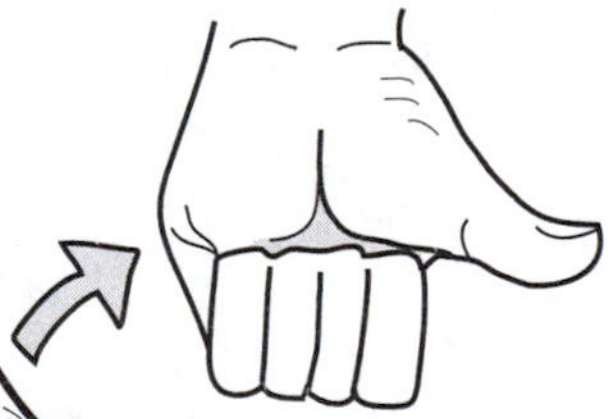

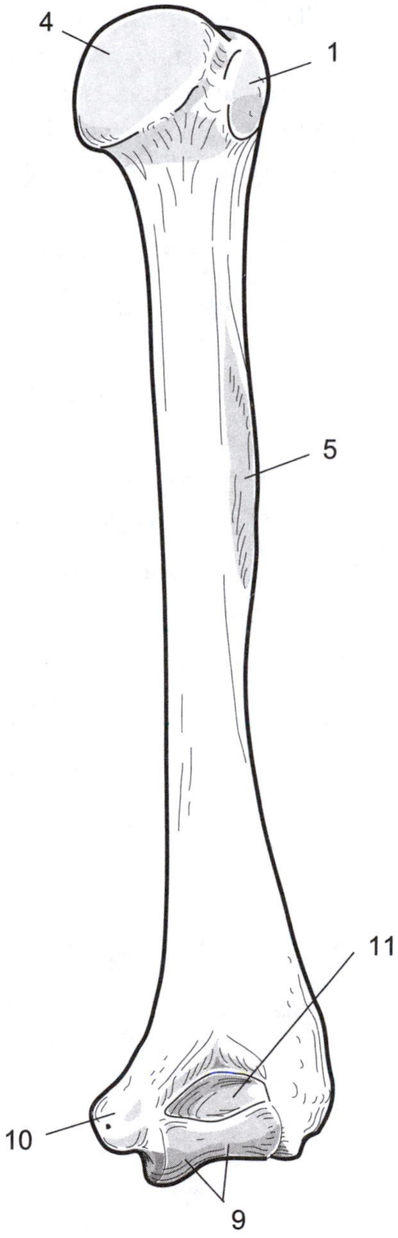

Right humerus
anterior view

Right humerus
posterior view

1. ______________________________
2. ______________________________
3. ______________________________
4. ______________________________
5. ______________________________
6. ______________________________

7. ______________________________
8. ______________________________
9. ______________________________
10. ______________________________
11. ______________________________

Description

The **radius** and the **ulna** articulate with each other at both their proximal and distal ends. At the proximal end, the head of the radius pivots on the radial notch of the ulna. At the distal end the head of the ulna joins the ulnar notch of the radius. A fibrous sheet of connective tissue connects the **diaphyses** of both bones.

The distal end of the radius articulates with the **carpal bones** in the wrist. The **olecranon process** of the ulna articulates with the olecranon fossa of the humerus, while the **coronoid process** of the ulna articulates with the coronoid fossa of the humerus. The head of the radius articulates with the capitulum of the humerus.

Analogy

- The **ulna** resembles a **crescent wrench**.
- The **head of the radius** resembles a **warped hockey puck**.

Location

The radius and ulna are located in the antebrachial region (forearm)

Study Tips

- The proximal end of the ulna has a "U" shaped structure called the trochlear notch that identifies it as the **U**lna because "**U**" is the first letter in the word "Ulna."
- The head of the **R**adius is **R**ounded. Use the alliteration, "**R**ounded **R**adius' to distinguish the radius from the ulna.
- Notches and other structures are sometimes named after the bones with which they articulate:

  e.g. —radial notch of ulna

  —ulnar notch of radius

  —olecranon *process* of the ulna hooks into the olecranon *fossa* of the humerus

  —coronoid *process* of the ulna hooks into the coronoid *fossa* of the humerus

Key to Illustration

Radius
1. Neck
2. Radial tuberosity
3. Radial styloid process
4. Ulnar notch of radius *(not fully visible in diagram)*

Ulna
5. Coronoid process
6. Olecranon process
7. Trochlear notch
8. Ulnar tuberosity
9. Head of ulna
10. Ulnar styloid process
11. Radial notch of ulna

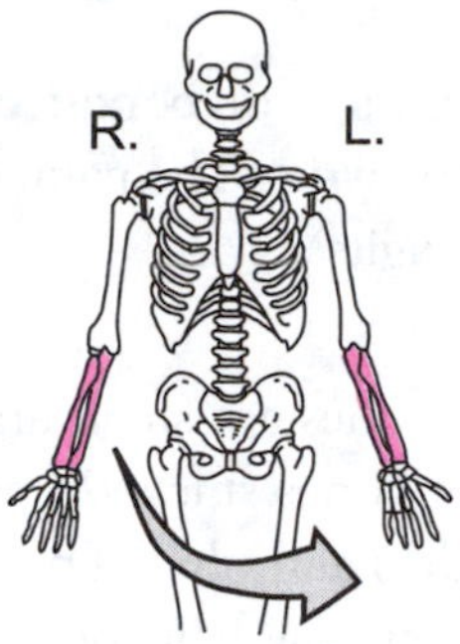
R.
L.

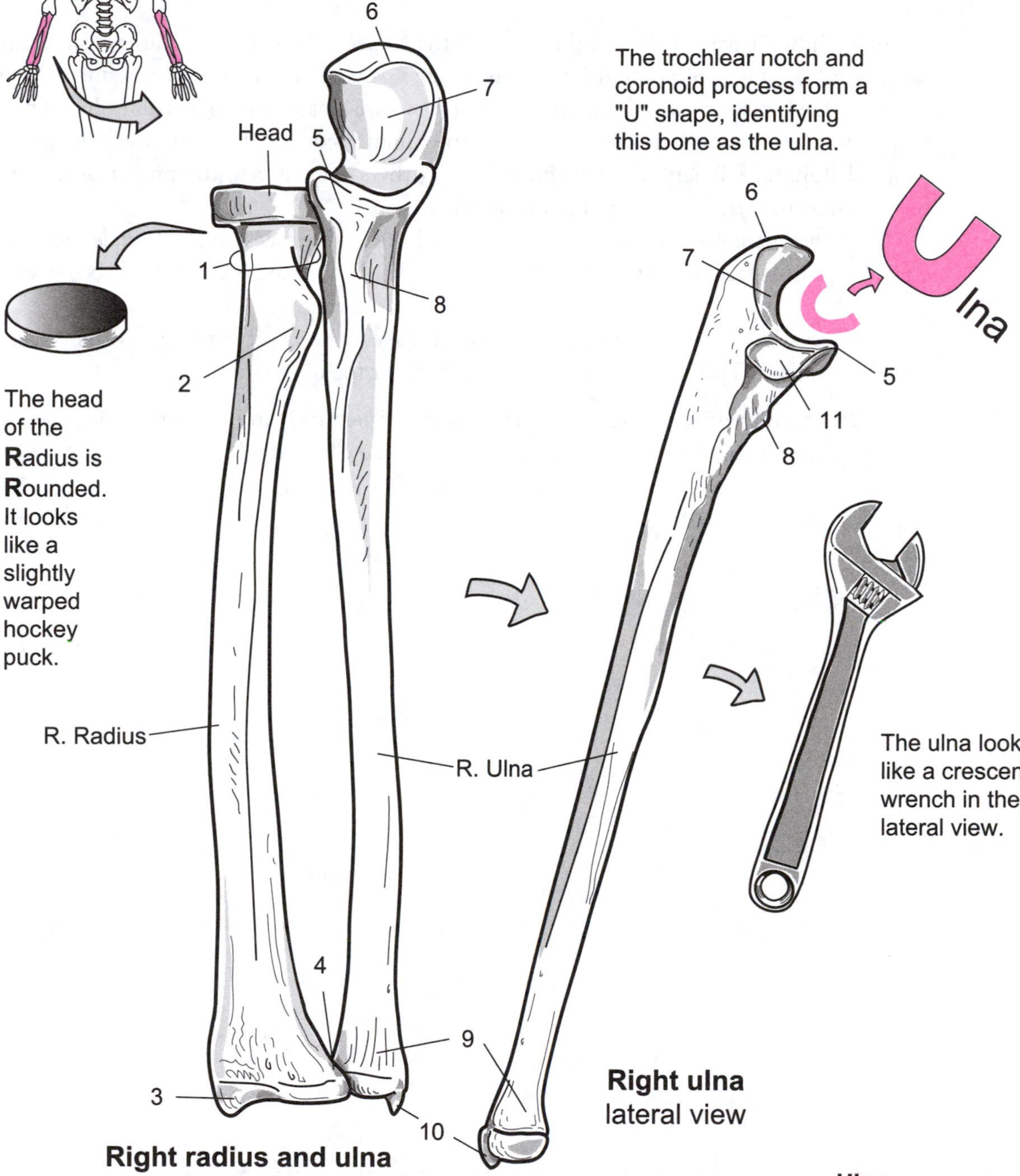
6
7
Head
5
1
8
2
The head of the **R**adius is **R**ounded. It looks like a slightly warped hockey puck.
R. Radius
R. Ulna
4
9
3
10
The trochlear notch and coronoid process form a "U" shape, identifying this bone as the ulna.
6
7
5
11
8
Ulna
The ulna looks like a crescent wrench in the lateral view.

Right radius and ulna
anterior view

Right ulna
lateral view

Radius

1. ______________________
2. ______________________
3. ______________________
4. ______________________

Ulna

5. ______________________
6. ______________________
7. ______________________
8. ______________________
9. ______________________
10. ______________________
11. ______________________

Description

Each hand contains a total of 27 bones and is divided into three groups of bones: **carpals** (8), **metacarpals** (5), and **phalanges** (14). The carpals are the small bones of the wrist. The palm of the hand contains the metacarpals, and the phalanges are located in the fingers or digits.

Study Tips

The most difficult part of the hand to learn is the *carpals*. These eight bones are small and clustered together in the wrist. Simplify it by viewing them as two equal rows of bones stacked one on top of the other. Each row has four bones in it. The first row is nearer the radius and ulna. The second row is next to the metacarpals. Beginning on the thumb side of the first row, the proper order is: **scaphoid, lunate, triangular, pisiform**. Following the same pattern, the proper order for the second row is: **trapezium, trapezoid, capitate, hamate.**

The *hamate* (*hamatum*, hooked) *bone* has a hook-like process on it (*visible only in the anterior view*). You can better recall this bone because the words ***hamate*** and ***hook*** both begin with the letter "***h***".

The *thumb* has only two phalanges instead of three. It has a proximal and distal phalanx but lacks a middle phalanx. This is the same pattern as the great toe of the foot.

- On many plastic models, the triangular (*triquetrum*) and pisiform bones are often fused together to appear as one bone. This causes some confusion for students.

Key to Illustration

Carpals

1. Scaphoid
2. Lunate
3. Triangular (*Triquetrum*)
4. Pisiform
5. Trapezium
6. Trapezoid
7. Capitate
8. Hamate

Phalanges

9. Proximal phalanx
10. Middle phalanx
11. Distal phalanx

Right hand
anterior view

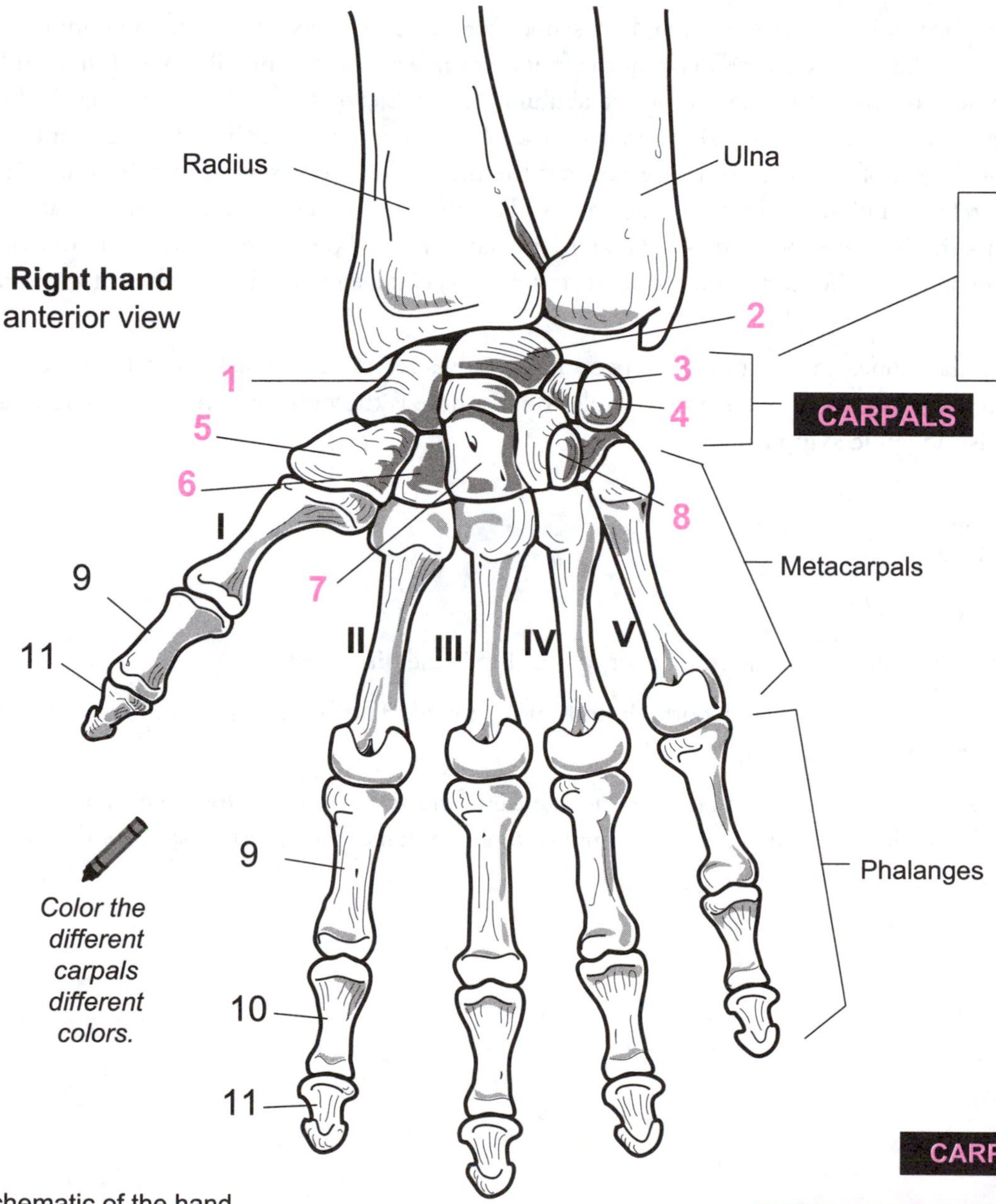

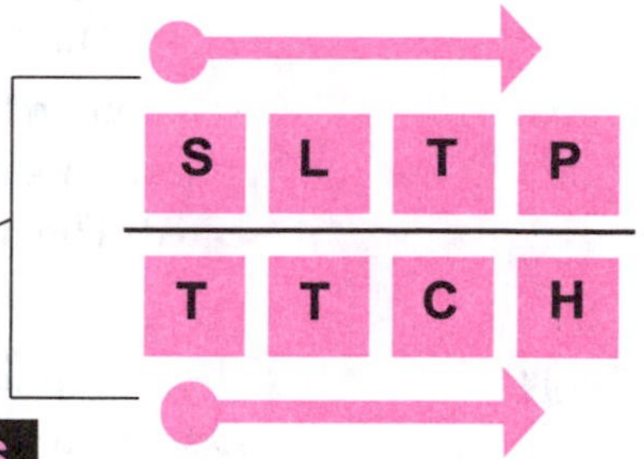

The carpals should be viewed as two rows of 4 bones each.

You can use the mnemonic,

"Sandy Left The Party To Take Cathy Home"

Move across the top row beginning at the thumb side. Then, proceed in the same direction for the bottom row.

Color the different carpals different colors.

Schematic of the hand

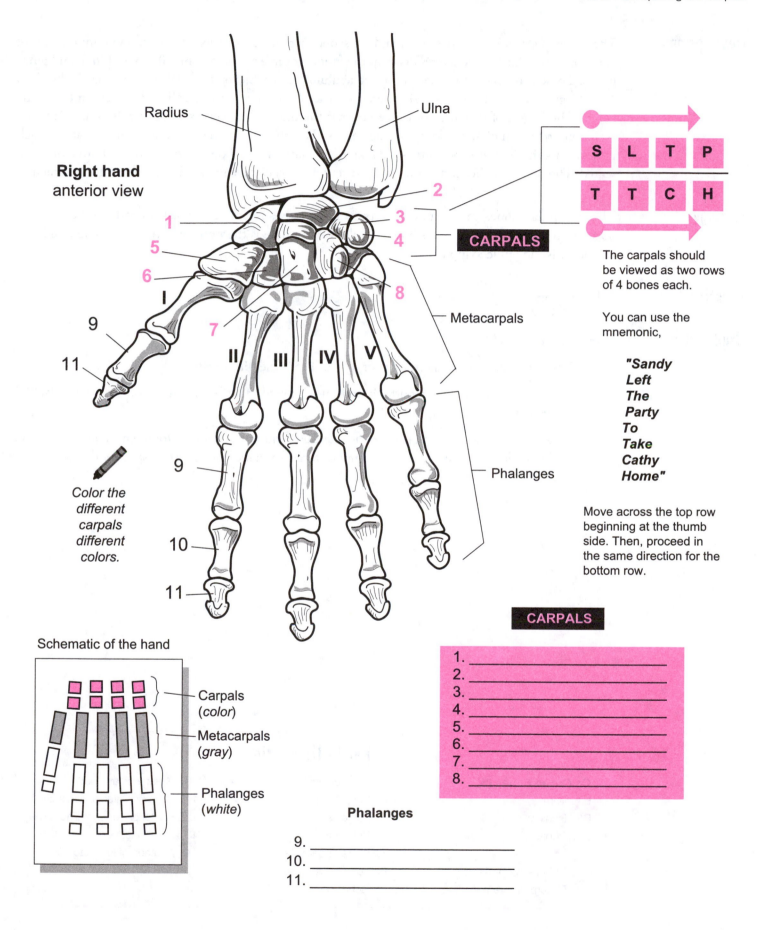

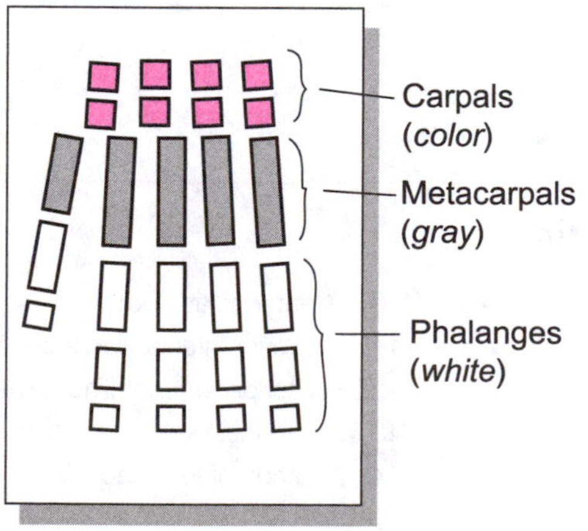

CARPALS

1. ______________________
2. ______________________
3. ______________________
4. ______________________
5. ______________________
6. ______________________
7. ______________________
8. ______________________

Phalanges

9. ______________________
10. ______________________
11. ______________________

Description

The **pelvis** (*hipbone*) is composed of four bones: one sacrum, one coccyx, and a left and right ossa coxae (sing. *os coxae*). Each os coxae results from the fusion of three separate bones: **ilium, ischium,** and **pubis** bones. These three meet together in the **acetabulum** (*means "cup of vinegar"*) to form what looks like a "peace sign." This deep depression is where the head of the femur forms a ball-and-socket joint with the pelvis. The largest of the three os coxae bones is the ilium. The broad surface of this bone is a major attachment for muscles, tendons, and ligaments. The strongest bone is the ischium. A large, rough projection—the **ischial tuberosity**—is a major landmark on this bone. It bears the body weight when seated. The pubis has the least bone mass and contains a very large foramen—the **obturator foramen.**

Analogy

The **two pubis bones** in the anterior aspect of the pelvis resembles a **mask** worn by someone. The mask itself represents both pubis bones, while the **eye hole** is the **obturator foramen**. The **middle of the mask** is the **pubic symphysis.**

Location

Pelvic region

Study Tips

Palpate (*feel by touch*):

- The curve along the superior aspect of your hip is the **iliac crest.**
- Place your hand on your buttock and push inward. The bony bump you feel is the **ischial tuberosity.**
- Good landmarks on a coxal bone: **acetabulum** and the **obturator foramen.** The acetabulum marks the lateral aspect, and the obturator foramen marks the anterior aspect of the bone.

Key to Illustration

Os Coxae Bones (B)

B1. Ilium
B2. Ischium
B3. Pubis

1. Pubic symphysis
2. Obturator foramen
3. Acetabulum
4. Sacroiliac joint
5. Iliac crest
6. Pubic crest
7. Ischial tuberosity
8. Lesser sciatic notch
9. Ischial spine
10. Greater sciatic notch
11. Posterior inferior iliac spine
12. Posterior superior iliac spine
13. Anterior superior iliac spine
14. Anterior inferior iliac spine

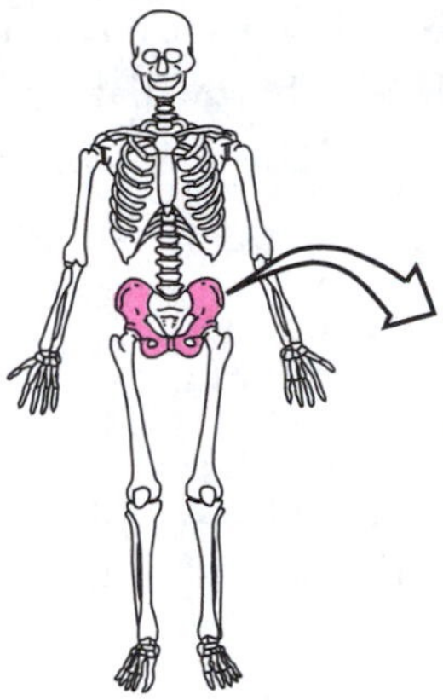

Note: The sacrum and coccyx are faded out since they are not part of the appendicular skeleton.

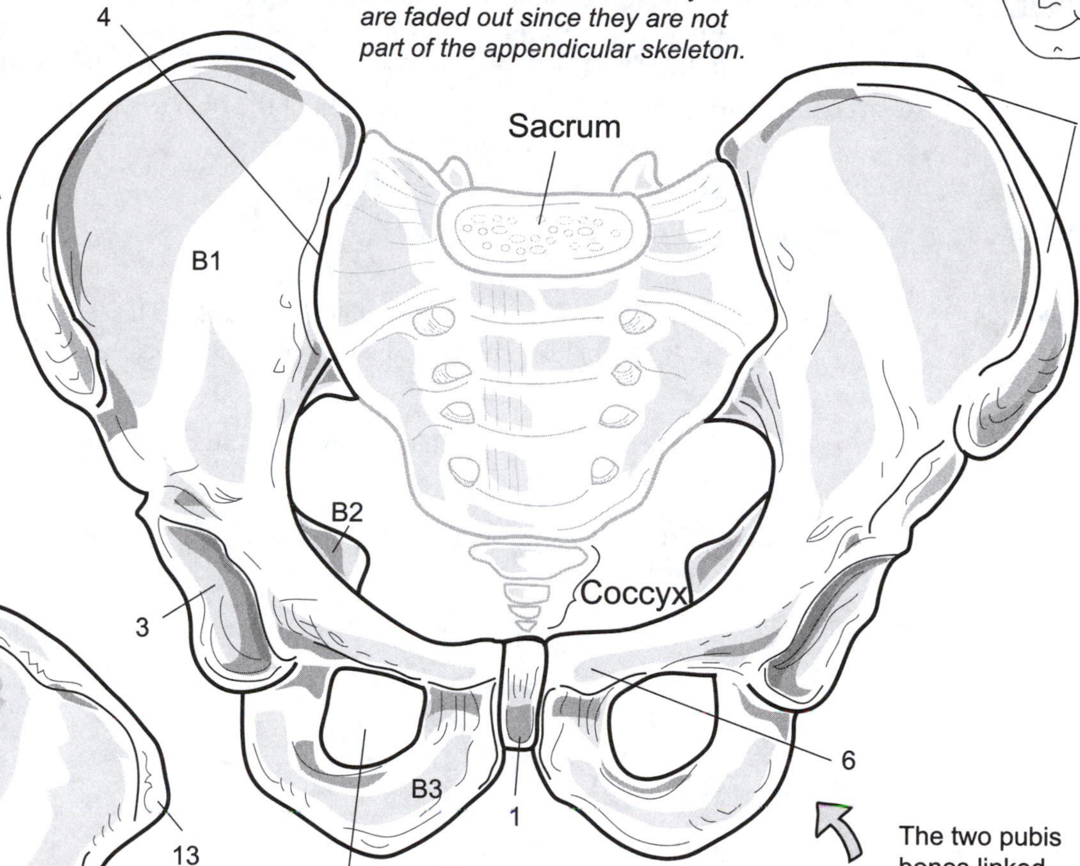

Male pelvis
anterior view

The two pubis bones linked together look like a mask.

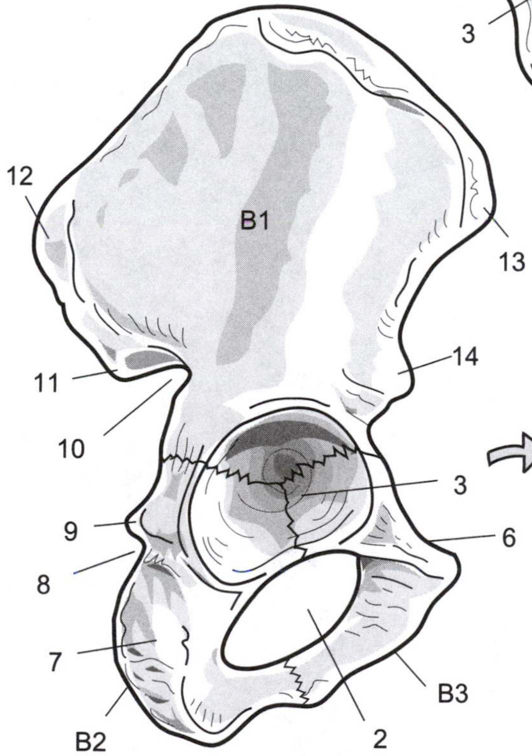

Right os coxae
lateral view

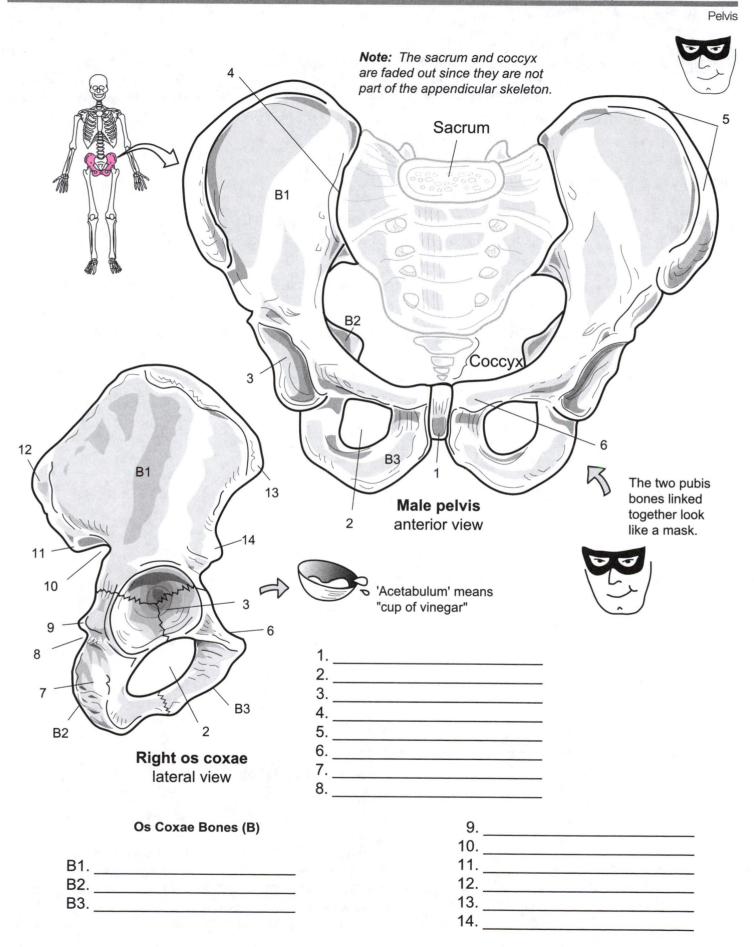

'Acetabulum' means "cup of vinegar"

1. ______________________________
2. ______________________________
3. ______________________________
4. ______________________________
5. ______________________________
6. ______________________________
7. ______________________________
8. ______________________________

Os Coxae Bones (B)

B1. ______________________________
B2. ______________________________
B3. ______________________________

9. ______________________________
10. ______________________________
11. ______________________________
12. ______________________________
13. ______________________________
14. ______________________________

Description

The **femur** is the longest long bone in the body and is commonly called the thigh bone. The **head** of the femur articulates with the acetabulum of the coxal bone to form the ball-and-socket joint in the hip. At the distal end, the patella (kneecap) covers the **patellar surface**. The **medial** and **lateral condyles** articulate with the proximal portion of the tibia to form the knee joint.

The **patella** is loosely held in place by ligaments.

Location

Femoral region (between hip and knee)

Study Tips

- Note that the **head** of the femur always points *medially*. This helps to distinguish a *left* femur from a *right* femur. It also helps to identify the medial condyle because it is on the same side as the head of the humerus.
- To identify the posterior view: **linea aspera** and **intercondylar groove** appear only in the posterior view.
- To identify the anterior from posterior in patella: The anterior surface is rough and the posterior surface is much more smooth.
- **LAP** = **L**inea **A**spera is **P**osterior
- For distinguishing the apex from the base of the patella, remember that apex is a general term meaning the *pointed tip* of a structure.

Palpate (*feel by touch*):

- You can easily feel your patella.

Key to Illustration

Patella

1. Base
2. Apex

Femur

3. Head
4. Neck
5. Greater trochanter
6. Lesser trochanter
7. Linea aspera
8. Patellar surface
9. Intercondylar groove
10. Medial condyle
11. Medial epicondyle
12. Lateral condyle
13. Lateral epicondyle

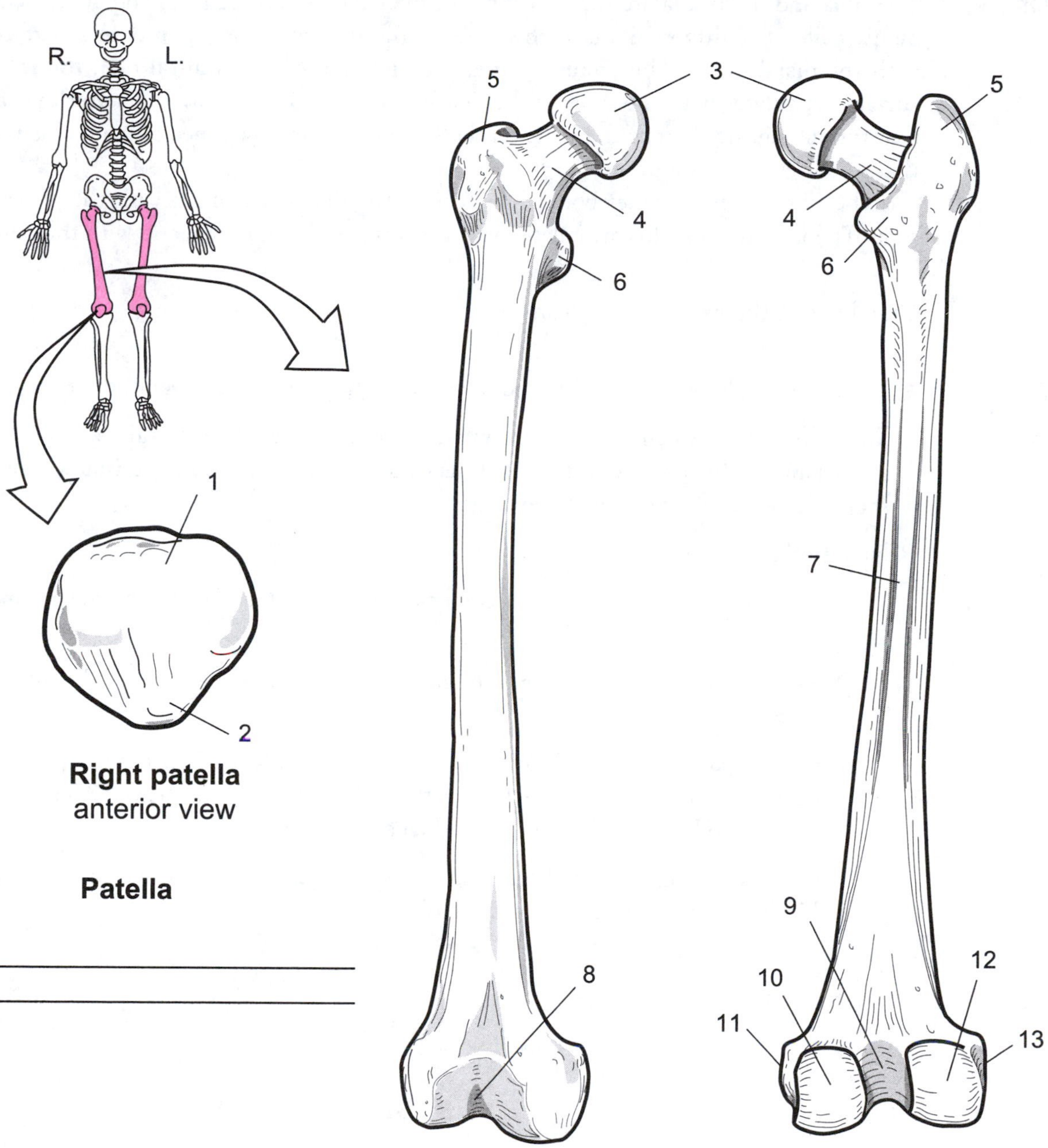

Right patella
anterior view

Right femur
anterior view

Right femur
posterior view

Patella

1. ______________________
2. ______________________

Femur

3. ______________________
4. ______________________
5. ______________________
6. ______________________
7. ______________________
8. ______________________
9. ______________________
10. ______________________
11. ______________________
12. ______________________
13. ______________________

Description

The **tibia** and the **fibula** are the two bones in the leg between the knee and ankle. The **tibia** is commonly called the shinbone and is the larger of the two bones. At its **proximal** surface it articulates with the **distal** end of the femur to form the knee joint. At its distal end, the **inferior articular surface** articulates with the talus in the foot to form the ankle joint. The **anterior margin** is a ridge that runs along the shaft of the bone on the anterior surface. The **medial malleolus** is a large process that stabilizes the ankle joint.

The **fibula** is the smaller bone. The **head** of the fibula articulates with the proximal end of the tibia. The **lateral malleolus** articulates with the distal end of the tibia and with the talus in the ankle.

Location

Crural region (between knee and ankle)

Study Tips

- To distinguish between the **tibia** and the **fibula**: The tibia is the larger of the two bones.
- To distinguish between the two different ends of the fibula: The **lateral malleolus** is more tapered and triangular in shape and is located near the ankle; the **head** of the **fibula** is more rounded in shape and is located near the knee.

Palpate (*feel by touch*):

- Touch the large bump on the medial side of your ankle. This is the **medial malleolus** of the tibia.
- Now touch the bony bump on the lateral side of your ankle. This is the **lateral malleolus** of the **fibula**.
- Feel your patella (*kneecap*), and slide your hand straight down toward your ankle. The first small bump you feel below the knee is the **tibial tuberosity** of the tibia. The soft spot between your knee and the tibial tuberosity is the **patellar ligament**.
- Feel the long ridge of bone beginning below the **tibial tuberosity** and running down toward the ankle. This is the **anterior crest** of the tibia.

Key to Illustration

Fibula
1. Head
2. Lateral malleolus
3. Inferior articular surface

Tibia
4. Lateral condyle
5. Medial condyle
6. Tibial tuberosity
7. Anterior crest
8. Medial malleolus
9. Inferior articular surface

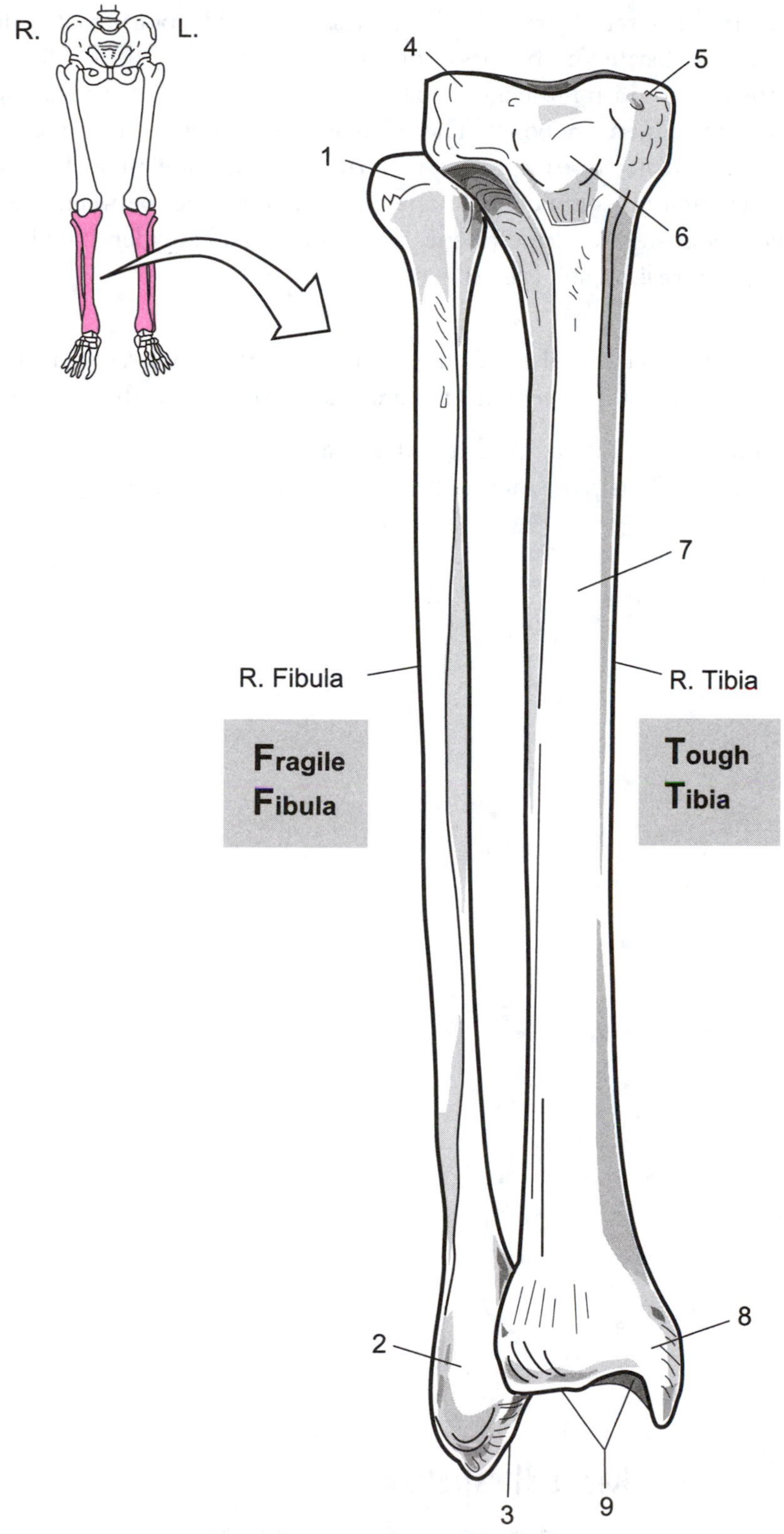

Right fibula and tibia
anterior view

Fibula

1. ______________________________
2. ______________________________
3. ______________________________

Tibia

4. ______________________________
5. ______________________________
6. ______________________________
7. ______________________________
8. ______________________________
9. ______________________________

Description

Each foot contains a total of 26 bones and is divided into three groups of bones: **tarsals** (7), **metatarsals** (5), and **phalanges** (14). The tarsals are the ankle bones. The middle of the foot contains the metatarsals, and the phalanges are located in the toes or digits.

The most difficult part of the foot to learn is the *tarsals*. These seven bones include the **talus, calcaneus, navicular, cuboid**, and three **cuneiform** bones. The **calcaneus** (*heel bone*) is the largest bone in this group. The talus is curved on its superior aspect to articulate with the tibia. Positioned between the calcaneus and the metatarsals is the **cuboid** bone. The **navicular** lies between the talus and cuneiforms. The **cuneiform** bones are the smallest bones in this group and are identified by their position within the foot (*medial*, *intermediate*, and *lateral*).

Study Tips

- The **talus bone** articulates with the tibia. It is also the tallest bone when the foot is viewed laterally. You can easily recall this because the words ***T****alus* and ***T****ibia* and ***T****allest* all begin with the letter "**t**."
- The **great toe** has only two phalanges instead of three. It has a proximal phalanx and a distal phalanx but lacks a middle phalanx. This is the same pattern as the thumb of the hand.

Key to Illustration

Tarsals

1. Calcaneus
2. Talus
3. Navicular
4. Medial **cuneiform**
5. Intermediate **cuneiform**
6. Lateral **cuneiform**
7. Cuboid

Phalanges

8. Proximal phalanx
9. Middle phalanx
10. Distal phalanx

Color the different tarsals different colors.

Schematic of the foot

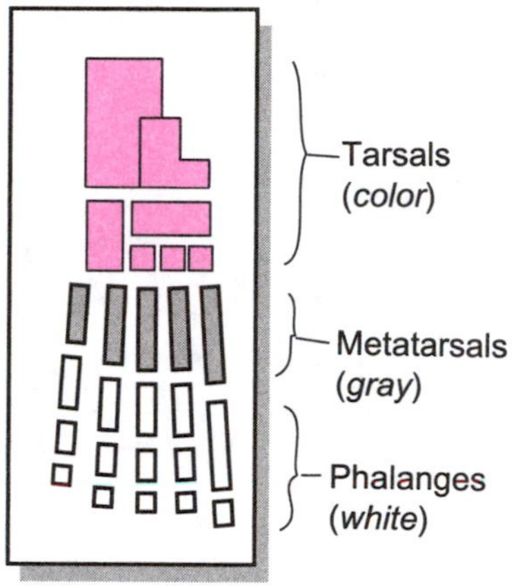

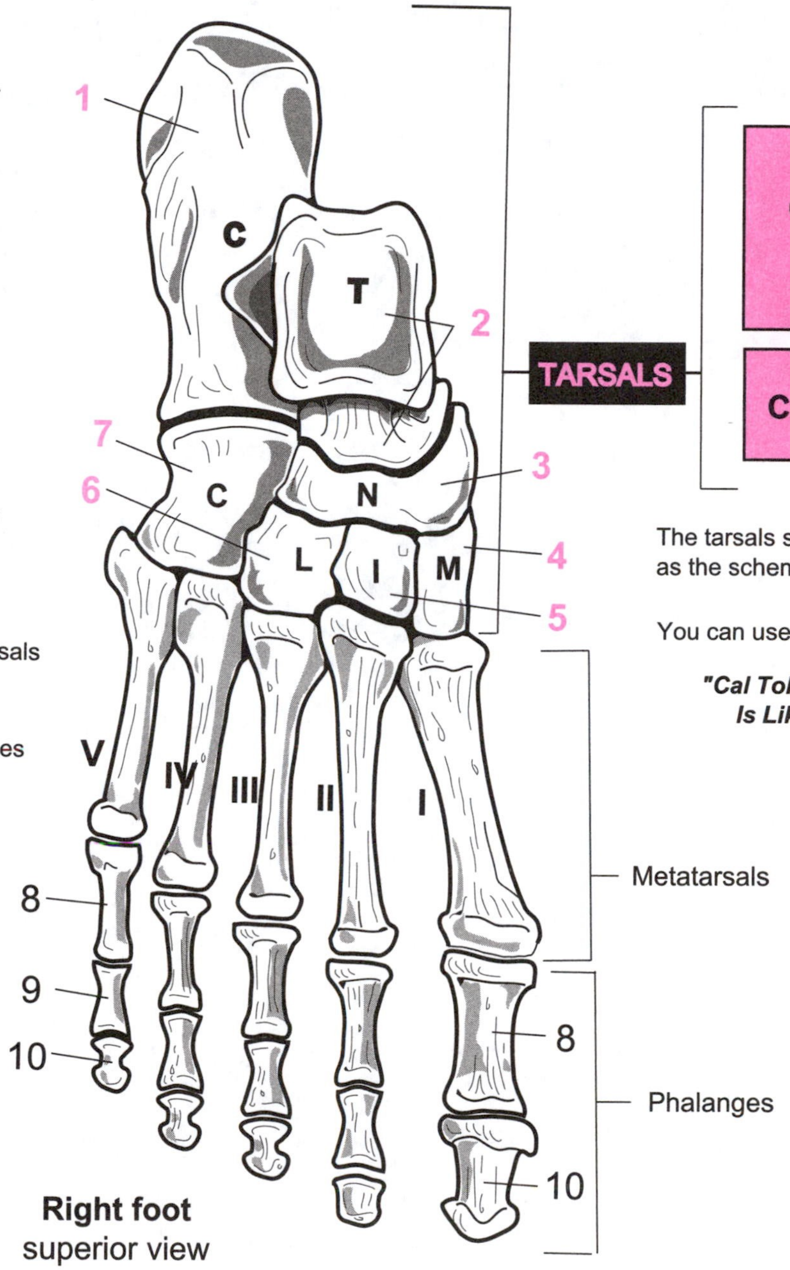

Right foot
superior view

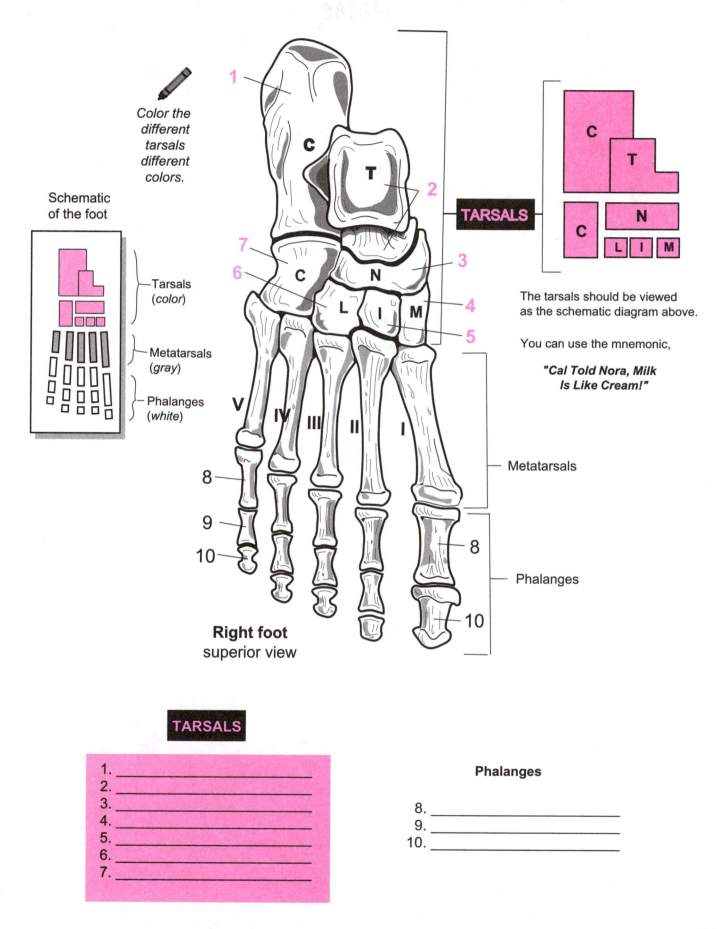

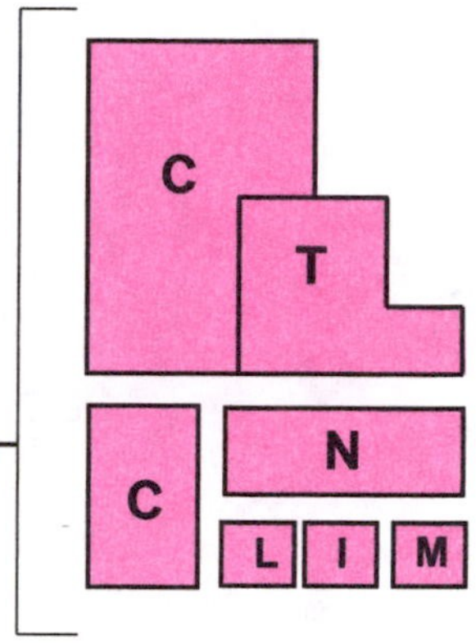

The tarsals should be viewed as the schematic diagram above.

You can use the mnemonic,

"Cal Told Nora, Milk Is Like Cream!"

TARSALS

1. ______________________
2. ______________________
3. ______________________
4. ______________________
5. ______________________
6. ______________________
7. ______________________

Phalanges

8. ______________________
9. ______________________
10. ______________________

Notes

Articulations

Description

Articulations (*joints*) are formed when two or more bones meet. Common examples that quickly come to mind are freely moving joints like the knee and elbow joints. Sutures in the skull are also joints but are often overlooked since they have no movement associated with them. Joints are classified according to either their structure or their degree of movement.

The structural classification system has three categories:

- **Fibrous joints**—only **fibrous connective tissues** anchor the joint together

  ex: sutures in the skull

- **Cartilaginous joints**—only **cartilage** anchors the joint together

  ex: pubic symphysis

- **Synovial joints**—structurally complex joints that contain a **variety of tissues/structures** such as articular cartilage, joint capsule, synovial fluid, and ligaments.

  ex: knee, shoulder, elbow, and hip joints

Here are the key structures in any **synovial joint**:

- **Articular cartilage**—smooth layer of hyaline cartilage that covers the ends of long bones
- **Joint cavity**—potential space filled with synovial fluid
- **Joint capsule**—double layered structure that surrounds the joint

  —outer layer = **fibrous capsule** (*dense irregular connective tissue*)

  —inner layer = **synovial membrane** (*loose connective tissue*)

- **Synovial fluid**—viscous, oily substance secreted by cells in the synovial membrane
- **Ligaments**—bands of fibrous connective tissue that connect one bone to another

Key to Illustration

1. Periosteum
2. Medullary cavity
3. Yellow marrow
4. Ligament
5. Fibrous capsule
6. Synovial membrane
7. Joint capsule
8. Joint cavity (*filled with synovial fluid*)
9. Articular cartilage

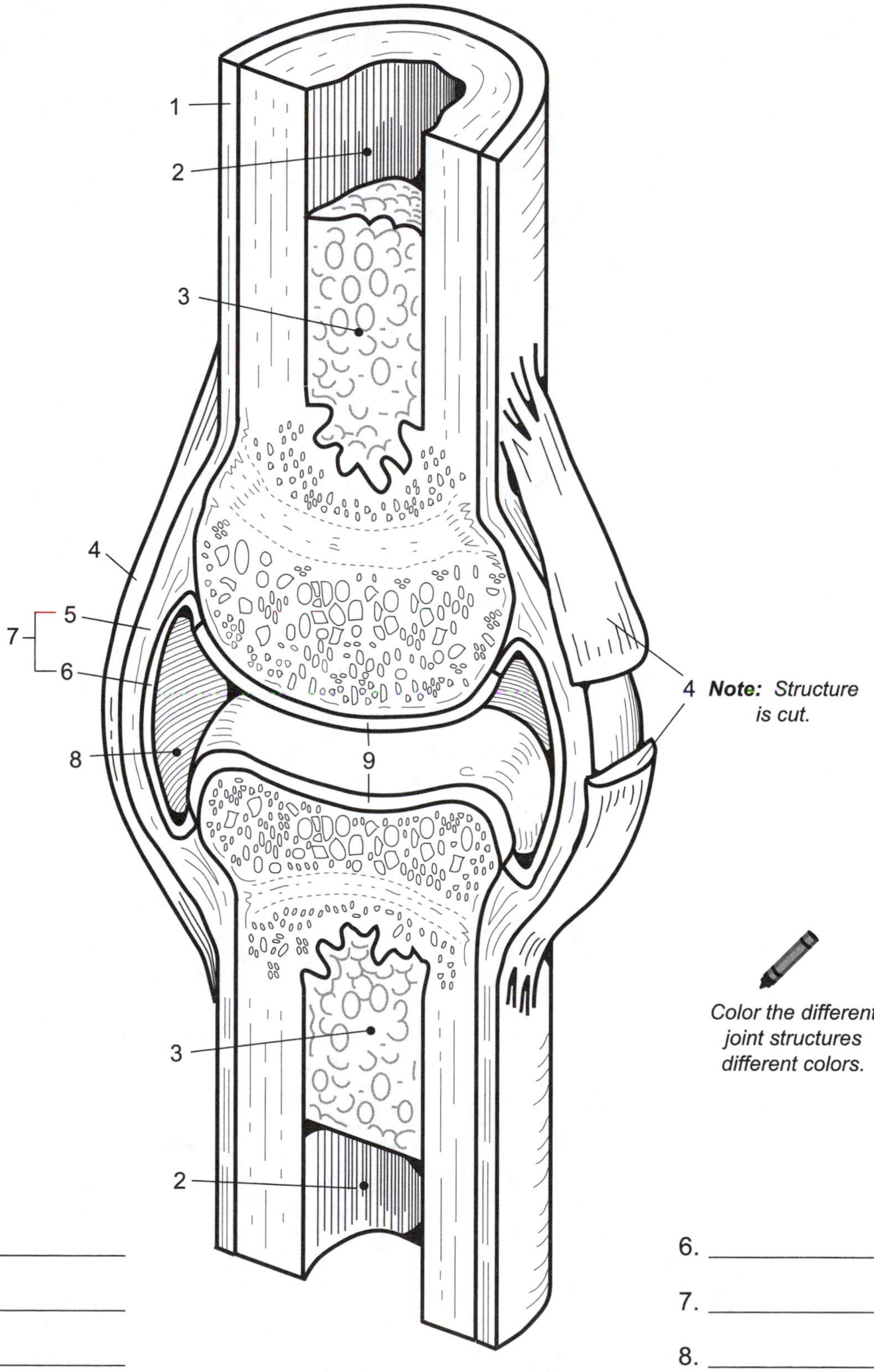

Color the different joint structures different colors.

1. ______________________
2. ______________________
3. ______________________
4. ______________________
5. ______________________

Typical Synovial Joint

6. ______________________
7. ______________________
8. ______________________
9. ______________________

Notes

Muscular System

Description

A whole skeletal muscle is packaged like a series of tubes within other tubes. A muscle first is subdivided into a bundle of long tubes called **fascicles.** Each fascicle is a bundle of skeletal muscle cells or fibers. Each skeletal muscle cell is a bundle of **myofibrils**. Each myofibril is composed primarily of two different protein filaments: **actin** and **myosin**. Because of their difference in size, myosin is called the thick filament and actin is called the thin filament. These proteins are part of a repeated unit called a **sarcomere**, which is the structural and functional unit for muscle contraction. The ends of a sarcomere are defined by the **Z-lines**, which are made of protein.

All of these different bundles of tube-like structures are held together with connective tissue. The **epimysium** is a tough, fibrous, connective tissue that completely surrounds the outside of a whole muscle. Within the whole muscle, the **perimysium** fills the space between fascicles. Surrounding each skeletal muscle cell is another connective tissue called the **endomysium**, which mainly serves to bind one skeletal muscle cell to another.

Analogies

Three analogies are given for structures at the level of the sarcomere.

1. An **actin** or thin filament is compared to a **double-stranded chain of pearls**. Each **pearl** is equivalent to **one molecule of actin**. (Note that this analogy does not include the troponin and tropomyosin proteins.)
2. The myosin filament has **heads** (cross bridges) branching off from it, which later attach to actin during muscle contraction. From the lateral view, these heads appear angled like the **tail feathers in an arrow**.
3. The **myosin heads** attach to actin and pull on it with a regular movement. To visualize this movement, imagine the **heads moving** like a **boat rower's oars**. Unlike the oar movements, however, the heads do not all move at the same time.

Location

All skeletal muscles in the human body (more than 600 in all)

Function

Contraction

Key to Illustration

1. Whole muscle (biceps brachii from the illustration)
2. Fascicle
3. Skeletal muscle cell (fiber)
4. Myofibril
5. Epimysium
6. Perimysium
7. Endomysium
8. Myosin heads
9. Myosin (*thick*) filaments
10. Actin (*thin*) filaments
11. Z-line

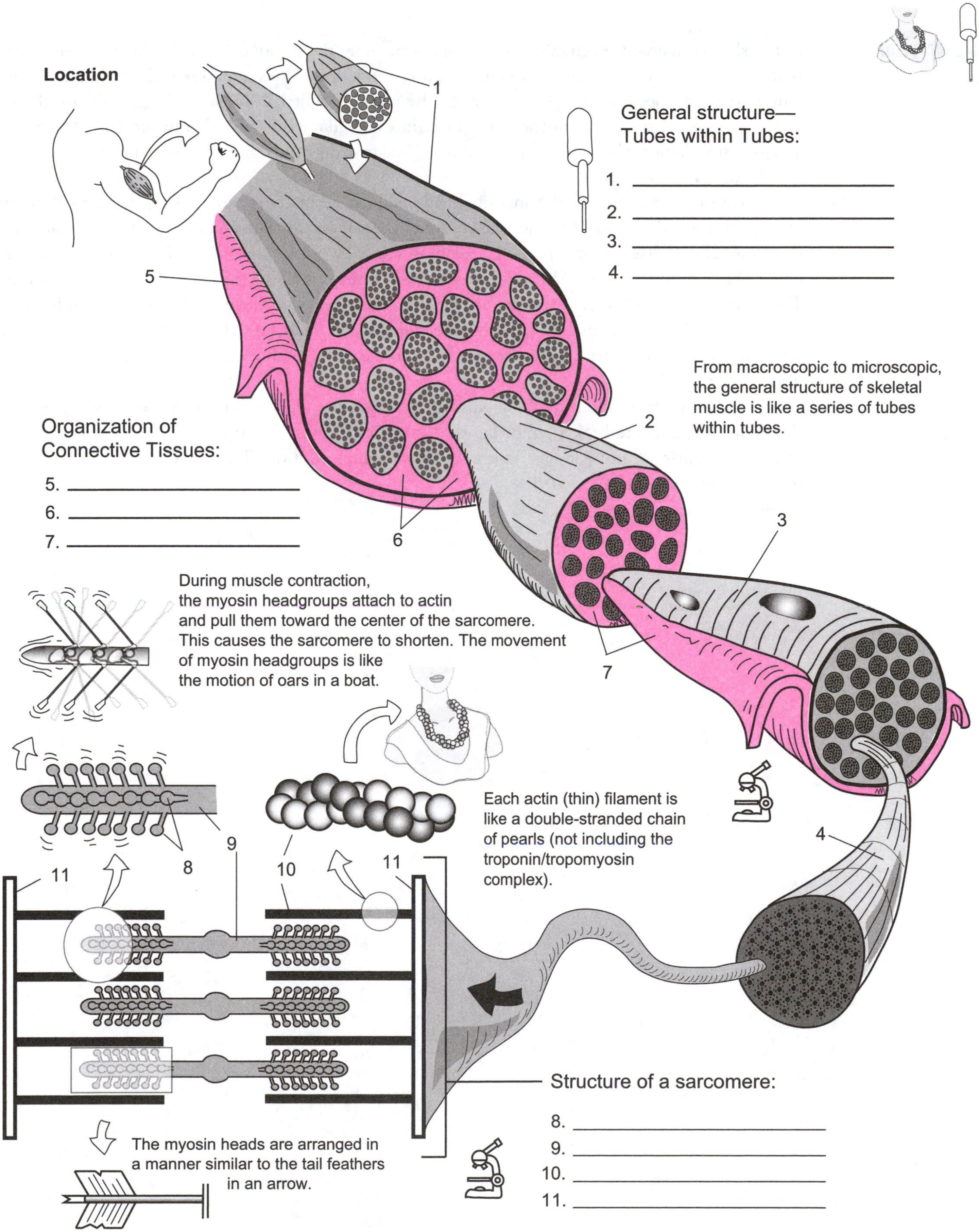

General structure—Tubes within Tubes:

1. ______________________
2. ______________________
3. ______________________
4. ______________________

From macroscopic to microscopic, the general structure of skeletal muscle is like a series of tubes within tubes.

Organization of Connective Tissues:

5. ______________________
6. ______________________
7. ______________________

During muscle contraction, the myosin headgroups attach to actin and pull them toward the center of the sarcomere. This causes the sarcomere to shorten. The movement of myosin headgroups is like the motion of oars in a boat.

Each actin (thin) filament is like a double-stranded chain of pearls (not including the troponin/tropomyosin complex).

The myosin heads are arranged in a manner similar to the tail feathers in an arrow.

Structure of a sarcomere:

8. ______________________
9. ______________________
10. ______________________
11. ______________________

Description

In the **sliding filament mechanism** of muscle contraction, the stimulus for skeletal muscle contraction is always a nerve impulse—no nerve impulse, no contraction. The **sarcomere** is the functional unit of muscle contraction, and its ends are marked by proteins called **Z-lines**. In the middle are the two proteins needed for muscle contraction: (1) **actin** (*thin*) **filaments**, and (2) **myosin** (*thick*) **filaments**. The myosin filaments have oar-like structures called **myosin heads** that stick out. Their function is to attach to and pull on actin.

During contraction, all the myosin heads pull on actin together, causing actin to slide past myosin as the actin is pulled toward the center of the sarcomere. The end result is that the sarcomere shortens. This shortening can be visualized most easily by noting that the Z-lines move toward each other. Also, note that neither the length of the myosin nor the actin changes during contraction. The thing that does change is the amount of *overlap* between the myosin and actin—which increases.

As goes the sarcomere, so goes the skeletal muscles cells and the whole muscle. When the sarcomeres contract, this causes the skeletal muscle cells to shorten, which shortens the whole muscle.

Relaxation occurs when nerve impulses stop. Then the linkages between the myosin heads and the actin filaments are broken, allowing the actin to slide past the myosin. This lengthening of the sarcomere continues until it has returned to its original position. These cycles of contraction and relaxation are repeated as needed.

Sliding Filament Mechanism of Muscle Contraction—Overview

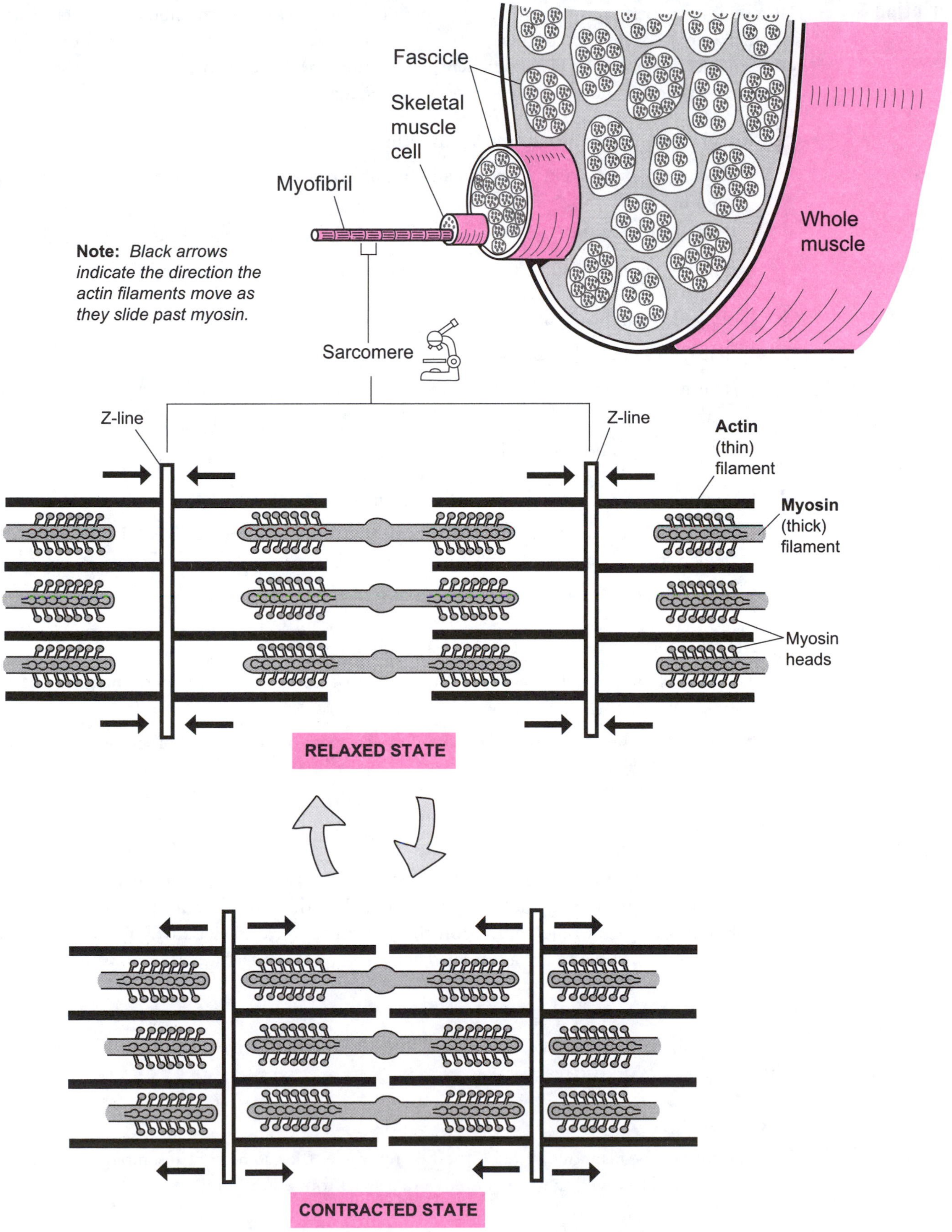

Description

All muscles can perform only one major function—contraction. At the microscopic level, it is the sarcomeres that shorten or contract. This process involves sliding actin filaments past myosin in a four-step contraction cycle: (1) **attachment**, (2) **pulling**, (3) **detachment**, and (4) **reactivation**.

A step-by-step summary of the contraction cycle is as follows:

(1) **Attachment**: Myosin head attaches to actin.

- The regulatory proteins **troponin** and **tropomyosin** already have shifted to expose the **myosin binding sites**.
- The products of **ATP hydrolysis—ADP** and **P**, are still bound to the myosin head.
- The myosin head attaches to the actin molecule (*peg in a hole analogy*).

(2) **Pulling**: Myosin head pulls on actin.

- The myosin head "cocks," which releases **ADP** and P.
- The myosin molecule bends, thereby pulling the actin filament toward the center of the sarcomere.
- This pulling action is called a *power stroke*.

(3) **Detachment**: Myosin head releases from actin.

- As a new **ATP** molecule enters the cycle and binds to a myosin head, it forces the head to detach from actin.
- The linkage between myosin and actin is broken.

(4) **Reactivation**: Myosin head reenergizes itself for another cycle.

- ATP hydrolysis occurs, producing a net release of free energy, which energizes the myosin head by "cocking" it again.
- The myosin head now is ready to attach to actin, and the contraction cycle is repeated.

Analogies

- **Peg in a hole analogy**

  The peg fits in the hole. In this case, the actin binding site on the myosin head is like the peg and the myosin binding site on the actin molecule is like the hole. In reality, it is a chemical bond rather than a peg and hole connection that allows the myosin head to bind to actin. But this analogy helps visualize the linkage formed between myosin and actin.

- **Cotton swab analogy**

  Each myosin molecule is like a double-headed cotton swab (*with one end cut off*) that is able to bend like a hinge in two places.

- **Rower analogy**

  As the myosin heads move through their 4-step cycle, they somewhat resemble the movement of oars in a boat. However, they do not move in a synchronized rhythm.

Contraction Cycle

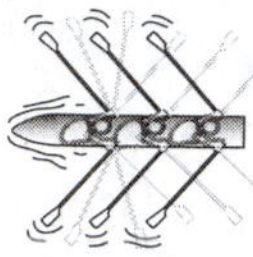

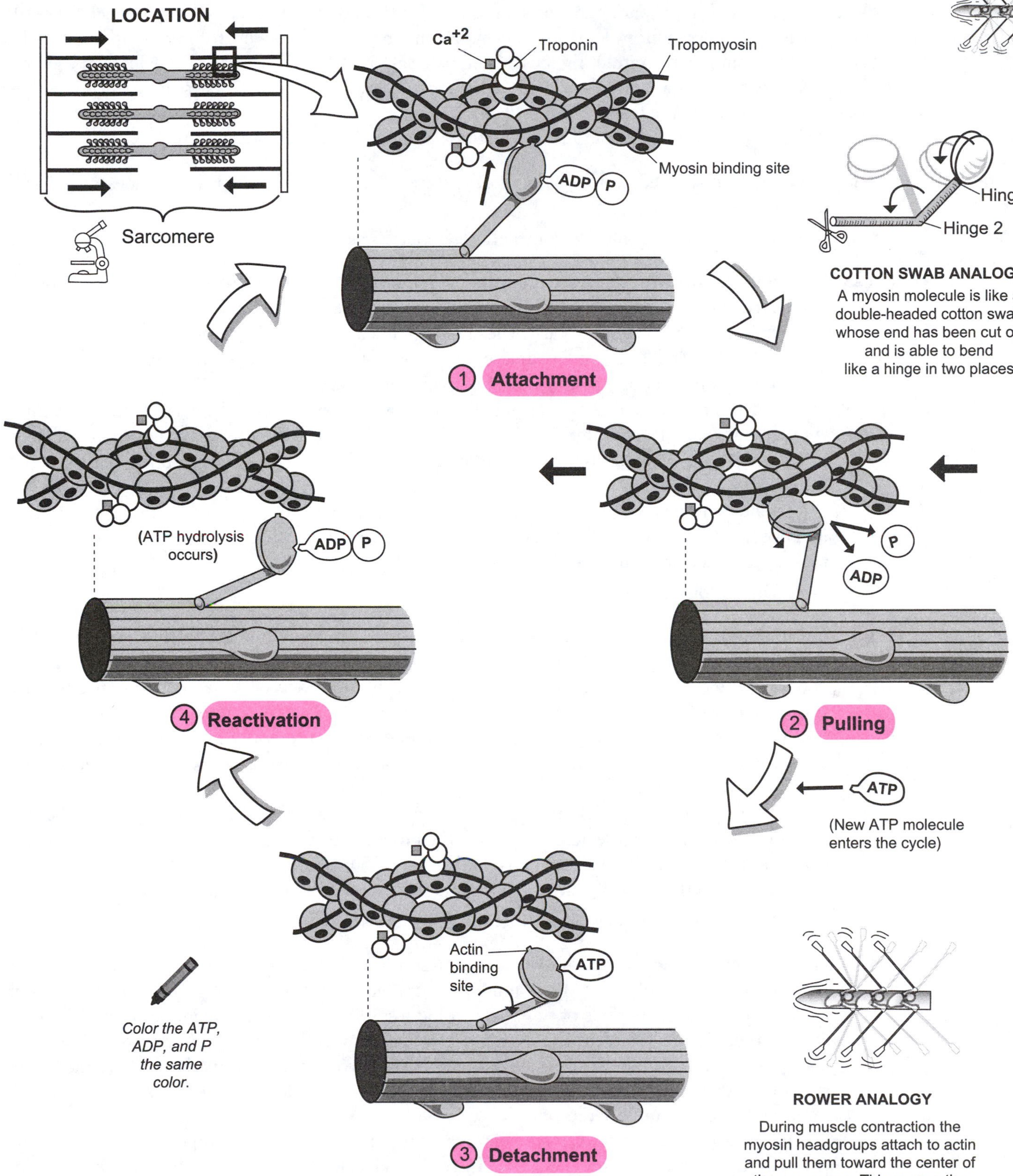

COTTON SWAB ANALOGY

A myosin molecule is like a double-headed cotton swab whose end has been cut off and is able to bend like a hinge in two places

Color the ATP, ADP, and P the same color.

ROWER ANALOGY

During muscle contraction the myosin headgroups attach to actin and pull them toward the center of the sarcomere. This causes the sarcomere to shorten. This movement of the myosin headgroups is rather like the motion of oars in a boat.

Description

ATP is the common currency that fuels cellular activities and the primary energy molecule used by all cells. Muscle contraction demands enormous amounts of ATP—even those that last only a few seconds. For example, a contracting muscle cell uses about 2 million molecules of ATP every second! Wow! Without a constant supply of ATP, muscle contraction ceases. This begs the question: Where does all of this ATP come from?

Here is a summary of the four major sources of energy for muscle contraction:

1. **ATP hydrolysis** (see p. 42): Muscle cells store a local supply of ATP, but it does not provide much energy.
   - Time allowed for muscle contraction: about 5 seconds.

2. **Phosphorylation from creatine phosphate**: While ATP is the *primary* energy molecule used by cells, **creatine phosphate (CP)** is the *secondary* energy molecule found *only* in muscle cells. Creating it is a two part process. First, the small chemical **creatine (C)** is produced by various organs including the liver and is delivered to muscle cells. Then, within the muscle cells, a phosphate group is added to C with the help of an enzyme.

   The function of CP is to transfer its phosphate group directly onto ADP to create more ATP. In a sense, CP "recharges" ADP to make more ATP. During the relaxed state, muscle cells make excess ATP, which quickly undergoes ATP hydrolysis, supplying a phosphate group to transform C into CP.

   In the contracting state, the levels of ADP are rising within the muscle cells, so CP transfers its phosphate group onto the excess ADP to form more ATP, which can be used to power more muscle contractions. But this process doesn't last long.
   - Time allowed for contraction: about 15 seconds.

3. **Fermentaton** (see p. 410): Fermentation is the production of two **lactic acid** molecules from the breakdown of a single **glucose** molecule when no oxygen is present in the cell. This process occurs in the cytosol and yields a net gain of **2 ATP**. **Glucose** is provided by either the blood or the breakdown of local **glycogen** within the muscle tissue. In **glycolysis**, a glucose molecule normally is converted into pyruvic acid. When oxygen is present, the pyruvic acid enters the mitochondria to begin aerobic respiration. But under conditions of no oxygen, the pyruvic acid is converted into lactic acid instead.

   The buildup of lactic acid in muscle cells causes soreness. It also presents a problem because it makes the local pH more acidic. To solve the problem, this waste product diffuses out of muscle cells and is transported into liver cells. Although 2 ATP is not much for each cycle of fermentation, it doubles the time of contraction compared to the previous two mechanisms.
   - Time allowed for muscle contraction: about 30 seconds.

4. **Aerobic respiration** (see p. 402): Aerobic respiration is the essential mechanism used to deliver a constant supply of ATP for muscle contraction. Without it, we couldn't sustain muscle contraction for long periods of time. In aerobic respiration, oxygen is required to produce more ATP through the **citric acid cycle** and **electron transport systems** inside the **mitochondrion**. The **pyruvic acid** produced in glycolysis enters the citric acid cycle in the mitochondrion instead of getting converted into lactic acid. The glucose needed for glycolysis comes from the same sources it did in fermentation.

   In addition to glucose, energy sources include **fatty acids** from adipose tissue and **amino acids** from the breakdown of proteins. Instead of producing only 2 ATP (as in fermentation), the new total is about **36 ATP** for every glucose molecule that is broken down. As an example, if you plan to run a marathon, you will be relying on aerobic respiration to fuel all those muscle contractions for running!
   - Time allowed for muscle contraction: hours.

ECF
Sarcolemma (plasma membrane of skeletal muscle cell)
ICF
1 ATP hydrolysis
P P P
A
ATP or ATP
A P P
ADP
+ P + Free energy (for muscle contraction)
P
2 Phosphorylation from creatine phosphate (CP)
ADP
excess ATP
relaxed state
C P
CP
C
C
contracting state
ADP
ATP or ATP
Color the ATP, ADP, and CP molecules.
3 Fermentation
GLYCOLYSIS
G G G G G
Glycogen (stored in muscle)
Glucose
Pyruvic acid
2 ATP
Lactic acid
KEY
C = Creatine
CP = Creatine phosphate
ATP = Adenosine triphosphate
ADP = Adenosine diphosphate
ECF = Extracellular fluid
ICF = Intracellular fluid
GLYCOLYSIS
4 Aerobic respiration
Glucose
Pyruvic acid
Glycogen (stored in muscle)
Amino acids
Fatty acids
Mitochondrion
CITRIC ACID CYCLE
ELECTRON TRANSPORT SYSTEM (ETS)
36 ATP + CO_2 + H_2O

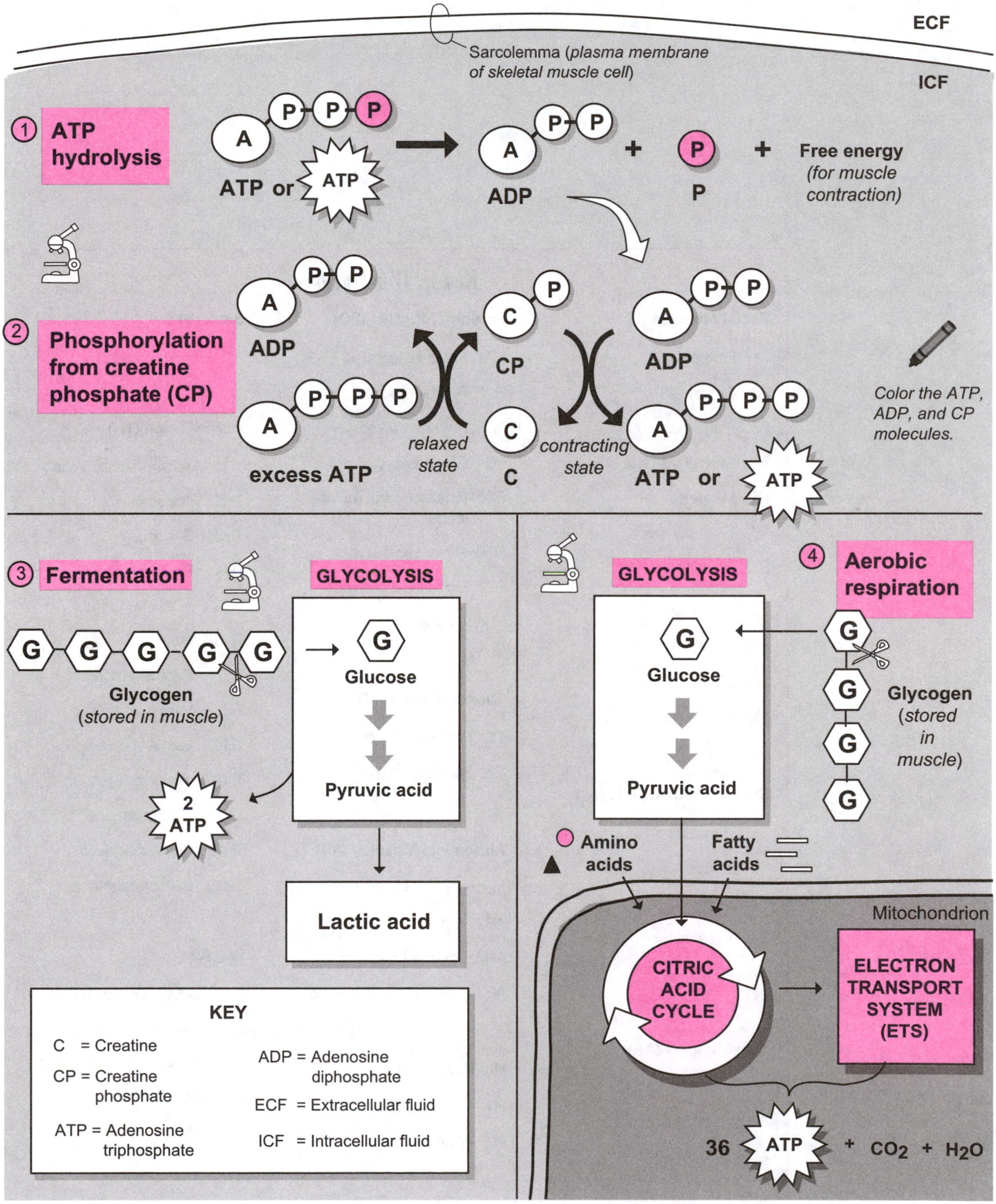

Key to Illustration

Facial Muscles (F)

F1. Frontalis
F2. Temporalis
F3. Platysma
F4. Orbicularis oculi
F5. Masseter
F6. Orbicularis oris
F7. Occipitalis

Neck Muscles (N)

N1. Sternohyoid
N2. Sternocleidomastoid

Shoulder Muscles (S)

S1. Trapezius
S2. Deltoid
S3. Infraspinatus
S4. Teres minor
S5. Teres major

Arm Muscles (AR)

AR1. Triceps brachii
AR2. Biceps brachii
AR3. Brachialis

Forearm Muscles (FO)

FO1. Pronator teres
FO2. Brachioradialis
FO3. Flexor carpi radialis
FO4. Palmaris longus
FO5. Extensor carpi radialis longus
FO6. Flexor carpi ulnaris
FO7. Extensor digitorum
FO8. Extensor carpi ulnaris

Thorax Muscles (T)

T1. Pectoralis major
T2. Serratus anterior

Abdominal Muscles (AB)

AB1. Rectus abdominis
AB2. External oblique
AB3. Internal oblique
AB4. Transverse abdominis

Hip Muscles (H)

H1. Gluteus medius
H2. Gluteus maximus

Back (B)

B1. Latissimus dorsi

Pelvis/Thigh (TH)

TH1. Tensor fasciae latae
TH2. Iliopsoas
TH3. Sartorius
TH4. Pectineus
TH5. Adductor longus
TH6. Gracilis
TH7. Rectus femoris
TH8. Vastus lateralis
TH9. Vastus medialis
TH10. Adductor magnus
TH11. Biceps femoris
TH12. Semitendinosus
TH13. Semimembranosus

Leg (L)

L1. Fibularis (*peroneus*) longus
L2. Tibialis anterior
L3. Extensor digitorum longus
L4. Gastrocnemius
L5. Soleus

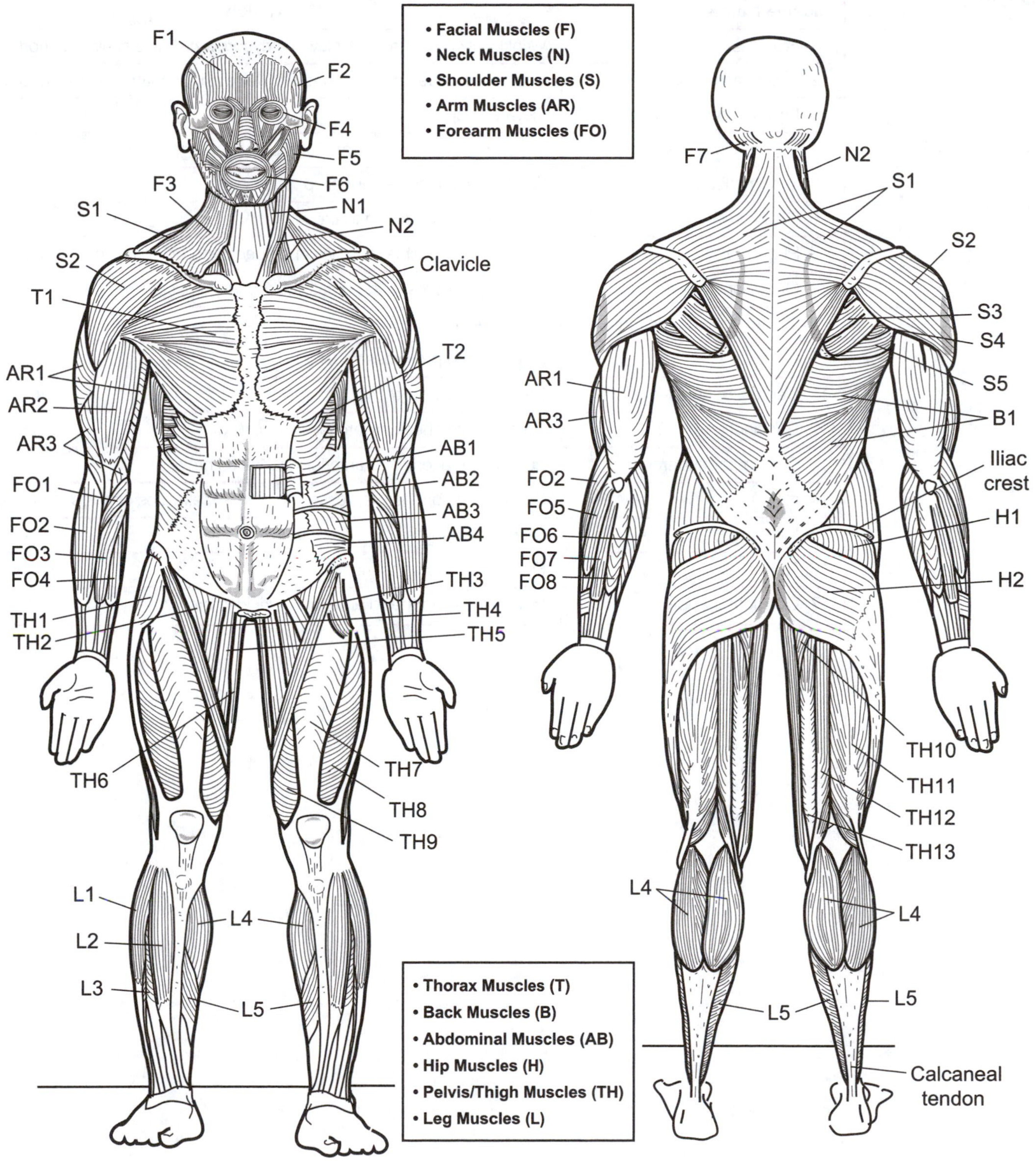
• Facial Muscles (F)
• Neck Muscles (N)
• Shoulder Muscles (S)
• Arm Muscles (AR)
• Forearm Muscles (FO)
F1
F2
F3
F4
F5
F6
F7
N1
N2
S1
S2
S3
S4
S5
Clavicle
T1
T2
AR1
AR2
AR3
AB1
AB2
AB3
AB4
B1
Iliac crest
H1
H2
FO1
FO2
FO3
FO4
FO5
FO6
FO7
FO8
TH1
TH2
TH3
TH4
TH5
TH6
TH7
TH8
TH9
TH10
TH11
TH12
TH13
L1
L2
L3
L4
L5
Calcaneal tendon
• Thorax Muscles (T)
• Back Muscles (B)
• Abdominal Muscles (AB)
• Hip Muscles (H)
• Pelvis/Thigh Muscles (TH)
• Leg Muscles (L)
Anterior view
Posterior view
Color the different muscles different colors.

Description

Muscle Name	Action
1. Frontalis	Wrinkles skin of forehead, elevates eyebrows; draws scalp anteriorly
2. Orbicularis oculi	Depresses upper eyelid, elevates lower eyelid, tightens skin around eyes
3. Levator labii superioris	Elevates upper lip
4. Zygomaticus major	Retracts and elevates corner of mouth
5. Zygomaticus minor	Retracts and elevates upper lip
6. Risorius	Draws corner of mouth to the side
7. Platysma	Depresses mandible, pulls lower lip back and down *(as in pouting)*
8. Orbicularis oris	Closes, protrudes, and purses lips *(kissing muscle)*
9. Depressor labii inferioris	Depresses lower lip
10. Mentalis	Elevates and protrudes lower lip
11. Depressor anguli oris	Depresses corner of mouth
12. Buccinator	Compresses cheek inward *(as in whistling)*
13. Corrugator supercilii	Pulls skin inferiorly and anteriorly; wrinkles brow
14. Nasalis	Compresses bridge, depresses tip of nose; elevates corners of nostrils
15. Masseter	Elevates mandible and closes jaw
16. Temporalis	Elevates and retracts mandible
17. Corrugator supercilii	Pulls skin inferiorly and anteriorly; wrinkles brow

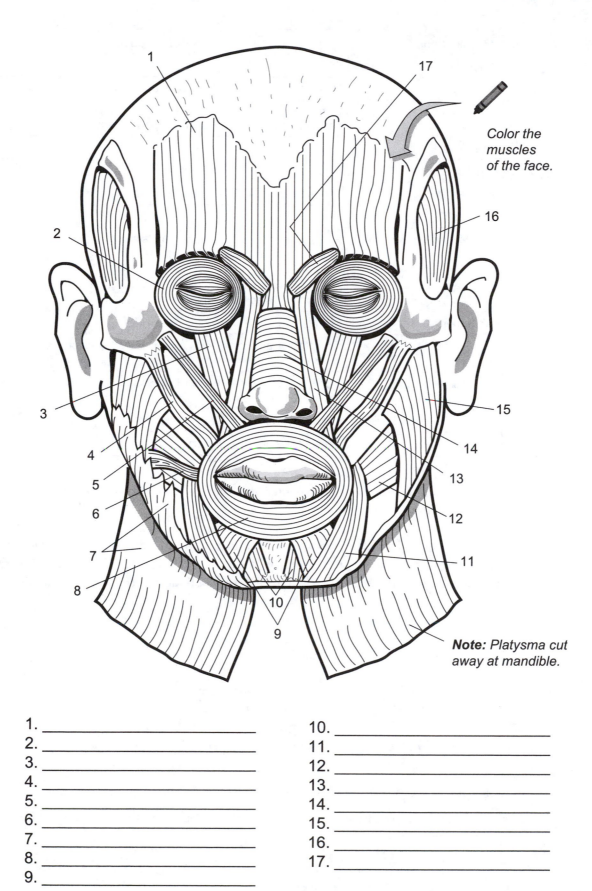

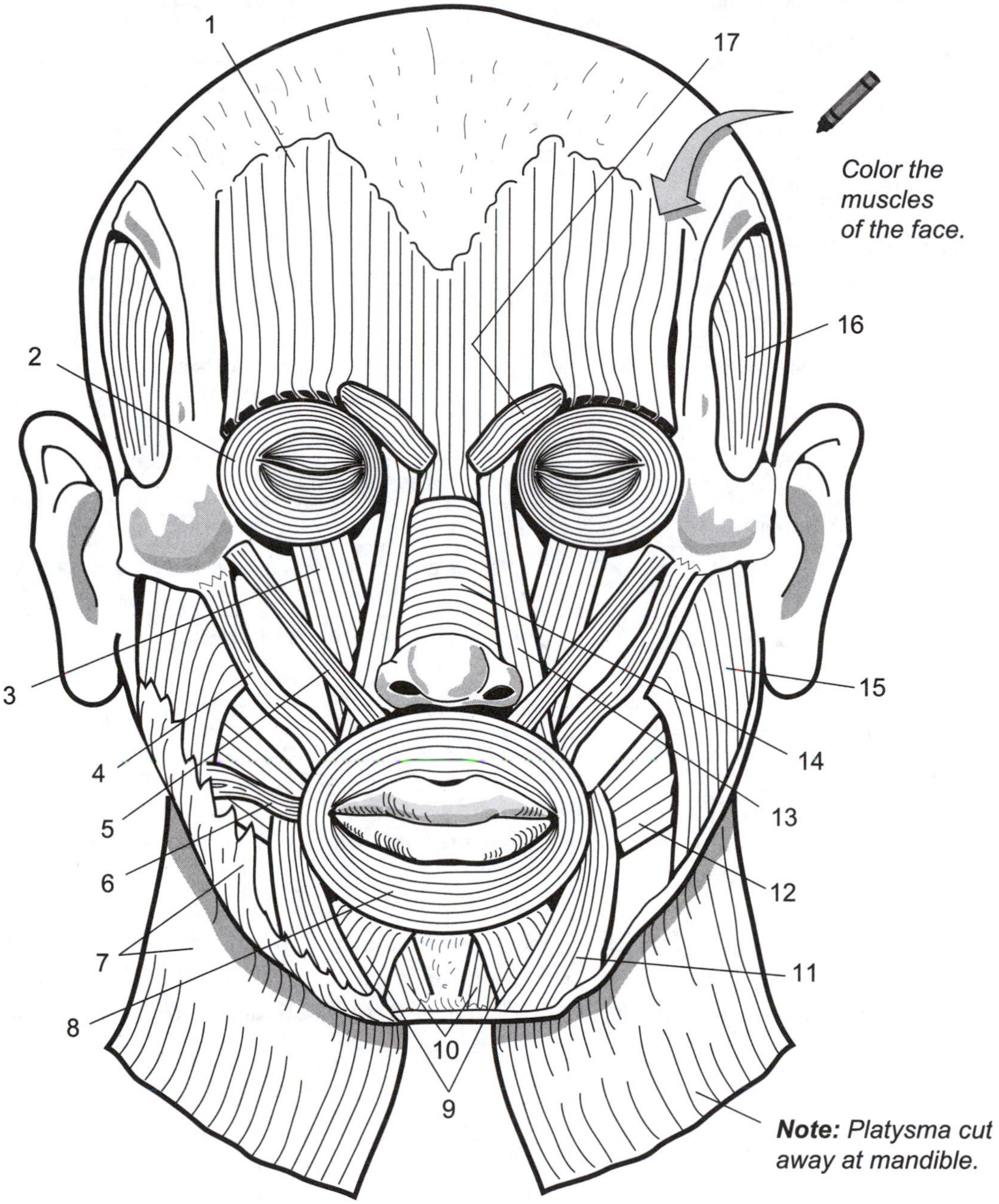

1. ______________________
2. ______________________
3. ______________________
4. ______________________
5. ______________________
6. ______________________
7. ______________________
8. ______________________
9. ______________________
10. ______________________
11. ______________________
12. ______________________
13. ______________________
14. ______________________
15. ______________________
16. ______________________
17. ______________________

Superficial and Deep Muscles of the Neck, Shoulder, Thorax, and Abdomen

Description

Muscle Name	Action
Neck (N)	
N1. Platysma	Depresses mandible; pulls lower lip back and down *(as in pouting)*
N2. Sternocleidomastoid	Simultaneous contraction of both heads: flexes neck. Individual action of each head: rotates head to shoulder on opposite side
Shoulder (S)	
S1. Trapezius	Extends neck; retracts scapula
S2. Deltoid	Flexes, extends, abducts; medially and laterally rotates arm
Thorax (T)	
T1. Pectoralis major	Primary muscle of arm flexion; adducts and medially rotates arm; with arm fixed, pulls chest forward *(as in a forced inspiration)*
T2. Pectoralis minor	Depresses and protracts scapula, elevates ribs
T3. Serratus anterior	Abducts and stabilizes scapula
Abdomen (A)	
A1. External oblique	Compresses anterior abdominal wall; flexes trunk; rotates trunk; depresses lower ribs
A2. Internal oblique	Compresses anterior abdominal wall; flexes trunk; rotates trunk; depresses lower ribs
A3. Transverse abdominis	Compresses anterior abdominal contents
A4. Rectus abdominis	Compresses anterior abdominal wall; flexes trunk

Analogies

- In the anterior, superficial view, the anterior portion of the **serratus anterior** looks like the blade of a **serrated knife.**
- The three abdominal muscles on the lateral surface of the abdomen—*external oblique*, *internal oblique*, and *transverse abdominis*—are layered on top of each other like a ham sandwich. The **internal oblique** is the **ham** (*in the middle*), and the **external oblique** and **transverse abdominis** are the layers of **bread.**

Study Tips

- Mnemonic for the 4 abdominal muscles: *"Really? Everything is terrible?"*

Really?	**R**ectus abdominis
Everything	**E**xternal oblique
Is	**I**nternal oblique
Terrible?	**T**ransverse abdominis

- *Rectus* means *straight*—this indicates the muscle fiber direction.
- Imagine you have your hands in your front pockets. Your fingers follow the fiber direction of the external oblique muscle.

Superficial and Deep Muscles of the Neck, Shoulder, Thorax, and Abdomen

Pectoralis major is a fan-shaped muscle.

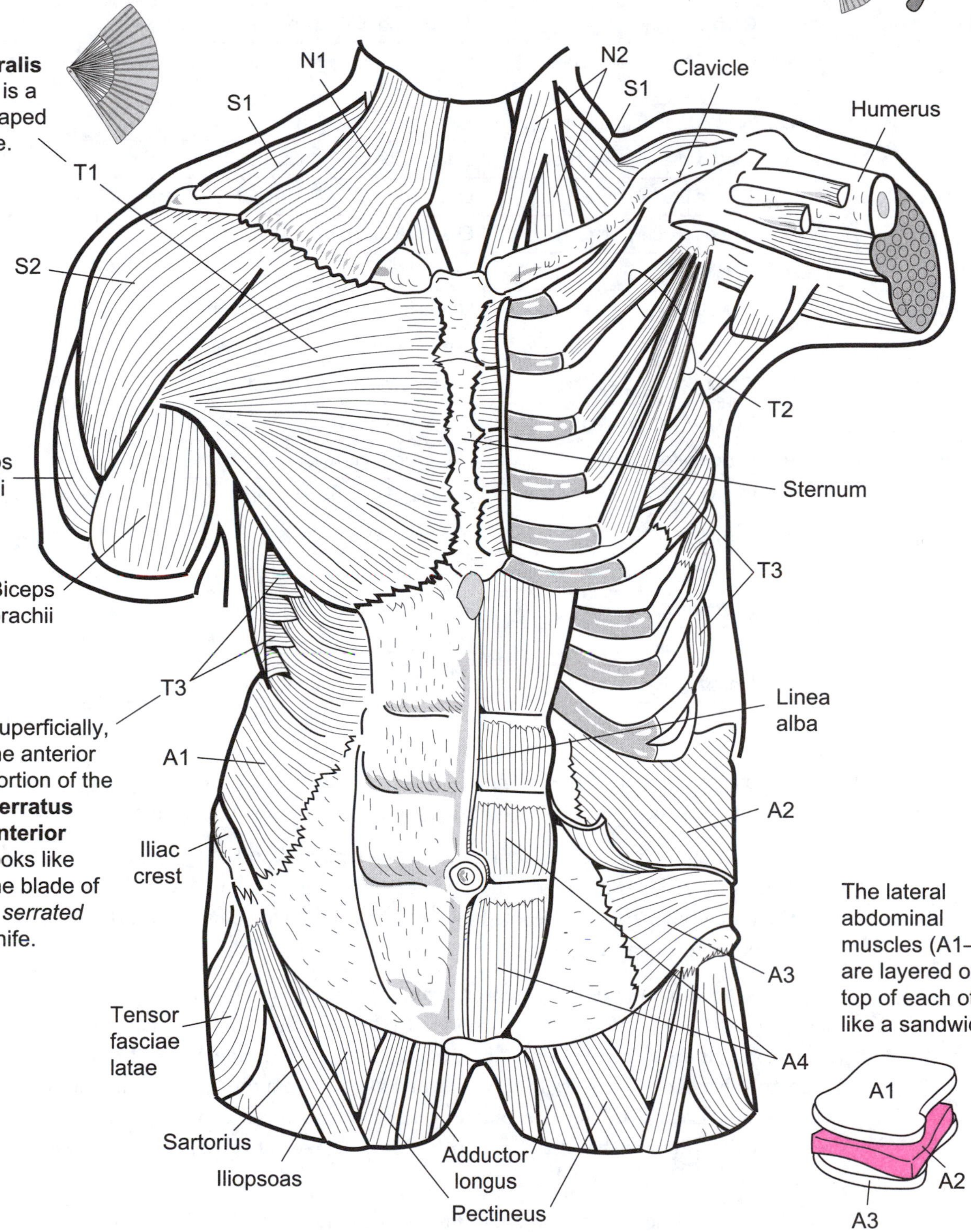

Superficially, the anterior portion of the **Serratus anterior** looks like the blade of a *serrated* knife.

The lateral abdominal muscles (A1–A3) are layered on top of each other like a sandwich.

Color the muscles in each regional area—neck, shoulder, thorax, abdomen—with different colors.

Superficial Muscles of the Arm, Anterior and Posterior Views

Description

Muscle Name	Action
1. Coracobrachialis	Flexion and abduction of arm/shoulder
2a. Biceps brachii, short head	Flexes and supinates forearm *(turns the corkscrew and pulls out the cork)*
2b. Biceps brachii, long head	Flexes and supinates forearm
3a. Triceps brachii, long head	Extends forearm; stabilizes shoulder joint
3b. Triceps brachii, medial head	Extends forearm
3c. Triceps brachii, lateral head	Extends forearm
4. Brachialis	Flexes forearm
5. Pronator teres	Flexes forearm; pronates forearm
6. Brachioradialis	Flexes forearm
7. Flexor carpi radialis	Flexes wrist; abducts hand
8. Palmaris longus	Weak wrist flexor
9. Flexor carpi ulnaris	Flexes and adducts hand
10. Anconeus	Adducts ulna during forearm rotation; weak forearm extensor
11. Brachioradialis	Flexes forearm
12. Extensor carpi radialis longus	Extends wrist; abducts hand
13. Flexor carpi ulnaris	Flexes and abducts hand
14. Extensor carpi ulnaris	Extends and adducts hand
15. Extensor digitorum	Extends hand, extends digits 2–5

Study Tips

- Most muscles that act as **flexors** are best seen in the anterior view of the upper limb, and muscles that act as **extensors** are best seen on the posterior view of the upper limb.
- In the anterior view, use the **brachioradialis** as a landmark in the forearm. It is the widest muscle on the lateral surface and inserts on the styloid process of the radius. Medially from the brachioradialis, use the mnemonic "**F**oolish **P**eople **F**ollow" for the following muscles: **F**lexor carpi radialis, **P**almaris longus, and **F**lexor carpi ulnaris.
- In the posterior view, use the **extensor digitorum** muscle as a landmark for the forearm. To correctly identify it, find the four tendons anchoring to the phalanges in all the fingers except the thumb. All these tendons are associated with this muscle. Then, use the mnemonic "**E**at **F**ood, **E**xcept **B**roccoli" to learn the adjacent muscles in the forearm. Laterally from the extensor digitorum is the first part of the phrase, "**E**at **F**ood," for the **E**xtensor carpi ulnaris and **F**lexor carpi ulnaris. Medially from the extensor digitorum is the last part of the phrase, "**E**xcept **B**roccoli" for the **E**xtensor carpi radialis longus and the **B**rachioradialis.
- The **PALM***aris longus* muscle anchors directly into the middle of the PALM.

Superficial Muscles of the Arm, Anterior and Posterior Views

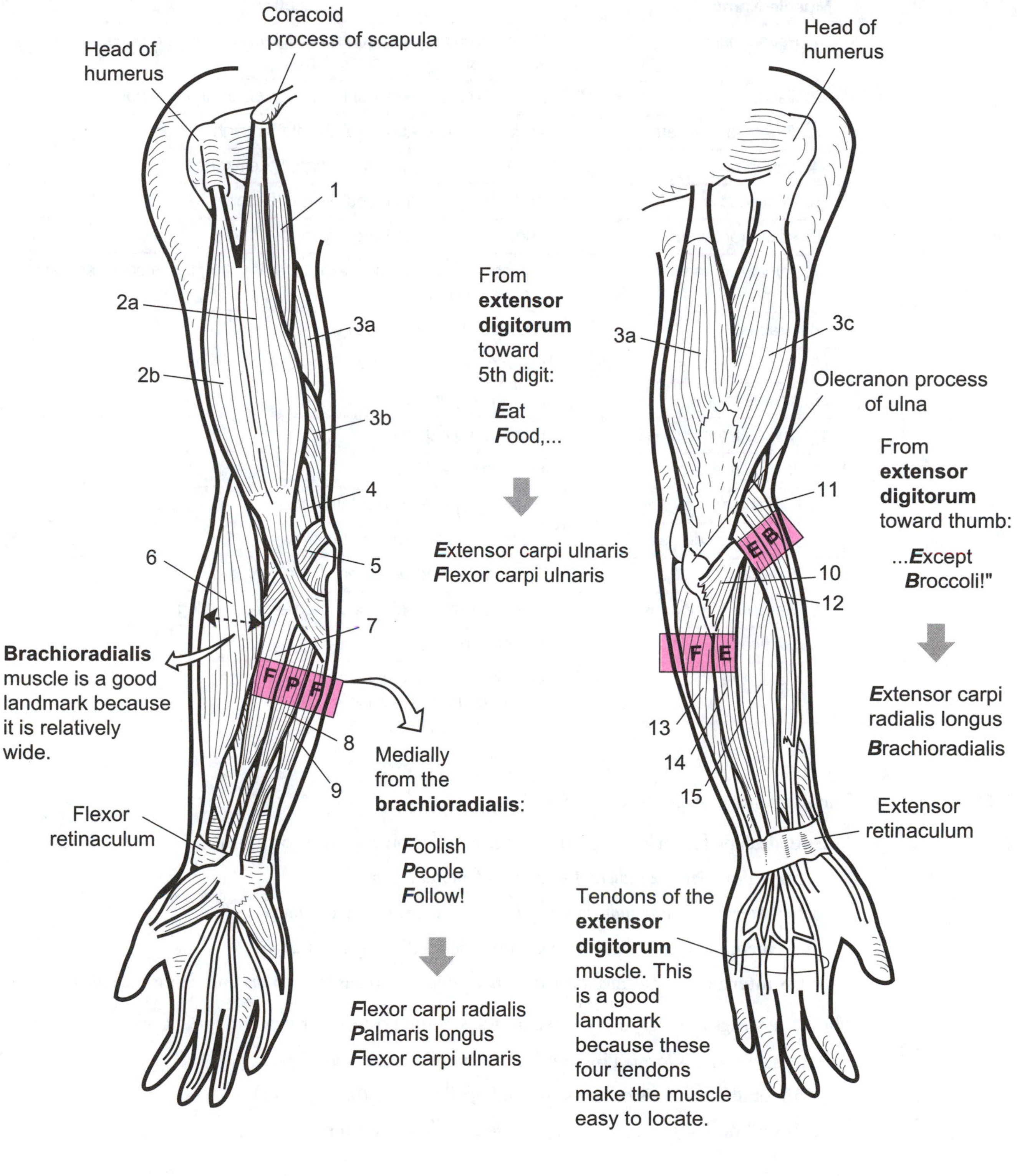

Right arm
anterior view

Right arm
posterior view

Muscles of the Thigh, Anterior and Posterior Views

Description

Muscle Name	Action
1. Psoas major	Flexes thigh or flexes trunk on thigh *(as in during a bow)*; also effects lateral flexion of vertebral column
2. Iliacus	Flexes thigh or flexes trunk on thigh *(as in during a bow)*
3. Tensor fasciae latae	Abducts, flexes, and medially rotates thigh
4. Sartorius	Flexes and laterally rotates thigh, flexes knee
5. Pectineus	Adducts, flexes, and medially rotates thigh
6. Adductor longus	Adducts, flexes, and laterally rotates thigh
7. Adductor magnus	Anterior part flexes and medially rotates thigh; posterior part extends and laterally rotates thigh
8. Gracilis	Adducts hip and flexes leg
9. Rectus femoris	Extends knee and flexes thigh at hip
10. Vastus lateralis	Extends leg at knee
11. Vastus intermedius	Extends leg at knee
12. Vastus medialis	Extends leg at knee
13. Gluteus medius	Abducts and medially rotates hip
14. Gluteus maximus	Extends and laterally rotates hip
15. Gracilis	Adducts hip and flexes leg
16. Semimembranosus	Extends thigh; flexes knee; medially rotates leg
17. Semitendinosus	Extends hip and flexes knee
18. Biceps femoris	Long head = extends hip and flexes knee Short head = flexes knee

Study Tips

- The **sartorius** muscle looks like a **sash**
- The **quadriceps femoris** is a group of four muscles on the anterior thigh:

  rectus femoris (*name indicates location—femoral region*)

  vastus **lateralis** (*name indicates location—on* ***lateral*** *aspect of thigh*)

  vastus **medialis** (*name indicates location—on* ***medial*** *aspect of thigh*)

  vastus **intermedius** (*located deep and* ***intermediate*** *to vastus lateralis and vastus medialis*)
- The **hamstrings** are a group of three muscles of the posterior thigh:

  Biceps femoris (*is located* **by** *itself on the lateral aspect of the thigh*)

  Semi**m**embranosus (*is the* **most medial** *of all the hamstring muscles*)

  Semi**tendin**osus (*is the hamstring muscle with the longest* **tendon**)

Muscles of the Thigh, Anterior and Posterior Views

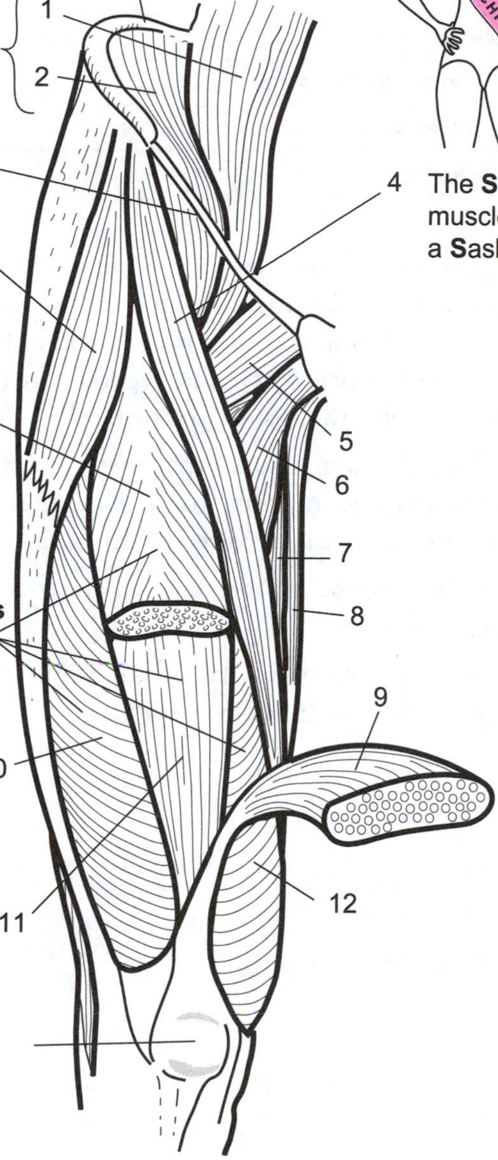

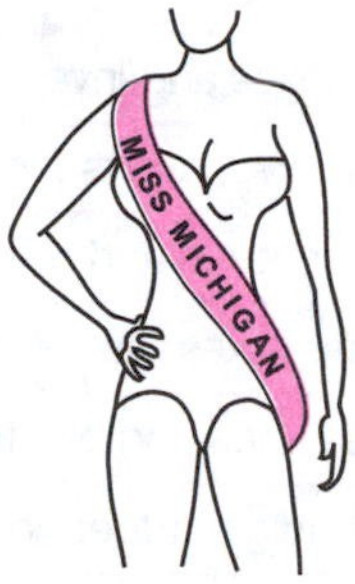

The **Sartorius** muscle looks like a **S**ash.

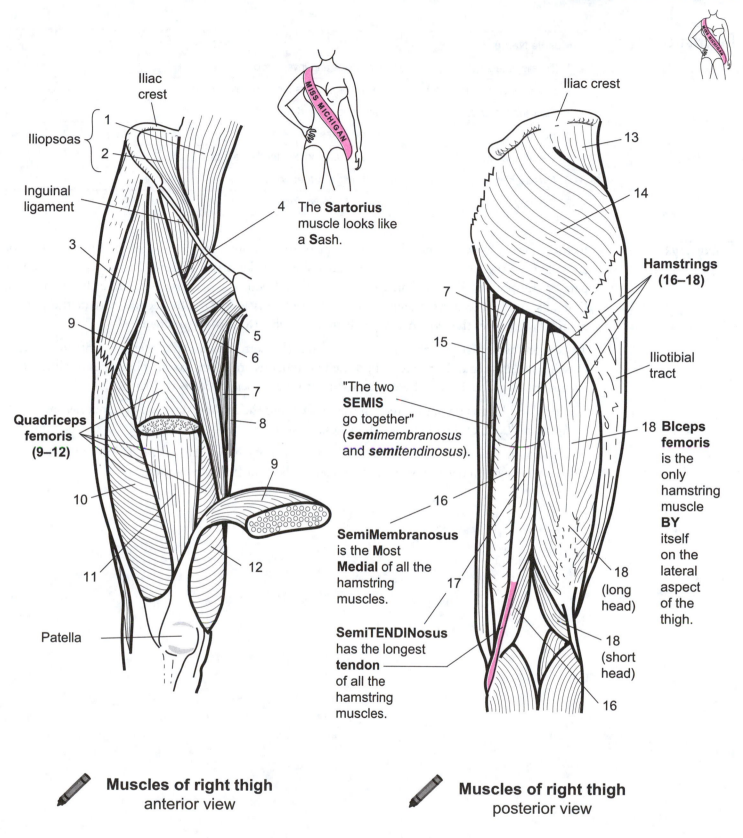

"The two **SEMIS** go together" (***semi****membranosus* and ***semi****tendinosus*).

SemiMembranosus is the **M**ost **Medial** of all the hamstring muscles.

SemiTENDINosus has the longest **tendon** of all the hamstring muscles.

BIceps femoris is the only hamstring muscle **BY** itself on the lateral aspect of the thigh.

Muscles of right thigh
anterior view

Muscles of right thigh
posterior view

Description

Muscle Name	Action
1. Tibialis anterior	Dorsiflexes and inverts foot
2. Extensor digitorum longus	Dorsiflexes, everts foot, and extends digits 2–5
3. Fibularis *(peroneus)* longus	Plantar flexes and everts foot
4. Soleus	Plantar flexes foot
5. Gastrocnemius	Flexes knee and plantar flexes foot
6. Extensor hallicus longus	Extends great toe; dorsiflexes ankle; everts foot
7. Plantaris	Plantar flexion of foot

Study Tips

- In the anterior view, use the **tibia** as a landmark to learn muscles in a sequence either medially or laterally from the tibia. Use the following mnemonic: **T**ake **E**than **F**ishing: **S**ounds **G**ood! The first part of the phrase, **T**ake **E**than **F**ishing, gives the sequence of muscles laterally from the tibia (**T**ibialis anterior, **E**xtensor digitorum longus, **F**ibularis longus). The second part of the phrase, **S**ounds **G**ood, gives the sequence medially from the tibia (**S**oleus, **G**astrocnemius).
- To double-check that you have identified the muscles correctly, use the **extensor digitorum longus** muscle as a landmark. To locate it, find the four tendons anchoring to the phalanges in all the toes except the great toe. Follow these tendons up into this muscle to correctly identify it.
- The **fibularis longus** has a long tendon that inserts into the 5th metatarsal. This tendon *loops around the lateral malleolus* of the fibula, which makes it easy to locate.
- The **soleus** (*soleus* = fish) is so named because it looks like a flat fish. It is located *deep* to the gastrocnemius like a flat fish would rest *deep* on the bottom of a body of water.

Superficial Muscles that Move the Ankle, Foot, and Toes, Anterior and Posterior Views

Laterally from the tibia:

***T**ake*
***E**than*
***F**ishing:*

→

***T**ibialis anterior*
***E**xtensor digitorum longus*
***F**ibularis longus*

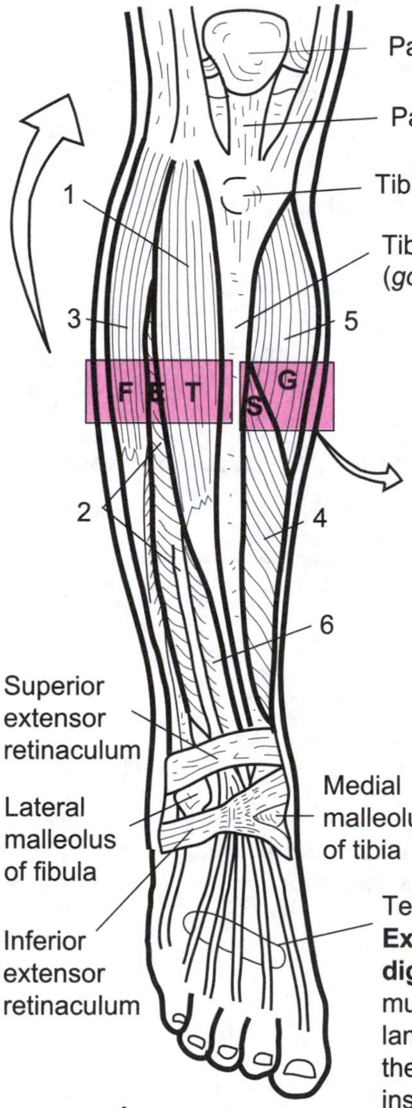

Tendons of the **Extensor digitorum longus** muscle. This is a good landmark muscle because these four tendons on its insertion point make it easy to locate.

Medially from tibia:

***S**ounds*
***G**ood!*

↓

***S**oleus*
***G**astrocnemius*

Right leg
anterior view

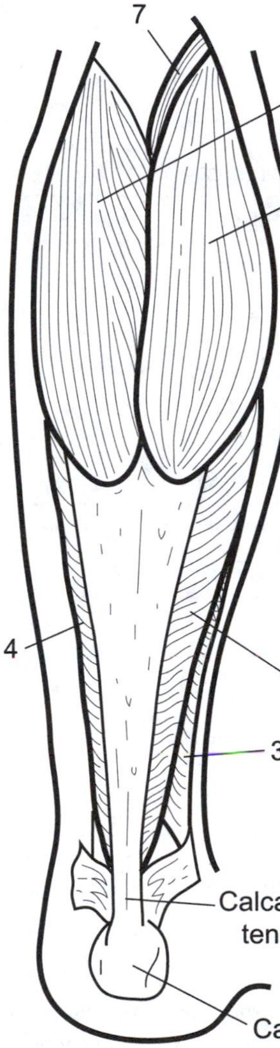

Right leg
posterior view

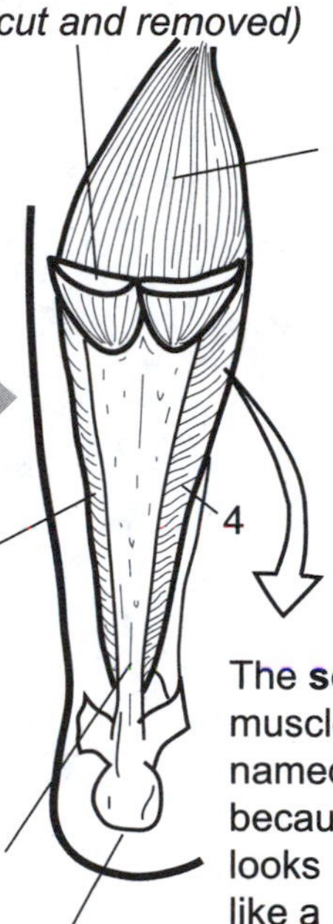

The **soleus** muscle is named because it looks like a **sole** flatfish.

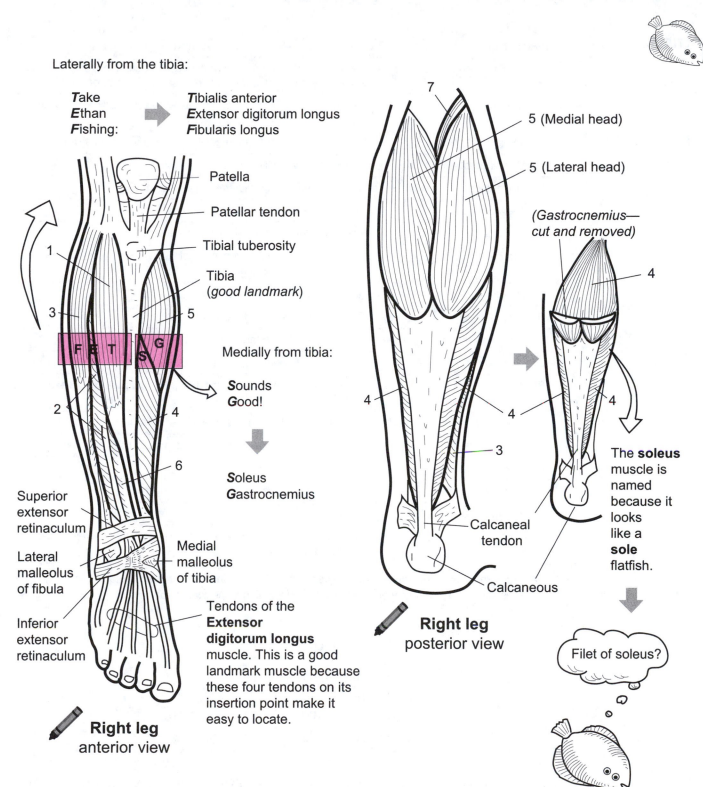

Description

Muscle Name	Action
1. Sternocleidomastoid	Simultaneous contraction of both heads: flexes neck forward. Individual action of each head: rotates head to shoulder on opposite side
2. Trapezius	Elevates, retracts, depresses, or rotates scapula upward; elevates clavicle; extends neck
3. Deltoid	Flexes, extends, abducts; medial and laterally rotates arm
4. Infraspinatus*	Abducts and laterally rotates arm
5. Teres minor*	Adducts, extends, and laterally rotates arm
6. Teres major	Extends, medially rotates, and adducts arm
7. Latissimus dorsi	Extends, adducts, and medially rotates arm
8. External oblique	Compresses anterior abdominal wall; flexes trunk; rotates trunk; depresses lower ribs
9. Gluteus medius	Abducts and medially rotates hip
10. Gluteus maximus	Extends and laterally rotates hip
11. Semispinalis capitis	Extends head and rotates to opposite side
12. Splenius capitis	Extends and hyperextends the head
13. Levator scapulae	Raises scapula and draws it medially; with scapula fixed, flexes neck to same side
14. Rhomboid minor	Retracts, adducts, and stabilizes the scapula
15. Rhomboid major	Adducts, retracts, elevates, and rotates scapula; stabilizes scapula
16. Supraspinatus*	Abducts arm; stabilizes shoulder joint
17. Serratus anterior	Abducts and stabilizes scapula
18. Serratus posterior inferior	Depresses last four ribs
19. Internal oblique	Compresses anterior abdominal wall; flexes trunk; rotates trunk; depresses lower ribs

Rotator cuff muscles:* **supraspinatus, **infraspinatus**, **teres minor**, and **subscapularis**. The only one of these muscles not shown in either the table or the illustration is the **subscapularis**.

Study Tips

- The rotator cuff = the "SITS" muscle group.

S	Supraspinatus
I	Infraspinatus
T	Teres MINOR
S	Subscapularis

A pro baseball pitcher injured his rotator cuff, so now he **SITS** down in the **MINOR** leagues (*minor* indicates teres *minor* instead of teres *major*)

Superficial and Deep Muscles of the Neck, Shoulder, Back, and Gluteal Region

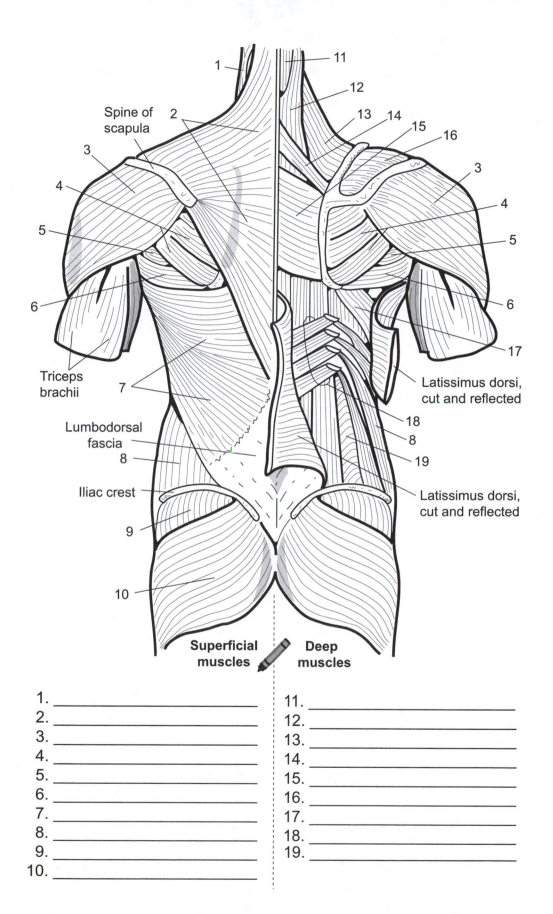

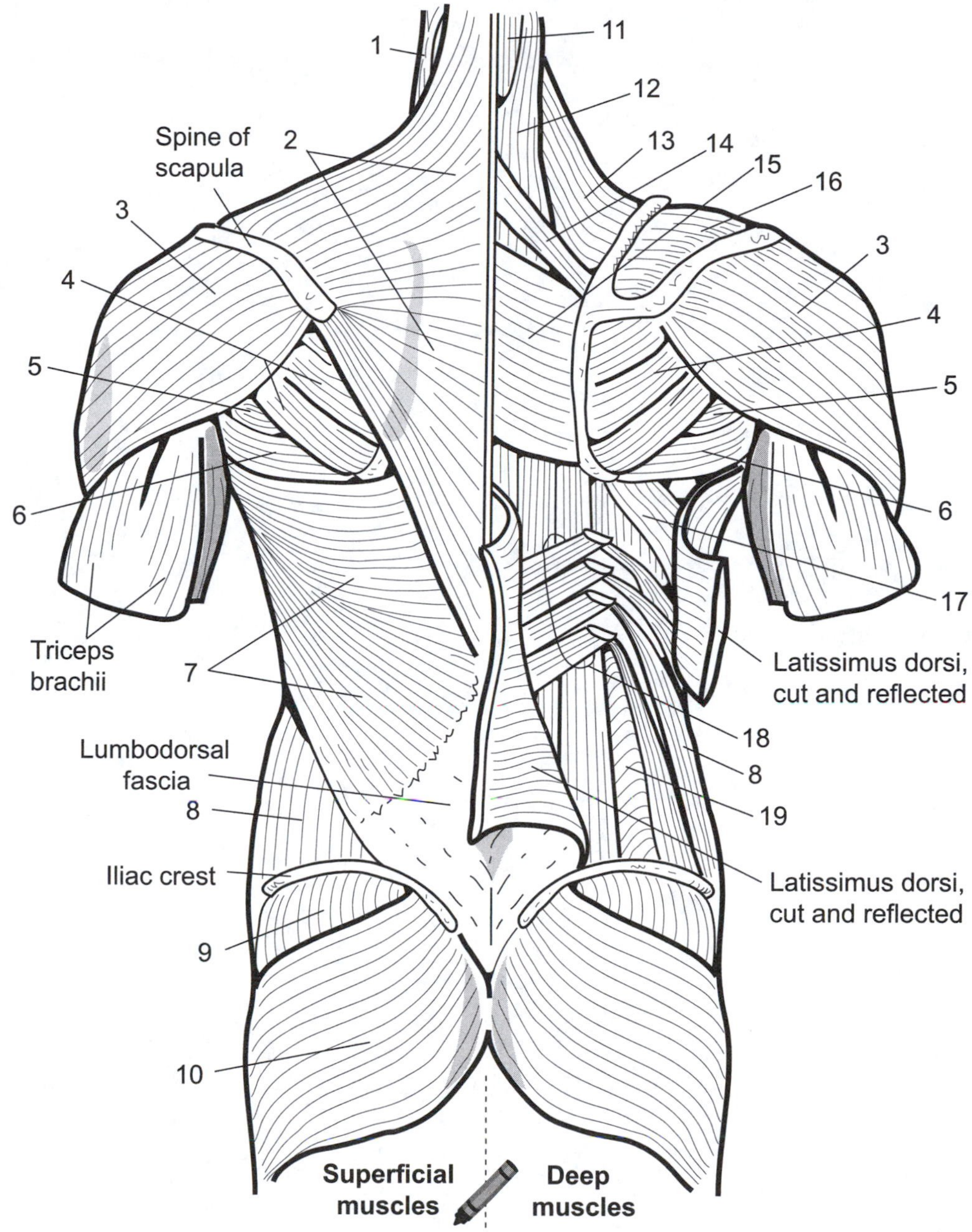

1. ______________________________
2. ______________________________
3. ______________________________
4. ______________________________
5. ______________________________
6. ______________________________
7. ______________________________
8. ______________________________
9. ______________________________
10. ______________________________

11. ______________________________
12. ______________________________
13. ______________________________
14. ______________________________
15. ______________________________
16. ______________________________
17. ______________________________
18. ______________________________
19. ______________________________

Notes

Nervous System

Description

This module provides an overview of the nervous system. Along with the endocrine system, the **nervous system** is one of the important regulators of the body. The nervous system helps maintain our internal stable states. For example, by sensing changes in blood pressure or body temperature, it can trigger the appropriate responses to keep our vital functions running smoothly. Without the nervous system, we literally would be disconnected from our inner and outer worlds. Imagine not being able to sense anything—no sights, no sounds, no smells! What if all 600+ muscles in your body were paralyzed? They would be unable to move because you would have nothing to stimulate them to contract. This will give you an appreciation of the importance of the nervous system.

To illustrate further, consider how much nervous system activity is involved in the seemingly simple act of getting your morning cup of coffee. All the sights, sounds, smells, and tastes of the coffee must be detected by various **sensory receptors**. These receptors are connected to **sensory nerves** that carry the information to the **brain** like data along a computer cable. All this information must be routed to the correct interpretation center in the cerebrum of the brain. When you decide to take that first sip, you need to contract your biceps brachii to bring the cup to your mouth. This is motor output that begins as a nerve impulse in the brain, runs down the spinal cord, and is carried by a **motor nerve** to the muscle. In summary, without the nervous system, no sensory input would go to the brain and no motor output to muscles or glands.

Organization

The nervous system can be divided into two major divisions: **central nervous system (CNS)**, and the **peripheral nervous system (PNS)**. Here is a brief summary of each:

- Central nervous system (CNS): consists of the **brain** and **spinal cord**.
- Peripheral nervous system (PNS): consists of all the **sensory nerves** and **motor nerves** that travel through the body like electrical wiring in a house; more broadly, any nerve tissue outside the CNS.

• **THOUGHTS, MEMORIES:** Your **brain** can remember how coffee looks, smells, and tastes. In addition, it can make associations with other memories, such as that trip you took where you had your best cup of coffee ever!

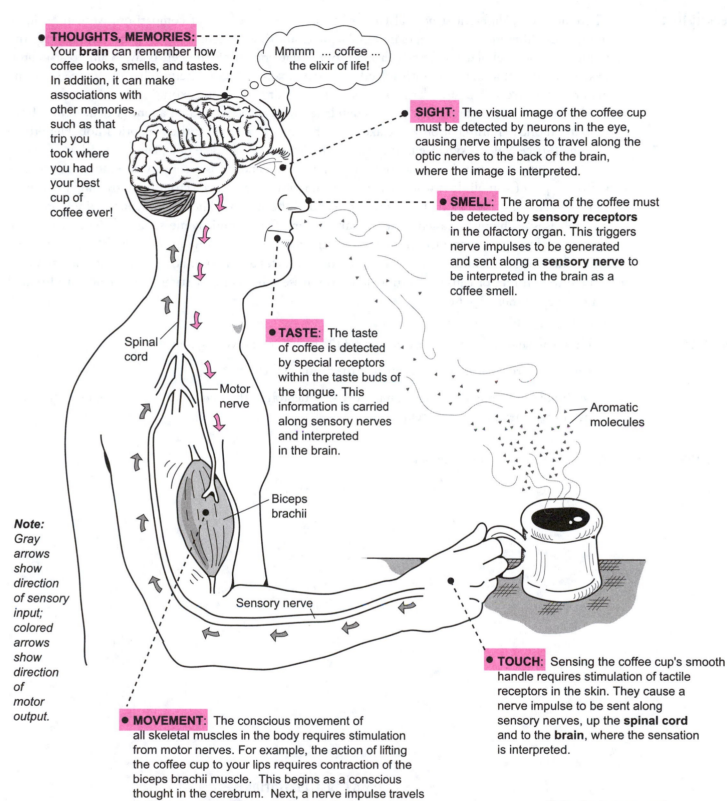

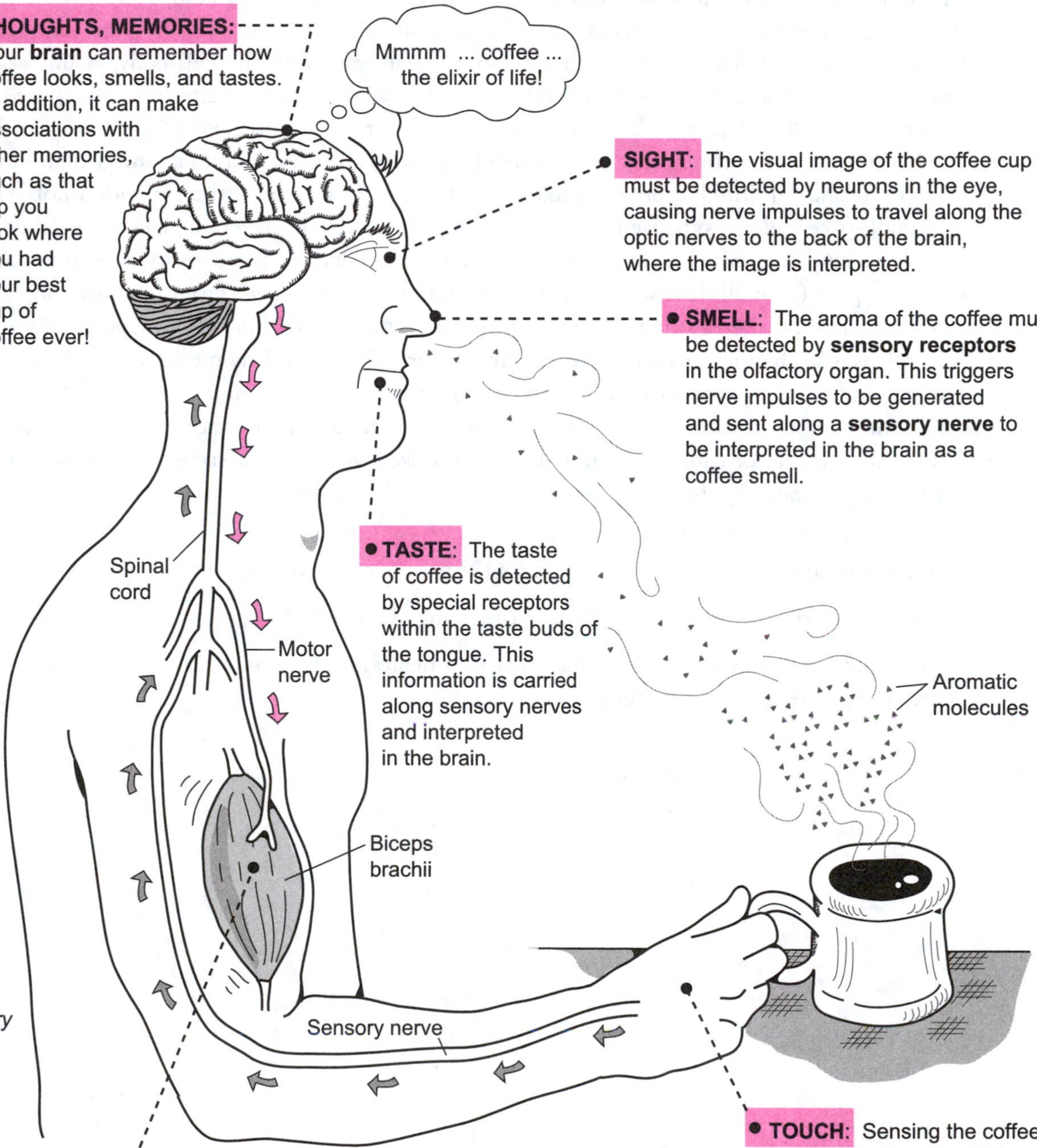

• **SIGHT:** The visual image of the coffee cup must be detected by neurons in the eye, causing nerve impulses to travel along the optic nerves to the back of the brain, where the image is interpreted.

• **SMELL:** The aroma of the coffee must be detected by **sensory receptors** in the olfactory organ. This triggers nerve impulses to be generated and sent along a **sensory nerve** to be interpreted in the brain as a coffee smell.

• **TASTE:** The taste of coffee is detected by special receptors within the taste buds of the tongue. This information is carried along sensory nerves and interpreted in the brain.

Note: *Gray arrows show direction of sensory input; colored arrows show direction of motor output.*

• **TOUCH:** Sensing the coffee cup's smooth handle requires stimulation of tactile receptors in the skin. They cause a nerve impulse to be sent along sensory nerves, up the **spinal cord** and to the **brain**, where the sensation is interpreted.

• **MOVEMENT:** The conscious movement of all skeletal muscles in the body requires stimulation from motor nerves. For example, the action of lifting the coffee cup to your lips requires contraction of the biceps brachii muscle. This begins as a conscious thought in the cerebrum. Next, a nerve impulse travels down the spinal cord and out through a motor nerve, stimulating the muscle to contract.

Color the sensory and motor nerves different colors.

Description

There are many different structural types of neurons. For the sake of comparison, we will examine only three different types: **unipolar**, **bipolar**, and **multipolar** neurons. They are named after the number of long cellular processes (*dendrite, axon*) that branch off the cell body. A unipolar neuron has one process that branches into a dendrite and an axon. A bipolar neuron has two processes—one dendrite, one axon. A multipolar neuron has many dendrites and one axon.

Every neuron has three basic parts: a **dendrite**(s), a **cell body**, and an **axon.** A neuron may have more than one dendrite but only one axon. At the end of the axon is the **synaptic knob**. Here there is a small space called a **synapse** that connects the neuron to a muscle or a gland or another neuron.

Nerve impulses travel in a one-way direction. The dendrite is a sensory receptor that receives various types of stimuli. It is where the nervous impulse is generated. From here, the impulse travels into the cell body, then along the axon to the end of the synaptic knob. The synaptic knob produces and releases a chemical called a **neurotransmitter.** This chemical messenger diffuses across the synapse and binds to a receptor in the muscle cell, glandular cell, or neuron to which it is connected. Once the neurotransmitter binds to its receptor it induces a response. For example, it may stimulate muscle to contract or cause cells in a gland to release a hormone or stimulate a neuron to fire and generate a nervous impulse.

Location

- **Unipolar neurons**—*ex:* sensory neurons of the peripheral nervous system
- **Bipolar neurons**—*ex:* photoreceptors in the retina of the eye
- **Multipolar neurons**—*ex:* most common type of neuron in the brain and spinal cord; motor neurons in the peripheral nervous system

Function

All neurons conduct nerve impulses.

Key to Illustration

1. Unipolar neuron	D = Dendrite
2. Bipolar neuron	C = Cell body
3. Multipolar neuron	A = Axon
	SK = Synaptic knobs

Color all the dendrites one color and the synaptic knobs a different color.

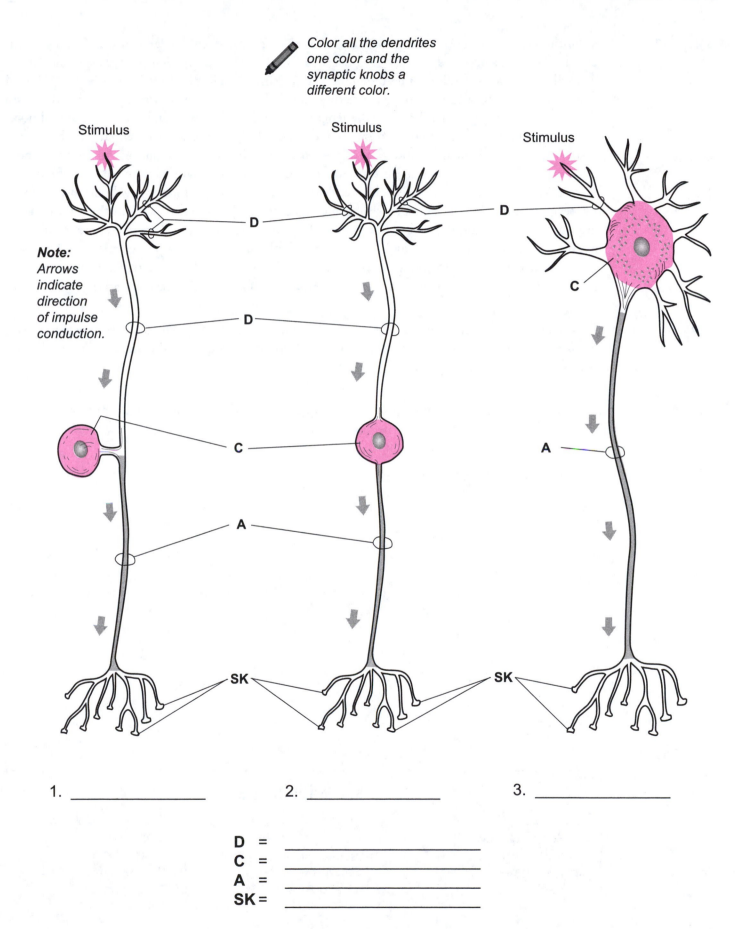

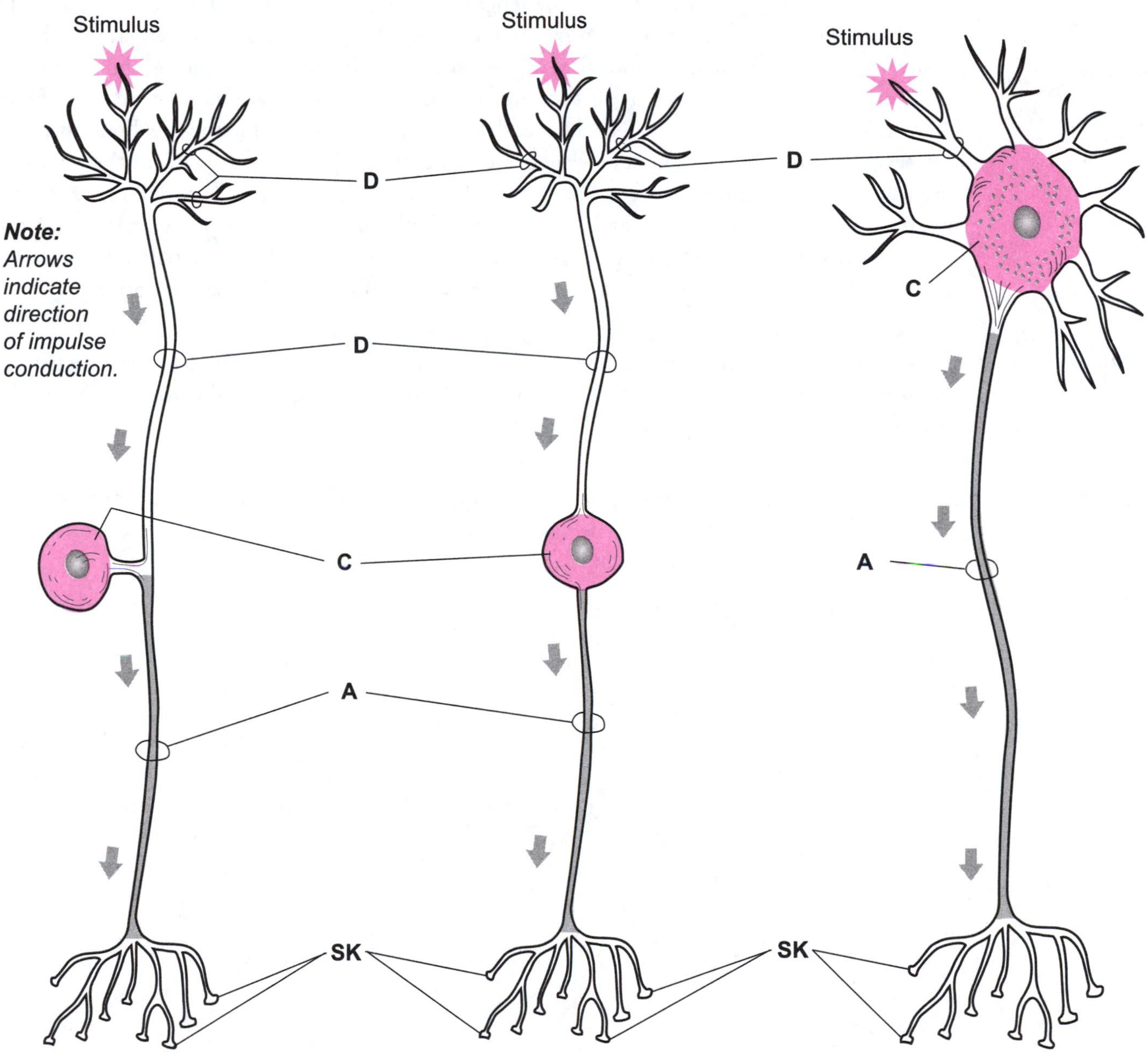

1. ________________ 2. ________________ 3. ________________

D = ______________________
C = ______________________
A = ______________________
SK = ______________________

Description

A multipolar neuron is one type of **neuron**. It has many processes called **dendrites**, which respond to stimuli. The **soma** (*cell body*) is the regional area that includes the nucleus, cytoplasm, and various organelles. The rough endoplasmic reticulum is found in large clusters called **Nissl bodies**. The cell body tapers off into a funnel-shaped structure called the **axon hillock**, which becomes the **axon**. Nervous impulses are conducted along the length of the axon. The axon ends in the **synaptic knob**.

Axons can be either **myelinated** or **unmyelinated**. In the process of **myelination**, the axon is wrapped in a sheath of lipid and protein that insulates the nervous impulse and speeds impulse conduction. This process begins in the fetus and continues into late adolescence. In the central nervous system (CNS) cells called oligodendrocytes are responsible for the myelination process. In the peripheral nervous system (PNS), **neurolemmocytes** (*Schwann cells*) perform this task by wrapping themselves around the axon many times. These layers of plasma (*cell*) membrane constitute the **myelin sheath**. During this process, the nucleus and other organelles are pushed to the outer surface of the neurolemmocyte (*Schwann cell*). This outer layer is called the **neurilemma**. Segments of unwrapped axon between neurolemmocytes (*Schwann cells*) are called **nodes of Ranvier**.

Analogy

Each neuron has only one **axon**. The axon is like an **electrical cord**. The axon conducts a nervous impulse like the copper wires in the cord conduct electricity. The **myelin sheath** serves to insulate the axon like the **plastic casing** around the electrical cord.

Location

Multipolar neurons are the most common type of neuron in the brain and spinal cord but also occur as motor neurons in the peripheral nervous system.

Function

All neurons conduct nerve impulses.

Key to Illustration

1. Dendrite
2. Soma *(cell body)*
3. Nucleus of neuron
4. Nissl bodies
5. Axon hillock
6. Axon
7. Neurolemmocyte (*Schwann cell*)
8. Nodes of Ranvier
9. Nucleus of neurolemmocyte (*Schwann cell*)
10. Neurilemma
11. Endoneurium
12. Synaptic knobs

Multipolar Neuron and Myelination

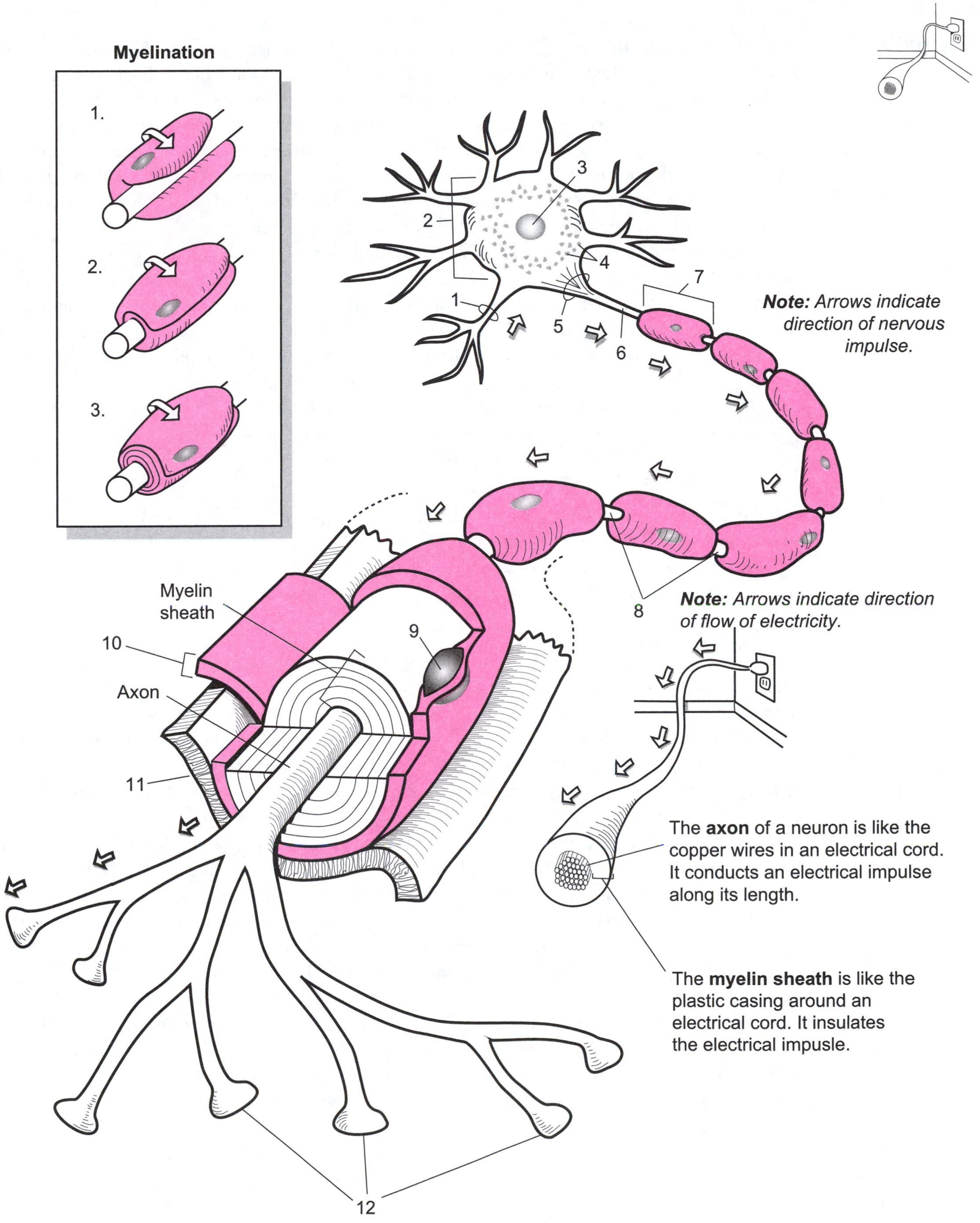
Myelination
1.
2.
3.
1
2
3
4
5
6
7
8
9
10
11
12
Myelin sheath
Axon
Note: *Arrows indicate direction of nervous impulse.*
Note: *Arrows indicate direction of flow of electricity.*
The **axon** of a neuron is like the copper wires in an electrical cord. It conducts an electrical impulse along its length.
The **myelin sheath** is like the plastic casing around an electrical cord. It insulates the electrical impulse.

Description

Your body has more than 600 skeletal muscles, which compose most of your body mass. More than 45 miles of nerves are "wired" through your body like electrical wiring in a house. Each nerve is a collection of **neurons**, or nerve cells. Motor neurons connect to muscle tissue to stimulate it to contract.

Neurons are the fundamental cells in nervous tissue. Though they are of various types, they all share certain features. Surrounding the nucleus of every neuron is a region called the **cell body**, where most of the organelles are found. This is the metabolic center of the cell. Branching out from the cell-like tentacles from an octopus are two types of processes—**dendrites** or **axons**. As a rule, each neuron has only one axon per cell but may have one or more dendrites.

The neuron in the illustration is called a multipolar (*multi* = many) neuron because it has *many* dendrites branching out of the cell body. The dendrites act as the sensory receptors of the cell, and the axon conducts the nerve impulse along the length of the cell like copper wires in an electrical cord. The doorknob-like structures called **synaptic knobs** located at the end of the cell contain chemical messengers called **neurotransmitters**. Any time a neuron is stimulated, it conducts a nerve impulse along its axon and usually responds by releasing neurotransmitters.

Nerve impulses flow through a neuron in a one-way direction:

dendrite → cell body → axon → synaptic knobs

First, a stimulus must arrive at the dendrite. The type of stimulus needed to cause a response depends on the type of neuron. For example, neurons in the retina of the eye respond to light as a stimulus. When dendrites detect a stimulus, it triggers a signal to be sent to the cell body, then along the axon, and finally to the synaptic knobs.

Neurons Connect To …

Neurons can connect to three different things: (1) **other neurons**, (2) **glands**, and (3) **muscles**. Let's look at each, in turn. In the nervous system, neurons connect to other neurons like paper dolls in a chain. When this occurs, the orientation of one cell to another always follows a predictable "head to tail"-type pattern. More specifically, the synaptic knobs ("tail") of the first cell always connect to the dendrites ("head") of the second cell. The neurotransmitters move between the two cells in a specialized junction called the **synapse**. For example, your brain is composed of billions of neurons that connect to each other.

Sometimes neurons connect to an endocrine gland to stimulate the release of other chemical messengers called **hormones**. For example, neurons are wired to your adrenal glands, located on top of your kidneys. As part of a normal response called the *fight-or-flight* response, neurotransmitters stimulate hormone-producing cells within the gland to secrete epinephrine, or *adrenaline*. It travels through the blood to specific target organs in the body. As one example, epinephrine targets the heart, resulting in an increase in heart rate, which may be needed to help you flee a dangerous situation.

Last, neurons that connect to muscle tissue (*skeletal, cardiac, and smooth*) are called motor neurons. Acetylcholine (ACh) is the name of the neurotransmitter released by motor neurons that stimulates all the skeletal muscles in your body to contract. As ACh is released, it crosses a synapse and docks with receptors in the skeletal muscle cells to induce a response—contraction!

ONE WAY
NEURON: FUNCTIONAL OVERVIEW
Arrows indicate direction of flow of electricity.
The axon of a neuron is like the copper wires in an electrical cord. It conducts an electrical current along its length.
Synaptic knobs
Dendrite
Nucleus
Neurotransmitters
Cell body
Axon
S
R
Neuron releases neurotransmitters
Neuron stimulated at a dendrite
ONE WAY
Information flows through a neuron in a one-way direction—from dendrite to cell body to axon to synaptic knobs.
KEY
Stimulus
Response
NEURONS CONNECT TO ...
1. Other neurons
Neurotransmitters
Synapse
Neurotransmitters
2. Glands
Hormone
To target
Blood
3. Muscles
Color the neurons.
Muscle contracts!
Synapse
Skeletal muscle

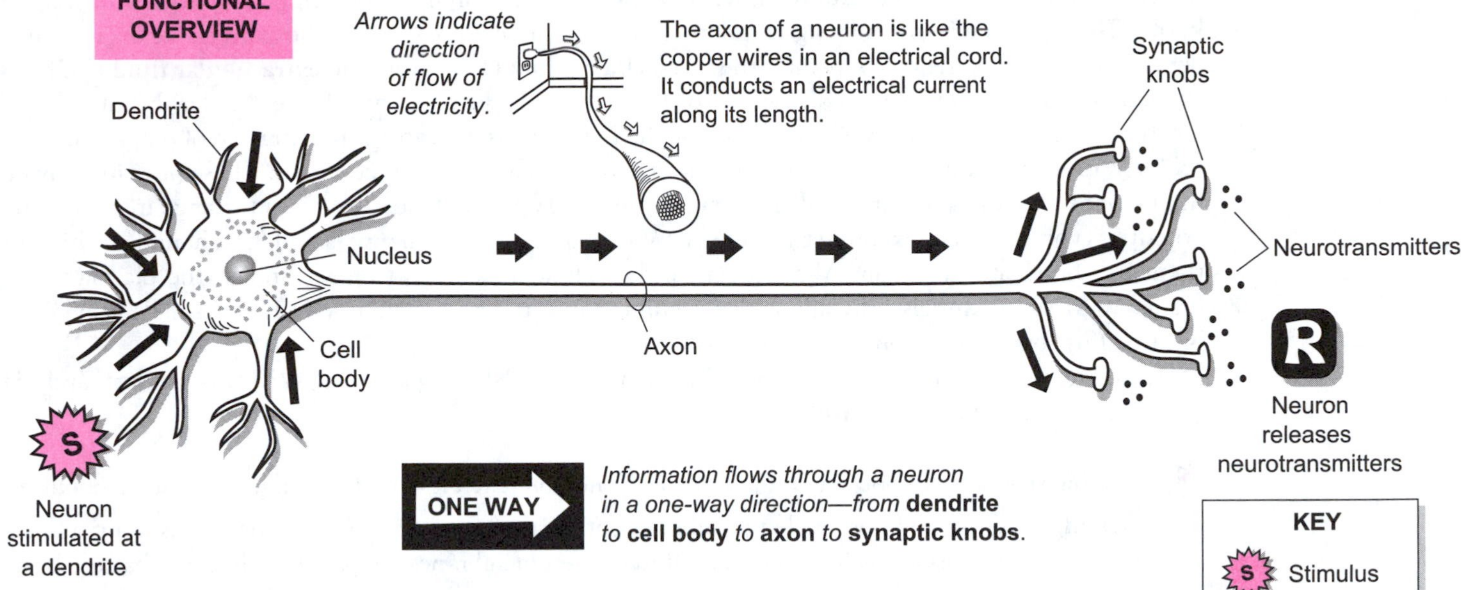

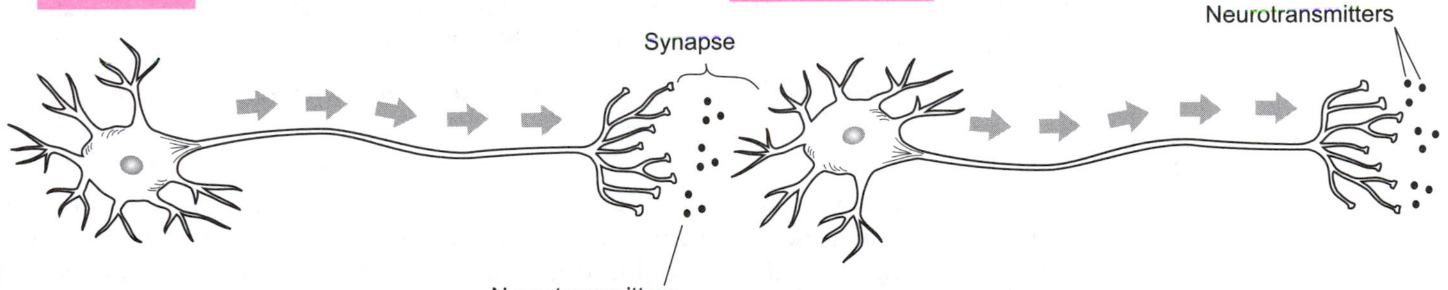

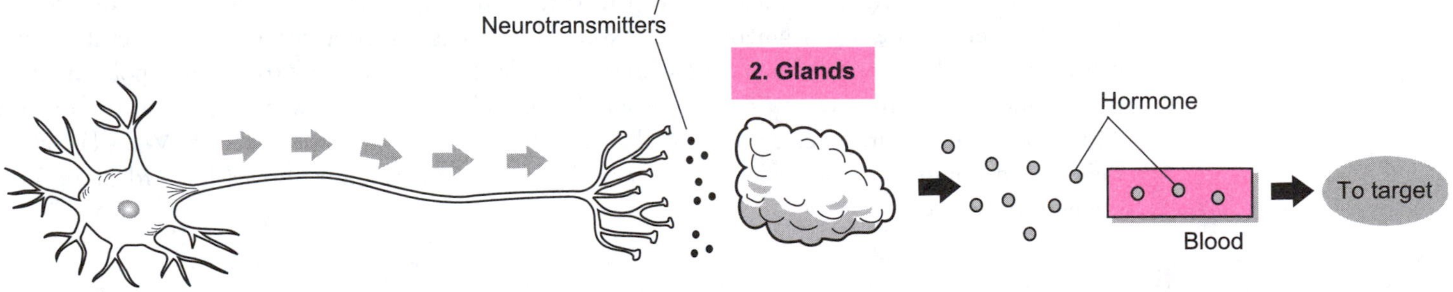

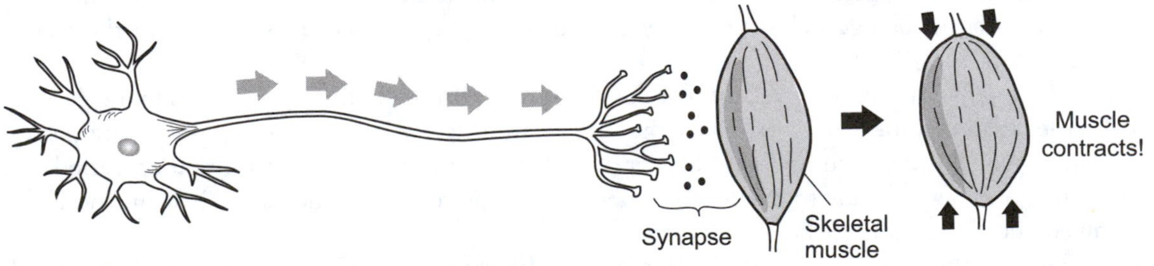

Description

Action potentials (*nerve impulses*) travel at great speeds along the long, slender **axons** of neurons. Like waves of electric current, they always move in a one-way direction from the cell body toward the synaptic knobs. Their purpose is to send a signal to other neurons, muscles, or glands. The axon is like a long cylinder of plasma membrane that separates the **intracellular fluid (ICF)** from the **extracellular fluid (ECF)**.

The chemical composition of these two fluids differs in several ways. For example, the ECF has a greater concentration of **sodium ions (Na^+)** while the ICF has a greater concentration of **potassium ions (K^+)**. Within the axon's plasma membrane are voltage-gated (V-G) ion channels. These proteins may be in one of two states—*open* or *closed*. They are triggered to open and close by a rapid change in membrane potential (voltage). Let's examine two different V-G channels: **Na^+ channels** and **K^+ channels.** Think of them like floodgates in a dam. When opened, they allow many ions to pass through the plasma membrane. V-G **Na^+ channels** only allow **Na^+** to diffuse into the axon, while V-G **K^+ channels** only allow **K^+** to diffuse out of the axon.

Conducting a nerve impulse involves three key steps: (1) *Resting potential*, (2) *Depolarization*, and (3) *Repolarization*. Let's look at each step.

(1) Resting (*membrane*) potential: *the state of the neuron when it is NOT conducting a nerve impulse.*

Resting potentials are also called **membrane potentials** (see p. 56). When a neuron is **not** conducting a nerve impulse, it is in this state. All living cell membranes are *polarized*. Just as a battery has a positive (+) end and a negative (−) end, the same is true for a cell membrane. The outside surface of the membrane is positive (+) and the inner surface is negative (−). Just as a battery stores a small voltage, the membrane also stores a voltage—normally about − 70 mV. This can be measured with a voltmeter that has one electrode in the ECF and the other in the ICF. This voltage is a type of potential energy that can used to conduct a nerve impulse. This potential energy comes from two ion gradients: (1) **Na^+ gradient** and the (2) **K^+ gradient**. These gradients are maintained by a protein pump not shown in the illustration called the **Na^+-K^+ pump** (see p. 58).

Action potential: ***the state of the axon when it is conducting a nerve impulse.***

(2) Depolarization

In order to conduct a nerve impulse, a threshold stimulus must arrive at the neuron. The stimulus is unique to the neuron in question. For example, a photoreceptor at the back of the retina would be responsive only to photons of light. Once this threshold is reached, it leads to the opening of V-G **Na^+ channels** in the first segment of the axon. This causes a sudden influx of **Na^+** resulting in a *depolarization*. This rapid shift of positive charges inside the axon reverses the normal polarity of the plasma membrane. In other words, the inside becomes more positive with respect to the outside. This sudden change in potential triggers the next segment of the axon to open its V-G **Na^+ channels** and a chain reaction ensues. Like a tidal wave, depolarization moves forward along each segment of the axon.

(3) Repolarization

In the wake of the depolarization "tidal wave," a *repolarization* occurs. The change in potential from the depolarization triggers V-G **K^+ channels** to open and **K^+** diffuses out of the axon. This rapid outflux of positive charge helps flip the polarity almost back to what it used to be in the resting potential.

After repolarization, more K^+ diffuses out of the axon through other special membrane proteins called **leakage channels** to re-establish the original resting potential. These proteins are passive channels that do not open and close. After a short resting period called a **refractory period**, the neuron is once again ready to conduct another nerve impulse. Lastly, the excess Na^+ in the ICF is pumped out of the axon by the Na^+-K^+ pump (see p. 58).

The cycle then repeats itself: *resting potential*, *depolarization*, *repolarization*. And on and on it goes!

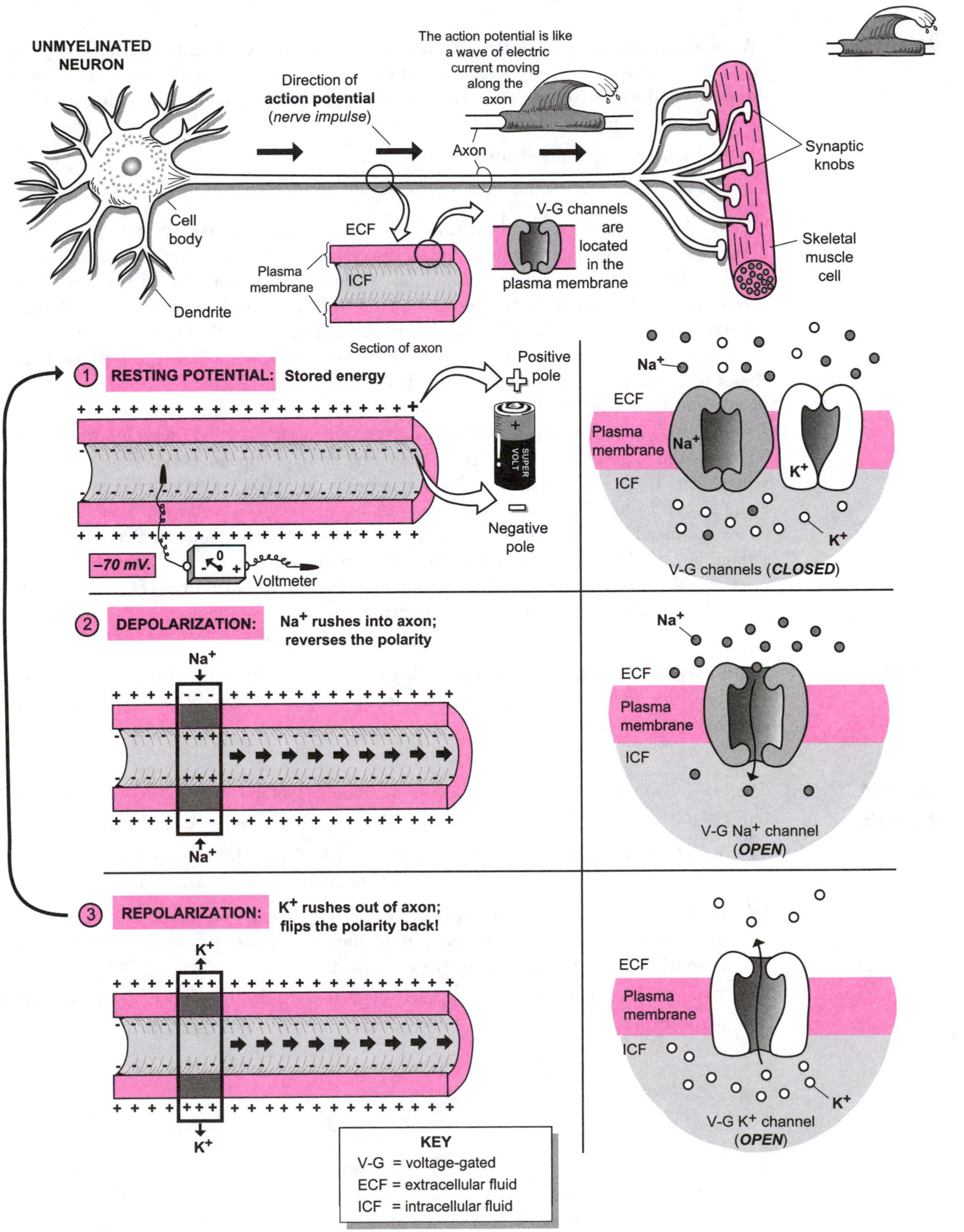
UNMYELINATED NEURON
Direction of action potential (nerve impulse)
The action potential is like a wave of electric current moving along the axon
Axon
Synaptic knobs
Cell body
Dendrite
ECF
Plasma membrane
ICF
Section of axon
V-G channels are located in the plasma membrane
Skeletal muscle cell
1 RESTING POTENTIAL: Stored energy
Positive pole
SUPER VOLT
Negative pole
–70 mV.
Voltmeter
ECF
Plasma membrane
Na+
K+
ICF
V-G channels (CLOSED)
2 DEPOLARIZATION: Na+ rushes into axon; reverses the polarity
Na+
V-G Na+ channel (OPEN)
3 REPOLARIZATION: K+ rushes out of axon; flips the polarity back!
K+
V-G K+ channel (OPEN)
KEY
V-G = voltage-gated
ECF = extracellular fluid
ICF = intracellular fluid

Description

Like an electric current travels through the copper wires in an electrical cord, nerve impulses travel along axons. The two general types of impulse conduction are: (1) continuous conduction, and (2) saltatory conduction (see p. 198). This module explains **continuous conduction**, which has the following key features:

- Occurs only in **unmyelinated neurons**.
- Is slower than saltatory conduction (about 2 m/sec).

Speed is determined mainly by myelination and diameter of the **axon**. Some axons are thick, and others are thin. The general rule is: "thicker is quicker." Therefore, thin, unmyelinated axons are much slower than thick myelinated axons. For example, in one second, impulses in thick myelinated axons can travel longer than the length of a football field while those in thin unmyelinated would not have covered enough ground to make a first down. That's a big difference in speed.

Unmyelinated neurons lack a myelin sheath around their axons. The **extracellular fluid (ECF)** comes into direct contact with the axon. This allows for easy exchange of ions across the plasma membrane of the axon.

Impulse conduction involves a repeated cycle of depolarization and repolarization all along the length of the axon (see p. 194). This is caused by the rapid opening and closing of two different voltage-gated (V-G) channels: (1) **Na^+ channels** and (2) **K^+ channels**. Like any gate, these channels can be in two states—either *open* or *closed*. They are triggered to open and close by a rapid change in membrane potential (volgate). First, V-G Na^+ channels are stimulated to open to allow sodium ions to diffuse into the **intracellular fluid (ICF)** within the axon. Then V-G K^+ channels open and K^+ ions diffuse out of the axon and into the **extracellular fluid (ECF)**. This flux of ions across the membrane creates an electrical current.

Analogy

Continuous conduction is like a domino effect in which each adjacent segment of the axon must be stimulated one after the other. The dominos are like the voltage-gated channels that must open and close in rapid succession.

Sequence of Events

The Na^+ channels are shown in the facing illustration. **Note:** K^+ channels are not shown in the illustration.

1. V-G **Na^+ channels** are stimulated to open and Na^+ ions diffuse into the axon resulting in a depolarization. This is followed by V-G **K^+ channels** opening to cause a repolarization.
2. In the adjacent part of the axon, V-G **Na^+ channels** again are stimulated to open, and Na^+ ions diffuse into the axon, resulting in another depolarization. This is followed by V-G **K^+ channels** opening to cause another repolarization.
3. This process of stimulating each adjacent segment of the axon continues along the entire length of the axon.

UNMYELINATED NEURON

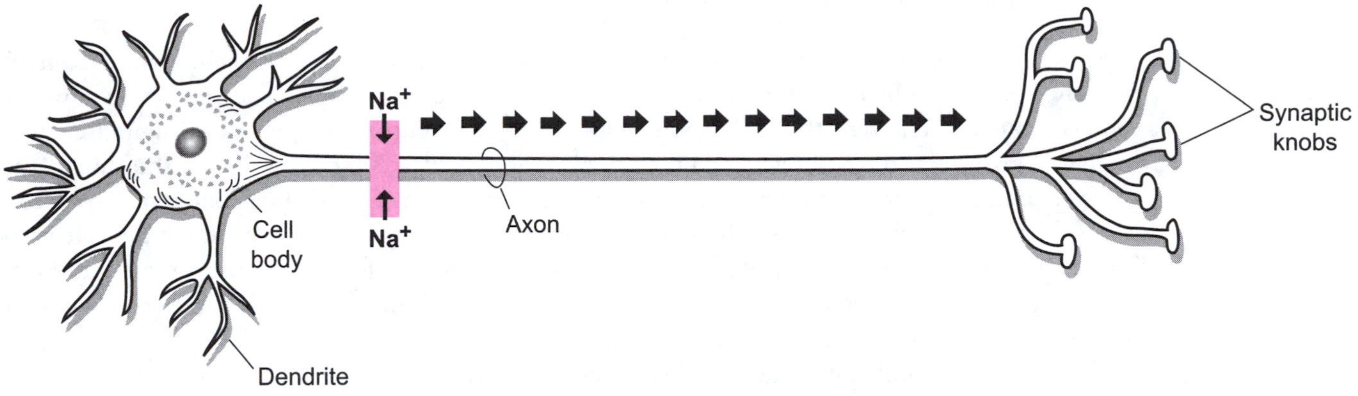

DOMINO EFFECT: Continuous conduction in an unmyelinated neuron is like a domino effect, in which each segment of the axon must be stimulated, one after the other.

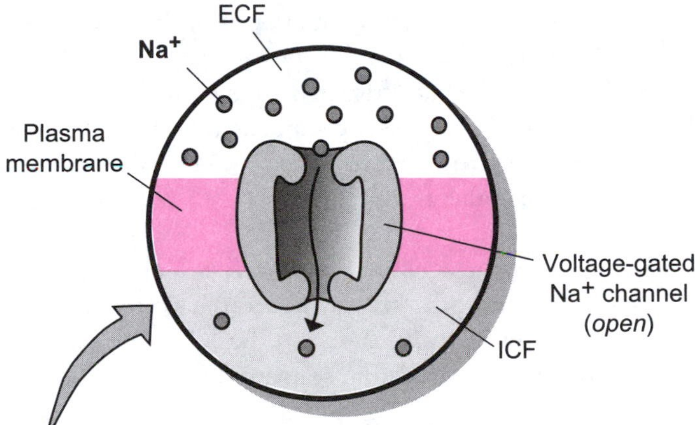

SEQUENCE of EVENTS

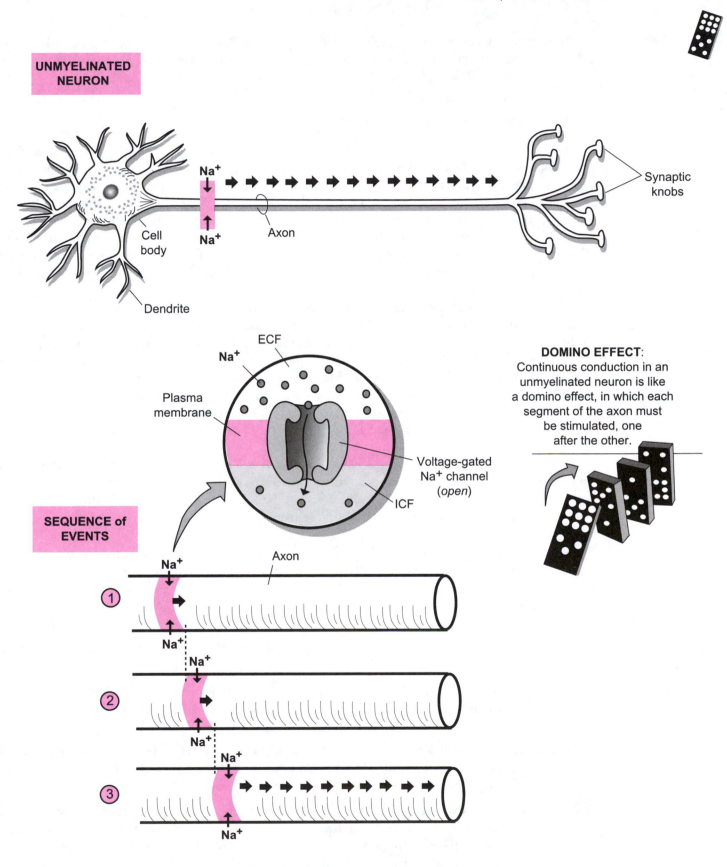

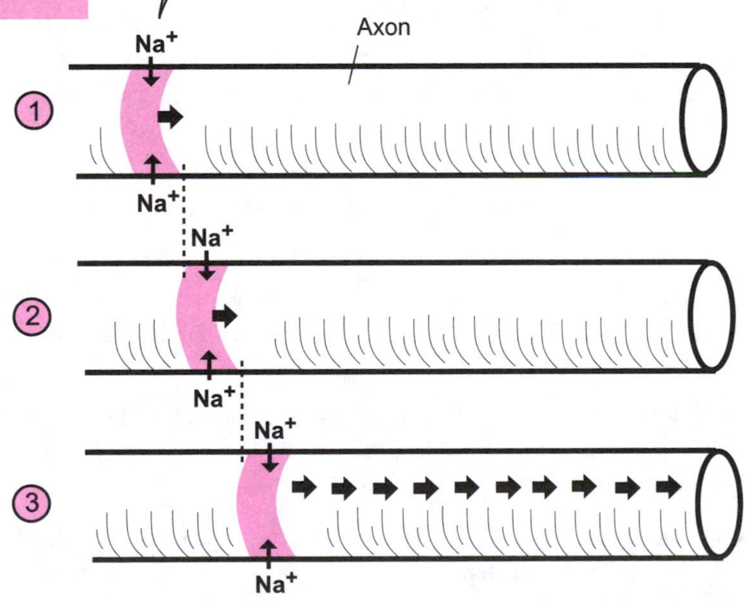

Description

The two general types of nerve impulse conduction are: (1) continuous conduction (see p. 196), and (2) saltatory conduction. This module explains the latter, **saltatory** (*saltare*, leaping) **conduction**. Saltatory conduction has the following key features:

- Occurs only in **myelinated neurons**.
- Is faster than continuous conduction (about 120 m/sec).

Speed of impulse conduction is mainly determined by myelination and diameter of the **axon**. Some axons are thick and others are thin. The general rule is: "thicker is quicker." Therefore, thick myelinated axons are much faster than thin unmyelinated axons. For example, in one second, impulses in thick myelinated axons can travel longer than the length of a football field while those in thin unmyelinated axons would not have covered enough ground to make a first down. That's a huge difference in speed.

Myelinated neurons in the peripheral nervous system have their axons covered by **neurolemmoctes** (*Schwann cells*). During development of the nervous system, these cells wrap themselves around the axon, forming layers of plasma membrane collectively referred to as the **myelin sheath**. This covering serves as insulation like the plastic coating around the copper wires in an electrical cord. Short segments of exposed axon between the neurolemmocytes are called **nodes of Ranvier**.

Impulse conduction involves a repeated cycle of depolarization and repolarization all along the length of the axon (see p. 194). This is caused by the rapid opening and closing of two different voltage-gated (V-G) channels: (1) **Na^+ channels** and (2) **K^+ channels**. Like any gate, these channels can be in two states—either *open* or *closed*. They are triggered to open and close by a rapid change in membrane potential (voltage). First, V-G Na^+ channels are stimulated to open to allow sodium ions to diffuse into the **intracellular fluid (ICF)** within the axon. Then V-G K^+ channels open and K^+ ions diffuse out of the axon and into the **extracellular fluid (ECF)**. This flux of ions across the membrane creates an electrical current.

In myelinated neurons, V-G Na^+ channels and V-G K^+ channels are located mostly at nodes of Ranvier. Why? This is the only part of the axon where ions can cross the plasma membrane because ions within the ECF come in direct contact with the axon. Because ions can't cross the barrier of the myelin sheath, voltage-gated channels are not needed in segments of the axon covered by myelin.

Analogy

Saltatory conduction is like skipping a stone across the surface of the water, as the impulse seems to "jump" from node to node. Strictly speaking, the impulse does not actually jump. But this is useful to illustrate that saltatory conduction is up to 60 times faster than continuous conduction, which is more like a domino effect.

Sequence of Events

The Na^+ channels are shown in the facing illustration. (**Note:** V-G K^+ channels are not shown in the illustration.)

1. V-G **Na^+ channels** at the **first node** are stimulated to open, and sodium ions diffuse into the axon, causing depolarization. The electric current that is generated is shuttled through the ICF of the axon to the next node of Ranvier.
2. The electric current triggers the V-G **Na^+ channels** at the **second node** to open, and sodium ions diffuse into the axon, causing another depolarization. The electric current that is generated is again shuttled through the axon to the next node of Ranvier.
3. The electric current triggers the V-G **Na^+ channels** at the **third node** to open, and sodium ions diffuse into the axon, causing another **depolarization**. The electric current that is generated is again shuttled through the axon to the next node of Ranvier. This process repeats itself until the nerve impulse is conducted along the entire length of the axon.

Study Tip

Remember the alliteration: <u>S</u>altatory is <u>S</u>peedy like <u>S</u>kipping a <u>S</u>tone.

MYELINATED NEURON

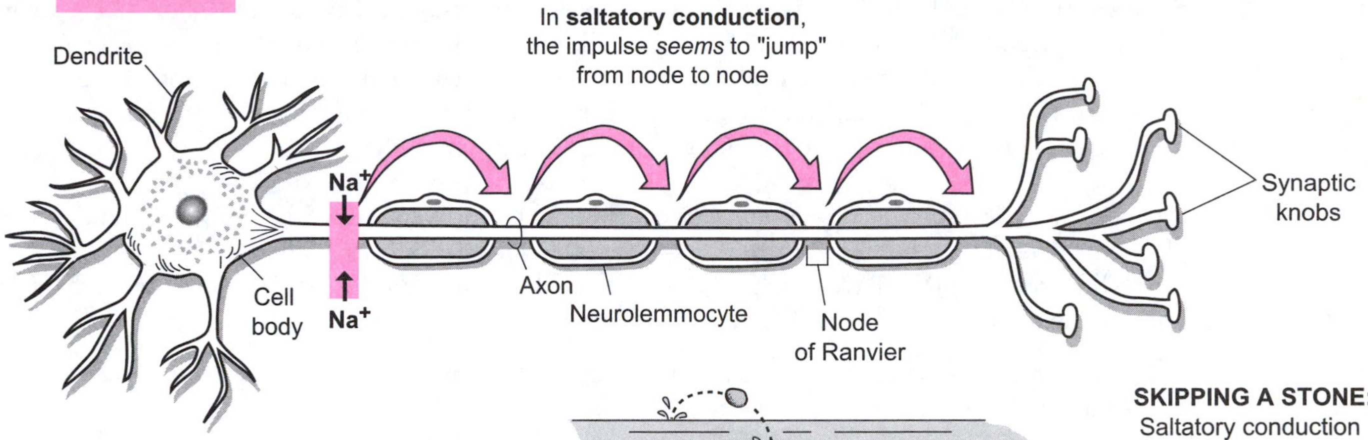

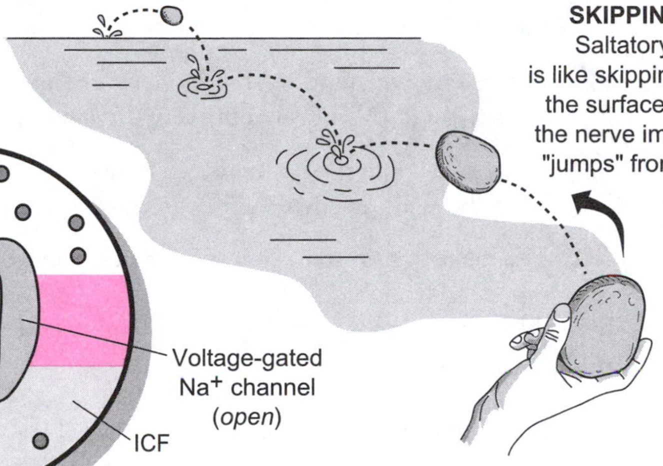

SKIPPING A STONE:
Saltatory conduction is like skipping a stone across the surface of the water as the nerve impulse seemingly "jumps" from node to node.

SEQUENCE of EVENTS

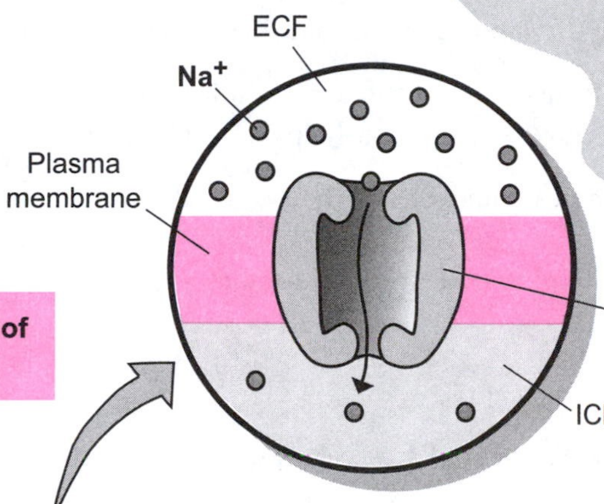

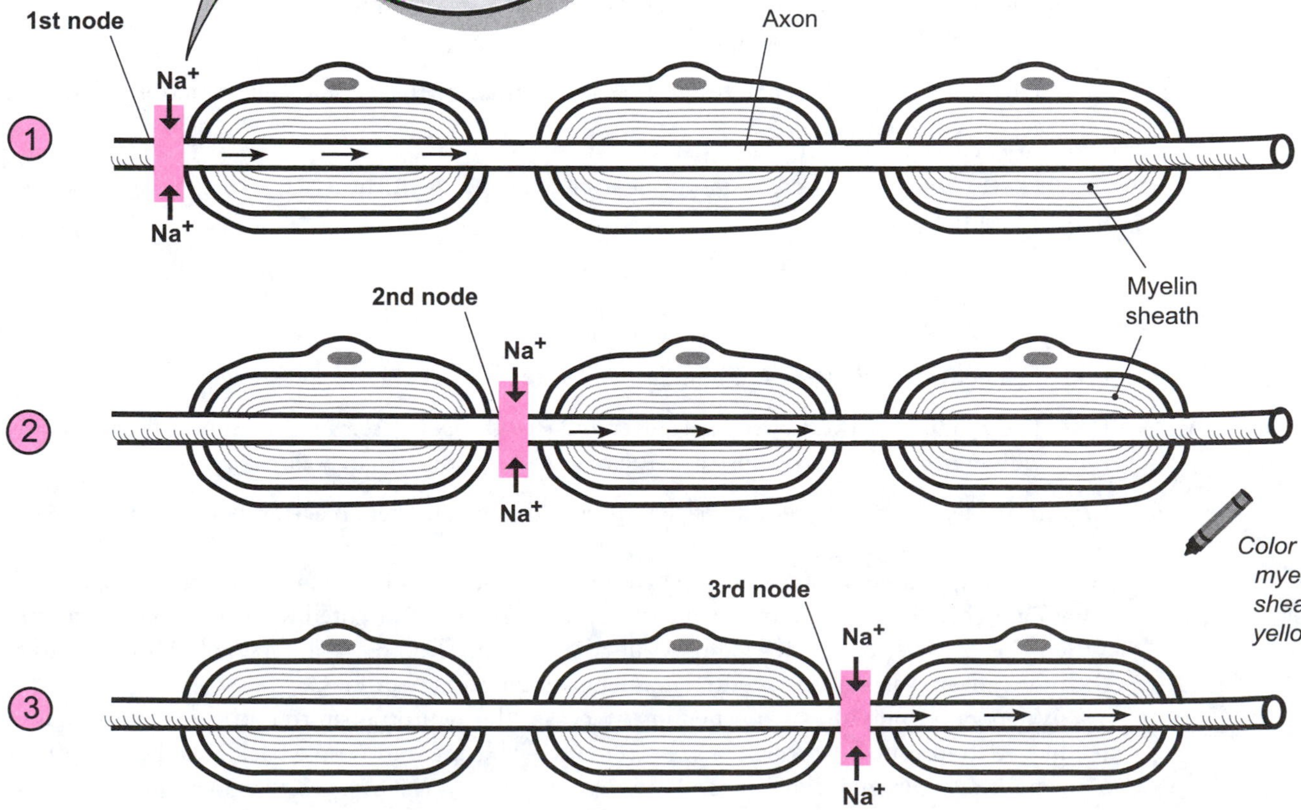

Color the myelin sheath yellow.

Description

Synaptic transmission is the method by which neurons communicate with other cells. The illustration shows the **synaptic knobs** of a **motor neuron** connecting to a **skeletal muscle cell** in the **neuromuscular junction** (NMJ), while the synaptic knobs of neuron #1 connect to the dendrites of neuron #2 in a **neuro-neuro junction** (NNJ). The **synapse** is the fluid-filled space between the synaptic knob and the muscle cell/nerve cell. In both cases, a chemical messenger called a **neurotransmitter** must cross this gap like a ferry boat crossing a channel, and dock at the receptor in the **plasma membrane** of the other cell. This may lead to either a stimulatory or inhibitory response in the other cell.

Steps #1–#4: *Secretion of neurotransmitters*. These steps are the same for both an NMJ and an NNJ. The only difference is the type of neurotransmitter(s) involved. Only **acetylcholine (ACh)** is released in the NMJ, whereas NNJs have dozens of different possible neurotransmitters. Here is a summary of this mechanism:

(1) Neurotransmitters are synthesized in the synaptic knob and stored within **vesicles** at high concentrations.

(2) When a nerve impulse travels along the axon to the synaptic knob, it triggers the opening of **voltage-gated (V-G) calcium (Ca^{++}) channels**. Note that these channels normally are *closed*.

(3) Because calcium is at a higher concentration in the extracellular fluid, it diffuses into the synaptic knob through the open V-G Ca^{++} channels, where it helps trigger exocytosis (see p. 54).

(4) The neurotransmitter is released into the fluid-filled **synapse** and diffuses across it.

Steps #5–#7: *Inducing a response in either muscle or nerve tissue*.

For the **NMJ**:

(5) **ACh** binds to the ACh-gated Na^+ channel (*closed state*).

(6) The binding of **ACh** induces the ACh-gated Na^+ channel to *open*, allowing only sodium ions **(Na^+)** to diffuse into the muscle cell.

(7) This influx of **Na^+** generates an action potential in the muscle cell, which eventually leads to a muscle contraction.

When nerve impulses from the motor neuron cease, **ACh** in the synapse is quickly broken down and muscle action potentials stop.

For the **NNJ**:

(5) The **neurotransmitter (NT)** binds to NT-gated Na^+ channel (*closed state*).

(6) The binding of the **neurotransmitter** induces the NT-gated Na^+ channel to *open*, allowing only sodium ions **(Na^+)** to diffuse into the neuron, resulting in a depolarization.

(7) If the depolarizing potential reaches threshold, it can trigger an **action potential** (*nerve impulse*).

Alternatively, the neurotransmitter may bind to a *different* NT-gated channel that is specific for either **potassium** (K^+) or chloride (Cl^-). The illustration shows a NT-gated K^+ gated channel as an example. When this channel opens, K^+ diffuses out of neuron #2, leading to a hyperpolarization. This inhibitory pathway prevents an action potential from being induced in neuron #2.

When nerve impulses from neuron #1 cease, the neurotransmitter in the synapse has two fates: (1) like ACh, it is broken down, or (2) it is re-used by being transported back into the synaptic knob or to a neighboring neuroglial cell.

Synaptic Transmission—Neuromuscular Junctions and Neuro-Neuro Junctions

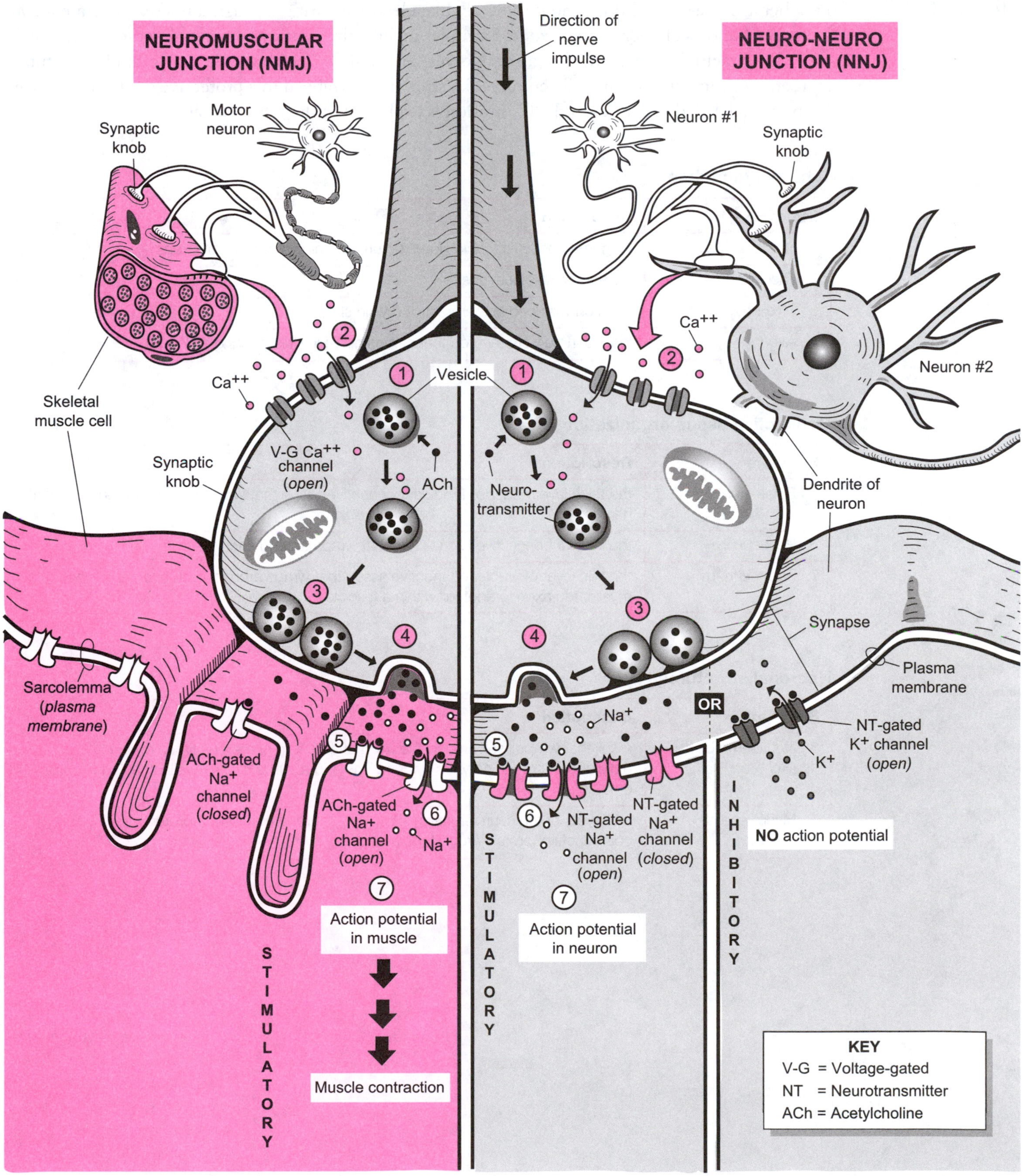

Description

Peripheral nerves spread throughout the body like electrical wiring through a house. They are a collection of bundles of microscopic axons. Examples include the sciatic and femoral nerves in the thigh and the brachial nerve in the arm. Nerves are visible to the naked eye and structured as a series of tubes within tubes. Each tube-like structure is wrapped in a protective connective tissue. Each of the labeled structures in the illustration is explained in the tables below.

Tubes within tubes

Structure	Description
1. Nerve	long, macroscopic, cable-like structure containing bundles of axons
2. Fascicle	a single bundle of axons
3. Myelinated axon	an axon wrapped in a protective myelin sheath
4. Axon	long, thin part of a neuron that carries nerve impulses along its length

Connective tissue organization

Structure	Description
5. Epineurium	the thick layer of dense irregular connective tissue that wraps around the outside of a nerve
6. Perineurium	the cellular connective tissue layer that wraps around each fascicle
7. Endoneurium	the thin layer of areolar connective tissue that wraps around each axon and binds one myelinated axon to another within a fascicle

Neuronal structures

Structure	Description
8. Neurolemmocyte (*Schwann cell*)	a cell that wraps itself around the axon of a neuron numerous times during development of the nervous system. The end result is that a myelin sheath is created
9. Myelin sheath	the layers of the neurolemmocyte's cell membrane that form a protective, insulting coating of lipoprotein around the axon

General structure—
Tubes within Tubes:

1. ____________________
2. ____________________
3. ____________________
4. ____________________

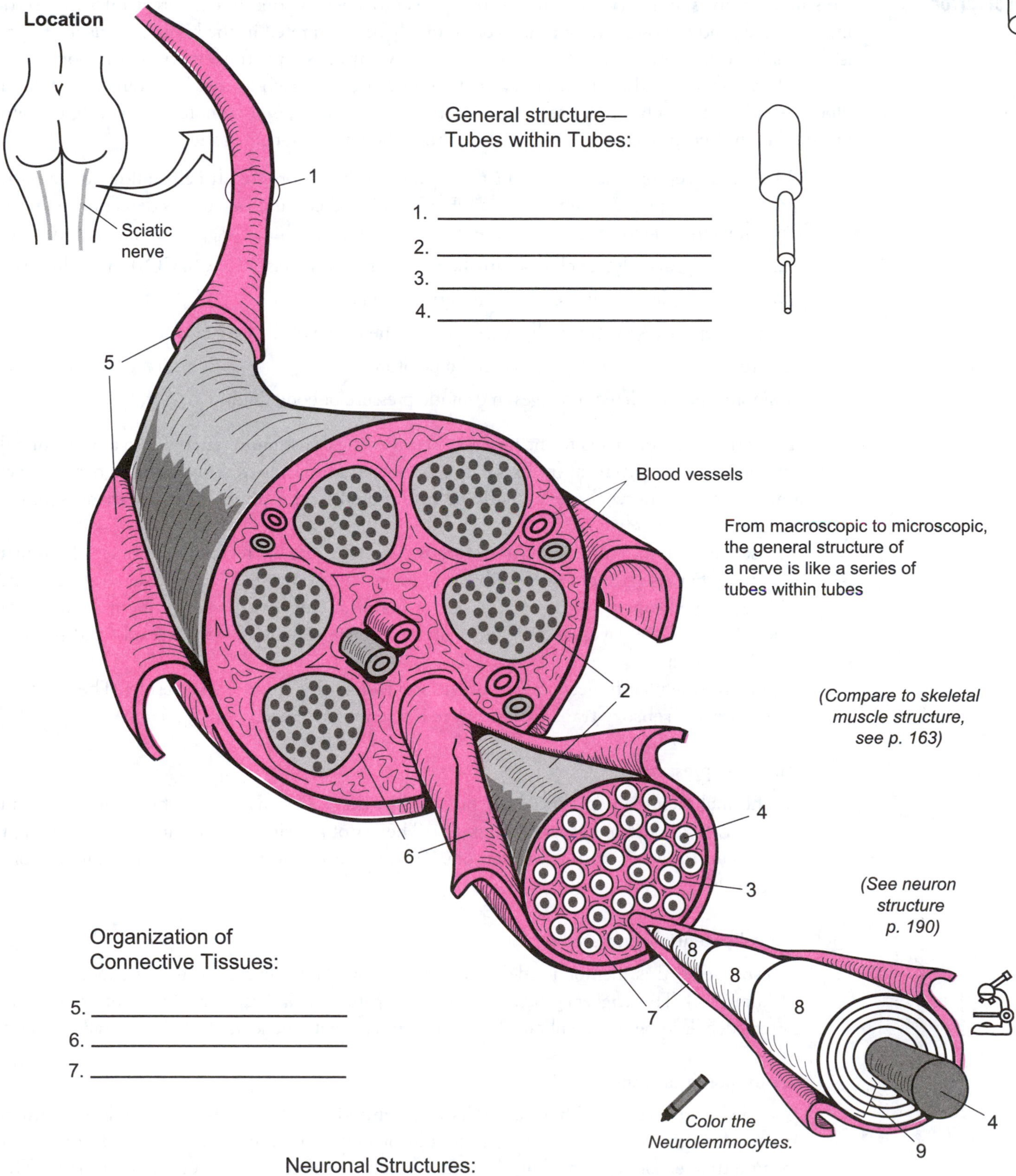

From macroscopic to microscopic, the general structure of a nerve is like a series of tubes within tubes

(Compare to skeletal muscle structure, see p. 163)

(See neuron structure p. 190)

Organization of Connective Tissues:

5. ____________________
6. ____________________
7. ____________________

Color the Neurolemmocytes.

Neuronal Structures:

8. ____________________
9. ____________________

Description

This module gives an overview of how sensory receptors detect stimuli, convert it into an electrical impulse, and send it along a **sensory neuron** to finally be interpreted in the brain. A **sensory receptor** is a specialized structure to detect a specific stimulus. Many are simply the dendrites of sensory neurons encapsulated by some other tissue. They are found in locations such as the skin, joints, tendons, and blood vessel walls. Each type of receptor responds to a single type of stimulus—touch, temperature, pressure, pain. Receptors can be organized into the following major categories:

- Mechanoreceptors: detect stimuli that compress, bend, or stretch cells; allow for sensations such as touch, pressure, and vibration. Ex: Lamellated corpuscles: mostly detect deep pressure; locations include lower dermis of the skin, joints, tendons, muscles, periosteum, and pancreas.
- Chemoreceptors: detect changes in chemical concentrations (H^+, O_2, CO_2) in a solution
- Nociceptors: detect pain resulting from tissue damage
- Photoreceptors: sensitive to light that strikes the retina of the eye
- Thermoreceptors: detect changes in temperature
- Osmoreceptors—detect changes in osmotic pressure of body fluids

Sensory transduction refers to the process of converting the stimulus energy into a nerve impulse. This conversion is essential, allowing all the sensory information to travel through the nervous system along sensory nerves to eventually be interpreted in the **brain**. Without it, no sensations would be able to be perceived.

Let's use the example of a **lamellated corpuscle** as our sensory receptor of choice. It primarily detects pressure, but also stretching and vibration. Its structure consists of a single dendrite surrounded by an oblong-shaped **capsule** of up to 60 layers of fibrous connective tissue, giving it a sliced-onion appearance. Between the layers is a viscous material. In the capsule's core is a fluid-filled space that surrounds the dendrite.

Imagine pressing your finger to a flat surface and feeling the pressure that results. The illustration shows the process of **sensory transduction**. Here is a summary of each step:

1. **Stimulus: pressure**

   The normal membrane potential is symbolized by the "+" signs outside the membrane and the "−" signs inside the membrane (see p. 56). The act of pressing serves as a stimulus to compress the concentric layers within the capsule. This, in turn, deforms the plasma membrane of the dendrite.

2. **Graded potential generated**

   In response to this membrane deformation, the membrane becomes more permeable to Na^+ (sodium ions). This triggers protein channels to open, and Na^+ rapidly diffuses into the dendrite, thereby depolarizing the membrane. The small current that is generated is called a **graded potential**.

3. **Action potential generated**

   A graded potential has different orders of magnitude. If it reaches a threshold or minimum strength—it can induce an **action potential** or nerve impulse in the first **node of Ranvier**. Because this sensory neuron is myelinated, it results in saltatory conduction (see p. 198). This is where the nerve impulse is shuttled from the first node, to the second, then third, and so on. Finally, the nerve impulse travels from the sensory nerve and up the **spinal cord**, where it is interpreted in the cerebral cortex.

Receptors and Sensory Transduction

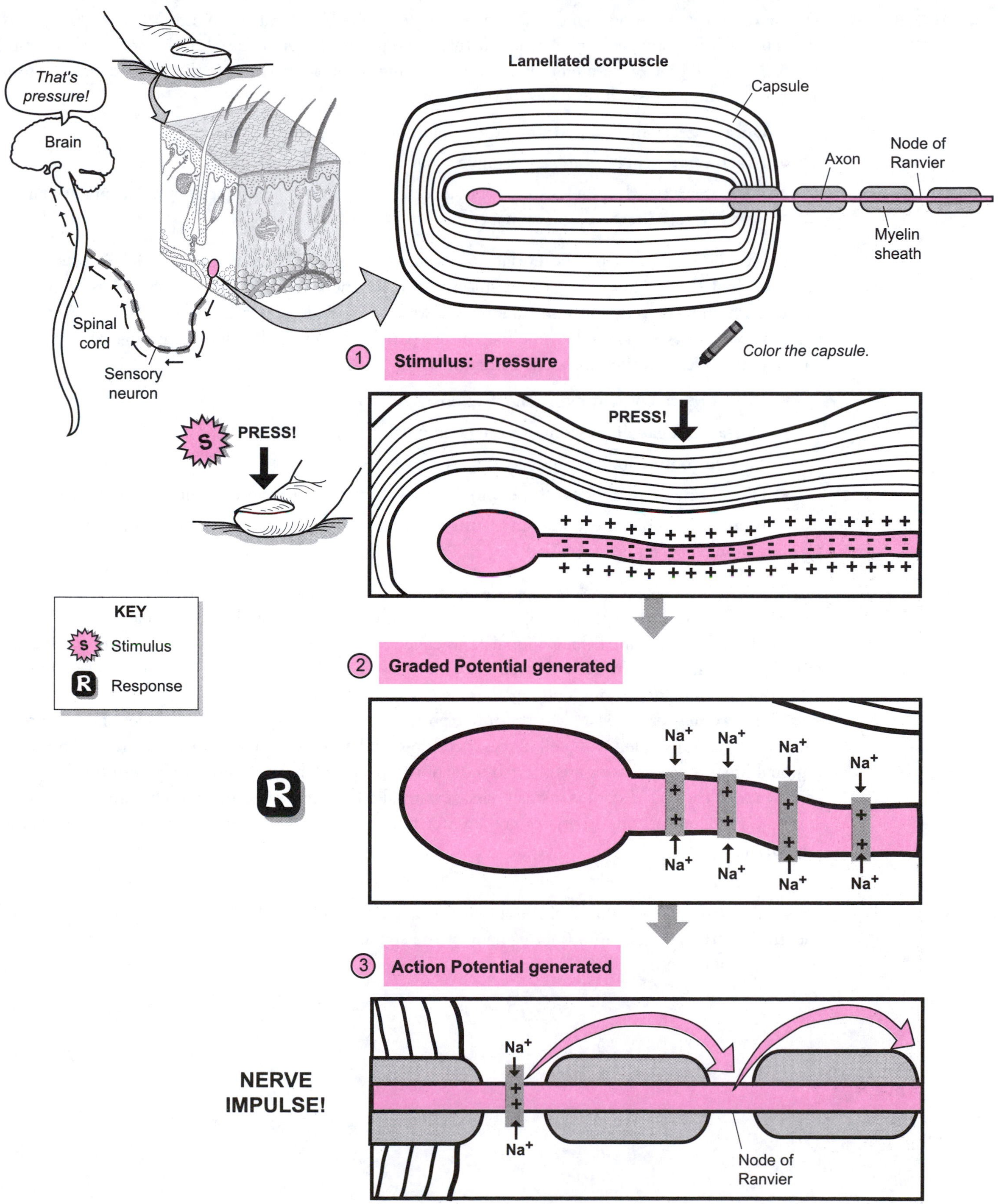

That's pressure!
Brain
Spinal cord
Sensory neuron
Lamellated corpuscle
Capsule
Axon
Node of Ranvier
Myelin sheath
Color the capsule.
1 Stimulus: Pressure
S
PRESS!
PRESS!
KEY
S Stimulus
R Response
2 Graded Potential generated
R
Na+
3 Action Potential generated
NERVE IMPULSE!
Na+
Node of Ranvier

Description

The **spinal cord** is a long, slender structure that links the body and the brain. Most of the cord is protected by the bony vertebrae because it runs through the vertebral canal of the vertebral column. Three layers of protective membranes called **meninges** surround the spinal cord and brain. From outermost to innermost, these are as follows:

- **Dura mater:** thickest and strongest; contains fibrous connective tissue
- **Arachnoid:** thin layer made of simple squamous epithelium; lacks blood vessels
- **Pia mater:** tightly adheres to the spinal cord and follows every surface feature; supplies many blood vessels directly to the spinal cord.

Below the arachnoid is a potential space called the **subarachnoid space**, which is filled with **cerebrospinal fluid**. This serves as a cushion to protect the spinal cord and functions as a medium through which to deliver nutrients and remove wastes. Extending laterally off the spinal cord are 31 pairs of **spinal nerves**. These become the various peripheral nerves that spread throughout the body. The spinal nerves and their associated structures are:

- **Dorsal root:** contains only *sensory* axons.
- **Dorsal root ganglion:** contains the neuron cell bodies of sensory neurons.
- **Ventral root:** contains only motor axons.
- **Dorsal ramus:** branches off the spinal nerves that innervate muscles and skin of the back.
- **Rami communicantes:** branch off the spinal nerves that contain axons related to the autonomic nervous system (ANS).

The spinal cord contains areas of **gray matter** and **white matter**. The white matter is located in the outer portion of the spinal cord and consists of myelinated axons that run along its length. It is divided into the following three regional areas called **funiculi** (sing. *funiculus*): a **posterior funiculus**, **lateral funiculi**, and an **anterior funiculus**. The **white commissure** is a narrow band of white matter that connects the anterior funiculi together.

The **gray matter** is located in the inner portion of the spinal cord and includes short neurons called interneurons along with cell bodies, dendrites, and axon terminals of other neurons. The three regional areas of gray matter are referred to as horns: **posterior horn**, **lateral horn**, **anterior horn**. In the center of the spinal cord is a small passageway called the **central canal**, which contains cerebrospinal fluid. The horizontal band of gray matter that surrounds the central canal is called the **gray commissure**.

Analogy

The gray matter in the middle of the spinal cord looks like a butterfly. Note that the exact shape of the gray matter changes from one segment of the spinal cord to another so it does not always look exactly like a butterfly.

Spinal Cord Structure and Function

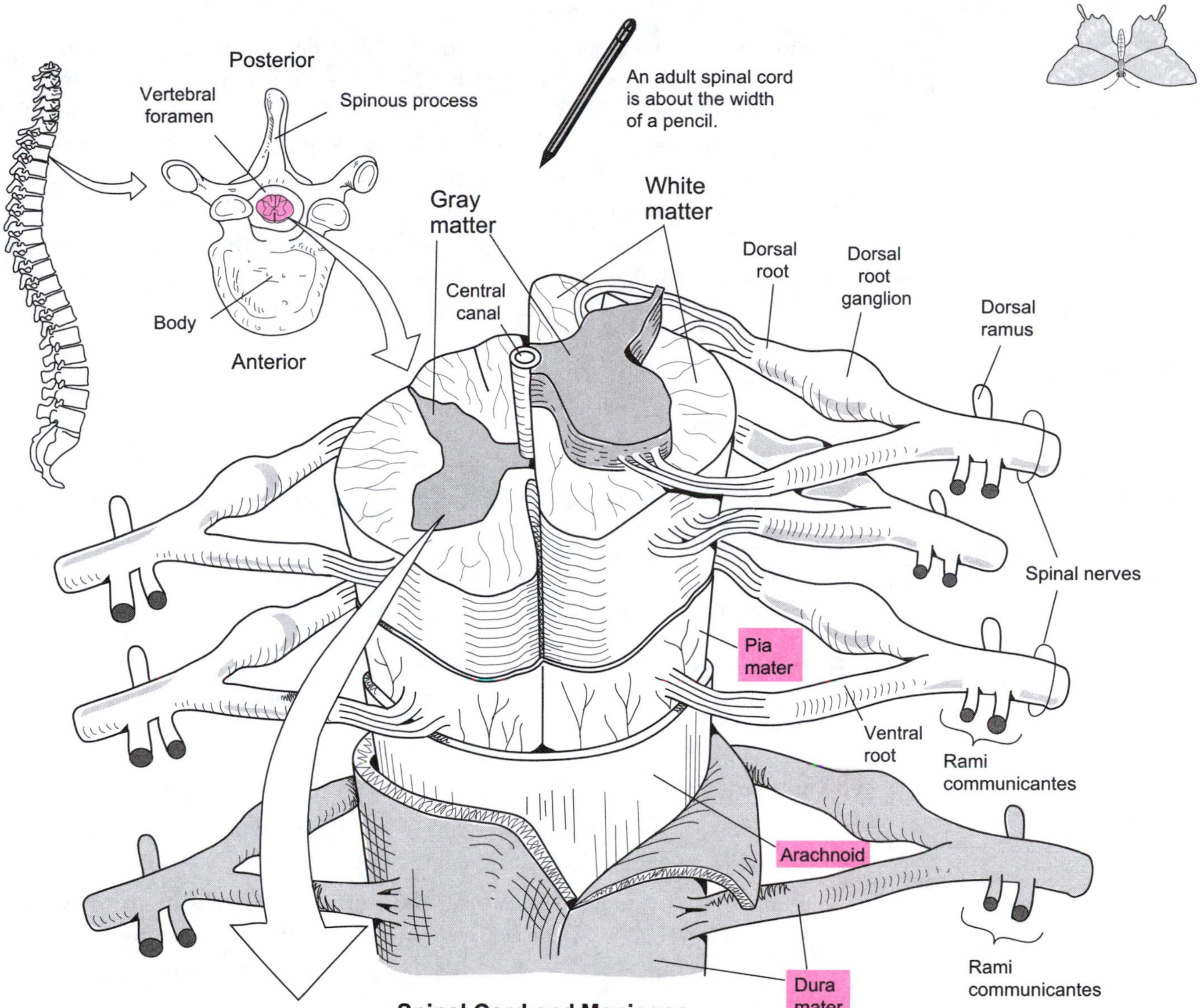

Spinal Cord and Meninges
anterior view

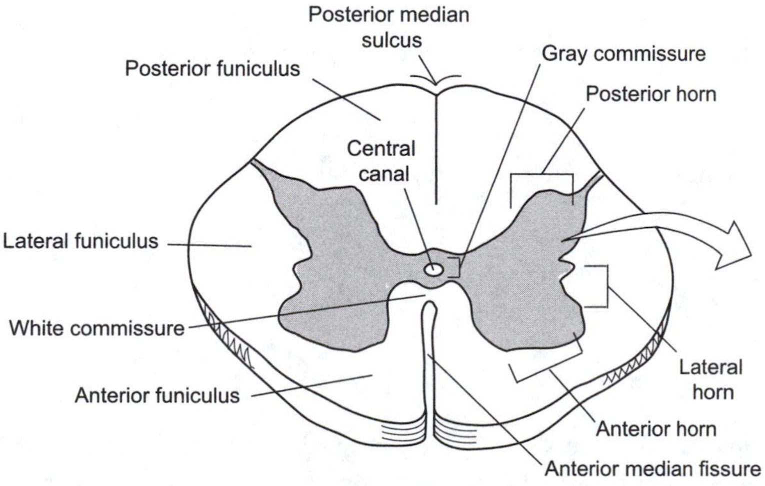

Spinal Cord
cross-section

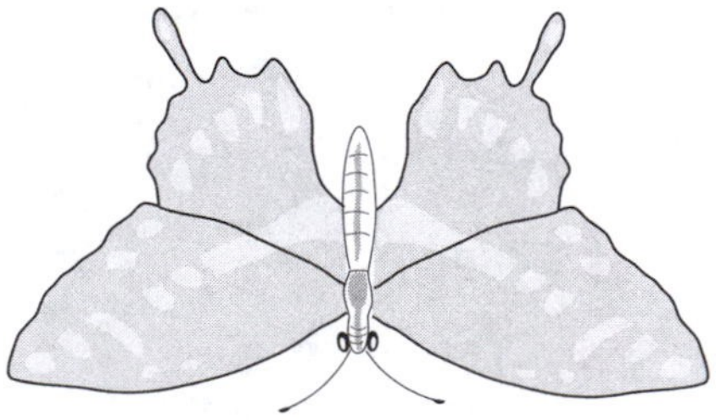

The gray matter in the middle of the spinal cord is shaped like a butterfly.

Description

A **reflex** is a rapid, involuntary response to a stimulus. A reflex arc refers to the neural pathway involved in a single reflex. The two general reflex categories are: somatic reflexes and autonomic reflexes. Only **somatic reflexes**—which result in the contraction of skeletal muscles—will be illustrated in this module. An example is the knee-jerk reflex. During a physical exam, a physician may strike your patellar tendon with a percussion hammer. This stimulates your quadriceps femoris muscle to contract, resulting in extension of the knee joint.

The autonomic reflexes are so named because they are part of the autonomic nervous system (ANS). These reflex arcs stimulate responses in smooth muscle, cardiac muscle, and glands, which allow for control of functions such as digestion, urination, and heart rate.

Somatic Reflex Arc

The parts of a typical somatic reflex arc are:

Input

1. **Receptors**
   —located at the end of sensory neurons.
   —respond to a specific stimulus (ex: touch, pressure, or pain stimuli).
   —different for different stimuli.
   In the illustration on the facing page, a thumbtack is shown penetrating the skin; this would be detected by pain receptors in the skin.

2. **Sensory Neuron**
   —conducts impulses from the receptor to the spinal cord (or brain).

Processing

3. **Integrating Center**
   —regions of gray matter in the spinal cord or brain that act as the integrating center; simple reflexes pass only through the spinal cord.
   A polysynaptic reflex is shown on the facing page; it always involves at least one interneuron that connects the sensory neuron to the motor neuron.
   —monosynaptic reflexes lack the interneuron and simply directly connect the sensory neuron to the motor neuron.

Output

4. **Motor Neuron**
   —conducts impulses from the integration center to the effector.

5. **Effector**
   —the last component in a somatic reflex arc; skeletal muscle.

Analogy

1. **Input**
2. **Processing**
3. **Output**

This is like typing the letter 'K' in a word processor on your laptop. Pressing on the keyboard is the input. This sends a message to the microchip—the brain of the computer or processing center. Then the microchip sends a signal for the final output—the appearance of the letter 'K' on the screen.

SOMATIC REFLEX ARC
S
OUCH!
1 Receptor
Skin
2 Sensory neuron
Interneuron
Gray matter
3 Integrating center
4 Motor neuron
Spinal cord
Effector 5
R
Skeletal muscle contracts
Color the sensory neuron, the interneuron, and the motor neuron different colors.
KEY
S Stimulus
R Response

1 INPUT
K
3 OUTPUT
CHIP
2 PROCESSING

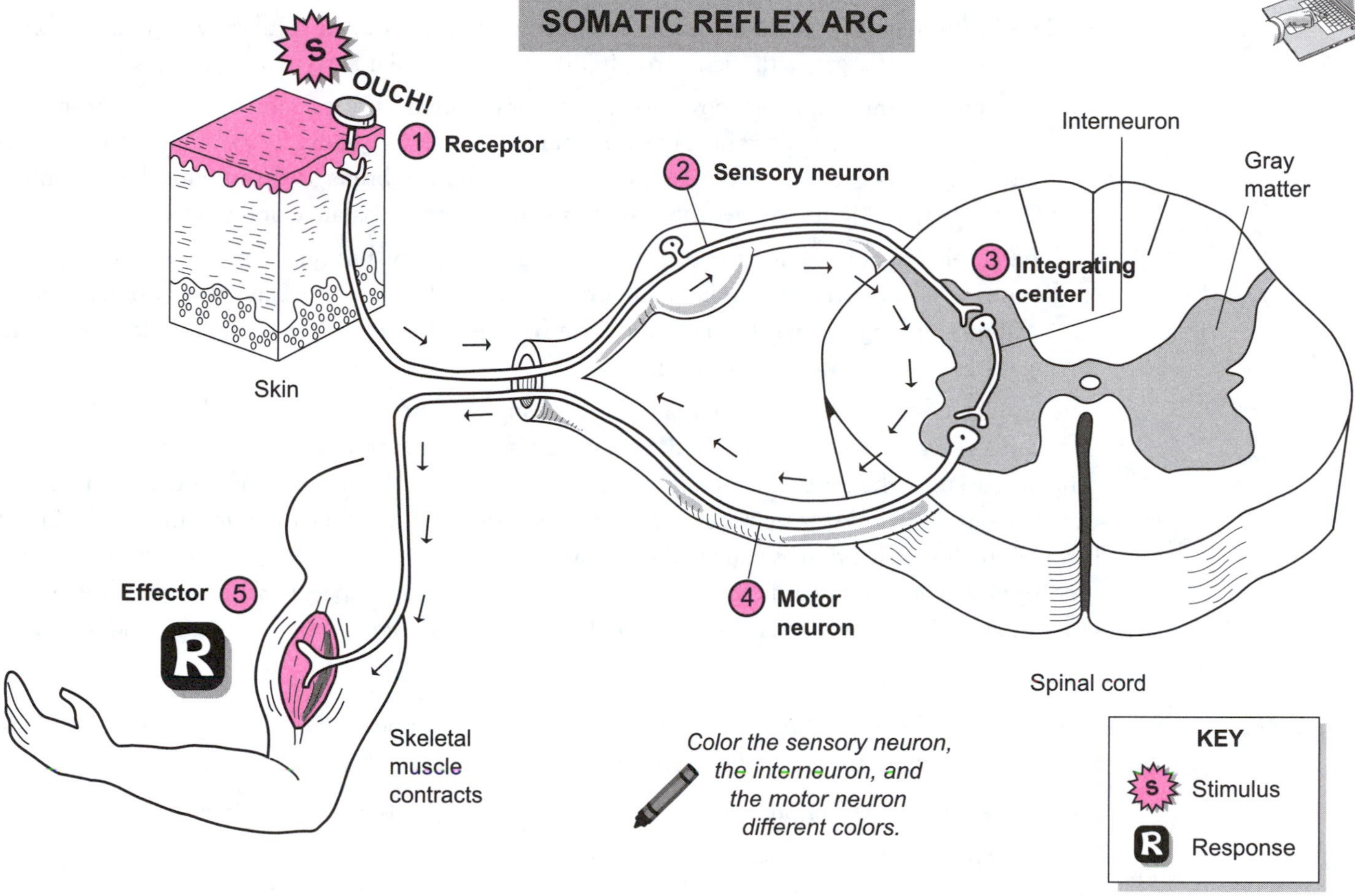

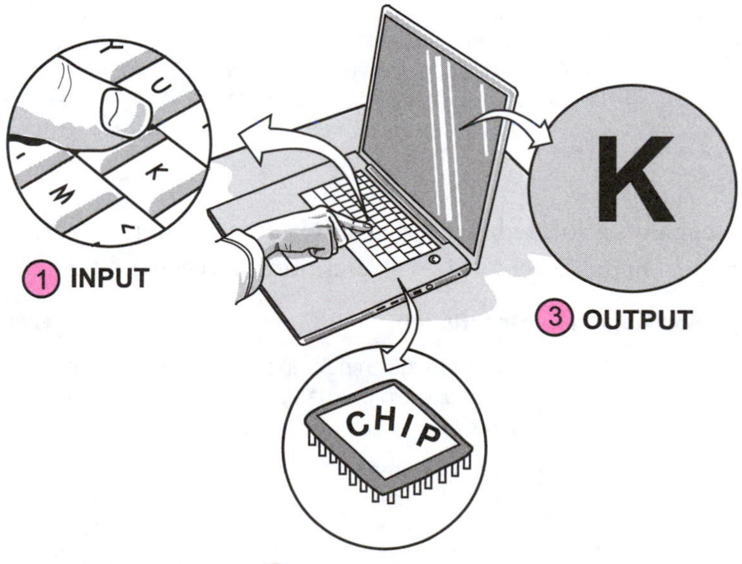

Description

The three *largest regions* of the brain are the **brain stem**, **cerebellum**, and **cerebrum**.

1. The **brain stem** is located at the base of the brain and contains regulatory centers to control things we take for granted, such as respiration, cardiovascular activities, and digestion.
2. The **cerebellum** is located posterior to the brain stem and inferior to the cerebrum. It is divided into two left or right halves, or **hemispheres**, and is extensively folded to increase surface area. Its general function is to work with the cerebrum to coordinate skeletal muscle movements, and it also allows the body to maintain proper balance and posture.
3. The **cerebrum** is the largest part of the brain and contains billions of neurons. Like the cerebellum, it is divided into two **hemispheres**. The deep division between these two hemispheres is called the **longitudinal fissure**. The term **fissure** indicates a deep groove or depression that separates major sections of the brain.

The surface of the cerebrum is not smooth but is folded into many little hills and gulleys. Each hill is called a **convolution** (or *gyrus*) and each valley is a shallow groove called a **sulcus**.

The cerebrum is the part of the brain associated with higher brain functions including planning, reasoning, analyzing, and storing/accessing memories. Ironically, without it, you would not be able to read and learn about the brain as you are doing now. It also perceives and interprets sensory information and coordinates various motor functions such as those involved in speech. The cerebrum is divided into four major lobes named after the bones that cover them: *frontal*, *parietal*, *temporal*, and *occipital*.

Brain Stem

The **brain stem** consists of three parts: *medulla oblongata*, *pons*, and *midbrain*. The table below gives a description and general function of each part.

Brain Stem Region	Description	General Functions
Medulla oblongata	Between spinal cord and pons	Respiratory control center; cardiovascular control center
Pons	Between medulla and midbrain; bulges out as widest region in brain stem	Controls respiration along with medulla; relays information from cerebrum to cerebellum
Midbrain	Between diencephalon and pons; includes corpora quadrigemina (*sensory relay station*) and cerebral aqueduct (*connects third and fourth ventricles; contains cerebrospinal fluid*)	Visual and auditory reflex centers; provides pathway between brain stem and cerebrum

Diencephalon

The **diencephalon** is located above the brain stem and contains three parts: *epithalamus*, *thalamus*,and *hypothalamus*. The table below gives a description and general function of each part.

Diencephalon Region	Description	General Functions
Epithalamus	Roof of third ventricle; includes pineal gland; choroid plexus (*forms cerebrospinal fluid*)	Pineal gland makes hormone melatonin, which regulates day–night cycles.
Thalamus	Two egg-shaped bodies that surround the third ventricle	Relays sensory information to cerebral cortex; relays information for motor activities; information filter
Hypothalamus	Forms floor of third ventricle; between thalamus and chiasm	Controls autonomic centers for heart rate, blood pressure, respiration, digestion, hunger center, thirst center, regulation of body temperature, production of emotions

Brain: Largest Regions, Brain Stem, and Diencephalon

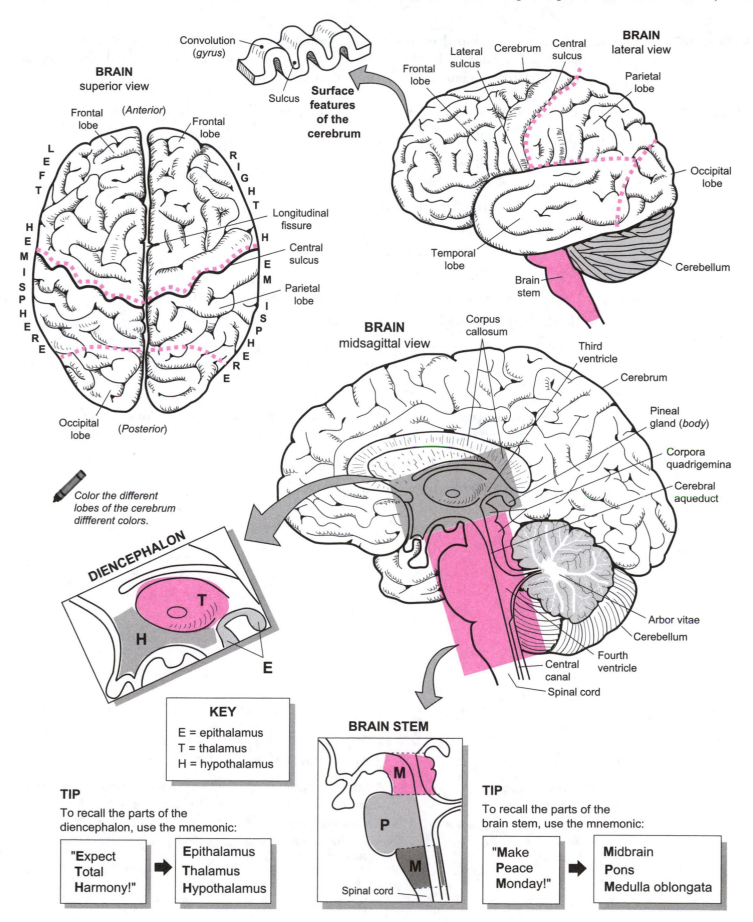

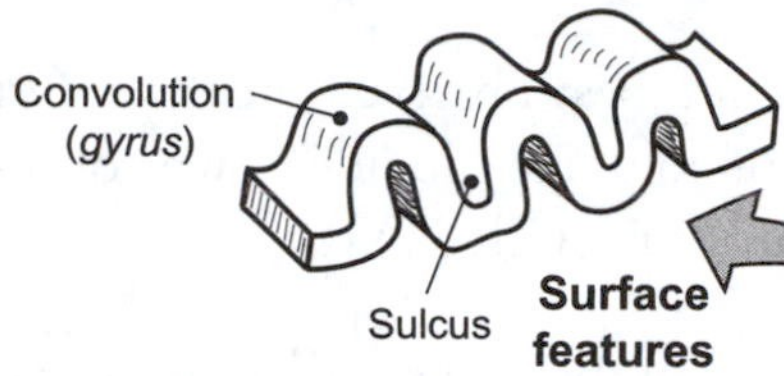

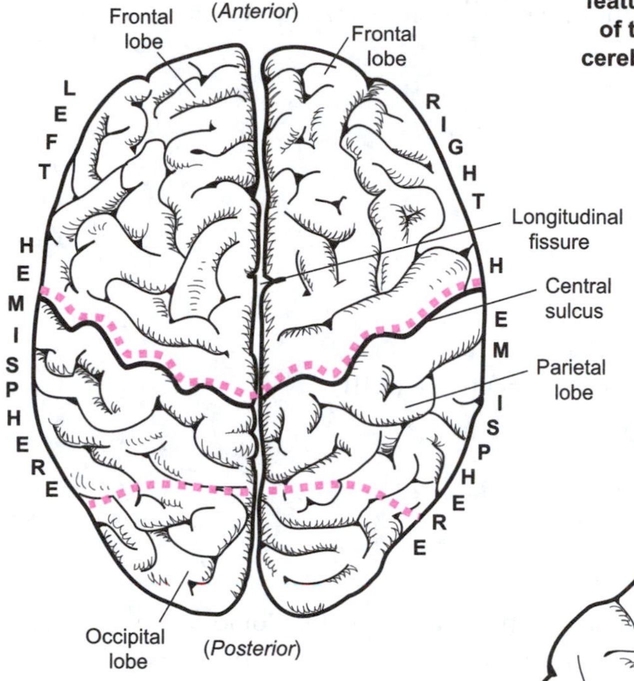

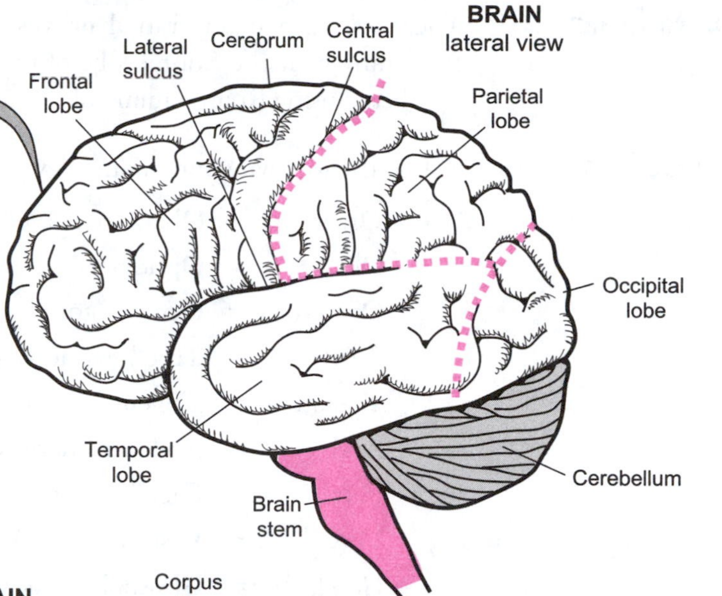

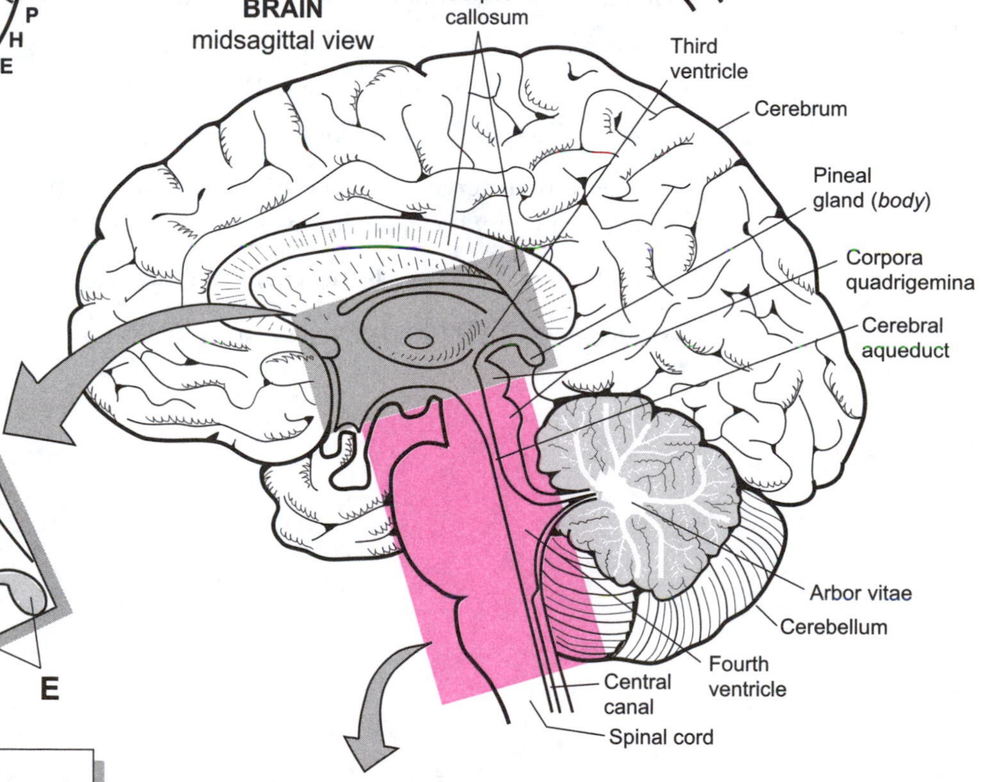

Color the different lobes of the cerebrum diffferent colors.

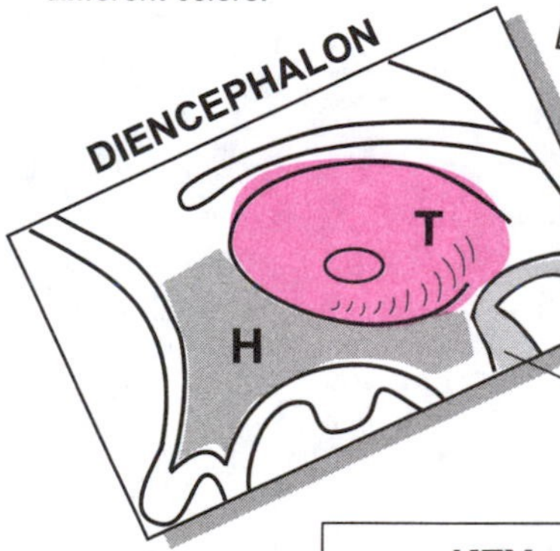

KEY

E = epithalamus
T = thalamus
H = hypothalamus

BRAIN STEM

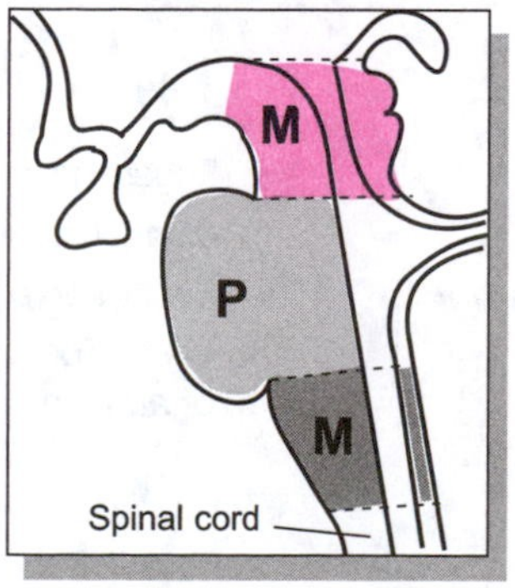

TIP

To recall the parts of the diencephalon, use the mnemonic:

"**E**xpect **T**otal **H**armony!" ➡ **E**pithalamus, **T**halamus, **H**ypothalamus

TIP

To recall the parts of the brain stem, use the mnemonic:

"**M**ake **P**eace **M**onday!" ➡ **M**idbrain, **P**ons, **M**edulla oblongata

Description

There are 12 pairs of **cranial nerves** that are best observed on the inferior view of a whole brain. Beginning near the frontal lobe of the cerebrum and moving down toward the spinal cord, they are numbered using Roman numerals from one (I) to twelve (XII).

Study Tips

- Use the following mnemonic device to recall the proper order of the cranial nerves:

Oscar's	=	Olfactory nerve (I)
Old	=	Optic nerve (II)
Ostrich	=	Oculomotor nerve (III)
Tasted	=	Trochlear nerve (IV)
Tomatoes	=	Trigeminal nerve (V)
And	=	Abducens nerve (VI)
Felt	=	Facial nerve (VII)
Very	=	Vestibulocochlear (*acoustic* or *auditory*) nerve (VIII)
Good,	=	Glossopharyngeal nerve (IX)
Vomited	=	Vagus nerve (X)
Any	=	Accessory nerve (XI)
How	=	Hypoglossal nerve (XII)

- Associate cranial nerves with specific landmarks on the brain—*ex:* **Oculomotor nerve** (III) is below the mamillary body, **Abducens nerve** (VI) is between the medulla and the pons
- The Thickest cranial nerve is the **Trigeminal nerve** (V)
- **Accessory nerve** (XI) runs parallel to the spinal cord

Key to Illustration

1. Olfactory nerve (I)
2. Optic nerve (II)
3. Oculomotor nerve (III)
4. Trochlear nerve (IV)
5. Trigeminal nerve (V)
6. Abducens nerve (VI)
7. Facial nerve (VII)
8. Vestibulocochlear *(acoustic* or *auditory)* nerve (VIII)
9. Glossopharyngeal nerve (IX)
10. Vagus nerve (X)
11. Accessory nerve (XI)
12. Hypoglossal nerve (XII)

To recall the cranial nerves from the first pair to the last pair, use the following mnemonic:

***O**scar's*
***O**ld*
***O**strich*
***T**asted*
***T**omatoes*
***A**nd*
***F**elt*
***V**ery*
***G**ood,*
***V**omited*
***A**ny*
***H**ow*

Color the matching pairs of cranial nerves.

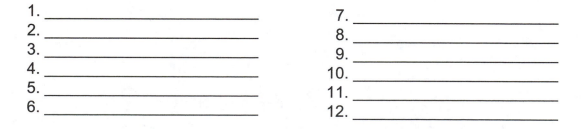

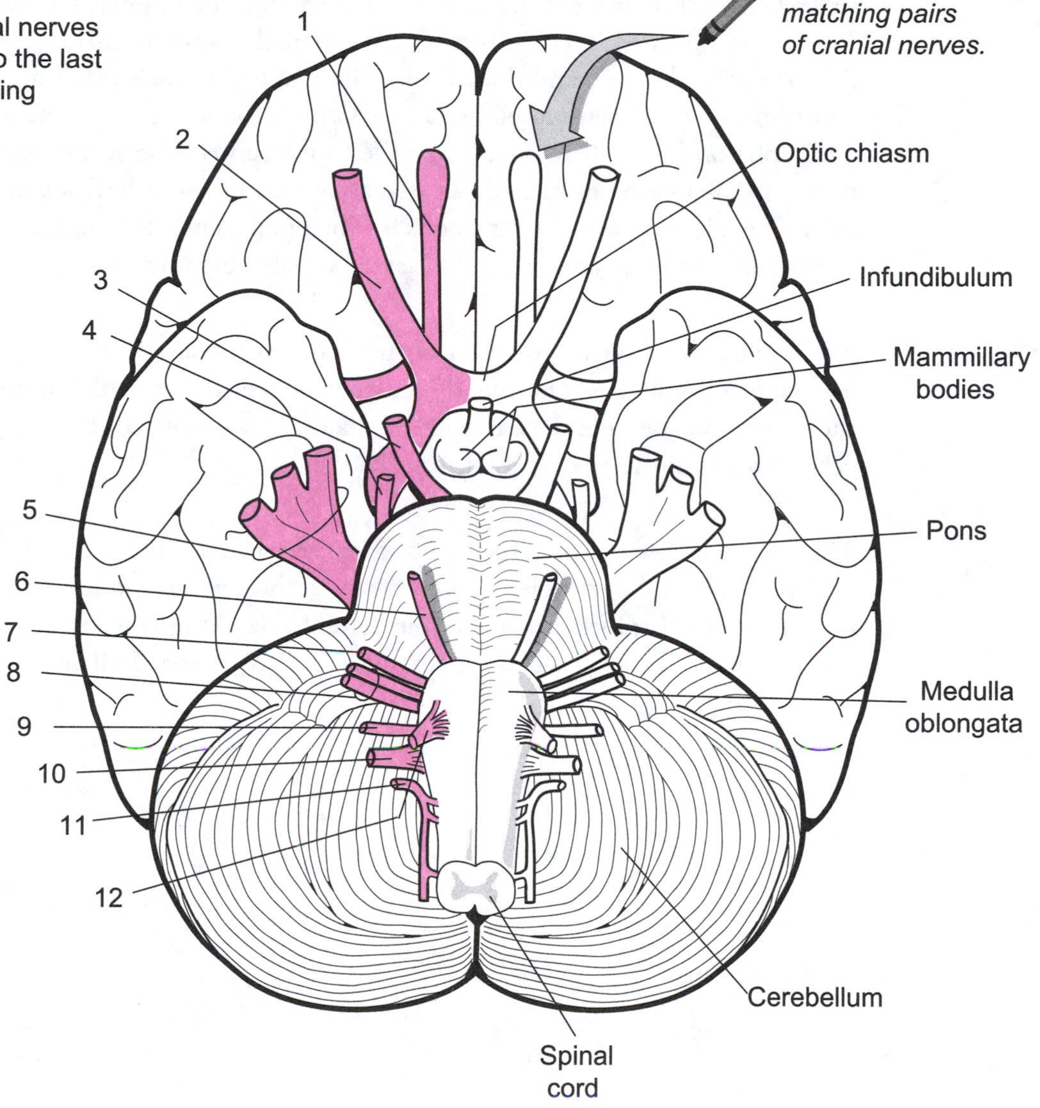

Cranial Nerves

1. ______________________
2. ______________________
3. ______________________
4. ______________________
5. ______________________
6. ______________________
7. ______________________
8. ______________________
9. ______________________
10. ______________________
11. ______________________
12. ______________________

Description

The heart contains ventricles that fill with blood, while the brain contains ventricles that are constantly filled with cerebrospinal fluid. In total, the brain has four ventricles inside it: *lateral ventricle* (of left hemisphere), *lateral ventricle* (of right hemisphere), *third ventricle*, and *fourth ventricle*. This entire network is referred to as the **ventricular system** in the brain. The lateral ventricles are the largest of the four and do not directly connect to each other as they are separated by a thin partition called the **septum pellucidum**. Both do connect to the third ventricle in the region of the diencephalon by small passageways called **interventricular foramina**. The third ventricle is connected to the fourth ventricle by a passageway called the **cerebral aqueduct** *(aqueduct of Sylvius)*. The fourth ventricle is located in the pons (of the brain stem) and the cerebellum. It communicates with a very narrow passageway called the **central canal**, which runs through the middle of the spinal cord.

Analogy

To visualize the relative positions of the ventricles, compare the whole **ventricular system** to the **hollow head of a ram**. The **fourth ventricle** is like the **neck of the ram**, the **third ventricle** is like the head, and the **lateral ventricles** are like the **two horns**. The ram's horns also follow the same general shape of the paired lateral ventricles.

Study Tip

The first and second ventricles are not numbered because they are the lateral ventricles. If you think of the two lateral ventricles as *first ventricle* and *second ventricle*, the numbering makes sense in relation to the **third ventricle** and **fourth ventricle**. Ah, the goofy things that anatomists do! As the saying goes, "you are not a good anatomist unless you know 87 different names for the same structure."

Key to Illustration

1. Lateral ventricles
1a. Anterior horns of lateral ventricles
1b. Posterior horns of lateral ventricles
1c. Inferior horns of lateral ventricles
2. Third ventricle
3. Cerebral aqueduct *(aqueduct of Sylvius)*
4. Fourth ventricle
5. Central canal

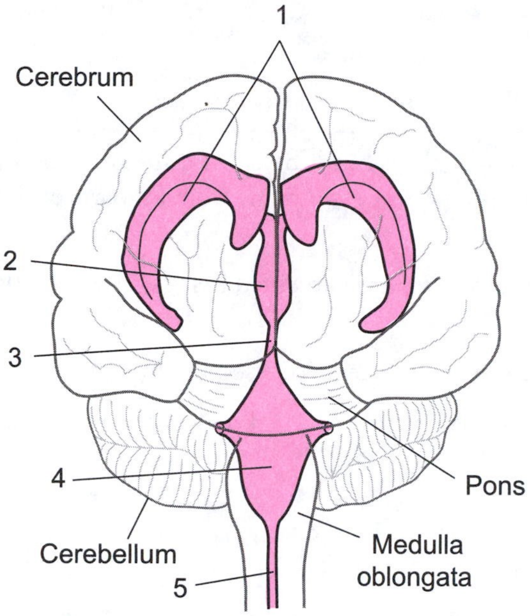

Anterior view

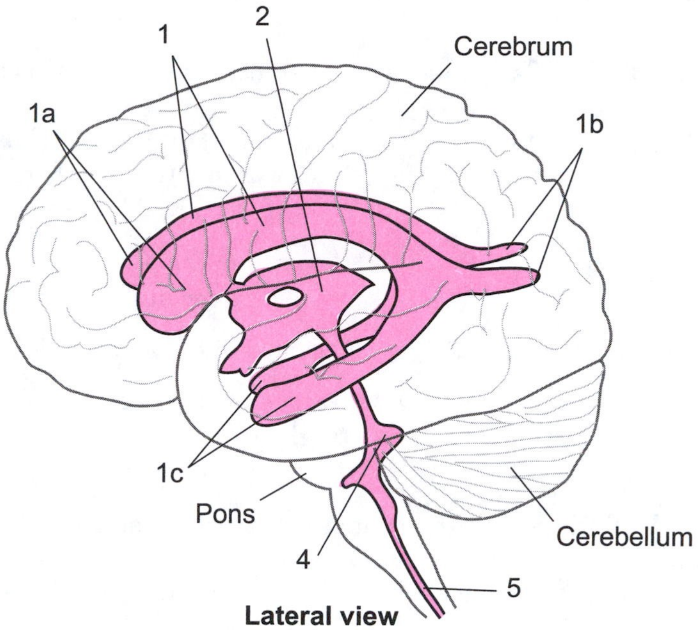

Lateral view

The ventricular system in the brain is like the neck, head, and horns of a ram if they were hollow.

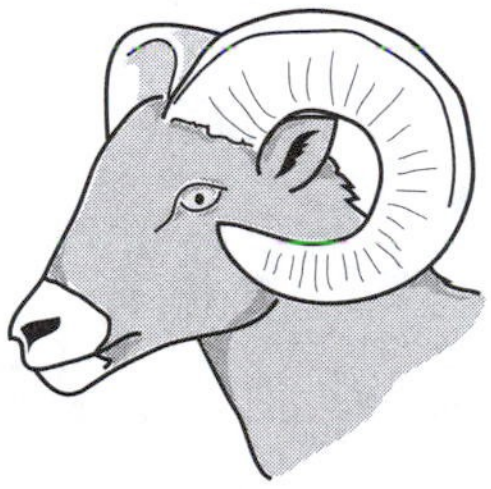

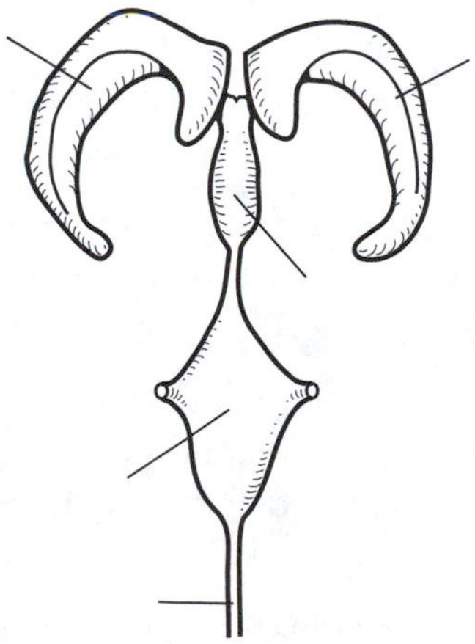

Anterior view

Color the ventricles and write labels on the illustrations.

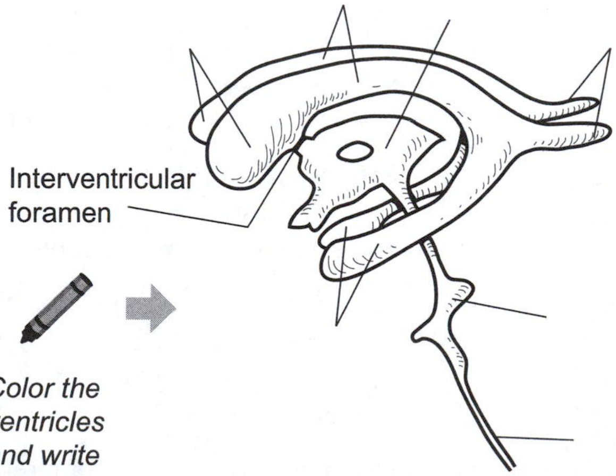

Lateral view

Functional Regions of the Cerebral Cortex

This module will describe some of the selected functional areas of the cerebral cortex. These areas have been divided into three general groups: **sensory areas**, **motor areas**, and **association areas**. Note that the words *cortex* and *area* are often used interchangeably.

SENSORY AREAS Control regions where sensations are perceived

1. **Primary somatic sensory cortex**	This important region is shown in dark gray behind the central sulcus. General sensory input (e.g. touch, temperature, pressure, and pain) from all parts of the body is perceived here.
2. **Gustatory cortex**	Located in the parietal lobe; taste sensations are perceived here, such as the flavors of the ice cream shown in the icon.
3. **Auditory cortex**	Located in the temporal lobe; auditory stimuli are processed by the brain here.
4. **Visual cortex**	Located in the occipital lobe; visual images are perceived here (like the star shown in the icon).

MOTOR AREAS Control centers for conscious muscle movements

1. **Primary motor cortex**	This important area is shown in color in front of the central sulcus. It controls voluntary muscle movements throughout the body, including those of the hands and feet, arms and legs, face and tongue.
2. **Premotor cortex**	This area serves as the "choreographer" for the primary motor cortex. It decides which muscle groups will be used and how they will be used prior to stimulating the primary motor cortex.
3. **Motor speech area** (*Broca's area*)	This area controls and coordinates the muscles involved in normal, fluent speech. Damage to this area can result in strained speech with disconnected words.
4. **Frontal eye field**	This area controls muscle movements of the eye, such as those needed to read this page.

ASSOCIATION AREAS Control regions—near sensory areas—involved in recognizing and analyzing incoming information

1. **Prefrontal area**	This area is most highly developed in humans and other primates. It regulates emotional behavior and mood and also is involved in planning, learning, reasoning, motivation, personality, and intellect.
2. **Somatic sensory association area**	This area allows you to *predict* that sandpaper is rough, for example, even without looking at it. It also stores memories about previous sensory experiences so you can determine when blindfolded, for example, that the object placed in your hand was a pair of scissors.
3. **Sensory speech area** (*Wernicke's area*)	This area seems to be an important part of language development—processing words we hear being spoken. It also appears important for children when they are sounding out new words. Damage to this area may result in deficiencies in recognizing written and spoken words.
4. **Auditory association area**	This area allows you to comprehend, interpret, analyze, and question what you are hearing. For example, it enables you to recognize a familiar song or disregard noise.
5. **Visual association area**	This allows you to associate the perceived image of the star with the letters "S-T-A-R". You connect the word "star" with the image of a star.

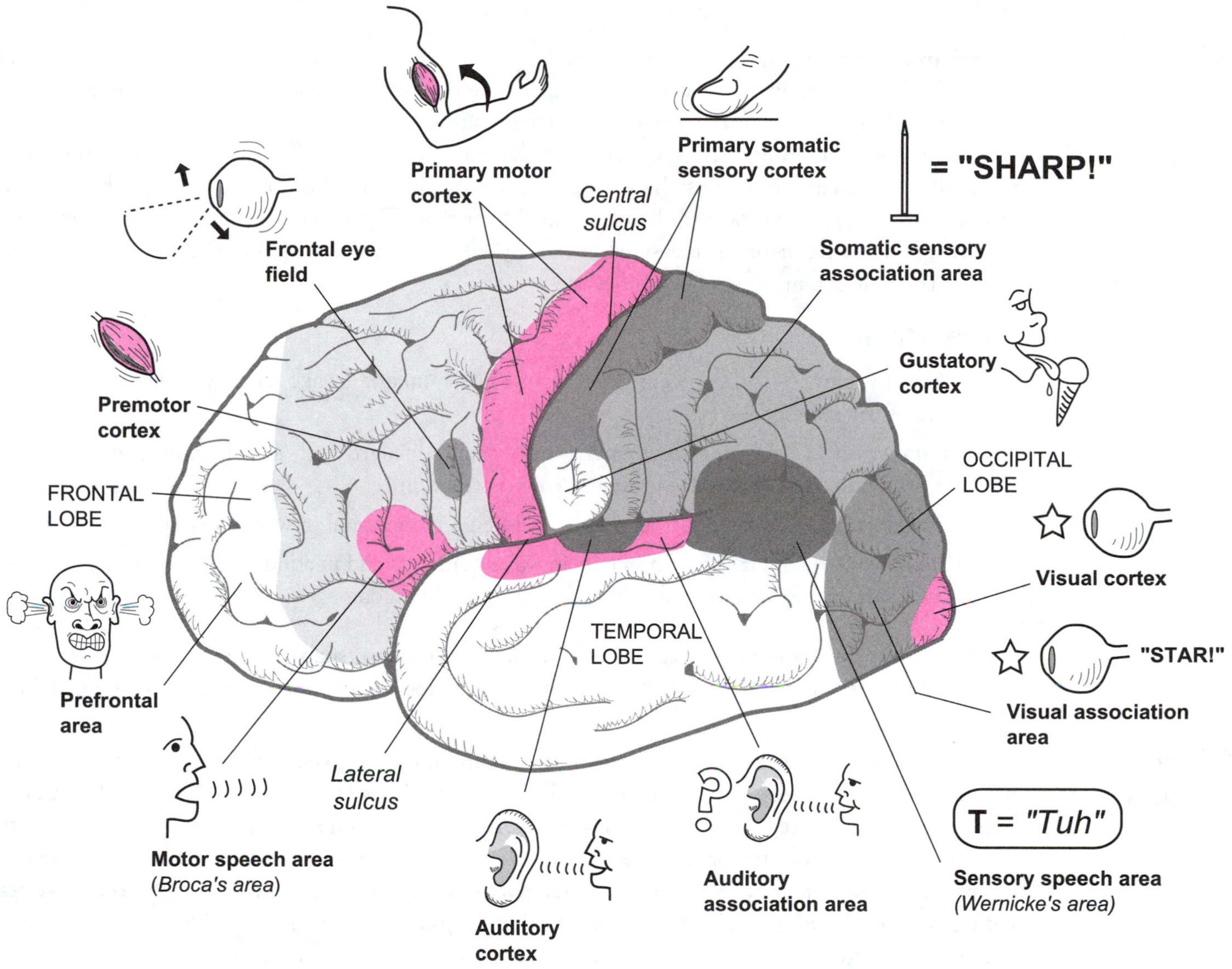

Left Cerebral Hemisphere
lateral view

Description

This module explains electroencephalogram (EEG), sleep/wakefulness, and reticular formation.

EEG

An **electroencephalogram (EEG)** is to the nervous system what an ECG is to the cardiovascular system (see p. 288). By detecting the electrical activity of the billions of neurons in the brain, an EEG measures normal brain function. **Electrodes** are placed around the scalp to detect the small voltage generated from these nerve impulses. Wires connect the electrodes to an EEG machine, where the voltage is amplified so it can be seen as characteristic waves. Each person has her own unique brain waves, displayed as a graph of **voltage** (millivolts or mV) over **time**. The height of the wave is the **amplitude** (in mV), and the frequency is measured in **hertz** (Hz), or cycles per second. The following is a summary of the four common types of brain waves:

Sleep/Wakefulness

- **Alpha** waves (8–13 Hz) occur when a person is calm and awake with eyes closed. This is like the restful state of the brain.
- **Beta** waves (14–30 Hz) occur when we are mentally alert, such as balancing your checkbook or studying a map to find your way to your destination.
- **Theta** waves (4–7 Hz) are common in children but abnormal in wakeful adults.
- **Delta** waves (< 4 Hz) occur normally in wakeful infants and in adults during deep sleep or while under anesthesia. In wakeful adults, they indicate brain damage.

EEGs also can be used to detect abnormalities in the brain, such as epilepsy, brain tumors, and traumatic injuries. The absence of any brain waves is the legal definition of what is called *brain dead*.

Reticular Formation

The brain never really "turns off." Neurons are sending impulses to each other constantly whether a person is awake or asleep, but the brain needs to have methods to control these states of consciousness. Sleep states are regulated through the hypothalamus. Wakeful states are controlled by the **reticular formation**—a cluster of neurons, centrally located and extending like columns through the brainstem—*midbrain*, **pons**, and **medulla oblongata**. One subset of these neurons—the **reticular activating system (RAS)** act like an alarm system that keeps the brain alert. The RAS sends stimulatory impulses through nerve pathways to other key parts of the brain, such as the cerebral cortex, thalamus, hypothalamus, cerebellum, and spinal cord. Sensory input is delivered to the RAS so it can filter out everything that is not important. Otherwise, we would go insane from sensory overload. Only the most important sensory input is routed to the cerebral cortex so we can be aware of it. For example, you probably are unaware of the sensation of your silky, long-sleeved shirt against your arm, but you would notice if someone suddenly were to pull on it.

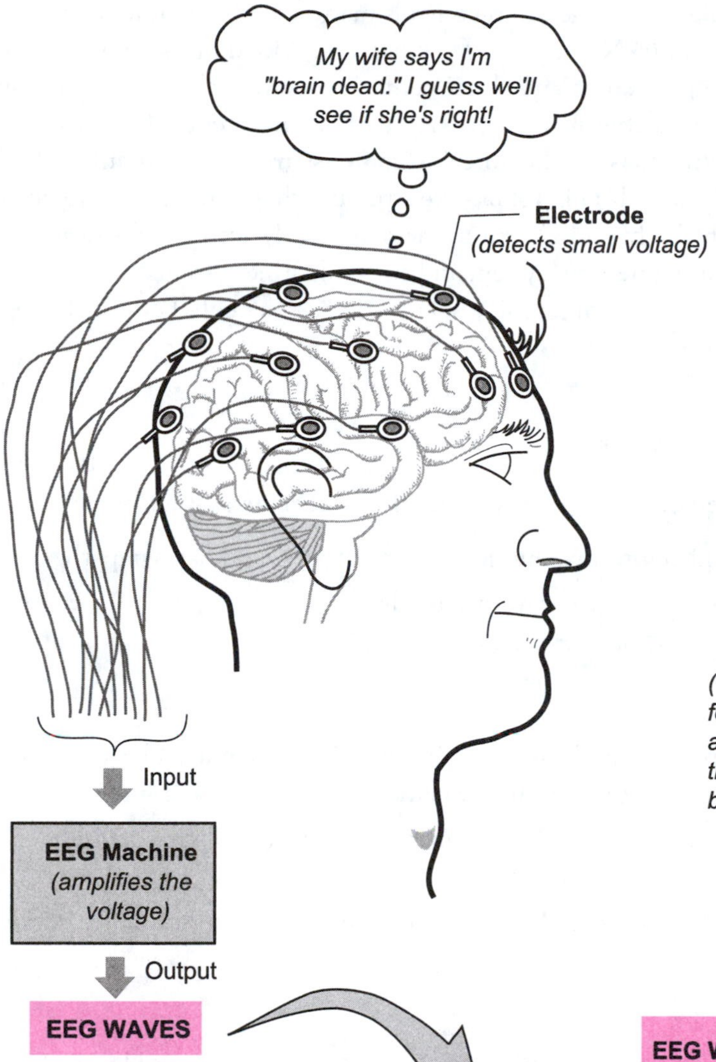
My wife says I'm "brain dead." I guess we'll see if she's right!
Electrode
(detects small voltage)
Input
EEG Machine
(amplifies the voltage)
Output
EEG WAVES

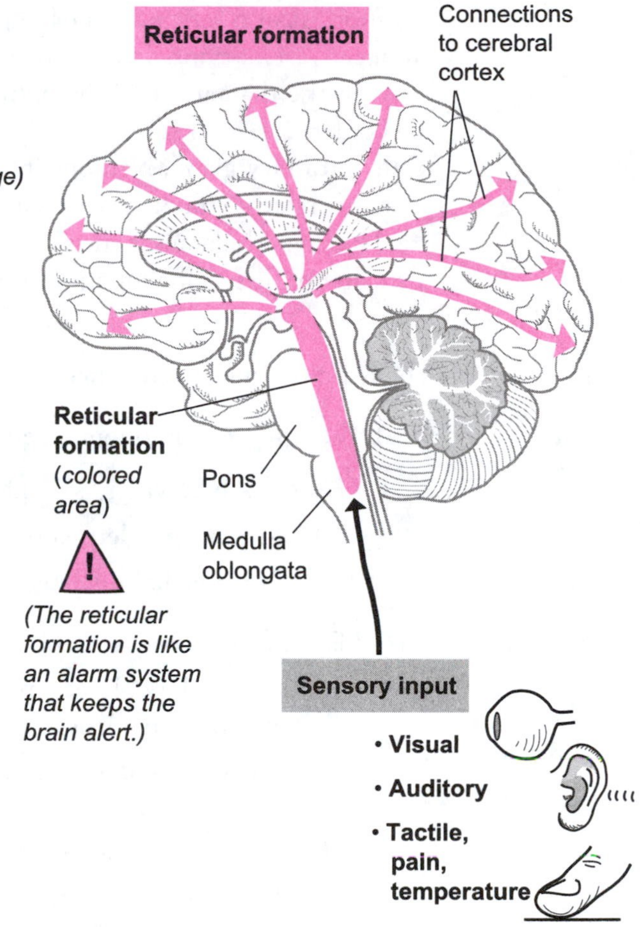
Reticular formation
Connections to cerebral cortex
Reticular formation
(colored area)
Pons
Medulla oblongata
(The reticular formation is like an alarm system that keeps the brain alert.)
Sensory input
• Visual
• Auditory
• Tactile, pain, temperature

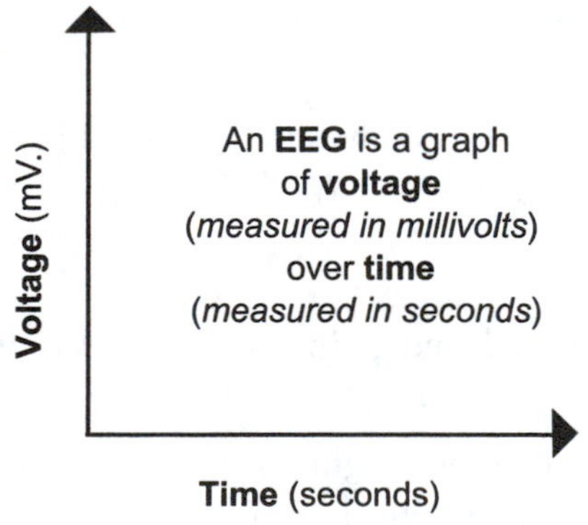
Voltage (mV.)
An EEG is a graph of voltage (measured in millivolts) over time (measured in seconds)
Time (seconds)

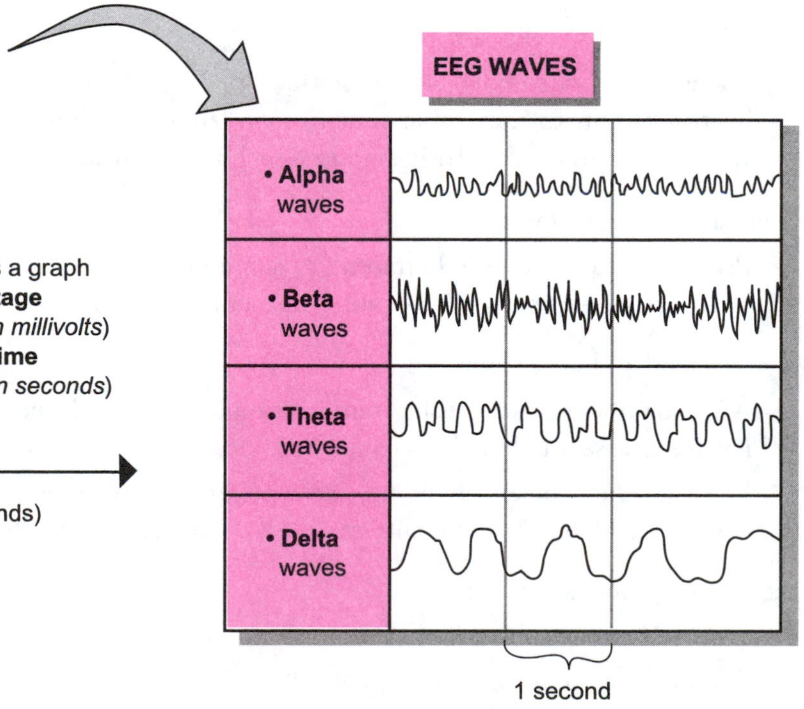
EEG WAVES
• Alpha waves
• Beta waves
• Theta waves
• Delta waves
1 second

Description

Though small in size, the **hypothalamus** in the **brain** controls many vital body functions. In general, it controls autonomic nervous system **(ANS)** centers for heart rate, blood pressure, respiration, and digestion. It also serves as our hunger center and thirst center, regulator of body temperature, and producer of emotions. It even plays a role in memory. As part of the diencephalon, it is located between the **thalamus** and the **optic chiasm**. The **anterior commissure** and **mammillary body** form the borders on either side. The **infundibulum** is a stalk-like structure that connects the hypothalamus to the **pituitary gland**. Neurons in the hypothalamus are wired directly into the **posterior pituitary** gland to create a vital link between the nervous system and the endocrine system.

The hypothalamus consists of many different nuclei—clusters of cell bodies. These nuclei correlate to the specific functions listed below.

Functions

The hypothalamus has five major functions:

1. Controls the autonomic nervous system **(ANS)** and the **endocrine system**.

   For the **ANS**, which is divided into two divisions—sympathetic and parasympathetic:
   - **Anterior nucleus** controls the parasympathetic division.
   - **Dorsomedial nucleus** controls the sympathetic division.

   For the **endocrine system**:
   - Notice the colored lines in the illustration, linking the paraventricular nucleus and the supraoptic nucleus to the posterior pituitary gland.
   - **Paraventricular nucleus** releases **OT** (*oxytocin*), the hormone that stimulates uterine contractions during labor and delivery.
   - **Supraoptic nucleus** releases **ADH** (*antidiuretic hormone*), important for maintaining water balance.
2. Regulates emotional behavior.
   - **Mammillary body** contains relay stations for the sense of smell. In addition, it links the hypothalalmus to the limbic system—our "emotional brain" (see p. 222). The limbic system helps us regulate basic emotions such as rage, fear, happiness, and sadness.
3. Regulate body temperature.
   - **Preoptic area** works as a thermostat to maintain our normal body temperature of 98.6° F by regulating shivering and sweating mechanisms.
4. Regulate food and water intake.
   - **Anterior nucleus** is the thirst center, checking dissolved solutes in the blood. As solute levels increase, fluid intake is stimulated.
   - **Ventromedial nucleus** is the hunger center, checking levels of glucose and other nutrients in the blood. As these nutrient levels decrease, food intake is stimulated.
5. **Regulates sleep/wake cycles.**
   - These cycles are called *circadian rhythms*.
   - **Suprachiasmatic nucleus** is the control center.

HYPOTHALAMUS FUNCTIONS

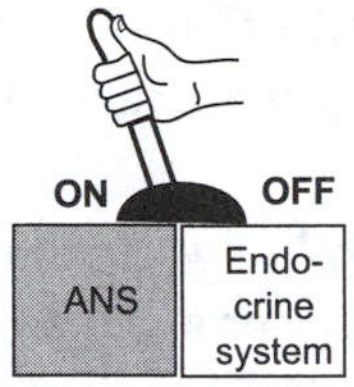

1. Regulates the ANS and the Endocrine system

   ANS
   - **Anterior nucleus**
   - **Dorsomedial nucleus**

   Endocrine system
   - **Paraventricular nucleus (OT)**
   - **Supraoptic nucleus (ADH)**

2. Regulates emotional behavior
   - **Mammillary body**

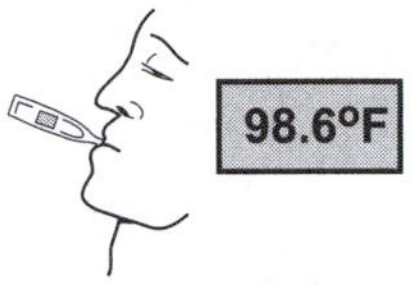

3. Regulates body temperature
   - **Preoptic area**

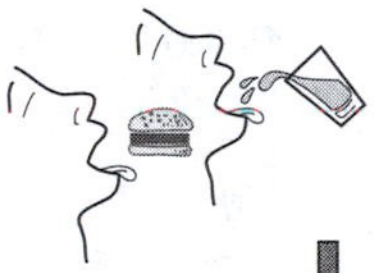

4. Regulates water and food intake
   - **Anterior nucleus - Thirst center**
   - **Ventromedial nucleus - Satiety center**

5. Regulates sleep/wake cycles
   - **Suprachiasmatic nucleus**

Color each of the different nuclei a different color.

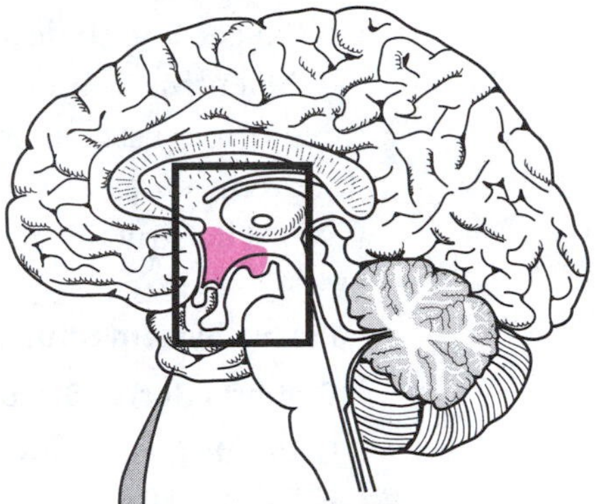

BRAIN, midsagittal view

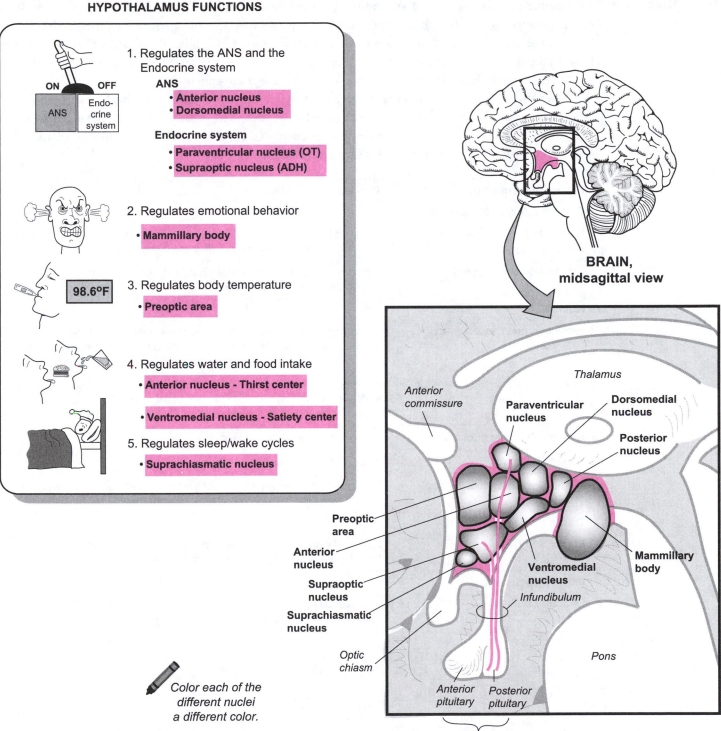

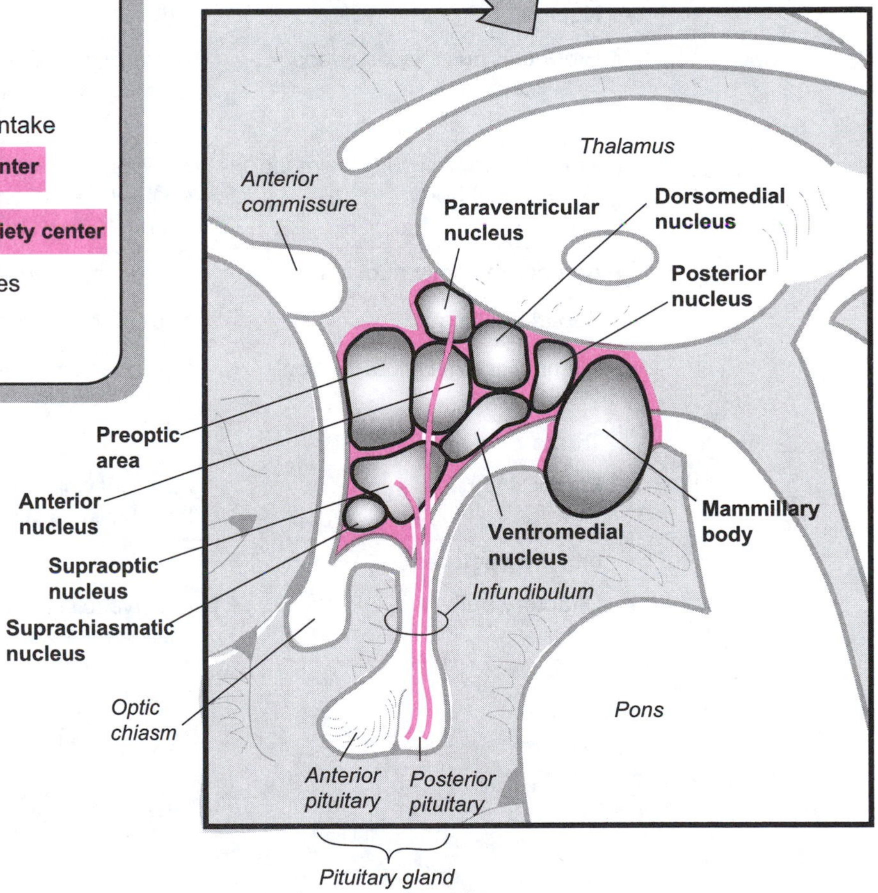

HYPOTHALAMUS, midsaggittal view

Description

The **limbic** (*limbus* = border) **system** is called our "emotional brain" because it processes our basic emotions such as fear, happiness, and sadness. To achieve this, it must integrate different brain centers such as sensory systems and conscious control centers. Physically, it is a collection of structures found mostly in the cerebrum and diencephalon (thalamus, hypothalamus, epithalamus). It is so named because these structures form a border around the diencephalon.

Scientists still do not universally agree on the structures that compose the limbic system. The table below lists the most commonly recognized structures, with similar structures grouped together. The term *nuclei* refers to clusters of cell bodies in the CNS, and *tracts* are bundles of axons in the CNS. The structures in the table are illustrated on the facing page.

Limbic System Structure	Description
Cerebral Cortex Structures	
Cingulate (*cingulum* = girdle, to surround) **gyrus**	A mass of cerebral cortex located above the corpus collosum; its name describes the fact that it surrounds the diencephalon.
Parahippocampal gyrus	A mass of cerebral cortex in the temporal lobe; works with the hippocampus to store memories.
Nuclei	
Hippocampus (*hippocampos* = seahorse)	Anatomists thought this nucleus looked like a seahorse; located above the parahippocampal gyrus and connects to the diencephalon through the fornix; important in learning and storing long-term memories.
Amygdaloid (*amygda* = almond shaped) **body**	Nucleus located in front of and connected to the hippocampus; directly involved in emotions such as fear; can store memories and correlate them with specific emotions.
Anterior thalamic nucleus	Located in the thalamus; relays visceral sensations to the cingulate gyrus.
Septal nucleus	Nucleus located near the anterior commissure.
Mammillary (*mammilla* = nipple) **body**	Nipple-like structures in the floor of the hypothalamus that serve as relay stations for reflexes related to sense of smell.
Tracts	
Fornix (arch)	A tract of white matter that connects the hippocampus to the hypothalamus; a portion of it is shaped like an arch.
Other Structures	
Olfactory bulbs	All serve to interpret various odors in the brain. Odors are linked to specific memories, so certain odors can trigger powerful emotions.

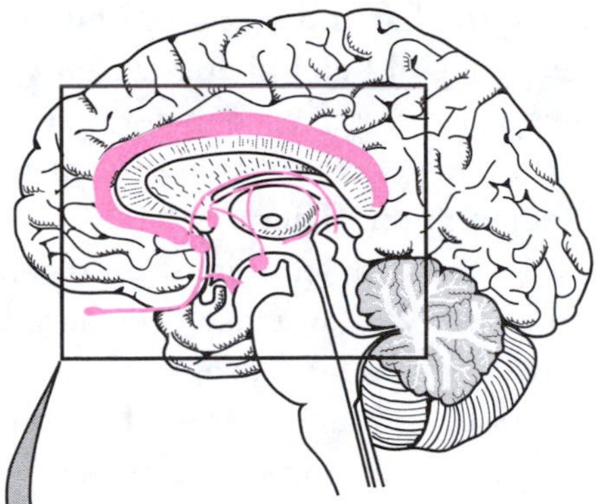

BRAIN
midsagittal view

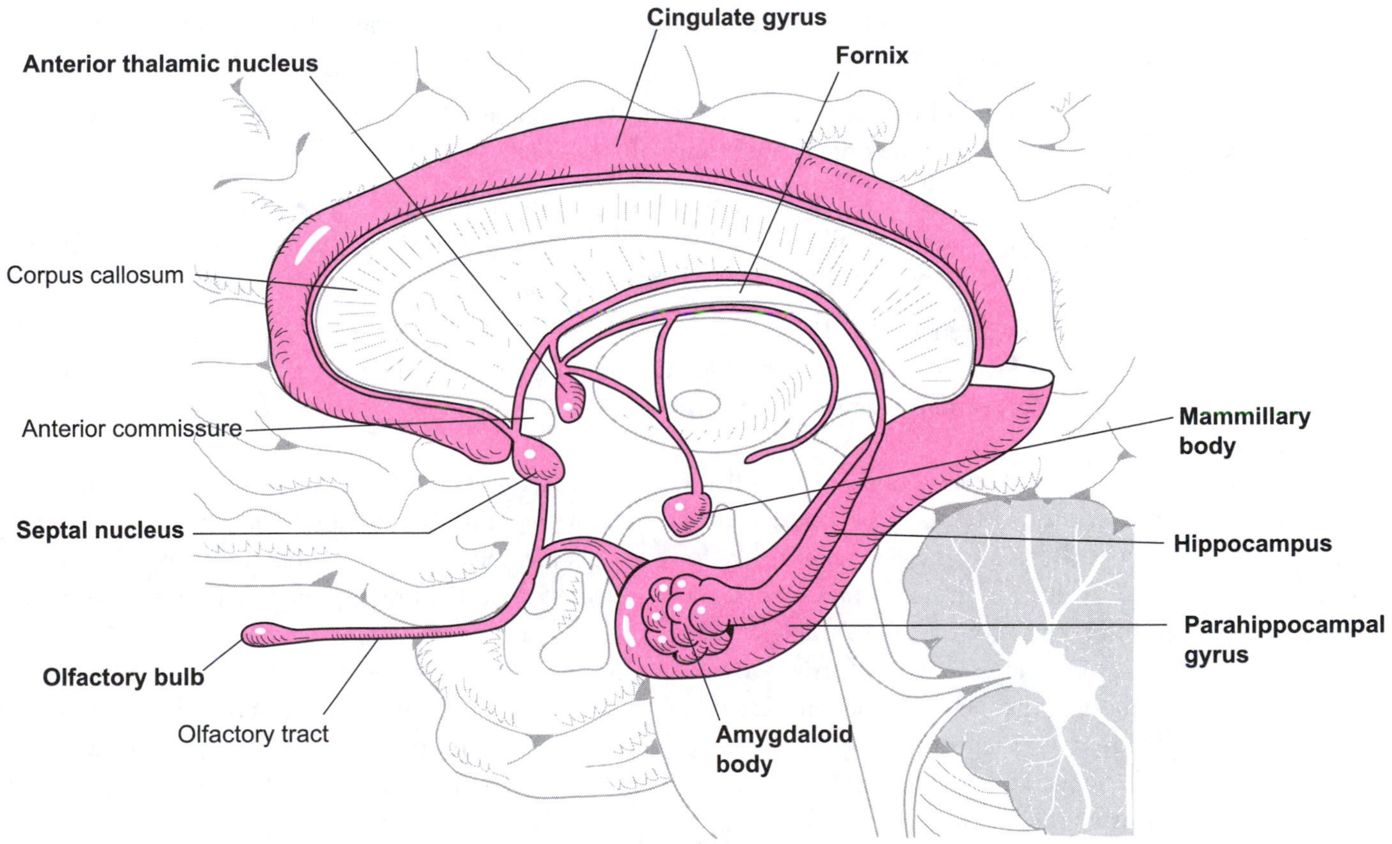

Midsagittal view
Note: *Major structures in the limbic system are shown in color.*

Description

This module explains the concept of lateralization between the left and right cerebral hemispheres. Looking at the illustration of the **brain**, the left and right cerebral hemispheres appear very similar. In fact, they are anatomically similar, and the two hemispheres work together for many functions. This is evidenced by the **corpus callosum**—a thick band of nerves connecting the left to the right hemispheres. Each hemisphere also has functional specialization. There is lateralization where certain functions are found only in one of the hemispheres. For example, in most people, the Broca's area for speech production is found only in the left hemisphere. We can make generalizations about the functional differences between the two hemispheres that apply to most people. Consider your left hemisphere to be your "analytical" hemisphere and your right to be your "creative" hemisphere.

The illustration lists the functional differences. Your left brain excels at language and logic. It deals with information in an organized, logical way as a scientist would. It helps you work with mathematical equations, write, and follow directions step by step. In contrast, your right hemisphere excels at musical and artistic abilities. It helps you understand shape and pattern relationships that are useful for facial recognition and drawing. It also is the seat of insight and inspiration.

But these generalizations are not set in stone. Here are some variables that are exceptions to the rule:

- **Individual differences.** Some individuals have one or more control centers in the hemisphere opposite from the one where it normally is found.
- **Gender differences**. Lateralization is greater in males than females. In typical females, a portion of the corpus callosum is thicker, indicating greater hemispheric integration. This means that both hemispheres work together more frequently.
- **Age differences.** Children can "re-wire" their brains more easily than adults. For example, if part of the brain is damaged or surgically removed in a child, the opposite hemisphere can take over and compensate.

One hemisphere doesn't actually dominate the other. Even so, the hemisphere that controls spoken and written language is designated as the *categorical* (or "dominant") hemisphere. As mentioned previously, this is the left hemisphere for most people and correlates to handedness. Because nerves cross over from one side of the body to the opposite side in the brain, motor activity on the right side of the body is controlled by the left hemisphere, and vice versa.

The same usually holds true for handedness. About 91% of the population is right-handed, and in most of these people the left hemisphere is the categorical one. Interestingly, the situation is a bit different for "lefties." In the majority of them, the left hemisphere is still their categorical one. In only about 15% is the right hemisphere categorical. In summary, although the two hemispheres work together all the time, they also specialize in specific functions.

Corpus callosum
Cerebral cortex
Pons
Medulla oblongata
Cerebellum
BRAIN, Coronal section
LATERALIZATION
Brain, Superior view
LEFT
HEMISPHERE
RIGHT
HEMISPHERE
Color the left hemisphere and the word "LEFT" one color; color the right hemisphere and the word "RIGHT" another color.
LEFT
RIGHT
• Speech, language ability
"Ask not what your country can do for you..."
• Musical ability
• Logical thought, mathematical ability
F = ma
E = mc2
a2 + b2 = c2
• Artistic ability
• Hearing vocal sounds
• Memory for shapes
• Orderly sequence
• Step 1
• Step 2
• Step 3
• Shape and pattern relationships
• Motor control—right side of body and right-handedness
• Motor control—left side of body and left-handedness

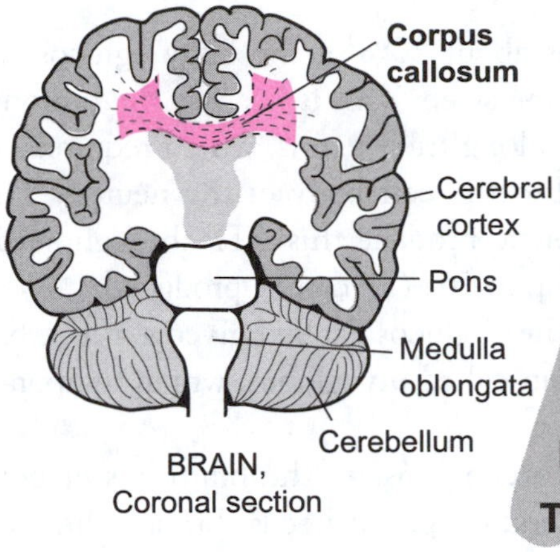

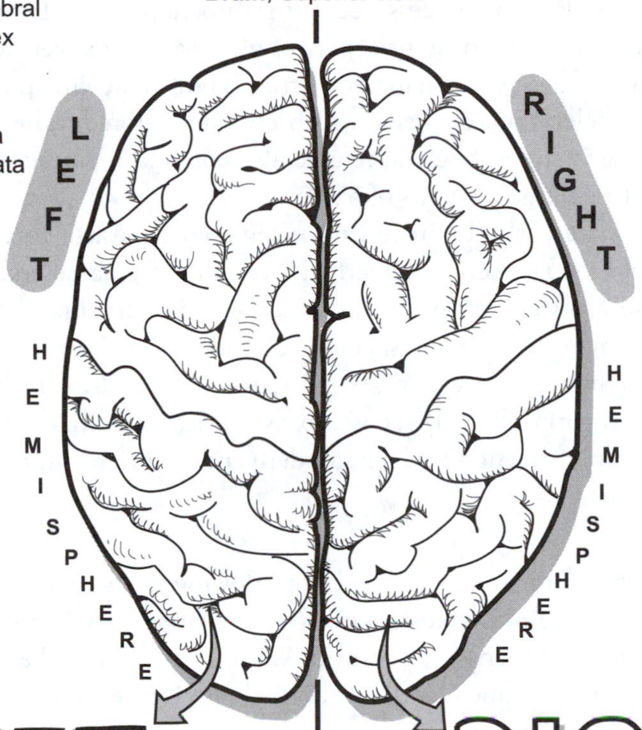

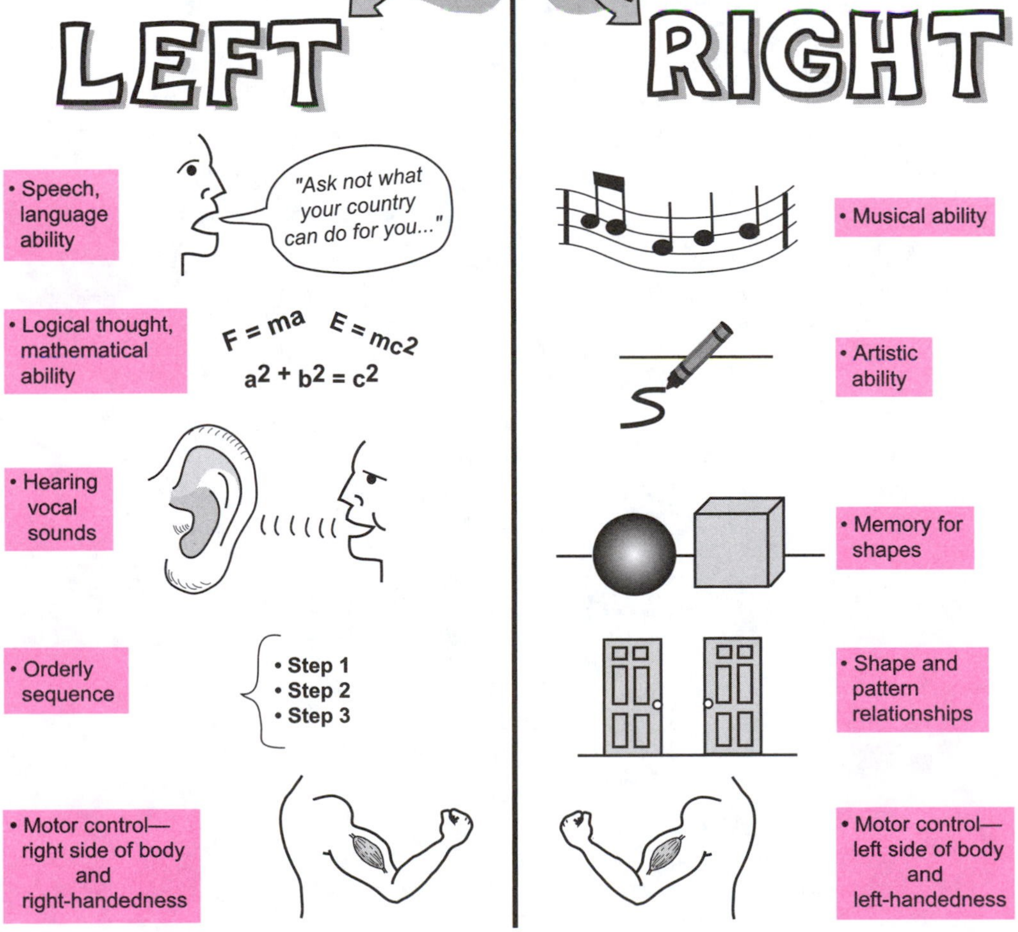

Description

Though the **brain** makes up only 2% of body mass, it uses about 20% of the body's glucose and oxygen. The brain never really shuts off; it's even active during sleep. This is because the billions of **neurons** in the brain are constantly sending nerve impulses along their axons, which requires large amounts of **ATP**. This process occurs in mitochondria within the **cell body** of the neuron. These hungry cells require a constant supply of **glucose** and **oxygen** to produce this ATP through **aerobic respiration**. For each glucose molecule that is broken down, up to 36 ATP can be produced.

Unlike skeletal muscle cells, which contain a stored form of glucose called **glycogen**, the brain has no glycogen stores. Skeletal muscle also contains a protein called **myoglobin**, which temporarily binds and stores large amounts of oxygen.

Unfortunately, the brain has no myoglobin deposits. To compensate, the brain has elaborate networks of blood vessels to directly deliver all the glucose and oxygen it needs through the blood. The **carotid arteries** transport oxygenated blood to the brain. But to protect vital neurons, a brain barrier system restricts what specific substances can be transported from the blood to the brain tissues. This prevents toxins that may enter the bloodstream from damaging neurons. Fortunately, this system is permeable to glucose, oxygen, carbon dioxide, sodium, potassium, and chloride. As a result, it doesn't present a problem for delivering nutrients to the neurons or eliminating wastes from them.

Cardiac Arrest

If a person has cardiac arrest, blood stops flowing to the brain. Of all the different cells in the body, neurons in the brain are most likely to die if blood flow is not restored. This is why **CPR** (cardiopulmonary resuscitation) is performed. Without it, neurons begin to die in about 3 to 4 minutes. In fact, this occurs any time blood flow to the brain is interrupted.

Brain Metabolism and Brain Function

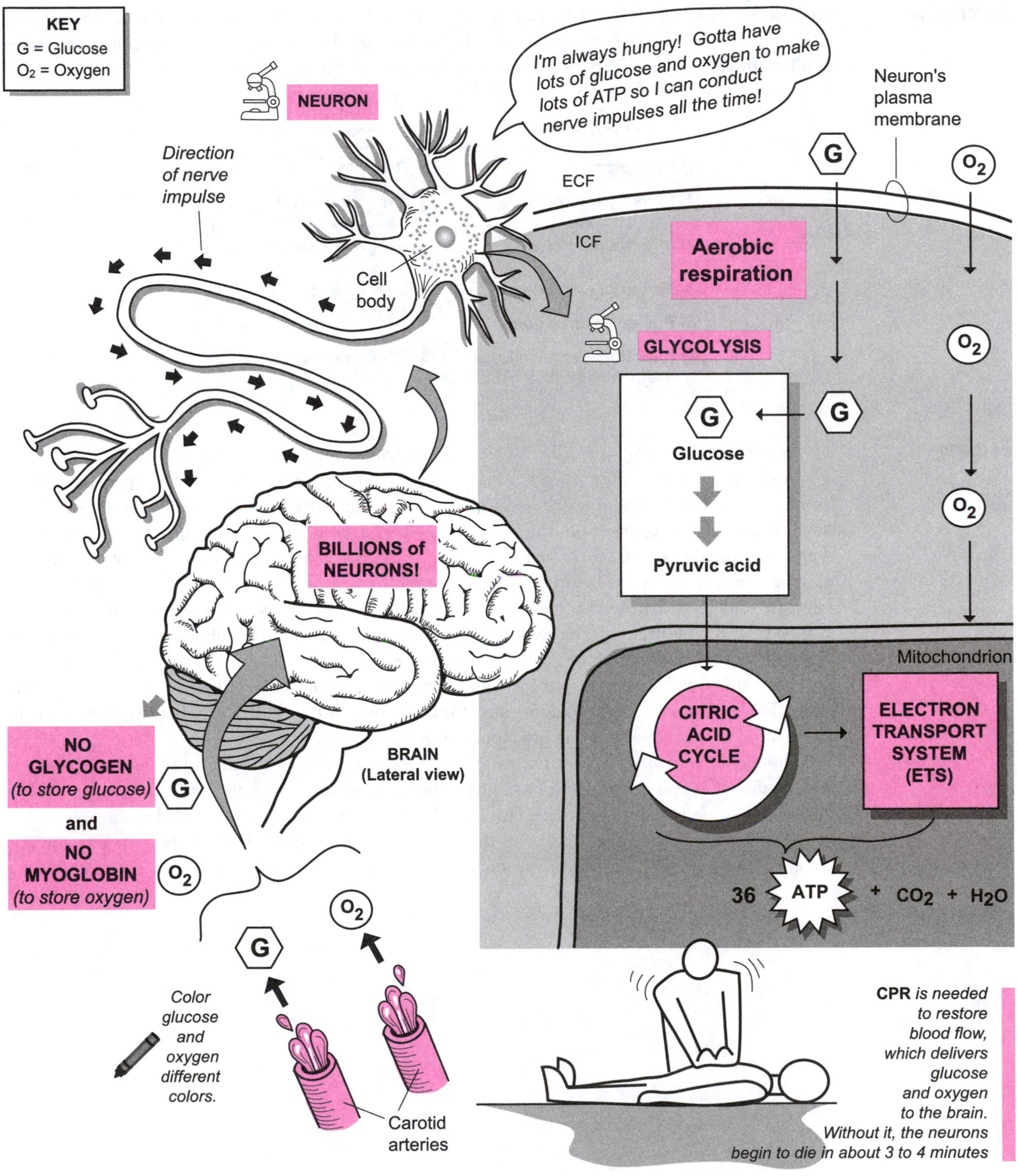

KEY
G = Glucose
O2 = Oxygen
NEURON
I'm always hungry! Gotta have lots of glucose and oxygen to make lots of ATP so I can conduct nerve impulses all the time!
Neuron's plasma membrane
Direction of nerve impulse
ECF
ICF
Cell body
Aerobic respiration
GLYCOLYSIS
Glucose
Pyruvic acid
BILLIONS of NEURONS!
Mitochondrion
CITRIC ACID CYCLE
ELECTRON TRANSPORT SYSTEM (ETS)
36 ATP + CO2 + H2O
NO GLYCOGEN (to store glucose)
and
NO MYOGLOBIN (to store oxygen)
BRAIN (Lateral view)
Color glucose and oxygen different colors.
Carotid arteries
CPR is needed to restore blood flow, which delivers glucose and oxygen to the brain. Without it, the neurons begin to die in about 3 to 4 minutes

Description

Somatic reflexes were covered previously (see p. 208). This module compares and contrasts somatic reflexes and autonomic reflexes. Somatic reflexes stimulate the contraction of skeletal muscle. **Autonomic reflexes** are used by the autonomic nervous system (ANS) to control many of the involuntary functions that we take for granted, such as digestion, urination, and heart rate.

	Autonomic Reflex	Somatic Reflex
Input	① **Receptor**	① **Receptor**
	② **Sensory neuron**	② **Sensory neuron**
Processing	③ **Integrating center**	③ **Integrating center**
Output	④ **Preganglionic fiber**	④ **Motor neuron**
	⑤ **Postganglionic fiber**	
	⑥ **Effector** (smooth muscle, cardiac muscle, or gland)	⑤ **Effector** (skeletal muscle)

Pathway

All reflexes involve a pathway called an arc, which includes input, processing, and output.

Input involves a receptor and a sensory neuron. In an autonomic reflex, the receptor typically is located on the organ/gland it is going to regulate. For example, the illustration shows a section of the small intestine. During digestion, the smooth muscle in its wall has to be stimulated to contract and thereby propel digested food through the intestine. This has to occur only when digested food is moving through the tract. In this example, a stretch receptor is located in the wall of the small intestine to detect when food is present. In contrast, most somatic receptors are located in the skin, as shown on the illustration. The sensory neuron has the same function in both reflexes: It conducts the impulse from the receptor to the integrating center.

Processing involves the integrating center, which is gray matter in either the spinal cord or the brain. This is where the input is connected to the output. In the illustration, both reflexes have an interneuron that connects the sensory input to the motor output.

The biggest difference between somatic and autonomic reflexes is seen in the output. The autonomic reflex uses two motor neurons—a preganglionic fiber and a postganglionic fiber, as opposed to the single neuron in a somatic reflex. This introduces an additional structure called a ganglion.

Finally, the effector is also different. In somatic reflexes skeletal muscle is the only effector. An autonomic reflex, has three different types of effectors: smooth muscle, cardiac muscle, or glands.

Autonomic Reflexes Compared to Somatic Reflexes

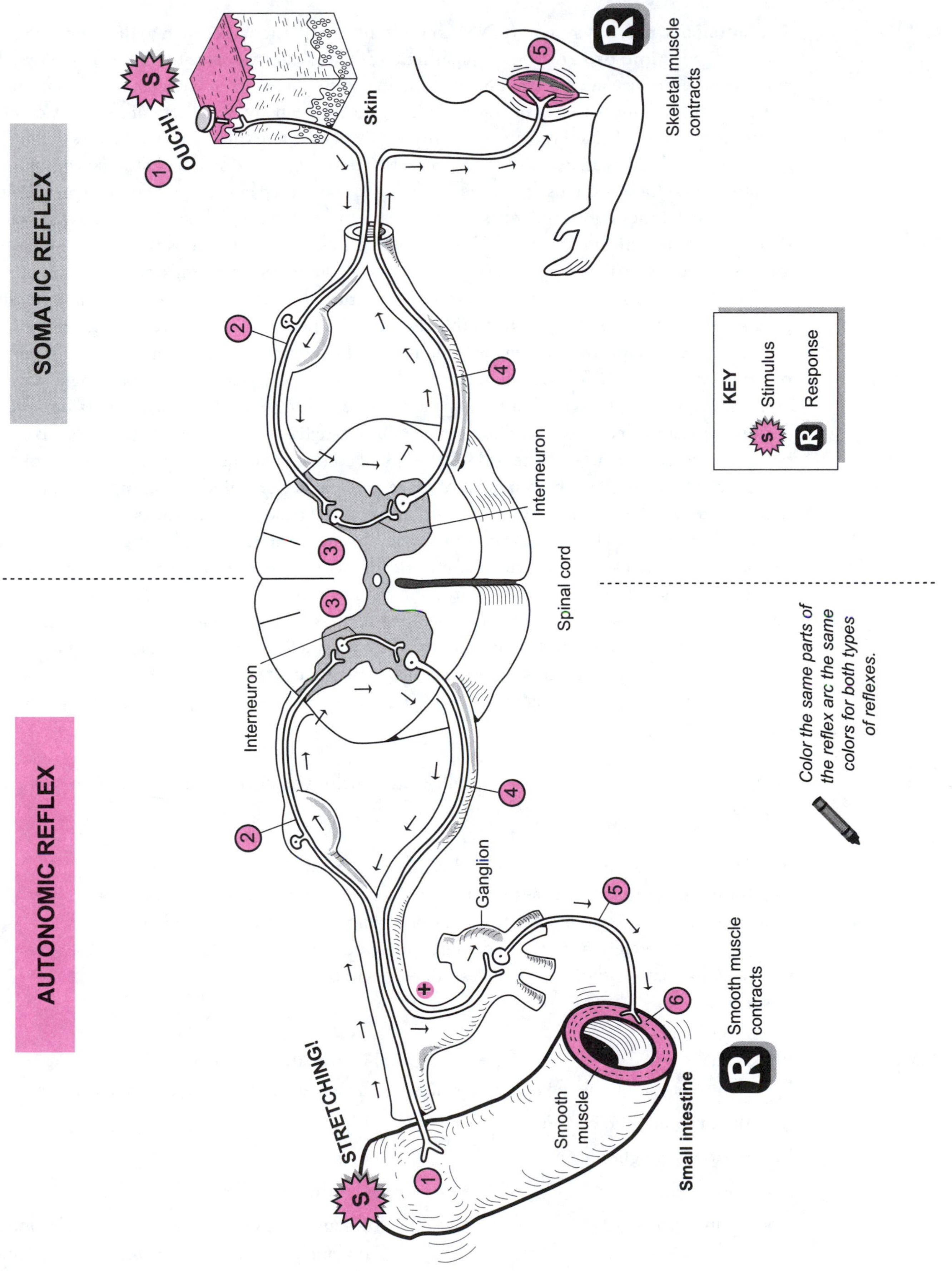

Color the same parts of the reflex arc the same colors for both types of reflexes.

Description

The **autonomic nervous system (ANS)** controls involuntary activities such as digestion and heart rate. It is subdivided into two divisions: sympathetic (SD) and parasympathetic (PD). These two divisions work together to regulate normal functions and maintain homeostasis. Each division has its own groups of nerves that connect to the major visceral organs in the body. But there are some exceptions. For example, the adrenal gland is controlled only by the SD because the PD has no nerves connected to it.

This module gives an overview of the SD, also referred to as the *thoracolumbar division*, as motor output arises from these sections of the spinal cord. Because it prepares the body for emergency situations, it is nicknamed the "fight-or-flight" system. Consider what happens when you get nervous before a test that you are not prepared to take. Your heart rate increases, your pupils dilate, your breathing rate increases, and your palms perspire. All these are a direct result of your sympathetic division in action.

The SD has a more complex anatomical structure than the PD. The illustration on the facing page shows only the motor output to the effector. The nerves involved are like the electrical wiring in a house. They consist of a **preganglionic neuron** linked to a **postganglionic neuron**. Preganglionic neurons arise from the T1–L2 segments of the spinal cord. Notice that the preganglionic axons are quite short, and the postganglionic axons are relatively long. Recall that ganglia (sing. *ganglion*) are clusters of neuron cell bodies. The **paravertebral ganglia**—two columns of ganglia like strings of beads that run along either either side of the spinal cord—are unique to the SD. As part of this chain are the cervical ganglia, which consist of three parts: **superior**, **middle**, and **inferior**. Postganglionic neurons are wired from these hubs to target organs in the head and thoracic cavity.

Three other ganglia—all located near the abdominal aorta and classified as the **prevertebral ganglia** are: (1) **celiac**, (2) **superior mesenteric**, and (3) **inferior mesenteric**. Postganglionic neurons arising from these hubs connect to all the target organs and glands below the diaphragm. The ANS has three possible effectors: smooth muscle, cardiac muscle, and glands.

To clarify the illustration, when you see the postganglionic neurons connecting to organs, they actually are connecting to muscle in the walls of these target organs. Notice that many of the postganglionic neurons pass through **plexuses** (sing. *plexus*), or neural networks, to link with their target organs.

Study Tip

Sympathetic = Stressful. This simple alliteration helps you distinguish the function of the SD from the PD.

Analogy

Regulating the visceral organs is like driving a car. Sometimes you have to step on the gas, and other times you have to step on the brake. Most of the time, the SD is like stepping on the gas. For example, sympathetic impulses sent along the accelerator nerve to the heart increase the heart rate. Unfortunately, it does not work this way with all organs/glands. There are exceptions to this rule.

Summary

Nickname	**Fight-or-flight system**
Location of preganglionic neuron cell bodies:	T1–L2 regions of spinal cord (*thoracolumbar*)
Length of preganglionic axon:	Short
Length of postganglionic axon:	Long
Location of ganglia:	Near the spinal cord
Neurotransmitters:	Preganglionic neurons secrete acetylcholine (ACh). Postganglionic neurons secrete norepinephrine (NE).

Sympathetic Division (SD) of the ANS

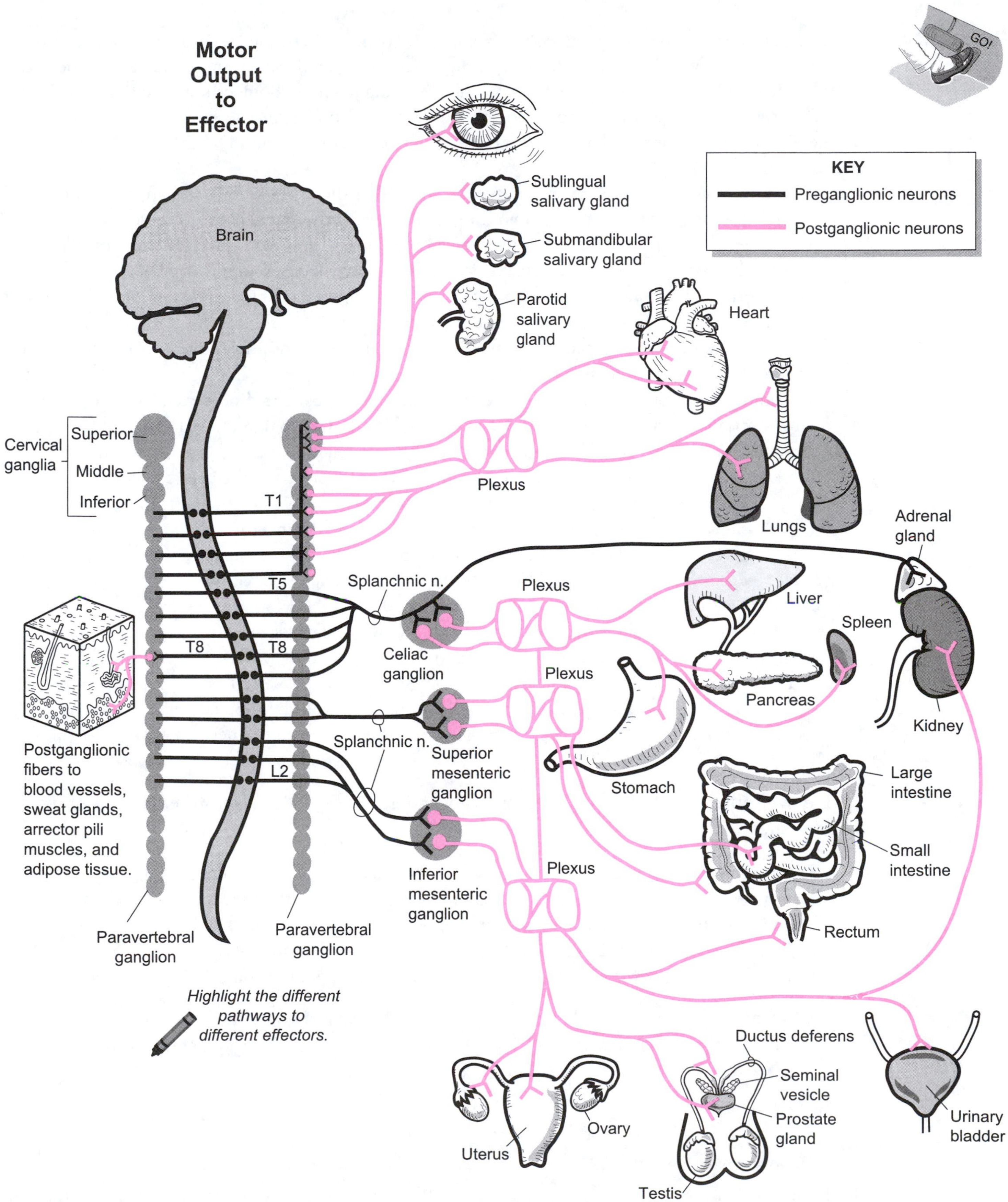

Highlight the different pathways to different effectors.

GO!
Motor Output to Effector
KEY
Preganglionic neurons
Postganglionic neurons
Brain
Sublingual salivary gland
Submandibular salivary gland
Parotid salivary gland
Heart
Lungs
Plexus
Cervical ganglia
Superior
Middle
Inferior
T1
T5
T8
T8
L2
Splanchnic n.
Celiac ganglion
Plexus
Adrenal gland
Liver
Spleen
Pancreas
Kidney
Plexus
Stomach
Splanchnic n.
Superior mesenteric ganglion
Large intestine
Small intestine
Inferior mesenteric ganglion
Plexus
Rectum
Postganglionic fibers to blood vessels, sweat glands, arrector pili muscles, and adipose tissue.
Paravertebral ganglion
Paravertebral ganglion
Highlight the different pathways to different effectors.
Ductus deferens
Seminal vesicle
Prostate gland
Urinary bladder
Uterus
Ovary
Testis

Description

The autonomic nervous system (ANS) controls involuntary activities such as digestion and heart rate. It is subdivided into two divisions: sympathetic (SD) and parasympathetic (PD). These two divisions work together to regulate normal functions and maintain homeostasis. Each division has its own groups of nerves that connect to the major visceral organs in the body. But there are some exceptions. For example, the lacrimal gland is controlled only by the PD, as the SD has no nerves that connect to it.

This module gives an overview of the PD—termed as the craniosacral division, as motor output arises from some *cranial* nerves and from the sacral region of the spinal cord. Because the PD is most active when the body is at rest, it is nicknamed the "resting and digesting" system. Consider what happens when you lie down on the couch after eating a heavy meal. Your digestive function increases, pupils constrict, heart rate decreases, breathing rate decreases, and blood pressure decreases. All of these illustrate your parasympathetic division in action.

The PD has a simpler anatomical structure than the SD. The illustration on the facing page shows only the motor output to the effector. The nerves involved are like the electrical wiring in a house and consist of a **preganglionic neuron** linked to a **post-ganglionic neuron**. Preganglionic neurons arise from the following cranial nerves (CN): **III** (oculomotor), **VII** (facial), **IX** (glossopharyngeal), and **X** (vagus) In the sacral region of the spinal cord, preganglionic neurons arise from **S2–S4** to become the **pelvic nerves**, which connect to target organs in the pelvic region.

As a general rule, the preganglionic axons tend to be quite long while the postganglionic axons are relatively **short**. Recall that ganglia (sing. ganglion) are clusters of neuron cell bodies. Four important ganglia are located in the head, where preganglionic neurons link to postganglionic neurons: (1) **ciliary**, (2) **pterygopalatine**, (3) **submandibular**, and (4) **otic**. These major hubs connect to target organs in the head, such as the eye, lacrimal gland, and salivary glands. The ANS has three possible effectors: *smooth muscle*, *cardiac muscle*, and *glands*. To clarify the illustration, when you see postganglionic neurons connecting to organs, they actually are connecting to muscle tissue in the walls of these organs.

The most important nerve in the PD is the **vagus nerve** (CN X) because it accounts for as much as 80% of all the preganglionic axons in the PD. It innervates all the thoracic organs and most of the abdominal organs. Notice that it runs through **plexuses** (sing. *plexus*) or neural networks to link with all its target organs.

Study Tip

Parasympathetic = **P**eaceful. This simple alliteration helps you distinguish the function of the SD from the PD.

Analogy

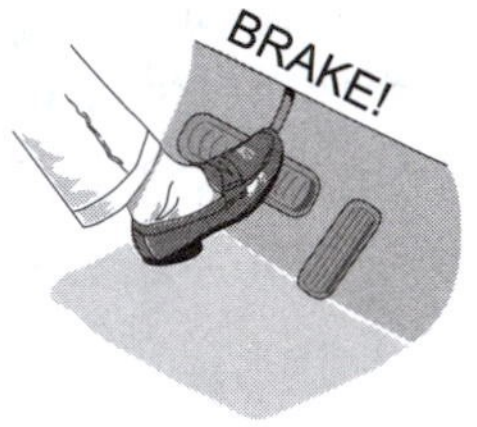

Regulating the visceral organs is like driving a car. Sometimes you have to step on the gas, and other times you have to step on the brake. Most of the time, the PD is like stepping on the brake. For example, parasympathetic impulses sent along the vagus nerve to the heart decrease the heart rate. Unfortunately, it does not work this way with all organs/glands. There are exceptions to this rule.

Summary

Nickname	**"Resting and Digesting" system**
Location of preganglionic neuron cell bodies:	Brainstem and S2–S4 regions of spinal cord (*craniosacral*)
Length of preganglionic axon:	Long
Length of postganglionic axon:	Short
Location of ganglia:	Usually near or within the wall of the target organ
Neurotransmitters:	Preganglionic neurons secrete acetylcholine (ACh). Postganglionic neurons secrete acetylcholine (ACh).

Parasympathetic Division (PD) of the ANS

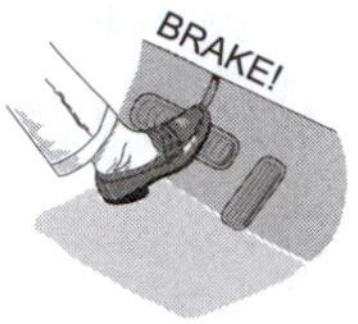

Motor Output to Effector

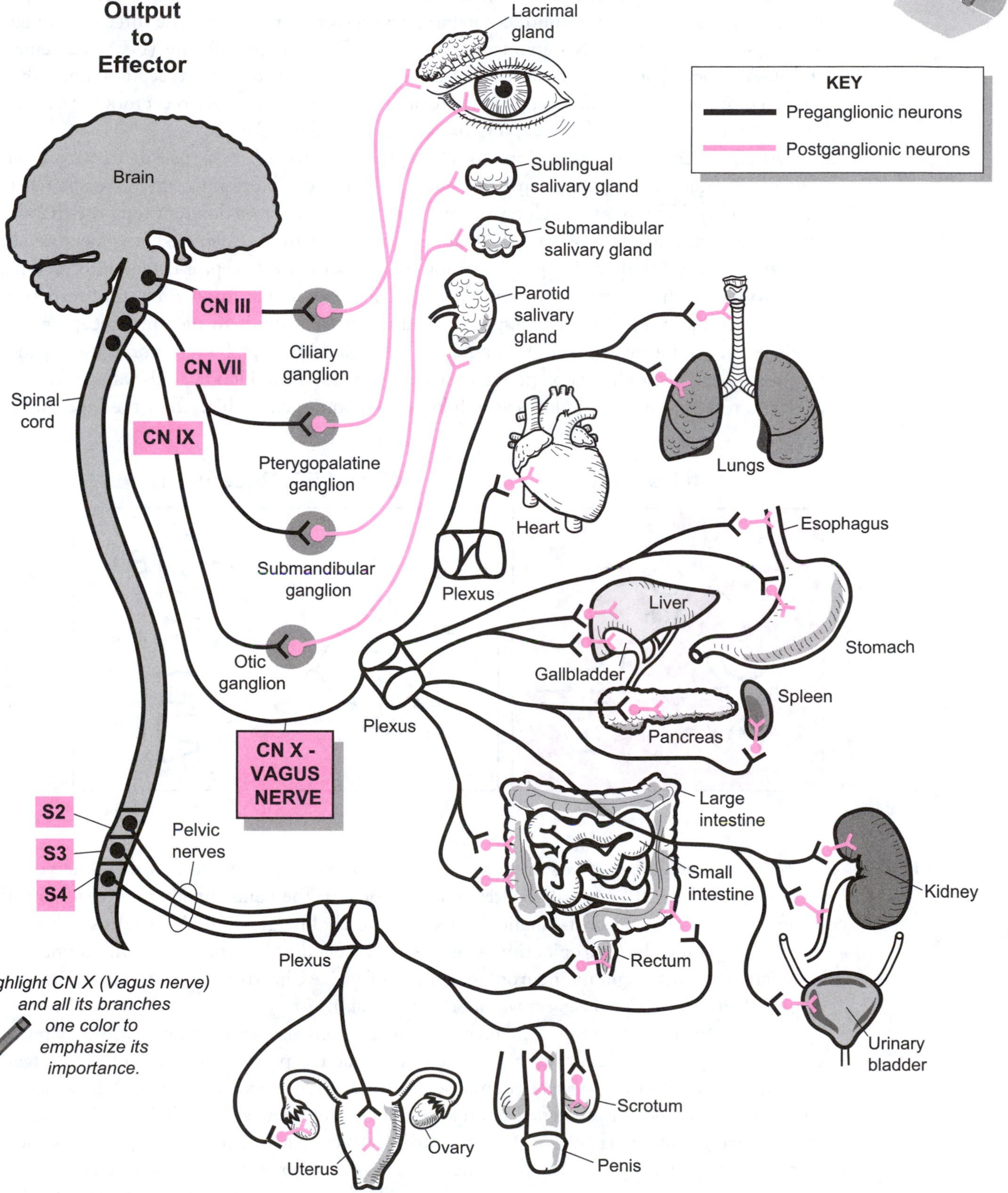

Highlight CN X (Vagus nerve) and all its branches one color to emphasize its importance.

Description

The autonomic nervous system (ANS) uses reflexes to regulate many of the unconscious activities we take for granted, such as digestive activities, heart rate, and pupillary responses. These responses are induced by specific neurotransmitters combining with specific receptors. The three major neurotransmitters secreted in the ANS are: **acetylcholine (ACh)**, **norepinephrine (NE)**, and **epinephrine (EPI)**. Acetylcholine binds to either **muscarinic (M)** or **nicotinic (N) receptors**, and NE and EPI bind to either alpha or beta receptors. There are two types of **alpha (∝) receptors** (*alpha 1* and *alpha 2*) and three kinds of **beta (β) receptors** (*beta 1*, *beta 2*, and *beta 3*).

When a neurotransmitter binds to a receptor, it induces a response in its target cells. The response is determined by which specific neurotransmitter combines with which specific receptor. For example, when **NE** binds to an alpha 1 receptor, it typically elicits a different response than when NE binds to a beta 1 receptor. Although there are exceptions to the rule, we can generalize about the different responses that are induced by various neurotransmitter/receptor combinations. When ACh binds to muscarinic receptors, for example, it sometimes results in an excitatory response. Other times, it results in an inhibitory response. And when ACh binds to nicotinic receptors, it usually produces an excitatory response. When NE (or EPI) binds to an alpha receptor, it usually induces an excitatory response such as constriction of smooth muscle within blood vessels walls. When NE binds to a beta receptor, it often induces an inhibitory response such as dilation of the bronchioles in the lungs.

ACh Receptors **NE and EPI Receptors**

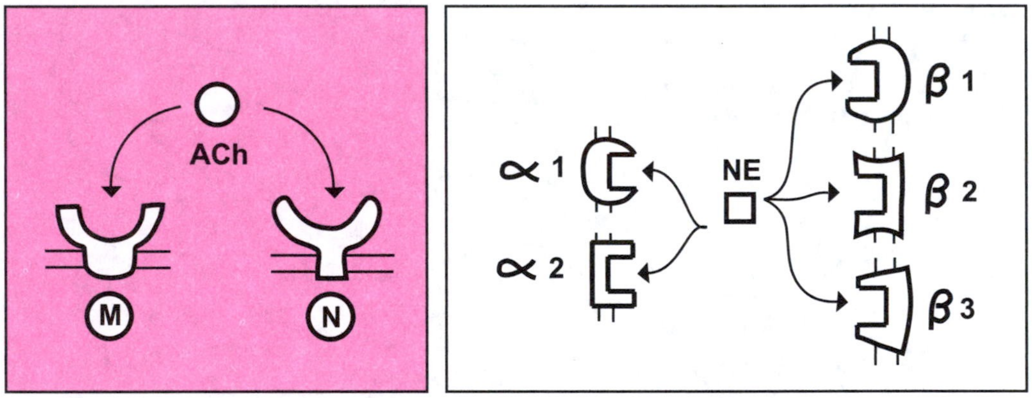

*Color the **neurotransmitters** and their **receptors** different colors.*

Cholinergic and Adrenergic Neurons

In the ANS, nerve fibers that secrete *only* ACh are called **cholinergic neurons**, and nerve fibers that secrete *only* NE or EPI are called **adrenergic neurons**. In the parasympathetic division, all the **preganglionic** neurons and **post-ganglionic** neurons are cholinergic. The ACh released from the preganglionic neurons binds to **nicotinic neurons** in the dendrites of the postganglionic neurons. This stimulates the **postganglionic neuron**, in turn, to release ACh, which binds to **muscarinic receptors** in the **effector** (*smooth muscle*, *cardiac muscle*, or a *gland*).

In the sympathetic division, the preganglionic neurons also are cholinergic, so they release ACh into the synapse. The ACh then binds to the nicotinic receptors in dendrites of the postganglionic neurons. Because all the postganglionic neurons are adrenergic, they release NE, which binds to alpha or beta receptors in the effector. In a special case, the sympathetic division also directly innervates hormone-producing cells in the inner medulla of the adrenal gland. The preganglionic neurons secrete ACh, which binds to the nicotinic receptors in the hormone-producing cells. This stimulates the secretion of both NE and EPI into the blood. These hormones travel through the bloodstream to their target cells, which contain either alpha or beta receptors.

ANS Neurotransmitters and Their Receptors

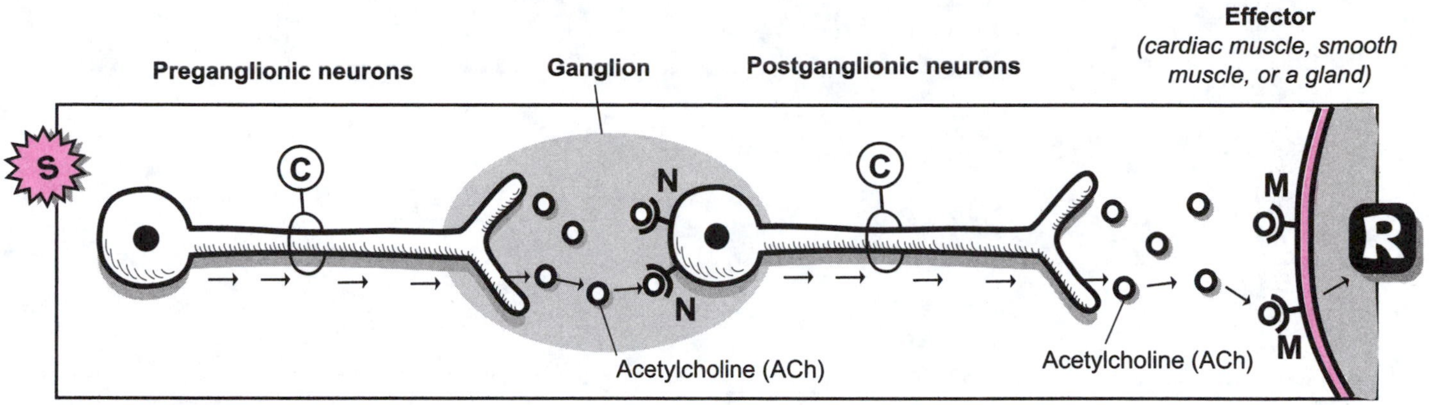

PARASYMPATHETIC Division

Color the preganglionic neurons, postganglionic neurons, and neurotransmitters different colors.

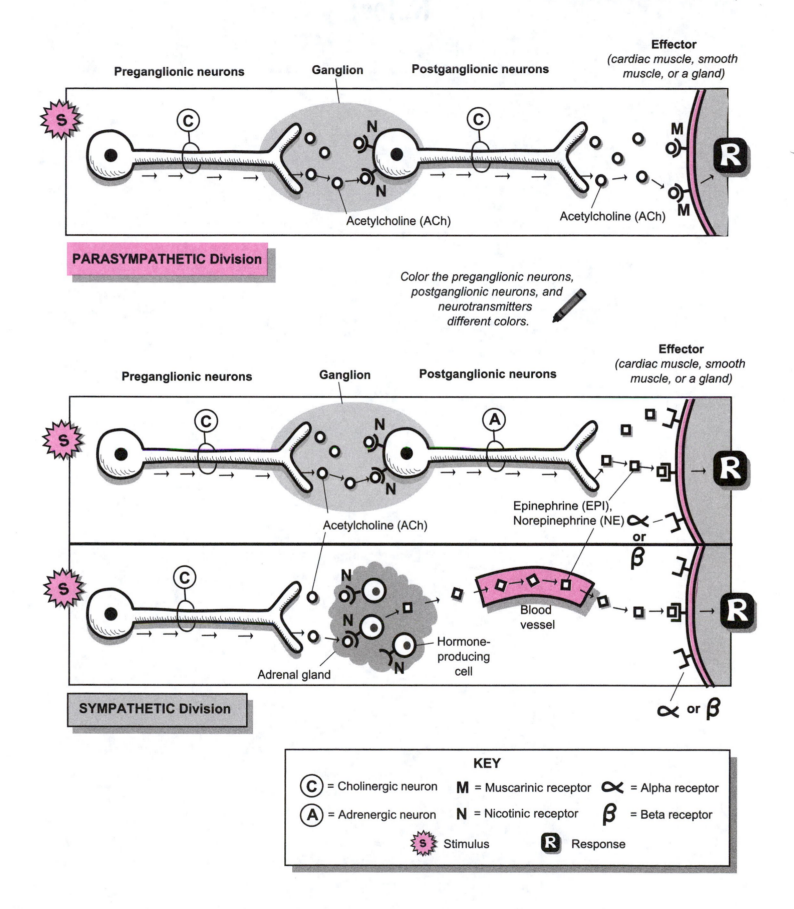

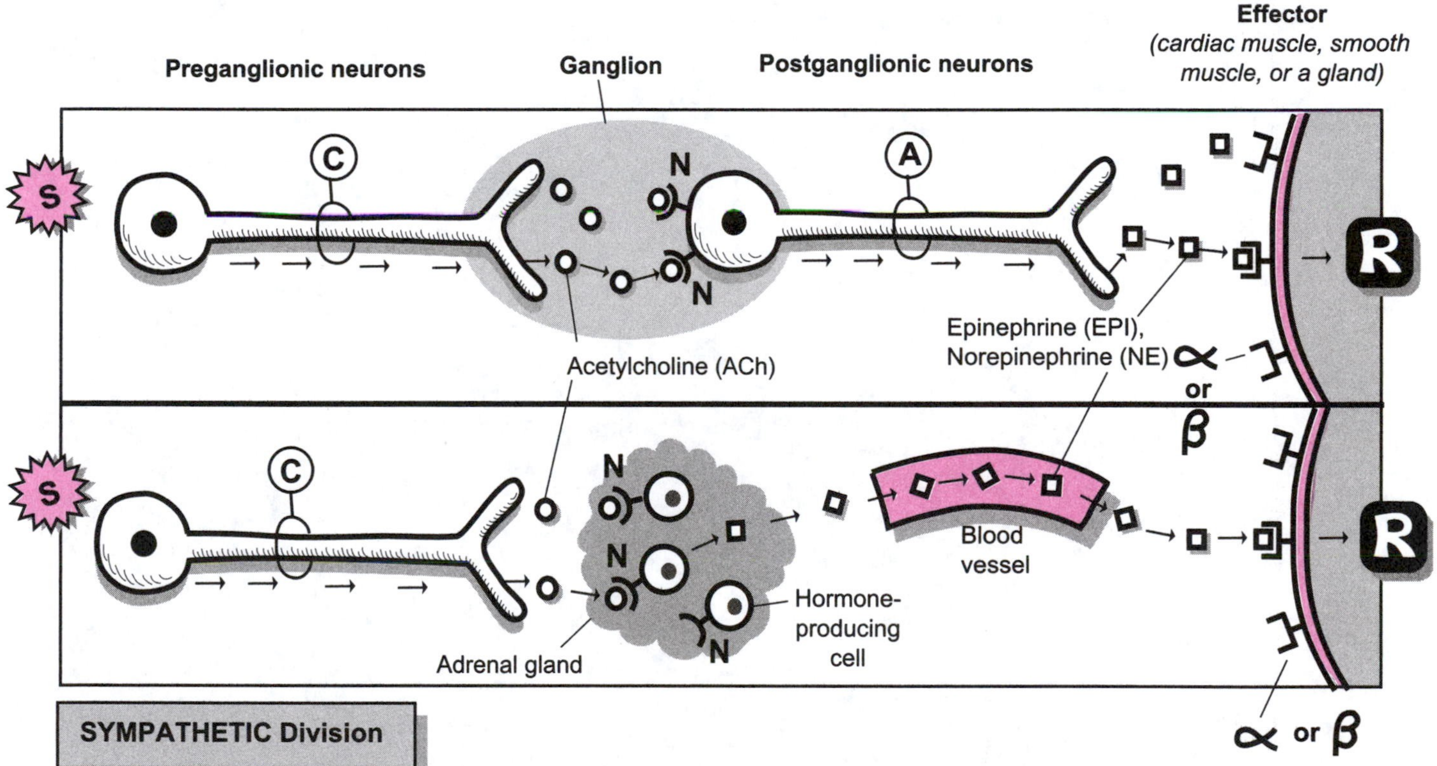

SYMPATHETIC Division

KEY

Ⓒ = Cholinergic neuron	M = Muscarinic receptor	α = Alpha receptor
Ⓐ = Adrenergic neuron	N = Nicotinic receptor	β = Beta receptor
Ⓢ Stimulus	Ⓡ Response	

Notes

Endocrine System

Description

Along with the nervous system, the endocrine system is one of the great regulators of activities in the human body. Its major purpose is to maintain homeostasis and to regulate various processes such as growth and human development. It consists of many different glands containing specific cells that synthesize and release chemical messengers called hormones. These hormones enter the bloodstream, where they travel to a specific cell called a target cell. This target cell has a receptor for that specific hormone. Once the hormone binds to the receptor, it induces a response in that cell. The general concept of how the endocrine system functions is illustrated below:

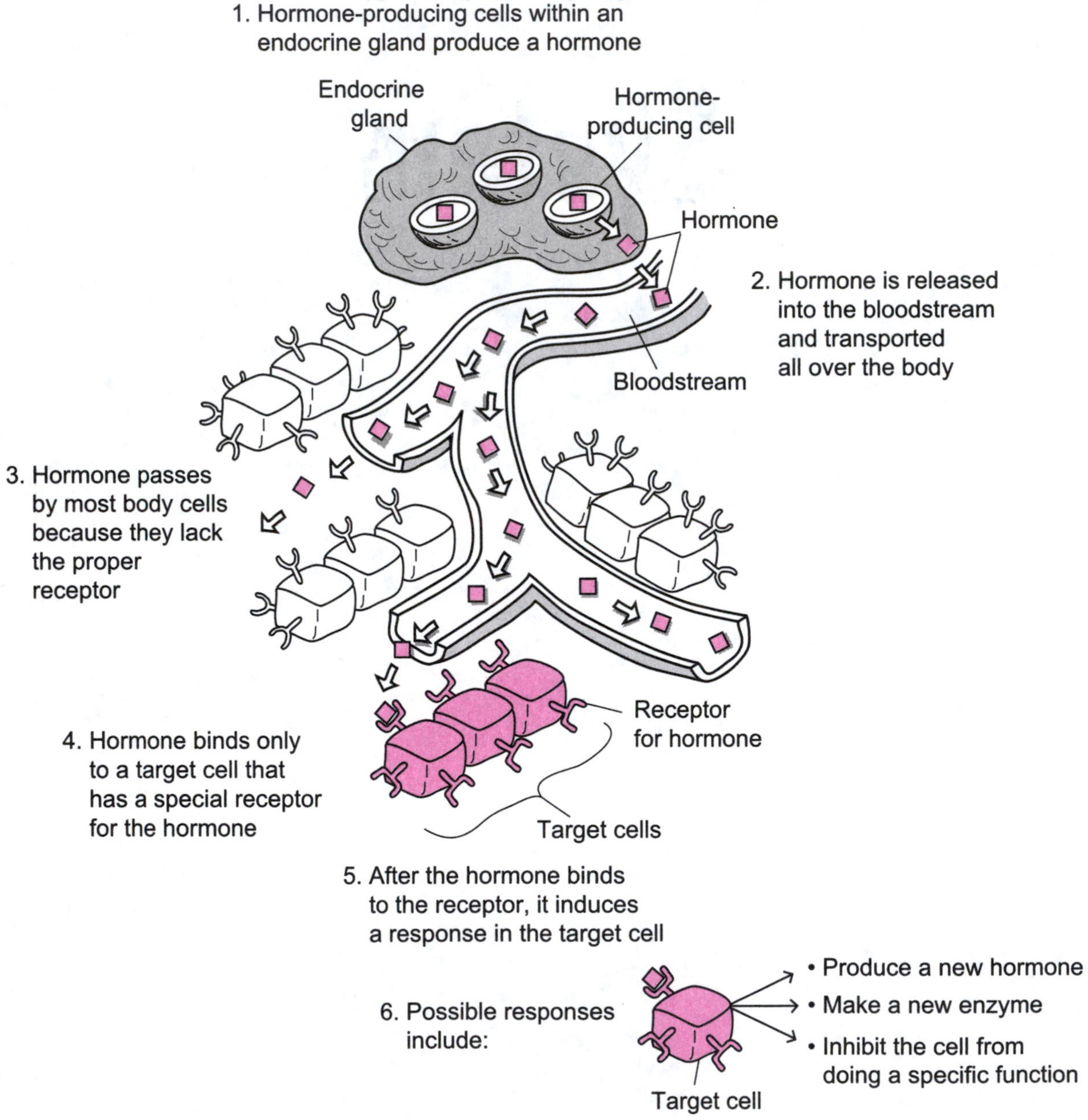
1. Hormone-producing cells within an endocrine gland produce a hormone
Endocrine gland
Hormone-producing cell
Hormone
2. Hormone is released into the bloodstream and transported all over the body
Bloodstream
3. Hormone passes by most body cells because they lack the proper receptor
Receptor for hormone
4. Hormone binds only to a target cell that has a special receptor for the hormone
Target cells
5. After the hormone binds to the receptor, it induces a response in the target cell
6. Possible responses include:
• Produce a new hormone
• Make a new enzyme
• Inhibit the cell from doing a specific function
Target cell

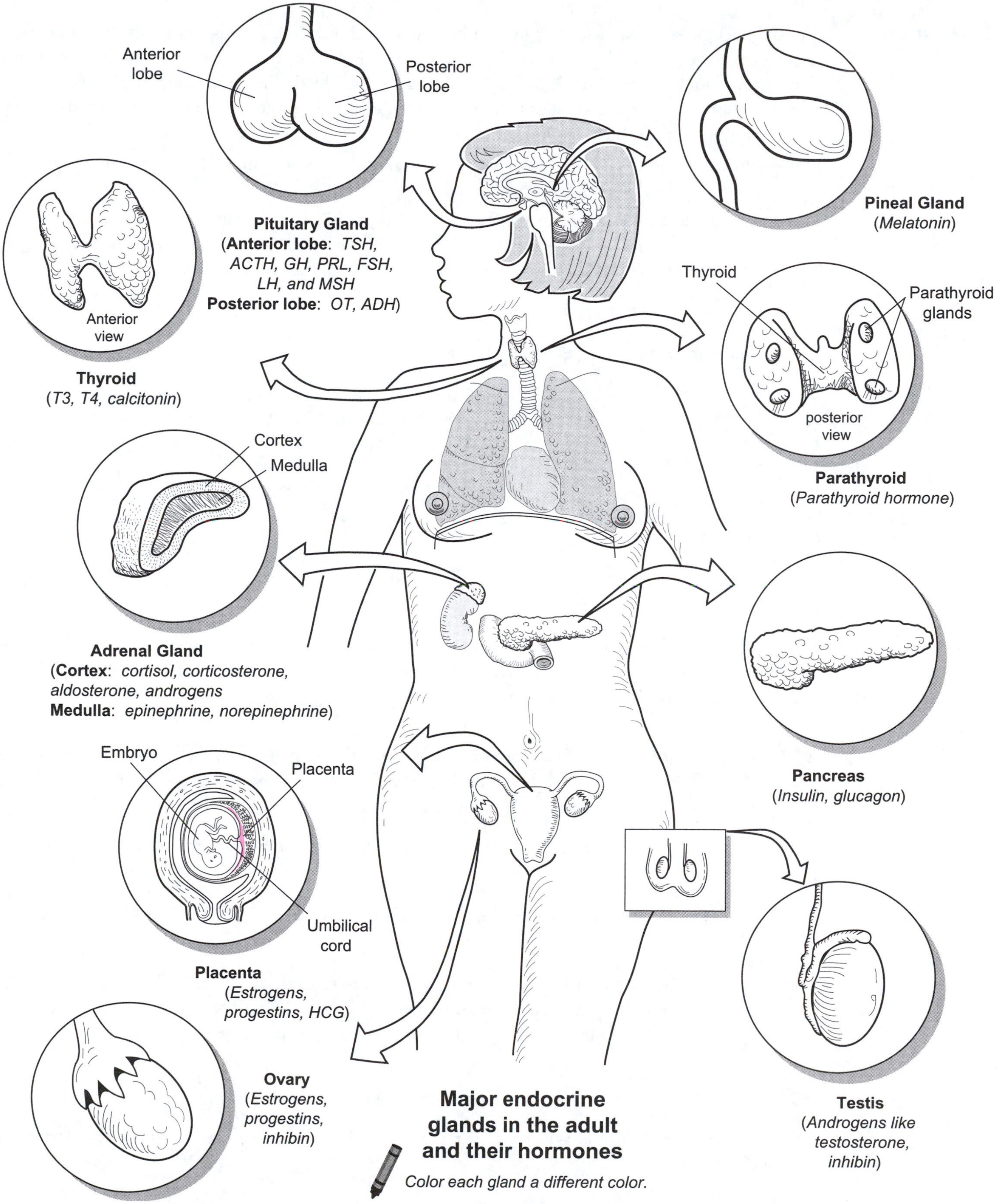

Major endocrine glands in the adult and their hormones

Color each gland a different color.

Description

Steroid hormones are lipid-soluble and synthesized from cholesterol. Because they must be transported through the watery blood, they need special transporters, called **protein carriers**. These carriers also prevent hormones from being degraded by enzymes in the blood. Once the hormones arrive at their target cells, they enter the nucleus to stimulate DNA to begin protein synthesis. The effects of any steroid hormone have the following key features:

- Slower action: They produce a response in hours or days after initial binding of hormone with its receptor—slower than non-steroid hormone.
- No amplification: The effects produced are proportional to the amount of hormone secreted; a greater amount of hormone is needed to induce a response, compared to a non-steroid, in which the effects of a hormone are amplified by a chain reaction.

Steroid Hormones

Examples of steroid hormones are given in the table below.

Steroid Hormones	Site of Secretion
• Aldosterone • Cortisol • Androgens (e.g., testosterone)	Adrenal glands (cortex)
• Testosterone	Testes
• Estrogens • Progesterone	Ovaries

Sequence

The following sequence of events is a simplified mechanism describing how steroid hormones induce a response in their target cells, illustrated on the facing page. Note that thyroid hormones (T_3 and T_4) use a similar mechanism of action.

1. The steroid hormone travels through the blood with the help of a protein carrier. After being released from this carrier, the hormone diffuses through the plasma membrane, into the **cytoplasm**, and into the **nucleus** through the **nuclear pore**.
2. Once inside the nucleus, the hormone binds with its **hormone receptor** to form a **hormone-receptor complex**.
3. The hormone-receptor complex binds to a **shape-specific receptor on DNA** like a key in a lock.
4. This stimulates DNA to begin the process of protein synthesis. First, DNA makes a copy of itself in the form of a mobile, single-stranded molecule called **mRNA** (messenger RNA). The smaller mRNA is able to leave the nucleus through the nuclear pore and it moves into the cytoplasm.
5. The mRNA molecule binds to the **ribosome**, the protein factory of the cell.
6. The message in mRNA is read by the ribosome, and it links free **amino acids** together to form a new protein.

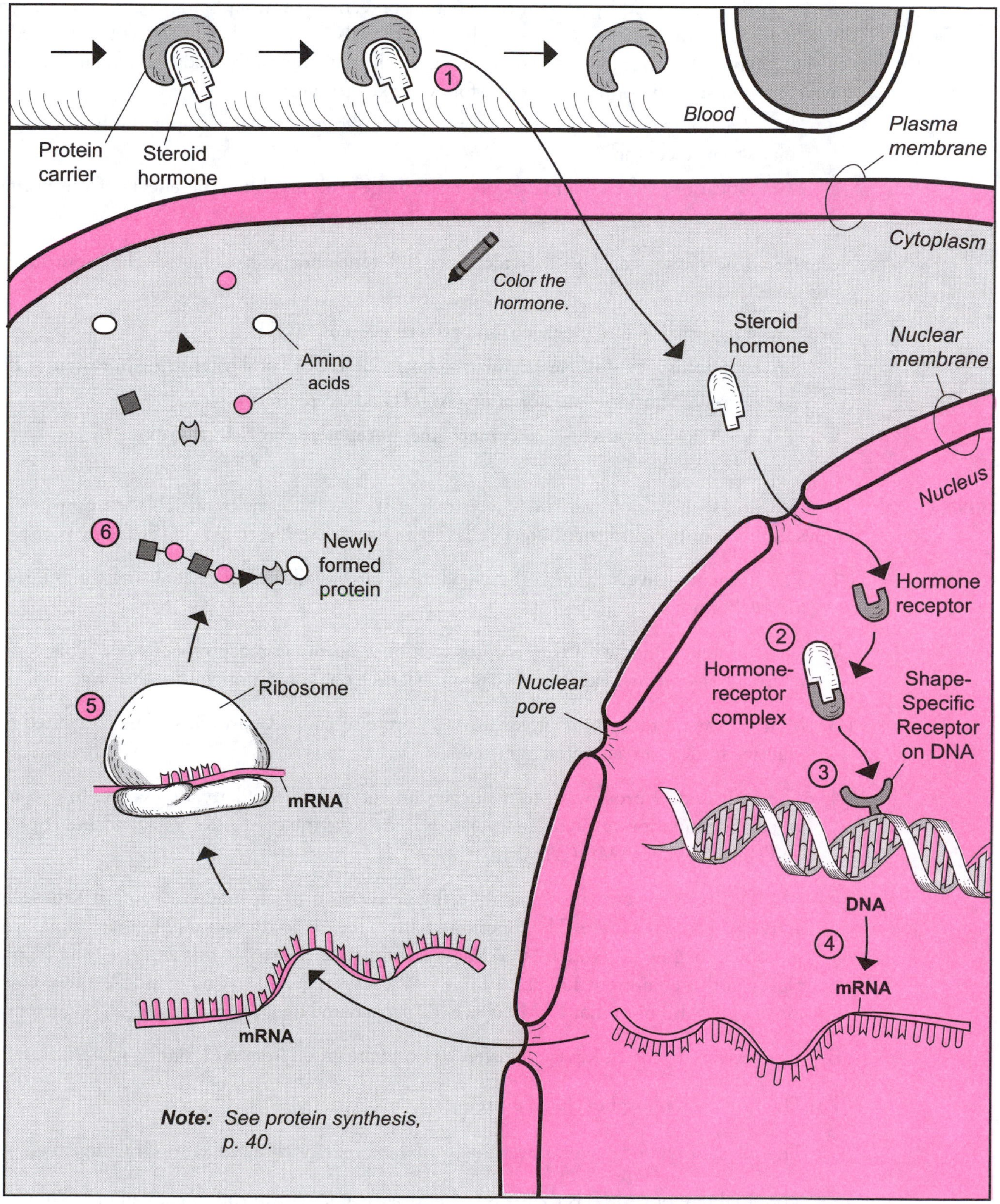
Protein carrier
Steroid hormone
1
Blood
Plasma membrane
Cytoplasm
Color the hormone.
Amino acids
Steroid hormone
Nuclear membrane
Nucleus
6
Newly formed protein
Hormone receptor
2
Hormone-receptor complex
Shape-Specific Receptor on DNA
Nuclear pore
5
Ribosome
mRNA
3
DNA
4
mRNA
mRNA
Note: See protein synthesis, p. 40.

Description

Non-steroid hormones are water-soluble, so they are easily transported through the blood. Many are protected and transported by protein carriers that prevent them from being degraded by enzymes in the blood. Once the hormones arrive at their target cells, they first bind at receptors located at the cell membrane. This triggers a cascade inside the cell that eventually leads to cellular changes. The effects of a non-steroid hormone have the following key features:

- Rapid action: They produce a response within *seconds* or *minutes* after the hormone initially binds to its receptor.
- Amplification: The chain reaction in the target cell amplifies the effects of the hormone; a little hormone produces a big response.

Non-steroid hormones can be subdivided into different chemical categories. The four major subgroups are:

- Proteins—ex: **insulin**, **glucagon**, and **growth hormone** (GH)
- Glycoproteins—ex: **follicle-stimulating hormone** (FSH), and **luteinizing hormone** (LH)
- Peptides—ex: **antidiuretic hormone** (ADH) and **oxytocin** (OT)
- Amino Acid Derivatives—ex: **epinephrine**, **norepinephrine**, and **thyroxine** (T_4)

Mechanism

The following sequence of events describes one of the mechanisms by which *some* non-steroid hormones induce a response in their target cells. These events are illustrated on the facing page.

1. The **hormone** travels through the blood to its target cell, which contains membrane **receptors** for the hormone.
2. The hormone binds with the receptor to form a **hormone-receptor complex**. This complex is referred to as the *first messenger* because it begins a chain reaction within the target cell.
3. Many hormone-receptor complexes trigger proteins called **G proteins** to be converted from an inactive form to an activated form.
4. The **activated G protein**, in turn, triggers an enzyme called **adenylate cyclase** to become activated. The function of adenylate cyclase is to catalyze the conversion of **adenosine triphosphate** (ATP) into **cyclic AMP (cAMP)**.
5. **cAMP** acts as an enzyme to catalyze the conversion of an **inactive protein kinase** into an **activated protein kinase**. The function of any kinase is to transfer a phosphate group from one substance to another. Note that cAMP is referred to as the *second messenger* because it's a critical player in this chain reaction and a threshold level is required to finally induce a response in the target cell. Also note that cAMP is not the *only* second messenger used within all target cells.
6. The **activated protein kinase** transfers a phosphate group from ATP onto a protein.
7. The result is a **phosphorylated protein**.
8. The phosphorylated proteins eventually produce cellular changes within the target cell.
9. The final fate of **cAMP** is that it is either deactivated within the target cell or it diffuses out of the cell. This ensures that the chain reaction stops.

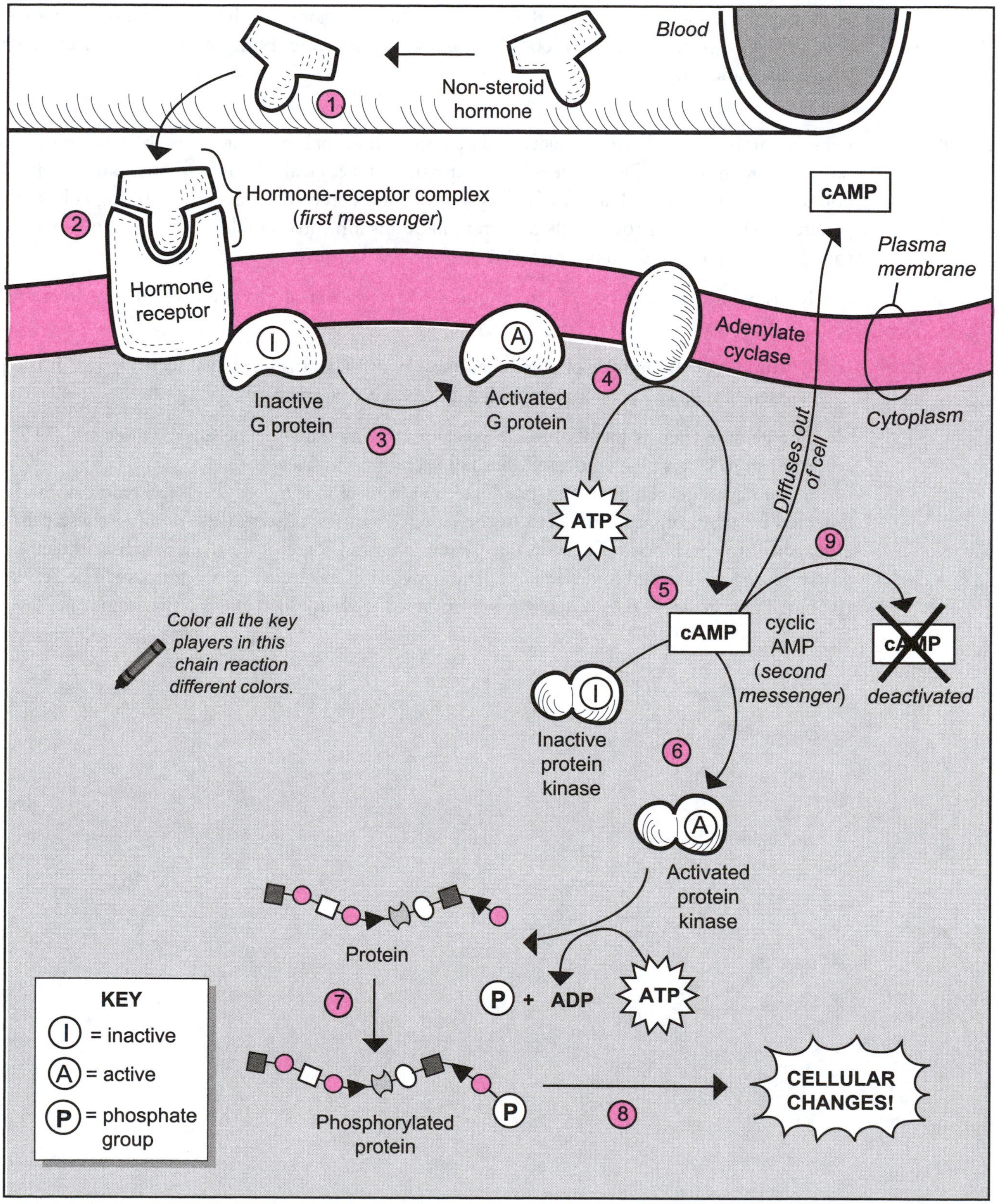
Blood
Non-steroid hormone
1
2
Hormone-receptor complex (first messenger)
Hormone receptor
cAMP
Plasma membrane
I
Inactive G protein
3
A
Activated G protein
4
Adenylate cyclase
Cytoplasm
Diffuses out of cell
ATP
5
9
cAMP
cyclic AMP (second messenger)
cAMP deactivated
Color all the key players in this chain reaction different colors.
I
Inactive protein kinase
6
A
Activated protein kinase
Protein
7
P + ADP
ATP
KEY
I = inactive
A = active
P = phosphate group
Phosphorylated protein
P
8
CELLULAR CHANGES!

Description

Regulation of blood glucose levels is controlled by two hormones: **insulin** and **glucagon**. These two hormones are antagonists because they have opposite functions. Insulin causes blood glucose levels to decrease while glucagon causes **blood glucose levels** to increase. Both hormones are produced by cells within the **pancreas**.

Mechanism

Let's examine the mechanism of action. The normal level of blood glucose is about 90 mg/100 ml and has to be maintained. Glucose levels typically rise after a meal rich in carbohydrates, such as baked potatoes or corn flakes. These high glucose levels are detected by insulin-secreting cells within the pancreas. This triggers these cells to secrete more insulin into the blood. As insulin travels through the blood, it induces two major responses:

1. stimulates the storage of excess glucose in the form of glycogen within the liver and some other tissues.
2. stimulates the uptake of glucose by most body cells. Insulin binds to receptors in the plasma membrane, which opens channels for glucose to pass through.

The glucose then is metabolized to produce large amounts of the energy molecule ATP. Insulin secretion stops when blood glucose levels fall back to normal levels.

If you haven't eaten anything for a long time, your blood glucose levels fall below normal. This is detected by glucagon-secreting cells in the pancreas. In response to this stimulus, these cells secrete **glucagon** into the blood. Glucagon targets the liver and some other tissues such as skeletal muscle, where glycogen is stored and stimulates the conversion of glycogen into **glucose**. The result is that the blood glucose level rises. Glucagon secretion stops when blood glucose rises to normal levels.

Hormonal Regulation of Blood Glucose Levels

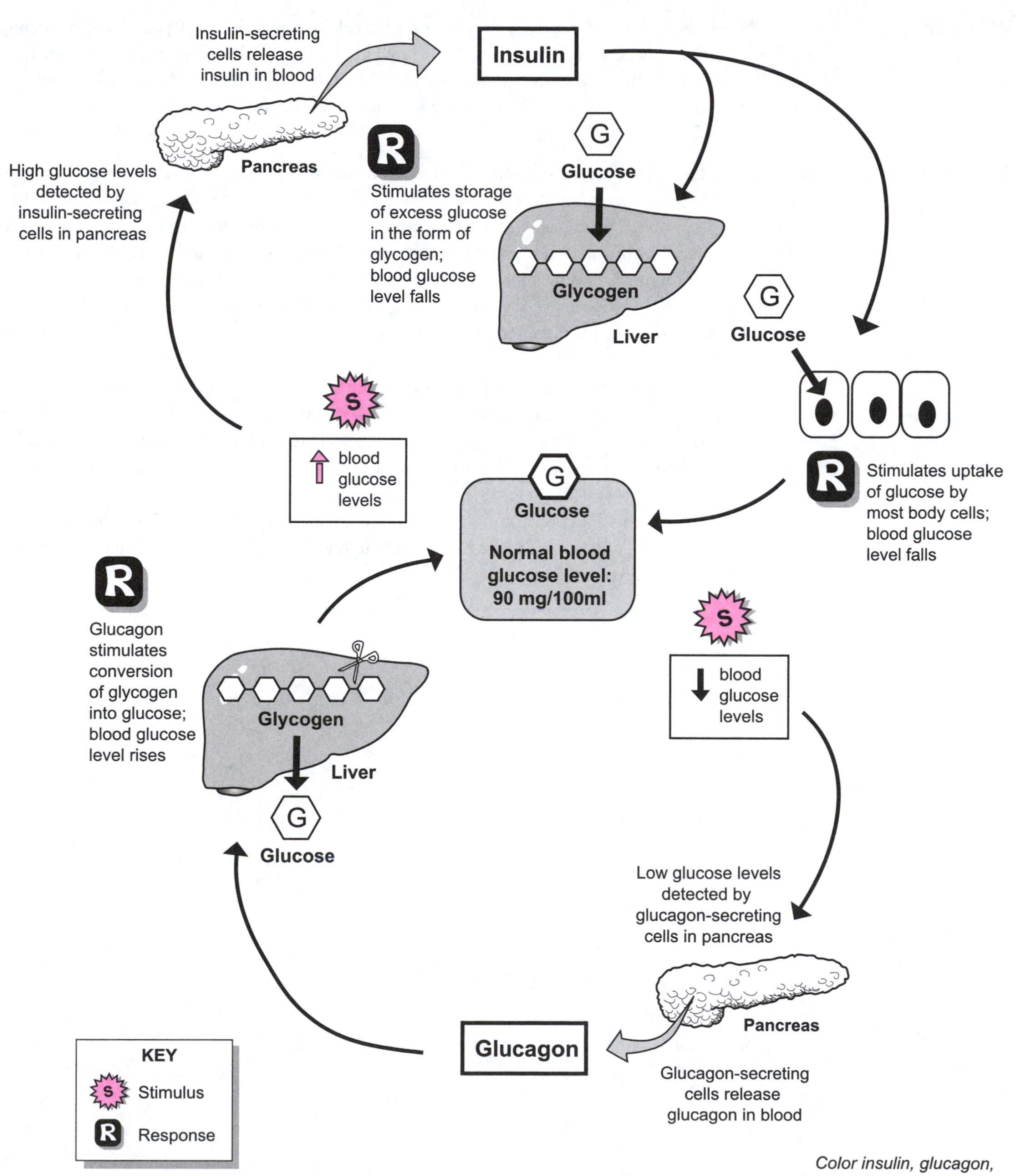
Insulin-secreting cells release insulin in blood
Insulin
Pancreas
High glucose levels detected by insulin-secreting cells in pancreas
R
Stimulates storage of excess glucose in the form of glycogen; blood glucose level falls
G
Glucose
Glycogen
Liver
G
Glucose
R
Stimulates uptake of glucose by most body cells; blood glucose level falls
S
blood glucose levels
G
Glucose
Normal blood glucose level: 90 mg/100ml
R
Glucagon stimulates conversion of glycogen into glucose; blood glucose level rises
Glycogen
Liver
G
Glucose
S
blood glucose levels
Low glucose levels detected by glucagon-secreting cells in pancreas
Pancreas
Glucagon
Glucagon-secreting cells release glucagon in blood
KEY
S Stimulus
R Response

Color insulin, glucagon, glucose, glycogen, and the pancreas.

Description

Blood calcium levels are regulated by two hormones: **calcitonin (CT)** and **parathyroid hormone (PTH)** or parathormone. CT is produced by the **thyroid gland**, and PTH is produced by the **parathyroid glands**. These two hormones are antagonists because they have opposite functions: CT causes blood calcium levels to decrease, and PTH causes blood calcium levels to increase.

Mechanism

Let's examine the mechanism of action. The normal level of blood calcium is in the range of 9–11 mg/100 ml, which has to be maintained. As calcium-rich foods such as milk, broccoli, and tofu are ingested, blood calcium levels rise. Calcitonin-producing cells within the thyroid gland sense this, and they secrete CT into the blood. CT targets the bone-forming cells called **osteoblasts** within the skeletal system. The plasma membranes in osteoblasts contain receptors for CT. Once CT binds to this receptor, it triggers the cells to deposit calcium salts into bone throughout the skeletal system. This causes the blood calcium levels to fall. CT stops being produced when blood calcium levels return to normal.

If you haven't eaten any calcium-rich foods or taken any calcium supplements for a long time, your blood calcium levels fall. This is detected by the two pairs of parathyroid glands located on the posterior surface of the thyroid gland, which are stimulated to secrete PTH into the blood. PTH targets **osteoclasts**—bone degrading cells within the skeletal system. Osteoclasts contain receptors for PTH.

Once PTH docks at its PTH receptor, it stimulates the osteoclasts to decompose bone and release calcium into the blood. Consequently, the blood calcium level rises. PTH stops being produced when blood calcium levels return to normal.

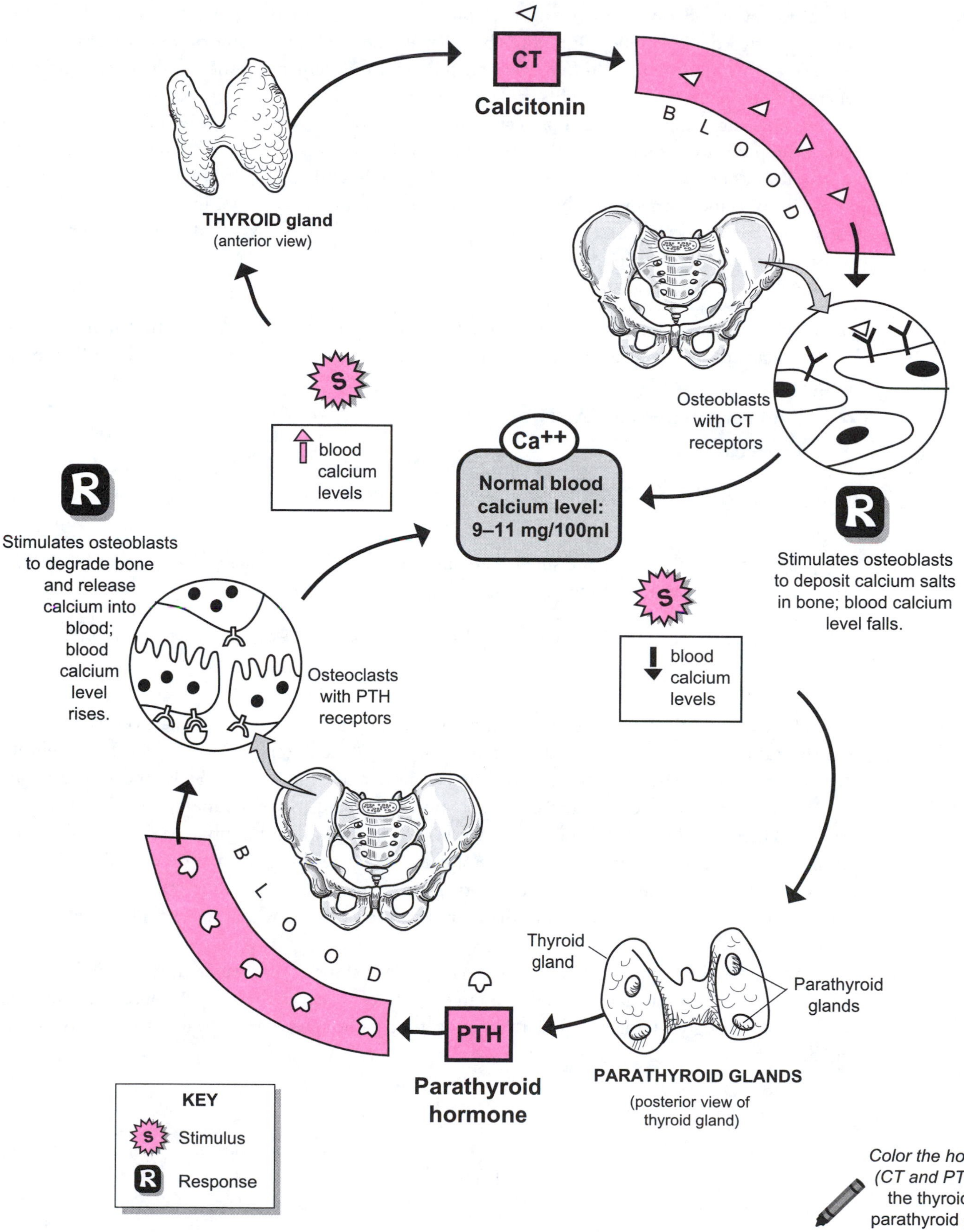

Color the hormones (CT and PTH) and the thyroid and parathyroid glands.

Description

The **pituitary gland** (*hypophysis*) is a small but powerful gland located at the base of the brain. It consists of two distinct lobes: anterior lobe (*adenohypophysis*) and **posterior lobe** (*neurohypohysis*). Because each lobe functions differently and produces different hormones, each is presented in a separate module. This one highlights the **posterior pituitary**.

Surrounding the pituitary is a protective pocket of bone called the sella turcica of the sphenoid bone. The pituitary is nicknamed the "master gland" because it makes many different hormones that control the other endocrine glands in the body. But the **hypothalamus** in the brain actually controls the pituitary gland with specific bundles of neurons. This provides a vital link between the nervous system and the endocrine system. The pituitary is connected to the hypothalamus by a slender stem-like structure called the **infundibulum** (*pituitary stalk*).

Hormones Produced

The cells in the posterior pituitary do *not* produce any hormones. Instead, neurons originating in the hypothalamus that extend into the posterior pituitary produce, store, and release the following two hormones:

- **Antidiuretic hormone (ADH)**
- **Oxytocin (OT)**

In short, both ADH and OT are released by **neurosecretion**.

Sequence

The posterior pituitary contains a simple **capillary network**. The following flowchart indicates blood flow through the posterior pituitary:

Inferior hypophyseal a. → capillary network → Posterior hypophyseal v.

Mechanism

Two different clusters of neurons in the hypothalamus produce hormones: the **paraventricular nuclei** and the **supraoptic nuclei**. The former produces OT, and the latter produces ADH. Each neuron in the cluster is a long, slender cell that has a kind of head, and a tail, the **cell body**, the **axon terminal**.

Hormones are produced in the cell body and then stored in membrane-bound chambers called vesicles. These vesicles travel through the neurons and into the axon terminals located in the posterior pituitary. Nerve impulses trigger the release of these hormones from the axon terminals and into the blood. ADH and OT diffuse into the capillary network and exit the pituitary gland via the posterior hypohyseal vein. As they travel through the blood, they bind to receptors in their target cells and induce a response.

For example, ADH targets cells in the kidney to reabsorb more water into the blood, thereby decreasing urine output. OT has multiple targets, one of which is the smooth muscle in the wall of the uterus. During childbirth, OT is released to trigger the muscular contractions needed for labor and delivery.

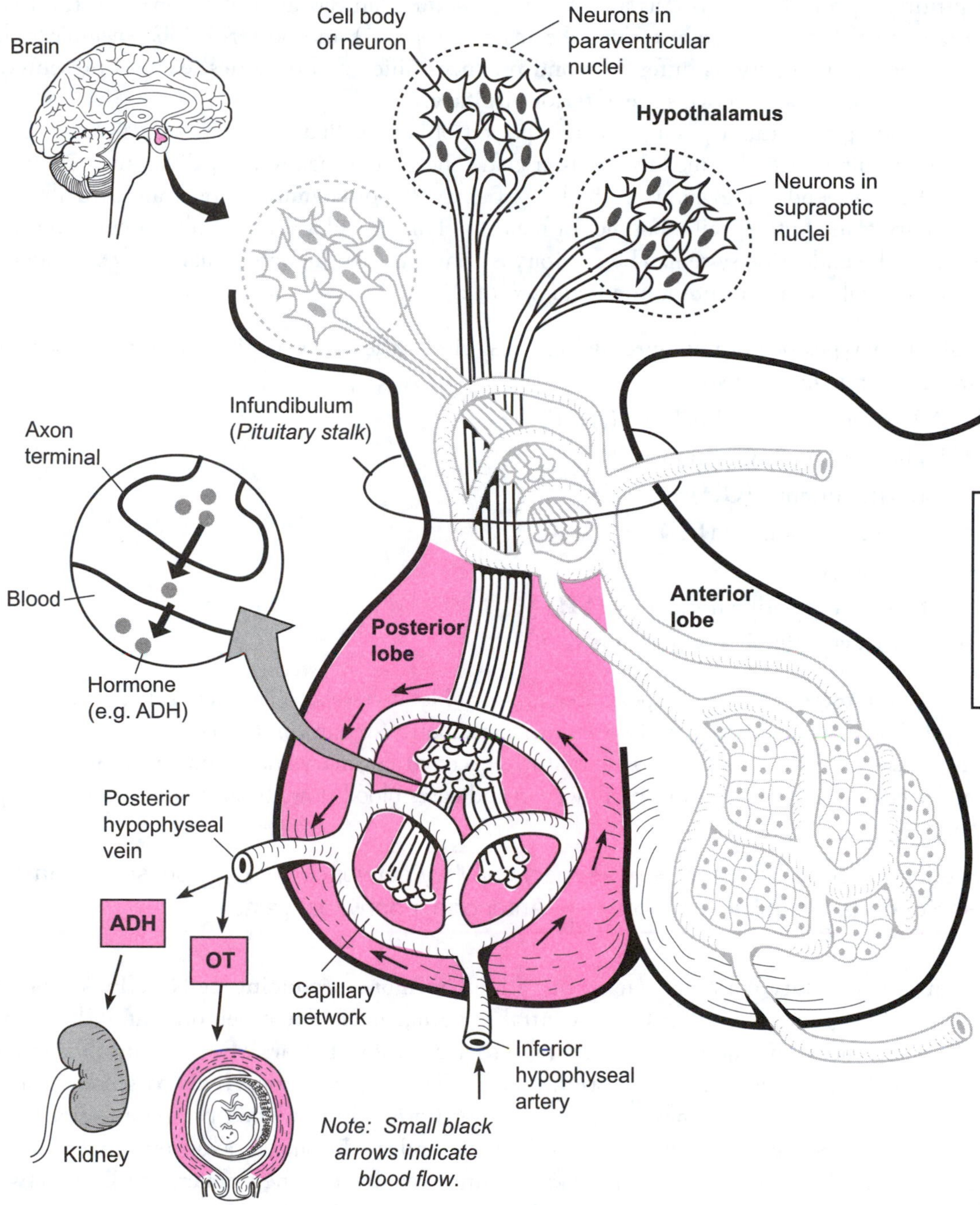

Neurosecretion
Is the key
To how the
Poster lobe
Works,
You see!

Color the neurons and blood vessels different colors.

Description

The **pituitary gland** (*hypophysis*) is a small but powerful gland located at the base of the brain. It consists of two distinct lobes: **anterior lobe** (*adenohypophysis*) and posterior lobe (*neurohypophysis*). Because each lobe functions differently and produces different hormones, each is presented in a separate module. This one covers the **anterior pituitary**.

Surrounding the pituitary is a protective pocket of bone called the sella turcica of the sphenoid bone. The pituitary is nicknamed the "master gland" because it makes many different hormones that control the other endocrine glands in the body. But the **hypothalamus** in the brain actually controls the pituitary gland with specific bundles of neurons. This provides a vital link between the nervous system and the endocrine system. The pituitary is connected to the hypothalamus by a slender stem-like structure called the **infundibulum** (*pituitary stalk*).

Hormones Produced

Five different types of **hormone-producing cells** within the anterior pituitary directly secrete the following seven major hormones:

- Adrenocorticotropic hormone (**ACTH**)
- Follicle-stimulating hormone (**FSH**)
- Growth hormone (**GH**)
- Luteinizing hormone (**LH**)
- Prolactin (**PRL**)
- Thyroid-stimulating hormone (**TSH**)
- Melanocyte-stimulating hormone (**MSH**)

Sequence

The anterior pituitary relies on a specialized group of blood vessels collectively called a portal system. The body has two important portal systems: hepatic portal system (*in the liver*) and the **hypophyseal portal system** (*in the anterior pituitary*). Portal systems have a **primary** and **secondary capillary network** linked by portal veins. The following flowchart indicates blood flow through the hypophyseal portal system:

Superior hypophyseal artery → Primary capillary network → Hypophyseal portal veins → Secondary capillary network → Anterior hypophyseal veins

Mechanism

The anterior pituitary contains a mass of various **hormone-producing cells (HPCs)**, which are controlled by a group of neurons in the **ventral hypothalamus**. These neurons can either stimulate or inhibit hormone production by secreting either **releasing factors (RFs)** or inhibiting factors, respectively. Let's examine the stimulation process. **RFs** stimulate the **HPCs** to secrete hormones. Different RFs are needed to make the seven major hormones produced by the anterior pituitary. These RFs are produced in the neuron's **cell body** and released from the **axon terminal**. After being triggered to be released by a nerve impulse, they diffuse rapidly into the **primary capillary network** of the portal system and then the hypophyseal portal veins quickly shuttle them into the secondary capillary network surrounding the HPCs. As the RFs diffuse out of the blood, they immediately bind to shape-specific receptors in the HPC's plasma membrane, which induces the production of specific hormones. These hormones then enter the secondary capillary network and exit the anterior pituitary via the anterior hypophyseal vein. As the hormones travel through the blood, they eventually bind to receptors in their target cells and induce a response.

For example, a thyroid-stimulating releasing factor is needed to stimulate the secretion of thyroid-stimulating hormone (TSH). After TSH leaves the anterior lobe via the **anterior hypophyseal vein**, it travels through the blood to all the different organs but binds only at those targets that have receptors for TSH. As its name implies, TSH targets the thyroid gland to induce the secretion of thyroid hormones such as triiodothyronine (T_3) and thyroxine (T_4).

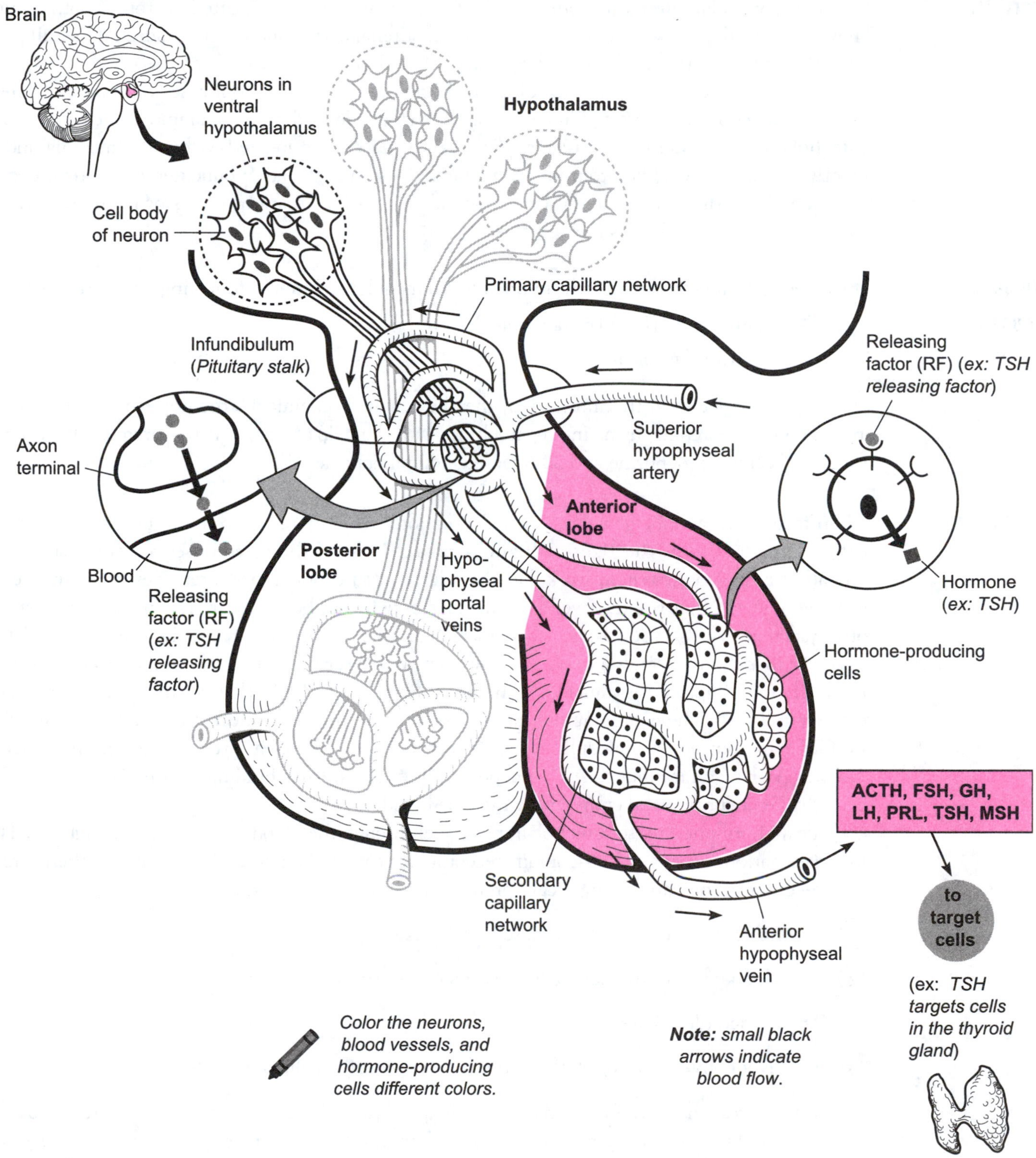

Color the neurons, blood vessels, and hormone-producing cells different colors.

Note: *small black arrows indicate blood flow.*

Description

The highly vascular **adrenal** (*suprarenal*) **glands** are small structures located on top of both kidneys. They are divided into two major regions: the *outer* adrenal cortex and the *inner* **adrenal medulla**. The adrenal cortex constitutes most of the mass of the adrenal glands. Each of these regions contains different types of cells that produce different hormones. This module focuses only on the adrenal medulla. **Catecholamines** are a group of chemicals that are derived from the amino acid called tyrosine. **Catecholamine-producing cells** (*chromaffin cells*), or **CPCs**, are located within the medulla and are directly stimulated by nerves from the sympathetic division of the autonomic nervous system. Because the sympathetic division helps the body respond to stressful situations on its own, the secretion of catecholamines simply boosts this response.

Hormones Produced

The catecholamine-producing cells in the adrenal medulla secrete the following two hormones:

- **Norepinephrine (NE)** (*noradrenaline*)
- **Epinephrine** (*adrenaline*)

As you can see from the illustration, they have similar chemical structures. In fact, the body is able to convert norepinephrine into epinephrine with the help of a converting enzyme. About 80% of the CPCs secrete epinephrine and 20% secrete norepinephrine.

Mechanism

When faced with an emergency or stressful situation, the body reacts in multiple ways that are collectively referred to as the **fight-or-flight** response. Most of these responses are the result of the sympathetic nervous system at work, but the release of catecholamines intensifies this response. For example, imagine that you discover a fire raging through your home. Your body needs to prepare you for physical activity so you can flee the situation and survive. As you recognize the fire as a danger, this conscious thought sends a nerve impulse from your **cerebral cortex** to your **hypothalamus** in the brain. Then an impulse is stimulated along nerve fibers in the brain that connect the hypothalamus to the spinal cord. The nerve impulse carried along the last nerve connects the spinal cord directly to the CPCs in your adrenal medulla. These sympathetic nerves release the neurotransmitter **acetylcholine (ACh)**, and it binds to ACh receptors in the CPCs. After binding, it stimulates the CPCs to secrete norepinephrine and epinephrine into the blood.

Like all hormones, these catecholamines travel through the blood like cars on the highway. Their final destination—**target cells**—contain receptors for catecholamines. The binding of the hormone to its receptor in the various target cells produces the following responses:

1. The heart rate increases (helps raise blood pressure)
2. The blood vessels constrict (helps raise blood pressure)
3. The bronchioles dilate.
4. The liver converts glycogen to glucose (to increase blood glucose levels).

All these responses are for the purpose of helping you flee the burning house. To run away from the fire, more blood will have to flow to your skeletal muscles to deliver more glucose and oxygen to the muscle cells. These nutrients will provide energy for muscle contraction as you flee. The following responses also occur (not included in the illustration):

- Metabolic rate increases.
- Urine output decreases.
- Blood is redirected from the digestive system toward the brain, skeletal muscles, and heart.

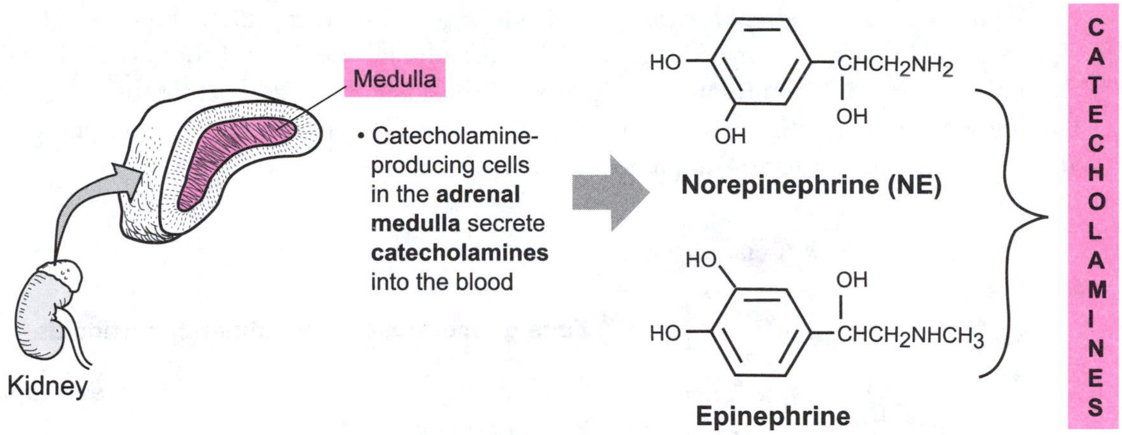
Medulla
• Catecholamine-producing cells in the **adrenal medulla** secrete **catecholamines** into the blood
Kidney
HO
OH
$CHCH_2NH_2$
OH
Norepinephrine (NE)
HO
HO
OH
$CHCH_2NHCH_3$
Epinephrine
CATECHOLAMINES

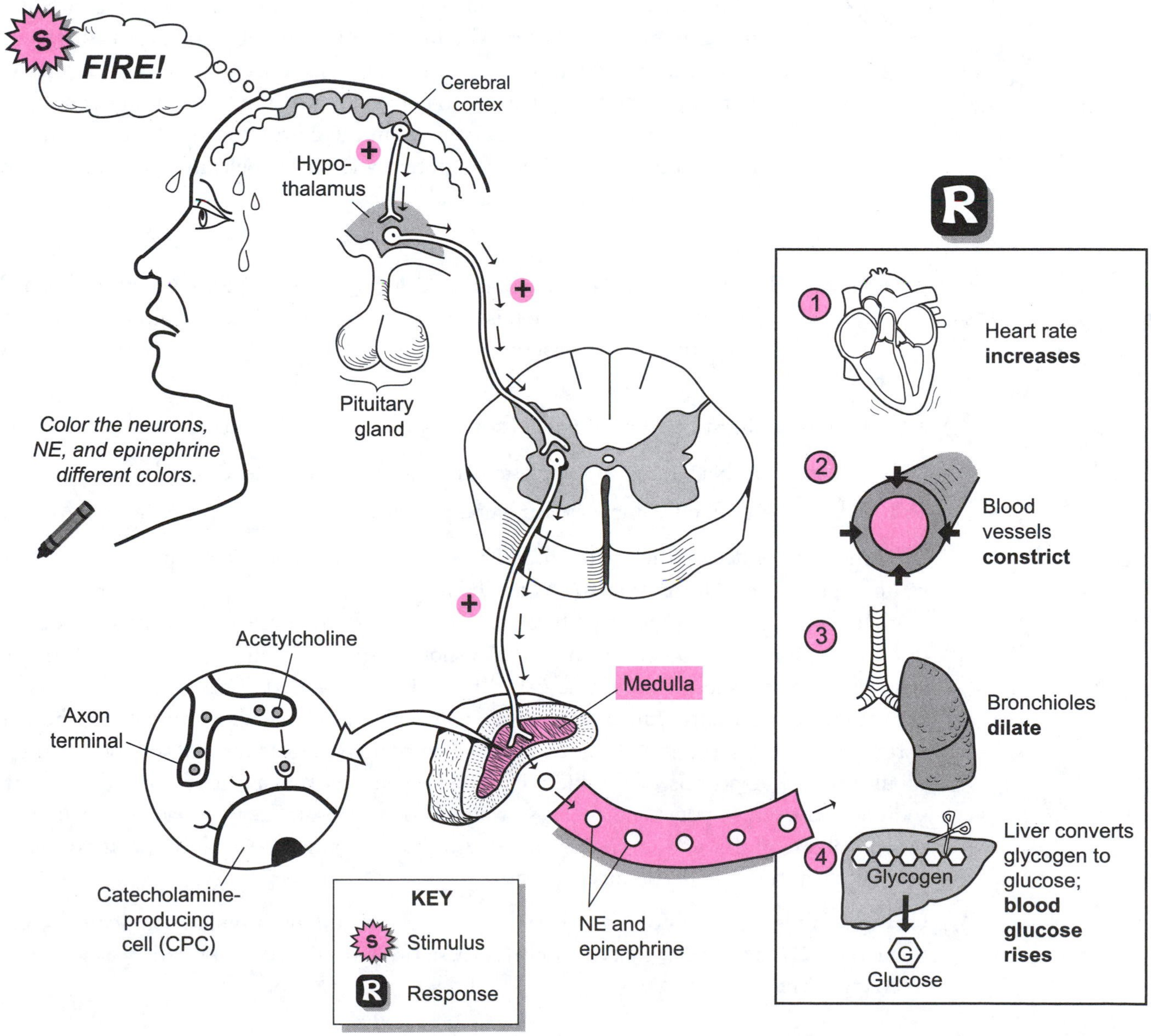
S
FIRE!
Cerebral cortex
Hypo-thalamus
Pituitary gland
Color the neurons, NE, and epinephrine different colors.
R
1
Heart rate **increases**
2
Blood vessels **constrict**
3
Bronchioles **dilate**
4
Glycogen
G
Glucose
Liver converts glycogen to glucose; **blood glucose rises**
Acetylcholine
Axon terminal
Medulla
Catecholamine-producing cell (CPC)
NE and epinephrine
KEY
S Stimulus
R Response

Description

The highly vascularized **adrenal** (*suprarenal*) **glands** are small structures located on top of both **kidneys**. They are divided into two major regions: the outer **adrenal cortex** and the inner adrenal medulla. The adrenal cortex constitutes most of the mass of the adrenal glands. Each of these regions contains different types of cells that produce different hormones. This module focuses on one aspect of the adrenal cortex, namely, its role in producing the hormone aldosterone.

Hormones Produced

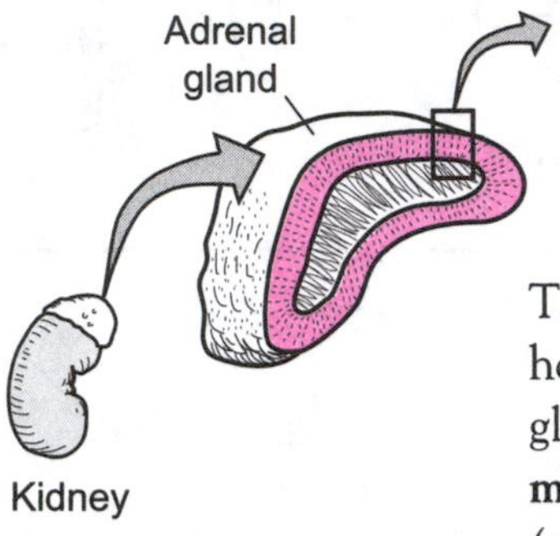

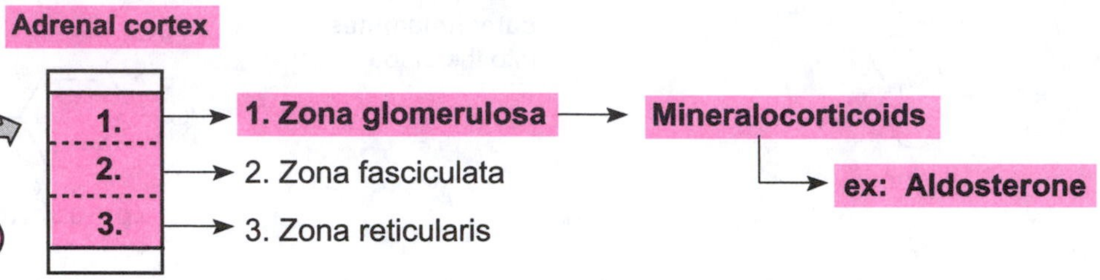

The **adrenal cortex** is divided into three different zones, each with its own hormone-producing cells. The first zone in the outer cortex is called the Zona glomerulosa. Various cells in this region secrete a group of hormones called **mineralocorticoids**. They are so named because they affect mineral homeostasis (such as Na^+ and K^+) and come from the cortex of the adrenal gland. Of this group, aldosterone is the most potent and physiologically significant, so we will focus on its mechanism of action.

Mechanism and Analogy

Aldosterone controls the levels of **Na^+** (sodium) and **K^+** (potassium) ions in extracellular fluids (such as the blood). The net result of its action is to reabsorb Na^+ ions into the blood and simultaneously excrete K^+ ions into the urine. Because "water follows the ions," as Na^+ is reabsorbed, water also is reabsorbed.

(1) As shown in the illustration, two types of stimuli stimulate aldosterone release:

1. a decrease in Na^+ levels in the plasma or an increase in K^+ levels, and
2. a decrease in **blood volume** and/or **blood pressure**.

The kidneys detect these stimuli and cause a chemical chain reaction referred to as the **renin-angiotensin-aldosterone (RAA) mechanism**.

This is the most important pathway to trigger aldosterone release of which the details are discussed in another module (see p. 442). In short, certain cells in the nephrons of the kidney are stimulated to release an enzyme called renin into the blood. This eventually leads to the activation of a hormone called angiotensin II, which stimulates secretory cells in the adrenal cortex to release aldosterone into the blood. The primary targets for aldosterone are the renal tubules in the nephrons of the kidneys. The response is that protein transporters reabsorb sodium and excrete potassium. Because water also is reabsorbed, this leads to an increase in blood volume and a corresponding increase in blood pressure. As blood pressure returns to normal, it shuts off the RAA mechanism.

(2) Changes in plasma levels of either Na^+ or K^+ also can directly stimulate the secretory cells in the adrenal cortex to release aldosterone. Of these two ions, the secretory cells are more sensitive to changes in K^+.

Adrenal gland
Adrenal cortex
Kidney
• Aldosterone-producing cells in the **adrenal cortex** secrete **aldosterone** into the blood
CH_2OH
C=O
O
H–C
HO
O
Aldosterone
Major Pathways for Release of Aldosterone
S
Na^+ in blood
and/or
K^+ in blood
1
S
blood volume
and/or
blood pressure
2
*Color **aldosterone**.*
KEY
Stimulus
Response
Kidney
Renin-angiotensin-aldosterone (RAA) mechanism
(see p. 442)
Adrenal cortex
Aldosterone (targets kidney tubules)
R
reabsorption of Na^+ and water,
K^+ excretion
blood volume
and
blood pressure
Kidney

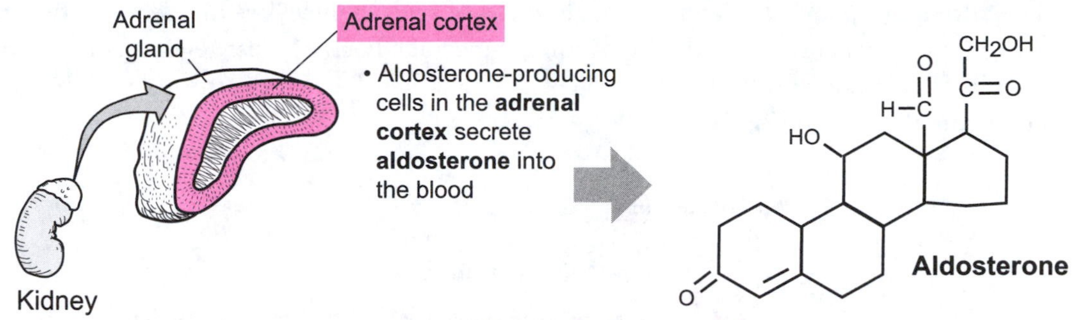

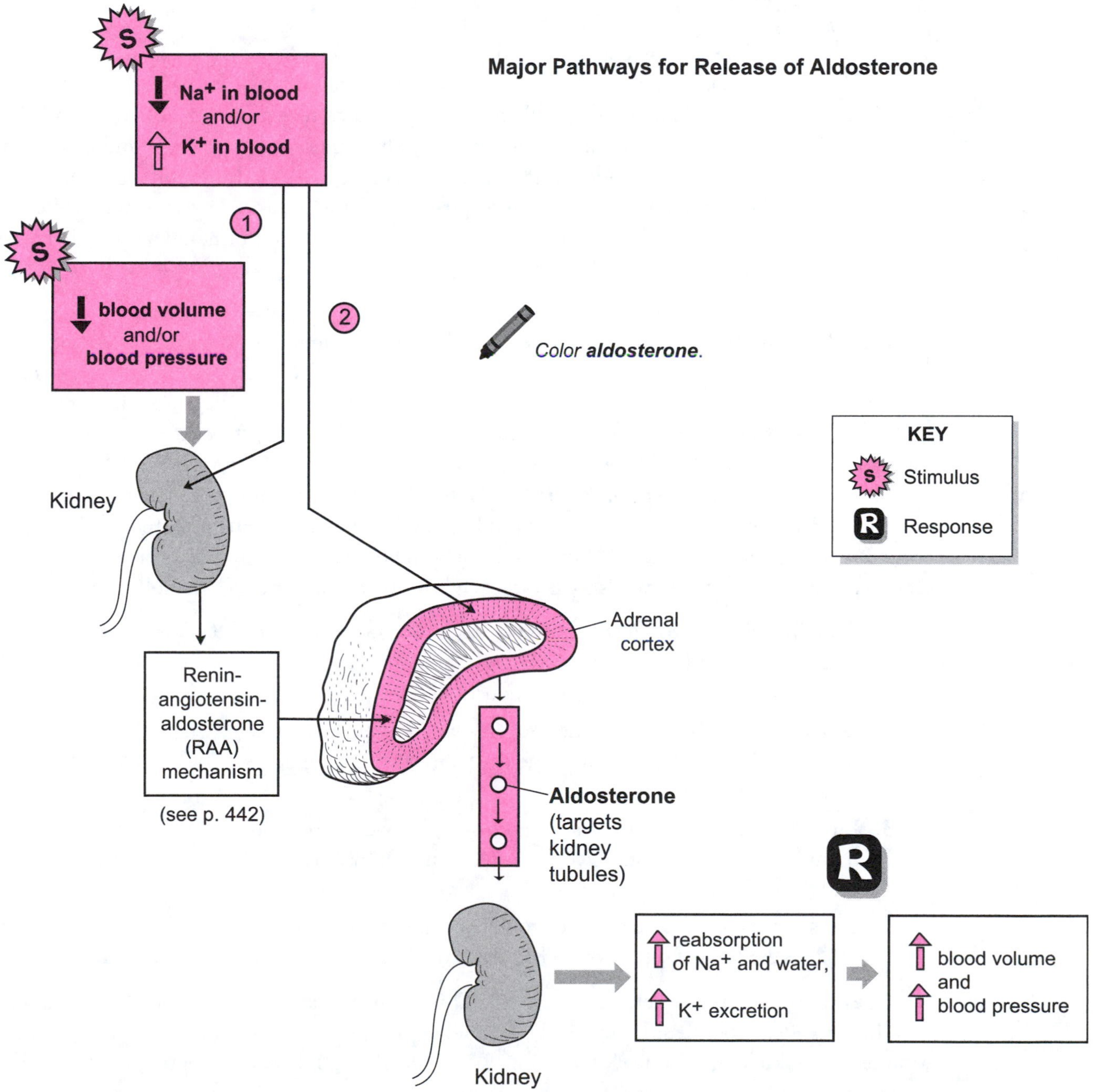

Description

The **adrenal** (*suprarenal*) **glands**, located on top of both kidneys, are divided into two major regions: the outer adrenal cortex and the inner adrenal medulla. Each of these regions contains different types of cells that produce different hormones. This module focuses on the role of the adrenal cortex, producing the hormone **cortisol**.

Hormones Produced

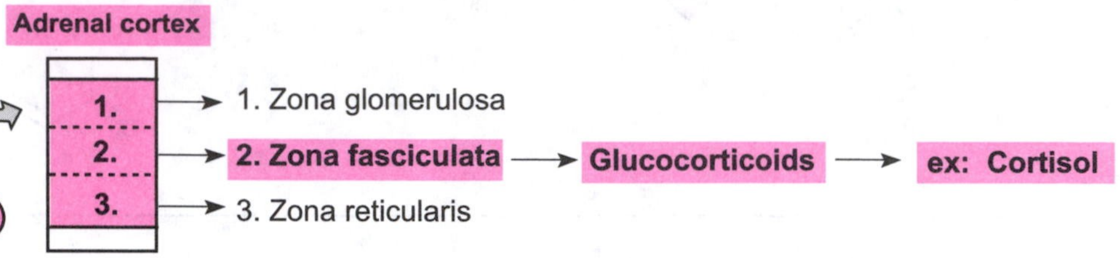

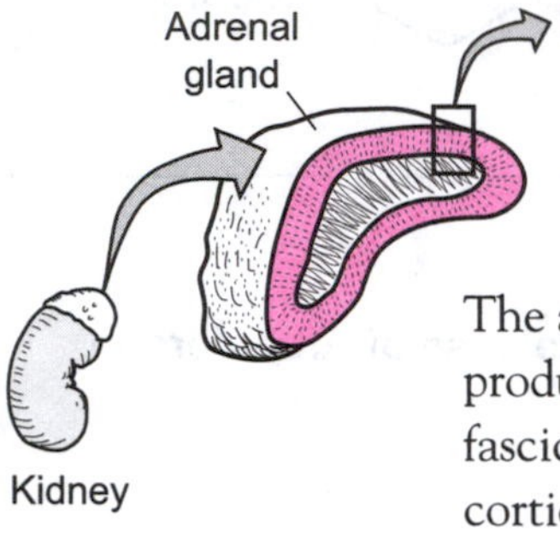

The **adrenal cortex** is divided into three different zones, each with its own hormone-producing cells. The second zone in the middle of the cortex is called the Zona fasciculata. Various cells in this region secrete a group of hormones called glucocorticoids—so named because they affect glucose homeostasis and come from the cortex of the adrenal gland. This group consists of the following three hormones: cortisol, corticosterone, and cortisone. Of these three, cortisol is the most abundant and potent. Because it accounts for nearly all the hormonal activity here, we will examine its mechanism of action.

Mechanism

We can't live without glucocorticoids, because they help us maintain normal blood glucose levels, normal blood volume, and resistance to stress—to name just a few functions. First let's examine how cortisol is secreted, then the responses it produces. Cortisol follows a predictable daily cycle, in which its levels peak after we awake in the morning and dip to their lowest point after we fall asleep at night. This cycle is regulated by a negative feedback loop. As blood levels of glucocorticoids fall, this stimulates neurons in the **hypothalamus** to secrete **CRH** (corticotropin-releasing hormone) into the blood. CRH targets cells in the anterior pituitary and stimulates them to secrete **ACTH** (adrenocorticotropic hormone). As ACTH travels through the blood, it targets cells in the adrenal cortex to secrete cortisol. As cortisol travels to its target cells in the body, it induces the following responses:

1. **Protein breakdown:** Proteins are broken down into **amino acids** (mostly in skeletal muscle). These amino acids enter the blood and travel to body cells, where they can be used to either make new proteins or be used to produce ATP.
2. **Gluconeogenesis:** This is the process that occurs in liver cells where a non-carbohydrate such as an amino acid or lactic acid is converted into glucose. The glucose then is used by body cells to produce ATP.
3. **Lipolysis:** This is the process of taking triglycerides stored in adipose tissue and converting them into fatty acids and glycerol. Cells can then use the glycerol and fatty acids to produce ATP.
4. **Stress resistance:** The glucose produced by liver cells in gluconeogenesis allows body cells to produce ATP. This supply of ATP enables the body to battle many different stresses.

Stress also triggers the release of cortisol. Chronic stress can lead to elevated levels of cortisol, and this can affect your health negatively. This leads to depression of the immune system. As a result, wounds may take longer to heal and you may be more susceptible to contracting a virus such as the common cold.

Adrenal gland
Adrenal cortex
Kidney
• Cortisone-producing cells in the **adrenal cortex** secrete **cortisol** into the blood
CH_2OH
C=O
HO
OH
O
Cortisol
S
The STRESS is killing me!
Cerebral cortex
+
Hypothalamus
CRH
Pituitary gland
Anterior lobe
Posterior lobe
ACTH
Adrenal cortex
Cortisol
Color ACTH and cortisol different colors.
KEY
Stimulus
Response
R
1.
Protein breakdown
Amino acid
2.
Gluconeo-genesis; blood glucose rises
G
Glucose
3.
Lipolysis
4.
Stress resistance
STR
ESS

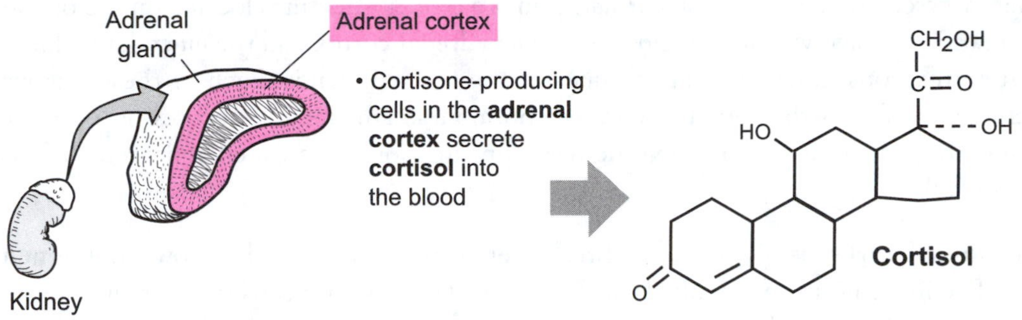

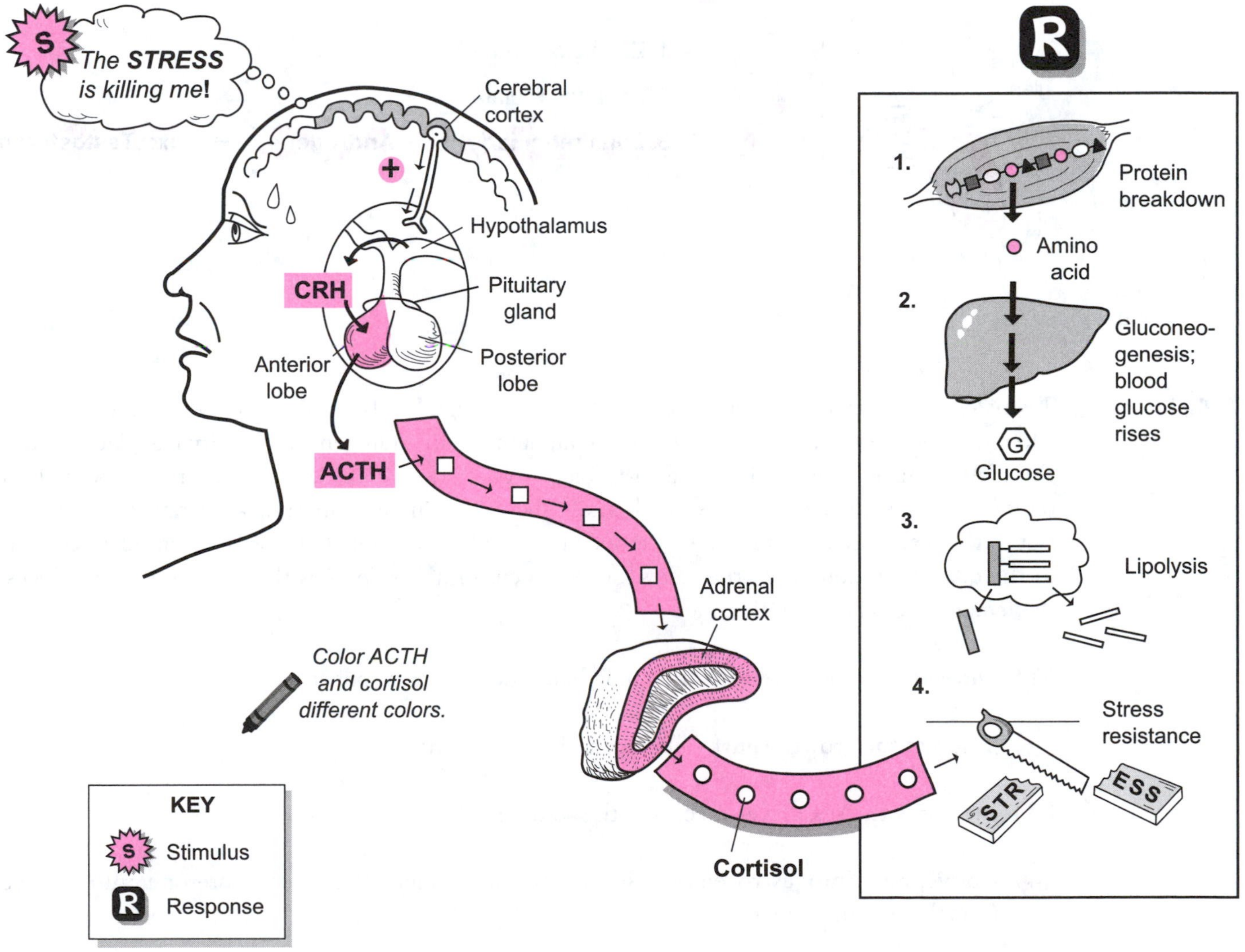

Description

The highly vascularized **adrenal** (*suprarenal*) **glands** are small structures located on top of both kidneys. They are divided into two major regions: the *outer* **adrenal cortex** and the *inner* adrenal medulla. The adrenal cortex constitutes most of the mass of the adrenal glands. Each of these regions contains different types of cells that produce different hormones. This module focuses on one aspect of the adrenal cortex—its role in producing the masculinizing hormones called androgens (*andros* = male human being).

Hormones Produced

- The **adrenal cortex** is divided into three different zones, each with its own hormone-producing cells. The innermost zone is called the Zona reticularis. Various cells in this region secrete a group of male sex steroids called androgens. The most potent type of androgen is testosterone. Like all steroidal hormones, it is derived from cholesterol.

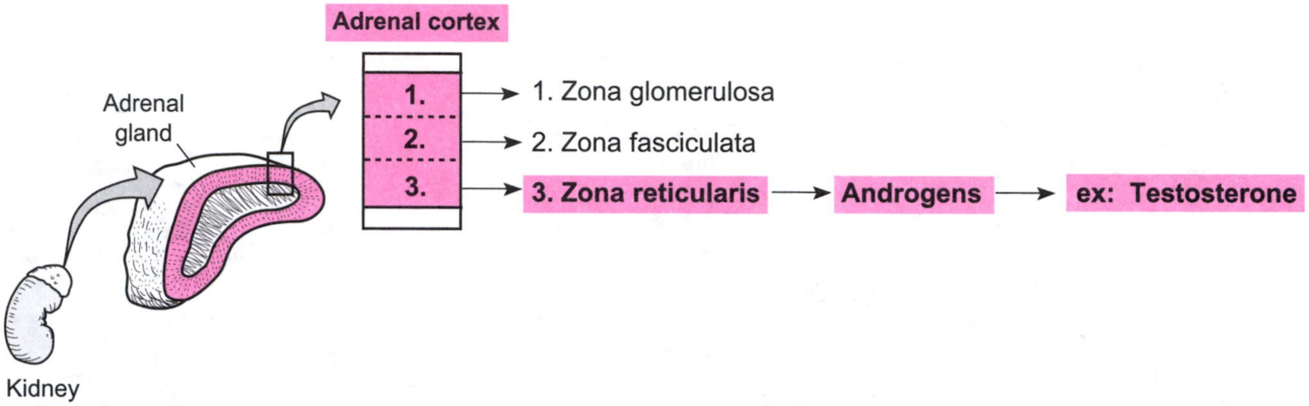

Mechanism

Testosterone is produced by the testes in the male at high levels during puberty and adulthood. The adrenal cortexes of both sexes produce small amounts of an androgen called **dehydroepiandrosterone** (**DHEA**). This is converted into **testosterone** by other tissue cells. At the onset of puberty, testosterone triggers similar responses in both sexes, such as an increase in under-arm hair and pubic hair. But in the male, the amount of testosterone produced by the adrenal glands is insignificant compared to the quantity produced by the testes. So let's focus on the responses that testosterone produces in the female throughout her lifetime:

1. **Development of underarm and pubic hair**—occurs before puberty

2. **Triggers the growth spurt**—also occurs before puberty

3. **Increases the sex drive**—occurs in the adult female

4. **Conversion into estrogens**—post-menopausal women convert testosterone into estrogens (female sex hormones).

Details of the mechanism by which the adrenal glands produce androgens is not well understood. Production seems to be stimulated by **adrenocorticotropic hormone** (**ACTH**), which is secreted by the **anterior pituitary**.

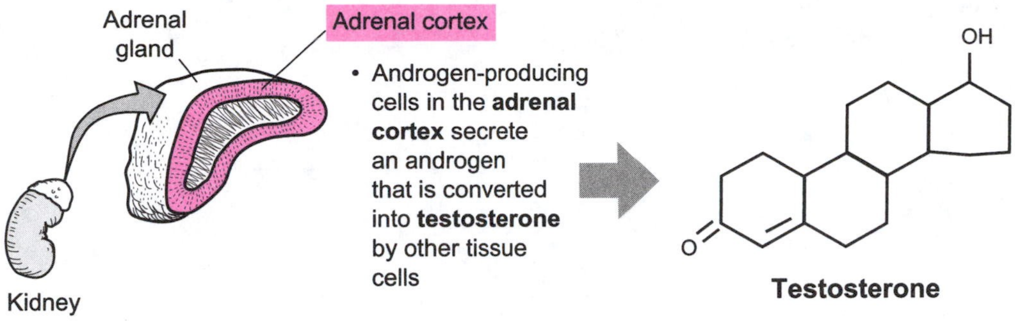
Adrenal gland
Adrenal cortex
Kidney
• Androgen-producing cells in the **adrenal cortex** secrete an androgen that is converted into **testosterone** by other tissue cells
OH
O
Testosterone

FEMALE

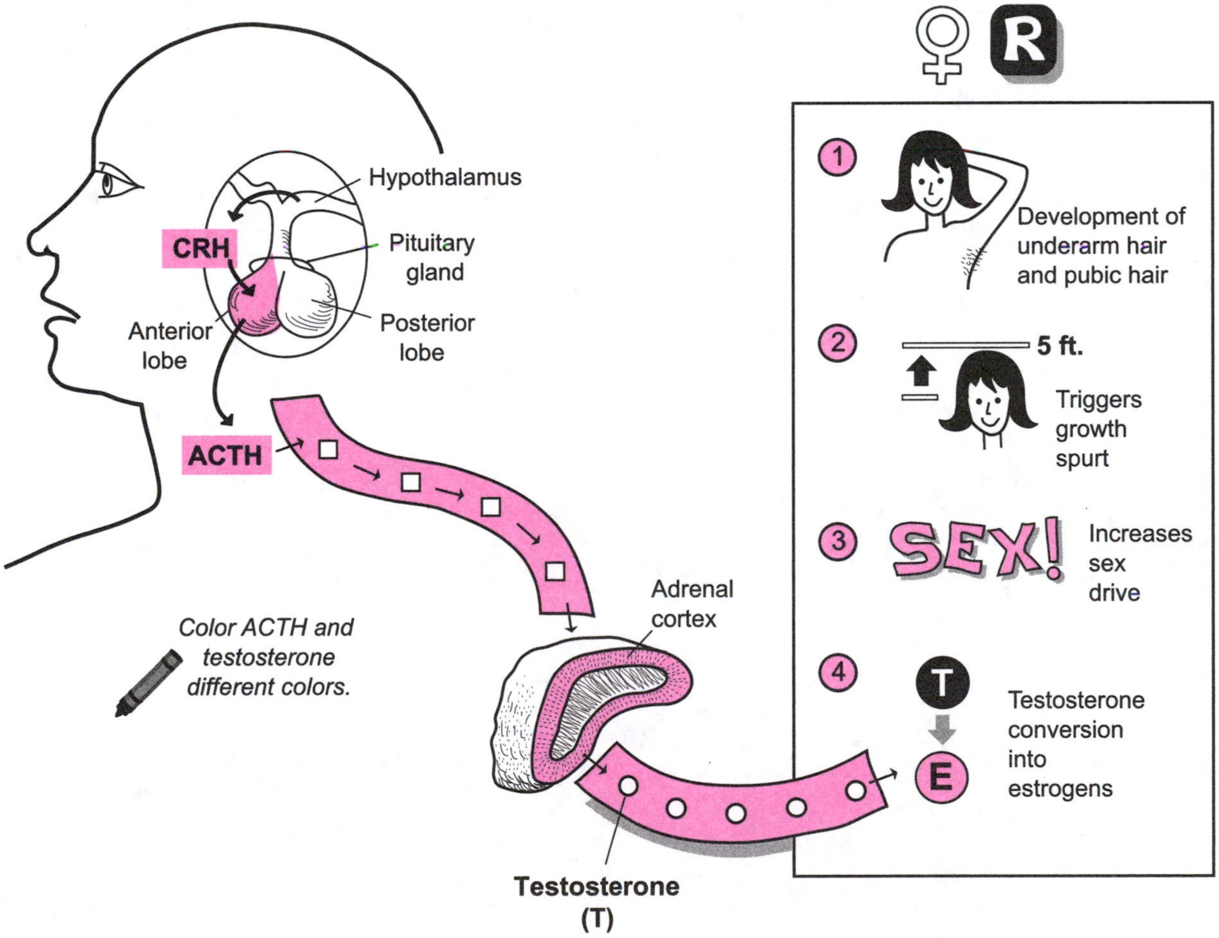
Hypothalamus
CRH
Pituitary gland
Anterior lobe
Posterior lobe
ACTH
Adrenal cortex
Color ACTH and testosterone different colors.
Testosterone (T)
R
1
Development of underarm hair and pubic hair
2
5 ft.
Triggers growth spurt
3
SEX!
Increases sex drive
4
T
E
Testosterone conversion into estrogens

Notes

Special Senses

Description

The eyeball rests in the orbit in the skull and is surrounded by fatty tissue. Six major muscles control the movement of the eyeball:

Muscle Name	Action
Superior rectus	Elevates eye
Inferior rectus	Depresses eye
Medial rectus	Moves eye medially
Lateral rectus	Moves eye laterally
Superior oblique	Depresses eye and laterally rotates eye
Inferior oblique	Elevates eye and laterally rotates eye

The lacrimal gland produces tears that are spread across the surface of the eye during blinking. The tears accumulate in a pooling area called the **medial canthus**. Then they enter openings called the **superior lacrimal punctum** and the **inferior lacrimal punctum**, which lead to passageways called the **superior lacrimal caniculus** and the **inferior lacrimal caniculus**. From here they travel to the **lacrimal sac** and down the **nasolacrimal duct** and are drained into the nose. This explains why your nose runs when you are crying heavily.

Study Tip

The **L**acrimal gland is located on the **L**ateral side of the eyeball. Use it as a landmark.

Key to Illustration

1. Superior lacrimal punctum *(opening)*
2. Superior lacrimal caniculus
3. Lacrimal sac
4. Inferior lacrimal punctum *(opening)*
5. Inferior lacrimal caniculus
6. Nasolacrimal duct
7. Opening of nasolacrimal duct
8. Superior oblique muscle
9. Superior rectus muscle
10. Lateral rectus muscle
11. Inferior oblique muscle
12. Inferior rectus muscle
13. Medial rectus muscle

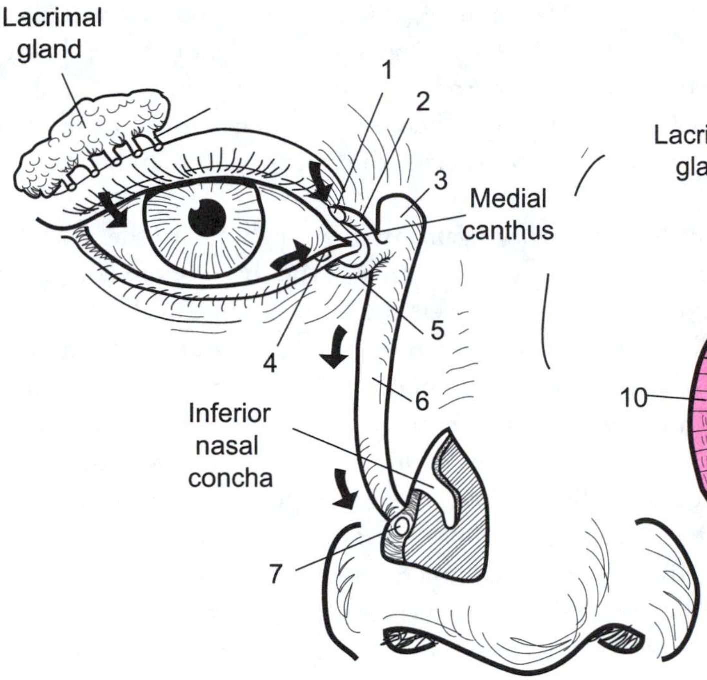

Lacrimal apparatus

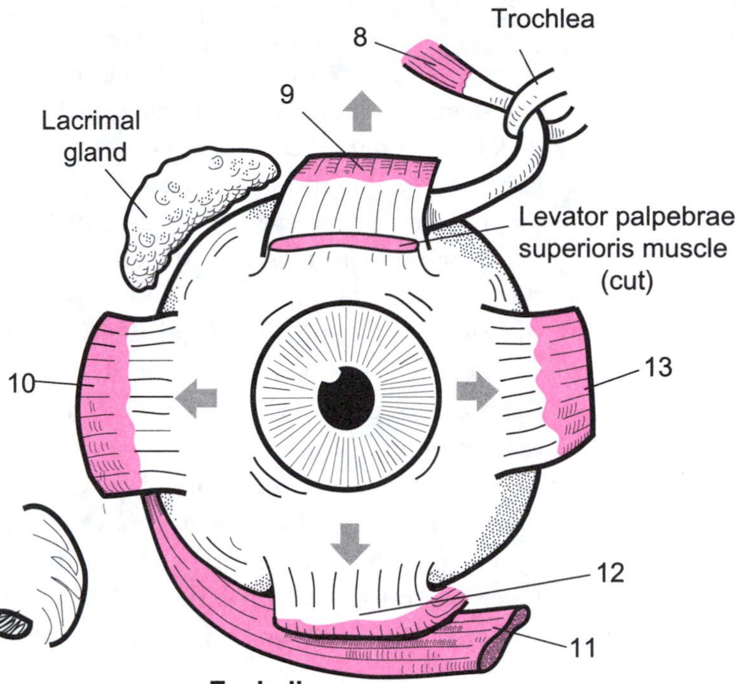

Eyeball
anterior view

Note: *Arrows indicate the direction the eye will move when a muscle is contracted.*

1. ______________________
2. ______________________
3. ______________________
4. ______________________
5. ______________________
6. ______________________

7. ______________________
8. ______________________
9. ______________________
10. ______________________
11. ______________________
12. ______________________
13. ______________________

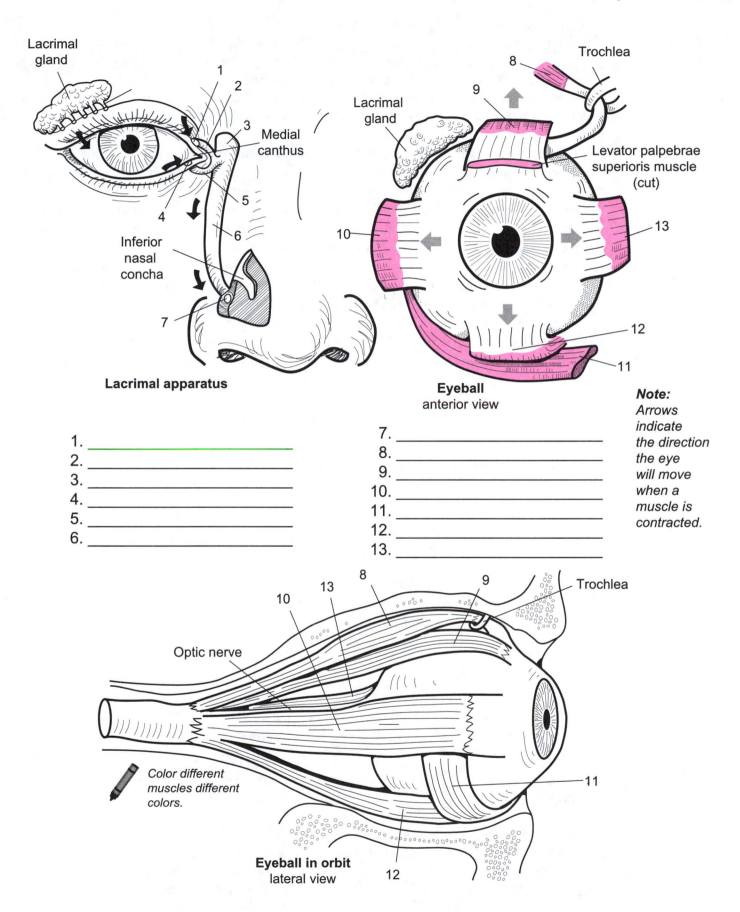

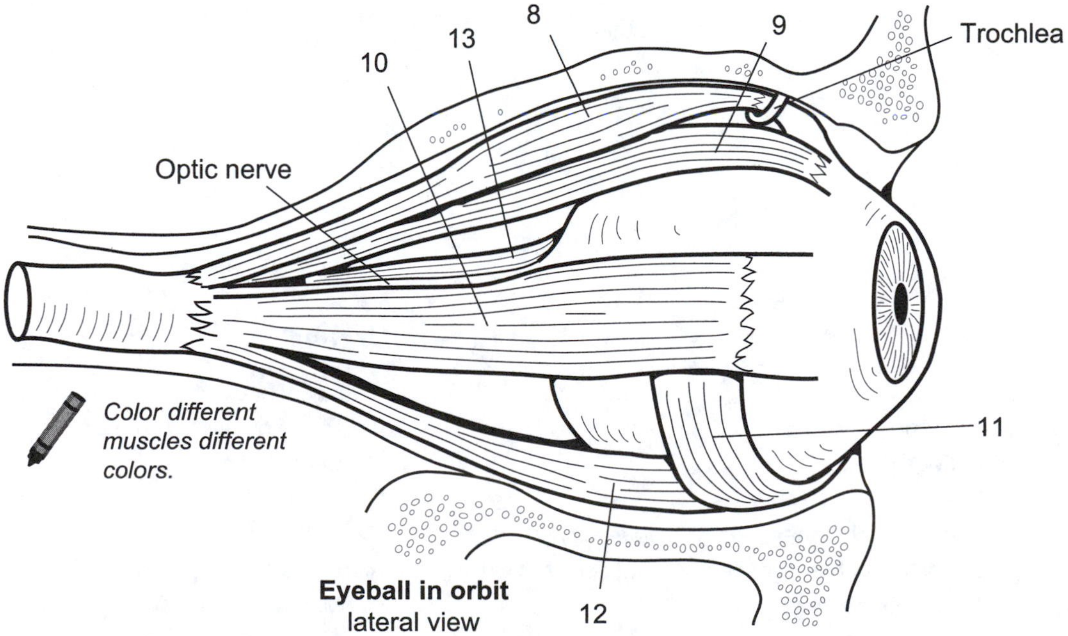

Color different muscles different colors.

Eyeball in orbit
lateral view

Description

The eye is divided into three tunics or sheaths: fibrous tunic, vascular tunic, and neural tunic. The **fibrous tunic** is a thick, tough layer of connective tissue and consists of the **sclera** (*white of the eyeball*) on its posterior portion and becomes the transparent **cornea** on its anterior portion. The cornea is cup-shaped and filled with a liquid called **aqueous humor**. The **vascular tunic** consists of the vascular, dark brown **choroid** coat on its posterior portion, which becomes the **ciliary body**, and the **iris** (*colored part of the eye*) on its anterior portion.

The lens of the eye is held in place by **suspensory ligaments** that anchor it to the ciliary body. When the smooth muscle in the ciliary body contracts, it can change the shape of the lens, allowing the eye to focus on near versus distant objects. The iris covers the front of the lens. In the center of the iris is a hole called the **pupil**, which allows light to enter the eye. Dilation and constriction of the pupil controls the amount of light that enters the eye. The **neural tunic** is the **retina** and covers only the posterior portion of the eye. This thin layer contains photoreceptors called **cones** and **rods**. Rods help us see in low-light situations, and cones help us see color and give us sharper, clearer images.

Filling the **posterior cavity** of the eye is a jelly-like substance called **vitreous humor**. It maintains the normal shape of the eyeball. On the back of the retina is a small disc called the **macula lutea**, which contains cones but no rods. At the center of this structure is a shallow depression called the **fovea centralis,** which has the highest concentration of cones and gives us our sharpest vision. In contrast, the one region of the retina that lacks photoreceptors is called the **optic disc** (*blind spot*). No images form here. It is located where the optic nerve leaves the eye.

Analogy

The **macula lutea** is like a **target** and the **fovea centralis** is like the **bullseye**.

Study Tip

To recall the order of the layers in the back of the eye, use the mnemonic: **S**cared **C**ats **R**un. This indicates the layers from outermost to innermost or **S**clera, **C**horoid, **R**etina.

Key to Illustration

1. Iris
2. Cornea
3. Edge of pupil
4. Lens
5. Anterior chamber *(filled with aqueous humor)*
6. Posterior chamber *(filled with aqueous humor)*
7. Anterior cavity
8. Suspensory ligaments
9. Ciliary body
10. Posterior cavity *(filled with vitreous humor)*
11. Central vein
12. Central artery
13. Optic nerve
14. Fovea centralis
15. Macula lutea
16. Optic disc (*blind spot*)

Eye—Internal Structure

Color all the different layers different colors.

To recall the layers on the posterior portion of the eyeball:

***S**cared **C**ats **R**un!*	=	***S**clera **C**horoid **R**etina*

The macula lutea is like the target and the fovea centralis is the bullseye.

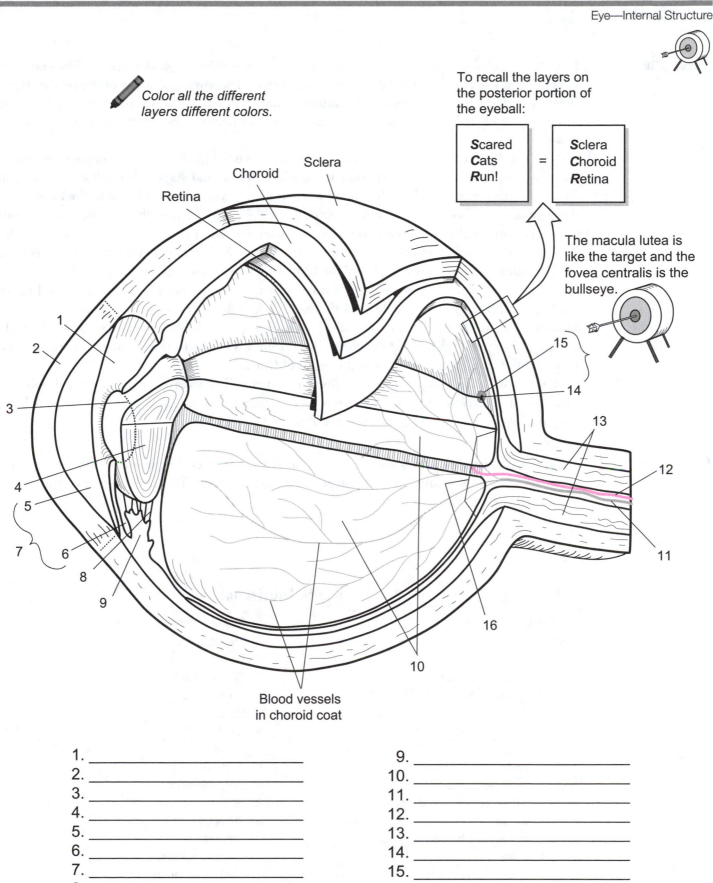

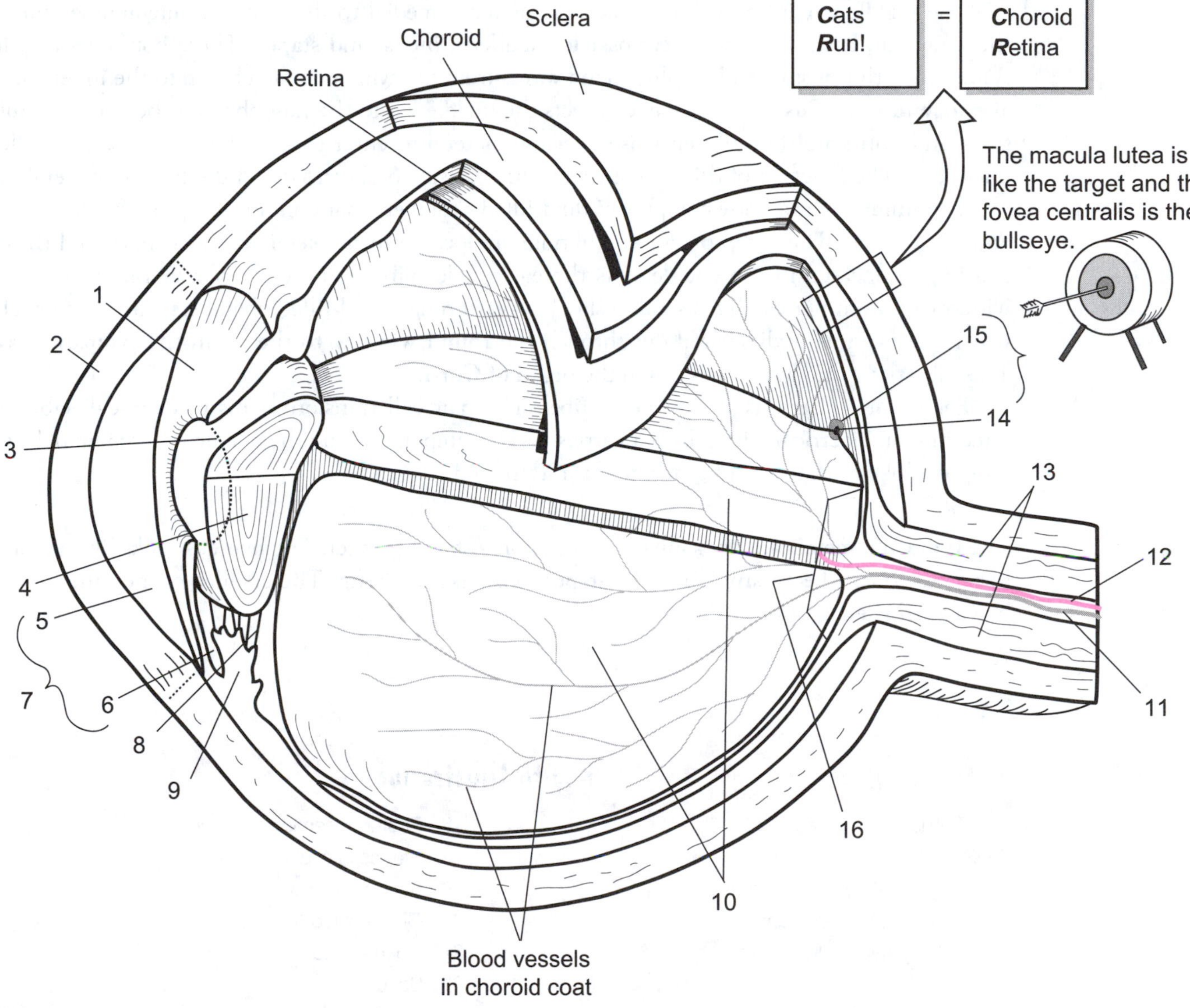

1. ______________________
2. ______________________
3. ______________________
4. ______________________
5. ______________________
6. ______________________
7. ______________________
8. ______________________
9. ______________________
10. ______________________
11. ______________________
12. ______________________
13. ______________________
14. ______________________
15. ______________________
16. ______________________

Description

The ear is divided into three regional areas: external ear, middle ear, and inner ear. The **external ear** extends from the ear itself to the tympanic membrane. The **auricle** is made of elastic cartilage and directs sound waves into the **external auditory meatus.** At the end of this passageway is the delicate **tympanic membrane** or eardrum. This cone-shaped connective tissue vibrates when sound waves strike it.

The **middle ear** is an air-filled space that extends medial to the tympanic membrane and up to the inner ear. It contains three ear **ossicles: malleus, incus**, and **stapes**. These bones act as a lever system to both transmit and amplify sound waves from the tympanic membrane to the inner ear. The **auditory tube**, or **Eustachian tube**, connects the middle ear to the nasopharynx. Because this tube is short and horizontal in children it is easier for bacteria to enter the middle ear, causing middle ear infections. The function of this tube is to equalize pressure on both sides of the tympanic membrane.

The **inner ear** is a bony complex of fluid-filled chambers that contain receptors for both hearing and equilibrium. The receptors for equilibrium are located in the **semicircular canals**, and those for hearing are located in the **cochlea**. As the ear ossicles vibrate, they transfer vibration to the **oval window** of the inner ear. This sends a shock wave through fluid-filled chambers within the cochlea, which are eventually dissipated out through the **round window** to the air in the **tympanic cavity**. This vibration stimulates hair cells in the **organ of Corti**.

Each hair cell is linked to a nerve fiber. The hair cell transforms this mechanical, vibrational force into an electrical stimulus that carries nervous impulses along the cochlear nerve branch to the temporal lobe in the cerebrum, where it is interpreted as sound.

Analogy

The **tympanic membrane** looks like the **cones** on a stereo speaker. The **malleus** looks like a **hammer**, the **incus** looks like an **anvil**, and the **stapes** looks like a **stirrup**. The **cochlea** of the inner ear looks like a **snail shell**.

Key to Illustration

1. Auricle
2. Lobule
3. External auditory meatus
4. Tympanic membrane
5. Tympanic cavity
6. Malleus
7. Incus
8. Stapes
9. Eustachian *(auditory)* tube
10. Semicircular canals
11. Vestibular branch of auditory nerve
12. Facial nerve
13. Cochlear branch of auditory nerve
14. Cochlea
15. Anterior semicircular canal
16. Semicircular ducts
17. Lateral semicircular canal
18. Posterior semicircular canal
19. Stapes in oval window
20. Round window
21. Organ of Corti
22. Cochlear duct
23. Saccule
24. Utricle
25. Vestibular duct
26. Cochlear duct
27. Tympanic duct
28. Tectorial membrane
29. Stereocilia
30. Outer hair cells
31. Supporting cells
32. Basilar membrane
33. Inner hair cell
34. Nerve fibers of cochlear nerve

Ear—General Structure

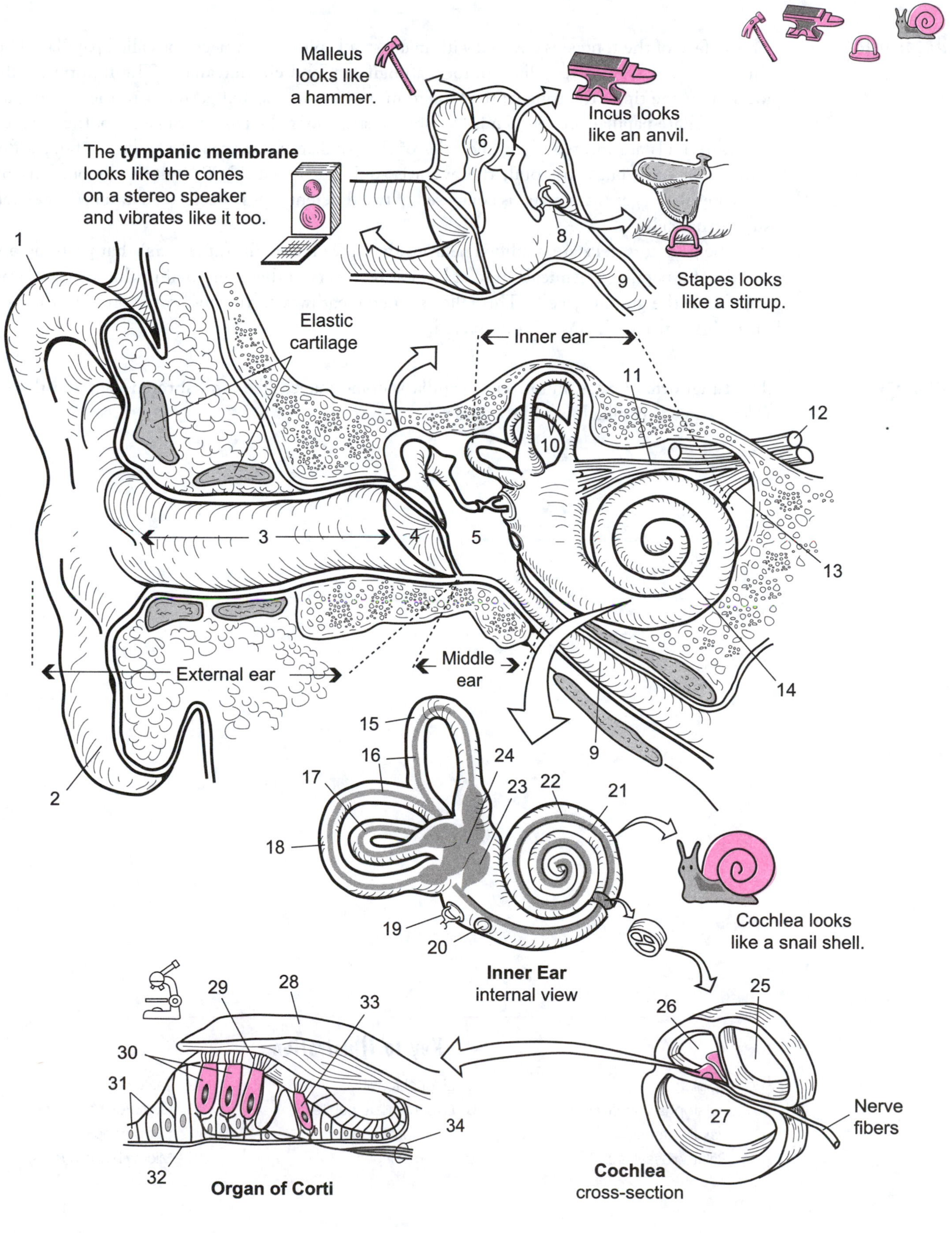

Malleus looks like a hammer.
Incus looks like an anvil.
The **tympanic membrane** looks like the cones on a stereo speaker and vibrates like it too.
Stapes looks like a stirrup.
Elastic cartilage
Inner ear
External ear
Middle ear
Cochlea looks like a snail shell.
Inner Ear
internal view
Organ of Corti
Cochlea
cross-section
Nerve fibers
1
2
3
4
5
6
7
8
9
10
11
12
13
14
15
16
17
18
19
20
21
22
23
24
25
26
27
28
29
30
31
32
33
34

Description

The surface of the tongue is covered with many small epithelial projections called **papillae**. There are three different types of papillae: **filiform**, **fungiform**, and **circumvallate**. The filiform papillae are located on the tip of the tongue; the fungiform papillae are located posterior to the filiform; and the **circumvallate papillae** are found in a "V" shaped strip along the posterior margin of the tongue.

Taste buds are located along the sides of the papillae. Each type of papilla contains a different number of taste buds. In total, the average adult has about 10,000 taste buds but this number decreases with age. A taste bud is composed of two different types of cells—**gustatory** (*taste*) **cells** and **supporting cells**.

The gustatory cells are modified neurons that have **microvilli** (*taste hairs*) that protrude onto the surface of a papilla. Chemicals in food bind at receptors in these microvilli, which trigger a nervous impulse in the gustatory cells. This follows a nerve pathway to the gustatory cortex in the cerebrum of the brain where the taste is interpreted.

Analogy

The **fungiform** (*fungus*, mushroom) **papilla** is dome-shaped like a **mushroom cap**, and the **filiform** (*filum*, thread) **papilla** looks like a **flame**.

Key to Illustration

1. Epiglottis
2. Palatopharyngeal arch
3. Palatine tonsil
4. Palatoglossal arch
5. Lingual tonsil
6. Circumvallate papilla
7. Fungiform papilla
8. Filiform papilla
9. Taste bud
10. Gustatory *(taste)* cells
11. Supporting cells
12. Microvilli *(taste hairs)*

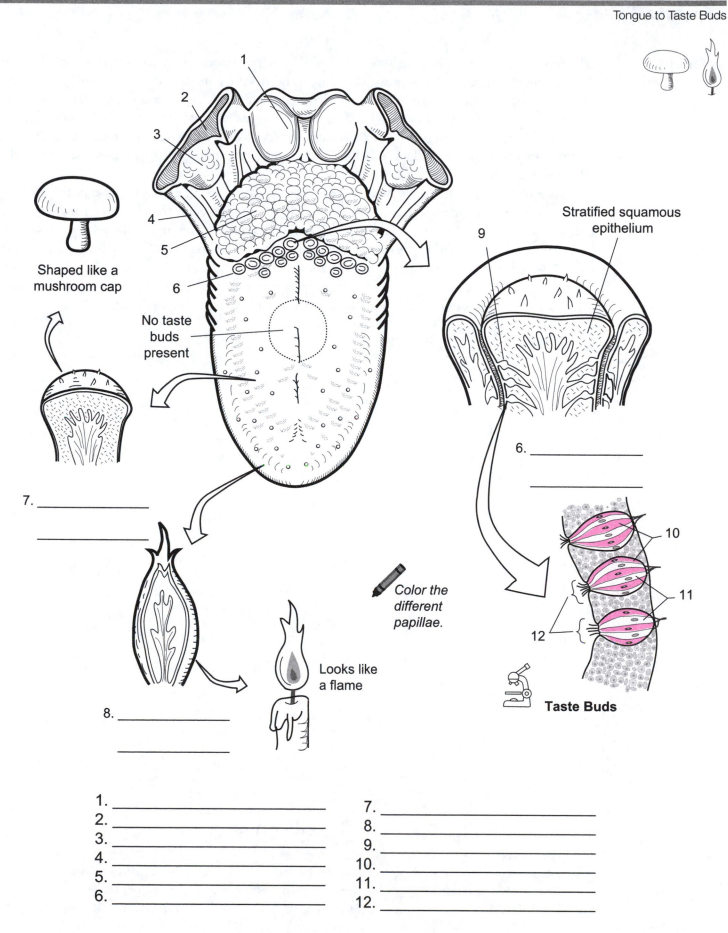

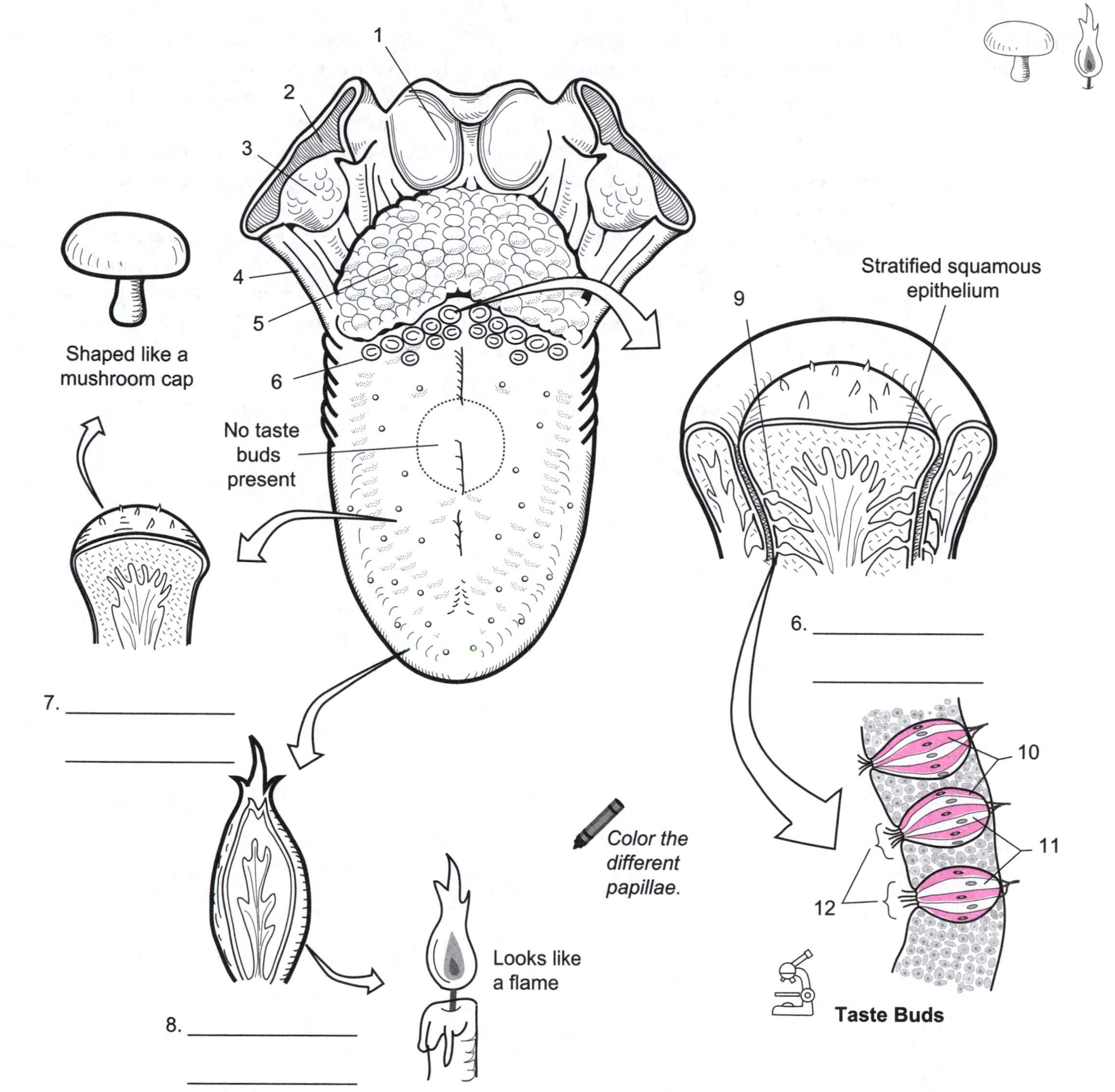

1. ______
2. ______
3. ______
4. ______
5. ______
6. ______
7. ______
8. ______
9. ______
10. ______
11. ______
12. ______

Description

Olfaction refers to the sense of smell. Let's follow what occurs when you smell a cup of coffee. As you inhale air through the nostrils it enters the nasal cavity and the nasal conchae produce turbulent airflow. This disperses the air and delivers aromatic molecules to the two **olfactory organs** located on the roof of the nasal cavity. These small organs are coated with a thick layer of mucus produced by the **olfactory glands**. The aromatic molecules diffuse through this mucus and bind to a receptor in the numerous **olfactory cilia**. These cilia are extensions of modified neurons called **olfactory receptor cells**.

Once the aromatic molecule binds to the receptor, it triggers a nervous impulse in the olfactory receptor cell. As this impulse travels along the cell, it passes through the **olfactory foramen** in the **cribriform plate** of the ethmoid. Then it reaches the **olfactory bulb,** which is the terminal portion of the first cranial nerve. Within the olfactory bulb, the olfactory receptor cell forms synaptic connections with other neurons. The impulse is transferred to these neurons and continues down the **olfactory tract**. Finally, the impulses are carried to the appropriate olfactory interpretation areas in the brain. These include regions in both the frontal and the temporal lobes of the cerebrum.

Key to Illustration

1. Olfactory bulb
2. Olfactory tract
3. Olfactory receptor cells *(neurons)*
4. Supporting cells
5. Olfactory foramen
6. Cribriform plate of ethmoid
7. Olfactory gland
8. Lamina propria
9. Olfactory epithelium
10. Mucous layer
11. Olfactory cilia

Sense of Smell

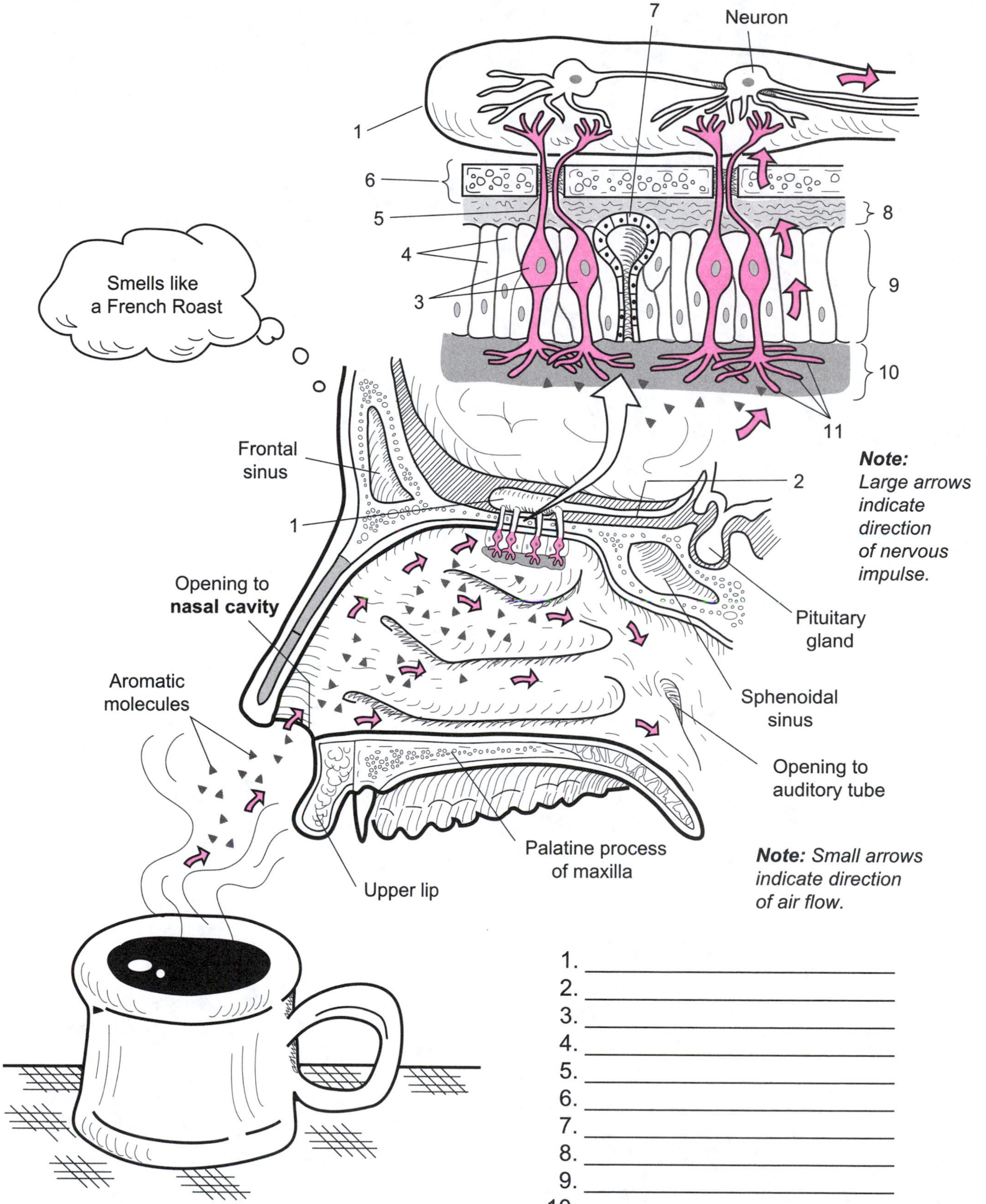
7
Neuron
1
6
5
8
4
9
3
10
11
Smells like
a French Roast
Frontal
sinus
2
1
Note:
Large arrows
indicate
direction
of nervous
impulse.
Opening to
nasal cavity
Pituitary
gland
Aromatic
molecules
Sphenoidal
sinus
Opening to
auditory tube
Palatine process
of maxilla
Upper lip
Note: Small arrows
indicate direction
of air flow.

1. ______________________
2. ______________________
3. ______________________
4. ______________________
5. ______________________
6. ______________________
7. ______________________
8. ______________________
9. ______________________
10. ______________________
11. ______________________

Notes

Blood

Description

Blood is a specialized type of connective tissue because it contains cells, fibers, and a liquid ground substance. It is composed of two major parts: **plasma** and **formed elements.** The **plasma** is a straw-colored fluid that contains mostly water, proteins, and other solutes. The **formed elements** consist of the following blood cells and cell fragments scattered in the plasma: **erythrocytes** (*red blood cells*), **leukocytes** (*white blood cells*), and **thrombocytes** (*platelets*).

Formed Elements of Blood

Cell	Description	Function
Erythrocytes *(red blood cells)*	Comprise 99.9% of all blood cells; biconcave discs; no nucleus in mature cell; filled with the protein hemoglobin	• Transport O_2 from lungs to body cells • Transport CO_2 from blood to lungs
Leukocytes *(white blood cells)*	Less than 0.1% of all blood cells; 5 different types; some have granules in cytoplasm, others do not; nucleus present in all types	• Fight against pathogens such as bacteria and viruses
Thrombocytes *(platelets)*	Less than 0.1% of formed elements; cell fragments; no nucleus; contain enzymes	• Involved in blood clotting

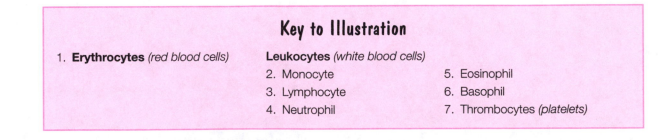

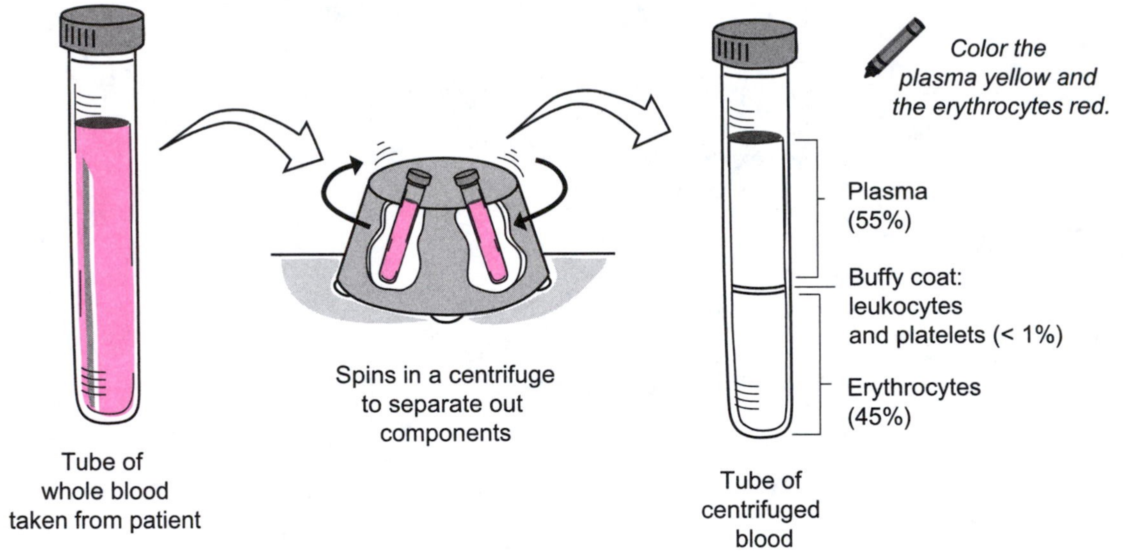

Key to Illustration

1. **Erythrocytes** *(red blood cells)*

Leukocytes *(white blood cells)*

2. Monocyte
3. Lymphocyte
4. Neutrophil
5. Eosinophil
6. Basophil
7. Thrombocytes *(platelets)*

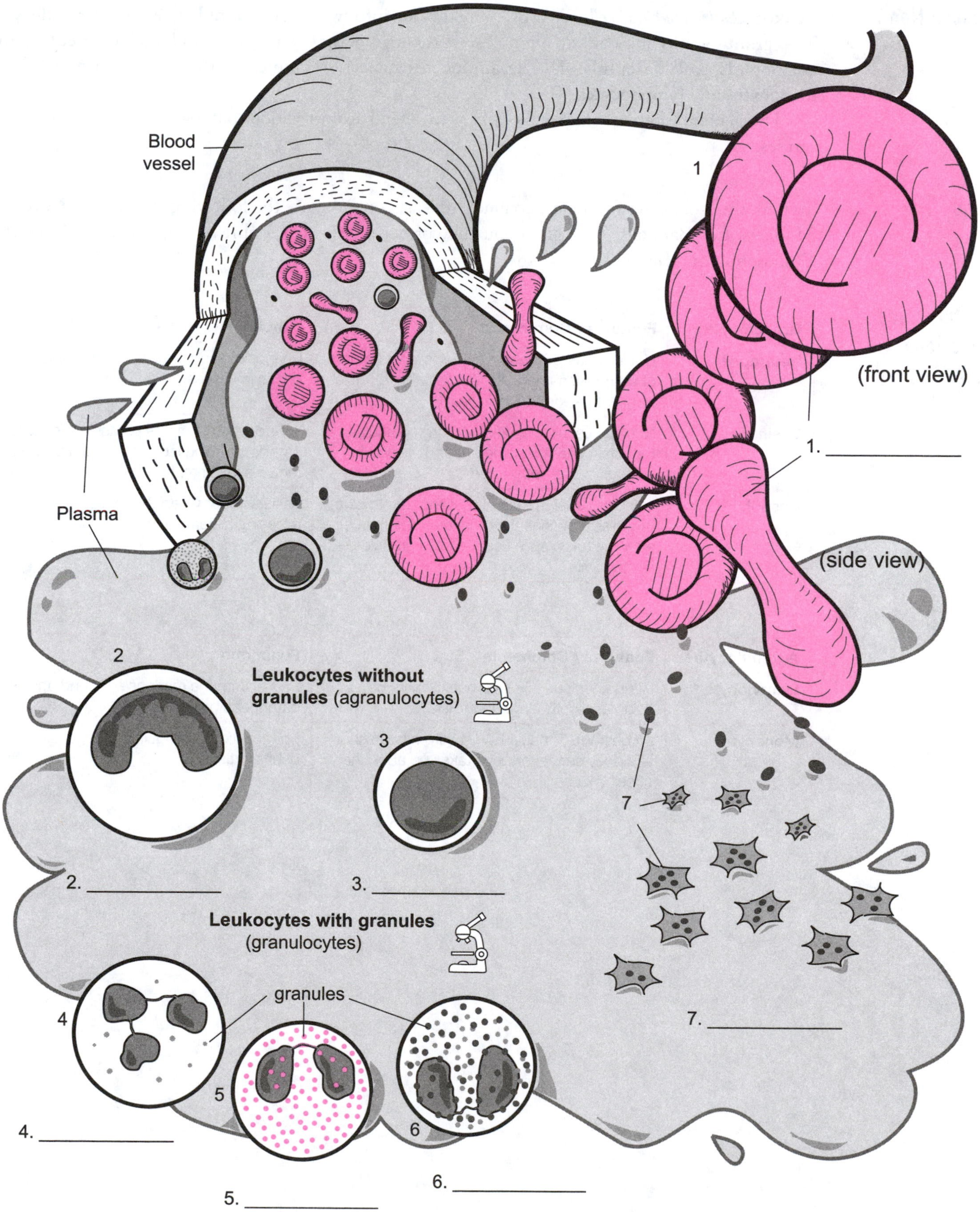
Blood vessel
1
(front view)
1. ______________
(side view)
Plasma
2
Leukocytes without granules (agranulocytes)
3
7
2. ______________
3. ______________
Leukocytes with granules (granulocytes)
granules
4
5
6
7. ______________
4. ______________
6. ______________
5. ______________

Description

Leukocytes (*white blood cells*, or *WBCs*) are divided into two groups: **granulocytes** and **agranulocytes**. The **granulocytes** all contain granules in the cytoplasm of the cell and include **neutrophils**, **eosinophils**, and **basophils**. The **agranulocytes** have no granules in the cytoplasm and include **monocytes** and **lymphocytes**.

The general function of all leukocytes is to defend against various pathogens such as bacteria and viruses.

Study Tip

To rank the leukocytes from most common to least common, use the following mnemonic: **N**ever **L**et **M**onkeys **E**at **B**ananas. This gives you the correct order: **N**eutrophils, **L**ymphocytes, **M**onocytes, **E**osinophils, and **B**asophils.

Features and Functions of WBCs

Granulocyte	Features / Comments	Functions
Neutrophil	Nucleus has 3–5 lobes; contains least amount of granules; named after the fact that granules are *neutral*—do not stain well.	Phagocytic cell; engulfs bacteria and debris in tissues.
Eosinophil	Nucleus usually has 2 lobes; large granules that stain brightly; named after dye used to stain granules—*eosin* dye.	Phagocytic cell; fights parasitic infections; engulfs anything labeled with antibodies; reduces inflammation.
Basophil	Nucleus usually masked by deep purple/blue granules; contains the most granules; named after the *basic* stain used to stain granules—hematoxylin.	Assists in damaged tissue repair by releasing histamine from granules.

Agranulocyte	Features / Comments	Functions
Lymphocyte	Round nucleus takes up nearly entire cell volume; slightly larger than RBC.	Part of immune response; defend against pathogens or toxins.
Monocyte	Largest WBC; nucleus varies from horseshoe to kidney shape; nucleus takes up about half of cell volume.	Phagocytic cell; engulfs pathogens and debris in tissues.

My Drawing of the WBCs

Neutrophil

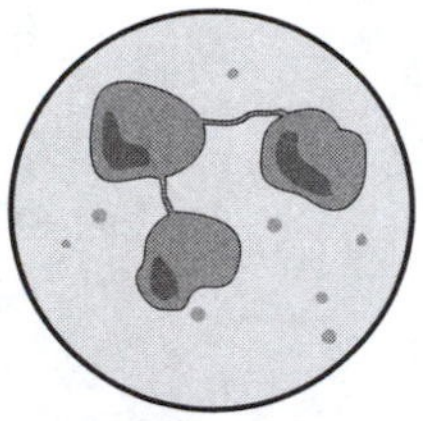

60–70 % of all WBCs

Lymphocyte

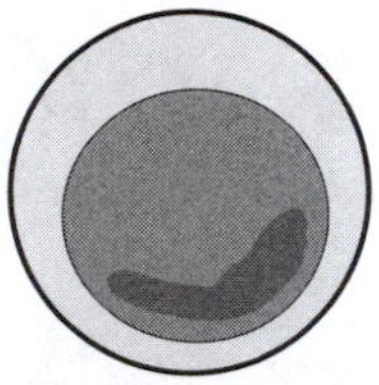

20–25 % of all WBCs

Monocyte

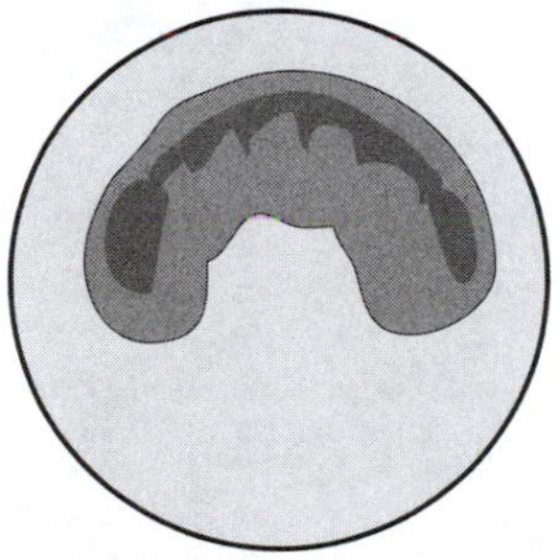

3–8 % of all WBCs

Eosinophil

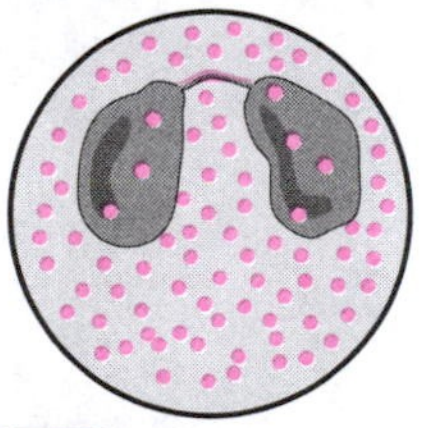

2–4 % of all WBCs

Basophil

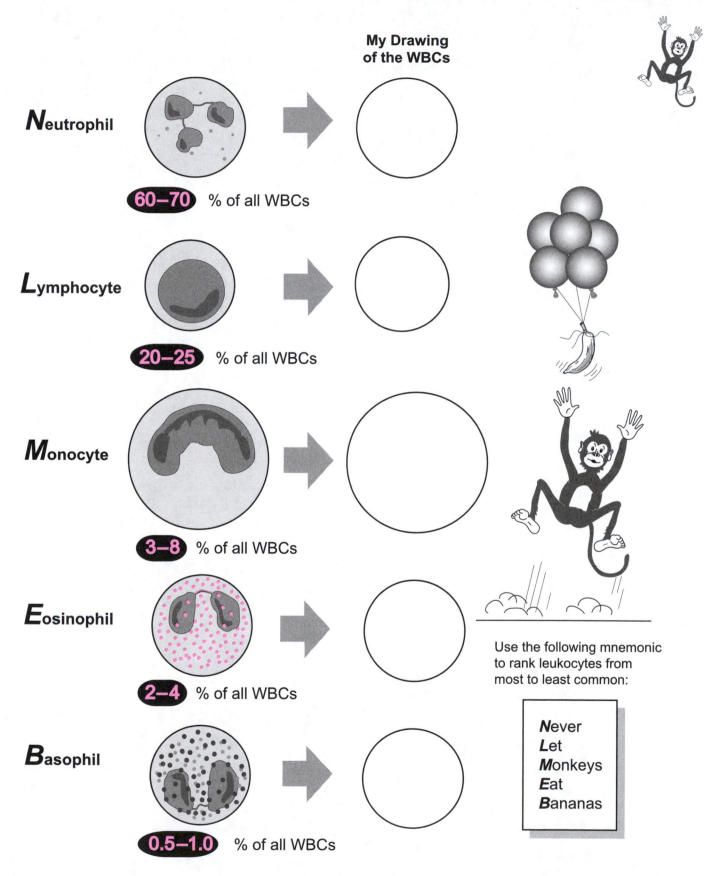

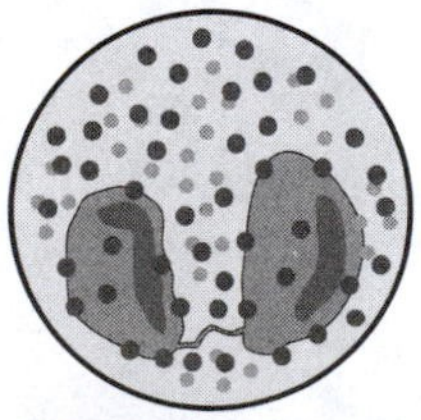

0.5–1.0 % of all WBCs

Use the following mnemonic to rank leukocytes from most to least common:

***N**ever*
***L**et*
***M**onkeys*
***E**at*
***B**ananas*

Notes

Cardiovascular System

Description

The **cardiovascular system** consists of the **heart** and all the **blood vessels**. Functionally, the heart is like a double pump. It consists of two receiving chambers called **atria** and two pumping chambers called **ventricles**. The left side of the heart always pumps oxygenated blood while the right side always pumps deoxygenated blood.

The illustration on the facing page shows blood flow through the heart, through the **pulmonary circuit** and through the **systemic circuit**. The pulmonary circuit refers to all the blood vessels that take deoxygenated blood from the right ventricle of the heart to the lungs and then return oxygenated blood to the left atrium. After this oxygenated blood is pumped from the left atrium to the left ventricle, it is pumped out to the rest of the body. The blood vessels that transport this oxygenated blood to the body are part of the systemic circuit.

All gas exchange occurs within **capillaries**. Capillaries are microscopic blood vessels that are only one cell layer in thickness. Their wall is made of **simple squamous epithelium**. These flat cells easily permit the diffusion of gases such as **oxygen** (O_2) and **carbon dioxide** (CO_2). Oxygen diffuses out of the blood and into body cells to be used in the process of cellular respiration. Carbon dioxide is a normal byproduct of cellular respiration and gradually builds up within body cells. Carbon dioxide diffuses from the body cells into the capillary.

Beginning in the right atrium, this is a flowchart for the blood flow:

Right atrium (1) → tricuspid valve (2) → right ventricle (3) →
pulmonary semilunar valve (4) → pulmonary trunk (5) →
pulmonary arteries (6) → lungs → pulmonary veins (7) → left atrium (8) →
bicuspid valve (9) → left ventricle (10) → aortic semilunar valve (11) →
ascending aorta (12) → aortic arch (13) → descending aorta (14) →
inferior vena cava (15) and superior vena cava (16) → right atrium (1)

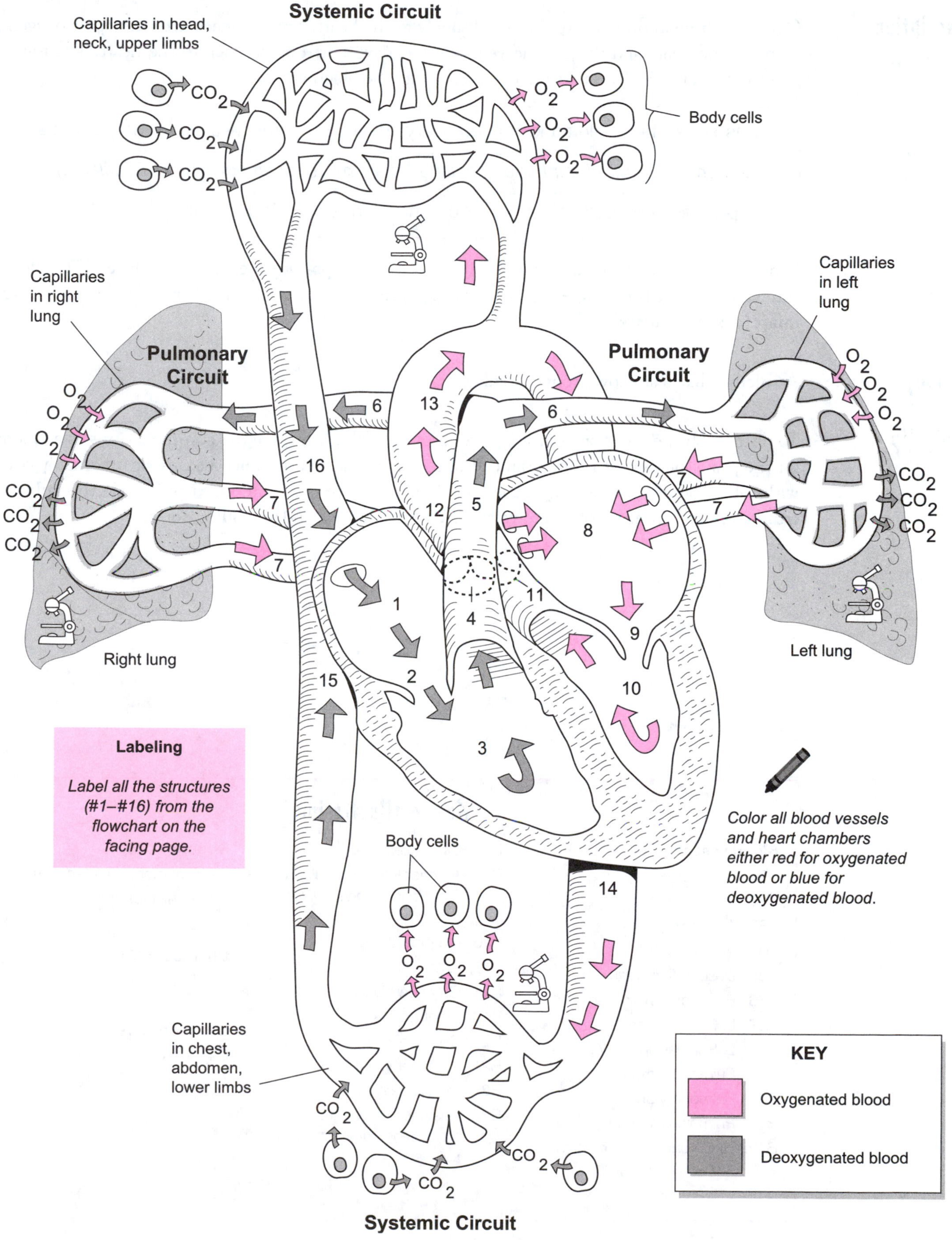

Labeling

Label all the structures (#1–#16) from the flowchart on the facing page.

Color all blood vessels and heart chambers either red for oxygenated blood or blue for deoxygenated blood.

Description

Coronary circulation refers to the blood supply to the heart. The **coronary arteries** supply oxygenated blood to the heart and the **cardiac veins** carry deoxygenated blood back to the heart. The following flowchart summarizes coronary circulation through the blood vessels.

Base of aorta → left and right coronary arteries → branches of coronary arteries (*circumflex a., anterior interventricular a., marginal a., posterior interventricular a.*) → capillaries → cardiac veins → coronary sinus → right atrium

When **coronary arteries** become blocked, the blood supply to the heart is reduced. This deprives cardiac muscle cells of oxygen. If this blockage persists over many years, it may lead to a **myocardial infarction** (*heart attack*).

Analogy

A **sulcus** is like a **shallow groove** or **gulley**.

Study Tip

To distinguish the anterior from the posterior view of the heart, use the **coronary sinus** as a landmark for the posterior view. It is often difficult to see on a dissected specimen because it is normally covered with a horizontal band of fatty tissue. Good landmarks for the anterior view include the **pulmonary trunk, anterior interventricular artery**, **circumflex artery**, and **ascending aorta**.

Key to Illustration

Blood vessels (B)

- B1. Superior vena cava
- B2. Inferior vena cava
- B3. Ascending aorta
- B4. Aortic arch
- B5. Descending aorta
- B6. Brachiocephalic trunk
- B7. L. Common carotid a.
- B8. L. Subclavian a.
- B9. Pulmonary trunk
- B10. Pulmonary arteries
- B11. Pulmonary veins
- B12. R. Coronary a. *(in r. anterior atrioventricular groove)*
- B13. L. Coronary a.
- B14. Circumflex a.
- B15. Anterior Interventricular a. *(in anterior interventricular sulcus)*
- B16. Marginal a.
- B17. Anterior cardiac v.
- B18. Great cardiac v.
- B19. Small cardiac v.
- B20. Coronary sinus
- B21. Posterior v. of l. ventricle
- B22. Middle cardiac v.
- B23. Posterior interventricular a. *(in posterior interventricular sulcus)*

Structures (S)

- S1. Ligamentum arteriosum
- S2. Apex *(tip)* of heart
- S3. Auricle

Chambers (C)

- C1. R. atrium
- C2. L. atrium
- C3. R. ventricle
- C4. L. ventricle

External Structure and Coronary Circulation

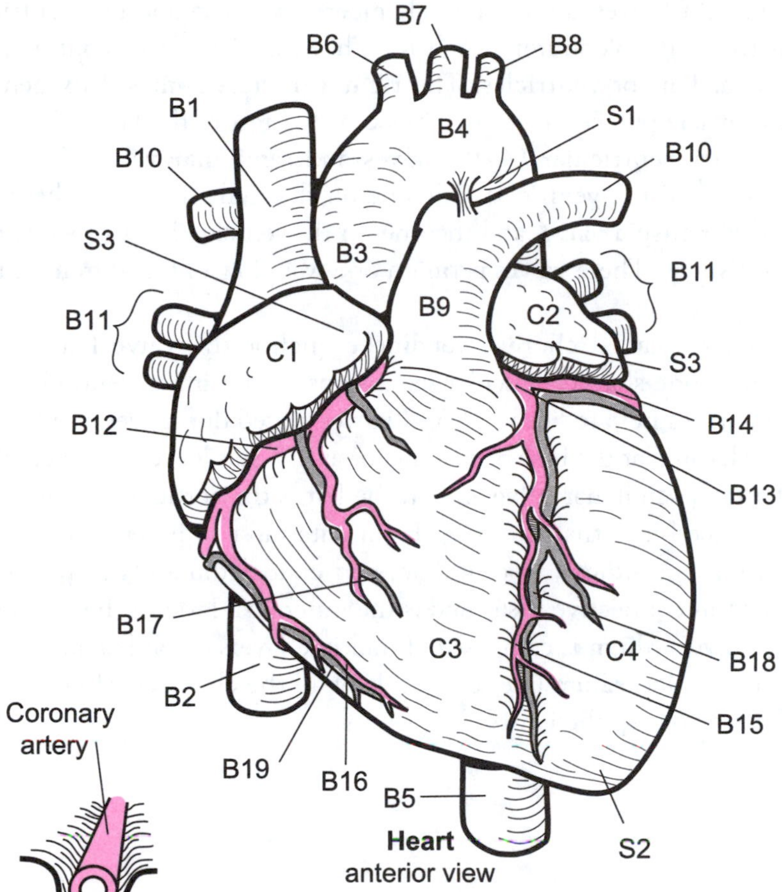

Heart
anterior view

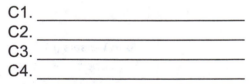

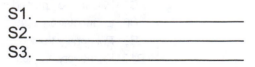

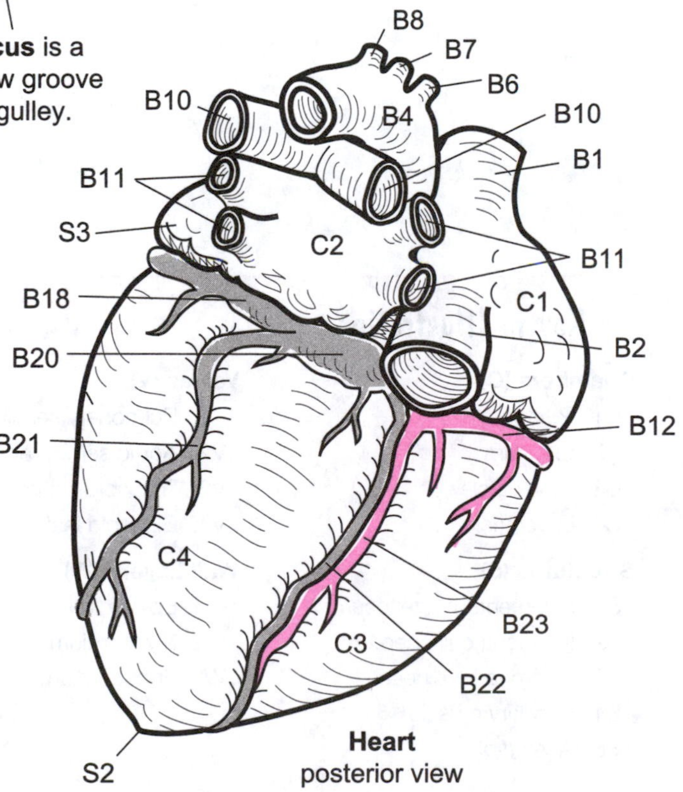

Heart
posterior view

Blood Vessels (B)

B1. ______
B2. ______
B3. ______
B4. ______
B5. ______
B6. ______
B7. ______
B8. ______
B9. ______
B10. ______
B11. ______
B12. ______
B13. ______
B14. ______
B15. ______
B16. ______
B17. ______
B18. ______
B19. ______
B20. ______
B21. ______
B22. ______
B23. ______

Chambers (C)

C1. ______
C2. ______
C3. ______
C4. ______

Structures (S)

S1. ______
S2. ______
S3. ______

Description

The heart is divided into left and right halves and has four chambers, two atria and two ventricles. The **atria** are the first chambers to receive blood from the body. They fill with blood, contract, and transfer blood to the pumping chambers or **ventricles**. The **right ventricle** pumps deoxygenated blood to the lungs and the **left ventricle** pumps oxygenated blood to the rest of the body. The heart has two different types of valves: **atrioventricular (A-V) valves** and **semilunar valves**. The A-V valves are located between the atria and the ventricles. The one on the right side of the heart has three valve flaps, so it is called the **tricuspid valve**, and the one on the left side has two valve flaps so it is called the **bicuspid** *(mitral)* **valve**. These valves permit a one-way flow of blood from atria to ventricles.

Long, fibrous, cord-like structures called **chordae tendineae** anchor the valve flaps to the **papillary muscles**, which are long, cone-shaped muscular extensions of the inner ventricles. The chordae tendineae and papillary muscles help to keep the A-V valves closed during ventricular contraction. The semilunar valves are located at the base of each major artery that leaves each ventricle.

On the right side is the pulmonary semilunar valve and on the left is the aortic semilunar valve. These valves prevent backflow of blood into the ventricles. From outermost to innermost, the wall of the heart is made of three layers: epicardium, myocardium, and endocardium. The **epicardium** (*visceral pericardium*) is made of fibrous connective tissue and is the innermost layer of the pericardial sac that surrounds the heart. The **myocardium** is composed of multiple layers of cardiac muscle and many blood vessels and nerves. The **endocardium** lines the inside of all the chambers along with all the valves and is made of simple squamous epithelium.

Key to Illustration

Blood Vessels (B)
- B1. Superior vena cava
- B2. Inferior vena cava
- B3. Ascending aorta
- B4. Aortic arch
- B5. Descending aorta
- B6. Brachiocephalic trunk
- B7. L. Common carotid a.
- B8. L. Subclavian a.
- B9. Pulmonary trunk
- B10. Pulmonary arteries
- B11. Pulmonary veins

Chambers (C)
- C1. R. atrium
- C2. L. atrium
- C3. R. ventricle
- C4. L. ventricle

Structures (S)
- S1. Ligamentum arteriosum
- S2. Interventricular septum
- S3. Chordae tendineae
- S4. Papillary muscle
- S5. Apex *(tip)*

Valves (V)
- V1. Pulmonary semilunar
- V2. Aortic semilunar
- V3. Tricuspid leaflet
- V4. Bicuspid leaflet

Wall Layers (W)
- W1. Epicardium
- W2. Myocardium
- W3. Endocardium

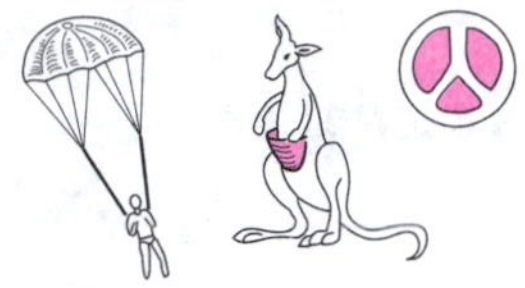

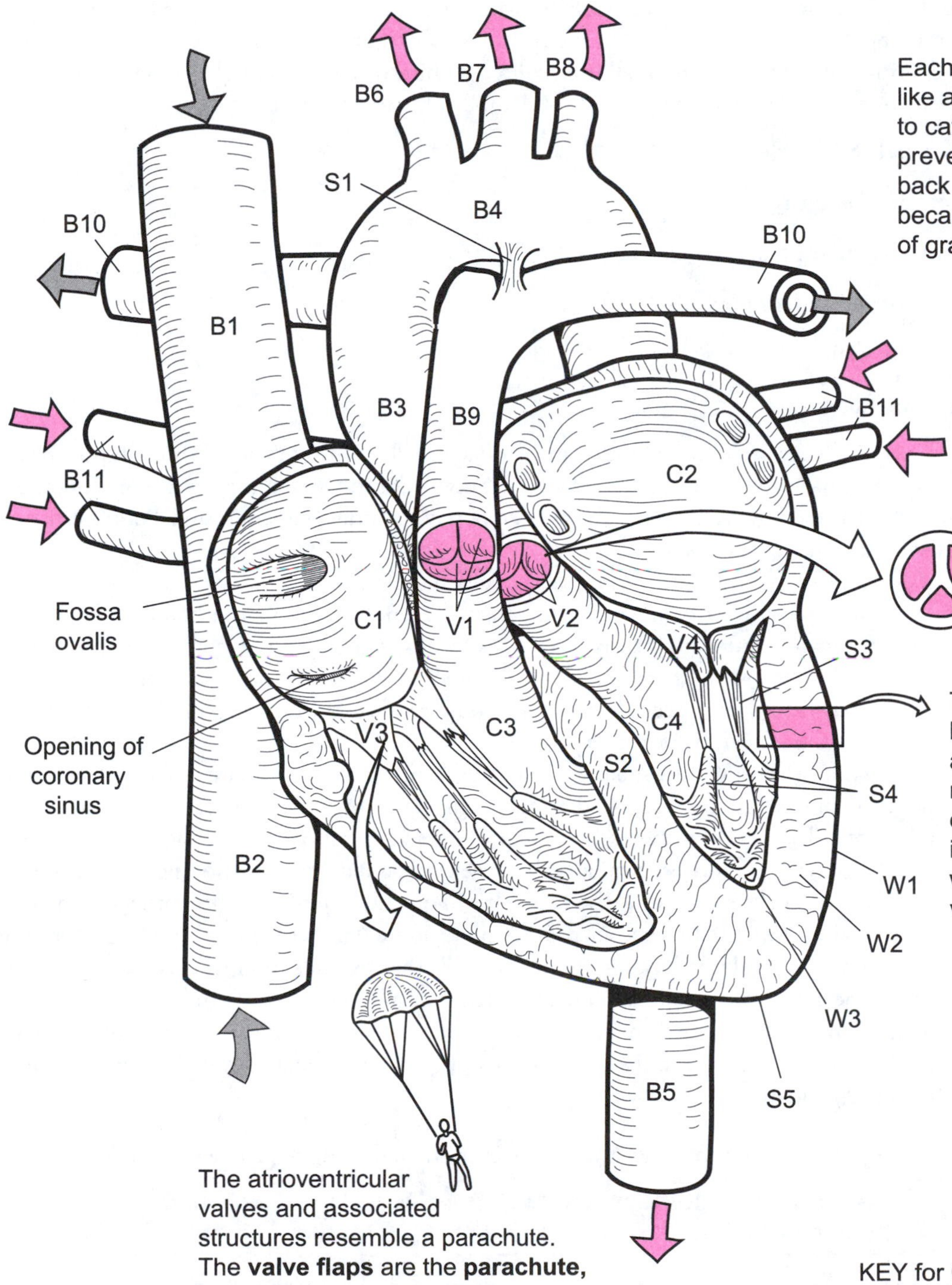

Each **valve flap** opens like a **kangaroo pouch** to catch blood and prevent it from falling back into the ventricle because of gravity.

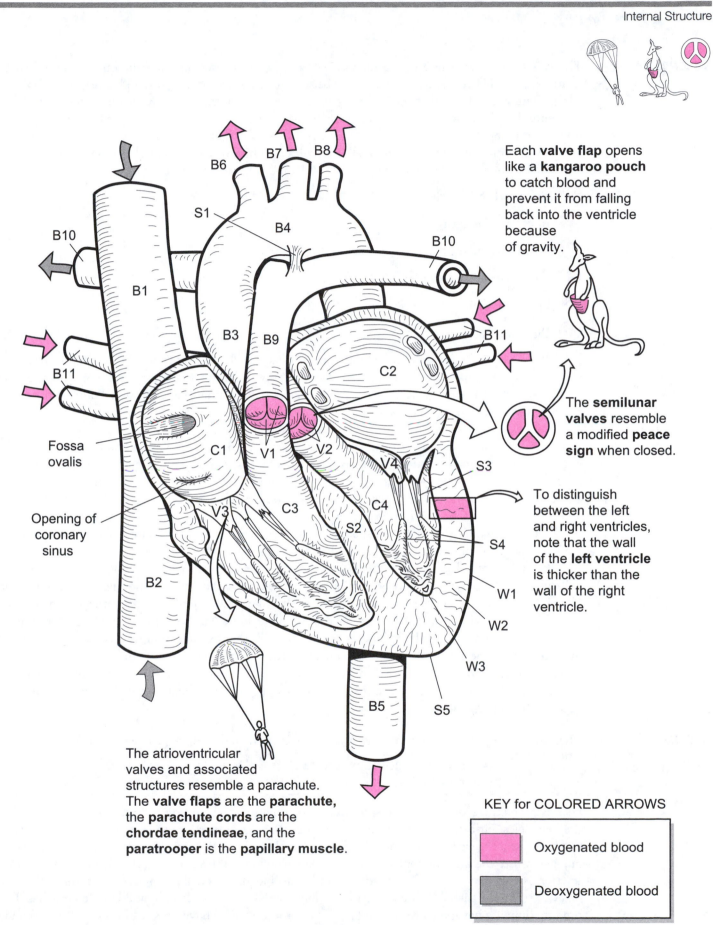

The **semilunar valves** resemble a modified **peace sign** when closed.

To distinguish between the left and right ventricles, note that the wall of the **left ventricle** is thicker than the wall of the right ventricle.

The atrioventricular valves and associated structures resemble a parachute. The **valve flaps** are the **parachute,** the **parachute cords** are the **chordae tendineae**, and the **paratrooper** is the **papillary muscle**.

KEY for COLORED ARROWS

(pink)	Oxygenated blood
(gray)	Deoxygenated blood

Description

The **heart** has its own internal regulation system to achieve two important functions: (1) creating the heartbeat, and (2) coordinating the timing between contraction of the atria and contraction of the ventricles. This control system is referred to as the **intrinsic conduction system**. Without this control system to ensure that the heart chambers completely fill with blood before contracting, the heart would be a very inefficient pump. Unlike most control systems in the body that are regulated by either the nervous system or the endocrine system, this system is all internalized within the cardiac muscle in the heart.

The intrinsic conduction system consists of the following six structures:

1. **sinoatrial (SA) node**
2. **internodal pathway**
3. **atrioventricular node**
4. **atrioventricular (AV) bundle**
5. **bundle branches**
6. **Purkinje fibers.**

This system is a network of interconnecting cardiac muscle cells that spread through the atria and the ventricles. Notice that this system can be structurally divided into two main parts: the **nodes** (SA and AV nodes) and the **pathway** (internodal pathway, AV bundle, bundle branches, and Purkinje fibers). Let's examine each, in turn.

The nodes consist of clusters of specialized cardiac muscle cells that lack the contractile proteins—myosin and actin—found in normal cardiac muscle cells. These cells are the body's only autorhythmic muscle cells; they manufacture their own beat. This means that instead of contracting, they serve as a kind of "sparkplug" to create the heartbeat. The **SA node** is located within the wall of the right atrium and is referred to as the *primary pacemaker*. The **AV node** is located in the septum between the two atria and is referred to as the *secondary pacemaker*.

A long network of specialized cardiac muscle cells—the pathway—ensures that the heart chambers are stimulated to contract in a coordinated manner. Like smooth muscle cells, cardiac muscle cells can stimulate adjacent cells. Think of the pathway like a series of dominoes. Knocking over the first domino will cause all the others to fall over. Similarly, because cardiac muscle cells interdigitate with each other like pieces of a jigsaw puzzle, stimulating the first cell will quickly stimulate all the others in the pathway.

The internodal pathway radiates from the SA node and extends to the left and right atria and AV node. The AV bundle is a relatively short segment of cells that extends from the AV node and penetrates into the top of the interventricular septum. From here, it splits into two pathways—the left and right bundle branches that extend through the interventricular septum toward the apex of the heart. The bundle branches extend into the walls of their respective ventricles and into the papillary muscles to become the Purkinje fibers.

Impulse Pathway

The flow of the impulse for contraction always moves in the following sequence:

SA node → internodal pathway → AV node → AV bundle →
bundle branches → Purkinje fibers

Let's examine this sequence with respect to the coordination of the filling and contracting of the atria and the ventricles. First, the autorhythmic cells within the SA node trigger the impulse to spread to the left and right atria through the internodal pathway. Simultaneously, the atria fill with blood and expand. The impulse causes the atria to contract, which forces blood into the ventricles. A delay in ventricular contraction is needed to allow the ventricles to fill with blood. This delay comes in the form of the time it takes to stimulate the AV node and send the impulse down the AV bundle and bundle branches. By the time the impulse has spread to the Purkinje fibers, the ventricles have finished filling with blood and the ventricles are stimulated to contract.

IMPULSE PATHWAY

1 ______________________

↓

2 ______________________

↓

3 ______________________

↓

4 ______________________

↓

5 ______________________

↓

6 ______________________

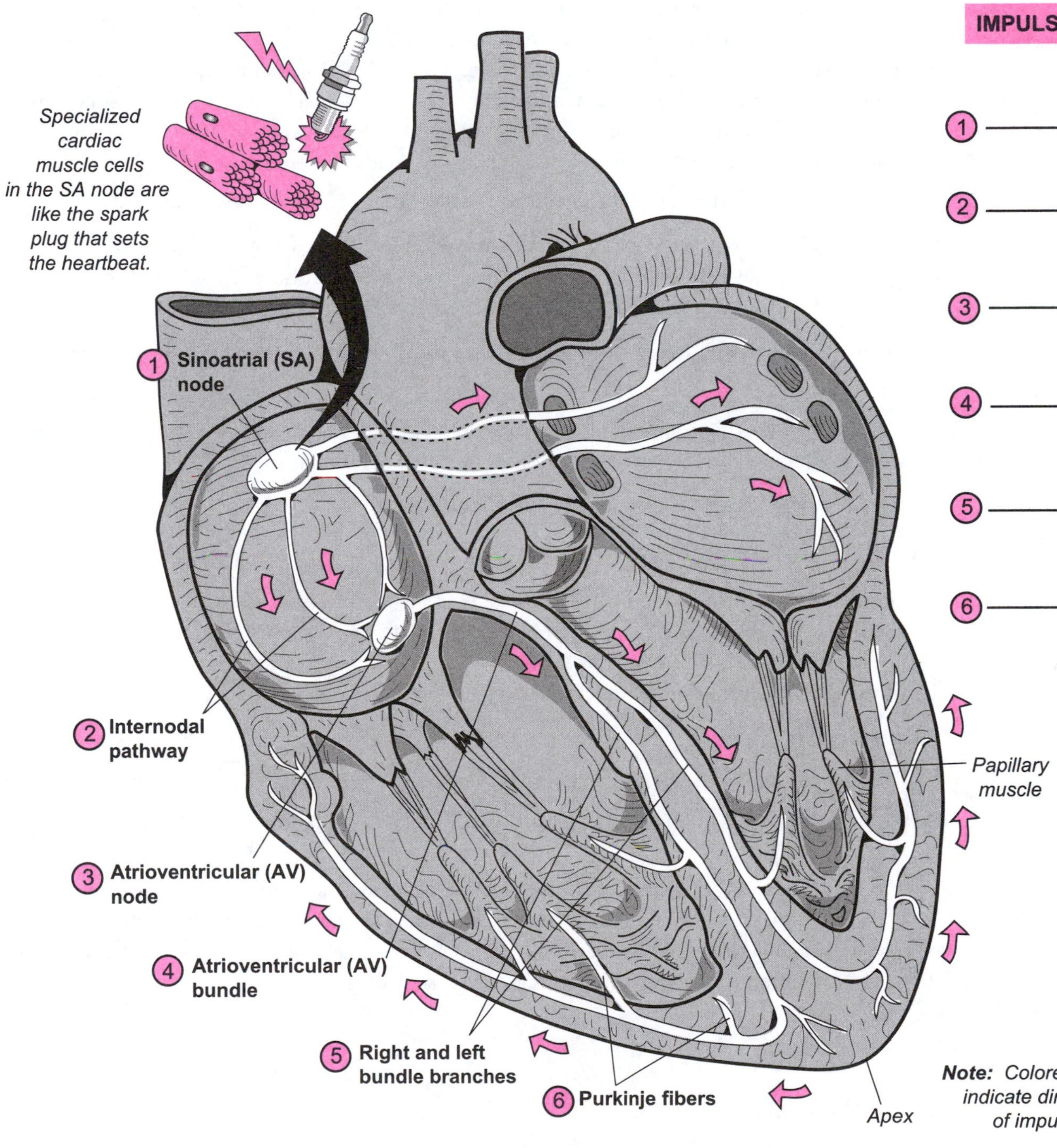

Note: *Colored arrows indicate direction of impulse.*

The cardiac conduction system is shown in white. Color it all one color.

Description

An **electrocardiogram** (**ECG** or **EKG**) is a graph of the heart's electrical activity as expressed in **millivolts** (**mV**) over time. The instrument used to obtain an ECG is called an **electrocardiograph**. A person is connected to the electrocardiograph with electrodes placed on the arms and legs (*limb leads*) and along the chest (*chest leads*). Each of the limb leads gives a slightly different picture of the heart's electrical activity. An ECG is used to detect if the electrical conduction pathway within the heart is normal and if any damage has been done to the heart.

In a typical lead II recording, three different waves appear: **P**, **QRS complex**, and **T**. Each wave represents an electrical event called a *depolarization* or a *repolarization*. These electrical events stimulate cardiac muscle within the heart wall to either contract or relax. Consequently, these events lead to the contraction and relaxation of the heart chambers—atria and ventricles.

- **P wave:** atrial depolarization—at the end of the P wave, both atria have depolarized
- **QRS complex:** ventricular depolarization—at the end of the QRS complex, both ventricles have depolarized. **Note:** Atrial repolarization also occurs during this period, but it is masked by the ventricular depolarization.
- **T wave:** ventricular repolarization—at the end of the T wave, both ventricles have relaxed.

Interpreting ECGs

Reading an ECG is an art that requires a lot of expertise. Even so, we can explain the general way in which they are interpreted. Two types of variations may signal abnormalities:

1. Variation in *wave height* (they may be *elevated* or *depressed*)
2. Variation in normal *time intervals*

Let's consider changes in wave height. For example, if a P wave is elevated, it may indicate atrial enlargement. If the QRS complex is elevated, it may indicate ventricular enlargement. A tall and pointed T wave may indicate myocardial ischemia.

The following three **normal time intervals** are examined on any ECG:

- **P–R interval** = time from beginning of P wave to start of QRS complex; 0.2 sec.
- **S–T segment** = time from end of S wave to beginning of T wave; 0.1 sec.
- **Q–T segment** = time from beginning of QRS complex to end of T wave; 0.4 sec.

Any time interval that is either longer or shorter than normal also may indicate an abnormality. For example, a P–Q interval that is much longer than normal indicates a blockage in the normal conduction pathway. This could be caused by scar tissue resulting from a childhood infection of rheumatic fever. Another example is a longer than normal Q–T segment, which may indicate myocardial damage.

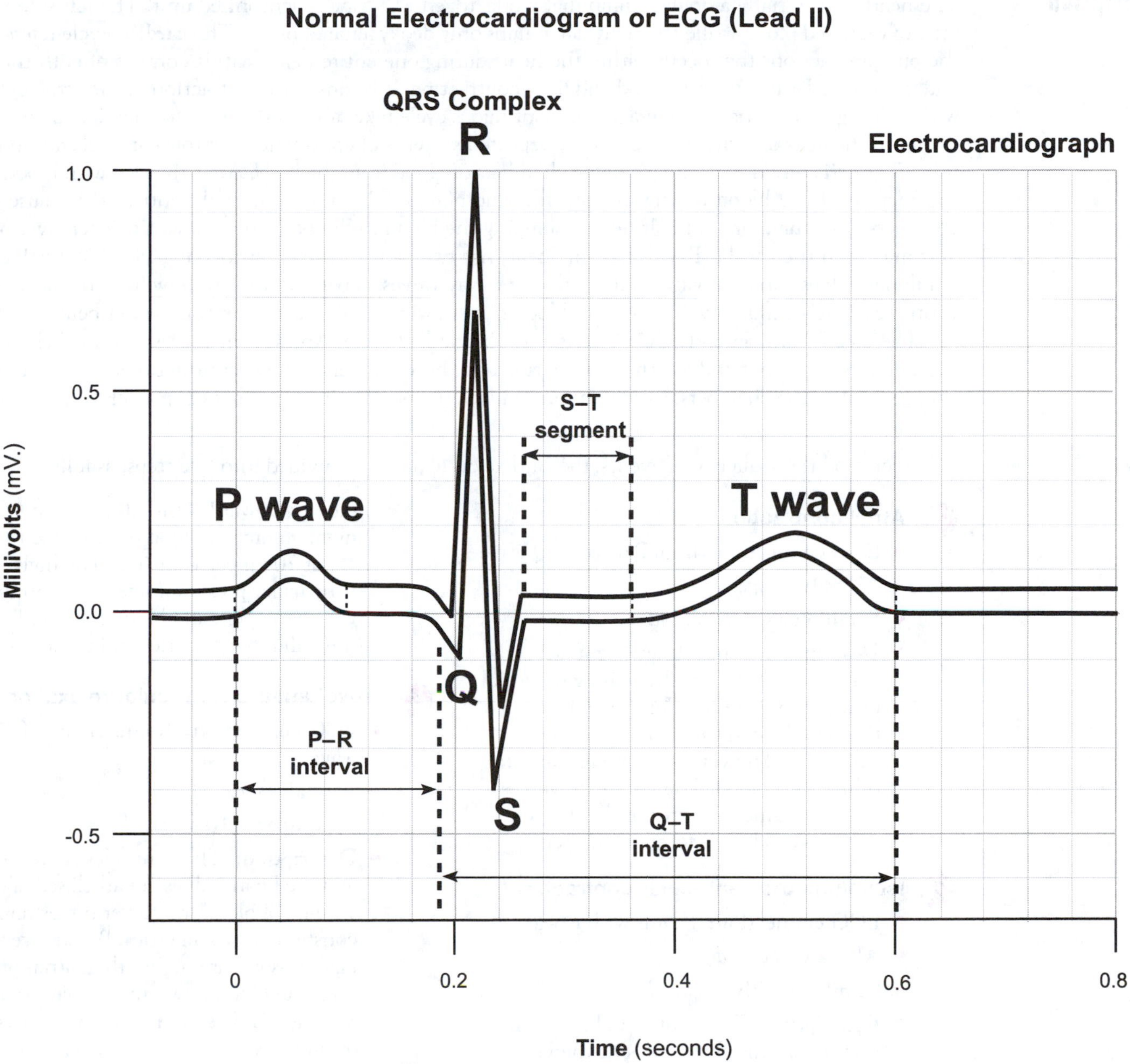

Color the P wave, QRS complex, and T wave different colors.

Description

The heart is essentially a double pump that works together as one synchronized unit. The left side pumps only oxygenated blood while the right side pumps only deoxygenated blood. The **cardiac cycle** refers to all the pumping actions that occur within the heart during one entire heartbeat. It consists of both the atria and the ventricles filling with blood and then contracting. It begins with contraction of the atria and ends with refilling of the atria. On average, this continuous cycle takes about 800 msec. to complete in an adult.

Both the atria and the ventricles have repeated patterns of contraction (systole) and relaxation (diastole). The atria function as the receiving chambers of blood from the body. After they fill with blood, they contract and force blood downward into the true pumps—the more powerful ventricles. Because these chambers have the more difficult task of pumping the blood to the body, they have a thicker layer of cardiac muscle in their walls. Recall that the heart has two pairs of valves: atrioventricular (AV) valves and semilunar valves. The AV valves function as one-way valves to permit blood to flow from the atria to the ventricles. The semilunar valves prevent blood from flowing back into the ventricles after being ejected.

The electrocardiogram (ECG) is related to the cardiac cycle. An ECG measures electrical changes in the heart muscle that induce the contractions of the chambers. These contractions lead to pressure changes within the chambers, which, in turn, induce the heart valves to either open or close.

Steps

To understand the sequence of events, the cardiac cycle has been divided into five steps, as follows:

1 Atrial contraction

- **ECG connection:** from P wave to Q wave
- **AV valves** *open*
- **Semilunar valves** *closed*
- **Description:** The left and right atria contract simultaneously, causing the atrial pressure to increase. This forces blood through the AV valves and into the ventricles. The ventricles are relaxed and filling with blood. The semilunar valves are closed since the pressure in the ventricles is too low to force them open.

2 Isovolumetric ventricular contraction

- **ECG connection:** begins with R wave
- **AV valves** *closed*
- **Semilunar valves** *closed*
- **Description:** The ventricles begin contracting in this phase, causing the ventricular pressure to increase. Higher ventricular pressure relative to the atria closes the AV valves. The volume of blood in the ventricles remains constant (isovolumetric). All heart valves are closed.

3 Ventricular ejection

- **ECG connection:** from S wave to T wave
- **AV valves** closed
- **Semilunar valves** open
- **Description:** The ventricles continue contracting during this phase. Like wringing out a wet rag, the ventricles wring the blood out, beginning at the apex and moving upward. This allows for the maximum volume of blood to be ejected—about 70 ml per ventricle. The high ventricular pressure forces the semilunar valves to open, and blood is ejected into the pulmonary arteries and the aorta.

4 Isovolumetric ventricular relaxation

- **ECG connection:** begins at end of T wave
- **AV valves** *closed*
- **Semilunar valves** *closed*
- **Description:** The ventricles are relaxing and the heart valves are all closed. The volume of blood in the ventricles remains constant (isovolumetric). Because ventricular pressures are higher than atrial pressures, no blood flows into the ventricles. Ventricular pressure is quickly decreasing during this phase.

5 Passive ventricular filling

- **ECG connection:** after T wave to next P wave
- **AV valves** *open*
- **Semilunar valves** *closed*
- **Description:** As blood flows into the atria, the atrial pressure increases until it exceeds the ventricular pressure. This forces the AV valves open, and blood fills the ventricles. This is the main way the ventricles are filled. At the end of this phase, the ventricles will be about 70% filled.

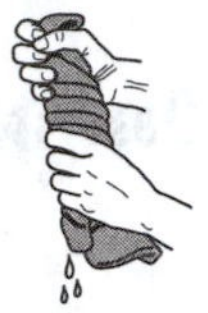

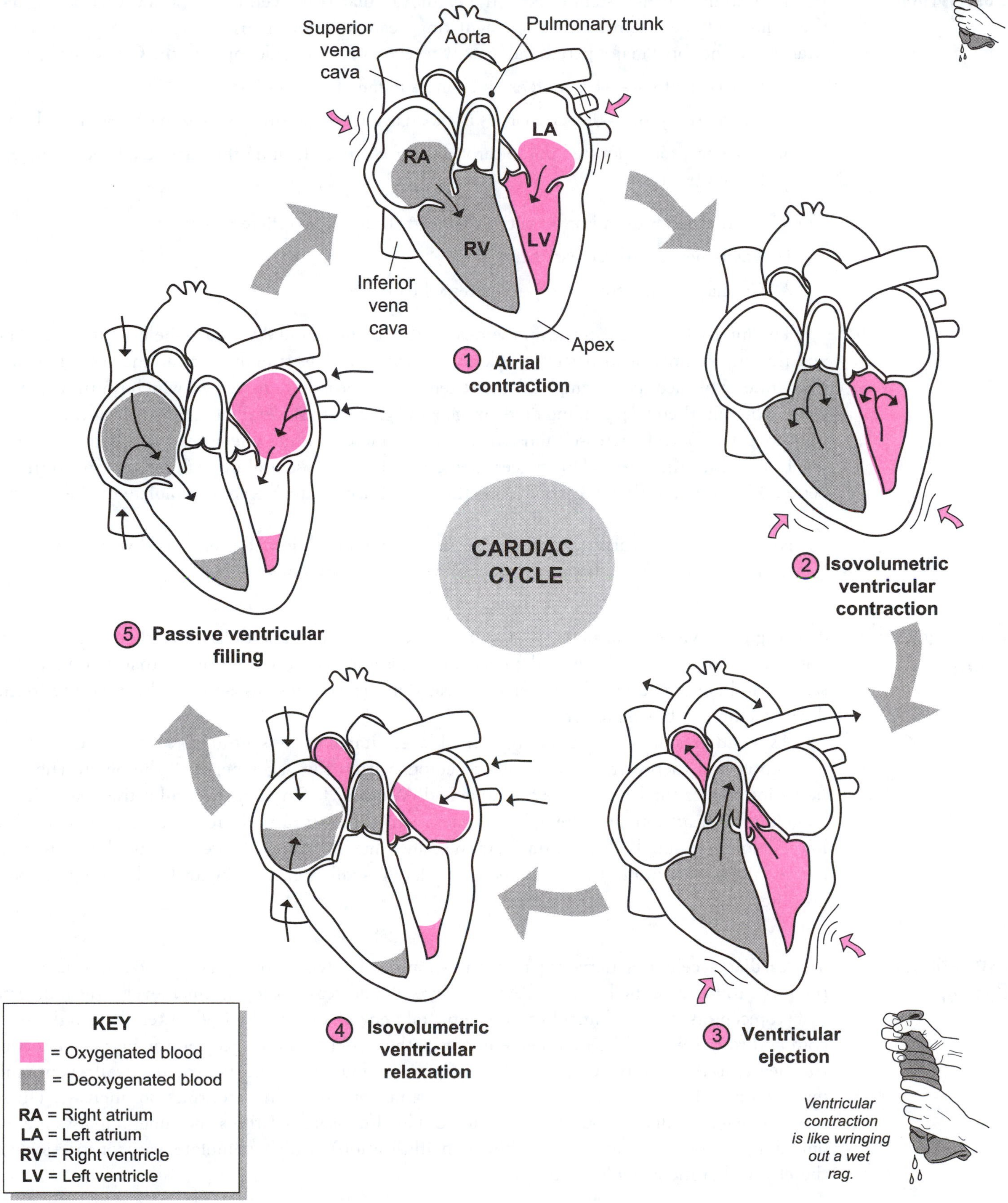
Superior vena cava
Aorta
Pulmonary trunk
LA
RA
RV
LV
Inferior vena cava
Apex
1 Atrial contraction
2 Isovolumetric ventricular contraction
CARDIAC CYCLE
3 Ventricular ejection
4 Isovolumetric ventricular relaxation
5 Passive ventricular filling
KEY
= Oxygenated blood
= Deoxygenated blood
RA = Right atrium
LA = Left atrium
RV = Right ventricle
LV = Left ventricle
Ventricular contraction is like wringing out a wet rag.

Description

In the central nervous system (CNS), the **cardiovascular (CV) center** in the **medulla oblongata** is the command-and-control center for regulating heart function. It uses reflex pathways to control heart rate. The three major peripheral sensory receptors that provide input to the CV center are:

1. **Proprioceptors**—measure tension changes in muscles and joints
2. **Chemoreceptors**—detect changes in blood acidity by sensing changes in CO_2 and H^+ levels
3. **Baroreceptors**—located in the carotid sinus, aortic arch, and other arteries, detect changes in blood pressure

The CV center sends its motor output to the heart through two different nerves:

1. **Vagus nerve**—to decrease heart rate
2. **Cardiac accelerator nerve**—to increase heart rate.

The internal pacemakers within the heart—the **SA node** and **AV node**—help set its normal rate and rhythm. But the heart has to be able to respond to various stimuli so it can increase or decrease heart rate as needed. For example, during exercise, the heart rate increases, which, in turn, elevates cardiac output, thereby providing more oxygen and glucose to skeletal muscle tissue. As you exercise, proprioceptors detect increased tension in skeletal muscles while chemoreceptors detect increased levels of CO_2 in the blood. This triggers nerve impulses to be sent along sensory neurons to the CV center. Motor output then is carried along the cardiac accelerator nerve to stimulate the heart rate to increase.

Now let's use the variable of blood pressure changes to follow the reflex pathway shown in the illustration for both the parasympathetic and sympathetic divisions.

Parasympathetic Control

The **vagus nerve** (cranial nerve X) is part of the parasympathetic division of the ANS and sends impulses to the heart to decrease the heart rate. It is a mixed nerve—meaning that it contains both sensory and motor neurons. Baroreceptors, located in the carotid sinus, aortic arch, and other arteries, detect changes in blood pressure.

As blood pressure increases above normal levels, impulses are sent along sensory neurons to the CV center and then carried along motor neurons within the vagus nerve to the heart. The vagus nerve innervates the heart at the SA node and AV node. It functions to inhibit the heart, thereby reducing the heart rate and strength of contractions. This, in turn, decreases cardiac output, which lowers blood pressure back to normal. At the same time, impulses also are sent out along vasomotor nerves (not shown in illustration), which stimulate vasodilation, thereby further lowering the blood pressure.

Sympathetic Control

The **cardiac accelerator nerve** is part of the sympathetic division of the ANS and sends impulses to the heart to increase the heart rate. As blood pressure decreases below normal levels, this is detected by baroreceptors that send impulses along a neural network over to the CV center, then to the spinal cord and to the heart. The final nerve in this pathway—the **cardiac accelerator nerve**—innervates the heart at the SA node, AV node, and cardiac muscle in the ventricular wall (ventricular myocardium). The response is that both the heart rate and strength of contractions increase. This, in turn, increases cardiac output, which increases blood pressure. At the same time, impulses are sent out along vasomotor nerves (not shown in illustration), which stimulate vasoconstriction and thereby further increase blood pressure.

SENSORY RECEPTORS
Proprioceptors—
tension
Chemoreceptors—
blood acidity
Baroreceptors—
pressure
INPUT
S
Change in
blood pressure.
Baroreceptors
(in carotid sinus)
Baroreceptors
(in aortic arch)
Cardio-
vascular (CV)
center
Glossopharyngeal n.
(Cranial n. IX)
Vagus n.
(Cranial n. X)
Medulla
oblongata
SA node
AV node
Ventricular
myocardium
Brain
Cardio-
vascular (CV)
center
Spinal
cord
Cardiac
accelerator n.
R
Either an
increase or
decrease
in heart
activity.
KEY
Sensory neurons
Motor neurons
Stimulus
Response
Highlight the parasympathetic
pathway in one color and
the sympathetic pathway
in a different color.

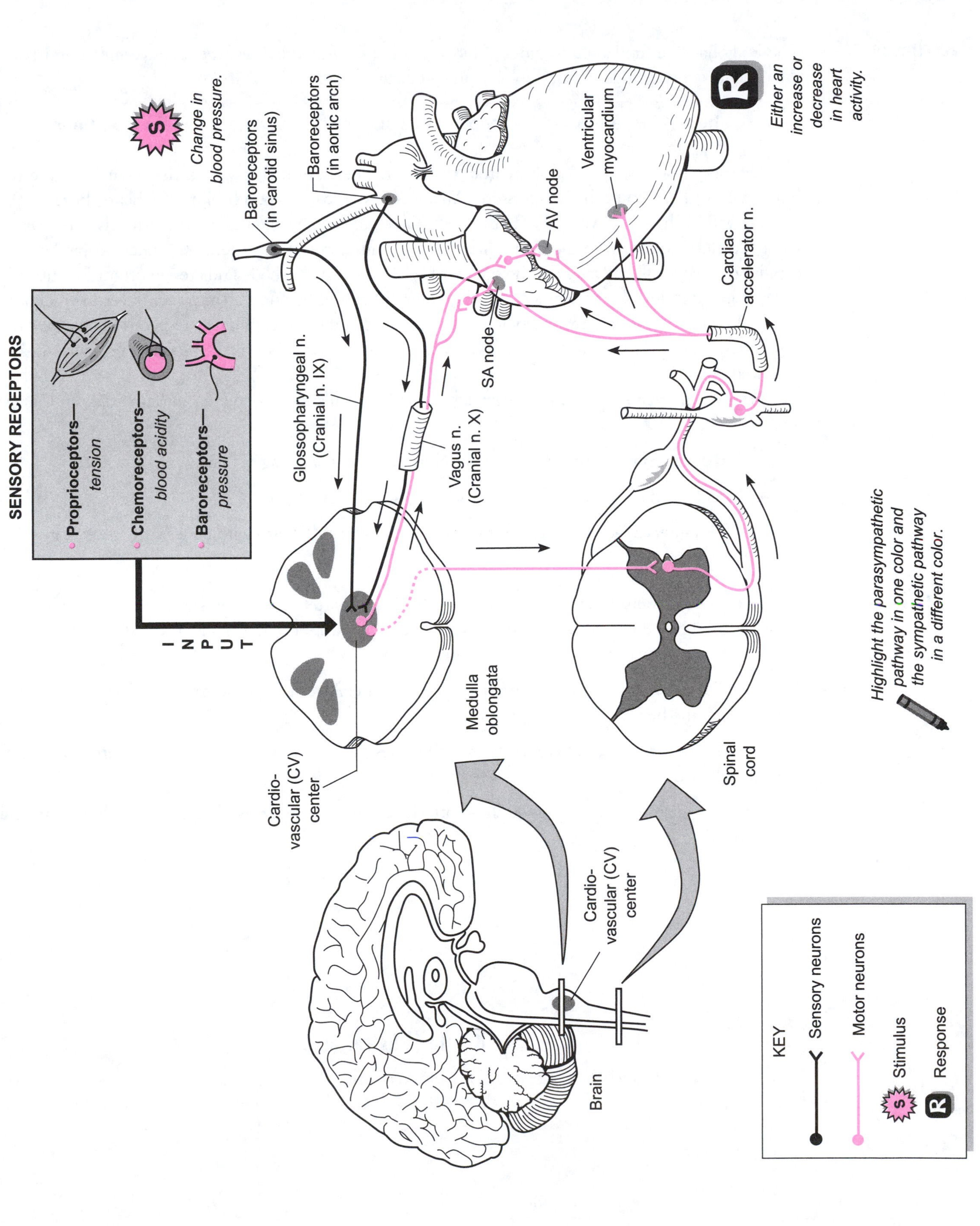

Description

The body has five fundamental types of blood vessels: *arteries*, *arterioles*, *capillaries*, *venules*, and *veins*. All of them connect together in the following predictable pattern:

heart → artery → arteriole → capillary → venule → vein → **heart**

Arteries always carry blood away from the heart. They are thicker-walled than veins because the blood within them is at a higher pressure. All **veins** always carry blood back to the heart. Because the pressure within them is lower, they are thinner-walled. Larger veins contain valves at regular intervals to assist the blood to return to the heart. Arteries and veins connect together at the microscopic level by capillary networks. **Capillaries** are the smallest blood vessels in the body and are important functionally because gas and fluid exchange occurs here. Their entire wall is often a single cell layer in thickness. **Arterioles** and **venules** are microscopic vessels that feed and drain capillaries, respectively. Nearest the capillaries, they are structurally similar to a capillary except they have small amounts of smooth muscle around them.

Arteries and veins have three major layers in their walls: *tunica externa*, *tunica media*, and *tunica interna*:

- **Tunica externa**—connective tissue layer made mostly of collagen fibers
- **Tunica media**—layers of smooth muscle and some elastic fibers
- **Tunica interna**—endothelial layer (simple squamous epithelium) with underlying loose connective tissue

Study Tips

- Don't confuse yourself by trying to distinguish between arteries and veins as to whether they carry oxygenated blood or deoxygenated blood. *This does not work* because some arteries/veins carry oxygenated blood and others carry deoxygenated blood.
- To recall the general function of arteries use the phrase: ***Arteries Away!*** **Arteries** always carry blood *away* from the heart.
- To recall one structural difference between arteries and veins: ***Veins have Valves*** (arteries do not have valves).
- Under the microscope, to distinguish an artery from a vein, **arteries** always have a **thicker tunica media**.

Key to Illustration

1. Tunica interna *(tunica intima)*
2. Tunica media
3. Tunica externa *(tunica adventitia)*
4. Lumen of blood vessel

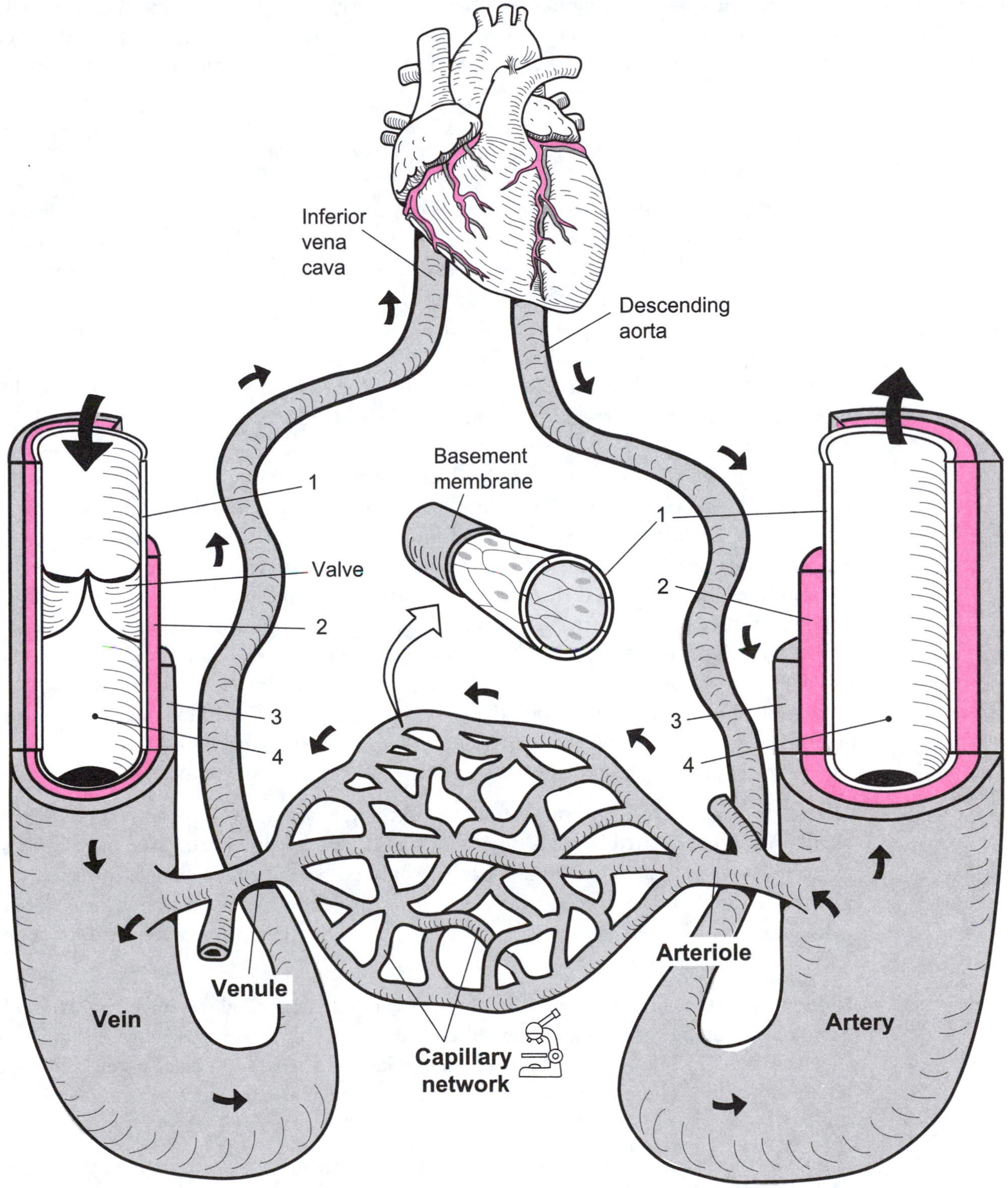
Inferior vena cava
Descending aorta
Basement membrane
1
Valve
2
3
4
1
2
3
4
Venule
Vein
Capillary network
Arteriole
Artery

1. ______________________________
2. ______________________________
3. ______________________________
4. ______________________________

Description

The illustration gives an overview of the general pattern of circulation. Blood always follows a predictable circuit through blood vessels. There are five fundamental types of blood vessels in the body: *arteries*, *arterioles*, *capillaries*, *venules*, and *veins*. All of them connect together in the following pattern:

Heart → artery → arteriole → capillary → venule → vein → heart

The schematic illustration on the facing page shows three of these five: *arteries*, *veins*, and *capillaries*. **Arteries** always carry blood away from the heart. They are thicker walled than veins because the blood within them is at a higher pressure. As distance from the heart increases, pressure decreases. All **veins** always carry blood back to the heart. Since the pressure within them is lower, they are thinner walled. Arteries and veins connect together at the microscopic level by capillary networks. **Capillaries** are the smallest blood vessels in the body and are very important functionally since gas exchange and fluid exchange occurs here. Oxygen exits the blood to be used by body cells and carbon dioxide enters the blood from cells. The liquid plasma is filtered out of the blood to become tissue fluid.

Let's follow the general pattern of circulation. Veins carrying low pressure, deoxygenated blood drain into the **vena cava**, which drains into the heart's **right atrium (RA)**. This receiving chamber fills with blood, contracts, and forces blood into the **right ventricle (RV)**. All this deoxygenated blood is then pumped out of the right ventricle to go to the lungs to get oxygenated. In the lungs, oxygen diffuses into the blood through the **pulmonary capillaries**. The oxygenated blood is then transported through veins to the **left atrium (LA)**. The LA fills with blood, contracts, and forces blood into the **left ventricle (LV)**. This oxygenated blood is then pumped out to the body via the **aorta**. The heart feeds its own cardiac muscle first through **coronary capillaries** so it can continue pumping blood every minute of every day. Arteries carry oxygenated blood above the heart to the capillaries in the brain, trunk, and upper limbs. Other arteries carry blood below the heart to the following major areas:

- **Digestive organs and spleen**—After gas exchange occurs at the **splenic** and **mesenteric capillaries**, deoxygenated blood is carried by veins to the **hepatic portal system** in the liver. Note that capillaries in this system are not for the typical purpose of gas exchange. Instead, these highly permeable capillaries are specialized for delivering nutrients absorbed by the digestive tract to liver cells. The liver cells serve as special processing centers that perform many functions. For example, they detoxify harmful substances.
- **Kidney**—Another unique group of permeable capillaries is the **glomerular capillaries**. Like the capillaries in the hepatic portal system, these are also not for the purpose of gas exchange. Instead, they are specialized to filter the blood plasma, place it in a separate tubular system, and process this liquid into urine. These capillaries lead into the **peritubular capillaries** where gas exchange does occur.
- **Gonads**—In the male, gas exchange occurs at capillaries in the testes whereas, in the female, gas exchange occurs at capillaries in the ovaries.
- **Liver, lower limbs**

Overview of the General Pattern of Circulation

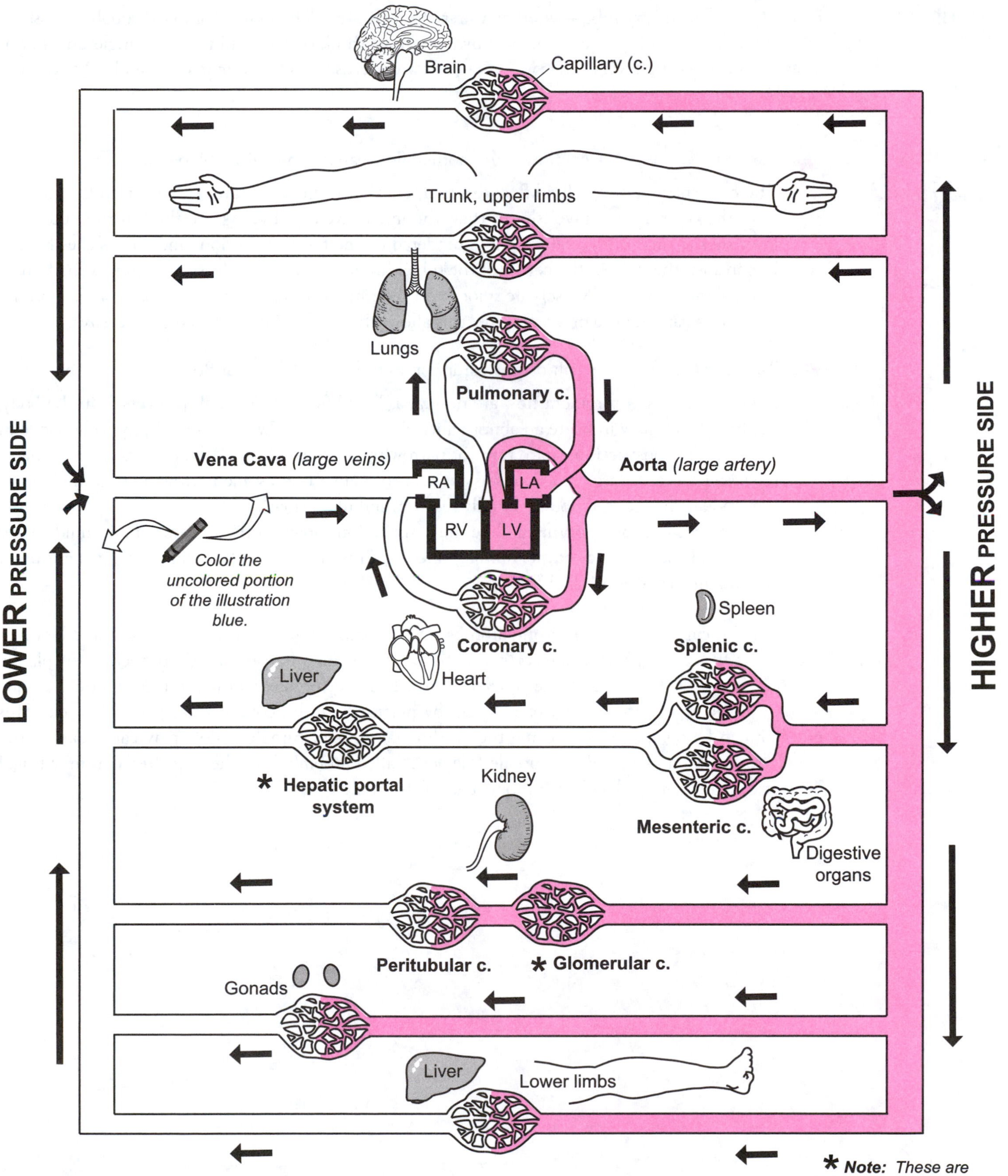

* **Note:** *These are specialized capillaries.*

Description

Fluid flow through any tube—whether water through a garden hose or blood through a vessel—follows a pattern. To understand **blood flow**, you must think of the fluid not as a single unit but as moving in concentric layers. The two types of flow patterns are: (1) **laminar flow**, and (2) **turbulent flow**.

Types of Blood Flow

(1) **Laminar** (*lamina* = layer) **flow** is the normal flow pattern in healthy blood vessels.

The concentric layers of fluid flow together, but not at the same rate. As shown in the illustration, the outer layers travel slower than the inner layers. Why? As the fluid in the outermost layer rubs against the vessel wall, it is hindered by the force of friction, and moves the slowest. In contrast, the fluid in the center of the tube is interfered with the least, so it moves the fastest. Normal, healthy blood vessels are smooth on their inner surface. Like the smooth surface on the inside of a garden hose or a drainpipe, this allows the fluid to flow through most easily.

(2) **Turbulent flow** is the result of any disruption in the normal, laminar flow.

Some low levels of turbulent flow are normal as blood flows through all the vessels in the body, but high levels may indicate an abnormality. What causes it? Two things: (1) physical change in a vessel (ex: constriction, sharp turn, or narrowing of a vessel) or (2) Disease state (ex: atherosclerosis). Because large blood vessels connect to smaller ones, some narrowing is normal, but it is a gradual change that doesn't usually result in turbulent flow.

Let's consider an abnormality like the constriction shown in the illustration. As fluid strikes the wall of the constriction, its normal, linear path is disrupted. The result is the generation of swirling little currents—like backflows.

As a clinical example, the carotid arteries are the major vessels that deliver oxygenated blood to the brain. If a person has atherosclerosis, these arteries may become blocked because of plaque deposits. This can lead to a stroke or even death. Because turbulent flow produces its own unique sounds, a physician can detect this problem by placing a stethoscope on the patient's neck. This problem can be compounded because the swirling currents of turbulent flow may cause part of the plaque to break off—becoming a mobile fragment called an embolus. The embolus then may travel downstream and cause a blockage in some part of the brain.

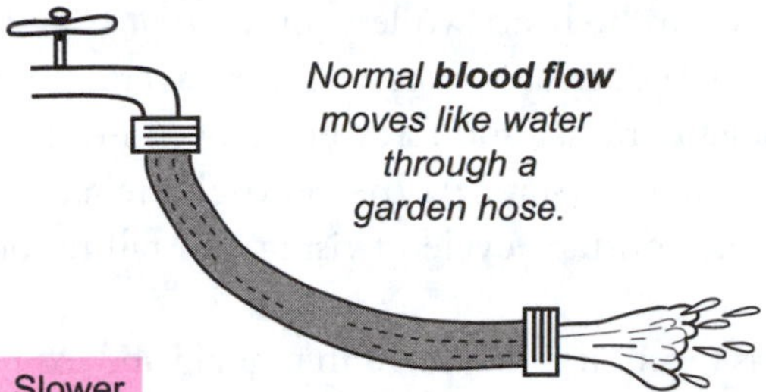

Normal ***blood flow*** *moves like water through a garden hose.*

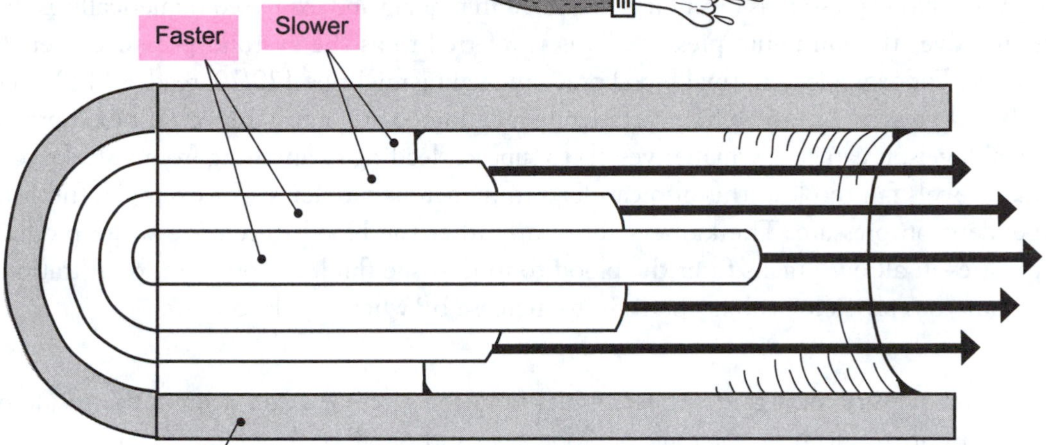

① **LAMINAR FLOW**

***L**aminar is **L**inear: Each fluid layer flows in a straight line.*

Blood vessel wall

Color each concentric ring of fluid a different color.

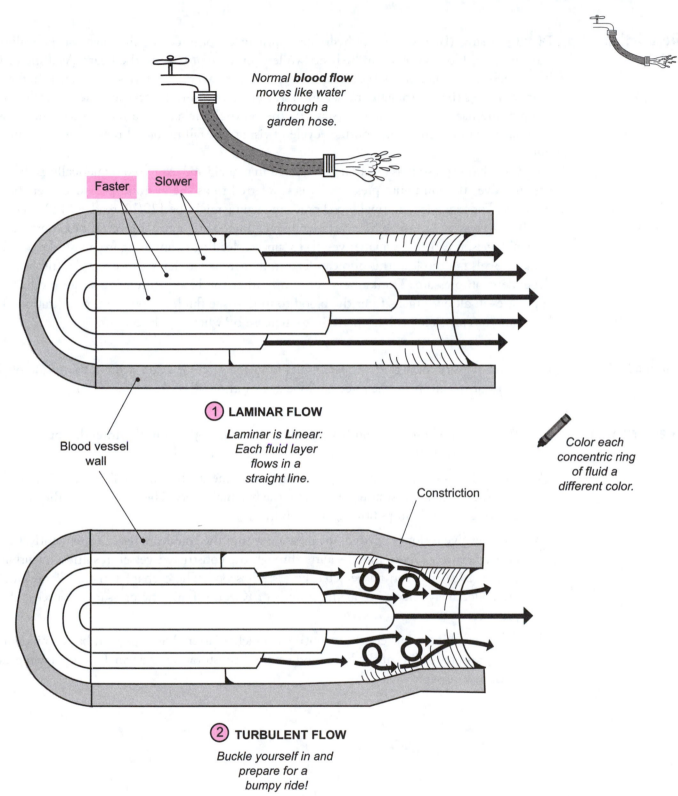

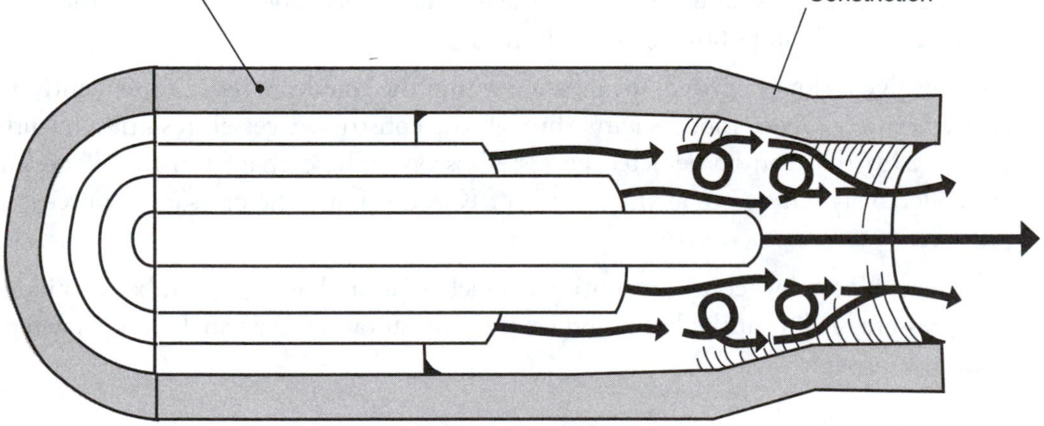

② **TURBULENT FLOW**

Buckle yourself in and prepare for a bumpy ride!

Description

Blood pressure (BP) is a type of **hydrostatic pressure**—or force of a fluid against the wall of a tube. Arteries carry blood away from the heart while veins carry it back to the heart. As distance from the heart increases, blood pressure decreases. Therefore, on average, arteries have much higher pressure than veins. As the ventricles contract and force blood out into the arteries, these vessels expand and the pressure rises to a *maximum* pressure. As the ventricles relax, the arteries recoil and pressure falls to a *minimum* pressure. This constant cycle of rising and falling blood pressure is the source of our pulse.

When blood pressure is taken, it is measured in mm Hg and expressed numerically as the maximum pressure over the minimum pressure. This is referred to as the **systolic pressure** over the **diastolic pressure**. For example, a normal blood pressure reading might be 120/70, read as "120 over 70."

Blood pressure has to be homeostatically maintained at a normal level. If blood pressure exceeds normal levels, it can cause smaller vessels to rupture, leading to anything from a stroke to even blindness. If levels fall too low, this also can lead to numerous problems. For example, filtration is totally dependent on pressure. The kidneys constantly filter the blood to remove waste products, and the capillaries in all our organs filter the blood to form tissue fluid. No pressure, no filtration. The body has numerous physiological mechanisms to increase BP when it falls too low.

Analogy

Like water pressure in a garden hose results from the force of water against the inside of the hose, blood pressure results from the force of blood against the wall of the blood vessel.

Measuring BP

1. When a blood pressure cuff is wrapped around the upper arm, the intended purpose is to compress the **brachial artery**, which delivers oxygenated blood to the arm.
2. The valve is closed on the blood pressure instrument, then the bulb is squeezed until the cuff pressure exceeds the systolic pressure in the brachial artery. The result is that the artery closes so blood temporarily stops flowing through it.
3. As the valve is slowly opened, the pressure within the cuff decreases. Consequently, the brachial artery begins to open. Blood squirts through the constricted vessel, resulting in turbulent flow (see p. 298) that can be detected by a stethoscope. These characteristic "whooshing" sounds are collectively referred to as the **sounds of Korotkoff** and the pressure at onset is the systolic pressure.
4. Turbulent blood flow continues until the brachial artery has expanded back to its normal size. The pressure reading at the last sound before the restoration of normal blood flow represents the diastolic pressure.

Blood pressure (BP) is a type of **hydrostatic pressure** (HP) or fluid force against the wall of a tube.

Like water pressure in a garden hose results from the force of water against the inside of the hose, blood pressure results from the force of blood against the wall of a blood vessel.

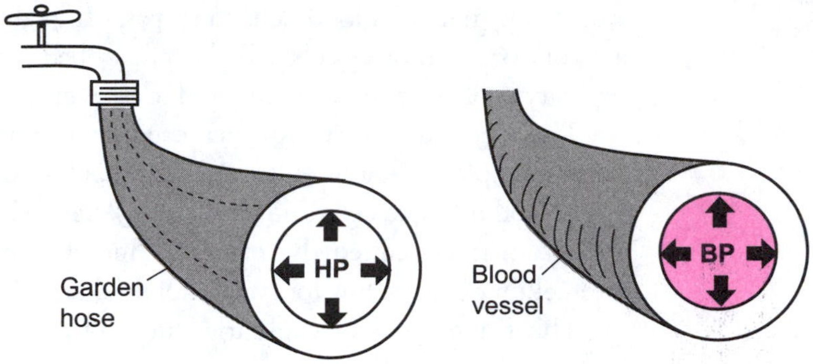

Measuring Blood Pressure

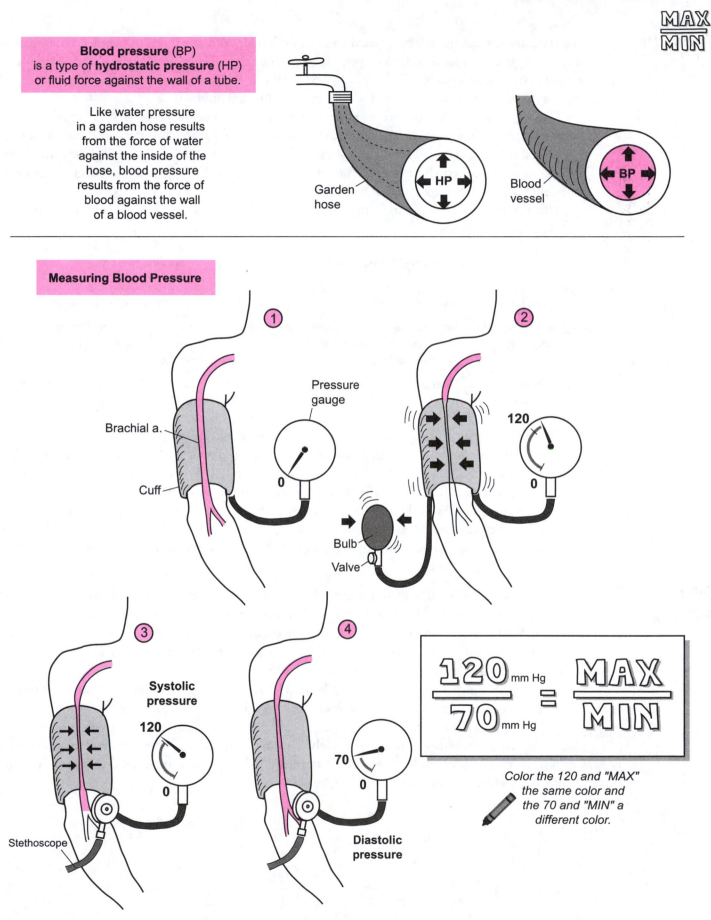

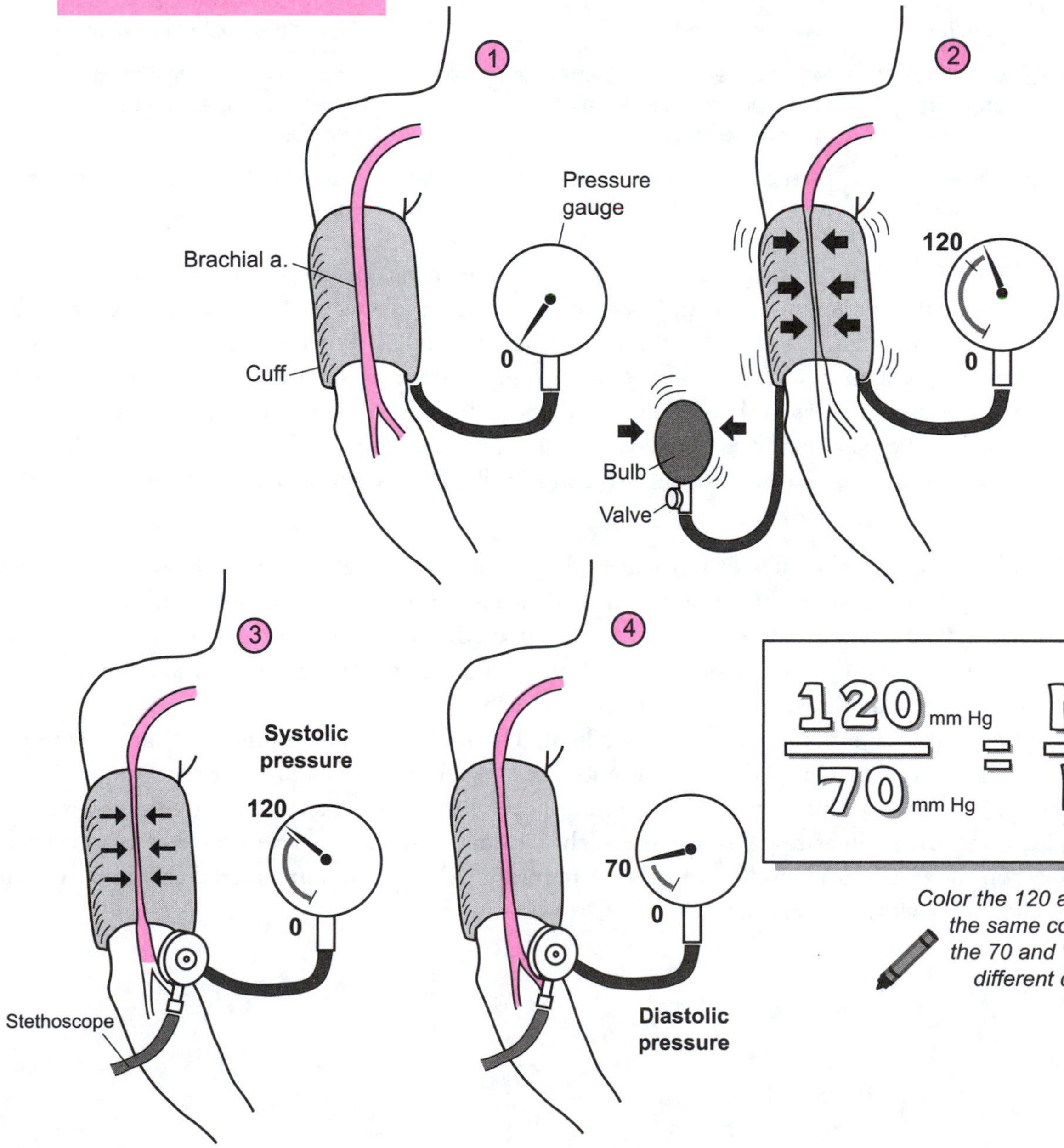

$$\frac{120 \text{ mm Hg}}{70 \text{ mm Hg}} = \frac{\text{MAX}}{\text{MIN}}$$

Color the 120 and "MAX" the same color and the 70 and "MIN" a different color.

Description

Capillaries, the most microscopic blood vessels in the body, join **arterioles** to **venules**. Their diameter is so small that red blood cells must pass through single file. These vessels form an interconnecting network or mesh of vessels called capillary beds. All the various tissues in the body depend on these capillary beds to remove wastes, and to deliver nutrients and fluid.

The tube-like structure of each capillary is simple. The wall of the vessel is made up of endothelial cells or simple squamous epithelial cells. Each of these cells is flat to allow for easy diffusion of solutes from blood to tissue cells and vice versa. Wrapped around the outside of these cells like a thin blanket of protein fibers called the basement membrane. Gaps between adjacent endothelial cells called intercellular clefts allow for passage of fluid and small solutes.

The three types of capillaries are: (1) continuous capillaries, (2) fenestrated capillaries, and (3) sinusoidal capillaries.

Type	Comments/Permeability	Location in Body
(1) **Continuous capillary**	• Most common • *Least permeable*	Skin, connective tissues, skeletal muscle, smooth muscle, and lungs
(2) **Fenestrated capillary**	• *More permeable.* Endothelial cells contain small holes called fenestrations, which increase permeability	Kidneys, small intestine, choroid plexuses in brain, and some endocrine glands
(3) **Sinusoidal capillary**	• *Most permeable*—has the largest fenestrations and the largest intercellular clefts.	Liver, red bone marrow, spleen, and some endocrine glands

Blood flow through a capillary bed is controlled by bands of smooth muscle called precapillary sphincters. When they are closed, blood bypasses the capillary bed by going through the **shunt**. When they are open, blood moves through the entire capillary bed (as illustrated). This allows the body to respond to different situations. For example, during exercise, the sphincters have to be opened in the capillaries in skeletal muscles to meet the muscle's increased oxygen demand and deal with increased production of carbon dioxide. During the digestion of a heavy meal, more blood has to be shunted to the capillaries in the digestive tract rather than to the skeletal muscles.

Solute Diffusion

Capillary beds are the sites of nutrient and waste exchange between the blood and tissue cells. Nutrients move from the blood into tissue cells while wastes move from tissue cells into the blood. For example, the respiratory gases—oxygen and carbon dioxide—diffuse across the wall of the capillary. Oxygen is a vital nutrient that diffuses out of the blood and into tissue cells while carbon dioxide is a waste product that diffuses from tissue cells into the blood.

Recall that diffusion depends on a gradient. The rule for simple diffusion is that a solute moves from a region of higher solute concentration to a region of lower solute concentration. Oxygen is carried in the blood primarily by the protein hemoglobin. As cells consume oxygen to metabolize glucose in the cellular respiration process, this ensures that the gradient for oxygen is maintained. Carbon dioxide is a byproduct of cellular respiration, so it maintains a gradient as it constantly accumulates within tissue cells.

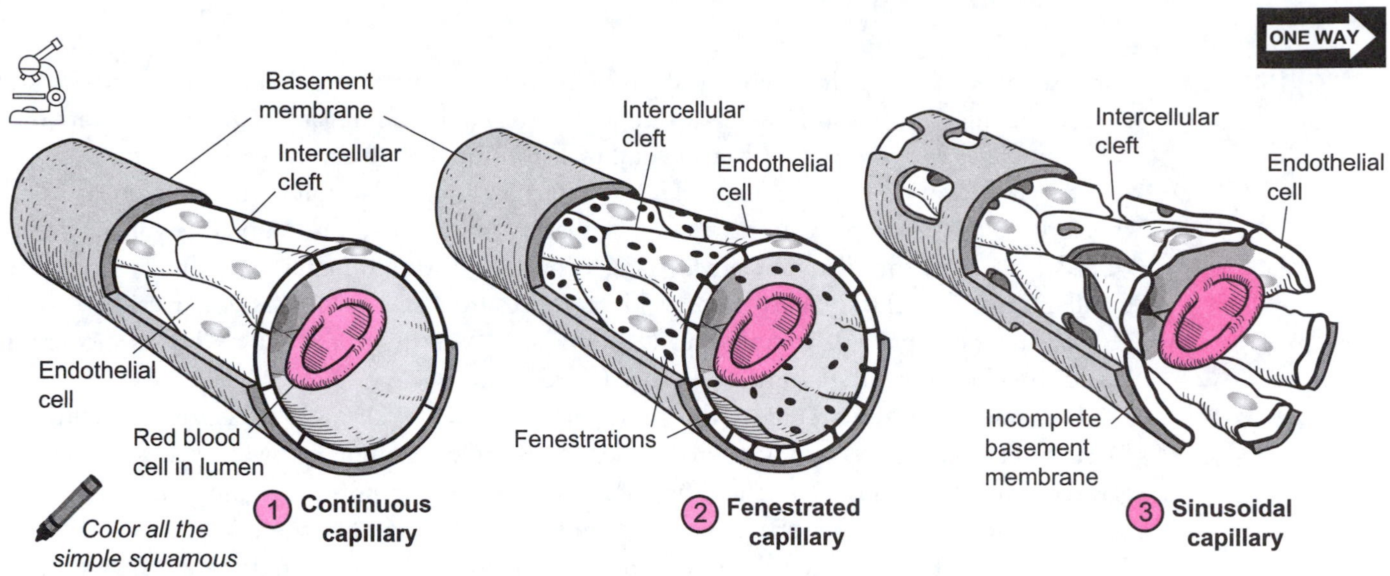

Color all the simple squamous epithelial cells the same color.

Three types of capillaries

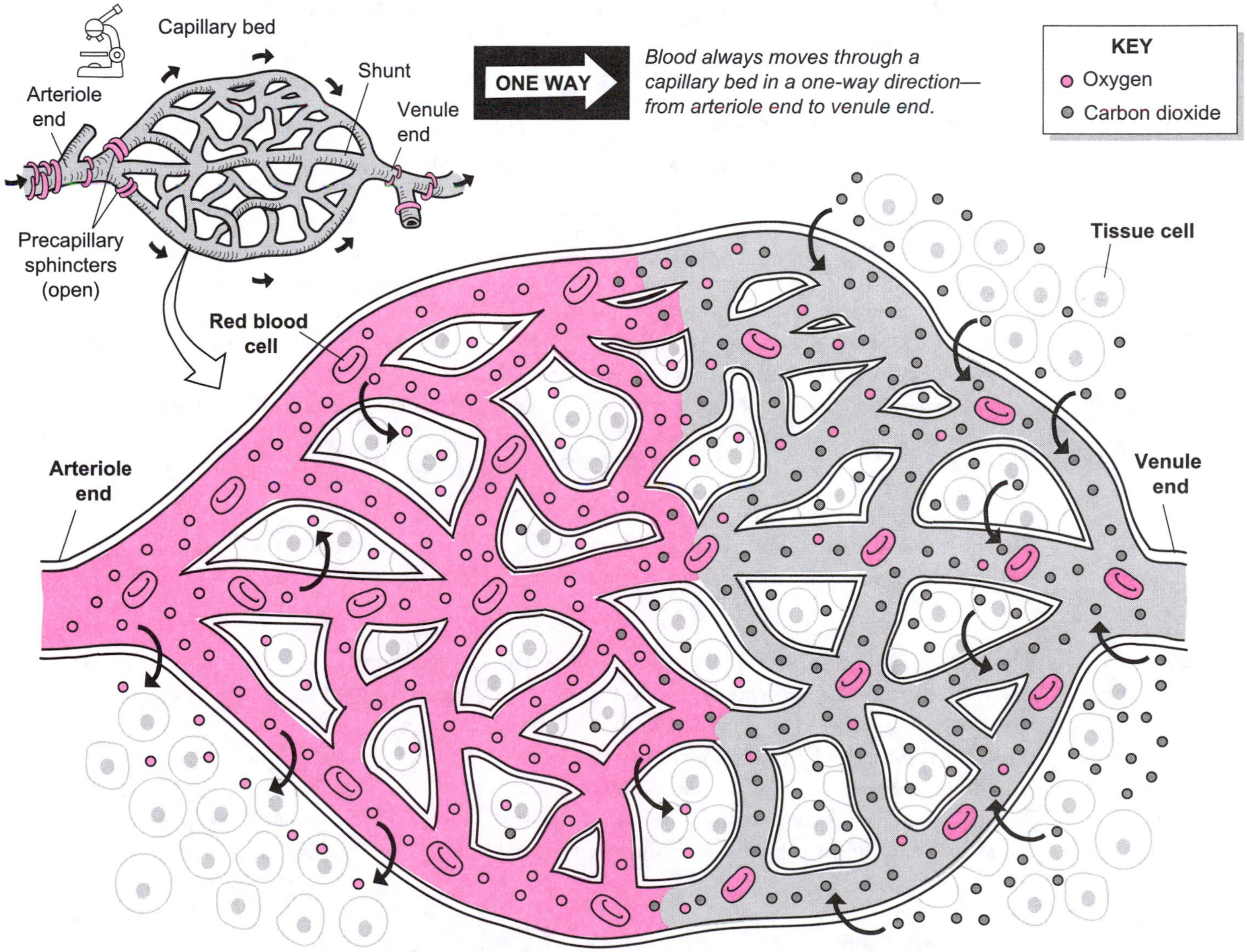

Diffusion of oxygen and carbon dioxide across a capillary bed

Description

Gas exchange occurs across capillary beds. This is how oxygen is delivered to your cells and how carbon dioxide is removed from tissues. In addition to diffusion of gases, two processes are occurring simultaneously with diffusion: **filtration** and **reabsorption**. Let's examine each, in turn.

The filtration process is completely dependent on a force. No force, no filtration. In capillaries, this force is the blood pressure inside the capillaries and is called the **capillary hydrostatic pressure (CHP)**. It drives the filtration process and is measured in millimeters of mercury (Hg). Across a capillary bed, the CHP quickly drops as blood moves from the *higher* pressure **arteriole end** (35 mm Hg) to the *lower* pressure **venule end** (18 mm Hg). Working against the **CHP** is a counter force called the **blood colloidal osmotic pressure (BCOP)**.

Focus on the last two words of this term: osmotic pressure. **Osmotic pressure** deals with the force from water's tendency to move across a semipermeable membrane toward the solution with the greater concentration of nonpenetrating solutes. The osmotic pressure is proportional to the solute concentration. The higher the solute concentration, the greater is the osmotic force. Because the concentration of nonpenetrating solutes in the blood is greater than the tissue fluid within the interstitial spaces, water's tendency is to move back into the blood.

Unlike the CHP, the BCOP remains constant across the capillary bed at 25 mm Hg. Why? Because the concentration of nonpenetrating solutes does not change from one end of the capillary bed to the other.

Formula

Filtration occurs *only* at the **arteriole end** of the capillary bed. Here is the formula to calculate the

net filtration pressure (NFP): $$\mathbf{NFP = CHP - BCOP}$$

substituting the numbers:
$$\mathbf{NFP = 35\ mm\ Hg - 25\ mm\ Hg}$$
$$\mathbf{= +10\ mm\ Hg}$$

This positive (+) number indicates a net movement of water and small solutes (such as sodium and chloride ions) out of the blood and into the interstitial spaces. This process of filtering the blood plasma is the only way the body has to create tissue fluid.

Reabsorption of water occurs *only* at the venule end. Let's calculate the NFP again:

$$\mathbf{NFP = 18\ mm\ Hg - 25\ mm\ Hg.}$$
$$\mathbf{= -7\ mm\ Hg}$$

The negative (−) number indicates net movement of water back into the blood. In short, some water is always reabsorbed at the venule end of the capillary because of osmosis.

Analogy

The capillary hydrostatic pressure (CHP) is the blood pressure inside the capillary. The CHP is like the water pressure inside a garden hose that has been turned on. Now imagine that you punctured some tiny holes in the hose. The fluid squirting out of the hose is like the solution being filtered out of the blood to create the tissue fluid within the interstitial spaces.

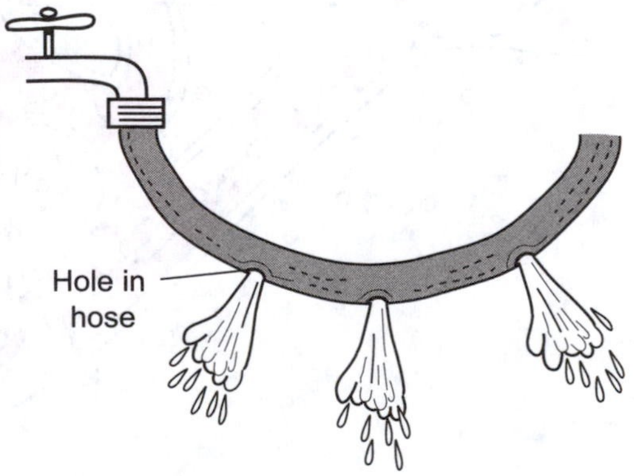

Capillary Function—Filtration and Reabsorption

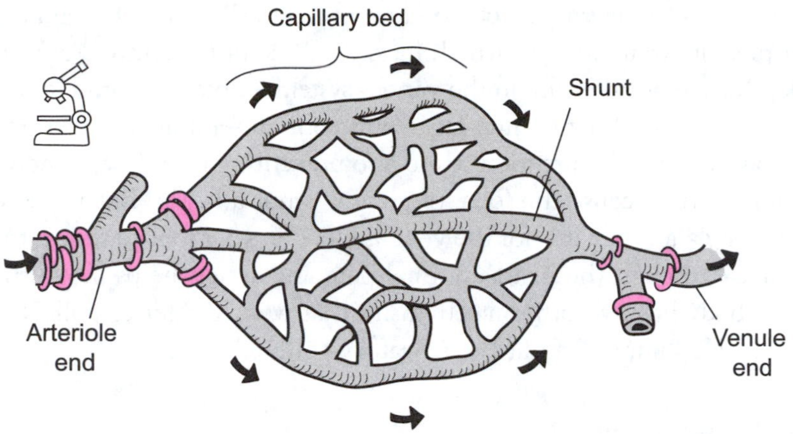

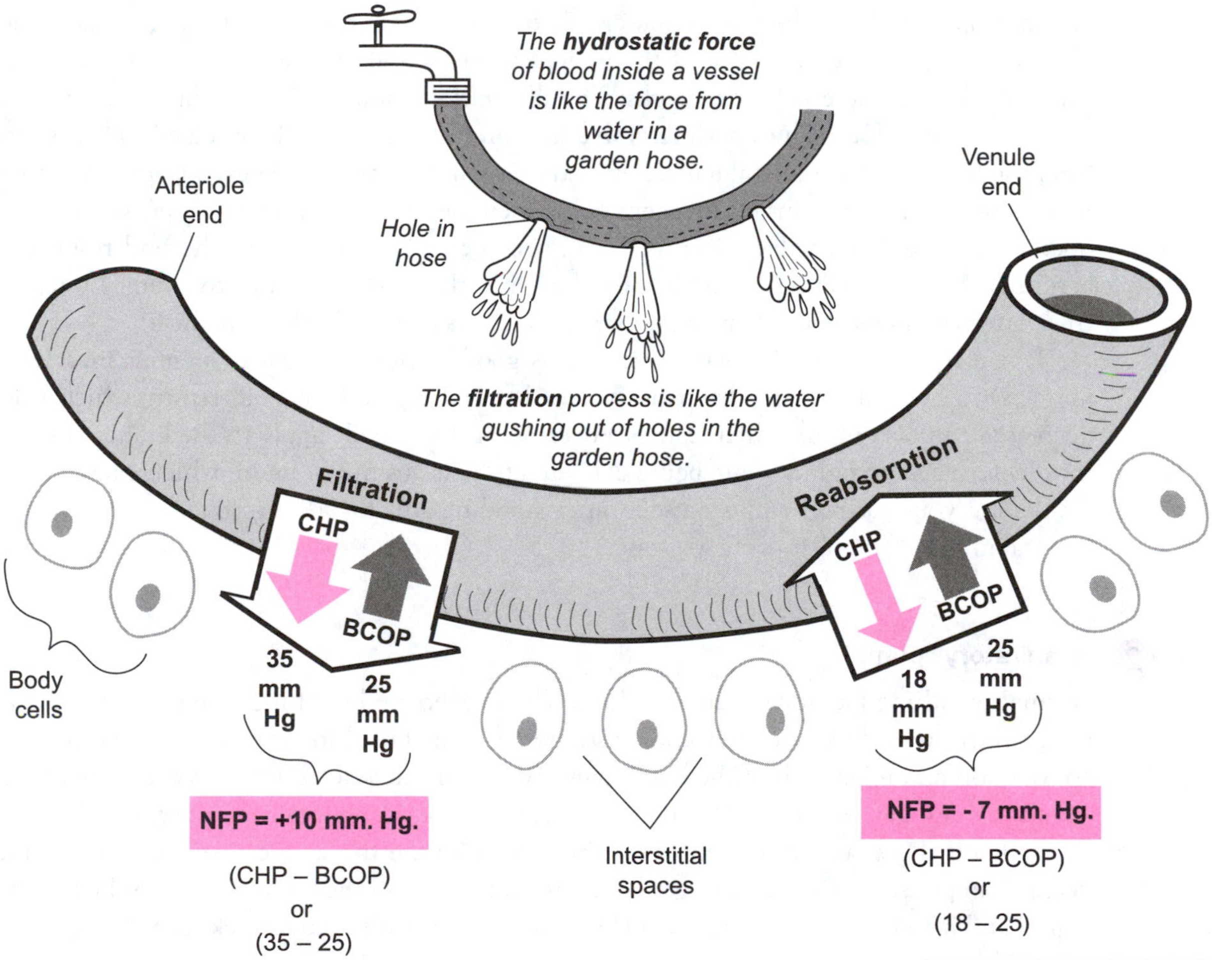

(see filtration, p. 28)

KEY

- **CHP** = Capillary hydrostatic pressure
- **BCOP** = Blood colloidal osmotic pressure
- **NFP** = Net filtration pressure

Description

Venous return refers to the volume of blood returning to the heart through the veins. Veins are under much lower pressure than arteries, which makes it difficult to return venous blood back to the heart. For example, the average pressure in the venous system is about 16 mm Hg, while the average pressure in the arterial system is about 60 mm Hg. Veins contain semilunar valves at regular intervals, which allow for blood flow in one direction while also preventing backflow. Venous return is easier when a person is lying down because the force of gravity is not a factor. But when someone is standing up, consider how long a journey it is for the venous blood in your foot to return to the heart while moving against gravity. Although the heart's normal pumping action helps move the venous blood back to the heart, the body has two other mechanisms that we will refer to collectively as venous "pumps": (1) **skeletal muscle "pump,"** and (2) **respiratory "pump."**

1 Skeletal muscle "pump"

When you are walking, the normal muscle contractions in your legs greatly aid in venous return. They actually squeeze blood out of the peripheral veins and up toward the heart in an action called "milking." Here is how it works: The illustration shows a short segment of a vein with two semilunar valves. The **proximal valve** is nearer the heart, and the **distal valve** is farther from the heart. As the skeletal muscle around this vein contracts, the vessel is squeezed, thereby increasing the pressure inside the vein. This increased pressure forces the proximal valve to open while the distal valve remains closed because of back pressure. The end result of this action is that some venous blood is moved up into the next section of the vein. By repeating this pattern, venous blood is given an extra push on its journey back to the heart.

If you have to stand for a long time, it is good to flex your gastrocnemius muscles (calf muscles) to prevent fainting. Fainting is caused by a decrease in venous return, which leads to decreased cardiac output, which causes a decrease in the blood supply to the brain. This means that less oxygen and glucose are being delivered to neurons in the brain, which causes fainting. By flexing your gastrocnemius muscles after standing still for a long time, you will put your skeletal muscle "pump" to work for you.

2 Respiratory "pump"

Normal breathing facilitates venous return. The deeper the breathing, the greater is the assistance in this regard. As shown in the illustration, normal breathing involves a repeating cycle of contraction and relaxation of the **diaphragm**. When this muscle contracts for a normal inspiration, it moves downward and compresses the contents of the **abdominal cavity**, which increases the abdominal pressure. Simultaneously, the volume in the **thoracic cavity** increases, causing a decrease in pressure. This pressure gradient forces blood out of the various veins in the abdominal region and into the inferior vena cava **(IVC)** to allow blood to move back into the right atrium of the heart.

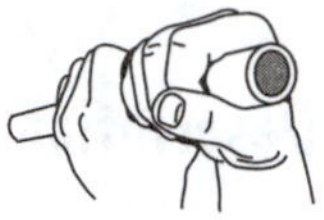

Venous "pumps"

The BIG SQUEEZE!

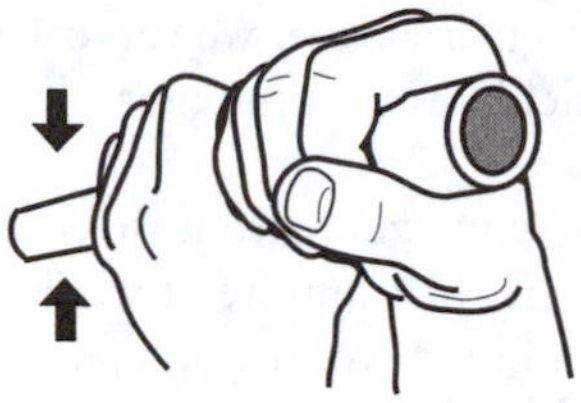

The venous "pumps" are external forces that act on veins like squeezing a hollow rubber tube with your hands.

① Skeletal muscle "pump"

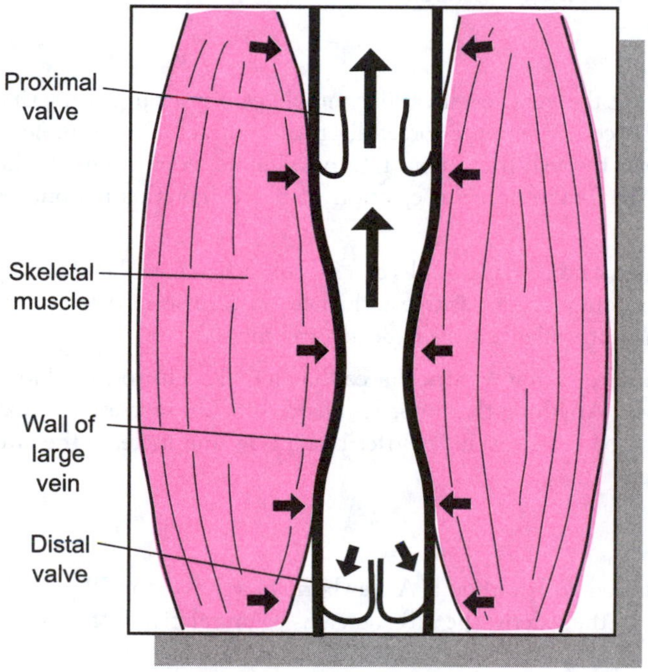

Color the blood in the large vein blue.

② Respiratory "pump"

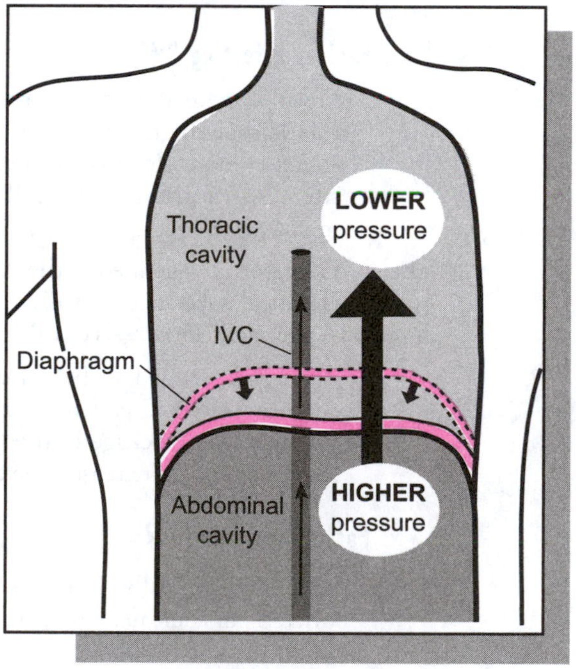

Which way does a pressure gradient go? It always moves from HIGH to LOW

Color the ovals for higher pressure and lower pressure different colors.

Description

Cardiac output (CO) is the amount of blood pumped out of each ventricle in a single contraction during one minute—expressed as **milliliters (ml)** of blood per minute. It is mathematically defined as the product of the **stroke volume** (SV) times the **heart rate** (HR). This is the mathematical equation: **CO = SV × HR**

A normal stroke volume is the amount of blood pumped out of each ventricle in one beat—about 70 ml for an adult heart. A healthy adult male has a resting heart rate of about 70 beats per minute (bpm). By plugging these numbers into the formula given above, we can calculate a normal CO. As shown in the equation at the bottom of the illustration, it calculates to approximately 4,900 ml of blood per minute. If converted to liters, it is 4.9 L per minute.

The illustration shows this by imagining that we could collect blood continually (from only one ventricle) in a 6-liter glass beaker. At the end of 1 minute, it would have filled the beaker to 4.9 L of blood. The CO can be measured by taking a person's blood pressure and pulse, then doing some calculations.

For example, let's say your blood pressure was measured as 118/80 and pulse was 70 bpm. From this, calculate the pulse pressure (systolic pressure minus diastolic pressure). In this case, 118 − 80 = 36. Then plug the numbers into this equation:

$$\begin{aligned} CO &= 2\text{ml} \times \text{Pulse Pressure} \times \text{HR} \\ &= 2 \times 36 \times 70 \\ &= \mathbf{5{,}320\ ml/min} \end{aligned}$$

In general, factors that effect SV also affect HR. For example, during exercise, both SV and HR increase. Even so, let's consider the key factors that govern each of these two variables.

Factors affecting SV

- Preload: defined as the amount of tension in the ventricular cardiac muscle cells prior to contracting. The rule is that the greater the tension on the cells, the more forcefully they contract. For example, during exercise, more blood is returned to the heart through the venous return. As more blood enters the heart, it stretches cardiac muscles cells, resulting in a more forceful ejection of blood. For this reason, exercise increases SV.
- Contractility: defined as the degree to which cardiac muscle cells can shorten when stimulated by a specific chemical substance. For example, calcium ions have a positive influence on cardiac muscle contraction. Abnormally low levels of calcium result in an irregular heartbeat, decreasing the SV.
- Afterload: defined as the amount of force needed from ventricular cardiac muscle cells to eject blood from the ventricles and past the semilunar valves. Anything that impedes blood flow can increase afterload. For example, any blockage in the peripheral blood vessels would restrict blood flow and increase the afterload. As the afterload increases, the SV decreases.

Factors affecting HR

- Age: HR gradually decreases from childhood to adulthood. A newborn may have an HR of 120 bpm, whereas an adult male may have 70 bpm. In the elderly, the HR increases again relative to that of a young adult.
- Sex: On average, females have slightly higher resting HR's than males. The difference is about 5 bpm.
- State of activity: During certain phases of the sleep cycle, the HR decreases. But during exercise the HR temporarily increases.
- Endurance training: Marathon runners may have a resting HR of 50 bpm. This type of training increases heart size as well as SV. This allows for a normal CO with a lower HR.
- Stress, anxiety: Stress and anxiety increase HR.

The CO does not remain constant. It regularly rises and falls. Maintaining the CO in a normal range is primarily the job of the cardiovascular (CV) center in the medulla oblongata of the brain. Using reflex pathways in the autonomic nervous system, the CV center regulates the rhythm and force of the heart rate. Hormones also help to control cardiac activity. For example, epinephrine and norepinephrine are powerful cardiac stimulators. In short, the nervous system and hormonal regulators work together to regulate heart activity, thereby indirectly ensuring that a normal CO is achieved.

KEY

A = Aorta
PA = Pulmonary artery
RV = Right ventricle
LV = Left ventricle

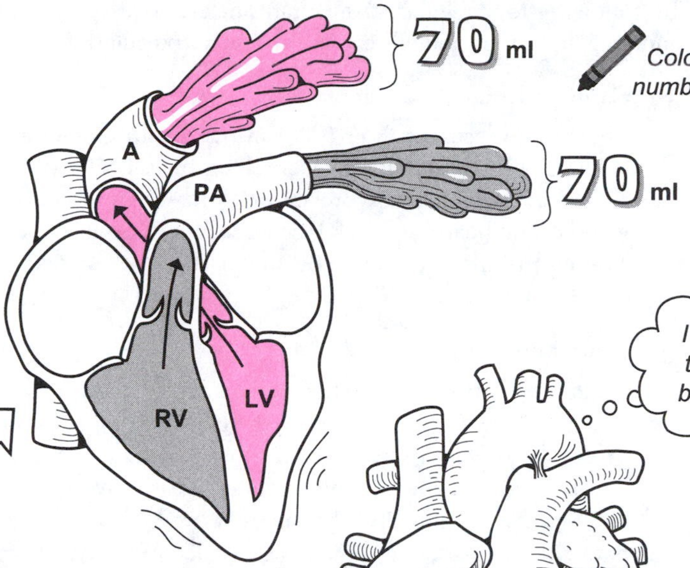

Color the number "70"

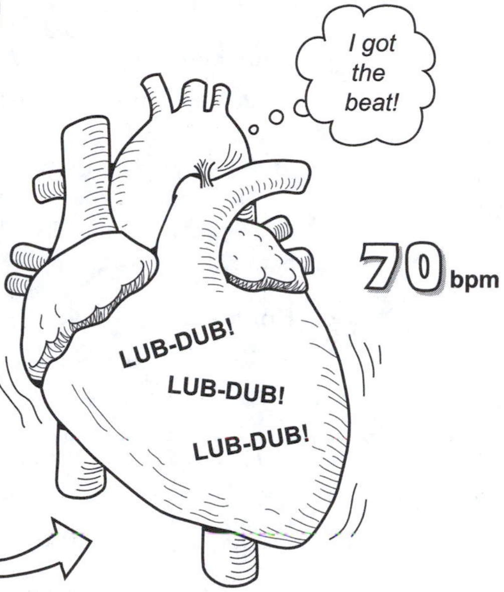

Color the equation for cardiac output.

Cardiac output (CO) (ml/min)

Stroke volume (SV) (ml)

Heart rate (HR) (bpm)

$$CO = SV \times HR$$

Imagine collecting blood from the left ventricle of a beating heart for 1 minute!

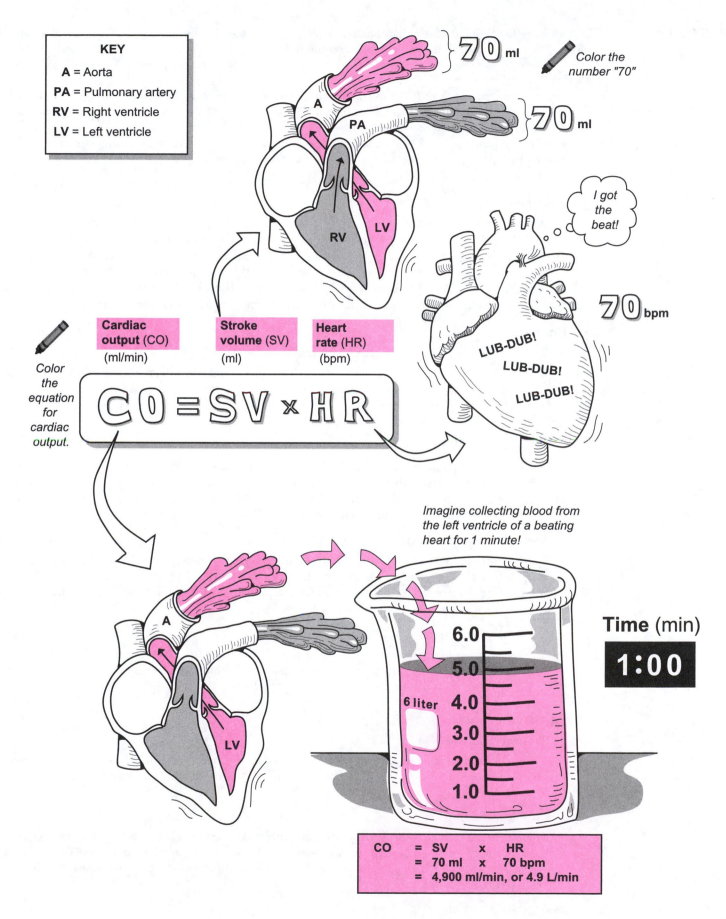

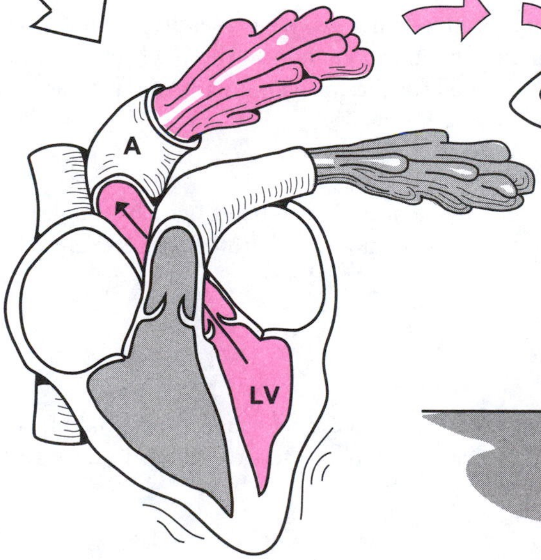

Time (min)

1:00

CO	**=**	**SV**	**x**	**HR**
	=	**70 ml**	**x**	**70 bpm**
	=	**4,900 ml/min, or 4.9 L/min**		

Description

The following four regulatory mechanisms are used to control **blood pressure (BP)**: cardiovascular center, neural regulation, hormonal regulation, and autoregulation.

(1) **Cardiovascular (CV) center**

The cardiovascular center (CV) in the **medulla oblongata** is the command-and-control center for controlling heart function. It receives peripheral sensory input from three major sources: **proprioceptors**, **chemoreceptors**, and **baroreceptors**. The CV responds by sending motor output to the heart and blood vessels. The heart rate **(HR)** either increases or decreases while blood vessels are stimulated to constrict. The end result is that the BP either increases or decreases back to normal levels. For details of this process, see p. 292.

(2) **Neural regulation**

Neural regulation uses a reflex pathway to respond to changes in BP. **Baroreceptors** located in the aortic arch and other arteries detect changes in blood pressure. When the BP rises or falls below a certain threshold, these sensors are stimulated to send a nerve impulse to the CV center. Then motor output is sent to the heart through different nerve pathways. One pathway triggers a decrease in heart rate, and the other pathway stimulates the opposite result. This change in heart rate leads to a change in cardiac output, which, in turn, either increases or decreases the BP. For details of this process, see p. 292.

(3) **Hormonal regulation**

- Renin-angiotensin-aldosterone **(RAA)** system.

  When blood pressure decreases, it stimulates the kidney to release the enzyme renin into the blood. This leads to the production of a powerful vasoconstrictor called angiotensin II. This, in turn, stimulates the adrenal cortex to produce the hormone aldosterone. The net result is an increase in blood pressure. For details of this mechanism, see p. 442.

- Epinephrine **(EPI)**/norepinephrine **(NE)**

  EPI and **NE** are produced by cells in the adrenal medulla in response to emergency or stressful situations. Two of the organs they target are the heart and blood vessels. The result is that heart rate increases and blood vessels constrict, respectively. Together, these two responses lead to an increase in blood pressure. For details of this mechanism, see p. 252.

- Antidiuretic hormone **(ADH)**

  Water loss from the blood, like what occurs during excessive sweating, stimulates the posterior lobe of the pituitary gland to produce **ADH**. This hormone targets the nephrons in the kidneys and causes more water to be reabsorbed into the blood thus restoring normal blood pressure. For the details of this mechanism, see p. 436.

- Atrial natriuretic peptide **(ANP)**

  As blood volume increases, this stretches the atria in the heart, which stimulates cells in the atria to produce **ANP**. By inducing blood vessels to dilate and water to be excreted in the urine, ANP decreases BP.

(4) **Autoregulation**

- **Physical changes**

  Changes in body temperature affect BP. For example, cold causes superficial blood vessels to constrict, which increases BP. In contrast, heat causes blood vessels to dilate, leading to a decrease in blood pressure.

- **Chemical changes**

  Different types of body cells, such as smooth muscles cells, endothelial cells, and macrophages, produce various chemicals that signal blood vessels to either dilate or constrict. For example, smooth cells produce lactic acid, which causes blood vessels to dilate. Other vasodilating chemicals include nitric oxide (NO), H^+, and K^+.

GOAL: Maintain Normal Blood Pressure
120 mm Hg
70 mm Hg
Color the 120 one color and the 70 a different color.
1 Cardiovascular (CV) center
• Proprio-ceptors - tension
• Chemo-receptors - blood acidity
• Baro-receptors - pressure
S
INPUT
Cardio-vascular (CV) center
OUTPUT
(see p. 292)
R
HR
HR
blood vessels constrict
2 Neural regulation
Baroreceptors
S
or
in BP
Cardio-vascular (CV) center
INPUT
Medulla oblongata
OUTPUT
or
in HR
R
(see p. 292)
3 Hormonal regulation
• Renin-angiotensin-aldosterone (RAA) system
• Epinephrine (EPI) /norepinephrine (NE)
• Antidiuretic hormone (ADH)
• Atrial natriuretic peptide (ANP)
Kidney
Adrenal medulla
Posterior lobe
Pituitary gland
RAA system
EPI, NE
ADH
ANP
BP
BP
BP
BP
(see p. 442)
4 Autoregulation
R
Physical changes
S
COLD
blood vessels constrict
HOT
blood vessels dilate
Chemical signals
R
Smooth muscle cells
Vasodilating chemicals
blood vessels dilate

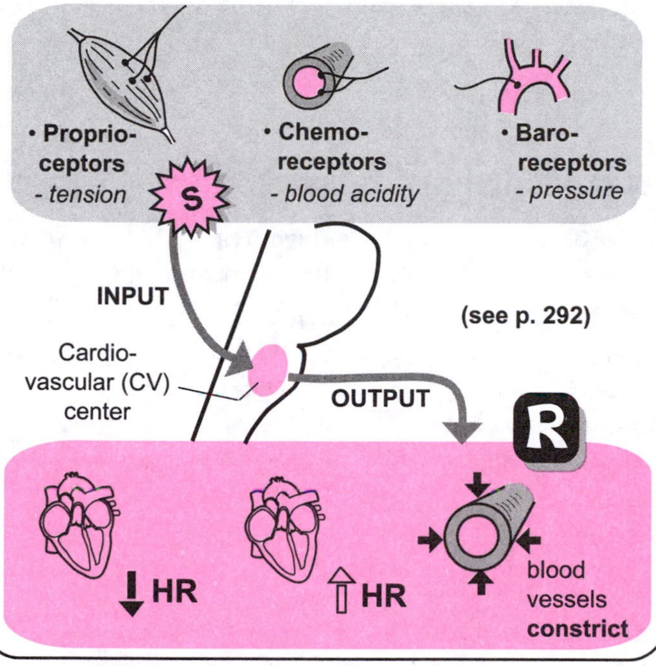

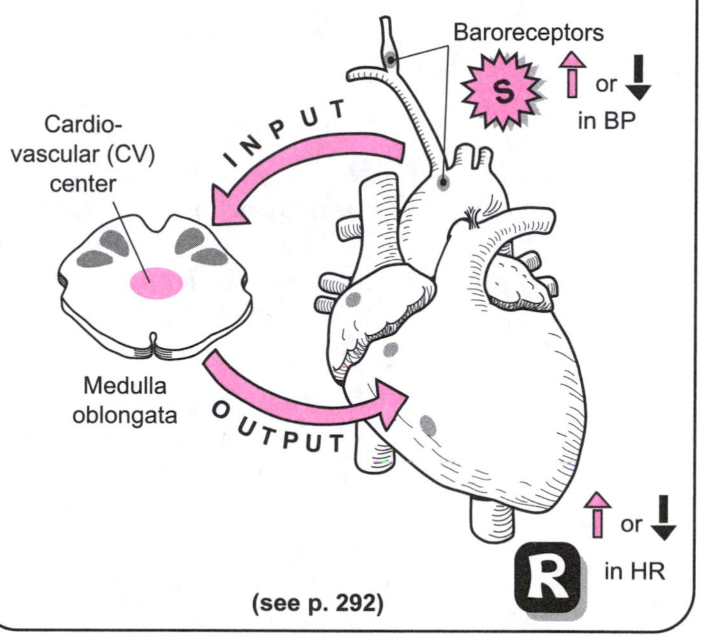

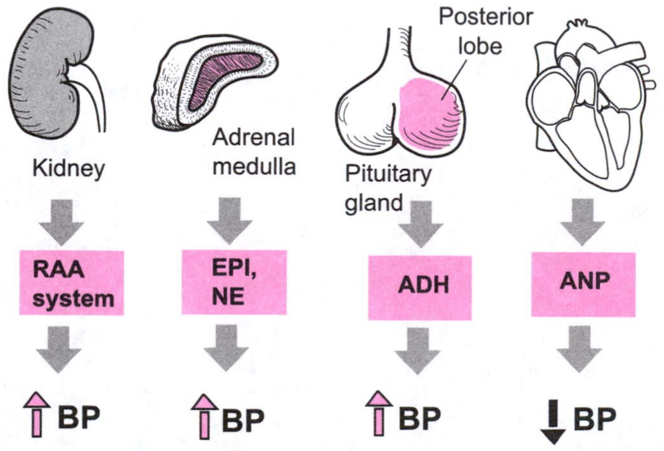

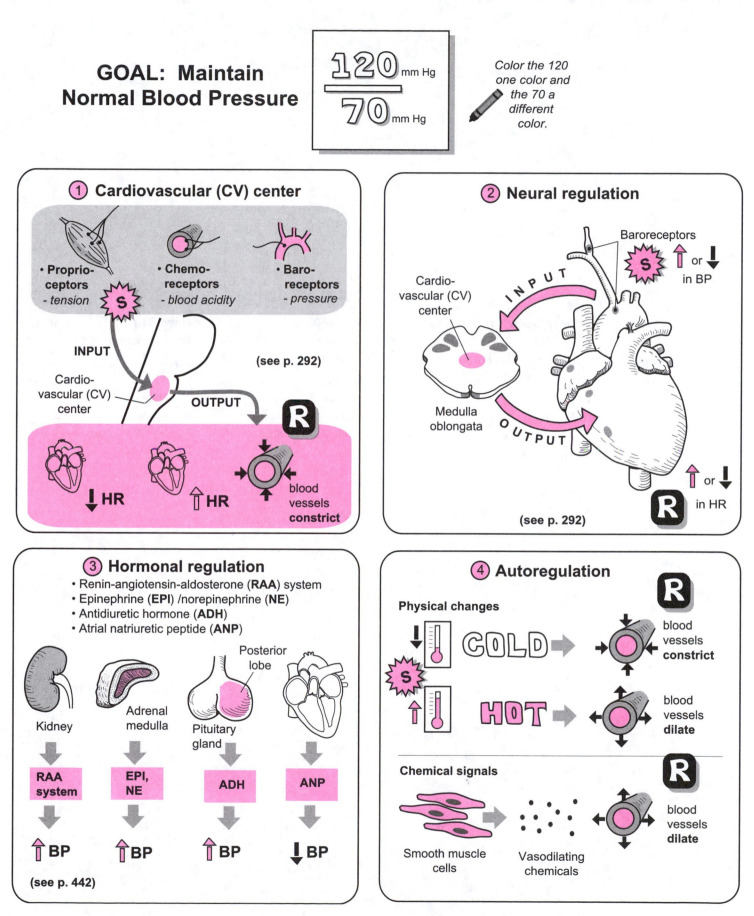

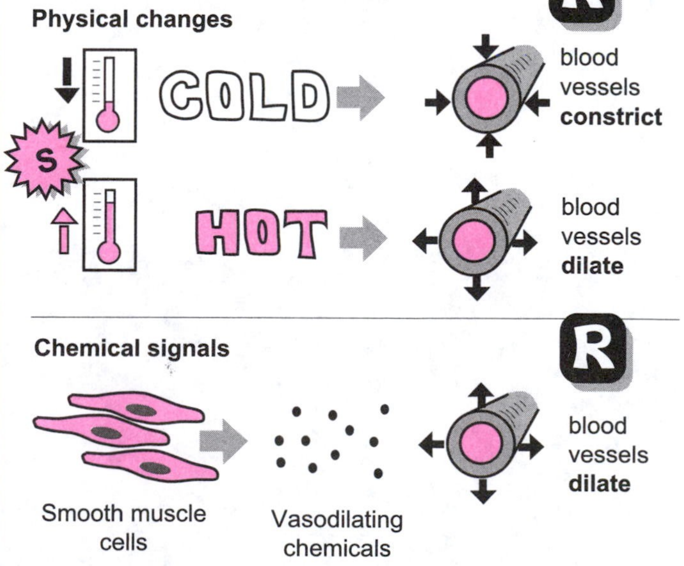

Key to Illustration

Head/Neck (H)

H1. Internal carotid a.

H2. External carotid a.

H3. Common carotid a.

H4. Subclavian a.

Shoulder (S)

S1. Axillary a.

Thorax (T)

T1. Aortic arch

T2. Pulmonary trunk

T3. Pulmonary a.

Arm (AR)

AR1. Brachial a.

Forearm (FO)

FO1. Ulnar a.

FO2. Radial a.

Abdomen (A)

A1. Abdominal aorta

A2. Celiac trunk

A3. Superior mesenteric a.

A4. Renal a.

A5. Gonadal a. *(testicular a. in males, ovarian a. in females)*

A6. Inferior mesenteric a.

A7. Common iliac a.

A8. Internal iliac a.

A9. External iliac a.

Thigh (TH)

TH1. Femoral a.

Leg (L)

L1. Popliteal a.

L2. Anterior tibial a.

L3. Fibular a.

L4. Posterior tibial a.

Major Arteries

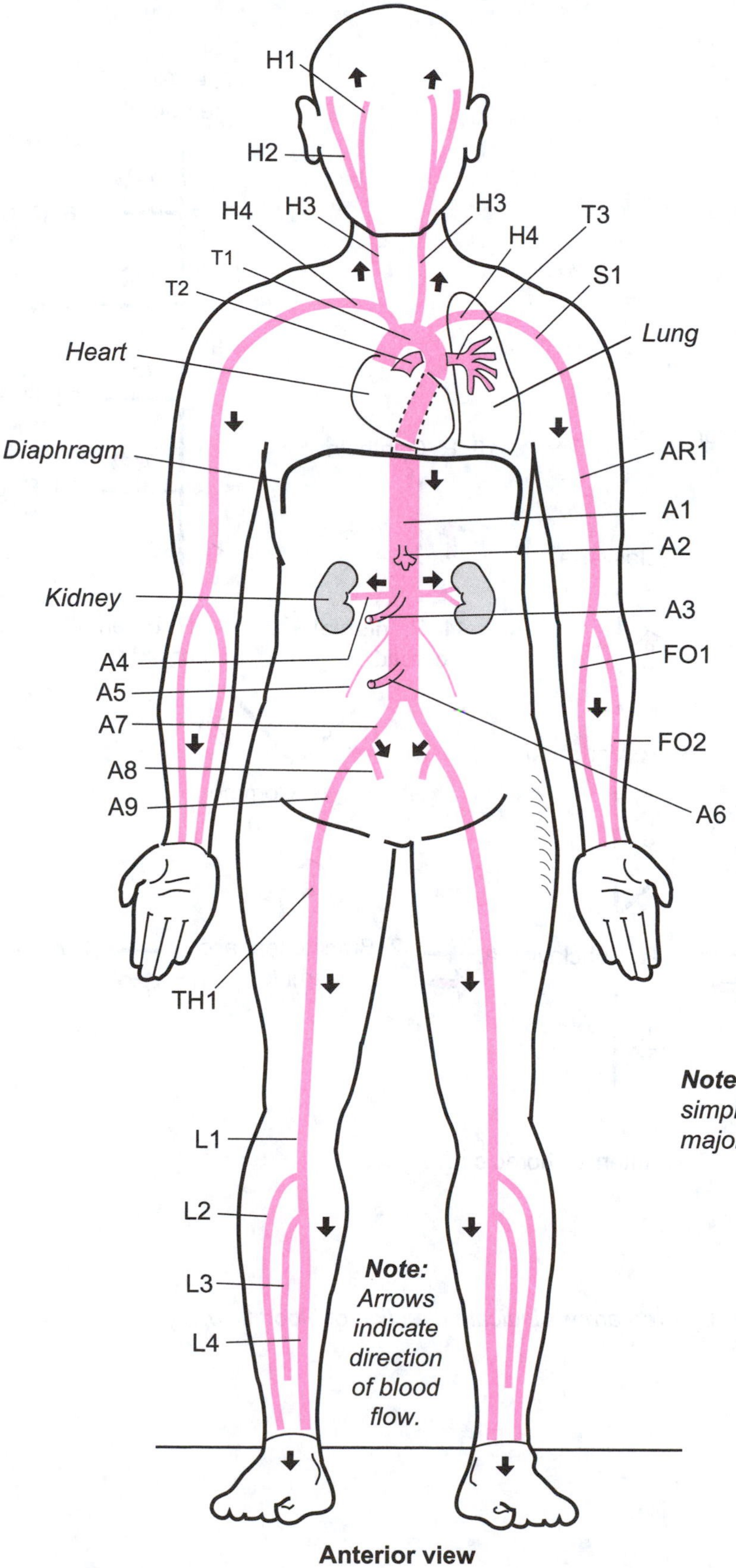

Anterior view

Arteries of:

- **Head/Neck (H)**
- **Shoulder (S)**
- **Thorax (T)**
- **Arm (AR)**
- **Forearm (FO)**
- **Abdomen (A)**
- **Thigh (TH)**
- **Leg (L)**

Note: *This illustration is simplified to show only the major arteries.*

Key to Illustration

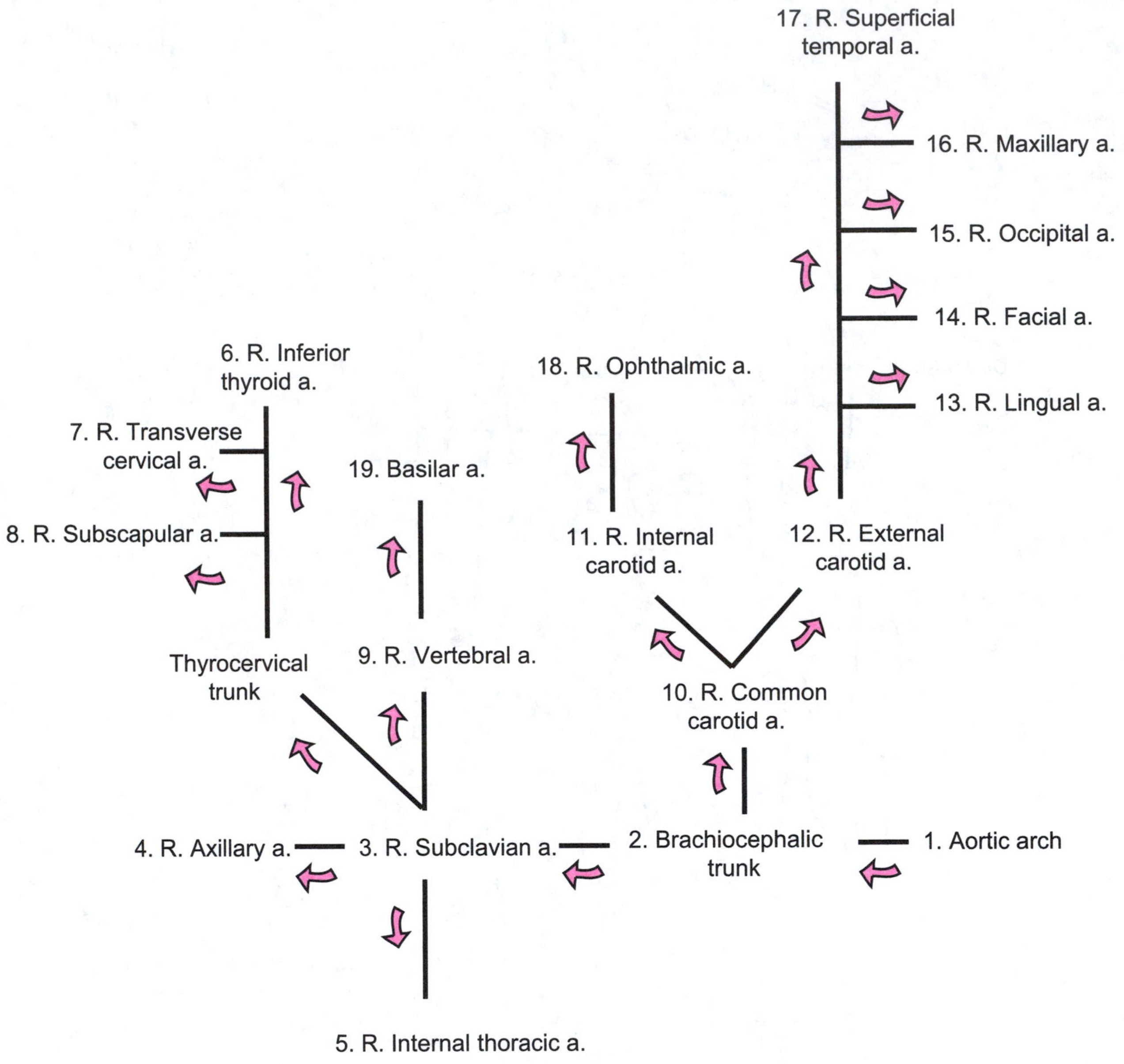

Note: *Colored arrows indicate direction of blood flow.*

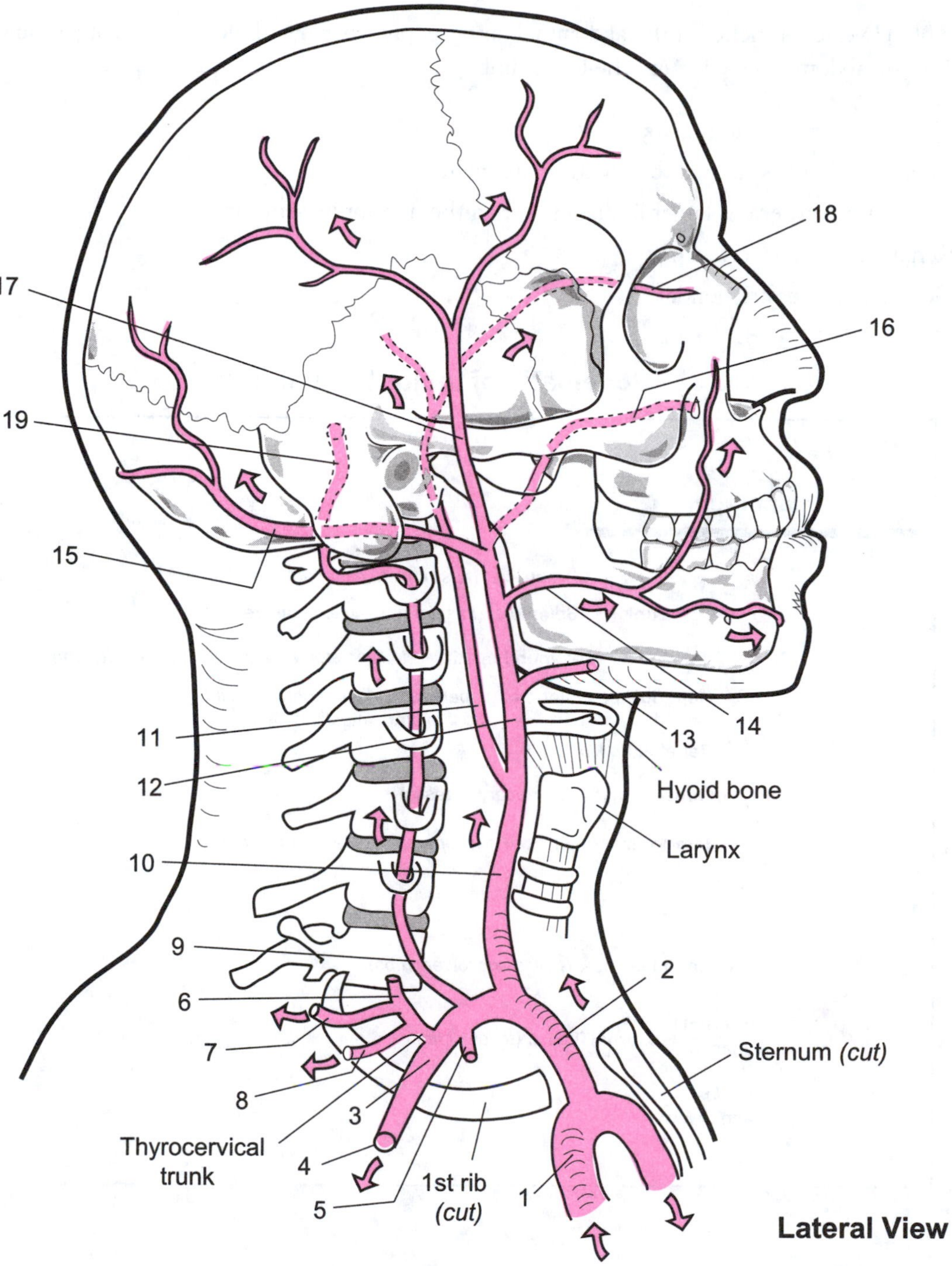
18
17
16
19
15
14
13
11
12
Hyoid bone
Larynx
10
9
2
6
Sternum *(cut)*
7
8
3
Thyrocervical trunk
4
1st rib *(cut)*
1
5
Lateral View
1.
2.
3.
4.
5.
6.
7.
8.
9.
10.
11.
12.
13.
14.
15.
16.
17.
18.
19.

Description

The blood vessel branches of the abdominal aorta supply oxygenated blood to the organs and structures within the abdominal cavity and the lower limbs.

Study Tips

- Use diaphragm as a landmark
- Celiac trunk is first branch below the diaphragm
- Superior mesenteric is larger in diameter than the inferior mesenteric
- Renal and gonadal are paired
- Gonadal arteries are slender

Schematic of major branches

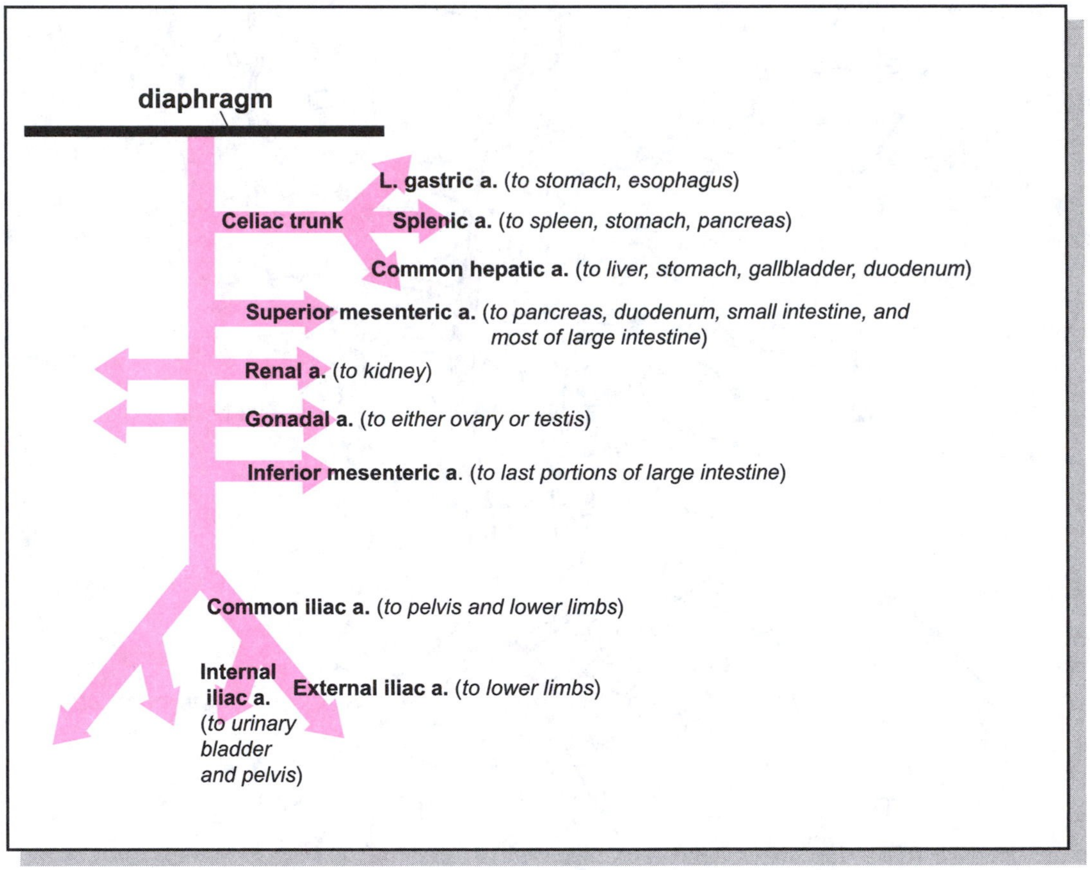

Key to Illustration

1. Celiac truck
2. Common hepatic a.
3. Left gastric a.
4. Splenic a.
5. Superior mesenteric a.
6. Renal a.
7. Gonadal a.
8. Inferior mesenteric a.
9. Common iliac a.
10. Internal iliac a.
11. External iliac a.

Schematic of major branches

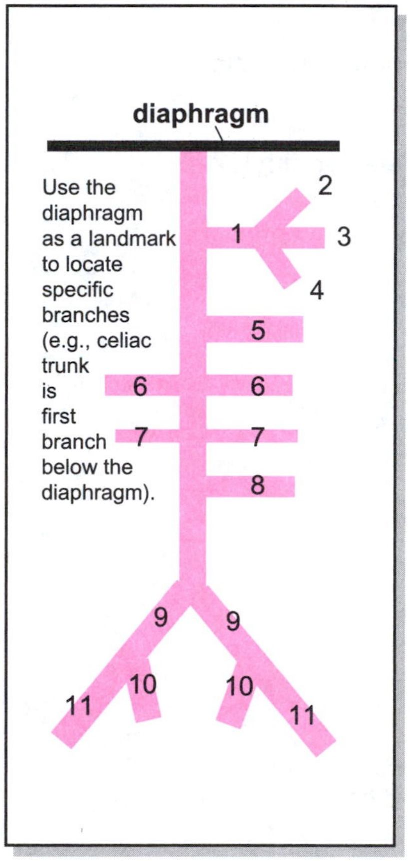

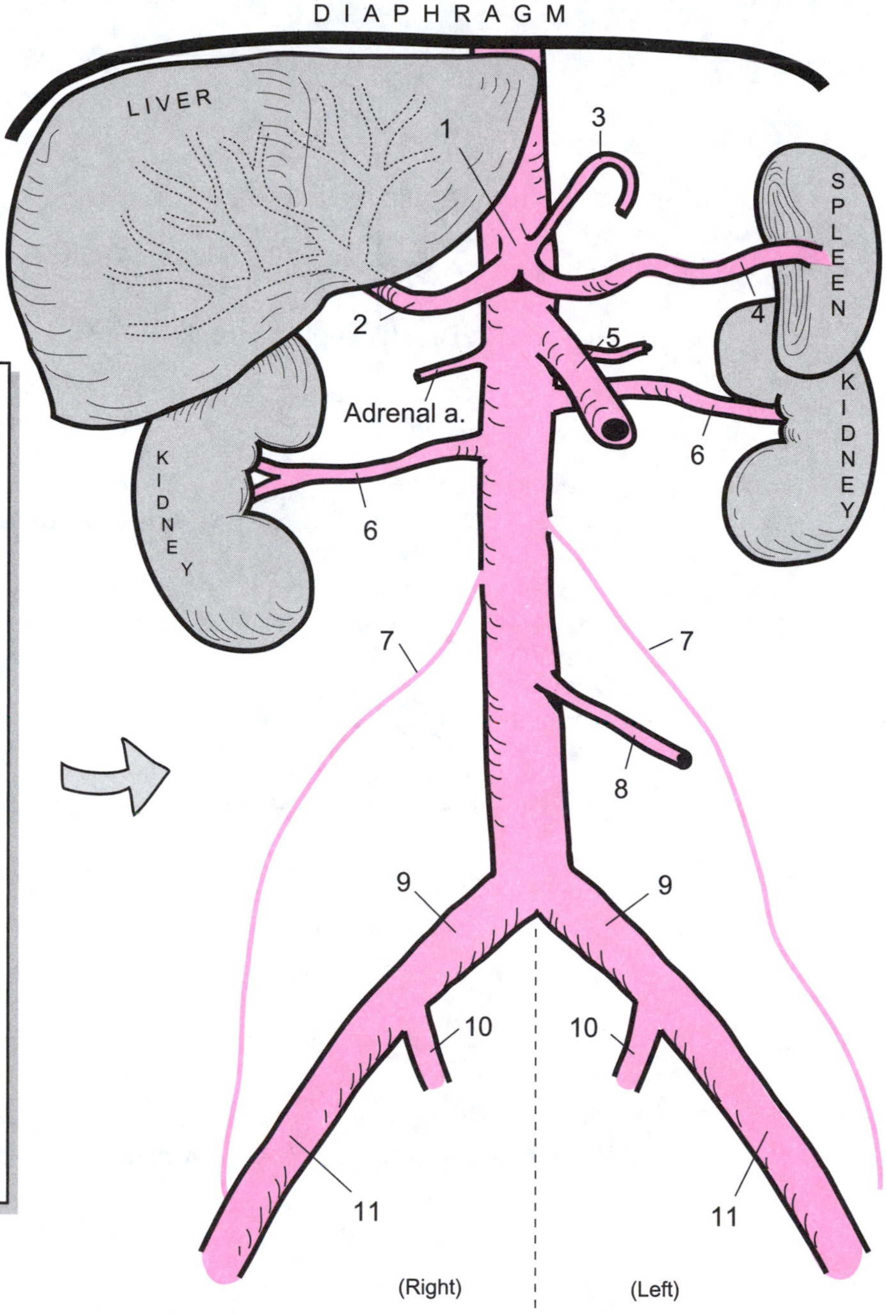

Anterior view

1. ______________________________
2. ______________________________
3. ______________________________
4. ______________________________
5. ______________________________
6. ______________________________

7. ______________________________
8. ______________________________
9. ______________________________
10. ______________________________
11. ______________________________

Key to Illustration

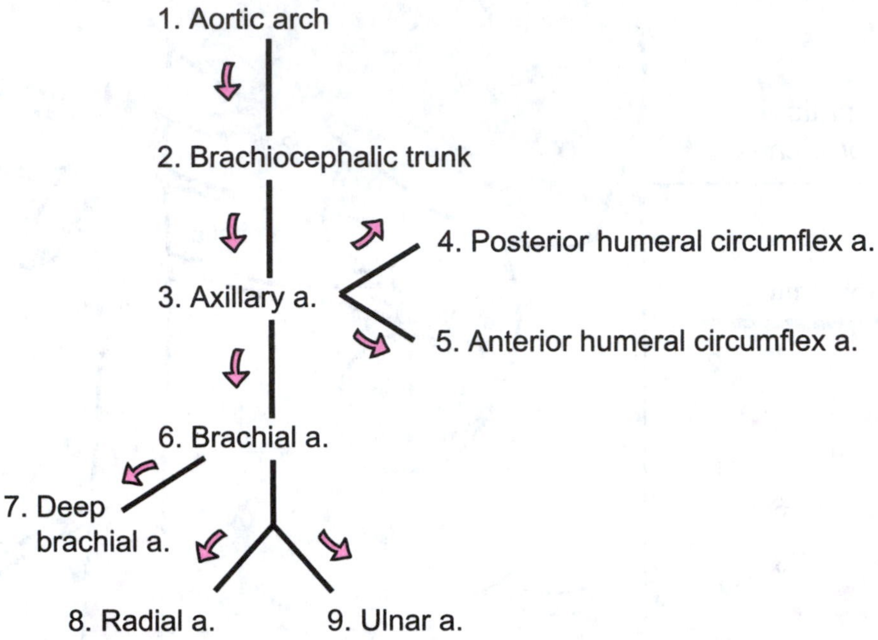

Note: *Colored arrows indicate direction of blood flow.*

Arteries of the Right Upper Limb and Thorax

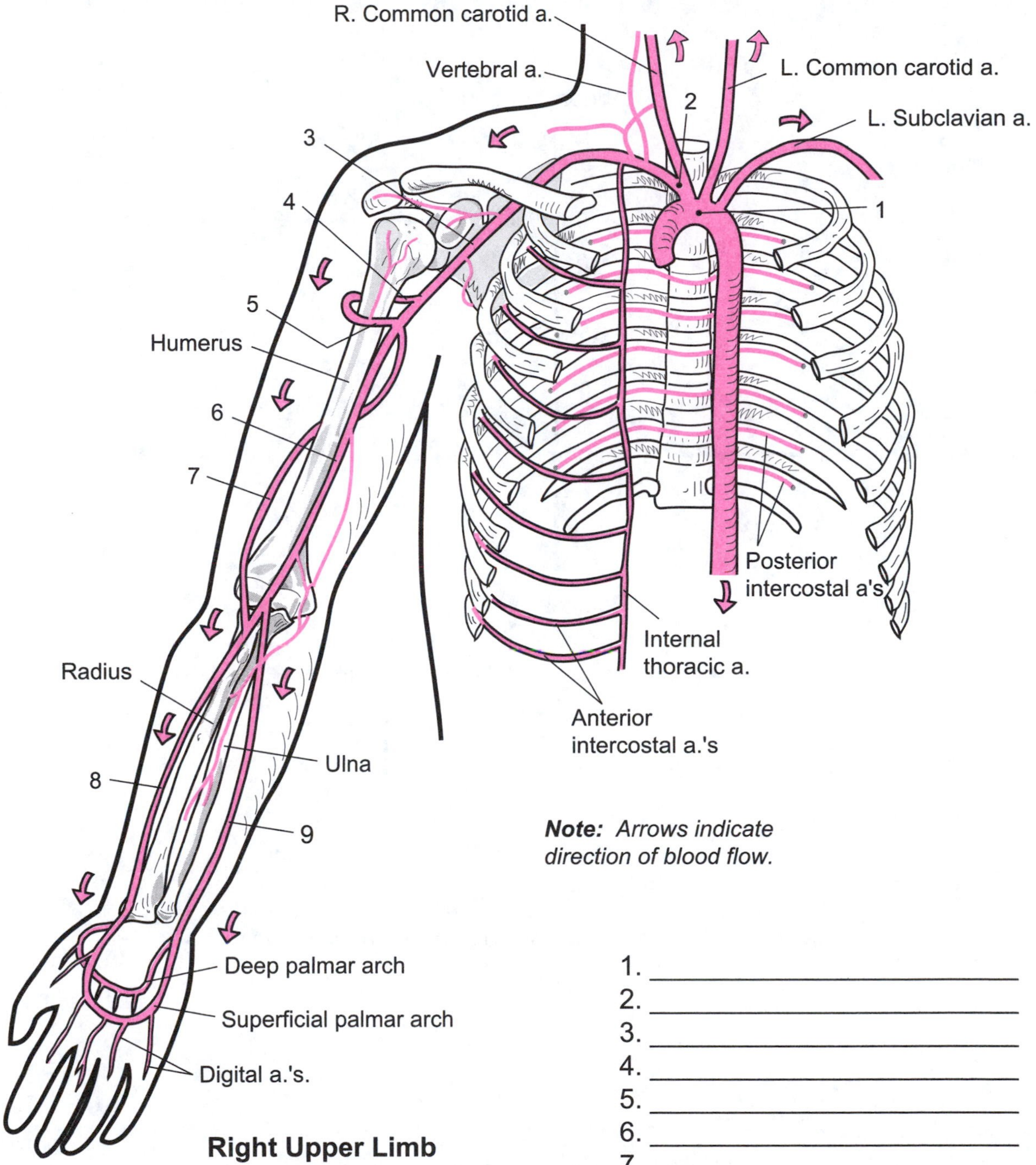

Right Upper Limb
anterior view

Note: *Arrows indicate direction of blood flow.*

1. ______________________
2. ______________________
3. ______________________
4. ______________________
5. ______________________
6. ______________________
7. ______________________
8. ______________________
9. ______________________

Major Arteries of the Right Pelvis and Lower Limb

Key to Illustration

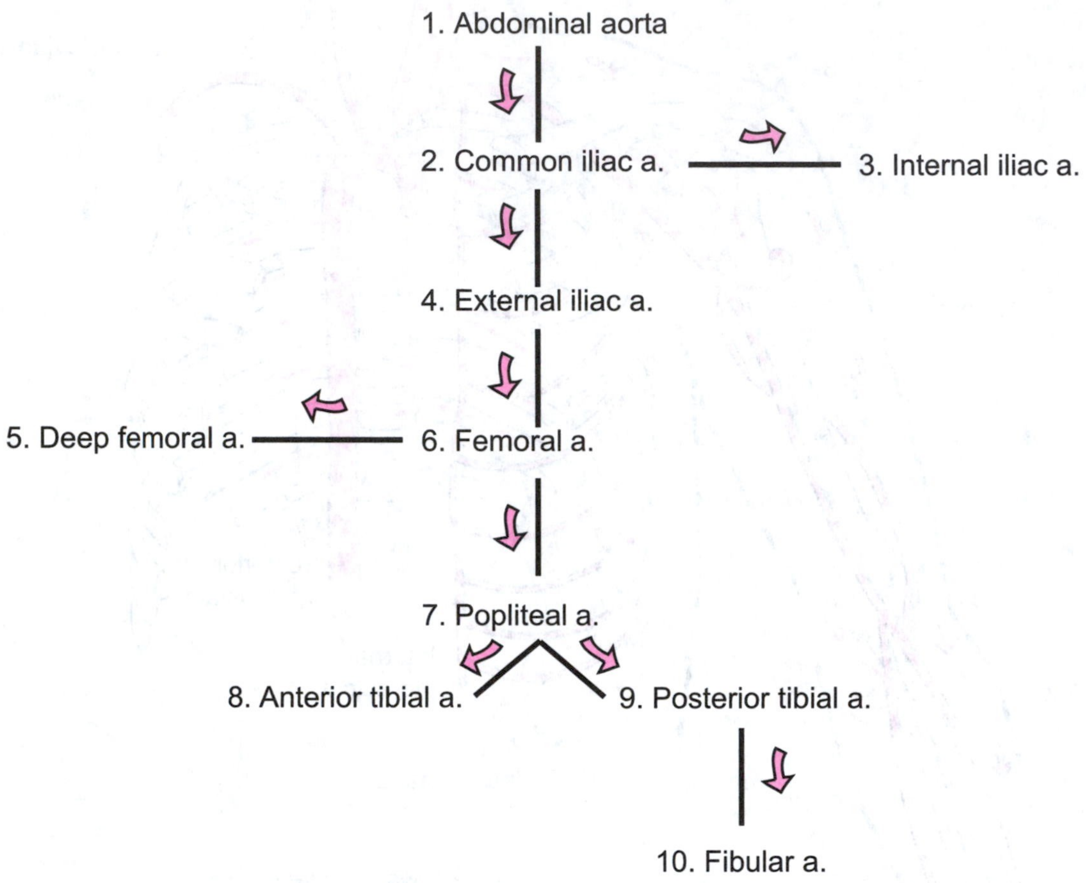

Note: *Colored arrows indicate direction of blood flow.*

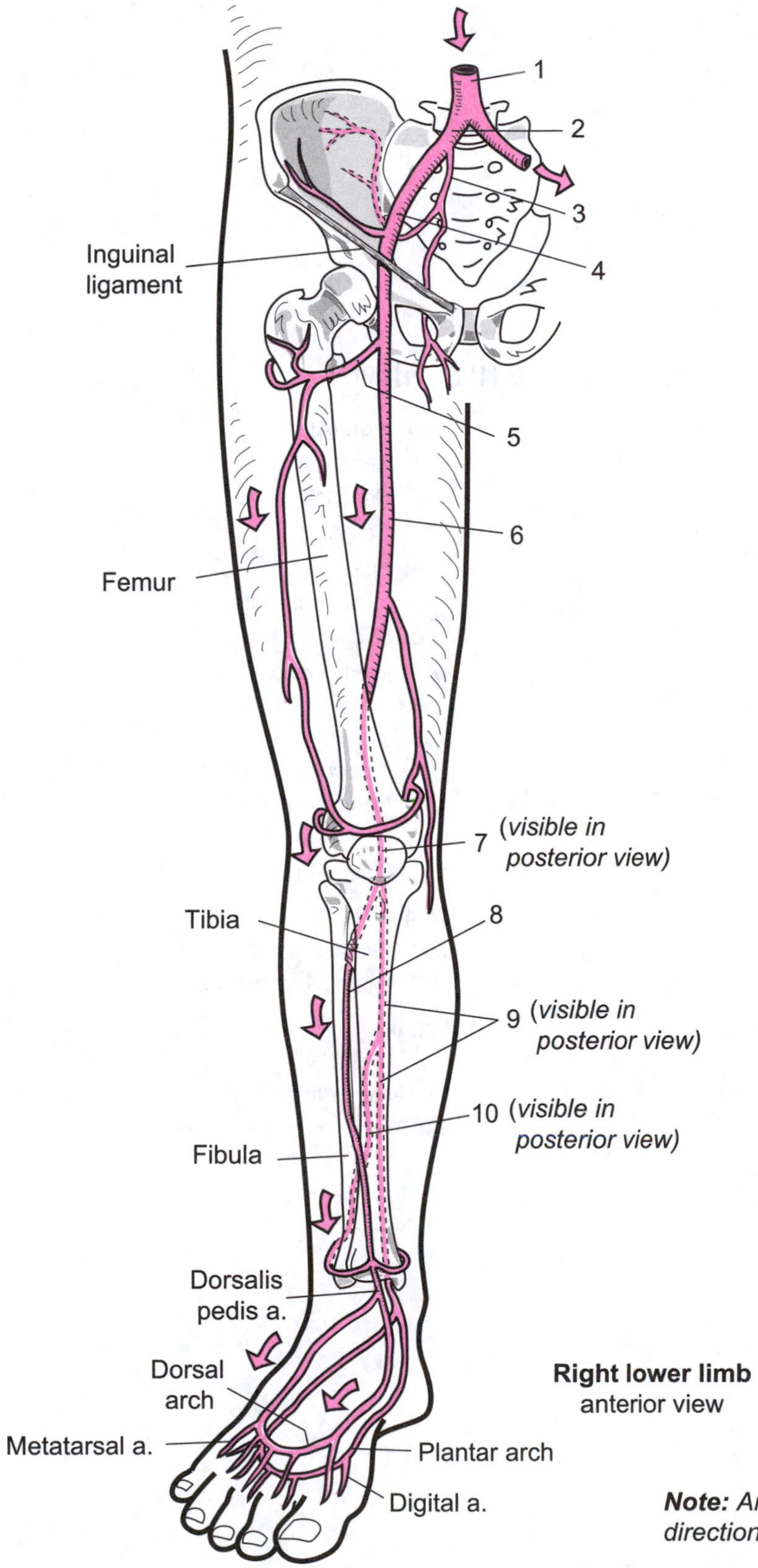

Right lower limb
anterior view

Note: *Arrows indicate direction of blood flow.*

Label major branches

1. ______________________
2. ______________________
3. ______________________
4. ______________________
5. ______________________
6. ______________________
7. ______________________
8. ______________________
9. ______________________
10. ______________________

Key to Illustration

Head/Neck (H)
H1. External jugular v.
H2. Internal jugular v.

Shoulder (S)
S1. Axillary v.

Thorax (T)
T1. Superior vena cava
T2. Brachiocephalic v.
T3. Subclavian v.

Arm (AR)
AR1. Cephalic v.
AR2. Brachial v.
AR3. Basilic v.

Forearm (FO)
FO1. Ulnar v.
FO2. Radial v.

Abdomen (A)
A1. Hepatic v.
A2. Hepatic portal v.
A3. Gastric v.
A4. Splenic v.
A5. Inferior mesenteric v.
A6. Superior mesenteric v.
A7. Inferior vena cava
A8. Renal v.
A9. Gonadal v.
(testicular v. in male, ovarian v. in female)
A10. Common iliac v.
A11. Internal iliac v.
A12. External iliac v.

Thigh (TH)
TH1. Femoral v.
TH2. Great saphenous v.

Leg (L)
L1. Popliteal v.
L2. Anterior tibial v.
L3. Posterior tibial v.

Major Veins

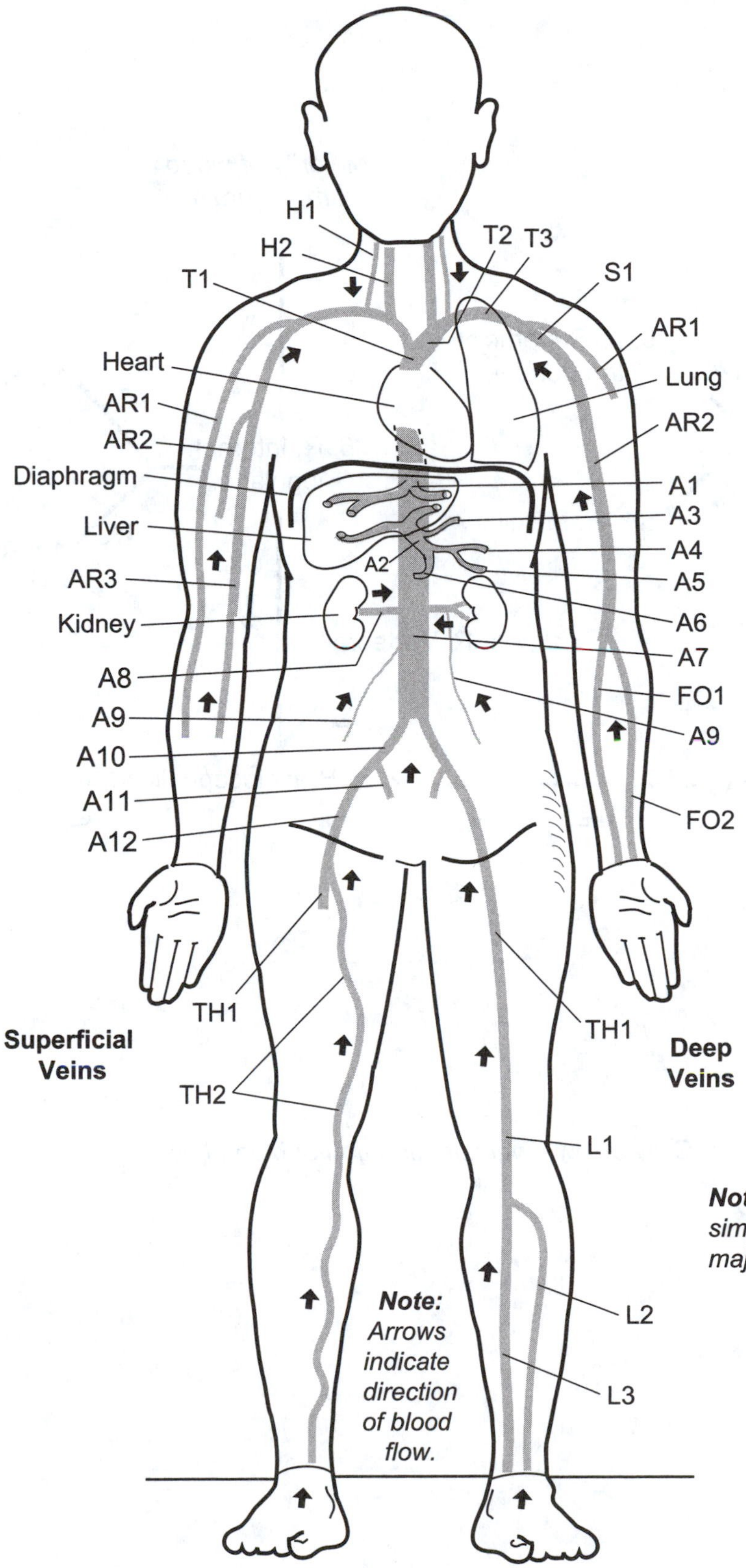

Anterior view

Veins of:

- **Head/Neck (H)**
- **Shoulder (S)**
- **Thorax (T)**
- **Arm (AR)**
- **Forearm (FO)**
- **Abdomen (A)**
- **Thigh (TH)**
- **Leg (L)**

Note: *This illustration is simplified to show only the major veins.*

Key to Illustration

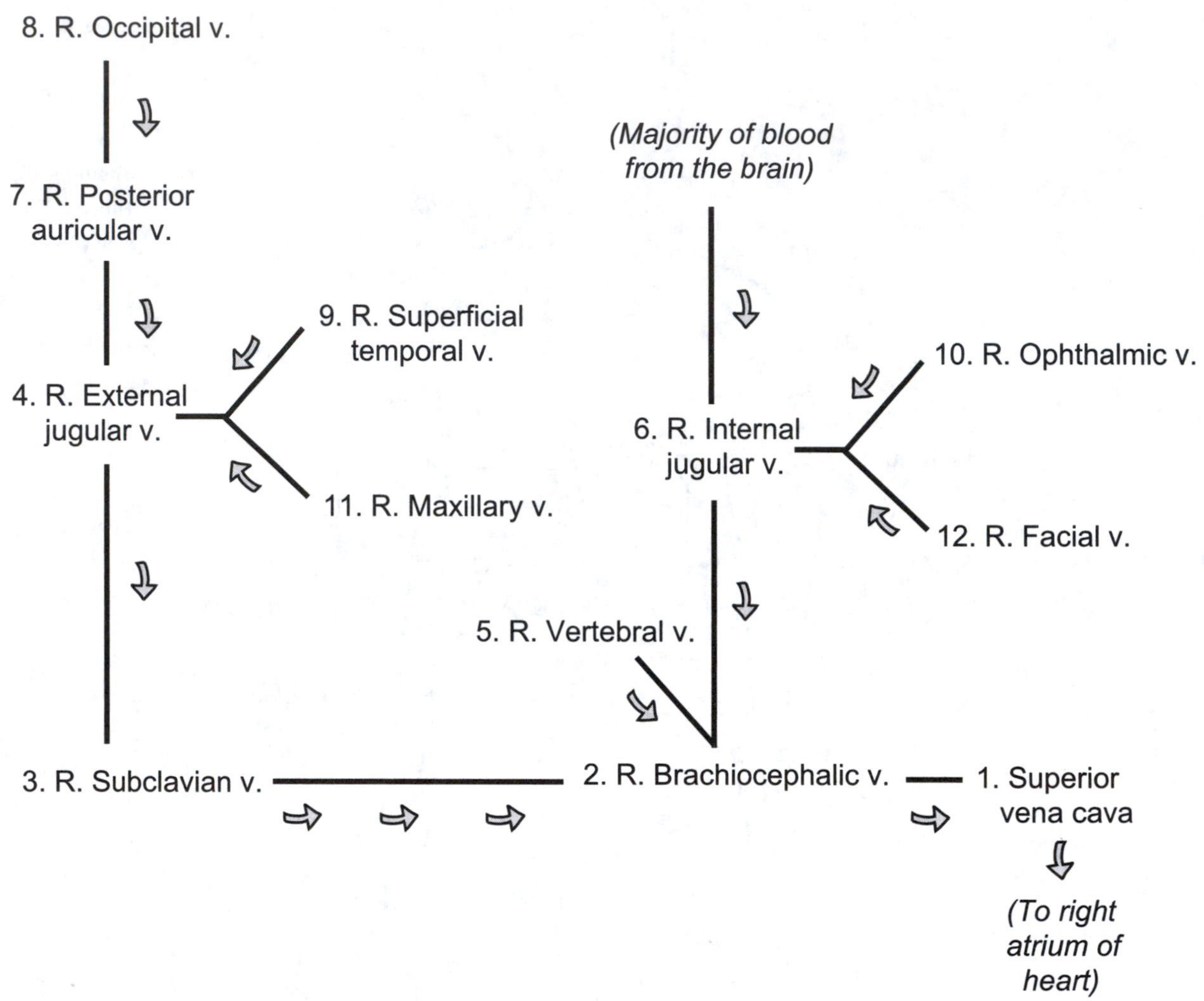

Note: *Gray arrows indicate direction of blood flow.*

Veins of the Head and Neck

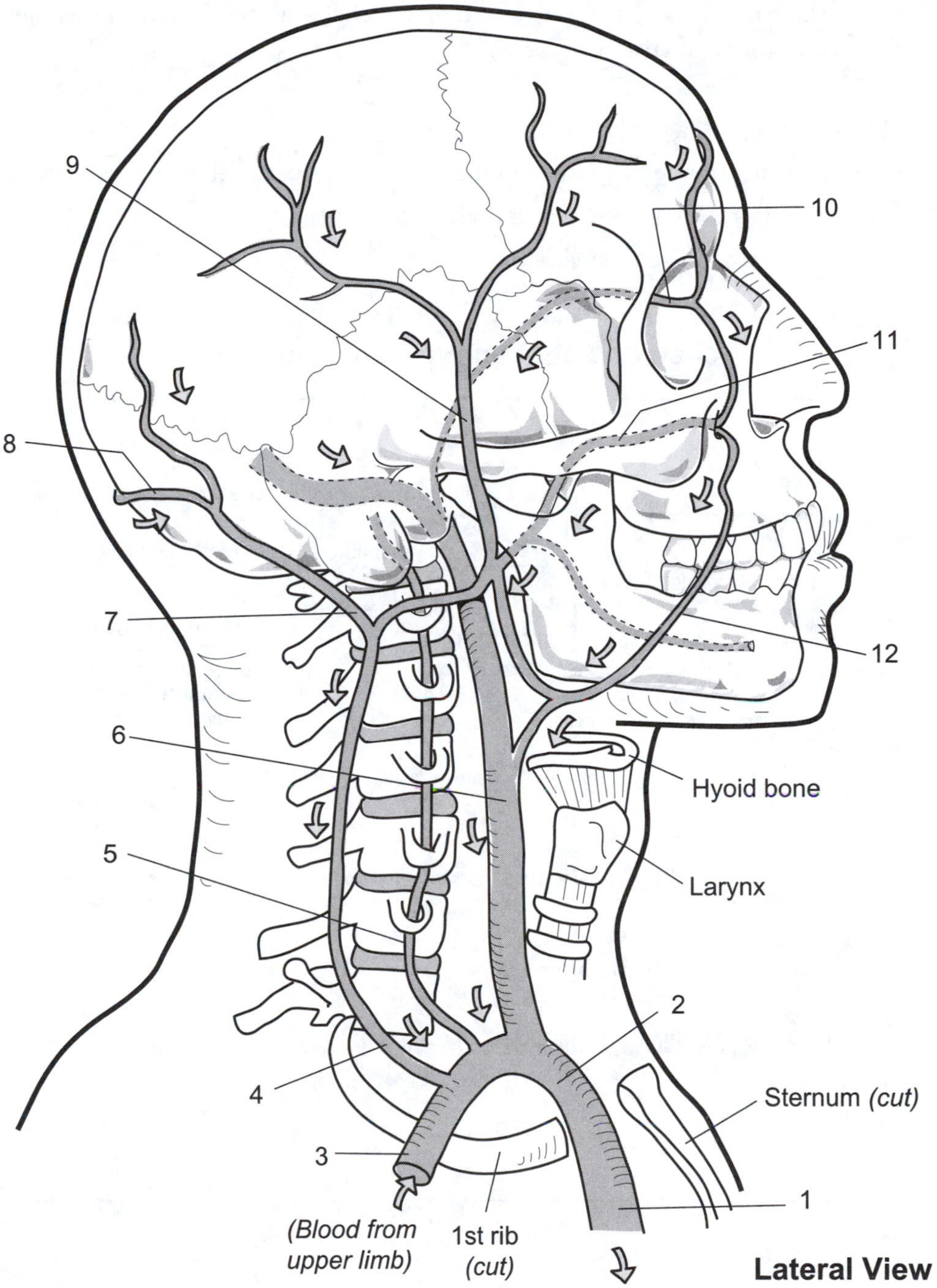

1. ______________________
2. ______________________
3. ______________________
4. ______________________
5. ______________________
6. ______________________
7. ______________________
8. ______________________
9. ______________________
10. ______________________
11. ______________________
12. ______________________

Description

Most of the blood vessels in the abdomen drain deoxygenated blood into the inferior vena cava, which empties into the right atrium of the heart.

Study Tips

- Use diaphragm as a landmark
- Hepatic veins are the first vessels of the inferior vena cava below the diaphragm
- **Renal veins** are the **widest** vessels of the inferior vena cava
- **Gonadal veins** are very **slender**

Schematic of Major Veins of Abdomen

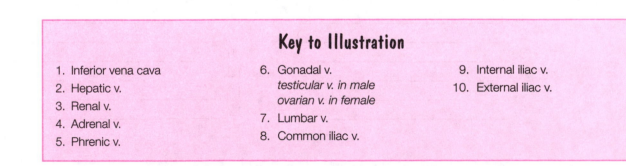

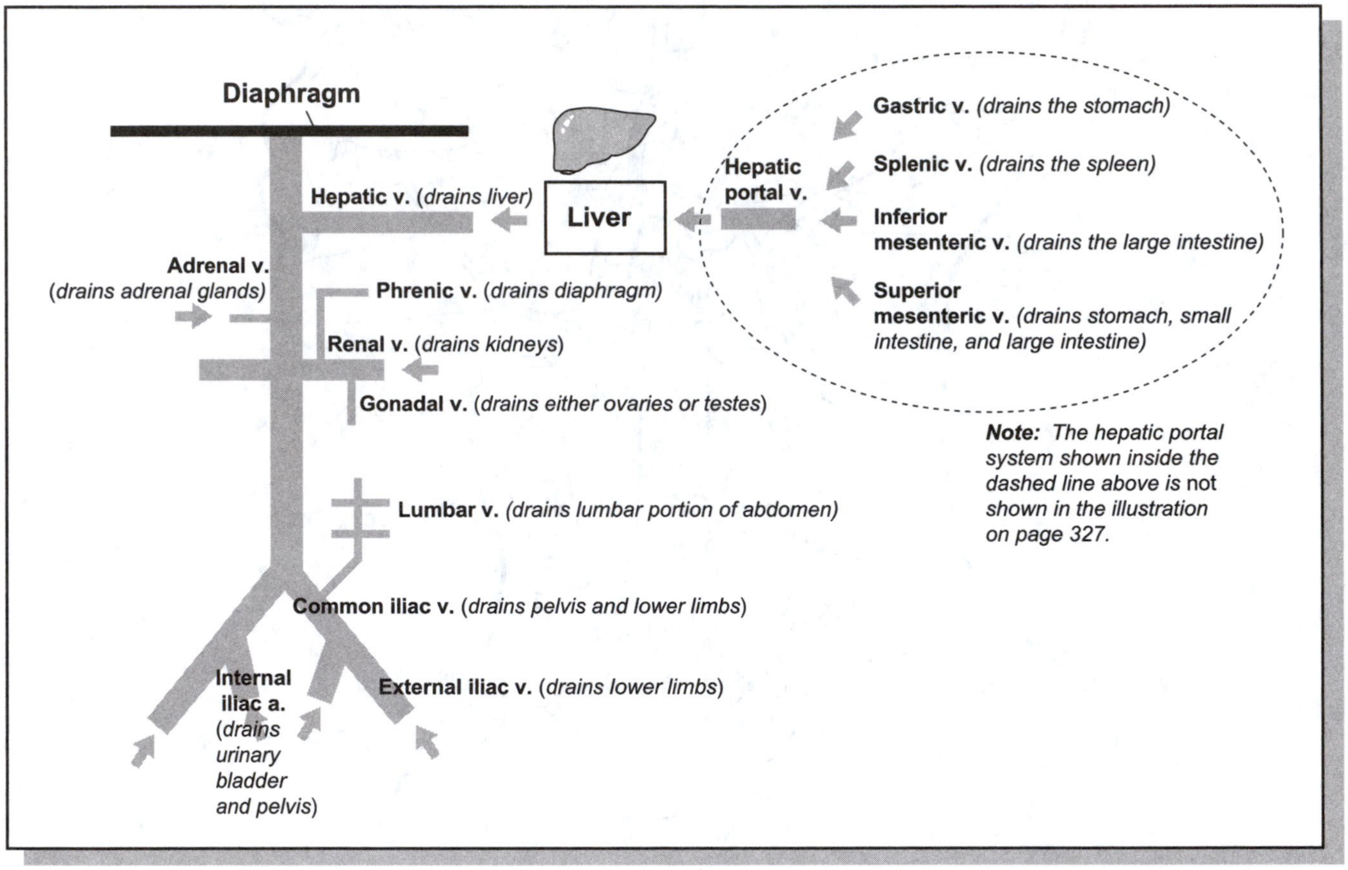

Key to Illustration

1. Inferior vena cava
2. Hepatic v.
3. Renal v.
4. Adrenal v.
5. Phrenic v.
6. Gonadal v.
   testicular v. in male
   ovarian v. in female
7. Lumbar v.
8. Common iliac v.
9. Internal iliac v.
10. External iliac v.

Major Veins of the Abdomen

Schematic of major branches

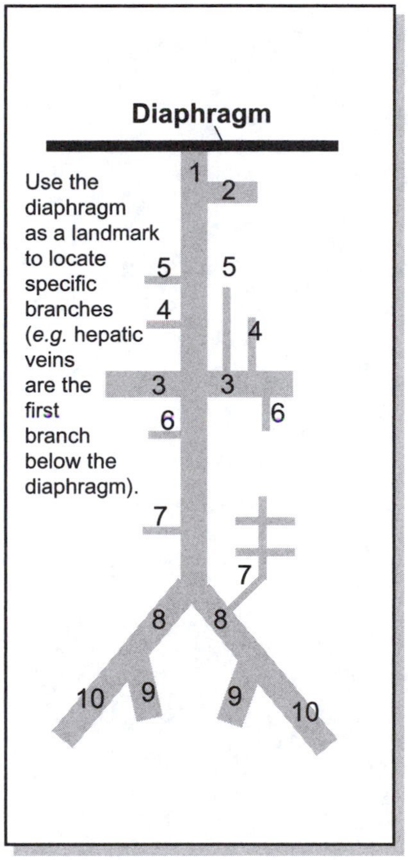

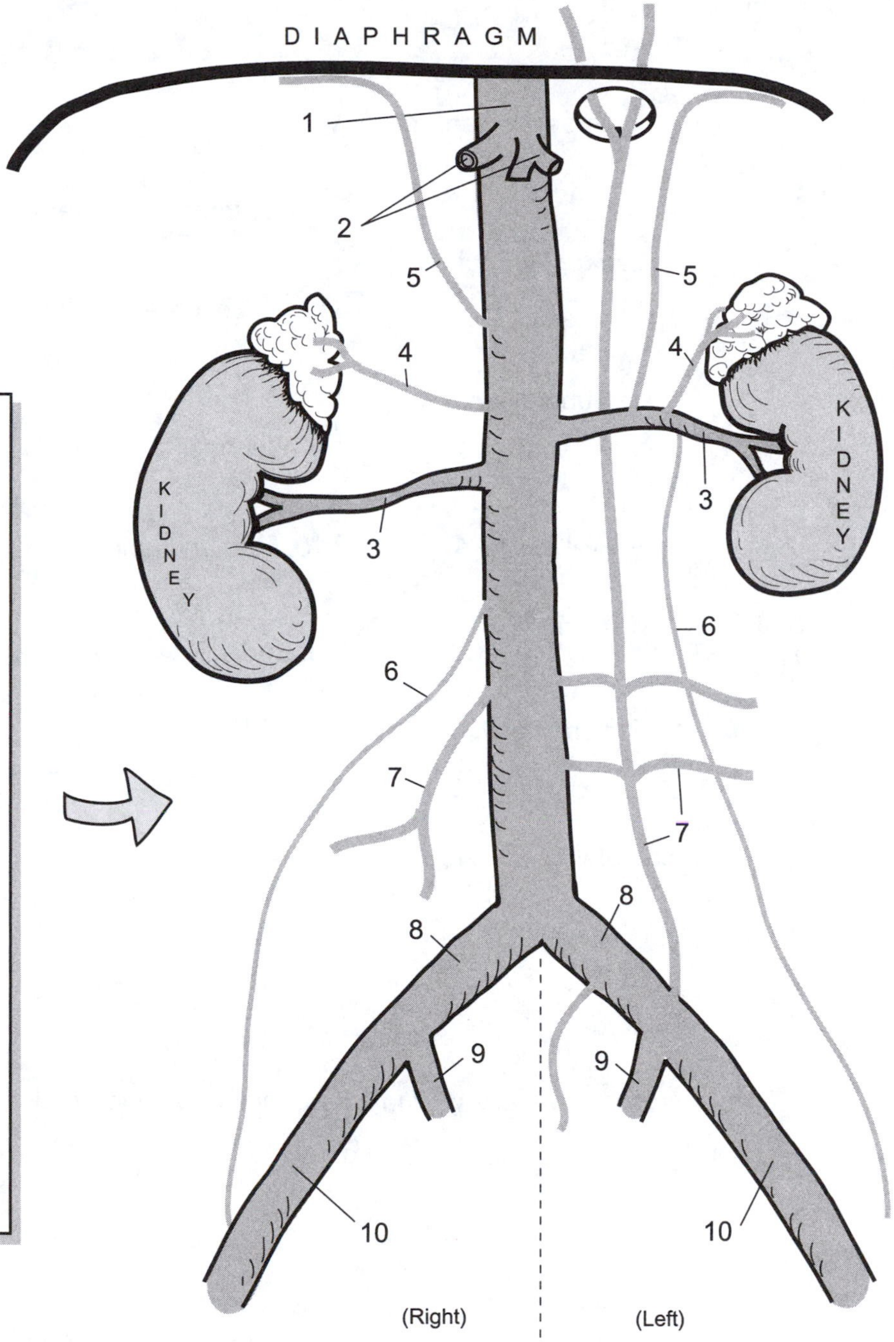

Anterior view

1. ______________________
2. ______________________
3. ______________________
4. ______________________
5. ______________________
6. ______________________
7. ______________________
8. ______________________
9. ______________________
10. ______________________

Key to Illustration

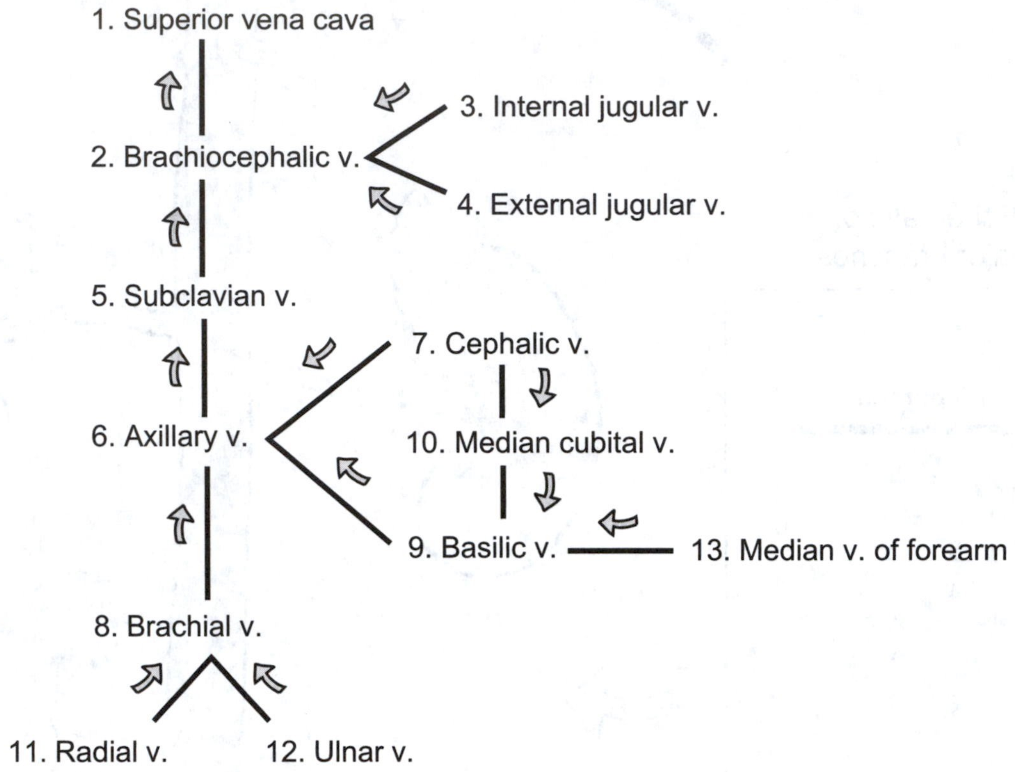

Note: *Gray arrows indicate direction of blood flow.*

Veins of the Right Upper Limb and Shoulder

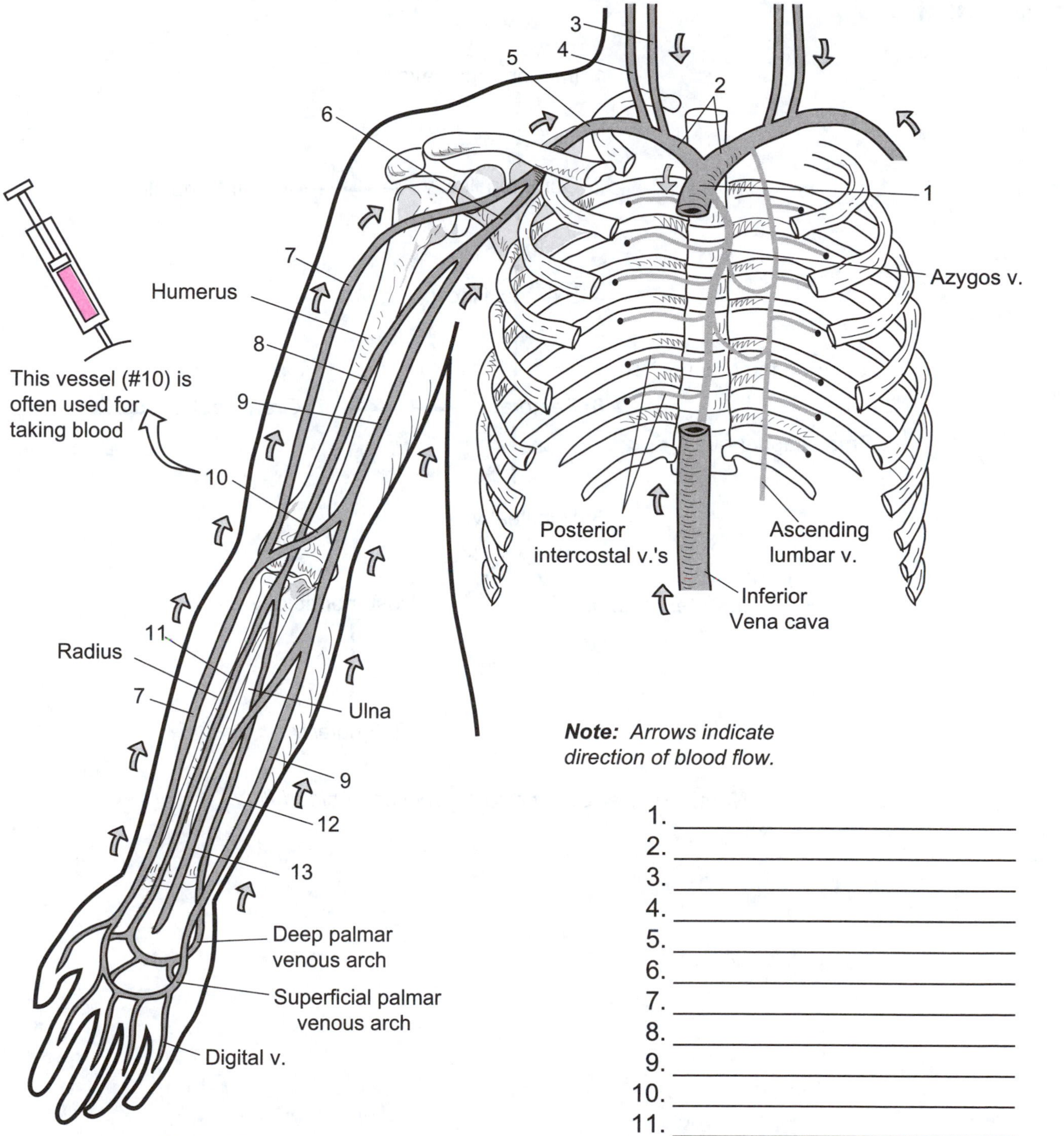

Right Upper Limb
anterior view

Note: *Arrows indicate direction of blood flow.*

1. ____________________
2. ____________________
3. ____________________
4. ____________________
5. ____________________
6. ____________________
7. ____________________
8. ____________________
9. ____________________
10. ____________________
11. ____________________
12. ____________________
13. ____________________

Major Veins of the Right Pelvis and Lower Limb

Key to Illustration

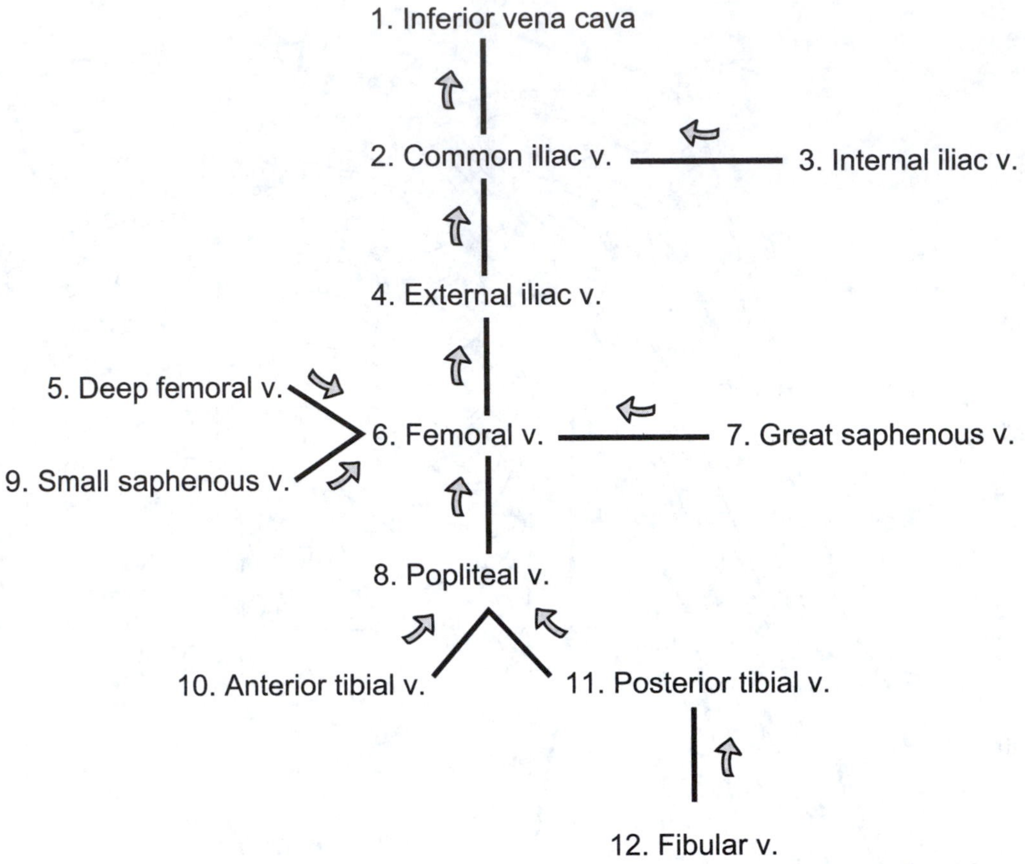

Note: *Gray arrows indicate direction of blood flow.*

Major Veins of the Right Pelvis and Lower Limb

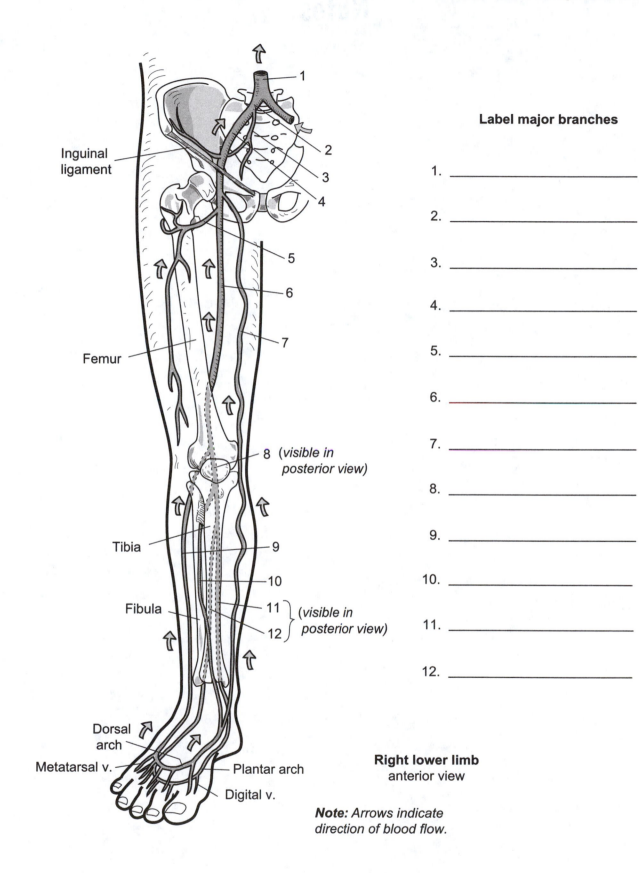

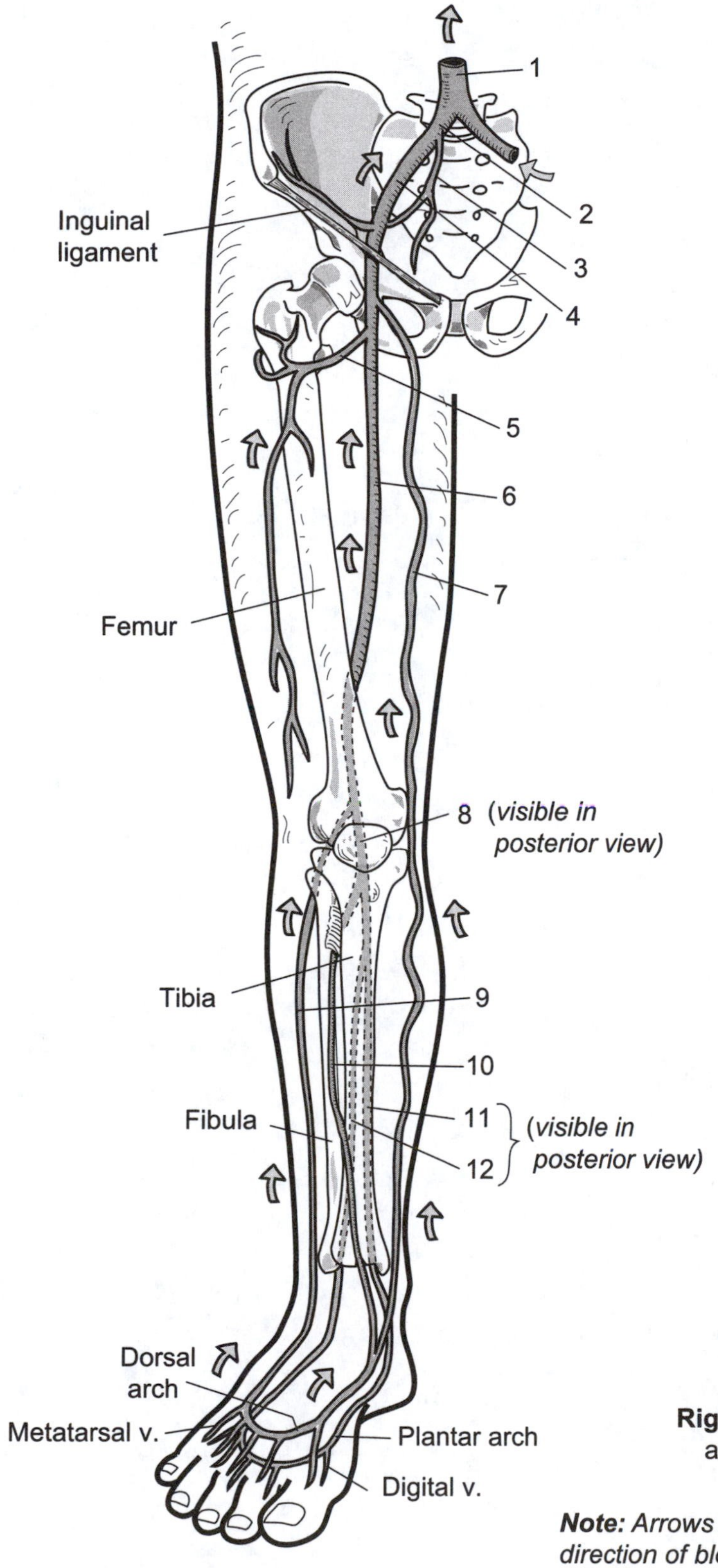

Right lower limb
anterior view

Note: *Arrows indicate direction of blood flow.*

Label major branches

1. ____________________
2. ____________________
3. ____________________
4. ____________________
5. ____________________
6. ____________________
7. ____________________
8. ____________________
9. ____________________
10. ____________________
11. ____________________
12. ____________________

Notes

Lymphatic System

Description

The cardiovascular system has a close relationship with the lymphatic system. Like the veins running through the body, the lymphatic system consists of a network of thin-walled vessels called **lymphatic vessels**. Like veins, they contain one-way valves (semilunar valves) that assist in circulating the lymph, which is under very low pressure.

Instead of carrying blood, the lymphatic vessels carry **lymph**—tissue fluid that was filtered from the blood. In the illustration, the gray areas indicate this filtration process. The composition of lymph is similar to plasma—mostly water along with some solutes such as salts.

Lymphatic vessels are connected to lymphatic capillaries and lymph nodes. **Lymphatic capillaries** are structurally similar to blood capillaries. Both are microscopic networks made of a single layer of simple squamous epithelium, but lymphatic capillaries contain flap-like structures that make them more permeable than blood capillaries. **Lymph nodes** are pea-sized structures that act as tiny filters to clean the lymph.

Like an oil filter cleans the motor oil in your car's engine, the lymph nodes filter the debris out of your lymph. These nodes contain macrophages that ingest and destroy pathogens such as bacteria. Lymph enters a lymph node through an **afferent lymphatic vessel** and leaves through an **efferent lymphatic vessel**. In short, "unclean lymph in, clean lymph out." The cleansed lymph is returned to the cardiovascular system via the subclavian veins.

Flow of Lymph

Here is a summary of the flow of lymph through the lymphatic system:

Lymphatic capillaries → afferent lymphatic vessel → lymph node → efferent lymphatic vessel → subclavian veins

Lymph is really nothing more than filtered blood plasma. All blood capillaries constantly filter the blood as a result of the force of blood pressure (see p. 300). This fluid, called **interstitial fluid**, fills the interstitial spaces between body cells, bathing them in fluid. Filtration occurs at the higher-pressure arteriole end of the capillary. Although some of this fluid is reabsorbed at the venule end, there is still an excess amount. As interstitial fluid pressure builds in the interstitial spaces, it is shunted into the nearby lymphatic capillaries, which act as a drain for the excess fluid. Although this fluid has not changed its chemical composition in any way, once inside the lymphatic capillary, it now is called **lymph**. Think of this **fluid cycle** as recycling of our plasma. This helps maintain normal fluid levels in the blood. As shown in the illustration: **Plasma** is filtered to become **interstitial fluid**, which becomes **lymph**, which becomes plasma once again. The cycle is complete!

Study Tip

The terms *afferent* and *efferent* apply to multiple organ systems. Here is a way to distinguish them: **A**fferent as in **A**pproach; **E**fferent as in **E**xit.

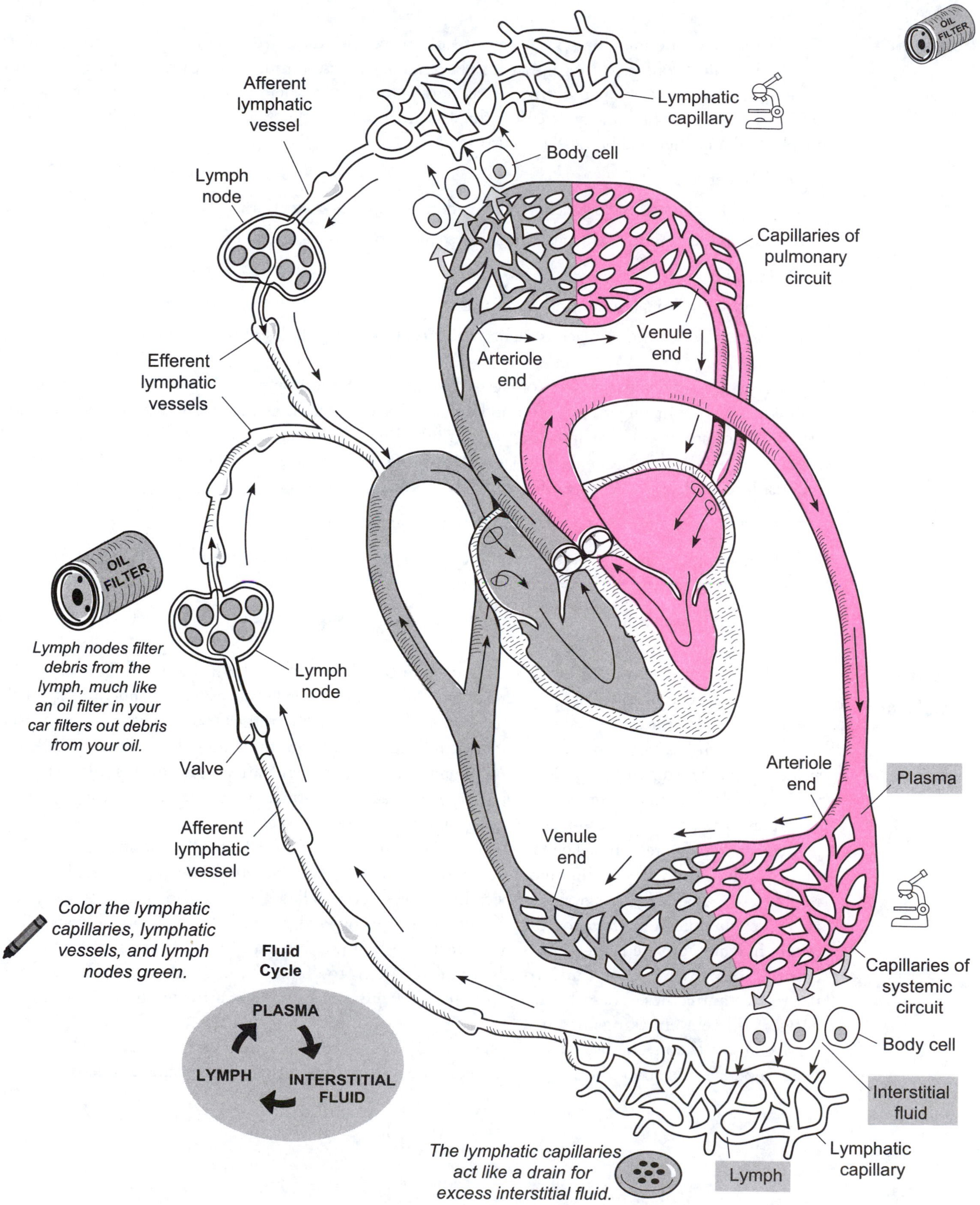
OIL FILTER
Afferent lymphatic vessel
Lymphatic capillary
Body cell
Lymph node
Capillaries of pulmonary circuit
Venule end
Arteriole end
Efferent lymphatic vessels
OIL FILTER
Lymph nodes filter debris from the lymph, much like an oil filter in your car filters out debris from your oil.
Lymph node
Valve
Arteriole end
Plasma
Afferent lymphatic vessel
Venule end
Color the lymphatic capillaries, lymphatic vessels, and lymph nodes green.
Fluid Cycle
PLASMA
LYMPH
INTERSTITIAL FLUID
Capillaries of systemic circuit
Body cell
Interstitial fluid
Lymphatic capillary
Lymph
The lymphatic capillaries act like a drain for excess interstitial fluid.

Description

This overview of the immune system has a focus on specific resistance. The **immune system** protects your body against foreign pathogens such as bacteria and viruses, using several lines of defense against these invaders. Think of a pathogenic invasion as an army of invaders attacking another army inside a medieval castle. The first line of defense is the wall around the castle. Similarly, your body has the following physical and chemical barriers as its first line of defense:

1. Physical barriers
   a. Skin: Thick layer of dead cells in the epidermis provides protection.
   b. Mucous membranes: Mucous film on these membranes traps microbes.
2. Chemical barriers
   a. Lysozyme in tears is an antibacterial agent.
   b. Gastric juice in the stomach is highly acidic (pH 2–3), which destroys bacteria.

The second line of defense consists of methods of nonspecific resistance that destroy invaders in a generalized way without targeting specific individuals. For example, if archers along the top of the castle were to shoot arrows into the invading army or if boiling oil were poured on them, this would help kill clusters of enemy soldiers. Similarly, your body has some general defenses for microbes that pass through the first lines of defense. A few examples are the following:

- Phagocytic cells ingest and destroy all microbes that pass into body tissues.
- Inflammation is a normal body response to tissue damage and other stimuli that brings more white blood cells to the site of pathogenic invasion.
- Fever inhibits bacterial growth and increases the rate of tissue repair during an infection.

Specific Resistance

The third line of defense deals with specific resistance, illustrated on the facing page. Think of these defenses like guided missiles that go after a specific target. In the medieval castle, they might be specially trained soldiers who act as assassins to kill the enemy's general. In short, they have a specific mission. Your immune system has "assassin" cells that attack microbes. Unlike in a war in which soldiers in different armies are wearing different uniforms, your immune system has a more difficult time distinguishing its own tissues ("self") from foreign microbes ("non-self"). To make this distinction, it relies on detecting **antigens**, which are specific substances found in foreign microbes. Most are proteins that serve as the stimulus to produce an immune response. The term "antigen" is coined from "**ANTI**-body **GEN**erating substances."

The illustration shows the immune response to an antigen. Once the antigen is detected, a dual response is activated by two groups of specialized lymphocytes called **T cells** and **B cells**. These cells are able to communicate with each other through chemical signaling. T cells typically are activated first. After they are activated, they can either directly destroy the microbes or use chemical secretions to destroy them. At the same time, T cells stimulate B cells to divide, forming other cells that are able to produce **antibodies**. These Y-shaped proteins circulate though the bloodstream and bind to specific antigens, thereby attacking microbes.

Details of the mechanisms that T cells and B cells use to attack microbes are examined in subsequent modules. Together, the T cells and B cells provide specific resistance to specific antigens.

Immune System: Overview of Specific Resistance
Direct attack by activated T cells
Let's get them!
Cell-mediated immunity (CMI) ➡ T Cells
I'm ready to fight but I'm going to need help from the B cells. Wake up, guys!
T cells get activated
Communication between T cells and B cells
Antibody-mediated immunity (AMI) ➡ B Cells
OK, OK. We're awake and ready to make antibodies!
B cells get activated
S
Antigen
antigens
R
Immune Response
Blood
Antibodies
We're ready to fight!
Blood
KEY
S Stimulus
R Response
Color T cells and B cells different colors.

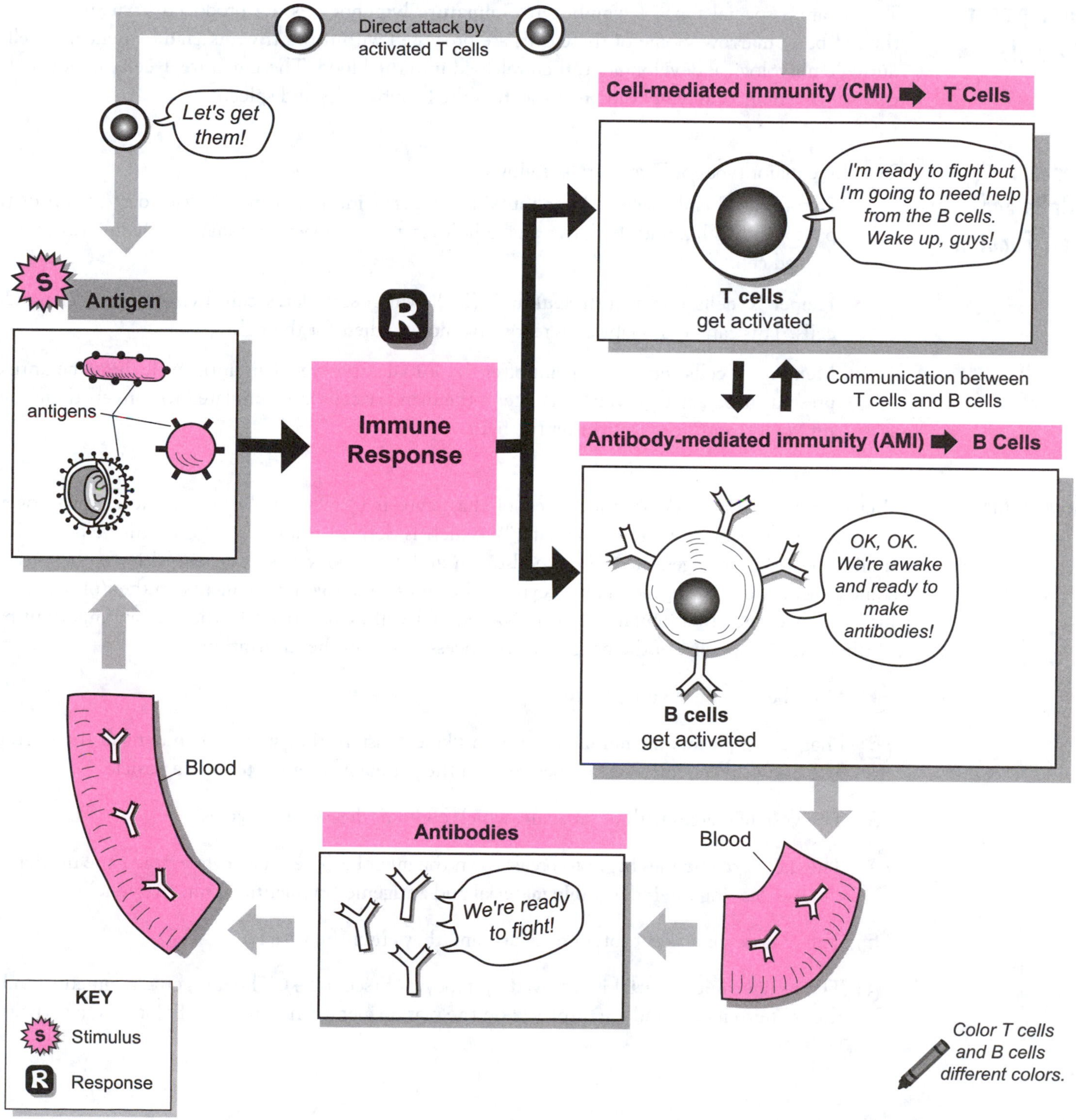

Description

Cell-mediated immunity (CMI) involves the activation of **T cells** (T lymphocytes) by a specific antigen. In total, the body contains millions of different T cells—each able to respond to one specific antigen.

Development of T cells

T cells are a special type of lymphocyte. **Immature lymphocytes** are produced from stem cells in the **red bone marrow**. Some of these cells are processed within the **thymus gland**—hence, **T cell**—during embryological development, then released into the blood. These mature T cells are located in the blood, lymph, and lymphoid organs such as the lymph nodes and spleen.

Common T Cells and Their Functions

The three major types of T cells are as follows:

- **Cytotoxic T cells** secrete **lymphotoxin** and **perforin**. The former trigger destruction of the pathogen's DNA, and the latter create holes in the pathogen's plasma membrane, resulting in a **lysed cell**.
- **Helper T cells** secrete **interleukin 2** (I-2), which stimulates cell division of T cells and B cells. This can be thought of as recruiting more soldiers for the fight.
- **Memory T cells** remain dormant after the initial exposure to an antigen. If the same antigen presents itself again—even years later—the memory cells are stimulated to convert themselves into cytotoxic T cells and enter the fight.

Phagocytosis

Phagocytes ("eater cells") use the process of **phagocytosis** ("cell eating") to ingest foreign pathogens. An example is a macrophage ("big eater"), which is derived from the largest white blood cells—monocytes. Macrophages leave the bloodstream and enter body tissues to patrol for pathogens. As some phagocytic cells engage in phagocytosis, they present antigenic fragments on their plasma membrane surface, thereby stimulating the activation of T cells. Consequently, they are an important part of the CMI. A summary of the phagocytosis process shown in the illustration is:

1. **Microbe** attaches to **phagocyte.**
2. Phagocyte's **plasma membrane** forms arm-like extensions that surround and engulf the microbe. The encapsulated microbe pinches off from the plasma membrane to form a **vesicle**.
3. The vesicle merges with a **lysosome**, which contains **digestive enzymes**.
4. The digestive enzymes begin to break down the microbe. The phagocyte extracts the nutrients it can use, leaving the **indigestible material** and **antigenic fragments** within the vesicle.
5. The phagocyte makes **protein markers**, and they enter the vesicle.
6. The indigestible material is removed by exocytosis (see p. 54). The antigenic fragments bind to the protein marker and are displayed on the plasma membrane surface. This serves to activate T cells.

Development of T cells
Red bone marrow
Immature lymphocytes
Thymus gland
(processing center)
T cell
(T lymphocyte)
Color the different T cells, different colors.
Common types of T cells and their functions
C
Cytotoxic T cell
lymphotoxin, perforin
Bacterium
Lysed cell
H
Helper T cell
interleukin 2 (I-2)
Stimulates cell division of T cells and B cells
M
Memory T cell
I will remember this antigen forever!
Years later...
Same antigen presents itself
M
C
Memory cells converted into cytotoxic T cells
Phagocytosis
1
Bacterium
Antigen
Plasma membrane
2
Vesicle
3
4
Lysosome
Digestive enzymes
Antigenic fragments
5
Protein marker
Endoplasmic reticulum
Indigestible material
Protein marker + antigenic fragment
6
Phagocyte

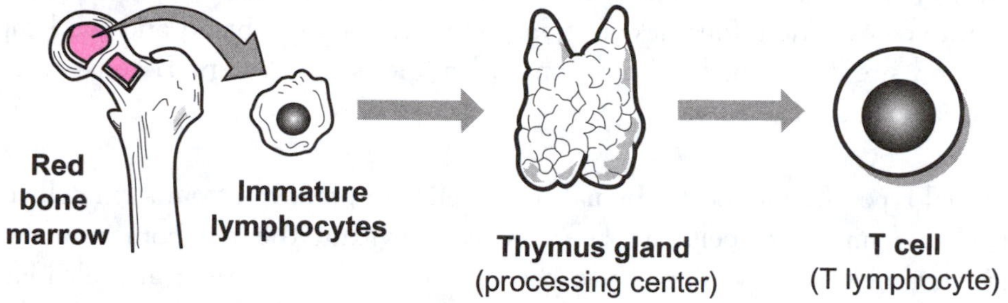

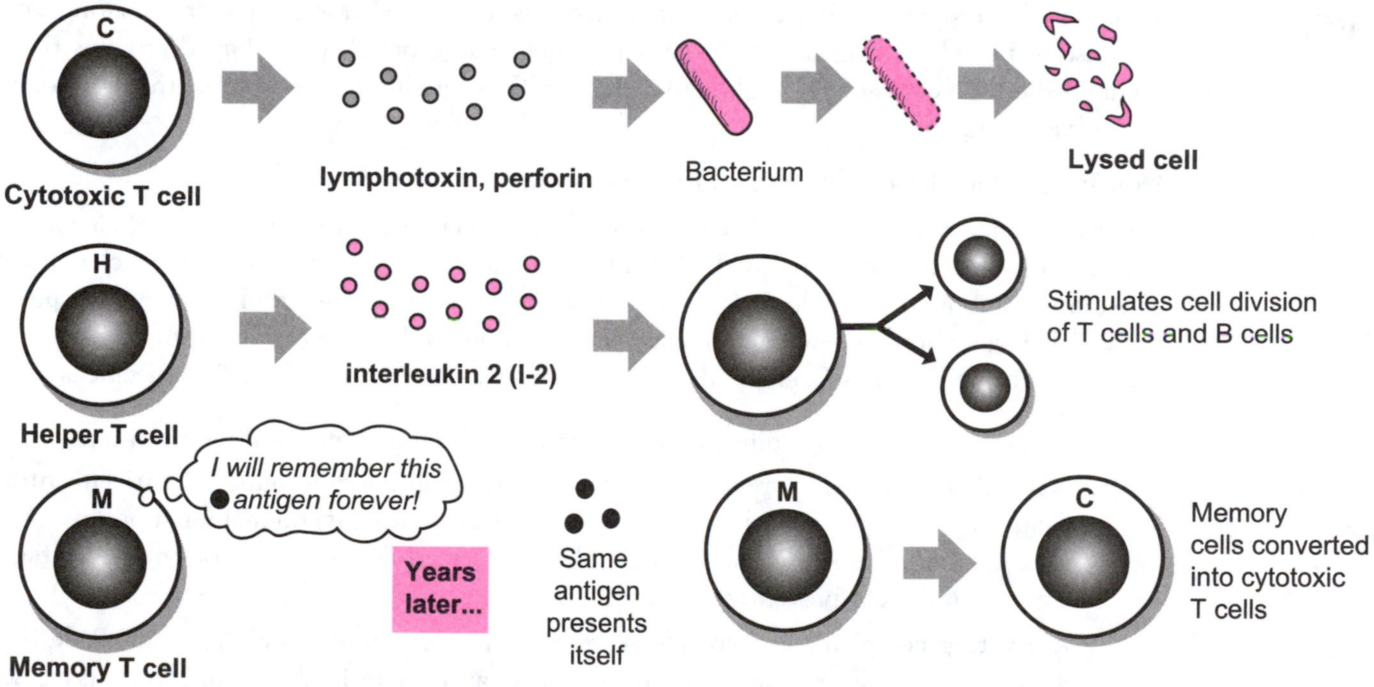

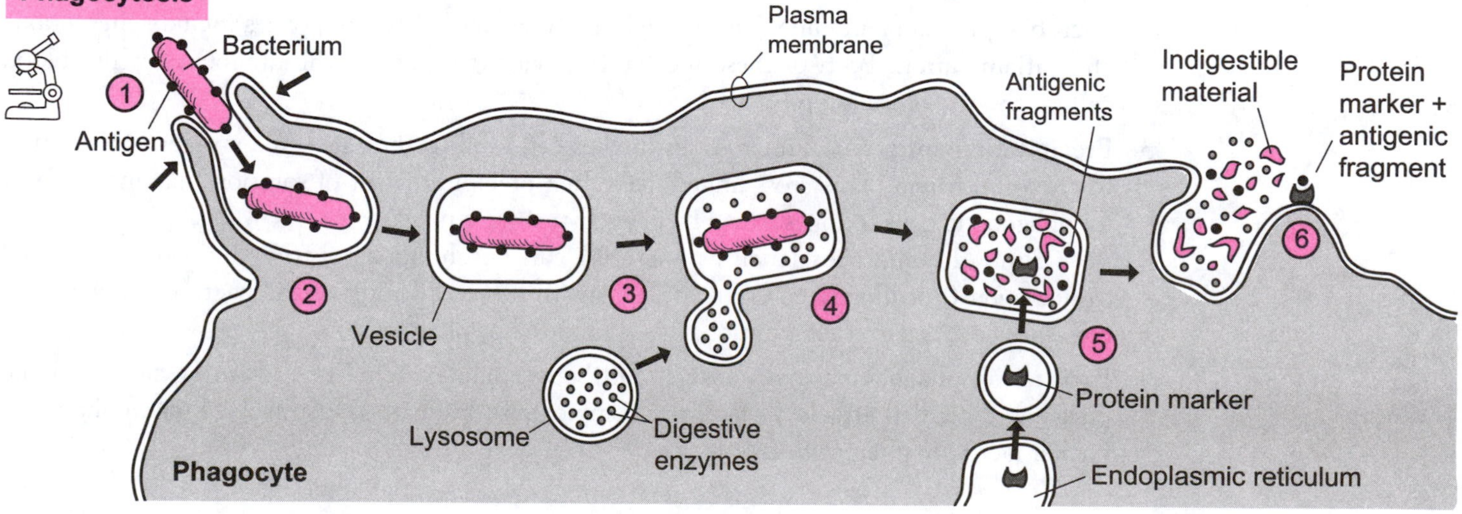

Description

Antibody-mediated immunity (**AMI**) involves the activation of **B cells** (B lymphocytes) by a specific antigen. This triggers the B cells to transform into **plasma cells**, which are able to secrete special proteins called **antibodies**. The antibodies are transported through the blood and the lymph to the pathogenic invasion site. In total, the body contains millions of different B cells—each able to respond to one specific **antigen**. Amazing!

Development of B Cells

B cells are a special type of lymphocyte. **Immature B cells** are produced from stem cells in the **red bone marrow**. These immature cells are later processed within the red **b**one marrow—hence, B cell—during embryological development to become mature B cells, then are released into the blood. The mature B cells are located in the blood, lymph, and lymphoid organs such as the lymph nodes and spleen.

Antibody Production

B cells can be stimulated to divide, forming two types of numerous cells: (1) **plasma cells**, which secrete **antibodies,** and (2) **B memory cells** which exist in the body for many years, ensuring a quick response to the same antigen. Antibodies (immunoglobulin, or Ig) are Y-shaped proteins that are subdivided into five classes: IgG, IgM, IgA, IgE, IgD. These are listed in order from the *most* common to the *least* common.

Study Tip: Mnemonic: *G**et** M**e** A**nother** E**xcellent** D**onut!***

The basic structure on an antibody consists of four polypeptide chains—two **heavy chains** and two **light chains**. Both heavy chains and both light chains are identical to the other, and each contains a **constant region** and a **variable region.** The constant region forms the "trunk" of the molecule, and the variable region forms the **antigen-binding site** on the antibody. Note that each antibody has two of these antigen-binding sites. Think of these like claws on a lobster used to "grab" its specific antigen.

How Do Antibodies Work?

Antibodies work through many different mechanisms, of which the following are major ones:

1. **Neutralizing antigen:** the **antibody** can bind to an **antigen**, forming an **antigen–antibody complex**. This forms a shield around the antigen, preventing its normal function. In this way, a toxin from a bacterium may be neutralized or a viral antigen may not be able to bind to a body cell thereby preventing infection.
2. **Activating complement:** "complement" refers to a group of plasma proteins made by the liver that normally are inactive in the blood. An **antigen-antibody complex** triggers a cascade reaction that activates these proteins to induce beneficial responses. For example, some of these activated proteins can cluster together to form a pore or channel that inserts into a microbial plasma membrane. This results in a **lysed cell**. Other responses include **chemotaxis** and **inflammation**. Both of these mechanisms serve to increase the number of white blood cells at the site of invasion.
3. **Precipitating antigens:** numerous antibodies can bind to the same free antigens in solution to cross-link them. This cross-linked mass then precipitates out of solution, making it easier for phagocytic cells to ingest them by phagocytosis (see p. 54).

   Similarly, microbes (such as bacteria) can be clumped together by a process called *agglutination* (not illustrated). The antigens within the cell walls of the bacteria are what are cross-linked. As with precipitation, this is followed by phagocytosis.
4. **Facilitating phagocytosis:** an **antigen-antibody complex** acts like a warning sign to signal phagocytic cells to attack. In fact, the complex also binds to the surface of **macrophages** to further facilitate phagocytosis.

Development of B cells
Red Bone marrow
Immature lymphocytes
Red bone marrow (processing center)
B cell (B lymphocyte)
Color the B cells, plasma cells, and antibodies different colors.
Antibody production
S
antigen
R
Activated B Cell
Plasma cells
Antibodies
Memory B cells
Antigen
Antigen-binding site
V
C
Light chain (gray)
C = Constant region
V = Variable region
Heavy chain (color)
Structure of an antibody
KEY
Stimulus
Response
How do antibodies work?
Color the antibodies all the same color.
1. Neutralizing antigen
Antigen
Antibody
Antigen-antibody complex
2. Activating complement
Activates complement
Bacterium
Lysed cell
• Inflammation
• Chemotaxis
3. Precipitating antigens
4. Facilitating phagocytosis
Attack!
Antigen-antibody complex acts like a warning sign
Macrophage

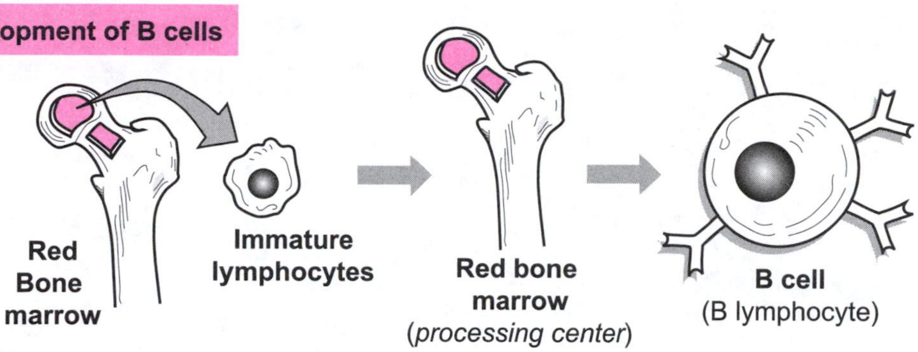

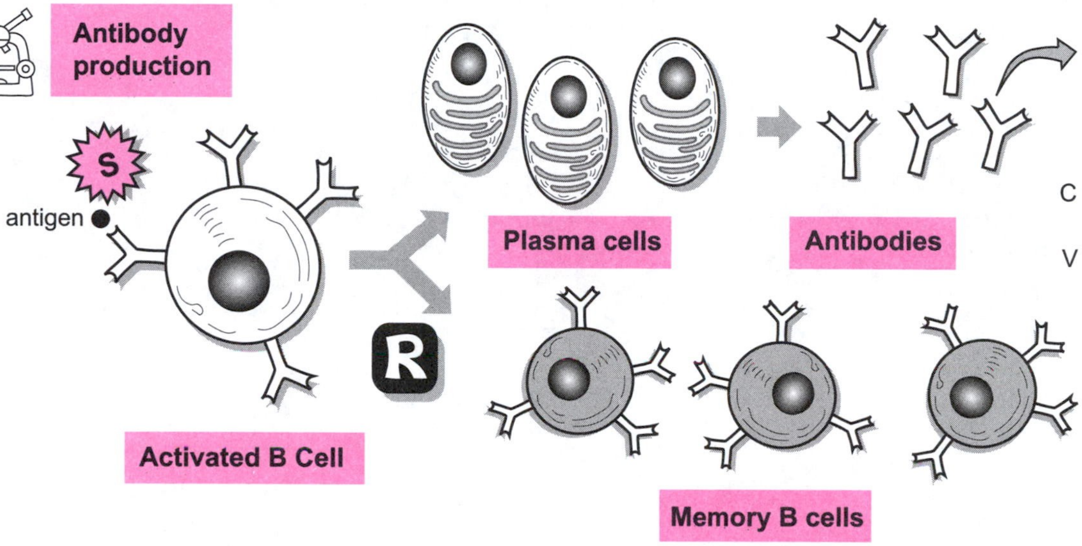

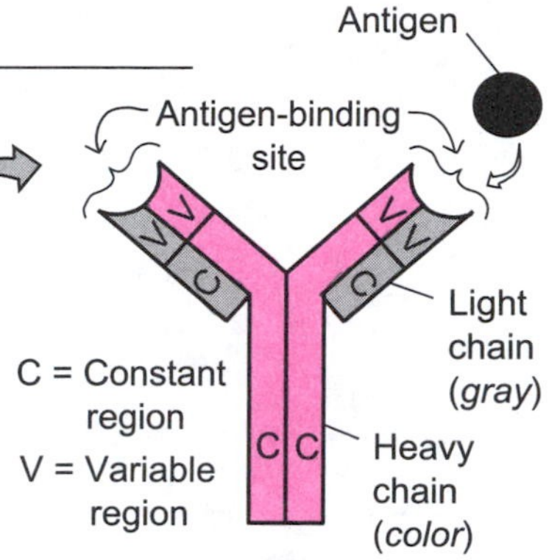

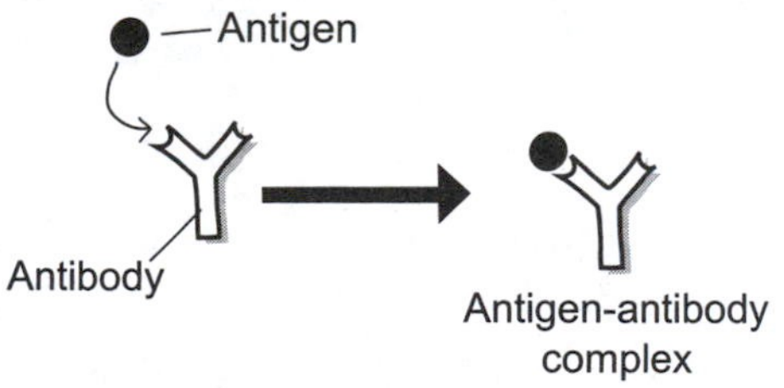

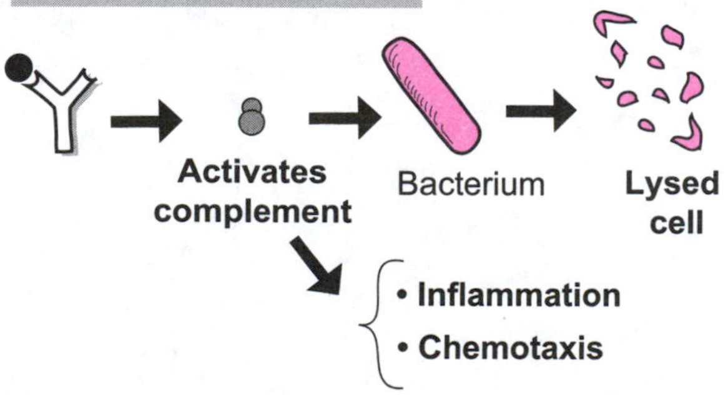

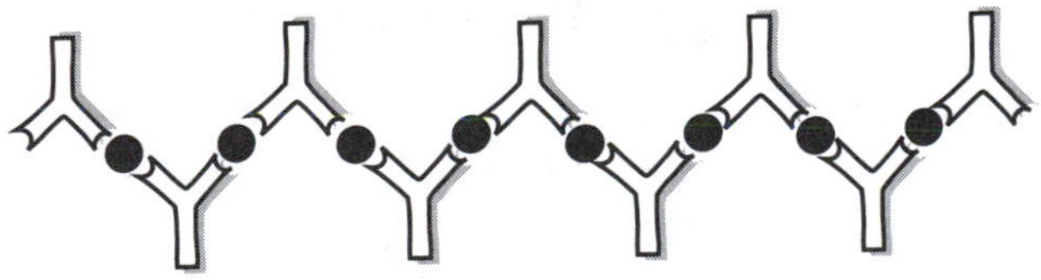

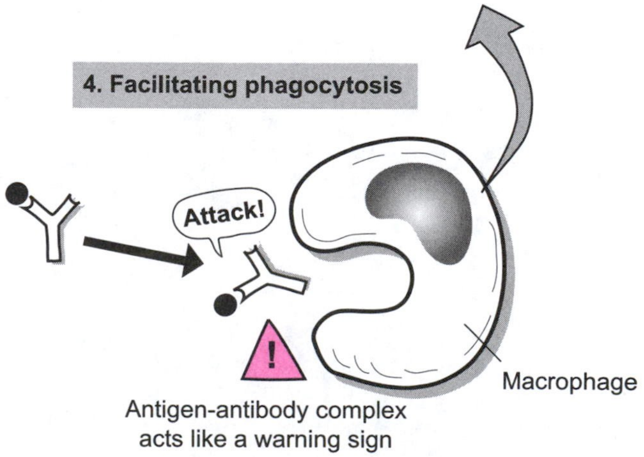

Notes

Respiratory System

Description

The respiratory system is divided into two major divisions: *upper respiratory system* and *lower respiratory system*. The **upper respiratory system** consists of the nose, nasal cavity, paranasal sinuses, and pharynx. The **lower respiratory system** consists of the larynx, trachea, bronchi, and lungs.

Let's trace the pathway of a molecule of oxygen (O_2) through the respiratory system to its final destination at a body cell. The O_2 molecule enters the **nasal cavity** through the **external nares** *(nostrils)*. As it passes to the back of this moist chamber, it enters the **nasopharynx**, then the **oropharynx**, and finally the **laryngopharynx.**

After passing the rigid, flap-like structure called the **epiglottis**, it enters the **larynx**, then passes through the slit-like opening between the vocal cords called the **glottis**. Next it moves through the long, rigid tube of the **trachea** until it reaches a split in this passageway. Following the passageway branching into the left lung, it enters the **left primary bronchus**, then enters the next split in the passageway, the narrower **secondary bronchus**. The next branch is the even narrower **tertiary bronchus**. Finally, the O_2 molecule enters a microscopic tube called a **bronchiole**.

This continues to branch into a **terminal bronchiole**, then a **respiratory bronchiole**, and finally terminates in an air sac called an **alveolus**. This delicate air sac is the end of the bronchial tree in the lungs. The wall of each alveolus is made of simple squamous epithelium, which allows for easy diffusion of the O_2 molecule out of the alveolus and into the bloodstream, where it will be delivered to a body cell.

Analogies

- The **larynx** looks like the **head of a snapping turtle**. The **turtle's head** is the **thyroid cartilage** and the **turtle's lower jaw** is the **cricoid cartilage**. The **neck of the turtle** is the **trachea**.
- The **clusters of alveoli** are like a **wad of bubble wrap** used in packaging. Both are small, sac-like structures filled with air.

Study Tip

Palpate (*feel by touch*):

On average, the larynx is slightly larger in males than in females, but you can easily feel a portion of it in either gender. The main structure you feel beneath the skin is the thyroid cartilage, commonly called the *Adam's apple*. The portion of the Adam's apple that protrudes most anteriorly is called the laryngeal prominence.

Key to Illustration

Larynx (L)

L1. Thyrohyoid membrane
L2. Thyroid cartilage
L3. Laryngeal prominence
L4. Cricothyroid ligament
L5. Cricoid cartilage

Upper Respiratory Tract (UP)

UP1. Frontal sinus
UP2. Sphenoidal sinus
UP3. External nares (nostrils)
UP4. Nasal cavity
UP5. Superior concha
UP6. Middle concha
UP7. Inferior concha
UP8. Pharyngeal tonsil
UP9. Nasopharynx
UP10. Palatine tonsil
UP11. Oropharynx
UP12. Lingual tonsil
UP13. Epiglottis
UP14. Laryngopharynx

Lower Respiratory Tract (LO)

LO1. Vocal fold
LO2. Trachea
LO3. Tracheal rings
LO4. Parietal pleura
LO5. Visceral pleura
LO6. Location of carina (*internal ridge*)
LO7. Primary bronchus
LO8. Secondary bronchus
LO9. Tertiary bronchus
LO10. Terminal bronchiole
LO11. Respiratory bronchiole
LO12. Alveoli
LO13. Simple squamous epithelium

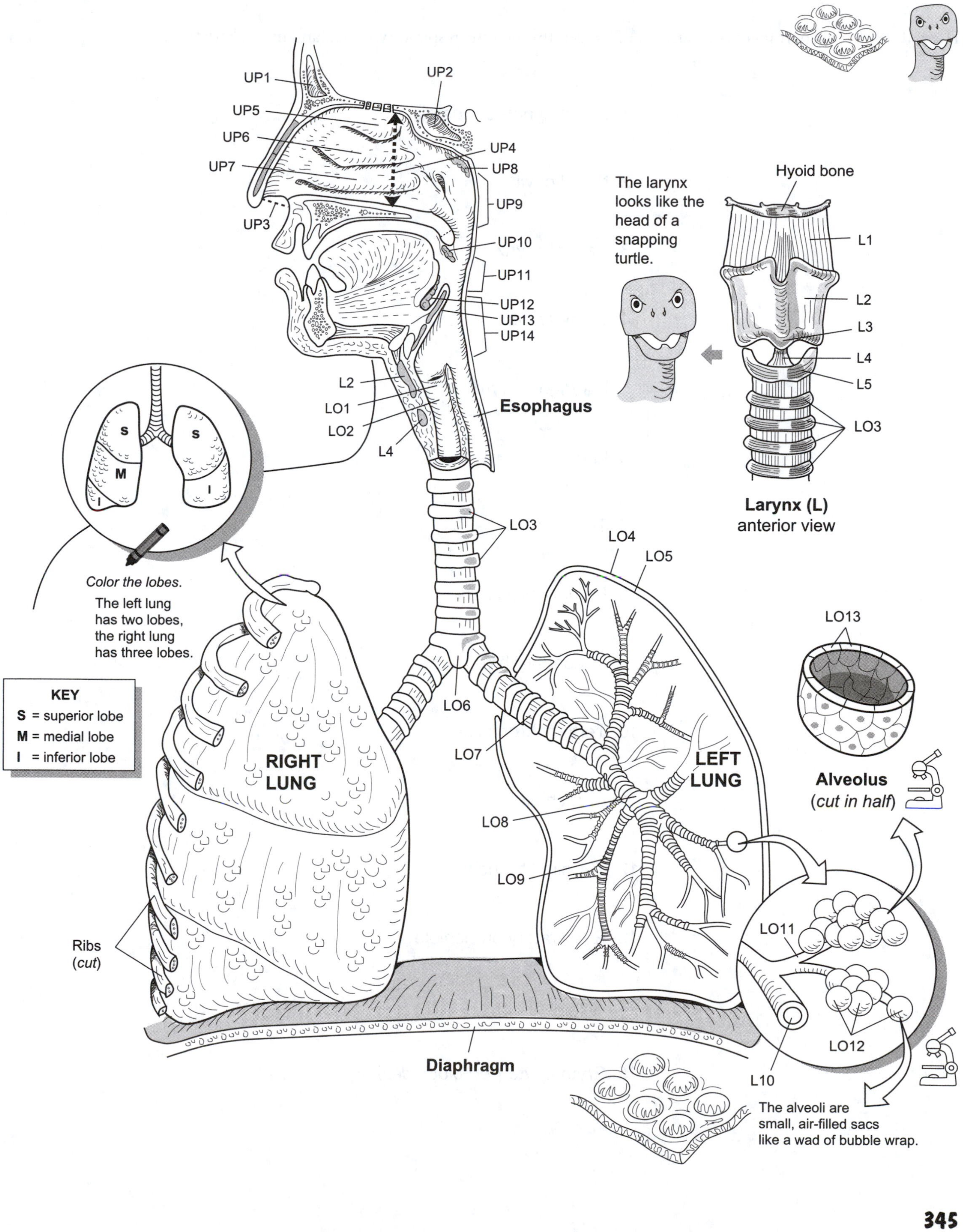
UP1
UP2
UP3
UP4
UP5
UP6
UP7
UP8
UP9
UP10
UP11
UP12
UP13
UP14
L2
LO1
LO2
L4
Esophagus
The larynx looks like the head of a snapping turtle.
Hyoid bone
L1
L2
L3
L4
L5
LO3
Larynx (L)
anterior view
S
S
M
I
I
Color the lobes.
The left lung has two lobes, the right lung has three lobes.
KEY
S = superior lobe
M = medial lobe
I = inferior lobe
LO3
LO4
LO5
LO6
LO7
LO8
LO9
RIGHT LUNG
LEFT LUNG
LO13
Alveolus
(cut in half)
LO11
LO12
L10
Ribs
(cut)
Diaphragm
The alveoli are small, air-filled sacs like a wad of bubble wrap.

Tracing the Pathway of an Oxygen Molecule

The following flowchart gives the pathway of oxygen through the respiratory system and into a body cell:

1. External nares *(nostrils)*

↓

2. Nasal cavity

↓

3. Nasopharynx

↓

4. Oropharynx

↓

5. Laryngopharynx

↓

6. Larynx

↓

7. Trachea

↓

8. Primary bronchus

↓

9. Secondary bronchus

↓

10. Tertiary bronchus

↓

11. Bronchiole

↓

12. Terminal bronchiole

↓

13. Respiratory bronchiole

↓

14. Alveolus

↓

15. Erythrocyte *(red blood cell)*

Tracing the Pathway of an Oxygen Molecule

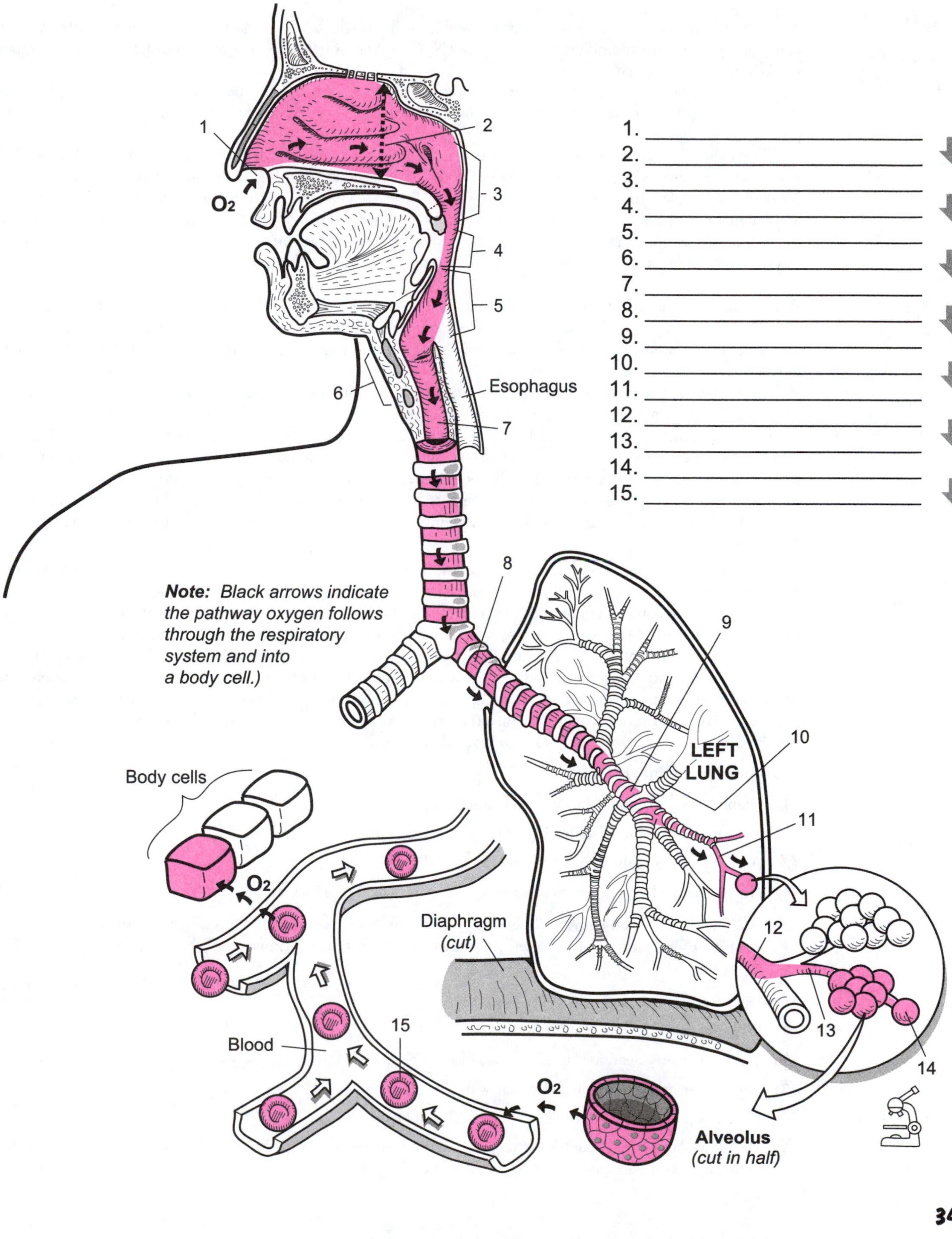

1. ______________________
2. ______________________
3. ______________________
4. ______________________
5. ______________________
6. ______________________
7. ______________________
8. ______________________
9. ______________________
10. ______________________
11. ______________________
12. ______________________
13. ______________________
14. ______________________
15. ______________________

Concept 1: Atmospheric Pressure (AP)

Atmospheric pressure (AP) is a force that combines the weight of all the gases in the air we breathe. Of the many gases, the most abundant are: **nitrogen (N_2)**, **oxygen (O_2)**, **carbon dioxide (CO_2)**, and **water** vapor **(H_2O)**. Each oof these gases has a different mass.

AP is the force on any surface that comes in contact with air. In the illustration, AP is indicated by the black arrows pushing against the external surface of the little boy's body. Even though he can't actually feel its presence, it is there.

The other illustration shows a **barometer**—a device used to measure the AP. A glass beaker is shown filled with liquid **mercury** (Hg). As the AP pushes down on the surface of this liquid, it forces the mercury into the open end of the glass tube so it rises to a specific height. At sea level, the standard measurement is **760 mm Hg**. This number changes as the elevation changes. The higher the elevation, the lower the AP. For example, a mountain climber and a person strolling in the park at sea level experience different AP. The mountain climber has less atmosphere above him, so he has less AP than the person in the park.

Concept 2: Boyle's Law

Boyle's law is a gas law stating that *volume and pressure have an inversely proportional relationship for a gas held at a constant temperature*. The illustration shows a clear, hollow sphere containing five gas molecules. These molecules have kinetic energy, so they randomly bounce around against the wall of the sphere. This is the source of the force of pressure inside the sphere.

Consider what would happen if the small sphere were to increase in size. Has the number of gas molecules changed? No. Five gas molecules are still present. The only thing that has changed is the inner volume of the sphere. If the volume increases, what do you predict will happen to the pressure inside? That's right. It will decrease.

If we refer to the volume and pressure in the small sphere as V_1 and P_1 and use V_2 and P_2 for the enlarged sphere, Boyle's law can be expressed as this **equation:** $V_1 \times P_1 = V_2 \times P_2$. The product of the volume and pressure in the first sphere should be equivalent to the product of the volume and the pressure in the second sphere. Fill in the given numbers, and plug them into the formula to see if this relationship holds true.

Application: Mechanics of Breathing

In addition to knowing atmospheric pressure and Boyle's law, we need to understand the terms **interpleural pressure** and **alveolar pressure**, so we can understand the mechanics of normal breathing. Interpleural pressure is the pressure inside the small, liquid-filled pleural space around the lungs. At rest, the normal value is **756** mm Hg. The pressure inside the lungs at any given moment is called the alveolar pressure. Its normal value at rest is the same as the normal AP, or **760** mm Hg. Normal, quiet breathing is a repeated cycle of **inhalations** and **exhalations**. Inhalation is an *active* process because it requires the contraction of muscles such as the **diaphragm** and the **external intercostal muscles** between the ribs. In contrast, an exhalation is a passive process because it mainly involves recoil of the elastic connective tissue in the lungs and thoracic wall. Let's summarize the steps involved in a normal inhalation and a normal exhalation.

Steps in a normal inhalation:

1. Diaphragm and external intercostal muscles contract
2. Thoracic cavity increases in size; lung volume expands
3. Alveolar pressure drops from 760 to 758 mm Hg
4. Air flows into lungs down its pressure gradient from **760** to **758** mm Hg

(**Note:** There is no such thing as "suction." We do not "suck" air into the lungs. Instead, air flows from a region of higher pressure to a region of lower pressure. This continues until the alveolar pressure is equal to the AP (760).)

Steps in a normal exhalation:

1. Diaphragm and exterior intercostals relax
2. Thoracic cavity decreases in size; lung volume decreases
3. Recoil effect from lungs and thoracic wall
4. Alveolar pressure increases from 760 to 762 mm Hg
5. Air flows out of the lungs down its pressure gradient from **762** to **760** mm Hg. This continues until the alveolar pressure is equal to the AP (**760** mm Hg).

CONCEPT 1: ATMOSPHERIC PRESSURE (AP)
AP
AP
AP
AP
I can't handle the pressure!
760 mm Hg = Atmospheric pressure (at sea level)
AP
AP
AP is a force that is exerted on any surface that comes in contact with air.
Open end of glass tube
Mercury (Hg)
Model of a Barometer
CONCEPT 2: BOYLE'S LAW
Gas molecule (ex: O_2 gas)
Hollow sphere
(increase size)
(decrease size)
Volume (V_1) = 2
Pressure (P_1) = 2
Volume (V_2) = 4
Pressure (P_2) = 1
Formula: $V_1 \times P_1 = V_2 \times P_2$
Color the gas molecules.
APPLICATION: MECHANICS of BREATHING
During inhalation, the ribs swing upward and outward like the handle on a pail.
Rib
Sternum
The recoil effect during exhalation is like the tension being released from a rubberband.
AP 760
Interpleural pressure
754
Thoracic wall
Alveolar pressure
758
Diaphragm
Diaphragm moves downward
External intercostal muscles
Diaphragm moves upward
AP 760
Interpleural pressure
756
Alveolar pressure
762
Diaphragm
Color the numbers for alveolar and interpleural pressures; Color the regional areas for alveolar and intrapleural pressures.
INHALATION (active process)
EXHALATION (passive process)

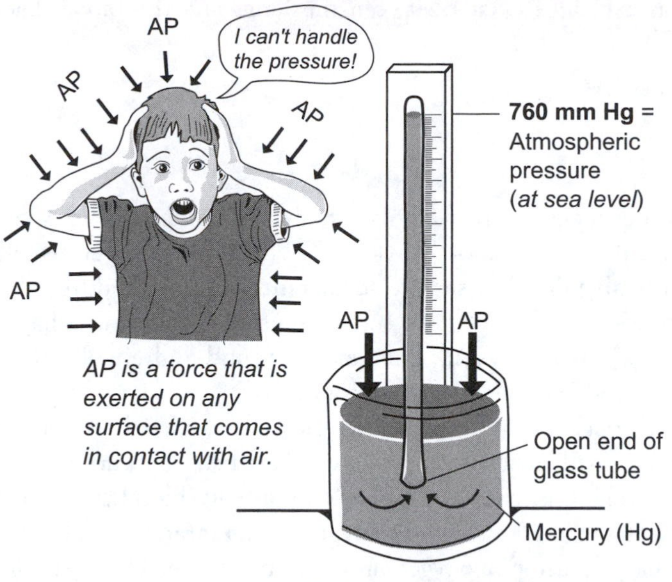

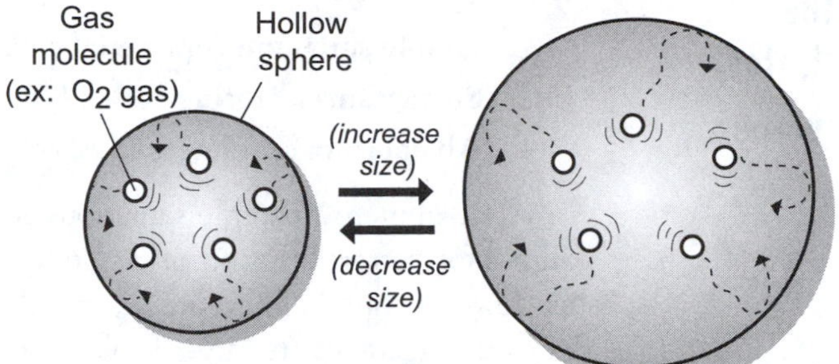

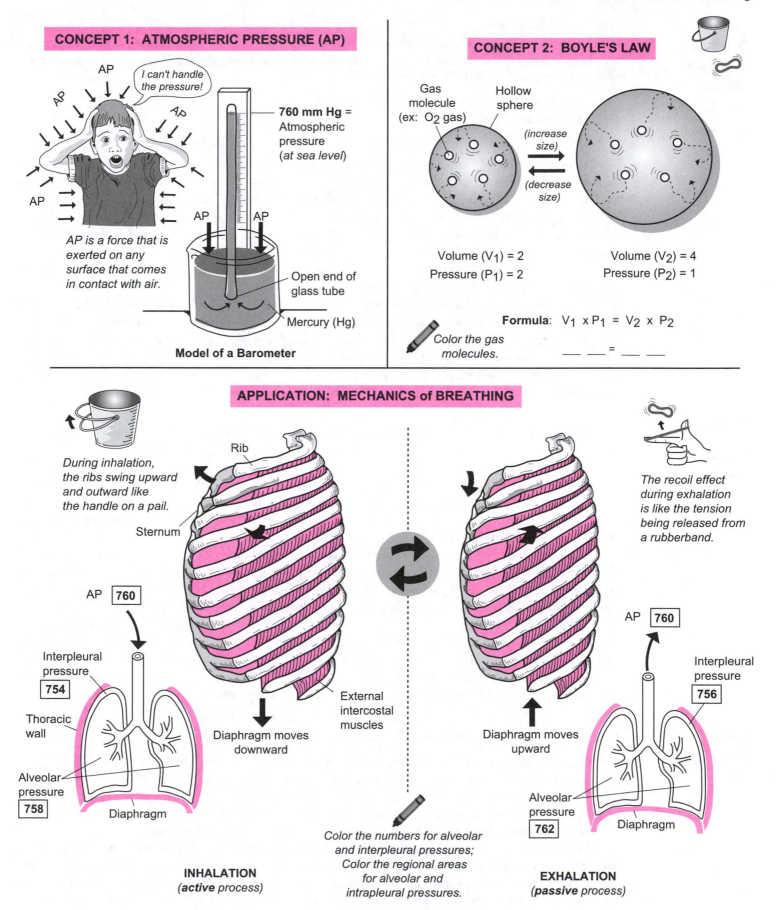

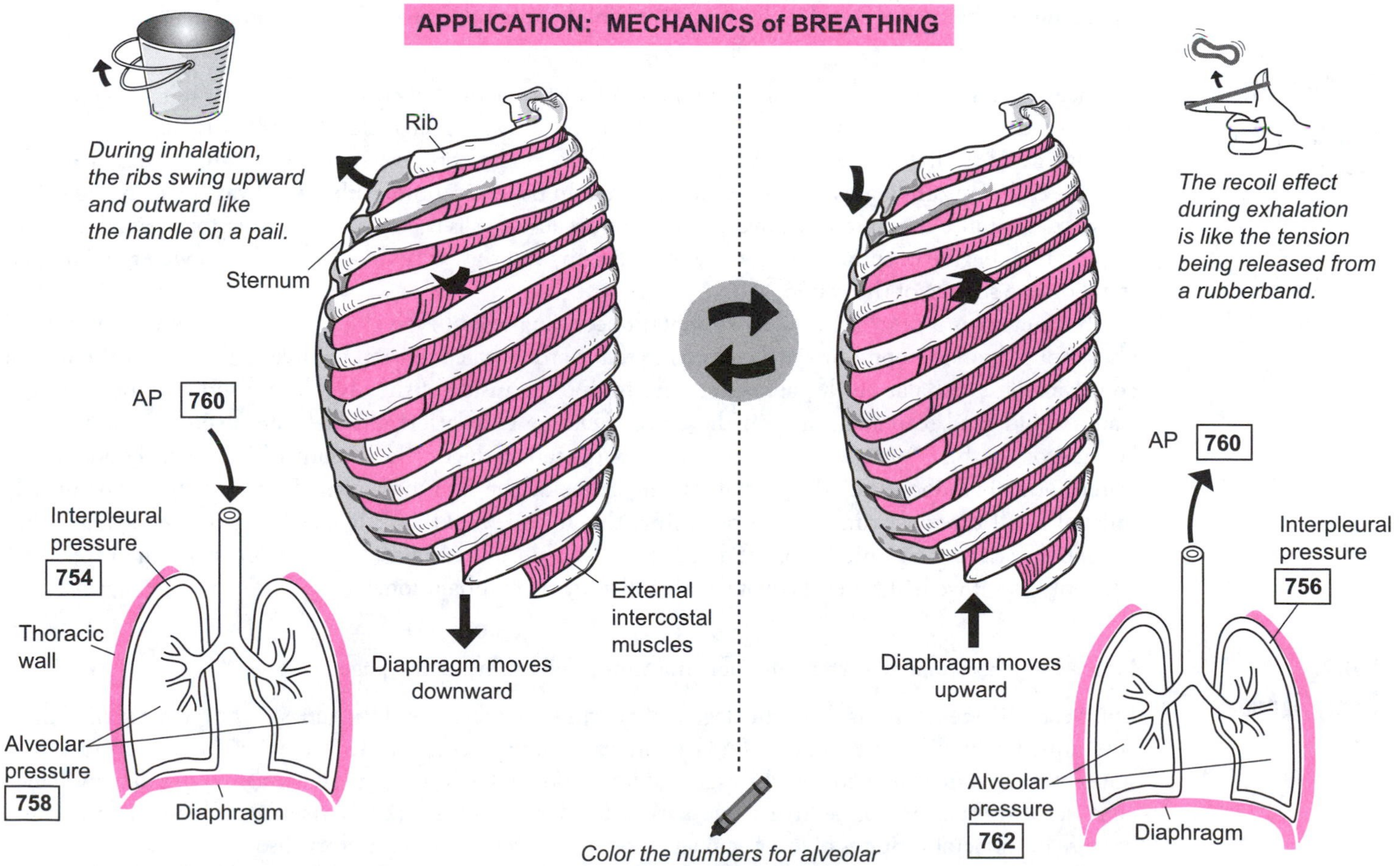

Alveolar Structure and the Respiratory Membrane

The lungs contain about 300 million alveoli. Like the bubbles in bubble wrap, the **alveoli** (sing. *alveolus*) are the delicate, microscopic air sacs where gas exchange occurs between the lungs and the blood. Each alveolus is composed of three different types of cells:

1. **Simple squamous epithelial cells** (*Type I cells*)
2. **Surfactant-secreting cells** (*Type II cells*)
3. **Alveolar macrophages** (*dust cells*).

The numerous simple squamous epithelial cells make up the wall of each alveolus, and their flat shape allows for better diffusion of respiratory gases (oxygen and carbon dioxide). The surfactant-secreting cells are fewer in number and are scattered within the alveoli. They secrete **surfactant**—an oily fluid film that lines the inside of the alveoli like a soap bubble and serves to reduce surface tension to prevent collapse. The alveolar macrophages are the "housekeepers" of the alveoli. They move around and engulf microorganisms, dust particles, and other debris to keep the lungs clean and free of disease.

Recall that the blood flows to each lung through the **pulmonary arteries** (left and right). Each of these vessels branches and becomes smaller until it finally reaches numerous capillary beds that surround the alveolar sacs—clusters of alveoli. This structural relationship is like a plastic mesh bag (blood capillaries) surrounding a cluster of grapes (alveolar sac) at the market. The **respiratory membrane** is called the "blood/air barrier" and is very thin to allow for easy diffusion of oxygen and carbon dioxide. This is the site where oxygen diffuses from alveolus to blood and carbon dioxide diffuses from blood to alveolus. This structure consists of the fusion of the alveolar and capillary walls. More specifically, it is where the **simple squamous epithelium** of the alveolus meets the **simple squamous epithelium** of the capillary.

Surfactant and Surface Tension

Surface tension is both a force and a property of water, attributed to the fact that water molecules are **polar**, or charged. Like mini-magnets, the oxygen end of any water molecule is slightly negative and the hydrogen end is slightly positive. When the negative end of one water molecule aligns itself with the positive end of another, a **hydrogen bond** is formed. In fact, water molecules are always rearranging themselves to maximize the number of hydrogen bonds because this is the most stable state for them. This gives water a high degree of surface tension and explains, by the way, why the insect called a **water strider** can stand on the surface of the water.

Surfactant is a fluid secreted by **surfactant-secreting cells** (Type II cells) that contains a mixture of phospholipids and lipoproteins. It functions to reduce the surface tension on alveoli to prevent them from collapsing after exhalation. If pure water were to line the inside of the alveoli, the surface tension in the water would tend to pull inward on the alveoli. During exhalation, the delicate alveoli would collapse because of the high surface tension on them. Surfactant reduces the amount of hydrogen bonds in the water normally found inside the alveoli, thereby reducing the surface tension. This keeps the alveoli partly inflated at all times. During normal breathing, the alveoli would look like slowly pulsating spheres that expand in size during inhalation and reduce in size during exhalation. This process is more energy-efficient than if we had to refill collapsed alveoli with every inhalation.

Lung Compliance

Note: Lung compliance is the only topic not illustrated on the facing page.

Lung compliance is a measure of the degree of effort needed to expand the lungs and thoracic wall. This is determined primarily by two factors: (1) elasticity of lung tissue, and (2) alveolar surface tension. Two states of compliance are mentioned—high and low. Normal, healthy lungs have a *high* degree of compliance because of the elastic connective tissue within the lungs and the reduced surface tension in the lungs because of surfactant. States of *low* compliance typically occur with respiratory disorders.

For example, a premature baby may lack the normal surfactant found in a newborn. This increases surface tension on the alveoli. Emphysema also leads to lower compliance because of destruction of the elastic connective tissue around the alveoli. In short, any disease state that increases alveolar surface tension and/or damages normal elastic connective tissue in the lungs will result in lowered compliance.

Alveolar Structure and the Respiratory Membrane

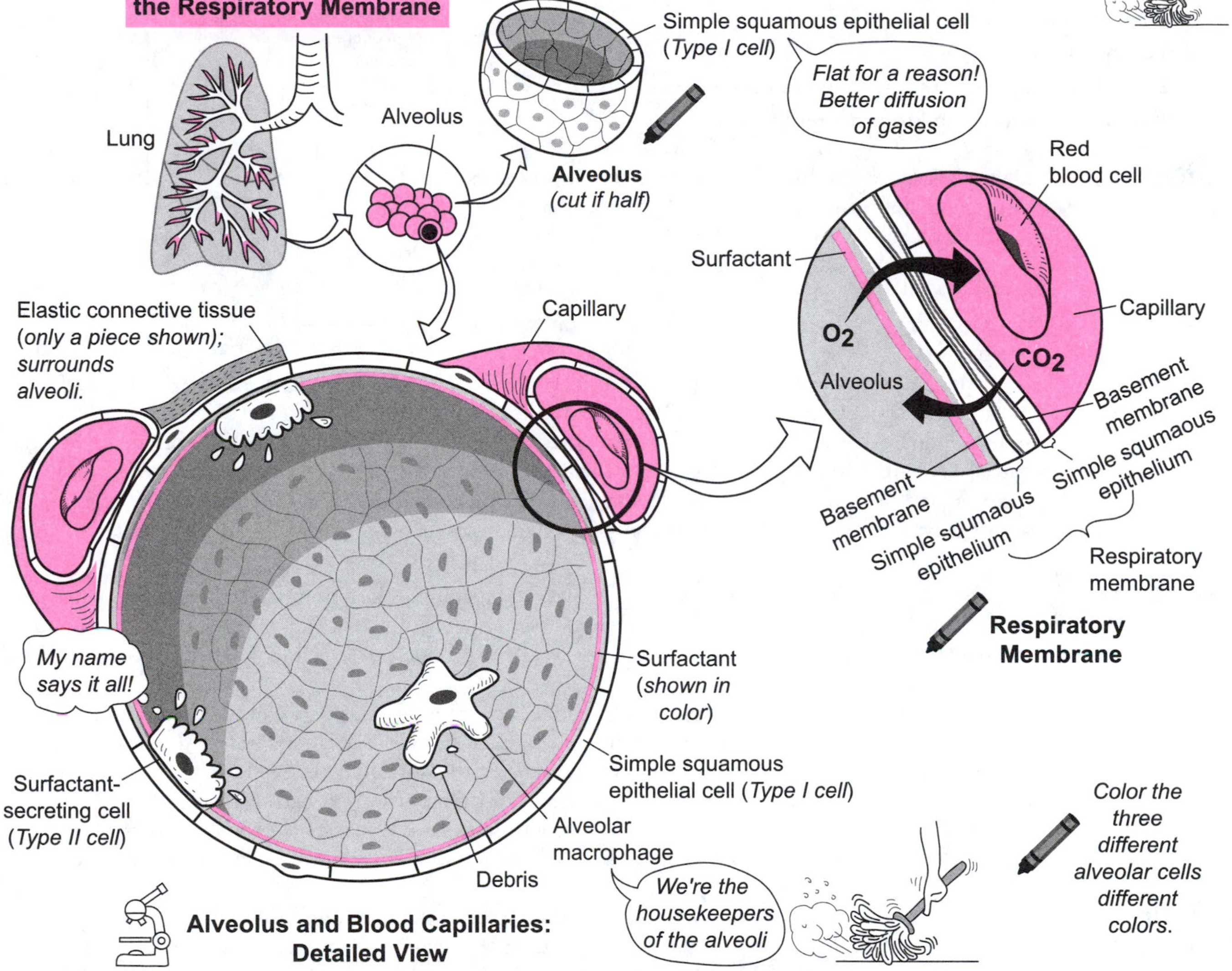

Surfactant and Surface Tension

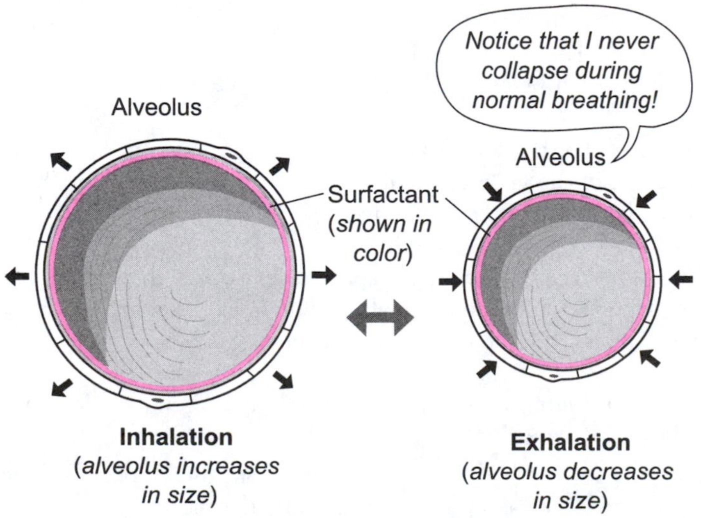

Surfactant *reduces* surface tension on alveoli which prevents them from collapsing

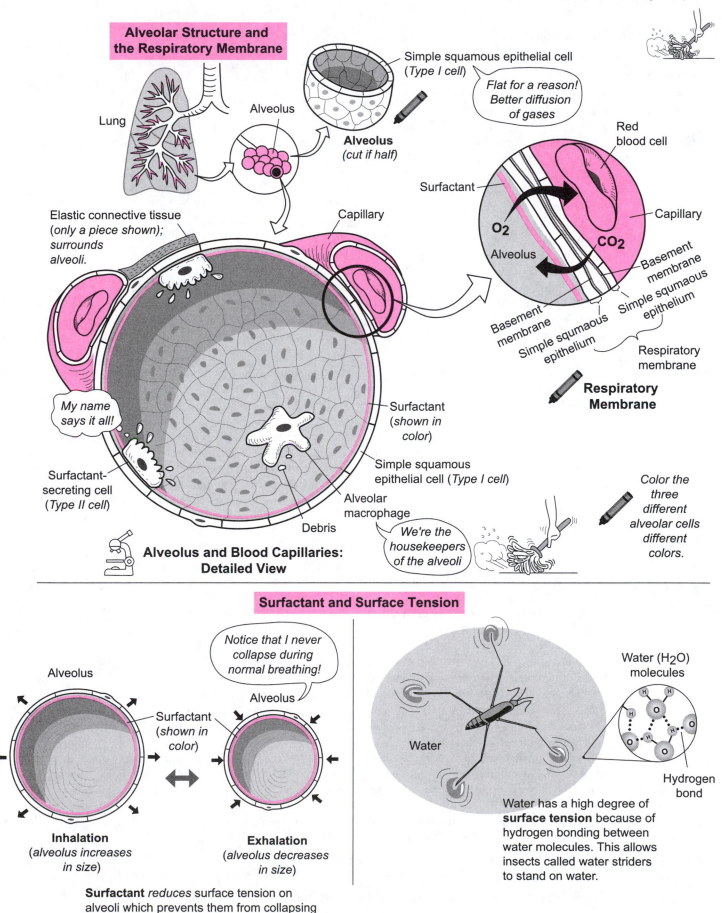

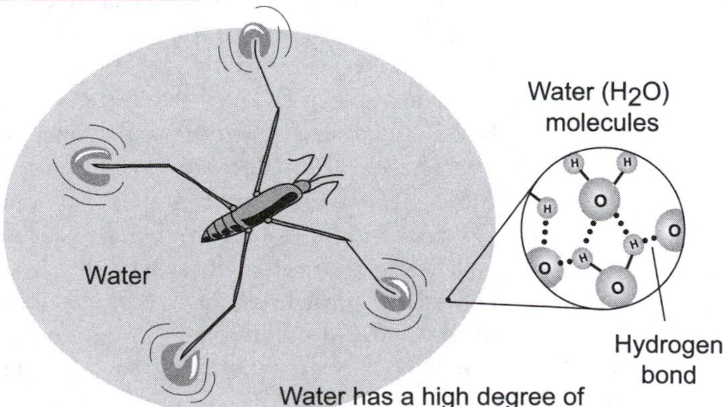

Water has a high degree of **surface tension** because of hydrogen bonding between water molecules. This allows insects called water striders to stand on water.

This module deals with the following concepts:

- **Dalton's law**
- **Henry's law**
- **Diffusion of O_2 and CO_2 into and out of the blood**
- **External respiration and internal respiration**

Dalton's Law

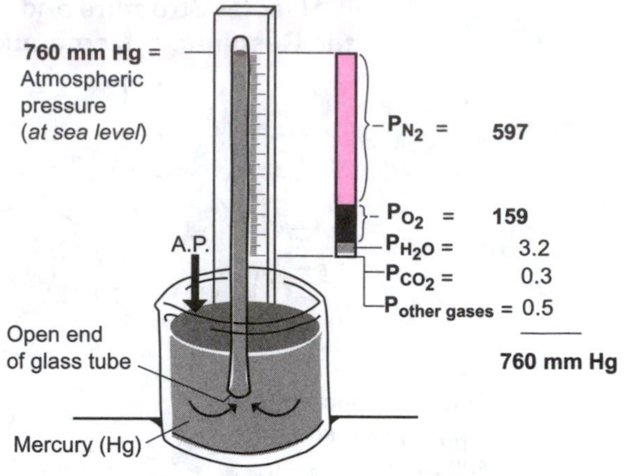

Model of a Barometer

Dalton's law is a gas law that deals with **partial pressures** of gases. It states that *in a mixture of gases (like the air we breathe), each gas exerts its own partial pressure*. Therefore, the sum of all the partial pressures is the total pressure exerted by the gas mixture.

Consider a barometer that measures atmospheric pressure (AP). At sea level, the normal AP is 760 mm Hg. Air contains the following gases: nitrogen (78%), oxygen (21%), water vapor (0.4%), carbon dioxide (0.04%), and other gases (0.06%). The partial pressure of nitrogen gas (P_{N_2}) is 78% of 760 or **597** mm Hg. The partial pressure of oxygen (P_{O_2}) is 21% of 760, or **159** mm Hg. Therefore, the combined partial pressures from nitrogen and oxygen account for **99%** of atmospheric pressure (756 of the 760 total). In the body, we have to consider the partial pressures of physiologically important gases such as O_2 and CO_2 in the alveolar air and the blood.

Henry's Law

Another gas law, Henry's law states that at a given temperature, the amount of gas dissolved in a liquid is directly proportional to the partial pressure of that gas. Gases normally move in and out of solution, with some being more soluble in liquids than are others. For example, carbon dioxide (CO_2) is highly soluble in water—much more so than oxygen (O_2).

As an example of a gas moving out of solution, let's consider soda pop. All sodas contain carbonated water—CO_2 dissolved in water. Although much of the CO_2 remains in solution, some is found in the air inside the bottle. If you open a soda and allow it to sit out at room temperature for a long time, all the CO_2 eventually will move out of solution and the soda will go "flat." If you shake a fresh bottle, you can see the CO_2 bubble out of solution, which builds up pressure inside the bottle. This increase in pressure causes the soda to shoot out when the bottle is opened. Similarly, in the lungs, CO_2 moves out of solution from the blood capillaries and into the alveolar air.

Diffusion of O_2 and CO_2 Into and Out of the Blood

The **blood P_{CO_2} level** is about 45 mm Hg as it approaches the lungs, and the P_{CO_2} in the alveoli is about 40 mm Hg. Because of the small gradient, CO_2 slowly diffuses out of the blood and into the alveoli. As the blood leaves the lungs, the P_{CO_2} level has dropped slightly, to about 40 mm Hg. Because tissues produce CO_2 as a byproduct of normal metabolism, it gradually diffuses back into the blood causing the P_{CO_2} levels to rise back to 45 mm Hg. Therefore, the P_{CO_2} levels in the blood have a narrow range, between 40–45 mm Hg. This stable amount of CO_2 forms bicarbonate in the blood, which helps buffer the blood's pH to keep it in the normal range of 7.35–7.45.

The **blood P_{O_2} level** is about 40 mm Hg as it approaches the lungs, while the P_{O_2} of the alveolar air is about 105 mm Hg. Because of the large gradient, O_2 quickly diffuses into the blood. Because the hemoglobin in the red blood cells is especially efficient at binding O_2, the blood P_{O_2} levels rise to 100 mm Hg—the typical level for oxygenated blood. When the oxygenated blood approaches tissues that are low in oxygen, it rapidly diffuses into tissues. In deoxygenated blood, the P_{O_2} drops to about 40 mm Hg. In short, the blood P_{O_2} levels vary widely, from 40 mm Hg in deoxygenated blood to 100 mm Hg in oxygenated blood.

External Respiration and Internal Respiration

- External respiration occurs between the alveoli in the lungs and the blood capillaries; O_2 moves from the alveoli into the blood and CO_2 moves from the blood into the alveoli.
- Internal respiration occurs between the blood capillaries and body cells; O_2 moves from the blood into body cells and CO_2 moves from the body cells (where it is produced as a waste product) into the blood.

Once oxygen is inside body cells, it is used in the important process of **cellular respiration** (see p. 406).

Diffusion of O_2 and CO_2

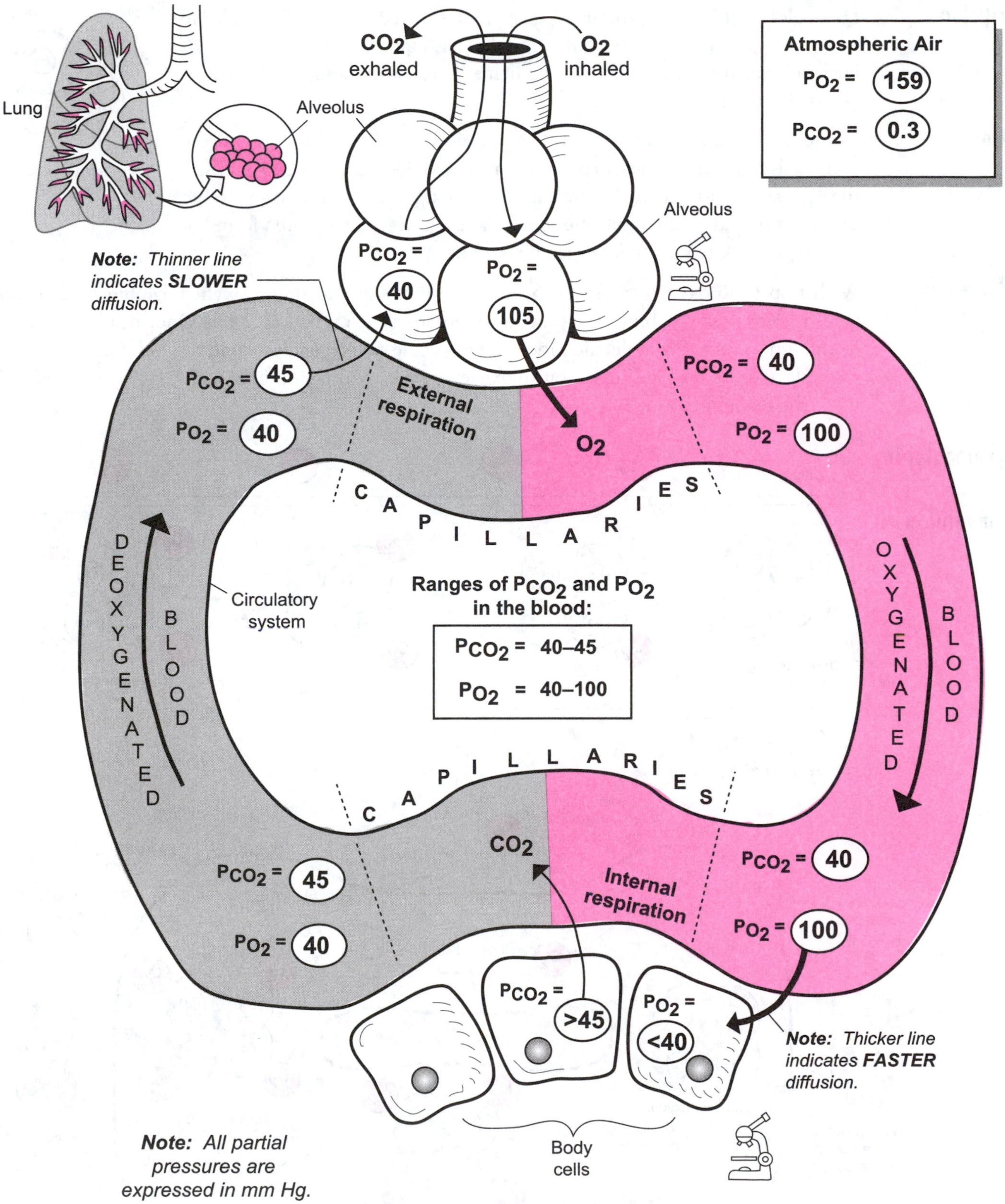

Note: All partial pressures are expressed in mm Hg.

Color the P_{O_2} numbers red and the P_{CO_2} numbers blue.

Description

Hemoglobin (Hb) is a protein pigment found in red blood cells. Each Hb molecule is composed of four subunits: alpha 1, alpha 2, beta 1, and beta 2. Each subunit has an iron-containing heme group where a molecule of oxygen is able to bind.

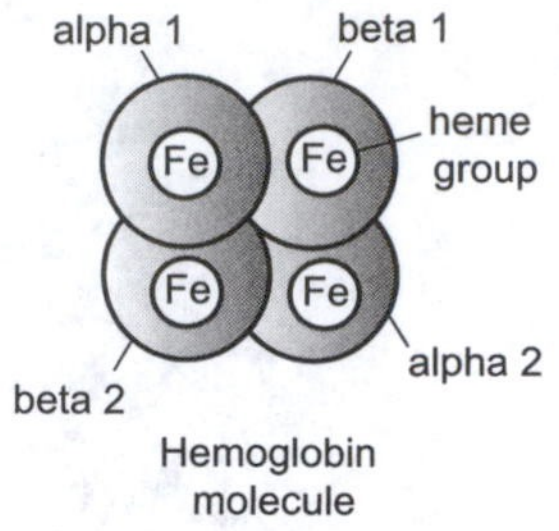

Hemoglobin molecule

Function

The primary function of Hb is to transport oxygen from the lungs to the body cells but it also transports some carbon dioxide from the body cells to the lungs. It is a pigment because its color changes depending on whether oxygen is bound to the molecule. It is bright red in its oxygenated form but changes to dark red in its deoxygenated form.

Two States for Hemoglobin

With respect to oxygen transport, hemoglobin exists in two states: **oxyhemoglobin** or **deoxyhemoglobin**. Deoxyhemoglobin is the state in which no oxygen is bound to Hb. Once inside the capillaries in the lungs, each Hb molecule binds one molecule of oxygen at a time to hold a maximum of four O_2 molecules. After binding its fourth oxygen molecule, it is saturated with oxygen and said to be in its oxyhemoglobin form.

Deoxyhemoglobin and Oxyhemoglobin

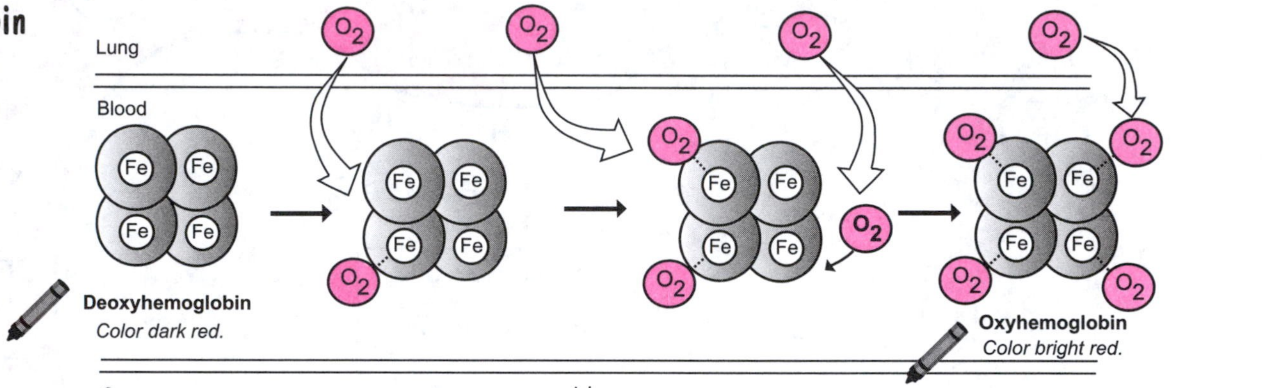

Oxygen transport by hemoglobin

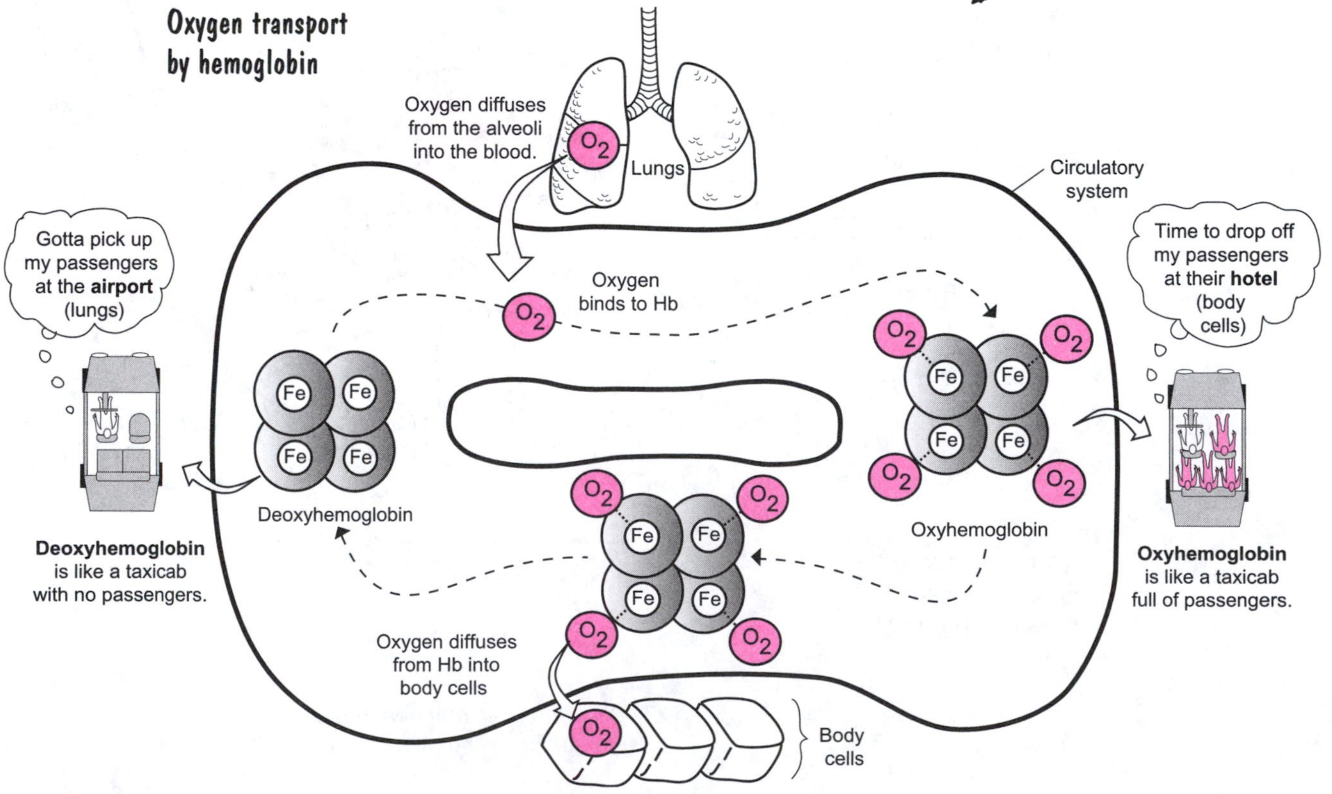

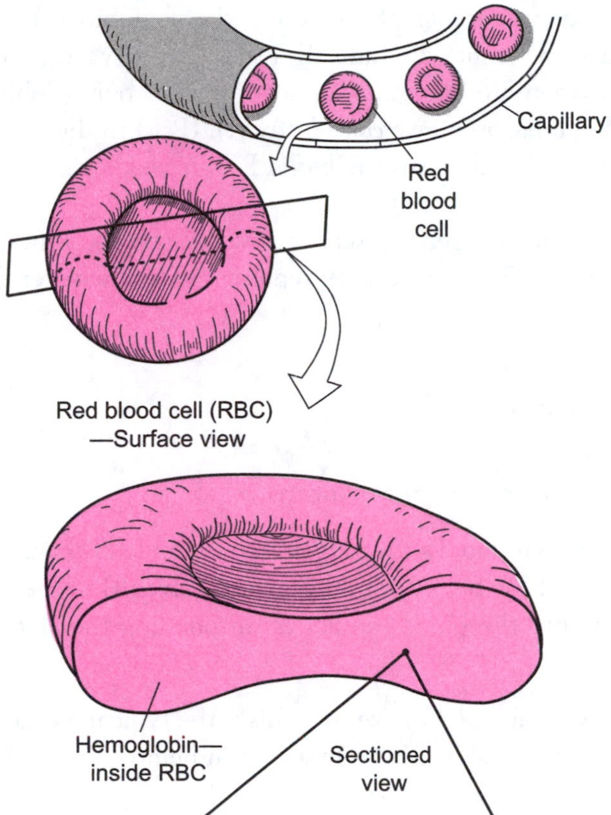

A molecule of Hb is functionally like a taxicab. Just as a taxicab picks up passengers at the airport and drops them off at their hotel, hemoglobin picks up oxygen from the lungs and delivers it to tissues. It also picks up some of the carbon dioxide generated from body cells and delivers it to the lungs. Just like a taxicab has seats for its passengers, an Hb molecule has four binding sites for oxygen.

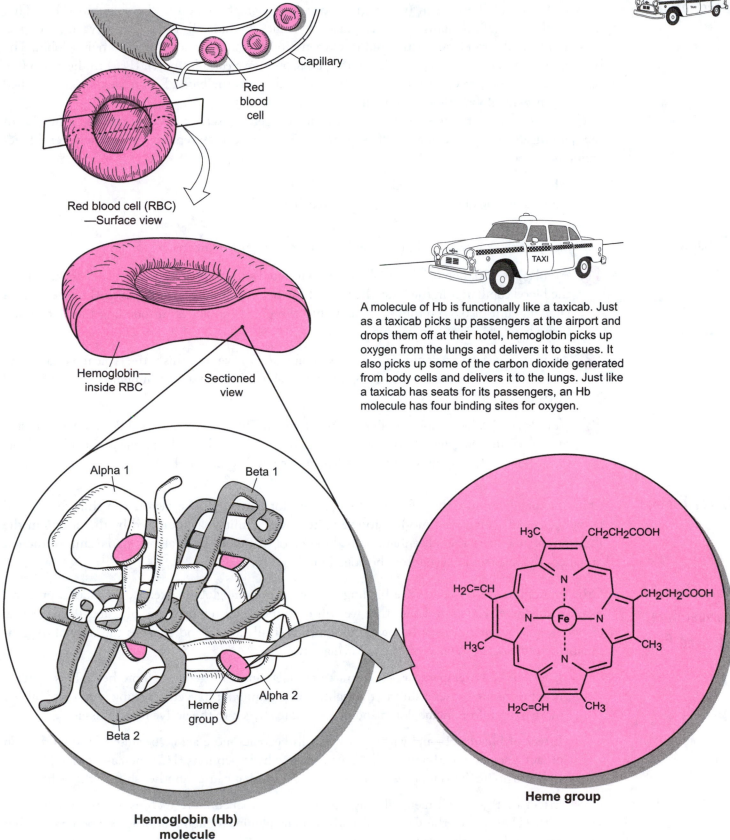

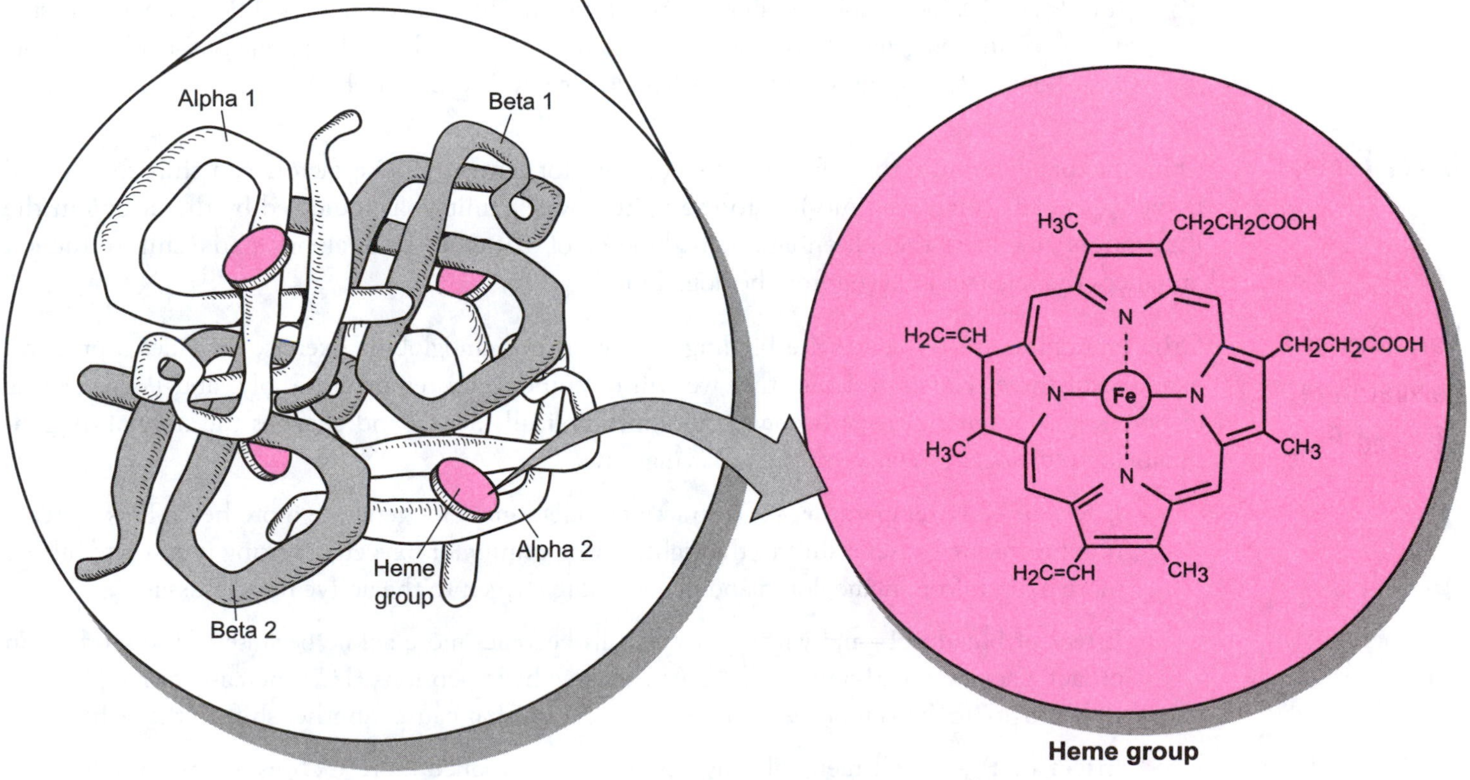

Hemoglobin (Hb) molecule

Heme group

Description

About 98% of all the oxygen (O_2) is transported through the blood on **hemoglobin (Hb)**. That means only about 2% remains to be released at the tissues. Because all tissues need oxygen to survive, it is vital to understand factors that aid in oxygen binding to, and releasing from, hemoglobin. The binding of O_2 to Hb depends primarily on the *partial pressure created by oxygen* (P_{O_2}) in the blood (see p. 352). When the percent saturation of Hb is plotted versus the blood P_{O_2}, it yields a graph called the **oxygen-hemoglobin dissociation curve**.

Reduced Hb is like a taxicab without any oxygen "passengers." When the taxi is filled with four passengers, it is oxyhemoglobin (see p. 354). This cycle can be represented by the reversible reaction given below:

Hb	+	O_2	↔	**Hb-O_2**
reduced hemoglobin				oxyhemoglobin

Examples

Let's interpret the curve according to the three different points numbered #1–3:

1. At a P_{O_2} of **100**, hemoglobin is about 98% saturated with oxygen. An example of this would be the blood capillaries in the lungs. Because of the high P_{O_2} in the air sacs (alveoli) in the lungs, a large amount of oxygen diffuses rapidly into the blood capillaries surrounding the alveoli and immediately binds to hemoglobin.

2. At a P_{O_2} of **40**, hemoglobin is about 75% saturated. On average, this is the typical blood P_{O_2} for tissues at rest. This means that 25% of the available O_2 is released from hemoglobin to be used by tissues at rest.

3. At a P_{O_2} of **20**, hemoglobin is about 35% saturated. This is a typical blood P_{O_2} for active tissues such as contracting muscle tissues during exercise. Notice that a large release of the available O_2—about 65%—comes with only a small change in P_{O_2} (20 mm Hg).

Other Factors

Other factors determine the affinity of hemoglobin for oxygen. These factors can shift the normal curve to the left (higher affinity) or to the right (lower affinity), as indicated by the arrows in the illustration. A shift to the left enhances the binding of oxygen to hemoglobin, and a shift to the left enhances the release of oxygen from hemoglobin.

Variables of Temperature, pH, and P_{CO_2}

Other variables also influence the binding of oxygen to hemoglobin. Exercise provides a practical example of three related variables that we will examine: body temperature, pH, and P_{CO_2}. During exercise, skeletal muscle tissue becomes more metabolically active and changes the normal oxygen-hemoglobin dissociation curve in the following ways:

- **Effect of body temperature**—Contracting skeletal muscles generate more heat. This increase in temperature causes a shift to the right (lower affinity for oxygen), making it easier to release more oxygen from hemoglobin and deliver more oxygen to the active muscle tissue.
- **Effect of blood pH**—pH within muscle cells becomes more acidic because of the production of lactic acid and carbonic acid. As a result, free hydrogen ions (H^+) increase. Some H^+ bind to hemoglobin. This induces a subtle shape change that causes another shift to the right.
- **Effect of P_{CO_2}**—All metabolically active tissues produce more CO_2 as a waste product. As with H^+, some of the CO_2 also binds to hemoglobin, inducing another shape change that causes a shift to the right.

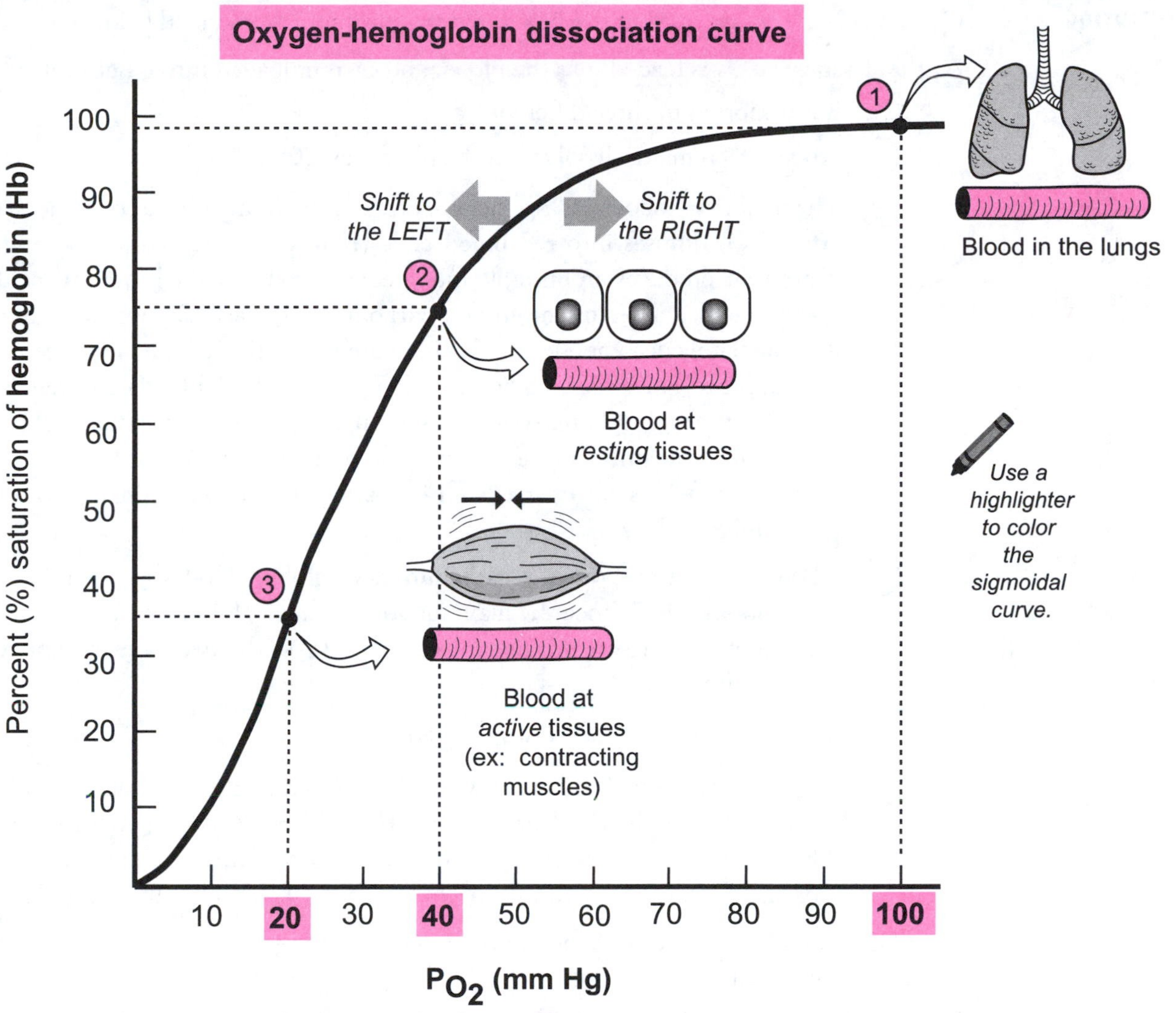
Oxygen-hemoglobin dissociation curve
Percent (%) saturation of hemoglobin (Hb)
100
90
80
70
60
50
40
30
20
10
1
2
3
Shift to the LEFT
Shift to the RIGHT
Blood in the lungs
Blood at resting tissues
Blood at active tissues (ex: contracting muscles)
Use a highlighter to color the sigmoidal curve.
10
20
30
40
50
60
70
80
90
100
P_{O_2} (mm Hg)

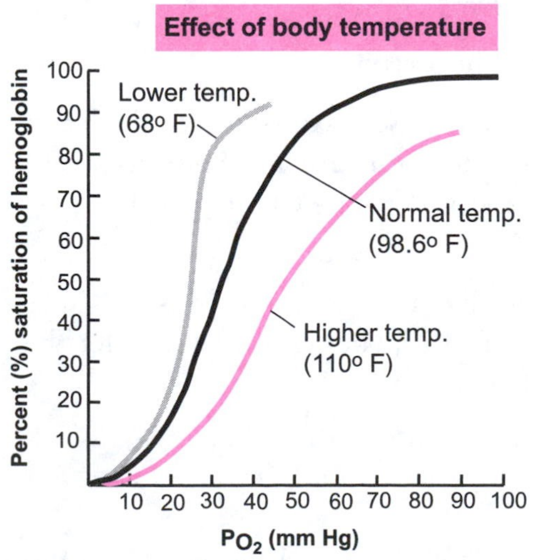
Effect of body temperature
Percent (%) saturation of hemoglobin
Lower temp. (68° F)
Normal temp. (98.6° F)
Higher temp. (110° F)
P_{O_2} (mm Hg)

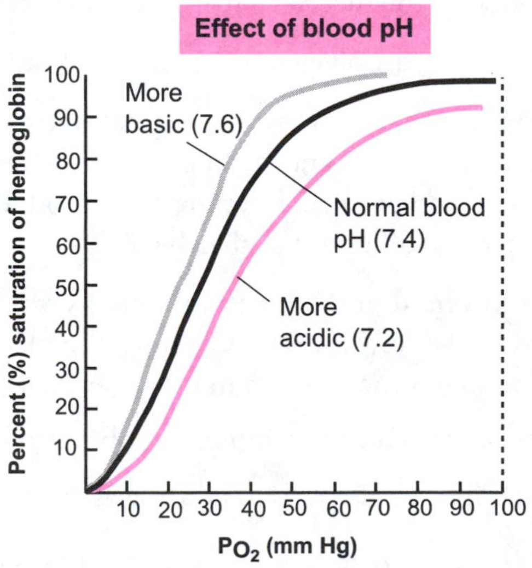
Effect of blood pH
Percent (%) saturation of hemoglobin
More basic (7.6)
Normal blood pH (7.4)
More acidic (7.2)
P_{O_2} (mm Hg)

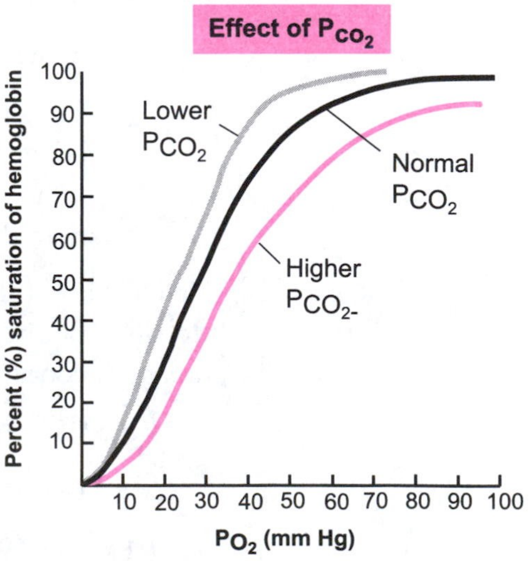
Effect of P_{CO_2}
Percent (%) saturation of hemoglobin
Lower P_{CO_2}
Normal P_{CO_2}
Higher P_{CO_2}
P_{O_2} (mm Hg)

Description

CO_2 and O_2 are transported in the blood in different ways, as described below.

At the Tissues: CO_2 is loaded into the blood, and O_2 is unloaded into tissues cells.

- **CO_2** is transported in three different ways:
  1. About 7% remains dissolved in the plasma as CO_2.
  2. The majority—about **70%**—is transported in the form of bicarbonate ions (HCO_3^-). How? After CO_2 diffuses into **red blood cells (RBCs)**, it undergoes a chemical reaction with water that produces an unstable intermediate product called carbonic acid (H_2CO_3). This reaction is fast, aided by the enzyme **carbonic anhydrase**. Because carbonic acid is unstable, it immediately decomposes into bicarbonate ions (HCO_3^-) and hydrogen ions (H^+).

     The levels of HCO_3^- quickly build up inside the RBC. The excess HCO_3^- has to be moved out into the plasma so more CO_2 can be converted into HCO_3^-. This is accomplished by the **chloride shift**, where a membrane protein exchanges a HCO_3^- ion for a Cl^- ion. Each ion has the same charge, so this does not result in any change in total charge across the membrane.
  3. About 23% is transported as **carbaminohemoglobin** (Hb-CO_2) when CO_2 loosely binds to hemoglobin (Hb). Since the main function of hemoglobin is to transport O_2, CO_2 binds to a different binding site on the molecule than O_2 so the two gases are not competing for the same binding site.
- **O_2** is unloaded to tissue cells in two different ways:
  1. Slightly more than 98% of the O_2 is transported in the form of **oxyhemoglobin** (Hb-O_2) when O_2 loosely binds to hemoglobin (Hb). Then oxyhemoglobin releases the O_2 and it diffuses tissue cells. Here it is used in cellular respiration. Some of the excess hydrogen ions (H^+) already present in the RBC temporarily bind to Hb to form H-Hb.
  2. Slightly under 2% of the O_2 is dissolved in the plasma as O_2. Some of this diffuses into tissue cells.

In the Lungs—CO_2 is unloaded into the alveoli, and O_2 is loaded into the blood.

- CO_2 is transported in the reverse as to how it was at the tissues.
  1. Some of the CO_2 dissolved in the plasma diffuses into the alveoli.
  2. The chemical reaction between CO_2 and water runs in reverse. In other words, the bicarbonate ions (HCO_3^-) and hydrogen ions recombine to form carbonic acid ($H_2CO_3^-$) again. Then, carbonic acid decomposes into CO_2 and water. The CO_2 that is liberated from this process diffuses into the **alveoli.**

The **chloride shift** is reversed in the lungs. In other words, a membrane protein transports a HCO_3^- into the RBC in exchange for moving a Cl^- into the plasma. This supplies lots of HCO_3^- for the purpose of continuing to run the reaction in the reverse direction.

  3. CO_2 is released from **carbaminohemoglobin** and it diffuses into the alveoli**.**

- **O_2** is transported in the reverse as to how it was at the tissues.
  1. Over 98% of the O_2 diffuses into red blood cells and loosely binds to **hemoglobin** (Hb) to form **oxyhemoglobin** (Hb-O_2). Some hydrogen ions (H^+) are produced as a byproduct.
  2. Less than 2% of the O_2 diffuses into the blood and remains dissolved in the plasma as O_2.

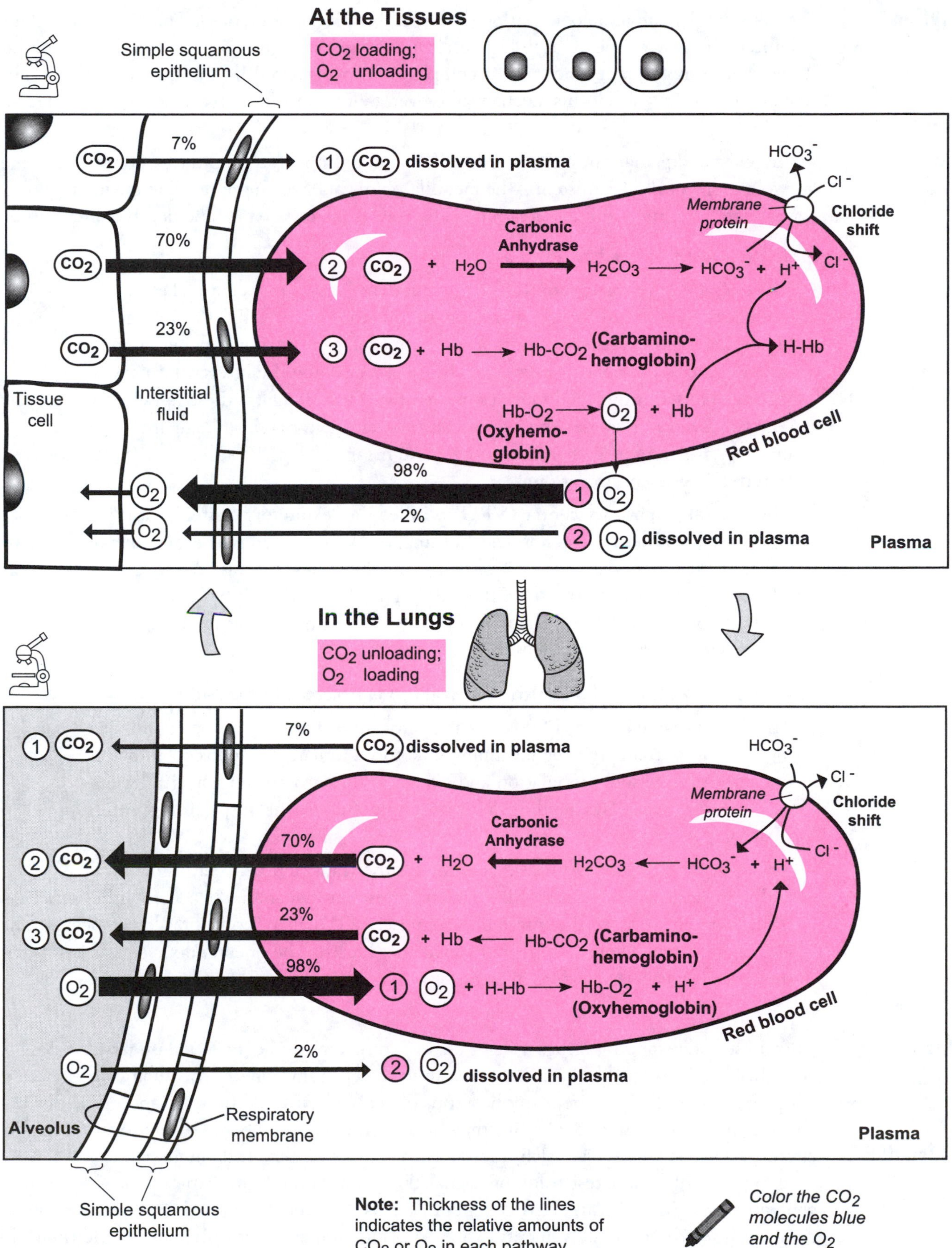

Note: Thickness of the lines indicates the relative amounts of CO_2 or O_2 in each pathway.

Color the CO_2 molecules blue and the O_2 molecules red.

Description

We take the rhythm and coordination of normal breathing for granted. The pattern of inhaling and exhaling constantly repeats itself. At rest, we take an average of 12–16 breaths per minute. This deceivingly simple process actually is complex and is regulated by the nervous system. The latest research has revealed that this mechanism is even more complex than previously thought.

Classic Explanation

In the classic explanation, clusters of specific neurons constitute respiratory control centers located in two regions of the brain stem—the **medulla oblongata** and the **pons**. The medulla oblongata sets the rate and rhythm of normal breathing. The pons regulates the rate and depth of breathing.

Medullary Respiratory Centers: DRG and VRG

- The **dorsal respiratory group (DRG)** is called the "inspiratory center" because it stimulates inhalations. Like a sparkplug, the neurons in the DRG fire in regular bursts. Each burst lasts about 2 seconds. As this happens, it simultaneously sends impulses along the **phrenic nerve** (to the diaphragm) and the **intercostal nerves** (to the **external intercostal muscles** of the ribs). This stimulates these muscles to contract, which increases the size of the thoracic cavity and pulls air into the lungs. A 3-second delay between consecutive firings of the DRG allows time for a passive exhalation to follow each inhalation.
- The **ventral respiratory center (VRG)** is called the "expiratory center," but it is active *only* during a *forced* exhalation. During normal breathing, the VRG is inactive because exhalation is a passive, recoil process that does not require stimulation of muscles. During a *forced* exhalation, though, the DRG sends impulses to stimulate the VRG. The VRG responds by sending impulses to the internal intercostal muscles and abdominal muscles, which then contract, decreasing the size of a thoracic cavity for a forced exhalation.

Pons Respiratory Centers: Pneumotaxic Center (PC) and Apneustic Center (AC)

- The **pneumotaxic center (PC)** is "the regulator." It is like a traffic cop coordinating the flow of traffic to keep it running smoothly. Similarly, the pons coordinates the transition between inhalation and exhalation. It also prevents overinflation of the lungs by always sending inhibitory impulses to the inspiratory center (DRG).
- The **apneustic center (AC)** also coordinates the transition between inhalation and exhalation by fine-tuning the medullary respiratory centers. It accomplishes this by sending stimulatory impulses to the inspiratory center (DRG), which result in a slower, deeper inhalation. This is necessary when you choose to hold your breath. The pneumotaxic center is able to inhibit the apneustic center so a normal breath is not too slow or too deep.

Changes in Naming and Conceptual Understanding

The latest understanding brings a slight change in naming and conceptual understanding. Let's look at each, in turn. As for the naming, a new term has been introduced: the *respiration regulatory center*. This consists of all the aforementioned control centers (DRG, VRG, PC, and AC). But instead of considering them as individual units, they now are thought of as being strung together in a loop or a circuit. This makes the classic definition of each control center a little more murky.

Conceptually, the latest thinking is that the respiration control center acts as a central pattern generator, or CPG. Like an electrical circuit, CPGs are neural circuits that generate periodic motor commands for rhythmic movements. The CPGs function automatically most of the time while also allowing voluntary control to override them.

Neural Control of Respiration

Respiration Regulatory Center

PC = pneumotaxic center
AC = apneustic center
VRG = ventral respiratory group
DRG = dorsal respiratory group

Color the PC, AC, VRG, and DRG different colors.

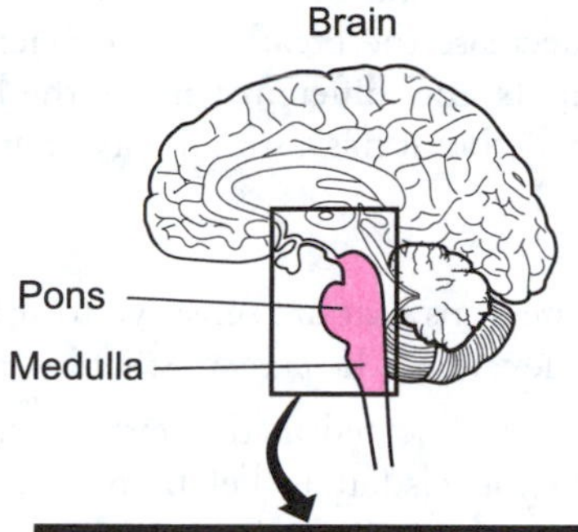

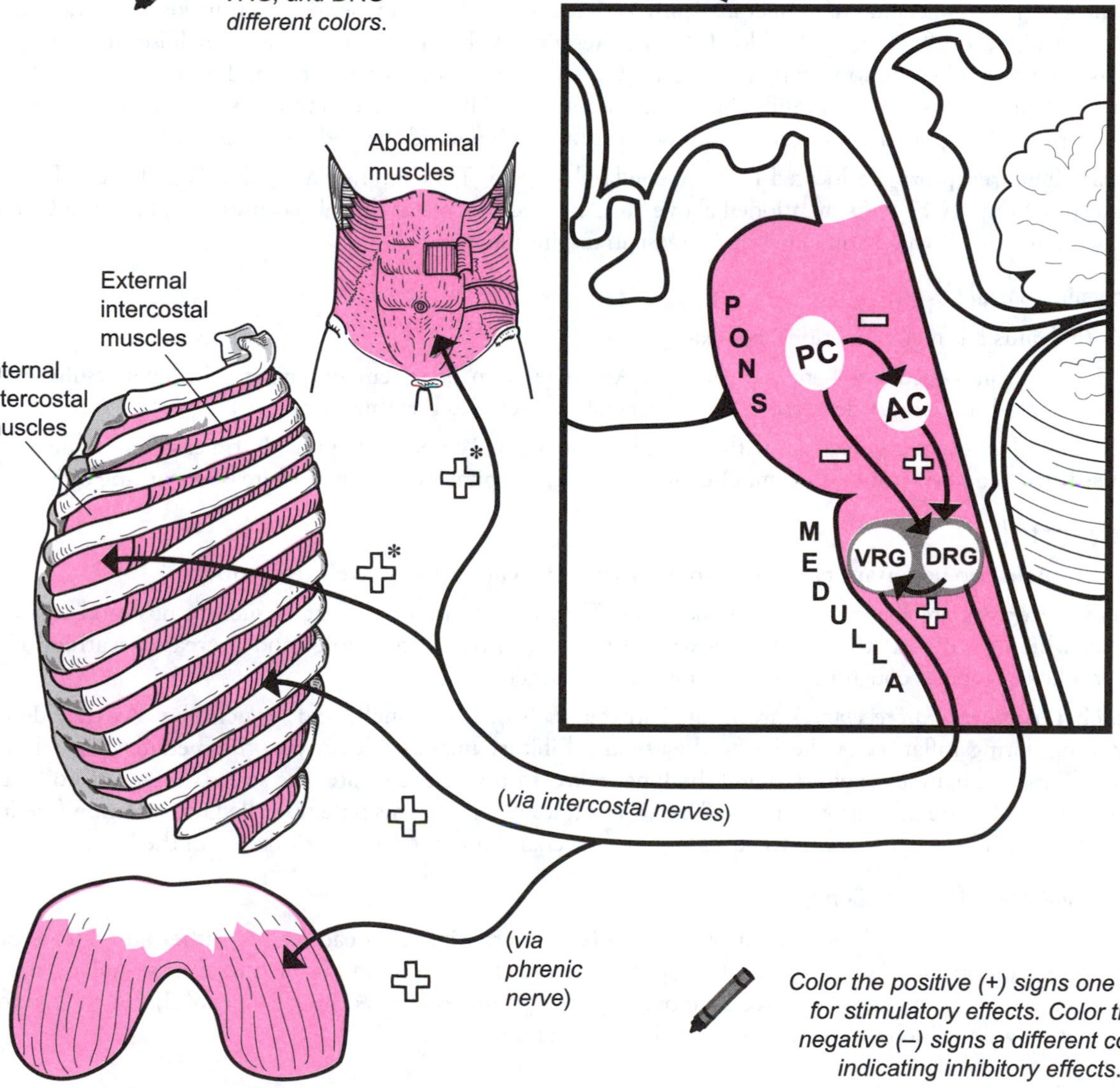

Color the positive (+) signs one color for stimulatory effects. Color the negative (–) signs a different color indicating inhibitory effects.

ANALOGIES

PC = Pneumotaxic Center

is like the traffic cop who keeps the flow of traffic (*breathing rhythm*) running smoothly.

DRG = Dorsal Respiratory Group

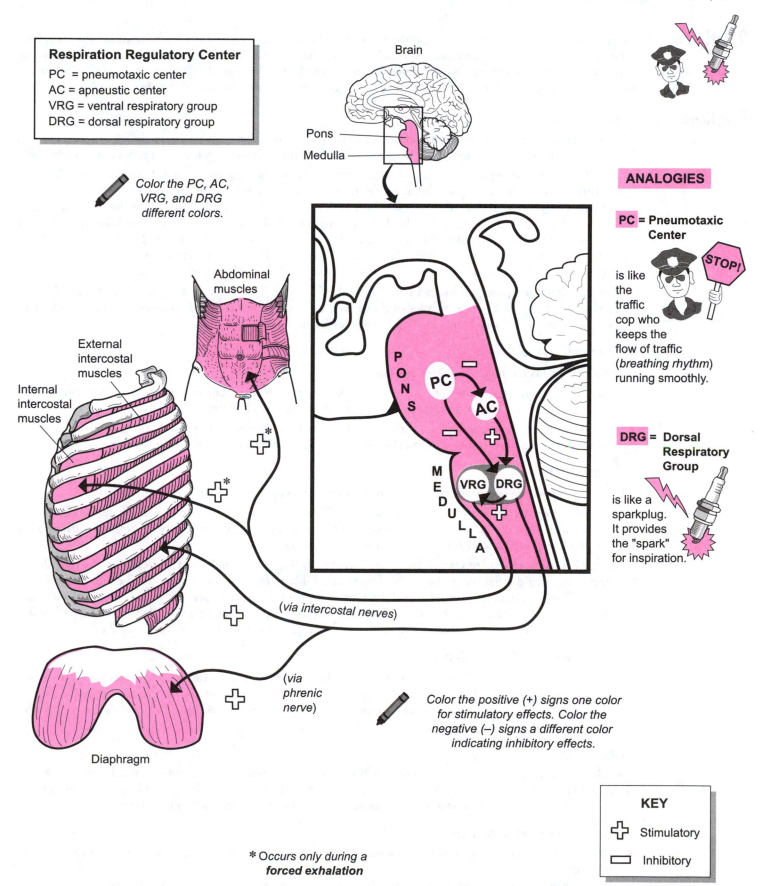

is like a sparkplug. It provides the "spark" for inspiration.

KEY

✚ Stimulatory

▭ Inhibitory

* Occurs only during a ***forced exhalation***

Description

Even though the medulla oblongata and the pons control the typical rate and rhythm of normal breathing, they must be able to increase or decrease the breathing rate when needed. To do so, these respiratory centers in the brain stem receive various inputs from different parts of the body. This module summarizes the different types of input received and indicates whether it has a stimulatory or an inhibitory influence on normal breathing.

Functions

Chemoreceptors

Chemoreceptors are sensory neurons that are sensitive to shifts in specific chemicals in the blood, such as CO_2, H^+, and O_2. Changes in CO_2 levels are the *primary* stimulus, and changes in O_2 levels act as a *secondary* stimulus.

- **Peripheral chemoreceptors** are located in the aortic arch and the common carotid arteries. Consider what happens during exercise: P_{O_2} levels fall as skeletal muscle cells consume more oxygen; P_{CO_2} levels increase as skeletal muscle cells become more metabolically active; and H^+ levels increase, which makes the blood more acidic. Why? CO_2 diffuses into the blood. In the presence of the enzyme carbonic anhydrase, it chemically reacts with water (H_2O) to form carbonic acid (H_2CO_3), which dissociates into H^+ and bicarbonate (HCO_3^-). In this way, an increase in CO_2 results in an increase in H^+. All of these conditions stimulate an increase in respiration rate to supply the skeletal muscle cells with more O_2 and rid the body of excess CO_2.
- **Central chemoreceptors** are located in the medulla oblongata. They monitor P_{CO_2} levels and the pH of the cerebrospinal fluid (CSF). As mentioned above, as CO_2 levels increase, this also causes the pH of the CSF to become more acidic, which stimulates an increase in respiration rate.

Hypothalamic Controls

The **hypothalamus** has many functions. For example:

- The hypothalamus regulates body temperature. An increase in body temperature from a fever results in an increased breathing rate. A decrease in body temperature decreases breathing rate.
- The hypothalamus has connections with the limbic system, our "emotional brain." In times of emotional stress such as fear or anxiety, stimulatory impulses are sent to the inspiratory centers to increase breathing rate.

Pulmonary Controls

The lungs contain two important types of receptors: **irritant receptors** and **stretch receptors.**

- **Irritant receptors** are located in the bronchioles. They detect airborne particles such as dust and cigarette smoke, which results in stimulating reflex pathways to constrict bronchioles and decrease breathing rate. Other reflexes, such as coughing and sneezing, may be triggered also.
- **Stretch receptors** (*baroreceptors*) are located in the walls of the bronchi and bronchioles. As they detect stretching during inflation of the lungs, they send inhibitory impulses directly to the medullary inspiratory center. This results in an expiration and the lungs relax. In this relaxed state, the stretch receptors no longer are stimulated, so the inhibition is lost and a new inspiration begins. This reflex is called the **inflation** (*Hering-Breuer*) **reflex**. The purpose seems to be a protective mechanism to prevent overinflation of the lungs.

Voluntary Controls—Cerebral Cortex

The **cerebral cortex** is the voluntary control over the body's activities. This can lead to either an increase or a decrease in respiration rate. For example: We may consciously choose to hold our breath. This stimulates motor neurons to send stimulatory impulses directly to our respiratory muscles for breathing. We may choose to go into a deep meditative state, which indirectly leads to a decrease in respiration.

Proprioceptors

Proprioceptors are located in muscles, tendons, and joints. They monitor the tension in muscles and the movement and position of joints and deliver the information to the brain. For example, at the beginning of exercise, as tension in muscles and movement in joints increases, it stimulates an increase in respiration rate.

Other Influences on Respiration

- **Pain**: Sudden, acute pain may trigger **apnea** (absence of breathing), and extended somatic pain increases respiration rate.
- **Blood pressure (BP)**: A quick increase in BP decreases respiration rate, and a drop in BP increases respiration rate.

Wedge of cerebral cortex
Voluntary control centers in cerebral cortex
Other influences on respiration (e.g., emotional stimuli such as fear or anxiety)
Hypothalamus
CO_2 ↑, H^+ ↑
Peripheral chemoreceptors
Pons
Medulla
Irritant receptors
Medulla
CO_2 ↑, H^+ ↑
Central chemoreceptors
Respiratory centers in medulla and pons
Stretch receptors
Proprioceptors in muscles, tendons, and joints
Color in the positive (+) signs one color for stimulatory effects; color the negative (–) signs a different color indicating inhibitory effects.
KEY
Stimulatory
Inhibitory
Stimulatory OR inhibitory

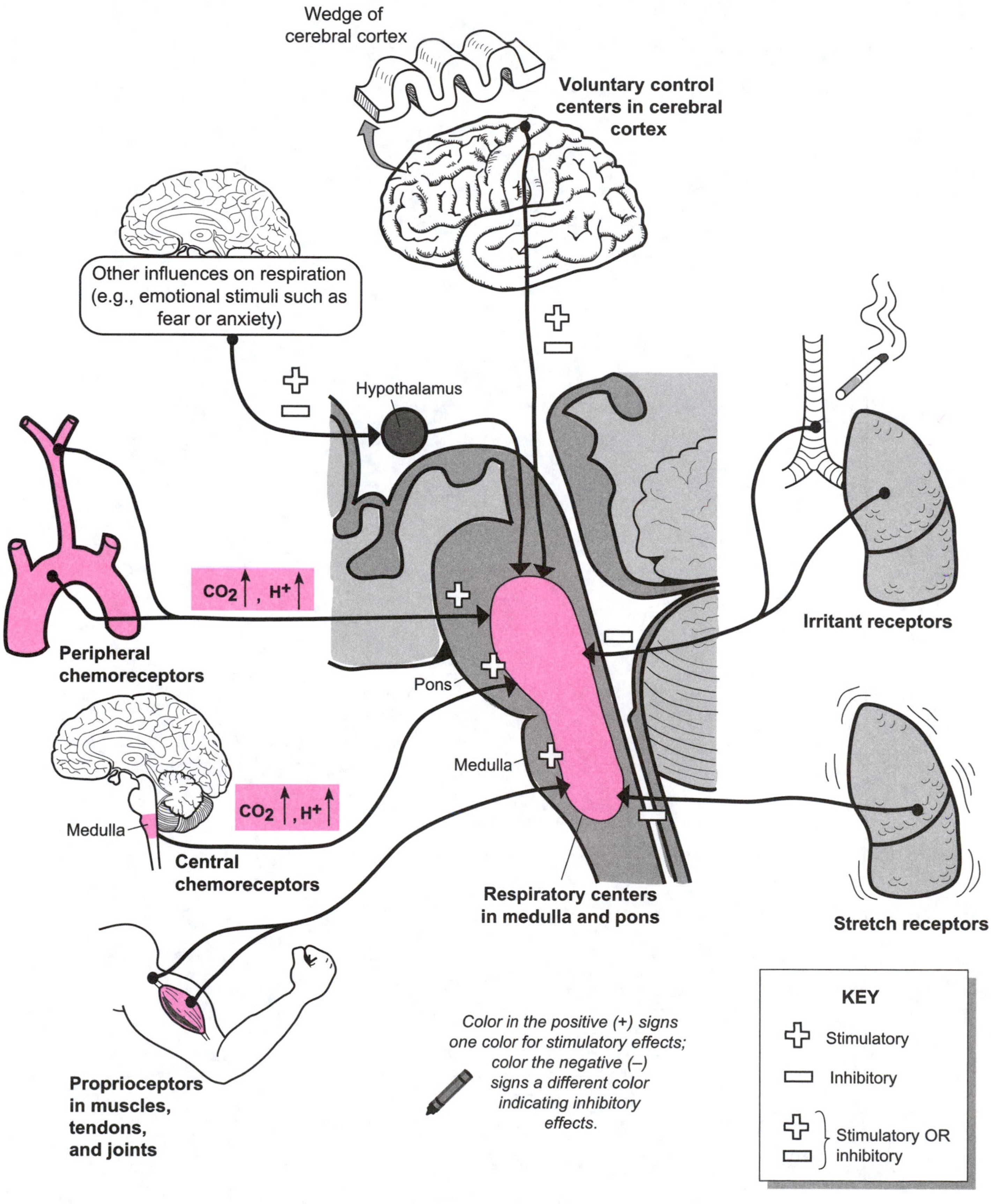

Notes

Digestive System

Overview of the Digestion Process

Let's look at the digestion process by examining what happens when eating a ham sandwich. The process begins in the mouth with mechanical digestion—the physical breakdown of food from chewing. The sandwich contains many different nutrients including starch—the complex carbohydrate in the bread, protein in the ham, and lipids (example: triglycerides) in the mayonnaise. Each of these macromolecules has to be broken down by chemical digestion into its fundamental building blocks so they can be used by body cells:

Complex carbohydrates	⟶	Monosaccharides
Proteins	⟶	Amino acids
Lipids	⟶	Monoglycerides and fatty acids

Overview of Chemical Digestion

The process is a series of catabolic reactions that breaks chemical bonds in the macromolecules with the help of digestive enzymes. Different enzymes are required for different macromolecules, shown on the illustration.

Note: The enzymes are shown in *italics*, and the icons indicate the location in the digestive system.

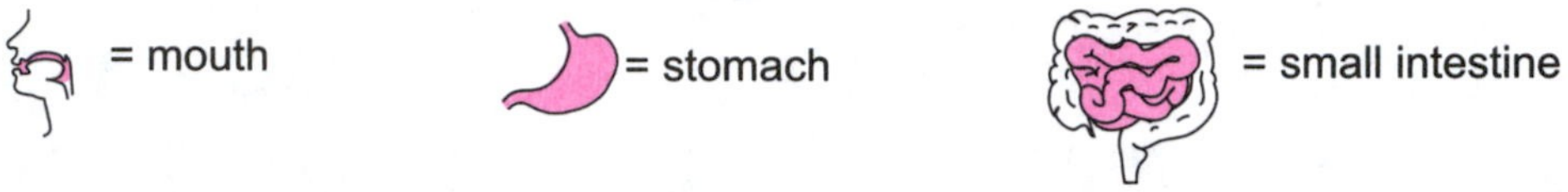

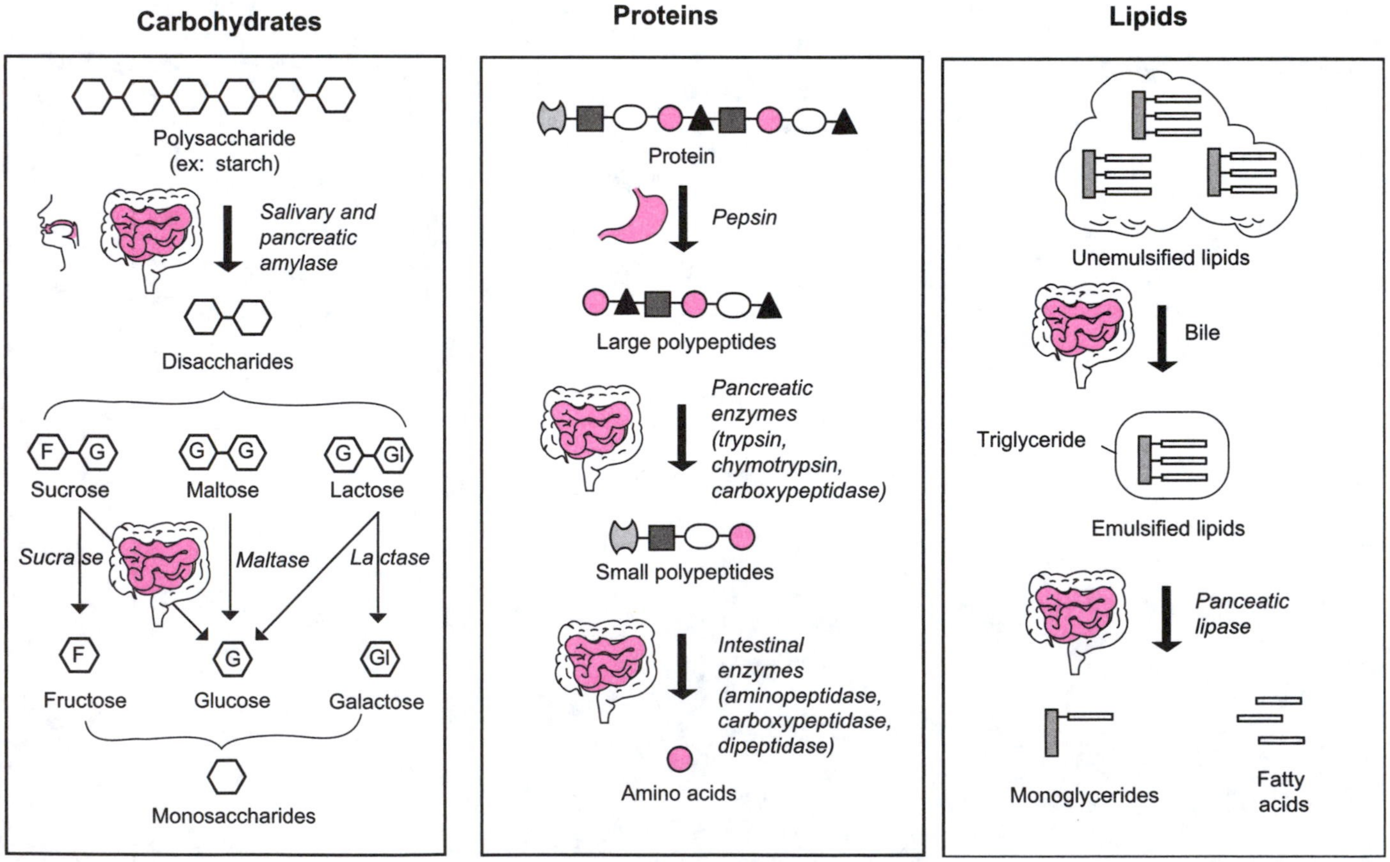

NUTRIENTS in a HAM SANDWICH
Bread
(contains complex carbohydrates like starch)
Starch
Ham
(contains protein)
Protein
Mayonnaise
(contains lipids like triglycerides)
Triglyceride
I love ham sandwiches!
Nasal cavity
Oral cavity
Parotid salivary gland
Tongue
Mandible
Epiglottis
Sublingual salivary gland
Submandibular salivary gland
Trachea
Primary bronchus
Color different organs different colors.
THE DIGESTIVE SYSTEM: OVERVIEW
Bolus of food
1.
2.
3.
4.
5.
Accessory glands
(salivary glands, liver, gallbladder, and pancreas)
Enzyme
(ex: amylase)
Digestive tract
(tube from mouth to anus)
Body cell
Circulatory system
Waste
Human body
Esophagus
Diaphragm
Liver
Pancreas
Spleen
Stomach
Gallbladder
Duodenum
Jejunum
Large intestine
Appendix
Ileum
Large intestine
Rectum
1. Macromolecules in food are ingested.
2. Enzymes are released from glands that aid in breaking down macromolecules into smaller products that the cells can use.
3. Final products are absorbed into the bloodstream.
4. Final products are transported through the bloodstream and finally delivered to cells, where they can be used as nutrients.
5. Waste products are eliminated from the body.

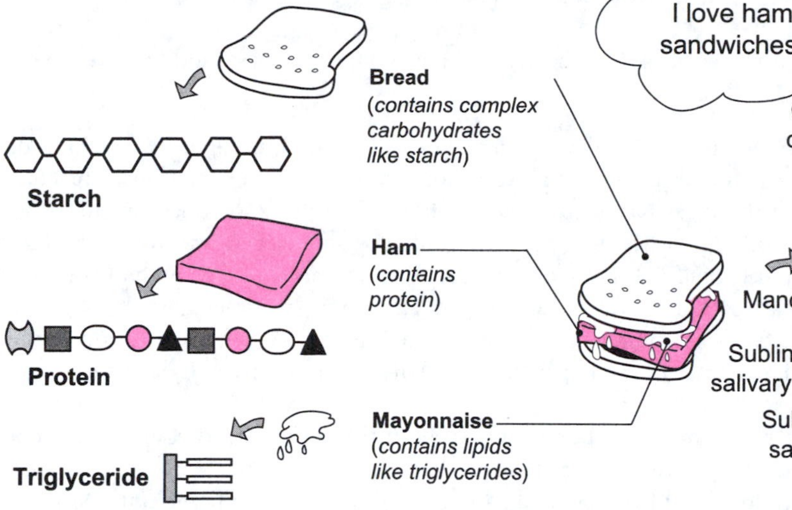

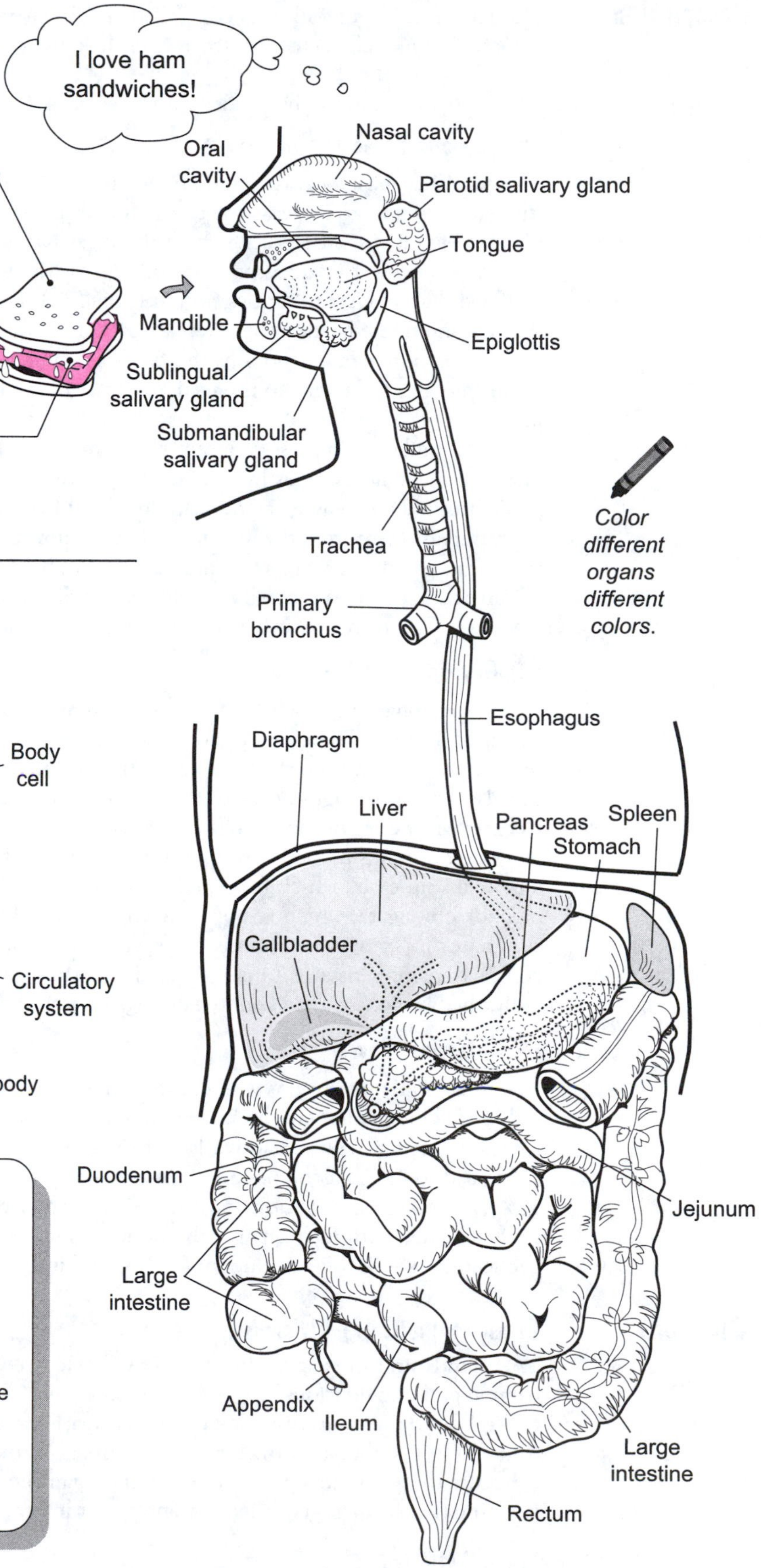

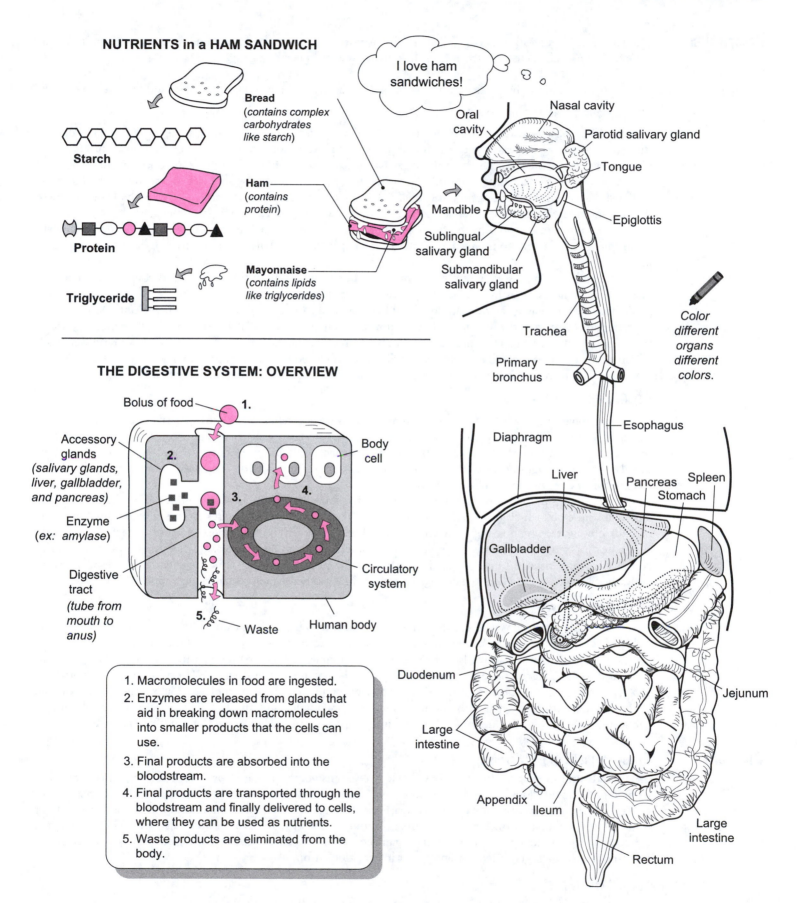

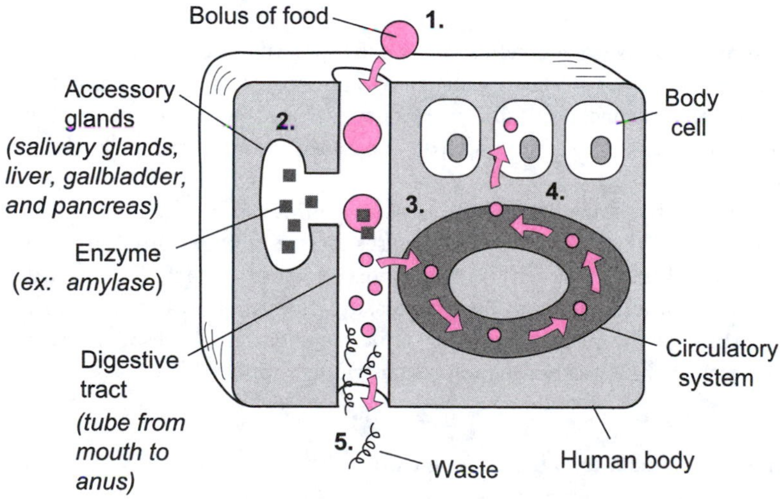

Description

The regulation of digestion is controlled simultaneously by the autonomic nervous system (ANS) and by hormones. This module focuses on the role of the nervous system in regulating digestion.

Phases

Control of digestion can be divided into three phases: (1) cephalic phase, (2) gastric phase, and (3) intestinal phase.

(1) Cephalic Phase

The **cephalic phase** begins before any food enters the oral cavity. The sight, smell, thought or first taste of food stimulates reflexes to increase the production of saliva, gastric secretions, and pancreatic secretions. The collective purpose of this response is to prepare the digestive tract to receive food. This phase produces about 10% of all gastric secretions. Key digestive control areas in the nervous system include the **cerebral cortex**, **hypothalamus**, and **medulla oblongata**. Various stimuli (from food) are sent through nerve impulses to the cerebral cortex, where they are interpreted. Then the cerebral cortex responds by sending impulses to the medulla.

The **hypothalamus** contains the feeding center. Specific stimuli, such as smells and tastes of food, stimulate nerve impulses to be sent to the feeding center in the hypothalamus. This causes the hypothalamus to send nerve impulses to the medulla oblongata.

As stimulatory impulses are sent to the medulla oblongata from both the cerebral cortex and the hypothalamus, the medulla responds by sending out nerve impulses that produce three key responses: (1) increased saliva production, (2) increased gastric secretions, and (3) increased pancreatic secretions (small amount). More specifically, impulses sent from the medulla oblongata down the **vagus nerve** stimulate both gastric and pancreatic secretions. The gastric secretions contain pepsinogen, hydrochloric acid, and mucus. The pancreatic secretions are rich in pancreatic enzymes that aid in carbohydrate digestion, triglyceride digestion, protein digestion, and nucleic acid digestion. Cranial nerves other than the vagus stimulate the salivary glands to produce saliva.

(2) Gastric Phase

The **gastric phase** begins as food enters the stomach. This triggers reflex pathways that stimulate about 80% of all gastric secretions. The two types of sensory receptors in the stomach wall are: stretch receptors and chemoreceptors. Upon entering the stomach, food stretches the stomach wall and triggers stretch receptors to send a nerve impulse along the vagus nerve to the medulla oblongata. The same response occurs when chemoreceptors detect a change in the pH of the gastric secretions.

After proteins in food enter the stomach, they buffer some of the normal acidity in the stomach. This leads to a more alkaline pH, which triggers the chemoreceptors to send out nerve impulses. These impulses are sent to the medulla oblongata, which responds by sending impulses back down the vagus nerve to the stomach. This further stimulates the gastric secretions of pepsinogen, hydrochloric acid, and mucus. It also stimulates the muscular contractions of peristalsis to move the chyme through the stomach and continue peristalsis in the esophagus. Finally, the hormone gastrin secreted during the gastric phase further stimulates gastric secretions.

(3) Intestinal Phase

The **intestinal phase** begins when chyme enters the **duodenum** after passing through the pyloric valve. Within the wall of the duodenum are the same two important sensory receptors as were found in the stomach: stretch receptors and chemoreceptors. Activation of these receptors stimulates reflex pathways through the medulla that initially lead to more gastric secretion, gastric motility, and emptying of the stomach. This phase accounts for about 10% of total gastric secretion and ensures that the digestion process is fully completed. Later, when the duodenum becomes filled with fatty chyme, the hormones secretin and cholecystokinin (CCK) are produced to inhibit gastric secretion and stomach emptying. This inhibitory response is needed to signal the end of the passage of a meal.

Divisions

Parasympathetic (P) Divisions

The glands and organs of the digestive system are extensively innervated by both the sympathetic and the parasympathetic divisions of the ANS. Consequently, the control of secretions and motility is under unconscious control by the nervous system. The **parasympathetic division (P)** controls stimulation of secretions and motility throughout the digestive tract primarily through action of the **vagus nerve** (cranial nerve X). The vagus nerve connects to all the major digestive system organs and structures except the salivary glands. It is a mixed nerve meaning that it carries both sensory and motor information within it.

Sympathetic (S)

The **sympathetic division** (S) inhibits secretion and motility in the digestive system.

1 CEPHALIC PHASE

The cephalic phase begins before any food enters the oral cavity. The sight, smell, thought, or first taste of food stimulates nerve impulses to increase the production of saliva, gastric secretions, and pancreatic secretions.

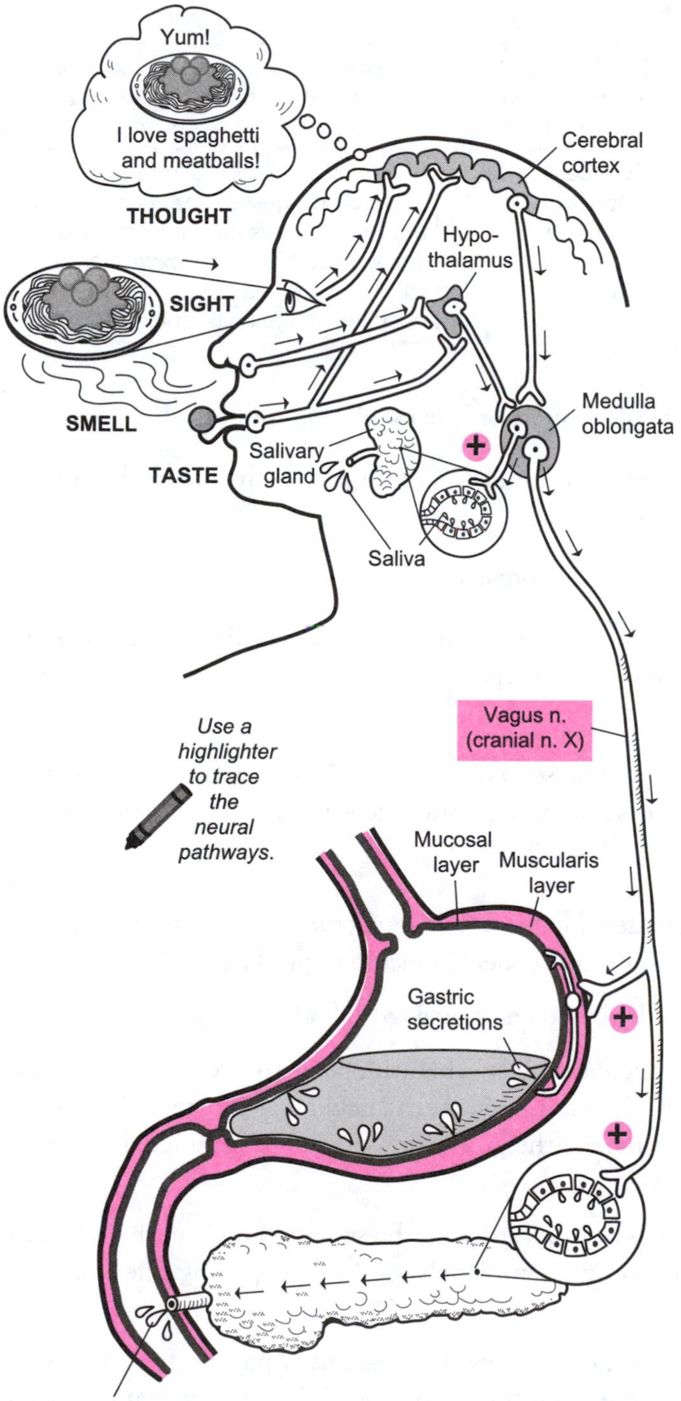

Use a highlighter to trace the neural pathways.

SYMPATHETIC (S) and PARASYMPATHETIC (P) CONTROL

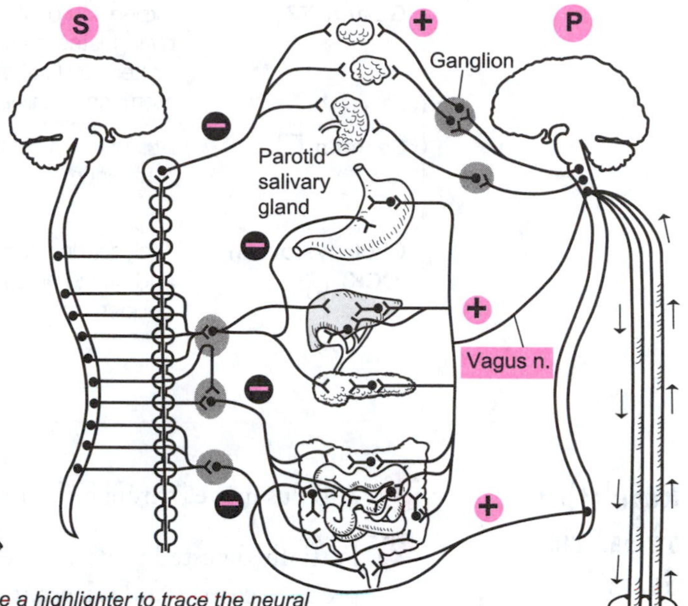

Use a highlighter to trace the neural pathways. Color the different digestive organs different colors.

2 GASTRIC PHASE

The gastric phase begins as food enters the stomach.

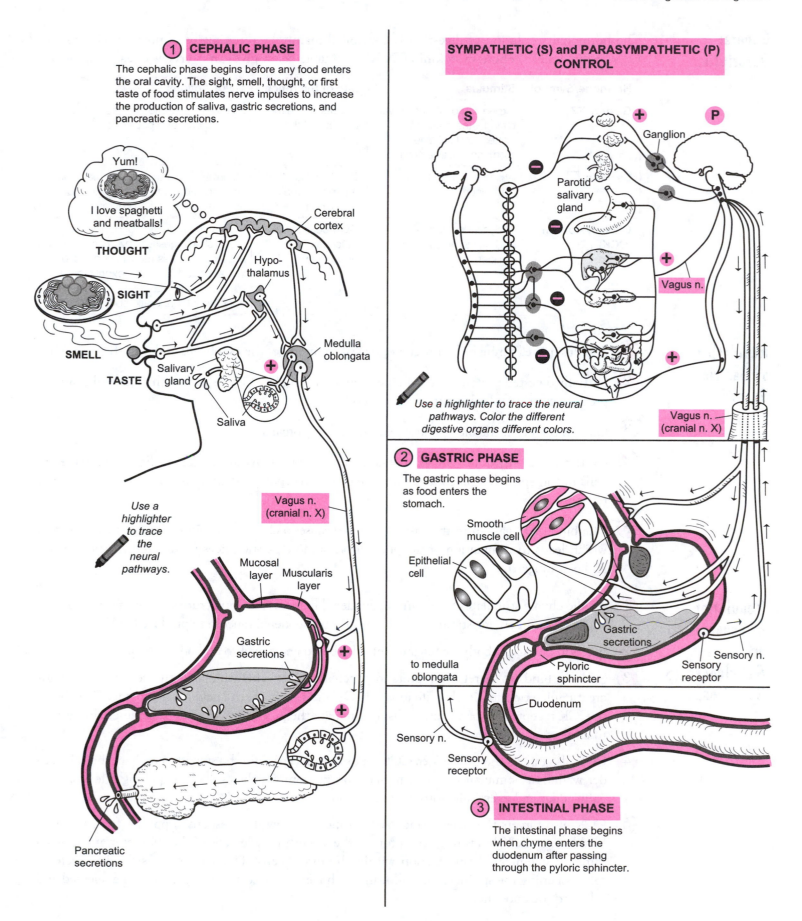

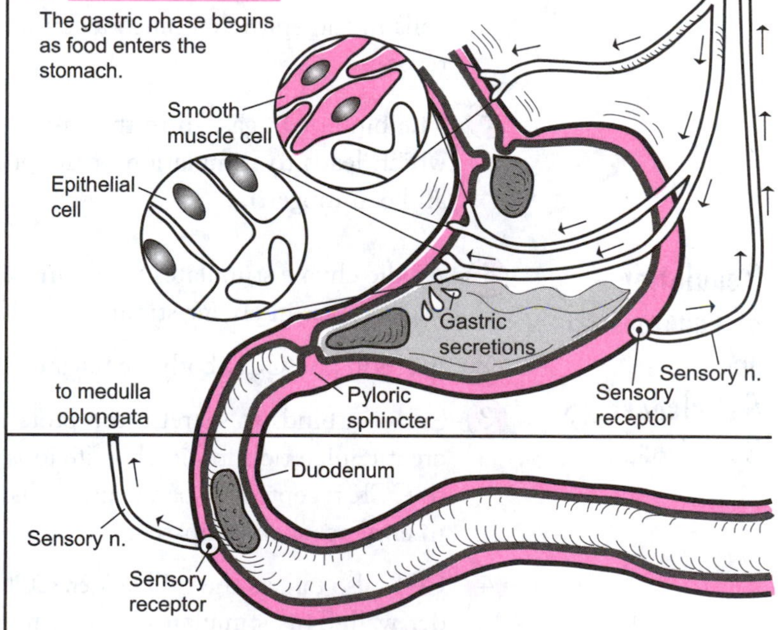

3 INTESTINAL PHASE

The intestinal phase begins when chyme enters the duodenum after passing through the pyloric sphincter.

General Description

The hormonal regulation of digestion is controlled mainly by three major hormones: (1) **gastrin**, (2) **secretin**, and (3) **cholescystokinin** (CCK). The table below gives an overview of each hormone:

Hormone/Symbol	Stimulus	Site of Secretion	Major Actions
Gastrin ▽	Expansion of stomach wall by ingested materials; protein and caffeine in stomach; alkaline chyme	**G cells** within the mucosa of the stomach	Stimulates secretion of gastric juice (HCl, mucus, and pepsinogen)
Secretin	Acidic chyme in the duodenum	**S cells** within the mucosa of the duodenum	Stimulates secretion of pancreatic juice rich in bicarbonate ions (HCO_3^-) to help neutralize acid in chyme
Cholecystokinin (CCK)	Triclycerides, fatty acids, and amino acids in the duodenum	**CCK cells** in the mucosa of the duodenum	Simulates production of pancreatic juice rich in digestive enzymes such as lipase; stimulates muscle contraction in wall of gallbladder to expel stored bile; stimlates relaxation of muscle to open hepatopancratic sphincter

Regulation of Gastric Secretion

1. A bolus moves through the lower esophageal sphincter of the esophagus and into the stomach.
2. Partially digested proteins and caffeine from food stimulate G cells within the mucosal lining to produce gastrin and release it into the blood.
3. Gastrin travels in the blood and spreads to all body organs.
4. Gastrin binds only to target cells that contain a receptor for gastrin. Because the various secretory cells in the epithelial mucosa contain the gastrin receptor, gastrin binds to them and induces a response.
5. The binding of gastrin to the gastrin receptor causes a chemical chain reaction within the cell, which leads to stimulation of the production of more gastric juices (hydrochloric acid, mucus, and pepsinogen).

Regulation of Pancreas and Gallbladder Secretions

1. Acidic chyme stimulates S cells in the duodenal mucosa to manufacture secretin; fatty acids and triglycerides in chyme stimulate CCK cells in the duodenal mucosa to produce CCK.
2. Secretin and CCK both are released into the blood and travel to all body organs.
3. Secretin binds to secretin receptors in pancreatic cells. This stimulates the cells to make pancreatic juice rich in bicarbonate ions (HCO_3^-) to neutralize the acid in the chyme. CCK binds to CCK receptors in pancreatic cells to stimulate the production of pancreatic juice that is rich in enzymes such as lipase.
4. CCK also targets the gallbladder. CCK binds to its receptors in the smooth muscle in the gallbladder wall. This stimulates smooth muscle contraction, which results in expelling bile from the gallbladder and eventually into the duodenum.
5. CCK also targets the hepatopancreatic sphincter, but with a response opposite from the gallbladder. As CCK binds to its receptors in the smooth muscle cells of the sphincter, it triggers a different chemical chain reaction within the target cells. This causes the smooth muscle to relax, opening the sphincter and allowing both pancreatic secretions and bile to be released and bile into the duodenum.

Three Major Hormones Regulate Digestion:

- Gastrin
- Secretin
- Cholecystokinin (CCK)

Color each hormone and its matching receptor the same color.

Site of Hormonal Secretion

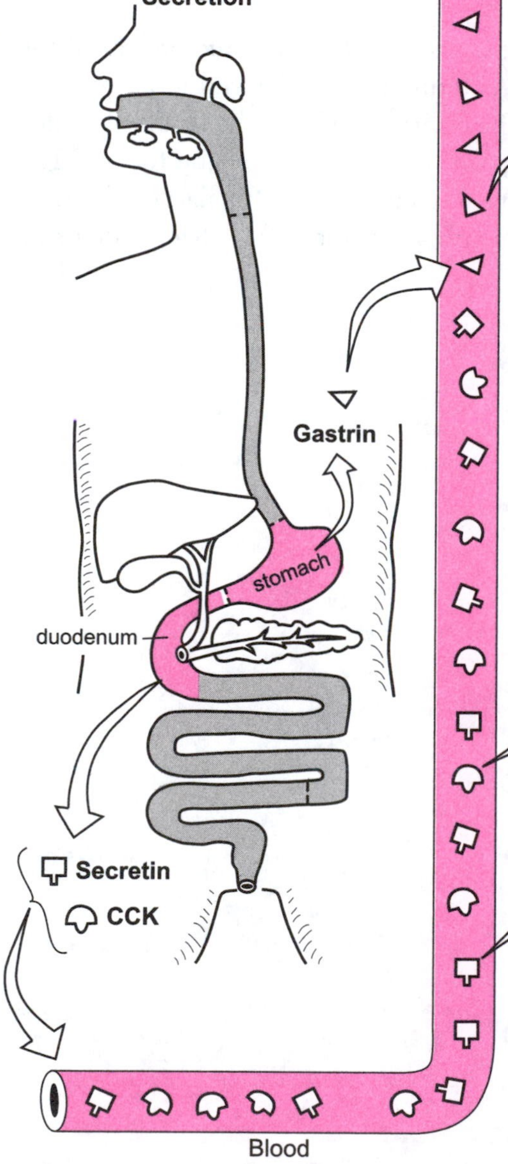

Regulation of Gastric Secretion by Gastrin

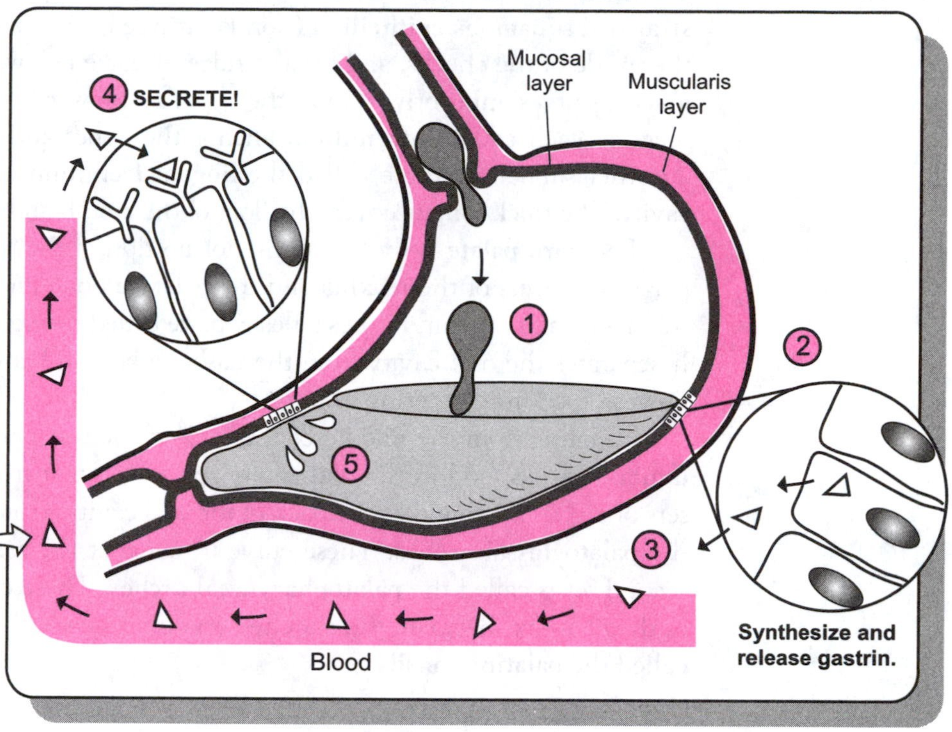

Regulation of Pancreatic and Gallbladder Secretions by Secretin and CCK

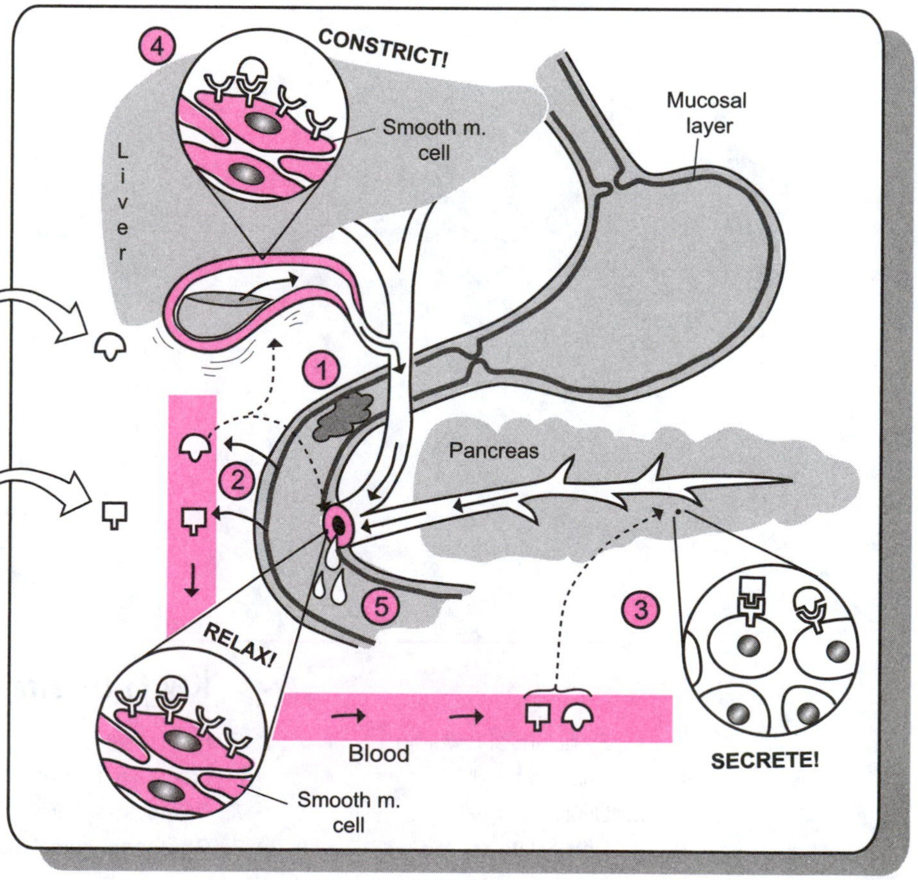

Description

The innermost lining of the **buccal cavity** (*oral cavity*) is called the **mucosa**. It is composed of **stratified squamous epithelium** (non-keratinized), which coats the roof and floor of the mouth, lines the inside of the cheeks, and forms a ridge of tissue below the teeth called the gums or **gingivae**. The mucosa curves anteriorly to form the upper and lower lips, or **labia**. Two separate, flattened plates of mucosa, each called a **frenulum,** anchor the upper and lower lips to the midline of the gingivae. Another similar structure called the **lingual frenulum** anchors the tongue to the floor of the oral cavity. The thick **tongue** covers the floor of the mouth and contains several different groups of muscles.

The **hard palate** is the most anterior portion of the roof of the oral cavity. It is composed of the palatine process of the maxilla and the palatine bone and functions to separate the oral cavity from the nasal cavity. Behind it lies a fleshy plate called the **soft palate**, which contains no bone within it. It separates the oral cavity from the nasopharynx and covers the nasopharynx during swallowing to prevent food from entering it.

Dangling from the middle of the back of the soft palate is a small, flap-like structure called the **uvula**. It holds food in the oral cavity and prevents it from entering the **oropharynx** too soon. Two sets of arches are found at the back of the oral cavity. From the front to the back, the first set is called the **palatoglossal arches**. These curve to connect the soft palate with the base of the tongue. The second set is called the **palatopharyngeal arches**. These curve to connect the soft palate to the sides of the pharynx. Between these arches, on either side of the mouth, are two masses of lymphatic tissue called the **palatine tonsils**.

Key to Illustration

1. Frenulum of lower lip
2. Palatine tonsil
3. Uvula
4. Soft palate
5. Hard palate
6. Frenulum of upper lip
7. Palatoglossal arch
8. Palatopharyngeal arch
9. Tongue
10. Gingivae

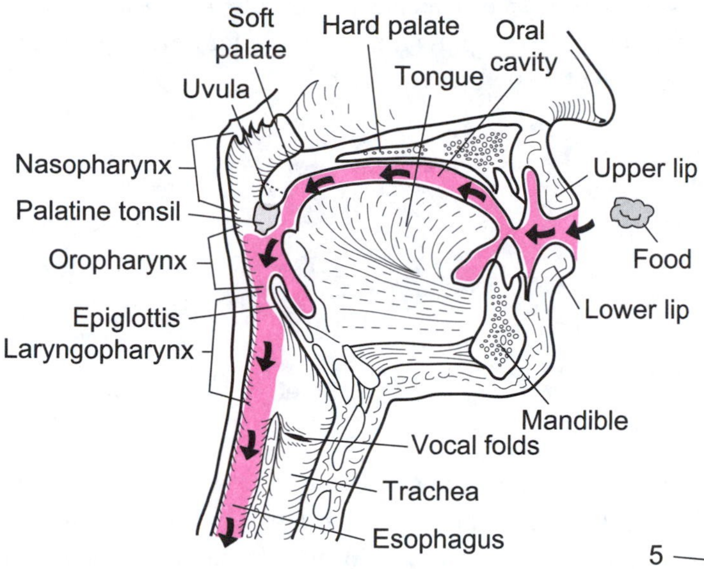

Note: *The colored section and the arrows show the pathway of food into the oral cavity and down through the esophagus as it moves toward the stomach.*

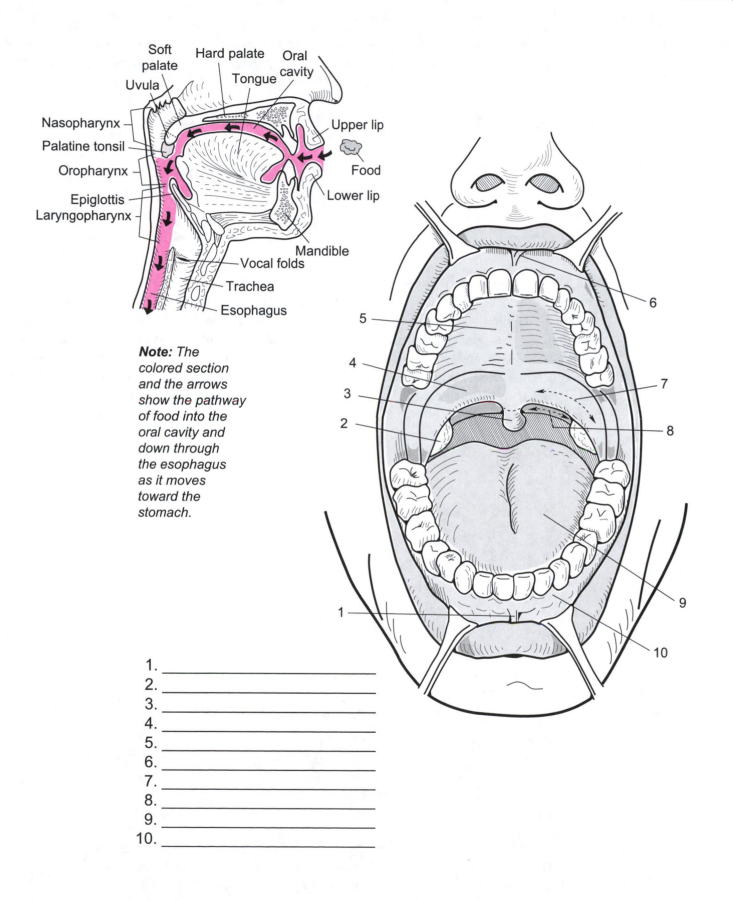

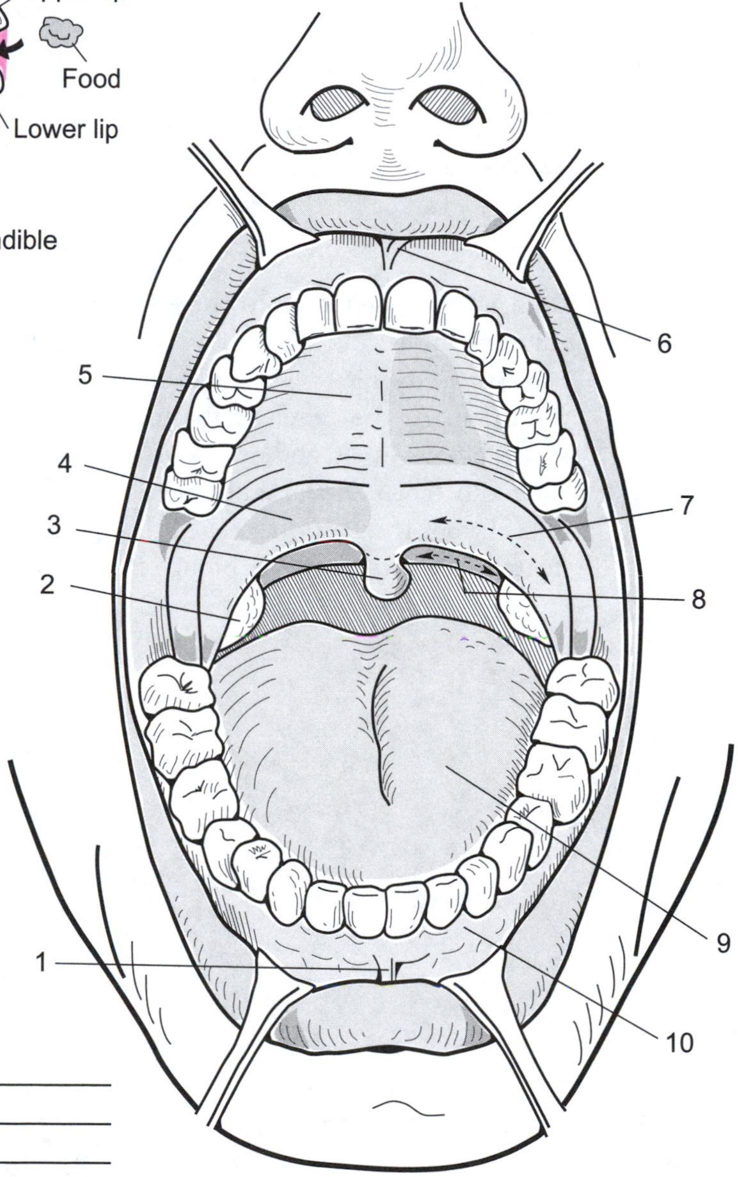

1. ______________________
2. ______________________
3. ______________________
4. ______________________
5. ______________________
6. ______________________
7. ______________________
8. ______________________
9. ______________________
10. ______________________

Description

There are four different types of teeth—incisors, cuspids (*canines*), bicuspids (*premolars*), and molars. Incisors are for cutting, cuspids are for tearing, and bicuspids and molars are for crushing and grinding food. The adult jaws can accommodate a total of 32 permanent teeth.

Tooth Type	Upper Jaw	Lower Jaw	Total
Incisor	4	4	8
Cuspid	2	2	4
Bicuspid	4	4	8
Molar	6	6	12
		Total	**32**

Each tooth is divided into three regions: **crown**, **neck**, and **root**. The crown is visible above the gumline and is covered with a calcified coating of **enamel**—the hardest substance produced in the body. Deep to this covering is a bone-like substance called **dentin**, which makes up the majority of the tooth. At the center of the tooth is a chamber called the **pulp cavity**, which contains spongy tissue, blood vessels, and nerves. A bony substance called **cementum** covers the dentin in the root of the tooth. The **periodontal ligaments** anchor the root of the tooth to the bone in the jaw. At the tip of each root is an opening called the **apical foramen** that allows the **dental artery**, **dental vein**, and **dental nerve**, to penetrate through the narrow **root canal** and up into the **pulp cavity**.

Key to Illustration

1. Incisors
2. Cuspids *(canines)*
3. Bicuspids *(premolars)*
4. Molars
5. Central incisors
6. Lateral incisor
7. Cuspid *(canine)*
8. 1st Premolar
9. 2nd Premolar
10. 1st Molar
11. 2nd Molar
12. 3rd Molar *(wisdom tooth)*
13. Crown
14. Neck
15. Root
16. Enamel
17. Dentin
18. Pulp cavity
19. Gingiva
20. Periodontal ligaments
21. Cementum
22. Bone
23. Root canal
24. Apical foramen
25. Dental artery
26. Dental nerve
27. Dental vein

1
2
3
4
(Upper jaw)
(Lower jaw)
Tooth types
5
6
7
8
9
10
11
12
Maxilla
Palatine bone
Upper jaw
Mandible
12
11
10
9
8
7
6
5
19
Lower jaw
16
17
18
13
14
15
19
20
21
22
23
24
25
26
27
Tooth anatomy

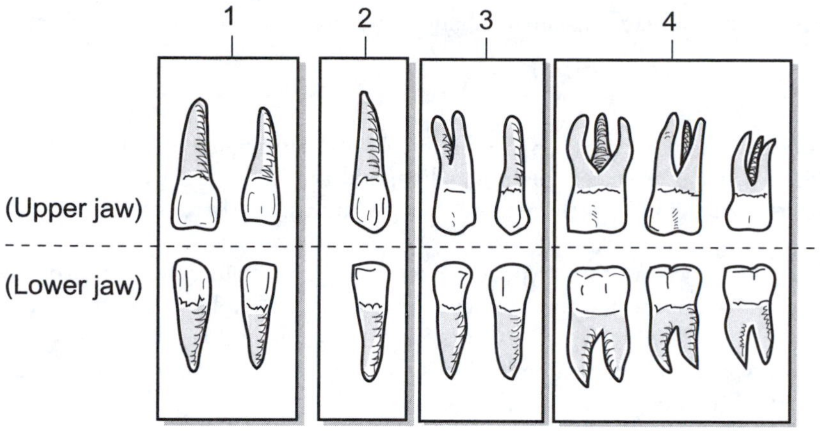

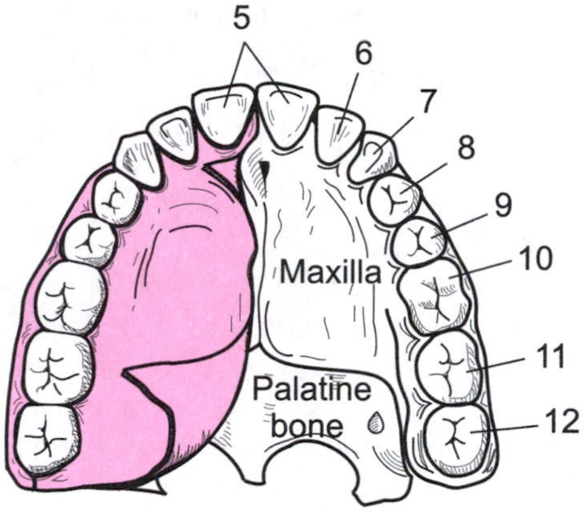

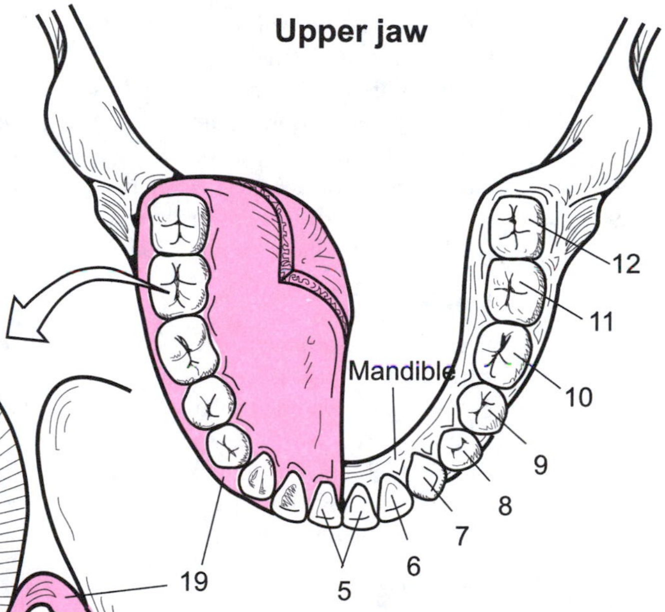

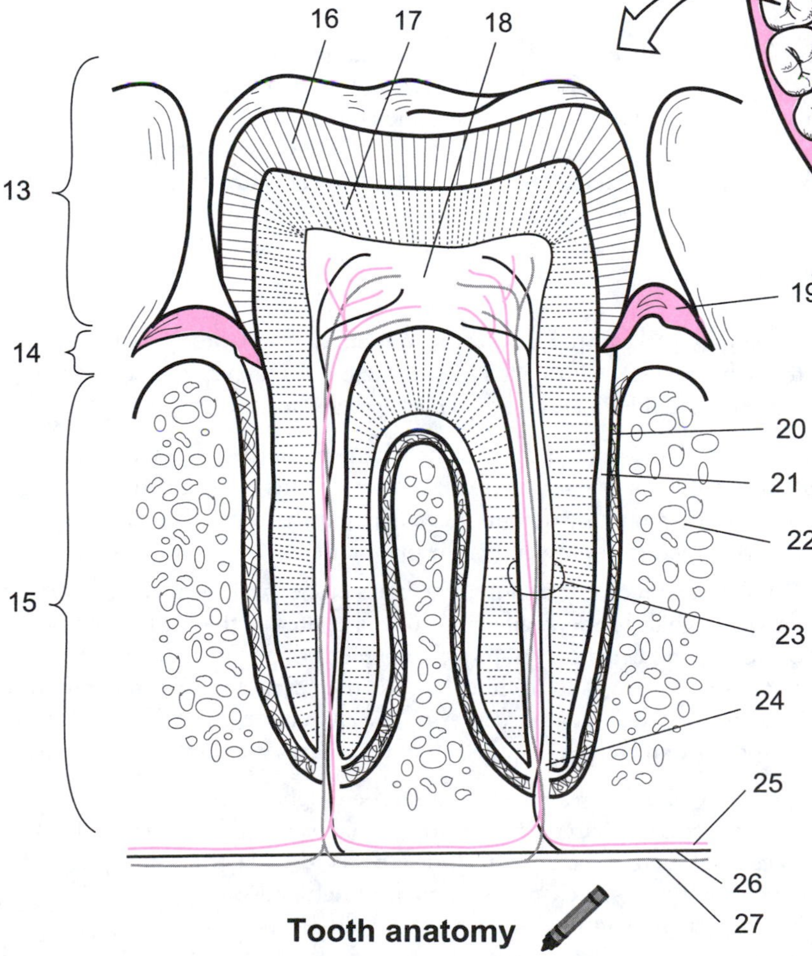

Description

Digestion in the mouth deals with producing saliva from salivary glands, chewing, and swallowing. Each of these processes is summarized here.

Salivary Glands

The body contains three pairs of **salivary glands**: **parotid**, **submandibular**, and **sublingual**. Of these three, the parotid is the largest. The submandibular gland produces the most saliva, and the sublingual gland produces the least. Salivary glands are classified as exocrine glands partly because they all have ducts that connect them to the oral cavity.

At the microscopic level, salivary glands contain two types of secretory cells: **mucous cells** and **serous cells.** The mucous cells secrete a protein called mucin, which forms mucus. The serous cells release a watery secretion containing electrolytes and salivary amylase. These collective secretions form saliva and release it into the oral cavity. Because secretion of saliva is controlled by the autonomic nervous system, salivary glands are innervated by the parasympathetic and the sympathetic divisions.

Saliva

Saliva is a mildly acidic (pH 6.4–6.8), watery secretion that has multiple functions. It contains mostly water (95%), as well as some important dissolved solutes (0.5%) including electrolytes such as sodium (Na^+), potassium (K^+), chloride (Cl^-), and others. **Mucus** is produced to lubricate the bolus to make swallowing easier. **Lysozyme** is an antibacterial agent that inhibits the growth of bacteria normally found in the oral cavity. **Antibodies** such as immunoglobulin A also control bacterial growth. The enzyme **salivary amylase** catalyzes the breakdown of complex carbohydrates into disaccharides. Like water spraying on cars in a carwash, saliva cleans oral cavity structures such as the teeth by constantly washing over them. Last, saliva serves as a fluid medium in which food molecules can dissolve to allow for chemical detection by taste buds. Without saliva, it is difficult to taste anything!

Chewing

Chewing (*mastication*) is the physical breakdown of food and the mixing of food with saliva. This is accomplished with the help of the teeth and the tongue. The four different types of teeth are: incisors, cuspids (canines), bicuspids (premolars), and molars. Each type is best suited for a specific task. Incisors have a flat, chiseled edge for cutting food like biting into an apple. Cuspids have a pointed edge for puncturing, tearing, and shredding food. Bicuspids and molars have larger, flattened surfaces, best for grinding and crushing food. The tongue moves during chewing to mix the pulverized food with saliva to form a pasty, compressed mass called a **bolus**. This mixing prepares the **bolus** for swallowing.

Swallowing

Swallowing (*deglutition*) is a complex process that requires many different muscles to move ingested substances from the oral cavity to the stomach. It typically is divided into three stages: oral, pharyngeal, and esophageal. Note that the same process applies to ingested liquids and solids. The stages in the process are:

1. **Oral** (or *buccal*) **phase**—is a voluntary phase that begins when substances are ingested. The tongue mixes these materials with saliva to form a moistened mass called a bolus. This bolus is compressed against the hard and soft palates in the roof of the mouth by upward movements of the tongue. Then the tongue pushes the bolus toward the oropharynx, which marks the end of the oral phase.
2. **Pharyngeal phase**—an involuntary process. It begins as the bolus pushes against the oropharynx, where it stimulates tactile receptors to send signals to the swallowing control center in the medulla oblongata. Three key events occur: (a) the soft palate and uvula move upward to prevent the bolus from entering the nasal cavity; (b) the bolus moves into the oropharynx; and (c) muscles raise the larynx forward and upward so the epiglottis closes over the glottis to prevent the bolus from entering the larynx or trachea. The pharyngeal phase ends after pharyngeal muscles constrict and force the bolus out of the pharynx and into the esophagus.
3. **Esophageal phase**—involuntary like the pharyngeal phase. The esophagus is a long, muscular tube that connects the pharynx to the stomach. This phase begins as the upper esophageal sphincter relaxes to allow the bolus to enter the esophagus. The rigid epiglottis springs back to its normal, upright position. Then the upper esophageal sphincter contracts to propel the bolus deeper into the esophagus. The muscular contractions of peristalsis (see p. 384) quickly move the bolus into the stomach. This entire phase takes less than 8 seconds!

DISCOVER for YOURSELF: Place your fingertips on the front of your larynx as you swallow. You can feel the larynx move forward and upward.

SALIVARY GLANDS

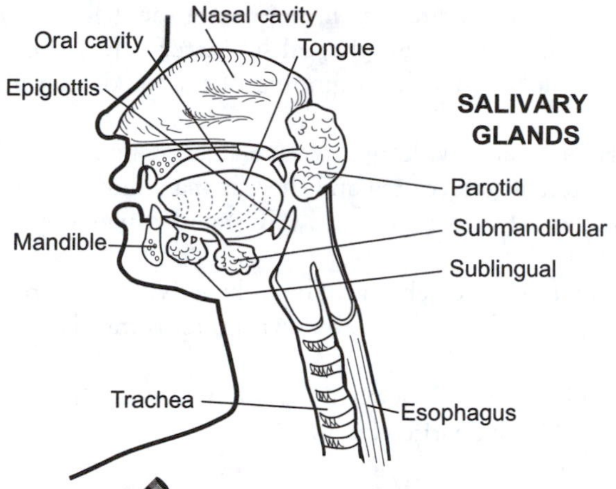

Color the three pairs of salivary glands the same color.

CHEWING

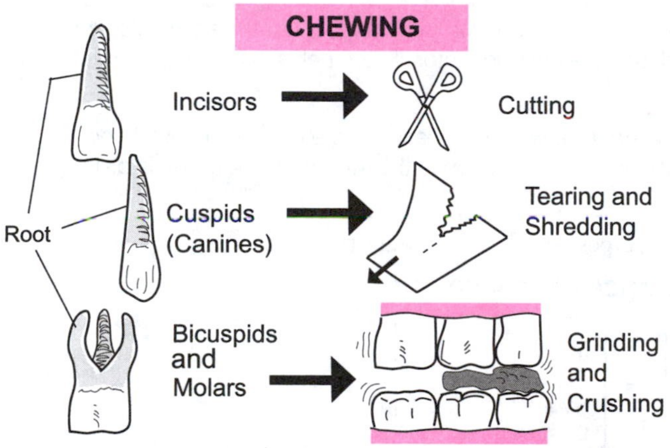

SALIVA

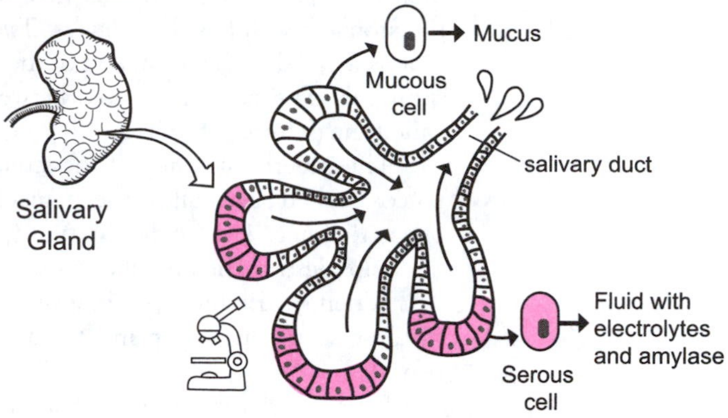

Functions of Saliva

- Lubrication

- Cleanses Oral Cavity Structures.

- Uses AMYLASE to catalyze digestion of carbohydrates.

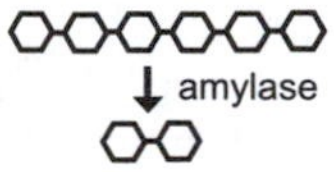

- Uses LYSOZYME and ANTIBODIES to inhibit bacterial growth.

- Dissolves FOOD MOLECULES to allow for taste.

Composition of Saliva

99.5%

0.5% SOLUTES

- Electrolytes (Na^+, K^+, Cl^-, etc.)
- Lysozyme
- Mucus
- Salivary amylase
- Immunoglobulin A

SWALLOWING

Color the bolus, larynx, uvula, and epiglottis different colors.

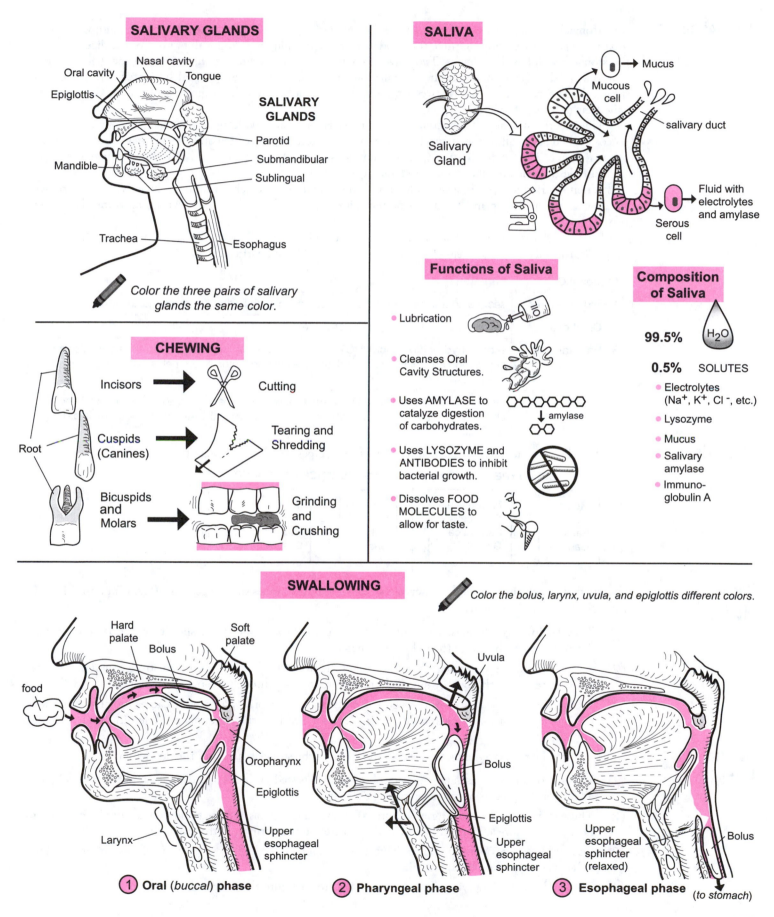

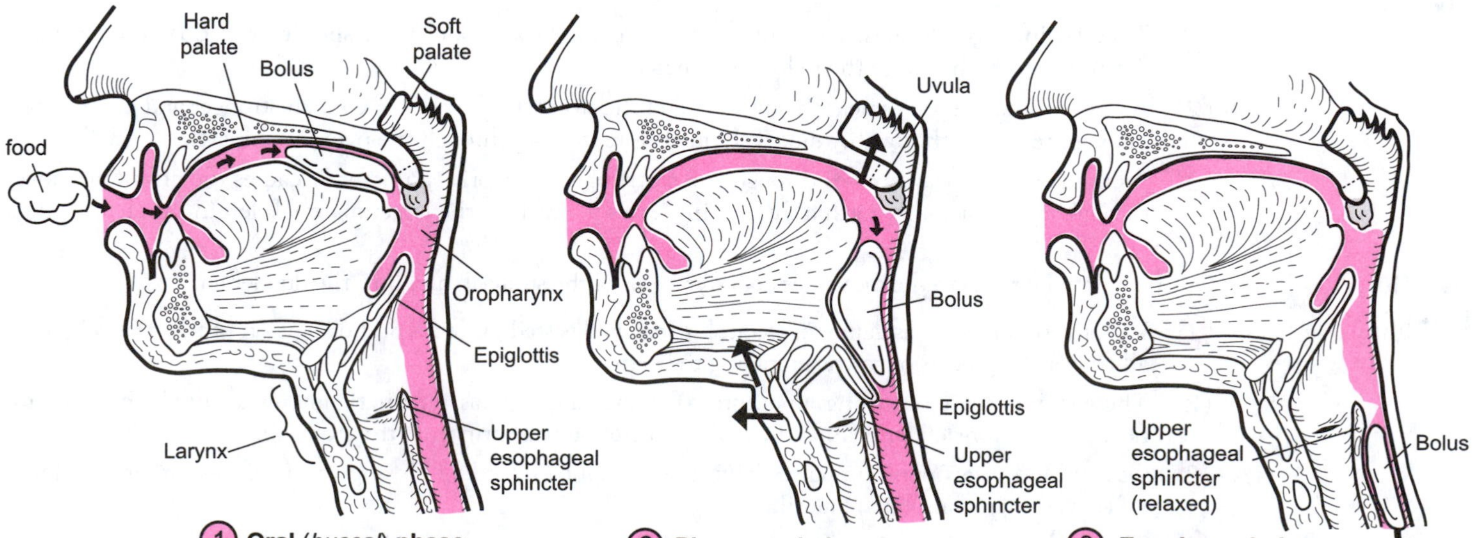

① **Oral (*buccal*) phase** ② **Pharyngeal phase** ③ **Esophageal phase** (*to stomach*)

Description

The **stomach** connects the esophagus to the duodenum. It is a specialized, muscular sac containing an acidic mixture of gastric secretions. Unlike the two layers of smooth muscle in the wall of most of the digestive tract, the stomach wall has three layers. Two valves called sphincters separate it from the rest of the digestive tract. One is located at the distal end of the esophagus and is called the **lower esophageal sphincter (LES)** or cardiac sphincter. The other is located between the end of the stomach and the beginning of the duodenum and is called the **pyloric sphincter**.

The innermost lining of the stomach—the **mucosa**—is coated with a protective alkaline mucus. Folds of mucosa called **rugae** allow the stomach to maximize its internal surface area and thereby maximize the amount of gastric juice. This acidic mixture (pH 2–3) contains mucus, hydrochloric acid (HCl) and enzymes. The low pH helps destroy the microbes present in food.

When gastric juice combines with food, it forms a mixture called **chyme**. The only active enzyme present in these secretions is **pepsin**. It initiates protein digestion by helping to break down proteins into large polypeptides.

Four different types of cells are used to produce secretions for the chemical digestion of food: *mucous cells*, *chief cells*, *parietal cells*, and *G cells*. The function of each cell is summarized below:

Mucosal Cell	Function
1. **Mucous cells**	secrete an alkaline **mucus** to protect the stomach lining from acidic gastric juice
2. **Chief cells**	secrete the inactive enzyme **pepsinogen**
3. **Parietal cells**	secrete **hydrochloric acid** (HCl), which helps convert pepsinogen into the active enzyme pepsin and kills microbes in food; secrete intrinsic factor, which helps with absorption of vitamin B_{12}
4. **G cells**	produce and secrete the hormone **gastrin**, which increases secretion from parietal and chief cells; it also induces smooth muscle contraction in the stomach wall for peristalsis.

Study Tip

Use the following mnemonics to recall the mucosal cell types and the substances they produce:

CELL TYPE			SUBSTANCE PRODUCED		
Men	=	**M**ucous cells	**M**en	=	**M**ucus
Can		**C**hief cells	**P**ass		**P**epsinogen
Pass		**P**arietal cells	**H**ot		**H**ydrochloric acid
Gas		**G** cells	**G**as		**G**astrin

Gastric Motility

Gastric motility is stimulated by the nervous system and by hormones such as gastrin. Waves of peristalsis move through the stomach regularly.

1. They begin as gentle muscular contractions near the **lower esophageal sphincter (LES)** and continue down the stomach toward the **pyloric sphincter**.
2. As the contractile waves near the distal end of the stomach, they become much stronger and more forceful. This results in thoroughly mixed **chyme** before it passes through the pyloric sphincter.
3. As the peristaltic wave passes through the partly opened pyloric sphincter, it causes the chyme to move through it in a back-and-forth fashion. This serves to break up the larger materials left in the chyme.

Gastric Emptying

Gastric emptying refers to the movement of **chyme** out of the stomach and into the duodenum.

1. Various stimuli, such as caffeine, alcohol, partially digested proteins, and distension of the stomach wall, all trigger the process of gastric emptying to begin.
2. These stimuli cause two things to occur: (a) an increase in parasympathetic impulses sent along the vagus nerve to the stomach, and (b) an increase in secretion of gastrin from the stomach.
3. Gastrin and vagus nerve impulses have the combined effects of closing the LES, opening the pyloric valve, and increasing gastric motility.
4. The final result is emptying of the chyme in the stomach into the duodenum.

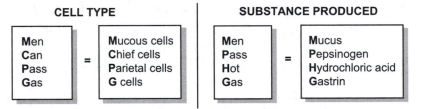

Structure and Function

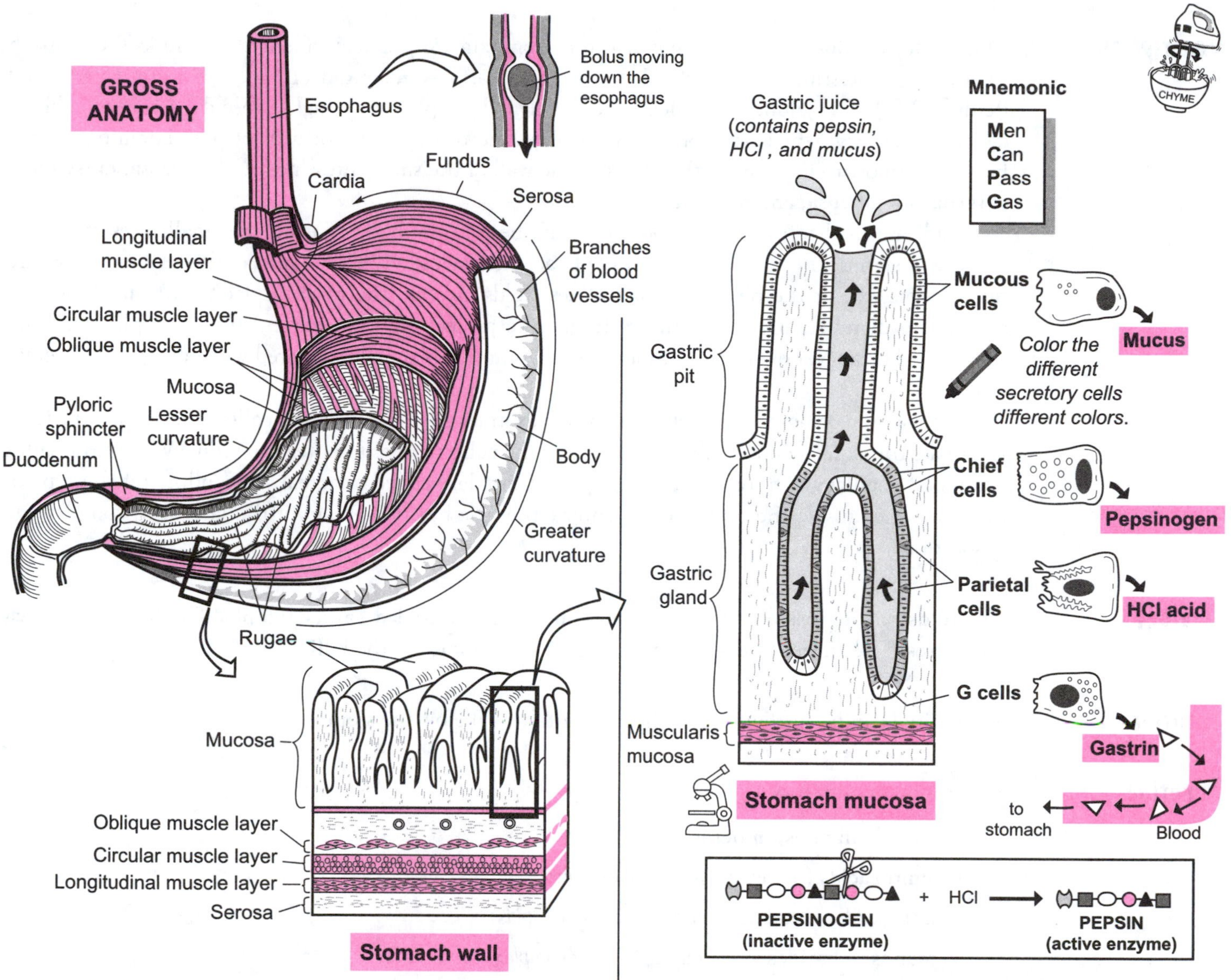

GASTRIC MOTILITY

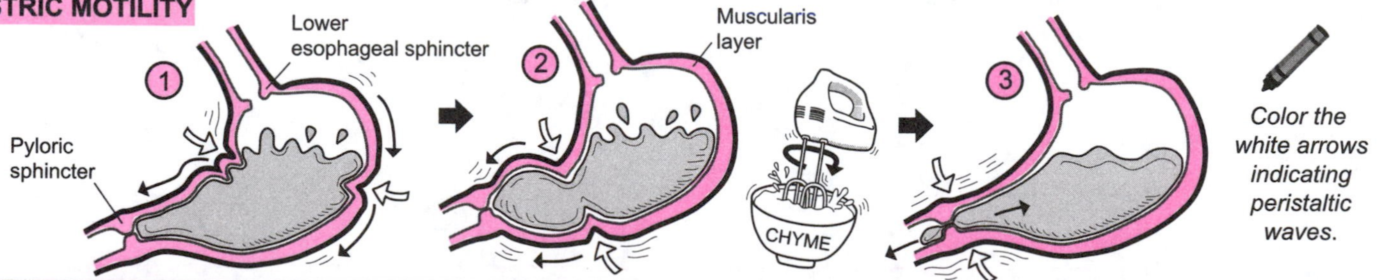

GASTRIC EMPTYING

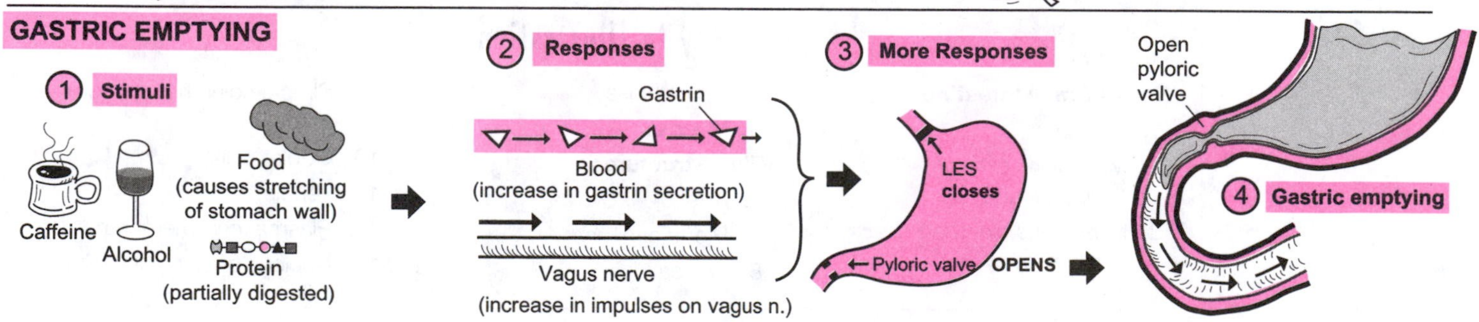

Description

The **small intestine** is a hollow, muscular tube approximately 20 feet in length that links the stomach to the large intestine. It is subdivided into three parts—**duodenum** (10 in.), **jejunum** (8 ft.), and **ileum** (12 ft.). The jejunum and ileum are loosely held in place by a highly vascular serous membrane called **mesentery**, which anchors the small intestine to the posterior wall of the abdominal cavity. From outermost to innermost, the layers in the wall of the small intestine are the **serosa**, **muscularis externa**, **submucosa**, and **mucosa**.

Folds are used to increase surface area anywhere in the body. Within the small intestine, there are three significant folded structures: **plicae circulares**, **villi**, and **microvilli**. All of these are microscopic except the plica circulares. These folds increase the total surface area to facilitate the absorption of nutrients. The **villi** are finger-like projections that extend from the plica and are surrounded by a layer of **simple columnar epithelium**. **Goblet cells** scattered within this tissue secrete mucus.

Each simple columnar epithelial cell has folds in its plasma membrane called microvilli. Nutrients must pass through the microvilli, then through the cell, in order to enter the villus. Within the villus is a **blood capillary** and a specialized lymphatic capillary called a **lacteal**. Depending on the type of nutrient, once inside the villus, it will enter either the blood capillary or the lacteal to complete its absorption.

Analogy

The **folds** on the inside of the **small intestine** are like a **folded carpet sample**. The **fold** is the **plica circulares**. The **carpet fibers** sticking out from this sample are the **villi**.

Location

Abdominal cavity; surrounded by the large intestine; below the stomach

Function

Absorption of nutrients

- **water, vitamins, minerals**
- **amino acids** (*from the digestion of proteins*)
- **fatty acids, glycerol** (*from the digestion of lipids*)
- **monosaccharides** (*from the digestion of complex carbohydrates*)
- **nucleotides** (*from the digestion of nucleic acids such as DNA*)

Key to Illustration

Wall of Small Intestine

1. Serosa
2. Muscularis externa *(longitudinal layer)*
3. Muscularis externa *(circular layer)*
4. Submucosa
5. Mucosa

Villus Structures

6. Arteriole
7. Blood capillary
8. Lacteal
9. Simple columnar epithelial cell
10. Goblet cell
11. Venule
12. Plasma *(cell)* membrane
13. Nucleus of one cell

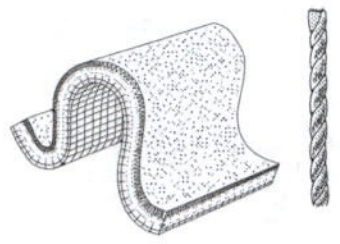

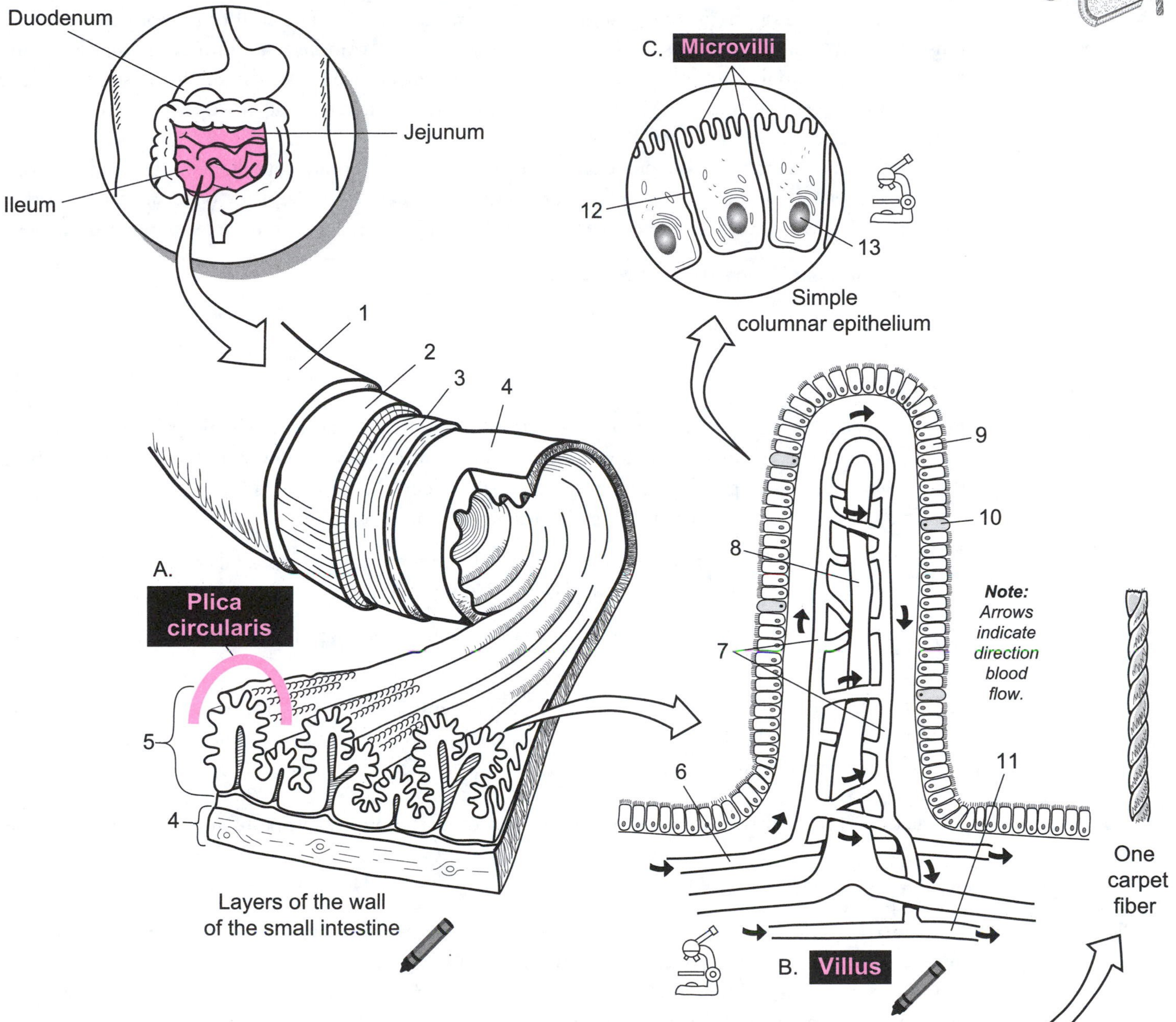

Imagine a folded carpet sample. The fold itself is like the large fold or **plica circularis** in the small intestine while each individual **carpet fiber** is like one **villus**.

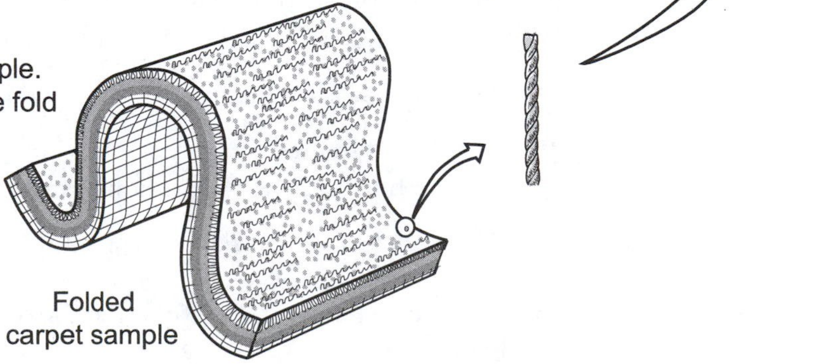

Description

The small intestine is a 20-foot-long muscular tube that connects the stomach to the large intestine. It is the major site of chemical digestion and nutrient absorption. The wall of this organ contains a muscularis layer of smooth muscle that is subdivided into two thin layers: an inner circular layer and an outer longitudinal layer. When contracted, the circular layer pinches the small intestine, and the longitudinal layer shortens it. All the materials from the stomach must move through this tube by different types of coordinated muscle contractions.

The small intestine uses the processes of peristalsis and segmentation to move undigested materials along its length. These processes are not unique to the small intestine; they occur in other parts of the digestive tract as well. The sites of action for these muscular movements are summarized below:

Type of Muscular Movement	Site of action
1. **Peristalsis**	Esophagus, stomach, small intestine, and large intestine
2. **Segmentation**	Small intestine and large intestine

Peristalsis

Peristalsis is a rhythmic wave of smooth muscle contraction that results in the propulsion of materials through the digestive tract. More specifically, it is the alternating, coordinated stimulation of both the circular and the longitudinal layers of smooth muscle in the muscularis layer.

The following is the series of events illustrated on the facing page:

1. The circular layer of smooth muscle contracts behind the mass of materials in the small intestine. The result is that the small intestine is pinched tighter, like rings becoming smaller in size.
2. A section of the longitudinal layer in front of the pinched region contracts, which shortens part of the small intestine.
3. This coordinated process of "pinching" and "shortening" continues to alternate at regular intervals along the length of the wall of the small intestine. Viewed in real time, it would look like a gradual, smooth undulating wave of muscle contraction.

Analogy

The action of peristalsis is like squeezing a tube of toothpaste so the toothpaste moves through the tube.

Segmentation

Segmentation is the pinching of the intestine into compartments and subsequent mixing of undigested materials with intestinal secretions. This ensures that the processes of chemical digestion and absorption are both completed. The circular layer of the muscularis is more active than the longitudinal layer. No net movement of materials results in this process as it did in peristalsis.

The following is the series of events illustrated on the facing page:

1. The circular layer pinches off the small intestine into similar-sized compartments or "segments."
2. It further pinches to even smaller compartments as the undigested materials are moved back and forth by muscle contractions like an agitator in a washing machine.
3. The undigested materials are mixed thoroughly with intestinal secretions.

Analogy

The action of segmentation is like the movement of a hand mixer combining all the ingredients in cake batter.

Movement Through the Small Intestine

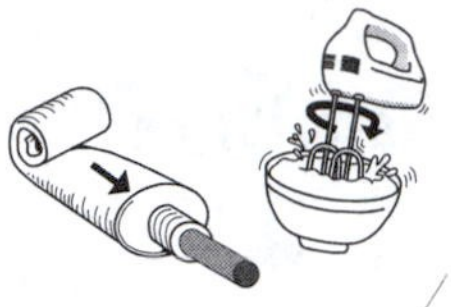

OVERVIEW

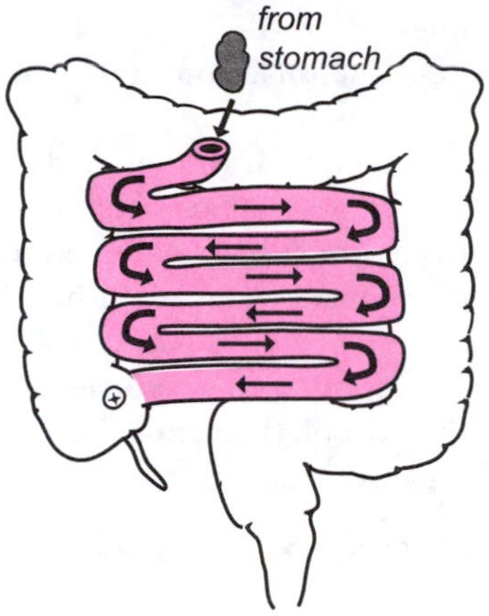

Black arrows indicate movement of undigested materials through the small intestine.

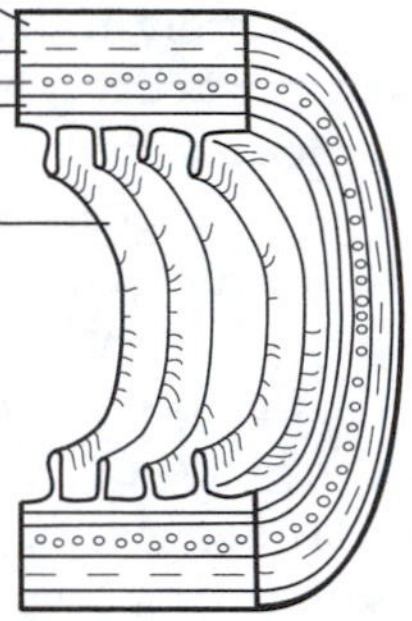

Layers of the Wall of the Small Intestine (midsaggital section)

Color each layer of the wall a different color.

PERISTALSIS

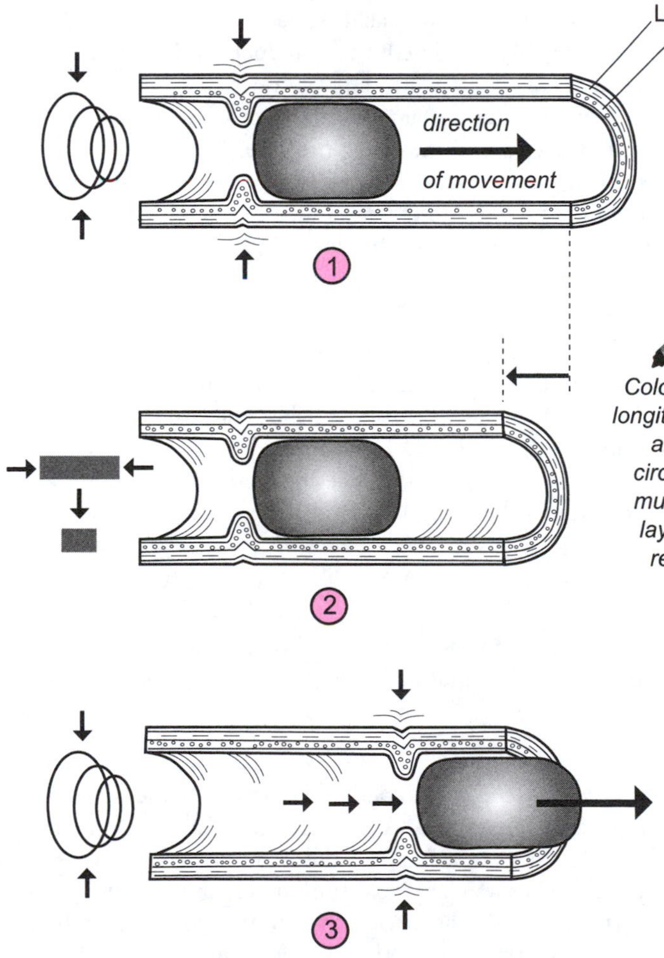

Color the longitudinal and circular muscle layers red.

SEGMENTATION

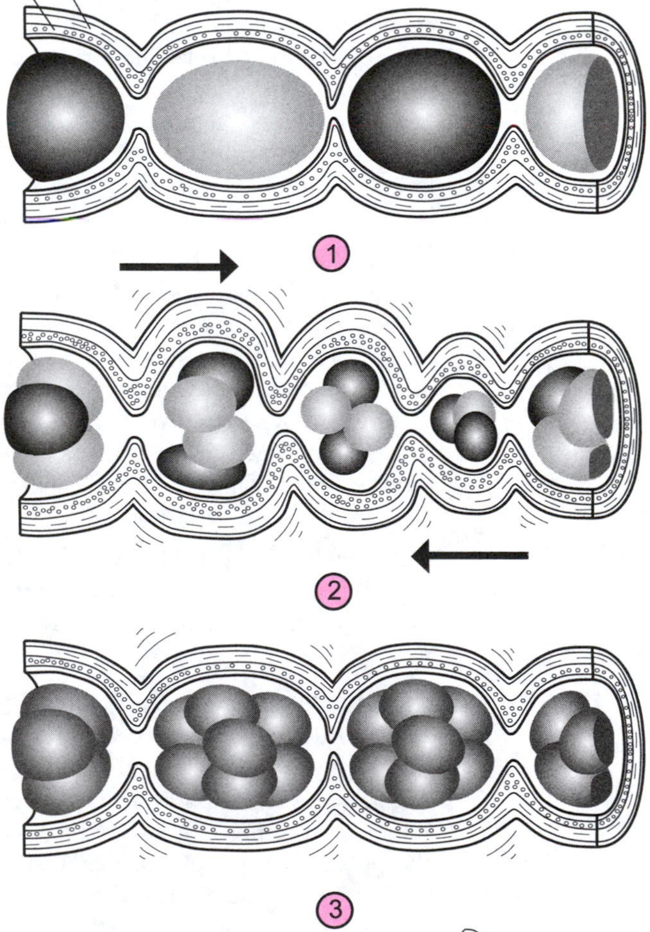

The purpose of peristalsis is to **propel** undigested materials through the digestive tract like squeezing a tube of toothpaste moves toothpaste through the tube.

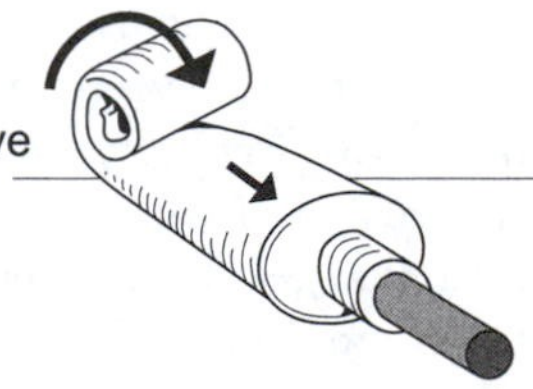

The purpose of segmentation is to **mix** undigested materials with enzymes and other secretions like a hand mixer combining all the ingredients in cake batter.

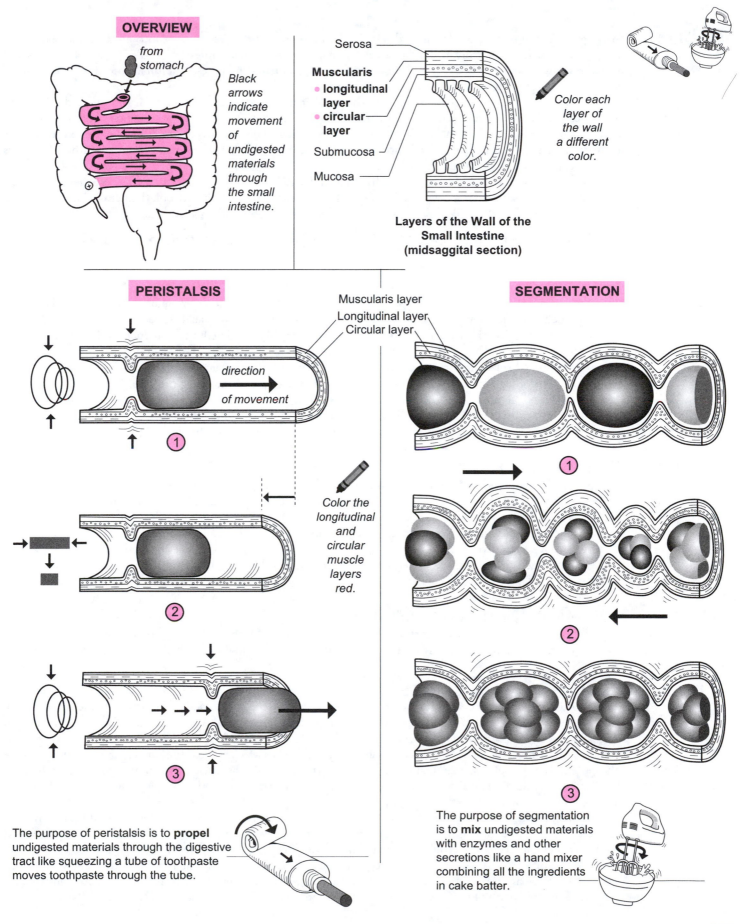

Description

Absorption is the movement of small, digested substances from the digestive tract into either the blood or the lymph. Almost all absorption in the digestive system occurs in the small intestine. Folds are used in organs and structures throughout the body to increase surface area for different purposes. The small intestine uses its many folds to maximize surface area for absorption of nutrients.

The three significant folded structures in the small intestine are the **plicae circulares**, **villi**, and **microvilli**. The plicae circulares are folds of the inner mucosal layer of the small intestine. Extending out from the plica are numerous finger-like projections called villi (sing. *villus*). Each villus is covered by a layer of **simple columnar epithelium**. The **goblet cells** scattered within this tissue secrete **mucus**. Each simple columnar epithelial cell has folds in its plasma membrane, called microvilli.

Each cell also has two surfaces through which nutrients must pass: **apical** and **basal**. Nutrients first pass through the apical surface that contains the microvilli, then move through the cytosol, then cross the basal surface that lines the inside of the villi. In summary: nutrients follow this absorption pathway:

Lumen (small intestine) → Epithelial cell: Apical surface → Cytosol → Basal surface → Blood or Lymph

Process

The purpose of chemical digestion is to break down macromolecules into their simplest building blocks. These building blocks include: monosaccharides from complex carbohydrates, amino acids, dipeptides, and tripeptides from proteins, and monoglycerides and fatty acids from triglycerides (lipids). These final products of digestion are small enough to be absorbed into the blood or the lymph and then pass directly into body cells where they can be used.

Some ingested substances that are already in their simplest form—such as the fructose in fruit juices—are partly absorbed through the wall of the stomach before they even reach the small intestine.

About 90% of all absorption occurs in the small intestine for the following reasons: (1) villi and other structures make it anatomically specialized for the process, (2) the final products of macromolecule catabolism are not even formed until they reach this site, and (3) the 20+ foot-long tube maximizes the time for absorption to take place.

Monosaccharides

Complex carbohydrates must be broken down into **monosaccharides** before they can be absorbed. The three types of monosaccharides are: glucose, galactose, and fructose. Fructose, found in fruit and fruit juices, is transported across both the apical and basolateral surfaces by **facilitated diffusion**. Glucose and galactose are transported by a different method—secondary active transport with sodium (Na^+). The sodium ion gradient is the driving force that helps move these two sugars across the apical surface and into the cell. Once inside the epithelial cell, both glucose and galactose cross the basolateral surface by facilitated diffusion. Fructose, glucose, and galactose all are absorbed directly into the blood capillary.

Amino acids Dipeptides Tripeptides

Most **amino acids** use **active transport** to cross the apical surface. Once inside the epithelial cell, the amino acids are actively transported again across the basolateral surface and then absorbed directly into the blood capillary.

Monoglycerides Fatty acids

Triglycerides are the most common type of lipids, so they will be considered here. When triglycerides are digested, two of their three fatty acid chains are broken off from their glycerol backbone, resulting in the formation of **monoglycerides** and **fatty acids**. In the chemical rule of "like goes into like," so these products of triglyceride digestion easily diffuse through the plasma membrane of the apical surface because they are chemically similar to the phospholipids in the plasma membrane. Some of the fatty acid chains are short-chain fatty acids, while others are long-chain fatty acids. The **short-chain fatty acids** diffuse directly across both the apical and basolateral surfaces and are absorbed directly into the blood capillary.

The pathway for monoglycerides and **long-chain fatty acids** is a bit more complex. After diffusing through the apical surface, the monoglycerides are digested further into glycerol and a single fatty acid chain. The glycerol then is reassembled back into triglycerides by linking with free fatty acid chains. The newly formed triglycerides are packaged in a protein coat to form a structure called a **chylomicron**. Chylomicrons leave the cell by exocytosis (see p. 54). They are too large to enter the blood capillary, so they enter the more permeable lacteal instead. As they travel in the lymph of the lacteal, they eventually are deposited into the blood via the left subclavian vein.

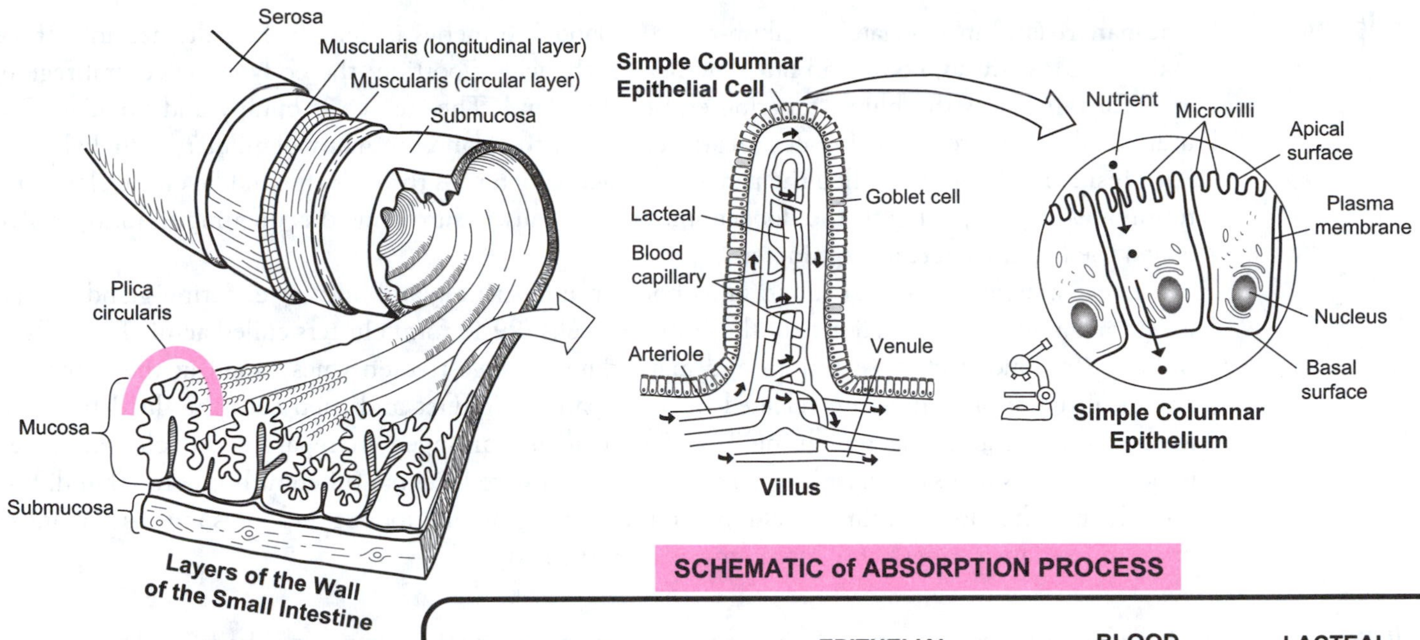
Serosa
Muscularis (longitudinal layer)
Muscularis (circular layer)
Submucosa
Plica circularis
Mucosa
Submucosa
Layers of the Wall of the Small Intestine
Simple Columnar Epithelial Cell
Lacteal
Blood capillary
Arteriole
Goblet cell
Venule
Villus
Nutrient
Microvilli
Apical surface
Plasma membrane
Nucleus
Basal surface
Simple Columnar Epithelium

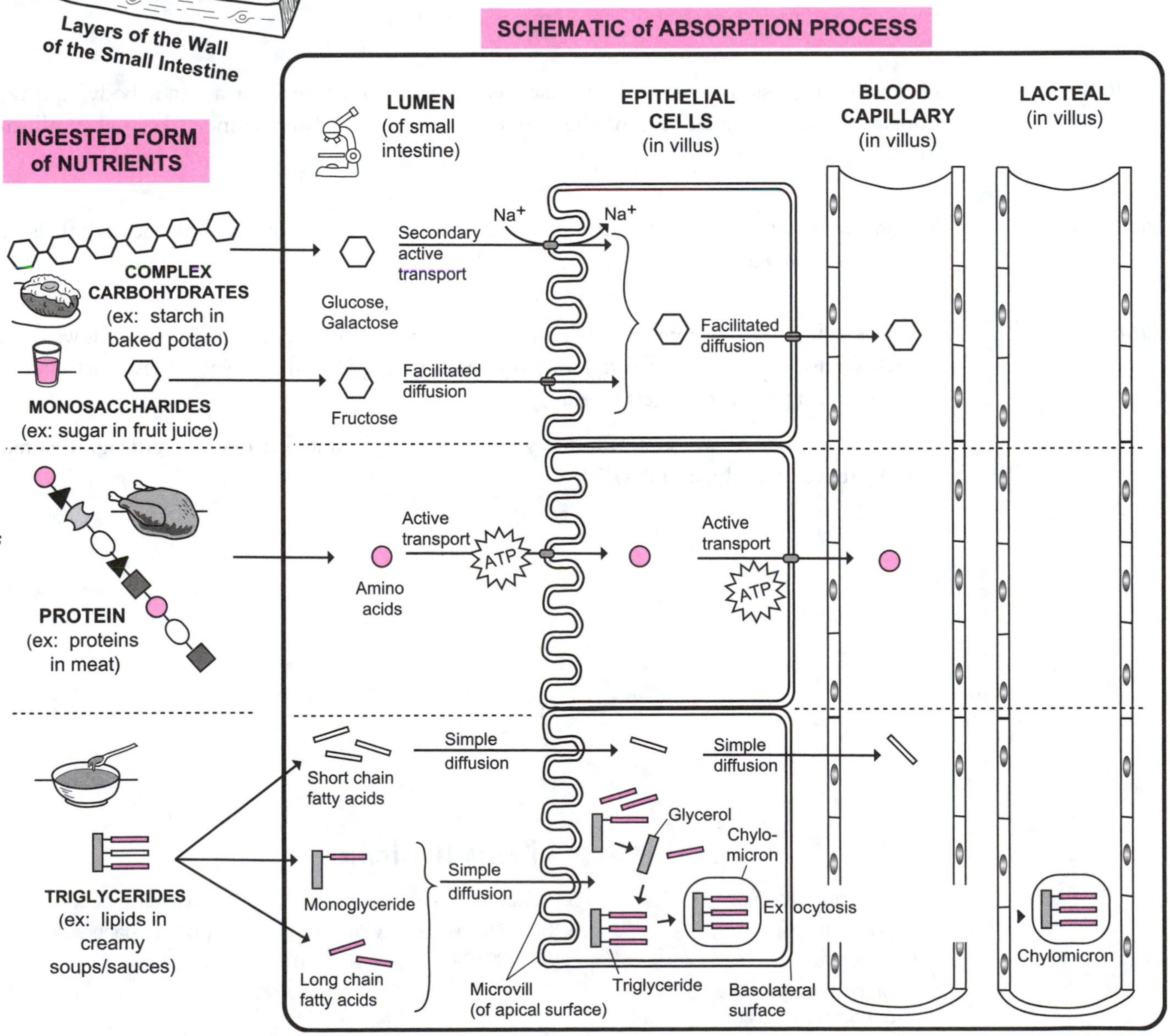
SCHEMATIC of ABSORPTION PROCESS
INGESTED FORM of NUTRIENTS
LUMEN
(of small intestine)
EPITHELIAL CELLS
(in villus)
BLOOD CAPILLARY
(in villus)
LACTEAL
(in villus)
COMPLEX CARBOHYDRATES
(ex: starch in baked potato)
MONOSACCHARIDES
(ex: sugar in fruit juice)
Na+
Na+
Secondary active transport
Glucose, Galactose
Facilitated diffusion
Fructose
Facilitated diffusion
PROTEIN
(ex: proteins in meat)
Active transport
ATP
Amino acids
Active transport
ATP
TRIGLYCERIDES
(ex: lipids in creamy soups/sauces)
Simple diffusion
Short chain fatty acids
Simple diffusion
Glycerol
Chylo-micron
Monoglyceride
Simple diffusion
Exocytosis
Long chain fatty acids
Microvill
(of apical surface)
Triglyceride
Basolateral surface
Chylomicron

Color the blood capillaries red, the lacteals green, the epithelial cells orange, and the ATP yellow.

Description

The **pancreas** is an elongated, pinkish-colored gland 5–6 inches in length. It is divided into three main sections—**head**, **body**, and **tail**. The head is the widest portion; the body is the central region; and the tail marks the blunt, tapering end of the gland. The surface is bumpy and nodular. The individual nodes are called **lobules**. Scattered within the gland are many glandular epithelial cells. Also inside is a long tube called the pancreatic duct, which runs through the middle of the gland and terminates in an opening to the duodenum. Smooth muscle surrounds this opening to form a valve called the **hepatopancreatic sphincter**.

The pancreas has a dual function as both an endocrine gland and an exocrine gland. As an exocrine gland, the vast majority of the epithelial cells (99%) form clusters called **acini**. The cells in the acini produce watery secretions and many different digestive enzymes to aid in the process of chemical digestion. These are released into the pancreatic duct and then into the duodenum. As an endocrine gland, the remaining 1% of its cells are arranged in separate cell clusters called pancreatic islets (*islets of Langerhans*). These various cells produce four different hormones that diffuse directly into the bloodstream, travel to various target organs, and induce responses in those organs in order to regulate a variety of different processes in the body.

Analogy

The general **gross anatomy of the pancreas** is like a **tadpole** with a head, body, and tail in one elongated shape. The **surface of the pancreas** has a nodular appearance like that of **alligator skin**. However, its texture is not as coarse.

Location

Abdominal cavity; the head of the pancreas is found near the duodenum and its tail is below and behind the stomach

Function

- For the **digestive system**—Cells produce a wide variety of enzymes in the following categories: **carbohydrases** *(digest carbohydrates)*, **lipases** *(digest lipids)*, **nucleases** *(digest nucleic acids such as DNA)*, and **proteases** *(digest proteins)*
- For the **endocrine system**—Produces the following hormones: **insulin**, **glucagon**, **somatostatin**, and **pancreatic polypeptide** (PP)

Key to Illustration

1. Common bile duct
2. Duodenal papilla
3. Duodenum
4. Droplet of either bile or pancreatic juice
5. Hepatopancreatic sphincter
6. Smooth muscle in wall of duodenum
7. Pancreatic duct
8. Body of pancreas
9. Lobules

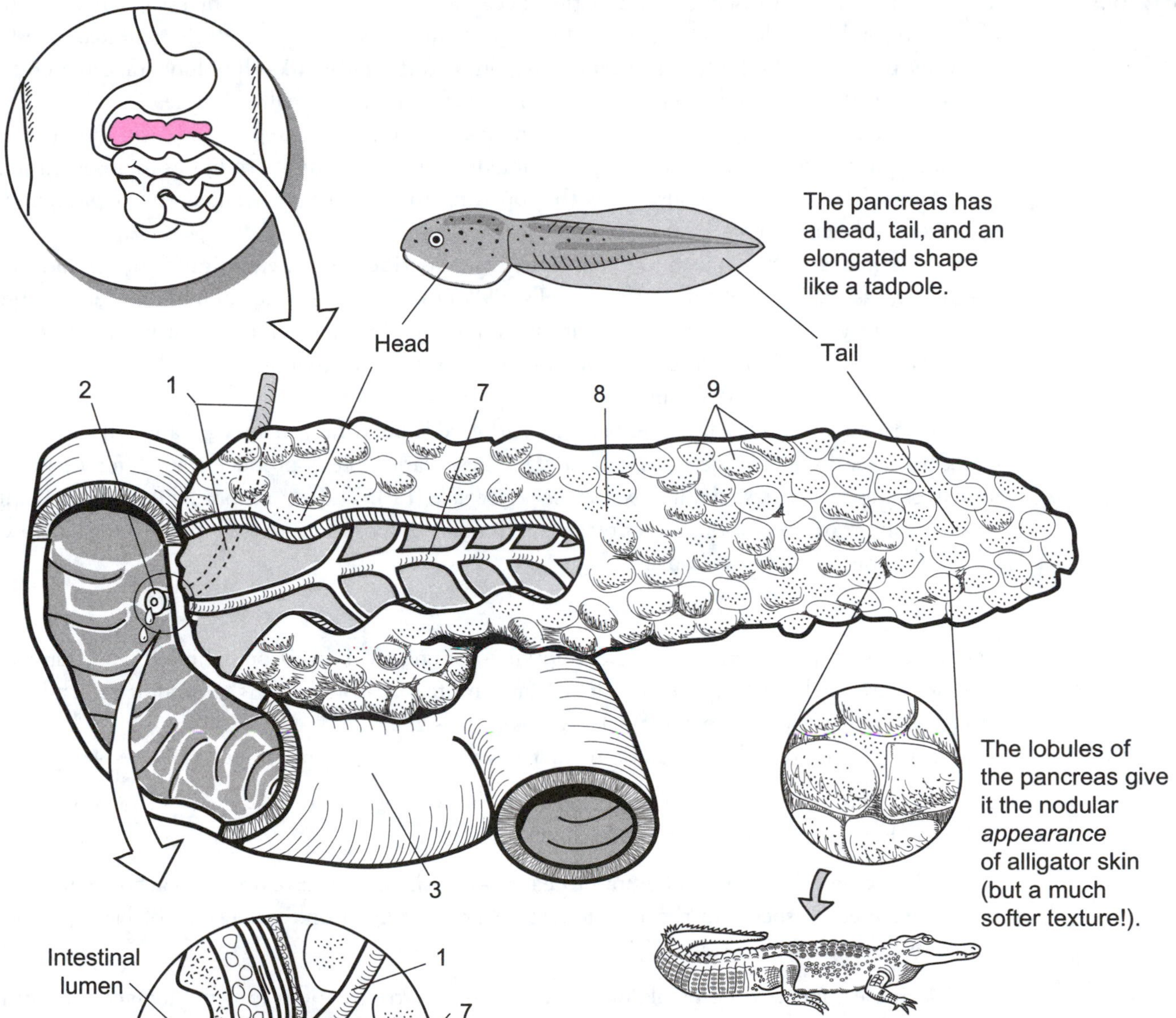
The pancreas has a head, tail, and an elongated shape like a tadpole.
Head
Tail
2
1
7
8
9
3
The lobules of the pancreas give it the nodular *appearance* of alligator skin (but a much softer texture!).

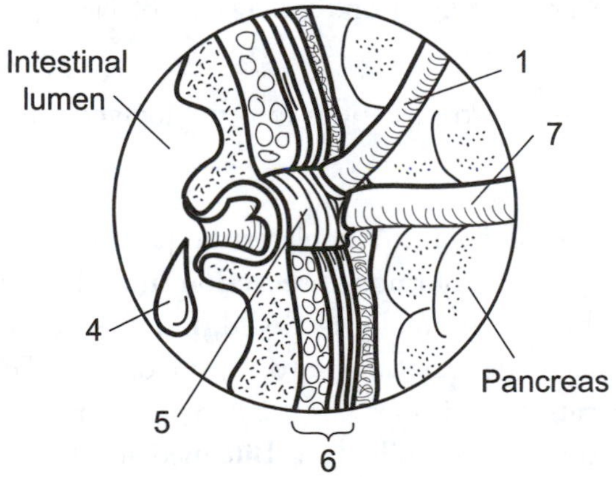
Intestinal lumen
1
7
4
5
6
Pancreas

Hepatopancreatic sphincter

1. ______________________________
2. ______________________________
3. ______________________________
4. ______________________________
5. ______________________________
6. ______________________________
7. ______________________________
8. ______________________________
9. ______________________________

Description

The **pancreas** is vital to chemical digestion because it produces so many digestive enzymes. It is an elongated, pinkish-colored gland divided into three main sections: *head*, *body*, and *tail*. The head is the widest portion; the body is the central region; and the tail marks the blunt, tapering end of the gland. The surface is bumpy and nodular. The individual nodes are called **lobules**.

Scattered within the gland are many glandular epithelial cells. Also inside is a long tube called the pancreatic duct, which runs through the middle of the gland and terminates in an opening to the duodenum. Smooth muscle surrounds this opening to form a valve called the **hepatopancreatic sphincter** (*sphincter of Oddi*).

The pancreas has a dual function as both an **endocrine** and an **exocrine** gland. As an exocrine gland, the vast majority of the epithelial cells (99%) form clusters called **acini**. The cells in the acini produce watery secretions including bicarbonate ions and many different digestive enzymes to aid in chemical digestion. These are released into the **pancreatic duct**, through the hepatopancreatic sphincter, and into the duodenum.

As an endocrine gland, the remaining 1% of cells are arranged in separate cell clusters called called **pancreatic islets** (*islets of Langerhans*). These various cells produce four different hormones: *insulin*, *glucagon*, *somatostatin*, and *pancreatic polypeptide* (PP). These hormones diffuse directly into the bloodstream, travel to various target organs, and induce responses in those organs to regulate a variety of different processes in the body.

Function

Pancreatic acini produce many different enzymes for chemical digestion. These include **amylase** (for complex carbohydrates), **lipase** (for lipids), and **proteases** (for proteins). In addition, the pancreas produces **nucleases** (for digesting nucleic acids such as DNA).

Let's examine the role of each of the first three enzyme types mentioned.

1. **Amylase** catalyzes the initial step in carbohydrate digestion, which is the conversion of a polysaccharide (ex: starch) into disaccharides.
2. **Proteases**: The three different proteases—**trypsin**, **chymotrypsin**, and **carboxypeptidase**—all are used to speed up the middle step in protein digestion: conversion of large polypeptides into small polypeptides.
3. **Lipase** catalyzes the breakdown of lipids such as triglycerides into **monoglycerides** and **fatty acids**.

As chyme passes from the stomach into the duodenum, it stimulates the pancreas to make and release digestive enzymes. This process is regulated by hormones **secretion** and **cholecystokinin (CCK)** (see p. 370). The hepatopancreatic sphincter normally is closed, but the hormone CCK causes it to open, which allows the release of pancreatic secretions into the duodenum. These secretions then mix with the chyme so the enzymes can do their jobs. As the hepatopancreatic sphincter opens, it also allows for the release of bile from the common bile duct. **Bile** mediates the first step in lipid digestion. It is needed to emulsify the lipids in the chyme so the lipases can work.

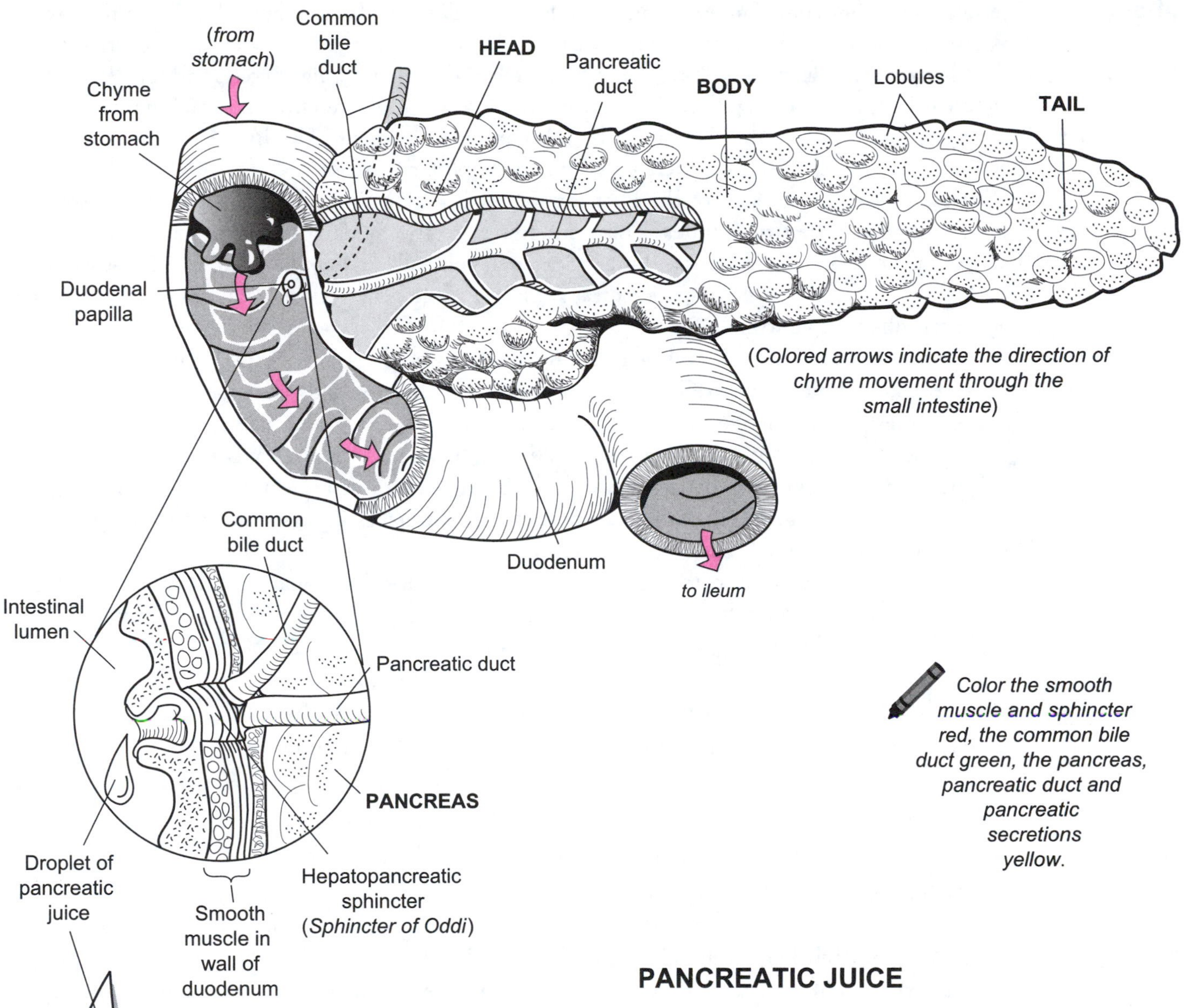

(Colored arrows indicate the direction of chyme movement through the small intestine)

Color the smooth muscle and sphincter red, the common bile duct green, the pancreas, pancreatic duct and pancreatic secretions yellow.

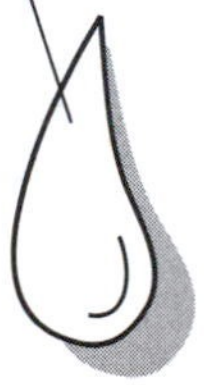

Pancreatic juice contains digestive enzymes such as **AMYLASE**, **PROTEASES**, and **LIPASE**.

PANCREATIC JUICE

AMYLASE for complex carbohydrates

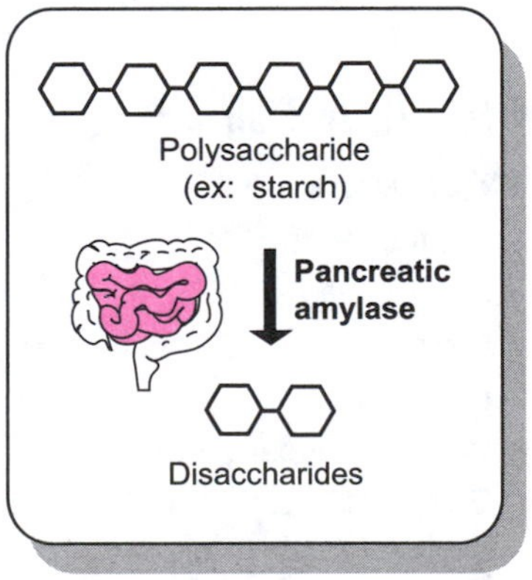

PROTEASES for proteins

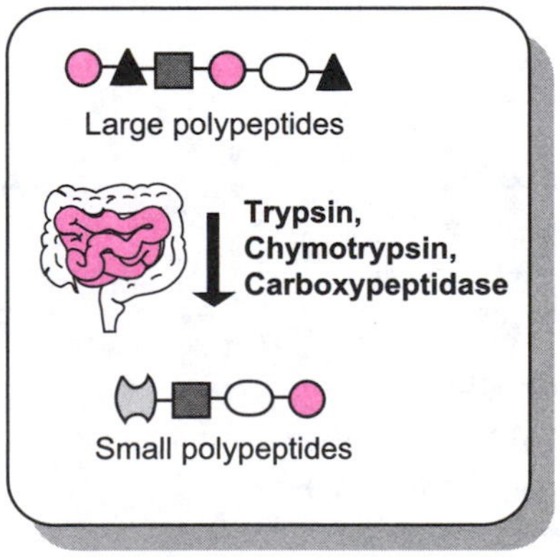

LIPASE for lipids

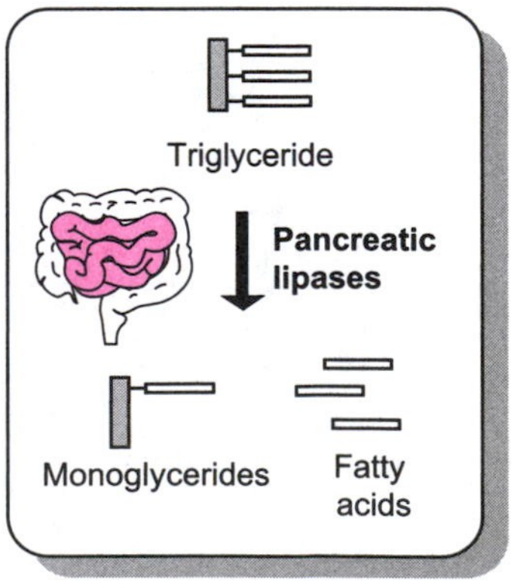

Description

The **liver** is the largest abdominal organ and is located below the diaphragm in the abdominal cavity. It is divided into two major lobes—left and right—that are separated by a sheet of connective tissue called the **falciform ligament**. Along the bottom edge of the falciform ligament is a fibrous, cable-like structure called the **round ligament** *(ligamentum teres)*, which is derived from the fetal umbilical vein. The liver is anchored to the diaphragm by the **coronary ligament**. The liver also has two minor lobes—**caudate** and **quadrate**—which are visible in the posterior view.

Microscopically, each lobe of the liver is composed of more than 100,000 individual hexagon-shaped units called **hepatic lobules**. These structures are the fundamental functional units of the liver. At each corner of a lobule is a cluster of three vessels called a **hepatic triad**, which consists of a hepatic artery, a branch of the hepatic portal vein, and a bile duct. At the center of each lobule is a long vessel called the **central vein.**

Radiating out from the central vein like spokes from a wheel are the permeable, blood-filled capillaries called **sinusoids**. Fixed to the inner lining of the sinusoids are phagocytic cells called **Kupffer's** cells, which remove most of the bacteria from the blood and digest damaged red blood cells. The hepatic artery and hepatic portal vein deliver blood into the sinusoids and it travels toward the central vein, then into the hepatic veins, and into the inferior vena cava. Most **hepatocytes** (*liver cells*) are located close to the blood in the sinusoids, which makes it easier to deposit products into the blood or screen materials out of it.

Bile is produced by hepatocytes and drains into vessels called **bile canaliculi**, then into the larger **bile ducts**, which transport the bile away from the liver and into the gallbladder for storage. Note that the direction of bile flow in the bile canaliculi is opposite to that of the blood in the sinusoids.

Functions

The liver has more than 200 different functions. Major functions of the liver are:

- **Synthesis** ex: plasma proteins, clotting factors, bile, cholesterol
- **Storage** ex: iron (Fe), glycogen, blood, fat-soluble vitamins
- **Metabolic** ex: convert glucose to glycogen and glycogen to glucose convert carbohydrates to lipids, maintain normal blood glucose levels
- **Detoxification** ex: alcohol and other drugs

Analogy

In cross-section, each **hepatic lobule** is like a simple **Ferris wheel** with six swinging chairs. Each **swinging chair** is a **hepatic triad**. The **hub** of the Ferris wheel is the **central vein**, and the **spokes** of the Ferris wheel are like the **sinusoids.**

Key to Illustration

Liver Lobes (L)

- L1 Left major
- L2 Right major

Ligaments (L)

- RL Round ligament (*Ligamentum teres*)
- FL Falciform ligament
- CL Coronary ligament

Lobule Structures

- CV Central vein
- HT Hepatic triad
- BC Bile canaliculus
- BD Bile duct
- HA Branch of hepatic artery
- HPV Branch of hepatic portal vein
- S Sinusoid
- H Hepatocyte
- K Kupffer's cell

Other

- GB Gallbladder

Structure and Function

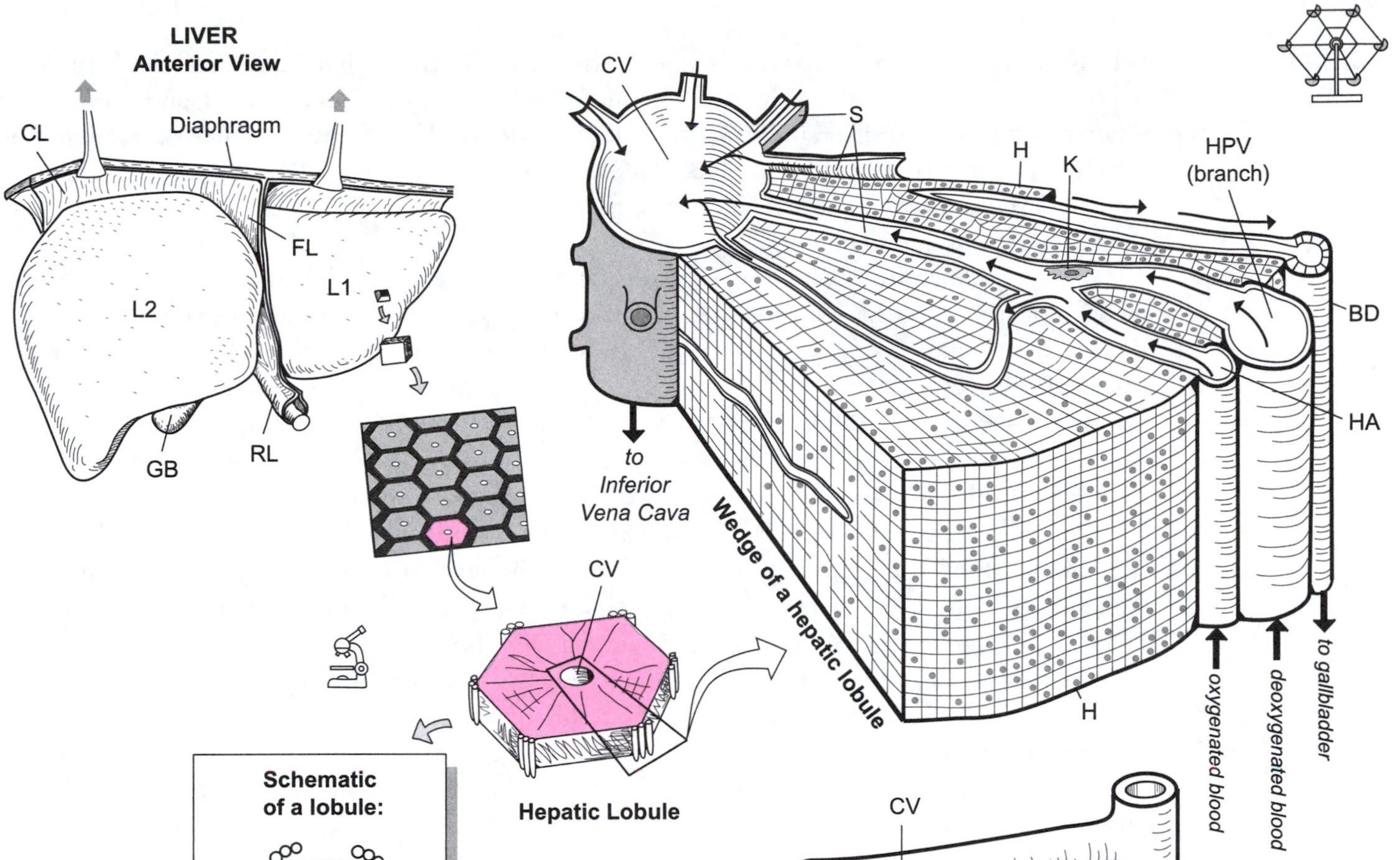

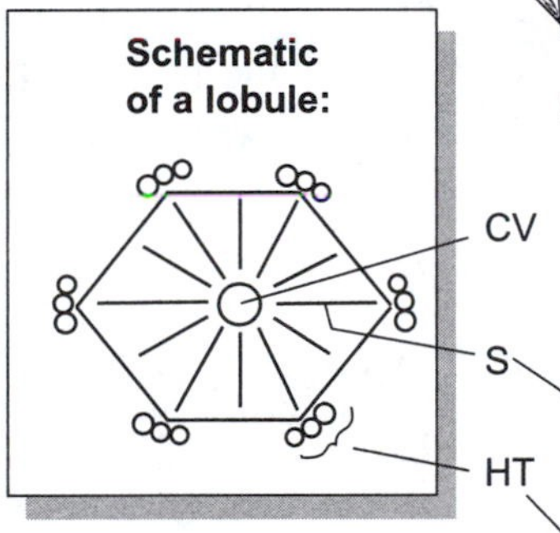

Color the bile canals green, the hepatic cells yellow, the sinusoids blue, the hepatic arteries red, and all the veins blue.

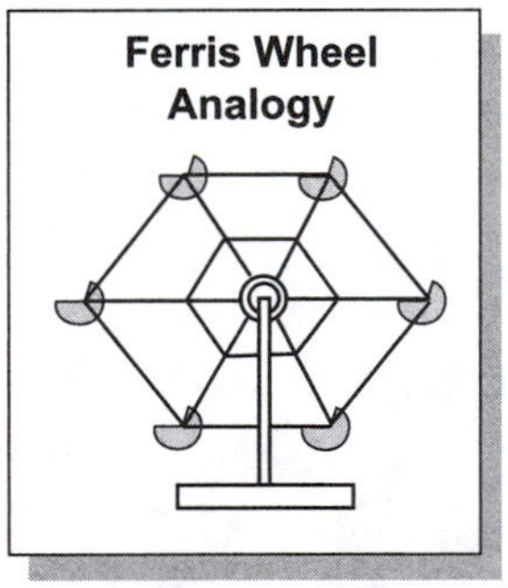

The schematic of a hepatic lobule resembles a Ferris wheel.

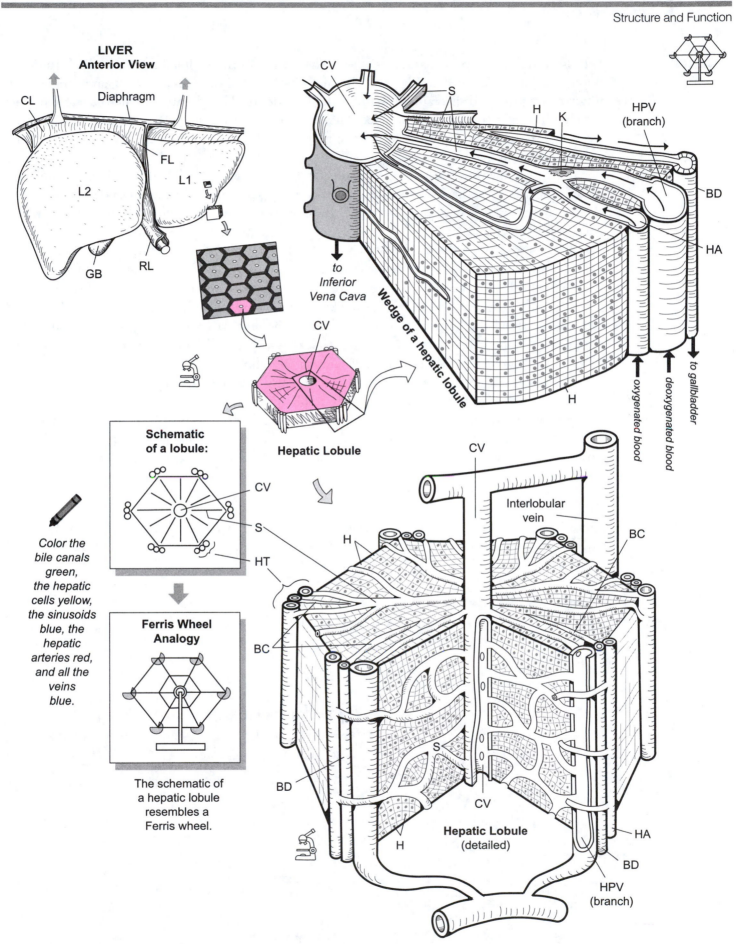

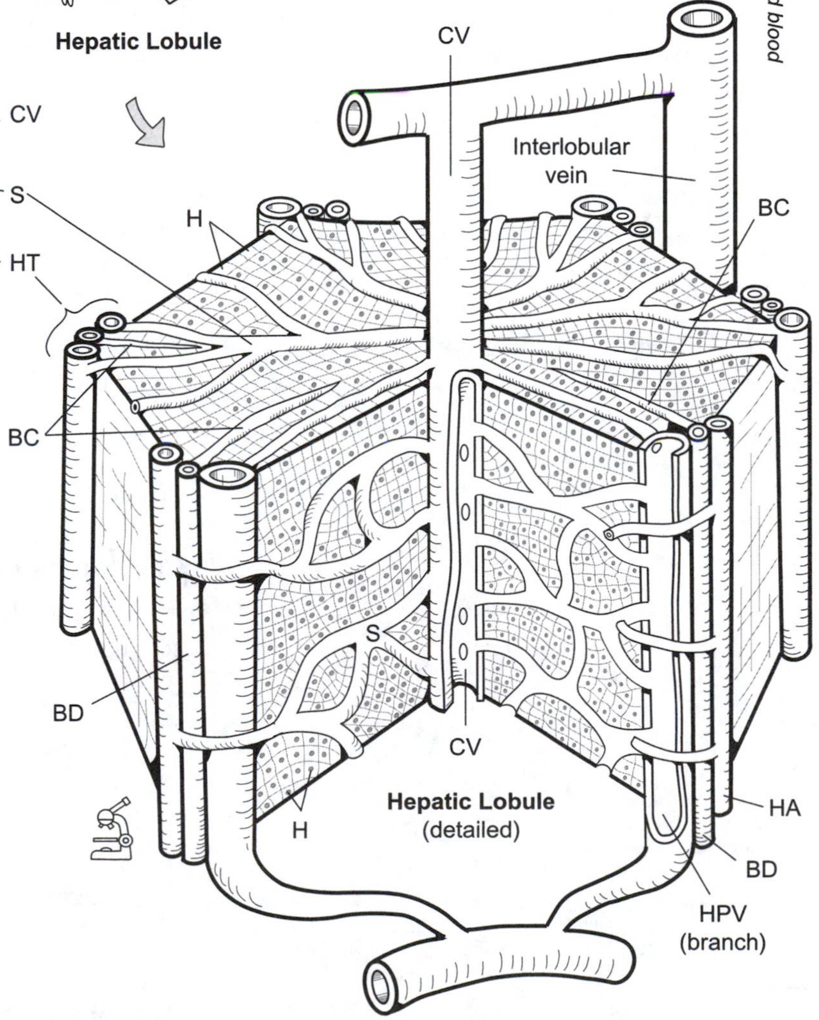

Bile Production, Storage, and Release

Description

Bile is produced by the **hepatocytes** (*liver cells*), stored in the **gallbladder**, and released into the **duodenum**. It is composed mostly of water, some ions, bilirubin (protein pigment), and various lipids. After being produced, it drains away from the hepatic lobule and enters a microscopic **bile duct**, which drains into a larger **hepatic duct**. The **left hepatic duct** drains bile from the left lobe of the liver, and the **right hepatic duct** drains bile from the right lobe of the liver. The two hepatic ducts then fuse to form a **common hepatic duct**.

Bile flows down the common hepatic duct and into the **common bile duct**, which terminates in an opening leading to the duodenum. Around this opening is a thickened band of smooth muscle called the **hepatopancreatic sphincter**. Normally the smooth muscle around this sphincter is contracted, causing it to be closed.

Bile is stored in the **gallbladder**, which is located under the right lobe of the liver. To fill the gallbladder with bile, the hepatopancreatic sphincter must remain closed. As bile drains from the liver, it backs up into the common bile duct and enters the **cystic duct**, which leads to the gallbladder.

Emptying of the gallbladder is regulated by a hormone called **cholecystokinin**, or **CCK**. As lipid-rich **chyme** (food and gastric juice) passes through the **mucosa** (inner lining) of the duodenum, it stimulates specific cells in the duodenum to release CCK. This hormone then travels through the bloodstream and targets the smooth muscle around the gallbladder.

These smooth muscle cells have receptors for CCK. Once CCK binds to these receptors, it induces the smooth muscle to contract. The result is that bile is expelled forcefully from the gallbladder. Simultaneously, CCK targets the smooth muscle of the hepatopancreatic sphincter and causes it to relax, opening the sphincter. With the sphincter open, bile moves out of the gallbladder, down the cystic duct, down the common bile duct, past the hepatopancreratic sphincter, and into the duodenum.

Function

Emulsification of lipids (fats) is the first step in the chemical digestion of lipids. Rather than breaking chemical bonds, bile physically breaks large masses of lipids into smaller lipid droplets. This makes it easier for enzymes such as **lipase** to facilitate chemical digestion.

Key to Illustration

1. Right hepatic duct
2. Left hepatic duct
3. Common hepatic duct
4. Cystic duct
5. Common bile duct
6. Pancreatic duct

Bile Production, Storage, and Release

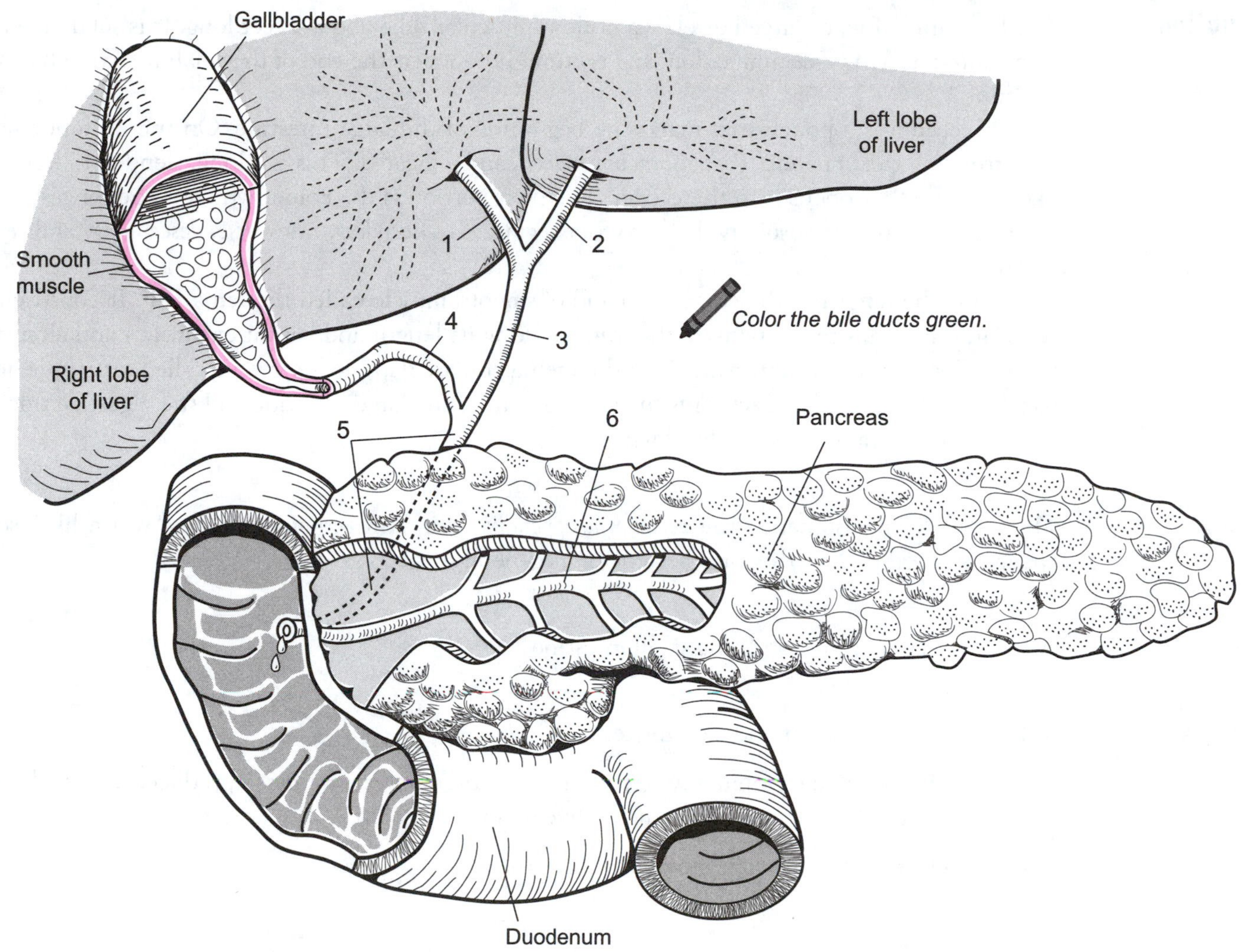

Color the bile ducts green.

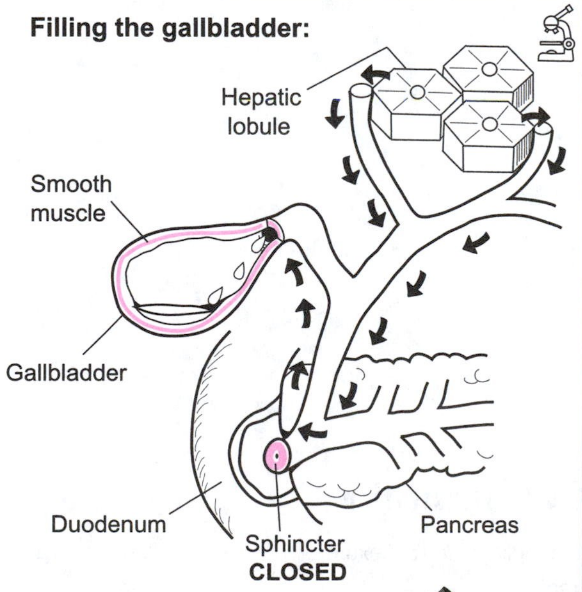

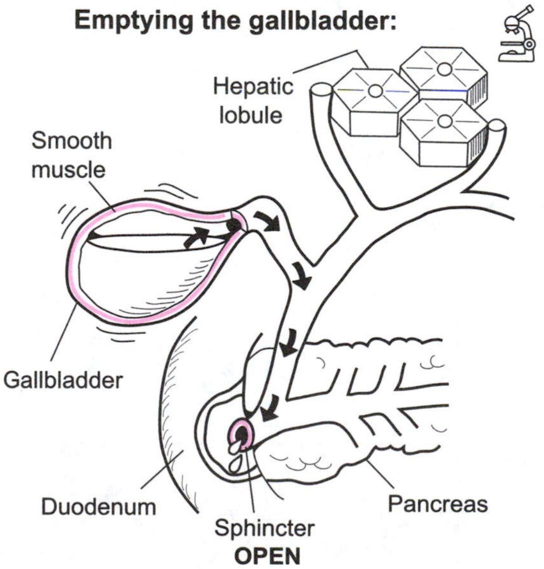

Use green to color the pathways indicated by the black arrows.

Description

The **large intestine**, or large bowel, is a hollow, muscular tube about 5 feet long. It is subdivided into three different parts—**cecum**, **colon**, and **rectum**. It connects the end of the small intestine (*ileum*) to the **anus**.

The **cecum** is a pouch that marks the beginning of the large intestine. On the posterior side of this structure, the **appendix** (*vermiform appendix*) can be located. This is a long, slender, hollow tube-like structure that opens into the cecum. The next portion is the **colon**, which is the longest part of the large intestine. It is subdivided into four segments—*ascending*, *transverse*, *descending*, and *sigmoid* colon.

Along the length of the colon is a band of smooth muscle called the **tenia coli**. It constricts the colon into pouches called **haustra** that run all along its length and permit expansion and elongation of the colon. At regular intervals along the taenia coli are flaps of fatty tissue called **fatty appendices** (*epiploic appendages*). The **rectum** is the last segment of the large intestine and the digestive tract and allows for temporary storage of fecal waste.

Analogy

The **appendix** is also called the *vermiform appendix*. The term *vermiform* means "worm-like" so the appendix is compared to a **worm** because of its general shape.

Location

Abdominal cavity; surrounds the small intestine

Functions

- **Reabsorption** of water and electrolytes
- **Absorption** of some vitamins (*e.g.*, vitamin K, B-complex vitamins) produced by the bacteria *Escherichia coli (E. coli)*, which naturally live within the colon.
- **Compaction and temporary storage** of fecal waste

Key to Illustration

1. Cecum
2. Right colic *(hepatic)* flexure
3. Haustra
4. Left colic *(splenic)* flexure
5. Tenia coli
6. Fatty appendages
7. Ileocecal valve
8. Rectum
9. Anus

Visualizing the General Structure

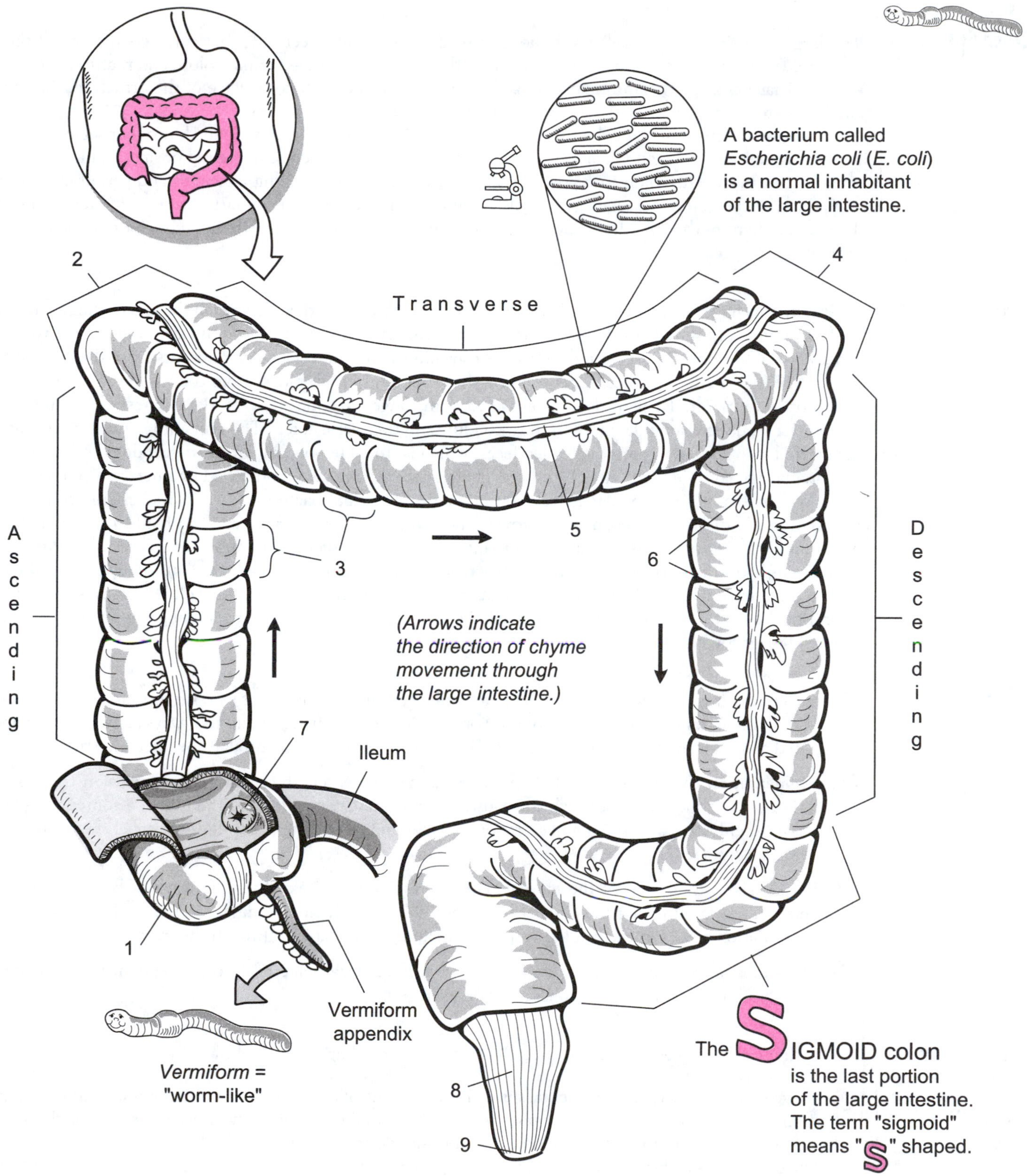
A bacterium called *Escherichia coli* (*E. coli*) is a normal inhabitant of the large intestine.
2
4
Transverse
Ascending
Descending
3
5
6
7
Ileum
(Arrows indicate the direction of chyme movement through the large intestine.)
1
Vermiform appendix
Vermiform = "worm-like"
8
9
The SIGMOID colon is the last portion of the large intestine. The term "sigmoid" means "S" shaped.

1. ______________________
2. ______________________
3. ______________________
4. ______________________
5. ______________________
6. ______________________
7. ______________________
8. ______________________
9. ______________________

Description

The **large intestine** (*large bowel*) is a hollow, muscular tube about 5 feet long that connects the end of the small intestine (*ileum*) to the anus. It is subdivided into three different parts—**cecum**, **colon**, and **rectum**. The cecum is a pouch that marks the beginning of the large intestine. On the posterior surface of this structure, the **appendix** can be located. This is a long, slender, hollow tube that opens into the cecum.

The next portion is the colon, the longest part of the intestine. It is subdivided into four segments—*ascending, transverse, descending,* and *sigmoid* colon. Along the length of the colon is a band of smooth muscle called the **tenia coli**. It bulges the colon into pouches called **haustra**, which run along its length and permit expansion and elongation of the colon. At regular intervals along the tenia coli are flaps of fatty tissue called **fatty appendices**. The last segment of the large intestine and the digestive tract is the rectum, which allows for temporary storage of fecal waste.

Process

Most nutrients have been digested and absorbed into the blood by the time they reach the end of the small intestine. The remaining undigested chyme passes through the **ileocecal sphincter** and into the large intestine. Two opposing processes—absorption and secretion—occur. **Absorption** is the movement of nutrients from the large intestine into the blood and then into cells. **Secretion** is the movement of substances from the blood to the large intestine for eventual excretion. Remaining to be absorbed in the large intestine are some electrolytes, water, and vitamins. Absorption and secretion take place in the ascending and transverse colon. The remaining undigestible material is compacted into a semisolid mixture called **feces**, which is stored temporarily in the descending and sigmoid colon and finally ejected from the body. The process of eliminating the feces is controlled by a defecation reflex.

We share a mutually beneficial relationship with the various **bacteria**—such as Escherichia coli (*E.Coli)*—that live inside the large intestine. We provide them with a warm environment and food to live and reproduce while they make vitamins for us that we can't make on our own, such as vitamin K and B-complex vitamins. But bacteria also contribue to *flatus*, or gas. As they metabolize any remaining carbohydrates, they produce hydrogen, carbon dioxide, and methane gases. Another gas produced by bacteria—hydrogen sulfide—gives off a "rotten egg" odor. But the bacteria are not entirely to blame because the major source of intestinal gas is air swallowed during the rapid ingestion of food.

Motility

The **muscularis layer** of the wall of the large intestine contains two layers of **smooth muscle**—an inner, circular layer and an outer, longitudinal layer. Contraction of this smooth muscle in different ways allows for movement of undigested materials inside the large intestine. The three types of motility through the large intestine are:

1. **Segmentation** (*haustral churning*). As undigested material enters the haustra, the expansion triggers muscularis layer contractions that pinch the haustra into separate compartments and turn the contents inside. This serves to enhance absorption and secretion of water and electrolytes, as well as to propel the contents to the next haustrum.

2. **Peristalsis**. Stimulation of the muscularis layer results in a weak, rhythmic wave of contraction that is a combination of pinching the circular layer and shortening the longitudinal layer. Together, this contraction combination results in the gradual propulsion of undigested material through the tract.

3. **Mass movement**. This strong muscle contraction forces fecal material into the rectum. It is regulated by a reflex and occurs two to three times a day on average.

Absorption and Secretion

Absorption follows a predictable sequence, illustrated on the facing page, as follows:

1. Levels of **sodium** (Na^+) and **potassium** (K^+) are controlled by the **sodium-potassium pump** (see p. 58), which transports these ions in opposite directions. Sodium is actively transported out of the lumen of the large intestine and into the blood, and potassium is transported into the large intestine. The net result is that sodium is reabsorbed into the blood and potassium is secreted into the large intestine to be removed in the feces. The result is that it causes a higher solute concentration in the blood relative to the lumen of the large intestine.

2. Because **water** (H_2O) always moves passively across a plasma membrane toward the higher solute concentration via osmosis (see p. 48), this is what occurs next: The body forces osmosis by actively transporting sodium out first.

Function of the Large Intestine

GROSS ANATOMY

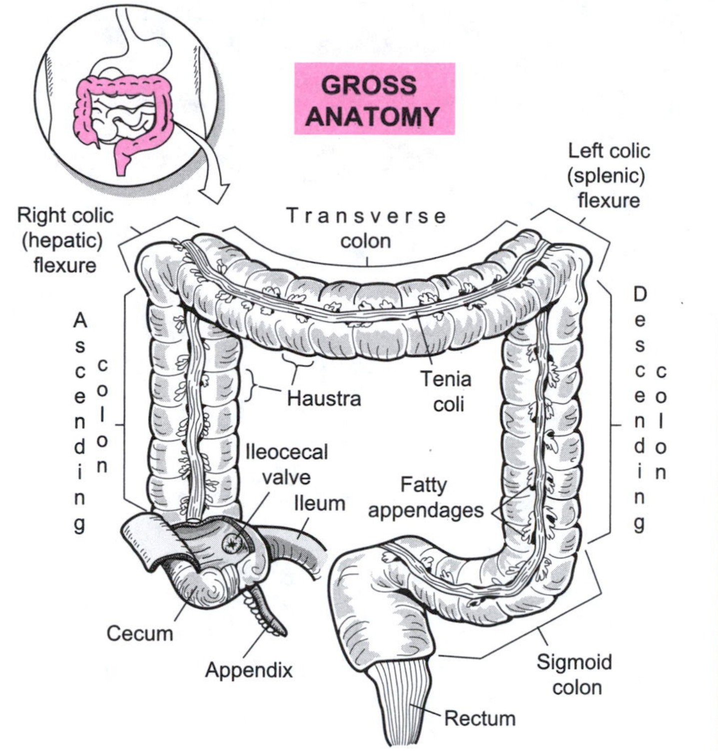

FUNCTIONAL OVERVIEW

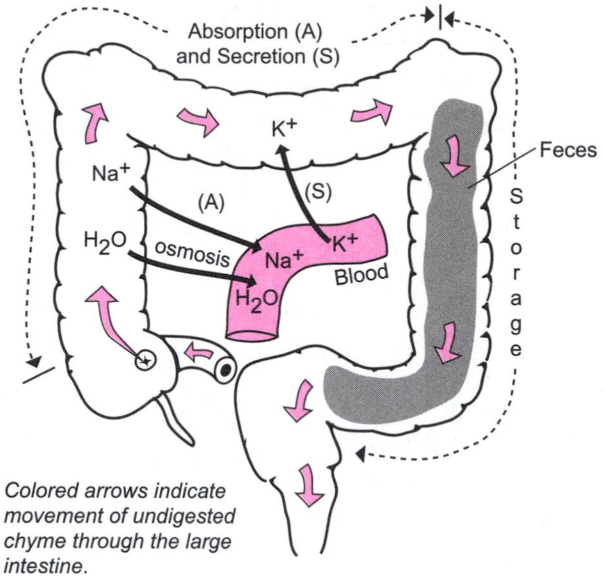

Colored arrows indicate movement of undigested chyme through the large intestine.

MOTILITY

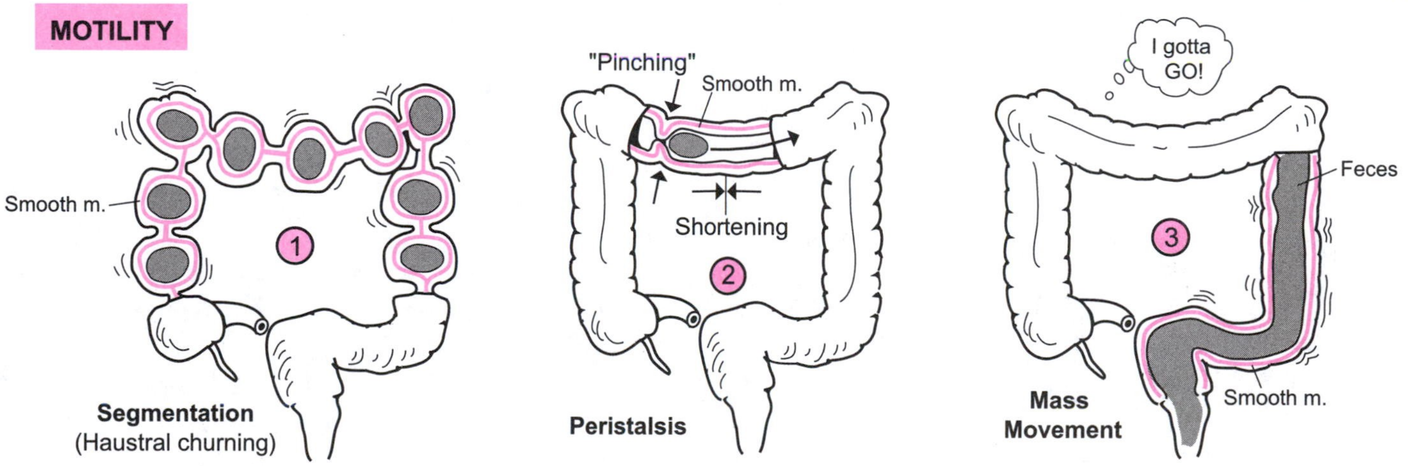

ABSORPTION (A) and SECRETION (S)

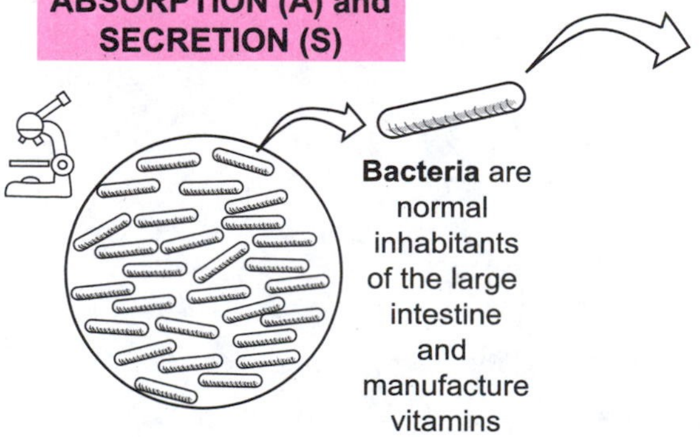

Bacteria are normal inhabitants of the large intestine and manufacture vitamins for the body

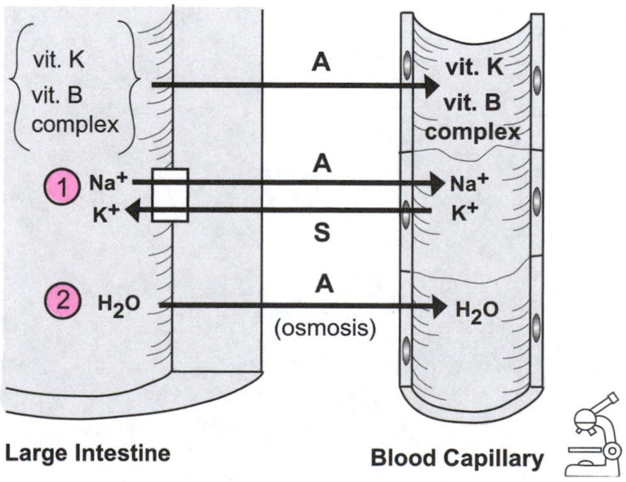

Color the large intestine brown and the blood capillary red.

Notes

Metabolic Physiology

Description

Metabolism is the sum total of all the chemical reactions within the body. It consists of two different processes—catabolic reactions and anabolic reactions. **Catabolic reactions** break down organic compounds into simpler building blocks (*e.g.*, lipids into fatty acids and glycerol) and **anabolic reactions** use these building blocks to synthesize more complex organic compounds (*e.g.*, amino acids into proteins). Energy metabolism focuses on the elements of metabolism that generate ATP. As a general overview of energy metabolism, let's examine it as having three main phases:

Phases

Phase 1: The release of nutrients in the digestive tract lumen

Let's examine the digestion of three major nutrients: **lipids**, **complex carbohydrates**, and **proteins**. As these macromolecules are ingested, they enter the digestive tract and are broken down gradually into their most fundamental building blocks:

- **Proteins** → **amino acids**
- **Complex carbohydrates** → **glucose** (and other simple sugars)
- **Lipids** (*e.g.*, triglycerides) → **monoglycerides** and **fatty acids**

This phase of catabolism is catalyzed by digestive enzymes in the digestive tract. These building blocks now are small enough for the cells to use. After being absorbed into the bloodstream through microscopic structures called **villi** on the inner lining of the small intestine, they are transported to various cells throughout the body.

Phase 2: The fate of nutrients within cells of various tissues

Glucose has three possible fates:

1. **Stored:** In the form of **glycogen**, this is a long chain of glucose molecules. Skeletal muscle cells and liver cells store lots of glycogen when excess glucose is present in the blood.
2. **Catabolized:** When the body requires energy, glucose is catabolized to produce ATP molecules. The three stages of glucose catabolism are: glycolysis, the citric acid cycle, and electron transport.
3. **Converted:** Glucose can be converted into several **amino acids**, which then are used to make proteins. Alternatively, when storage of glycogen is maximized, liver cells can convert glucose into **glycerol** and **fatty acids** to form **triglycerides**. The triglycerides then are deposited in fatty tissue.

Fatty acids and **glycerol** have one of two general fates:

1. **Anabolized:** Glycerol and fatty acids recombine to form lipids within liver cells and fat cells.
2. **Converted:**
   a. **Glycerol** can be converted into glyceraldehyde-3-phosphate—one of the compounds formed during the process of glycolysis. Then, depending on the ATP levels in the cell, it is converted into either **pyruvic acid** (via glycolysis) or **glucose** (via gluconeogenesis).
   b. **Fatty acids** follow a different pathway than glycerol. They can be converted into **acetyl CoA**, which then enters the citric acid cycle to produce ATP.

Amino acids have three different fates:

1. **Anabolized:** Amino acids may be converted back into **proteins**.
2. **Converted:** Some amino acids can be converted into **acetyl CoA** and enter the citric acid cycle to produce ATP. When excess amounts of amino acids are present, they can be converted into either **glucose** (via gluconeogenesis) or **fatty acids.**
3. **Deaminated:** Liver cells can remove the amino group from some amino acids to produce carboxylic acids and ammonia (NH_3), a process called *deamination*. Then the carboxylic acids can enter different stages of the citric acid cycle.

Phase 3: Aerobic catabolism of nutrients in the mitochondria of cells

Aerobic respiration refers to the two parts of cellular respiration that occur inside the mitochondrion—citric acid cycle (see p. 406) and the electron transport system (see p. 408). **Acetyl CoA** is a key chemical intermediate in metabolism and is generated from fatty acids, pyruvic acid, and some amino acids within the mitochondrion. Once acetyl CoA is produced, it may enter the citric acid cycle as part of normal catabolism.

Trace the pathway for lipids, carbohydrates, and proteins, using a different color for each.

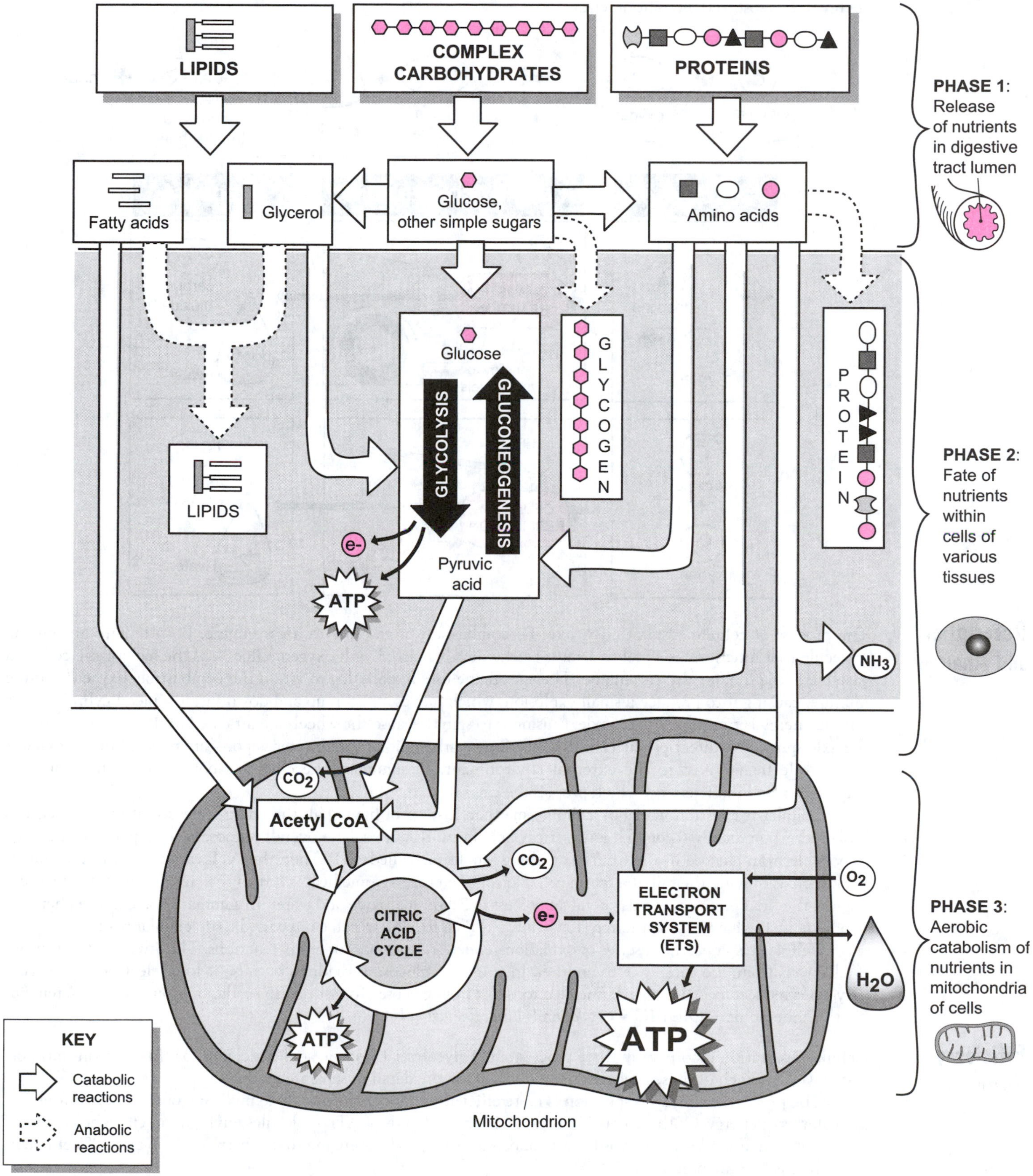

Equations

In the process of **cellular respiration**, one molecule of **glucose** reacts with six **oxygen** molecules through a series of chemical reactions to produce six **carbon dioxide** molecules, six **water** molecules, 36 molecules of **ATP** and heat energy that is lost to the environment.

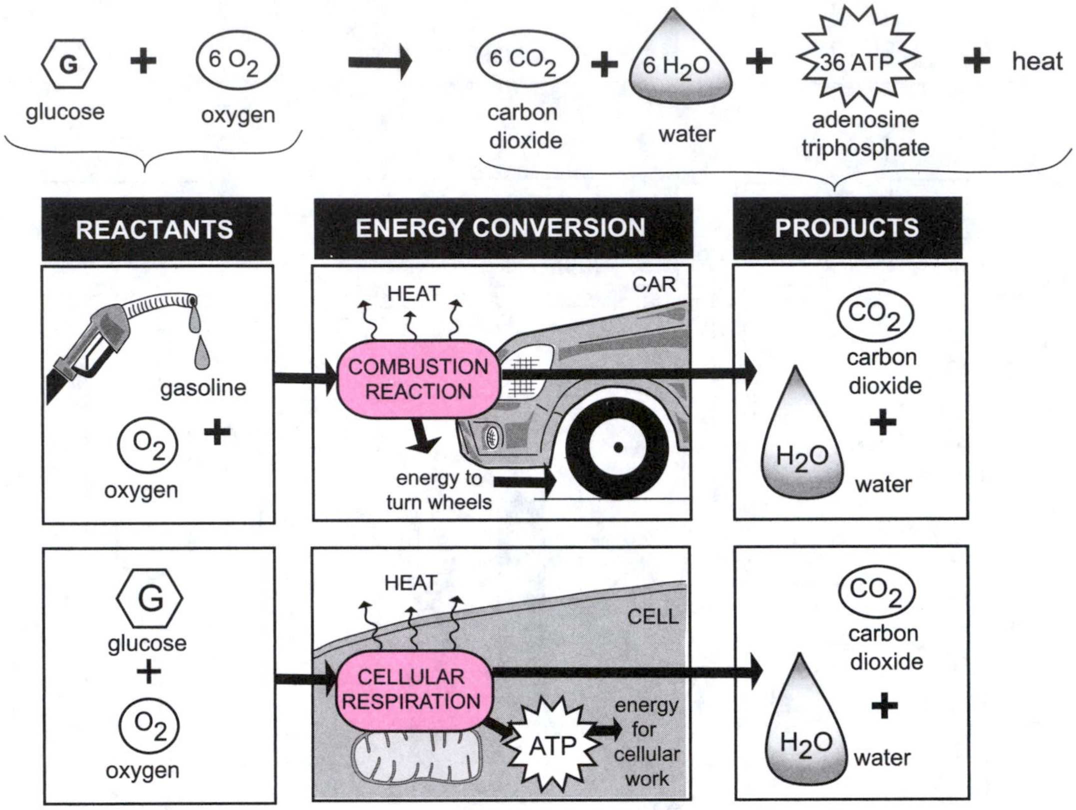

Description and Analogy

The process of cellular respiration is like the combustion of gasoline in a car engine. Both require an organic molecule that functions as a fuel that must be chemically reacted with oxygen. Glucose is the fuel for the cell, and gasoline is the fuel for the car engine. The car engine uses a sparkplug to ignite the combustible oxygen/gasoline mixture, causing it to produce a small explosion, which moves a piston up and down inside a metal cylinder. This kinetic energy is transferred to an axle, causing it to spin. Because the wheels are attached to the axle, they spin as the axle spins. The direct products of this reaction—carbon dioxide and water vapor—are released in the exhaust gases while heat is lost to the external environment. In summary, the chemical-bond energy in gasoline is converted into kinetic energy which propels the car.

Cellular respiration works in a similar manner. Instead of having all the energy released at once, glucose is broken down gradually through a long series of chemical reactions. The overall purpose is to trap as much energy as possible from glucose in the form of an energy molecule called ATP. Then the ATP can be used to do cellular work such as moving the tail of a sperm cell. During the process, some heat is lost to the environment. Like the car engine, the final products of this chemical process are carbon dioxide and water. In summary, some of the chemical bond energy in glucose is transferred slowly into ATP molecules, which can be used to do cellular work.

Cellular respiration is a series of oxidation-reduction reactions, or redox reactions. This involves the transfer of electrons from one substance to another. In this case, glucose is oxidized because it loses electrons to oxygen. Oxygen is reduced because it gains the electrons lost from glucose. To distinguish oxidation from reduction, remember this mnemonic: "OIL RIG = *Oxidation Is Loss, Reduction Is Gain.*"

Efficiency Rating

Cellular respiration incorporates three processes: (1) glycolysis, (2) citric acid cycle, and (3) the electron transport system (ETS). Each of these processes are discussed in more detail in separate modules.

The process of cellular respiration is more efficient than a car engine. One molecule of glucose contains 686 kilocalories of energy. Of this total, 278 is captured in the bonds of ATP molecules, giving it an efficiency rating of 41%. The remaining 59% is lost as heat. Compared to a typical car engine that is between 10% and 30% efficient, this rating is looking pretty good!

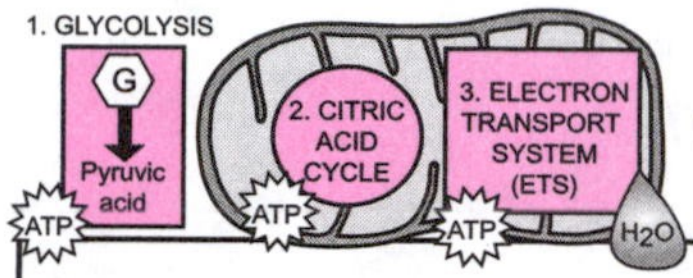

Roadmap

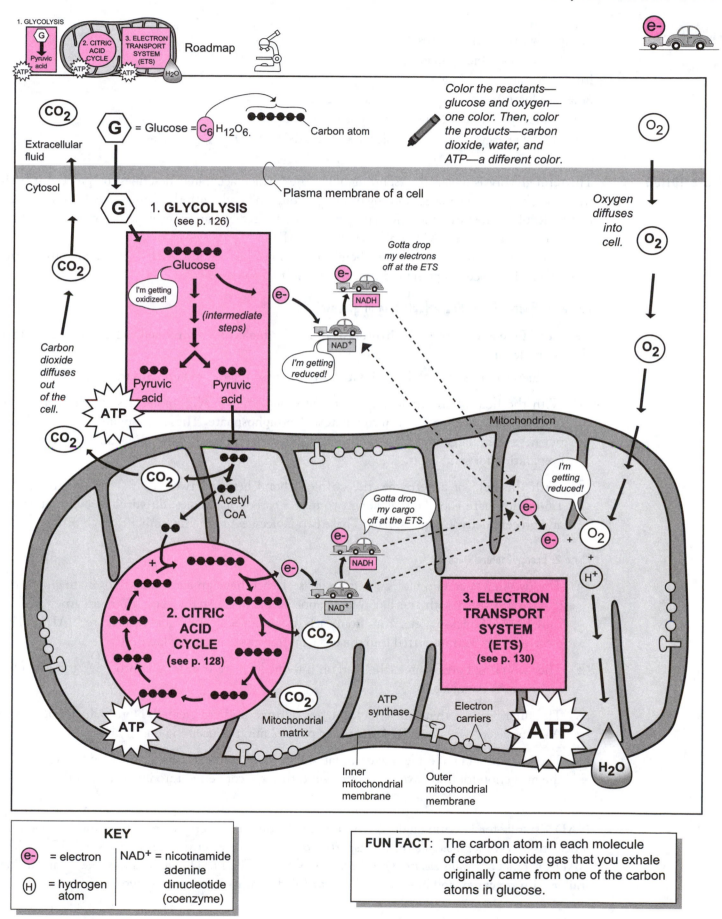

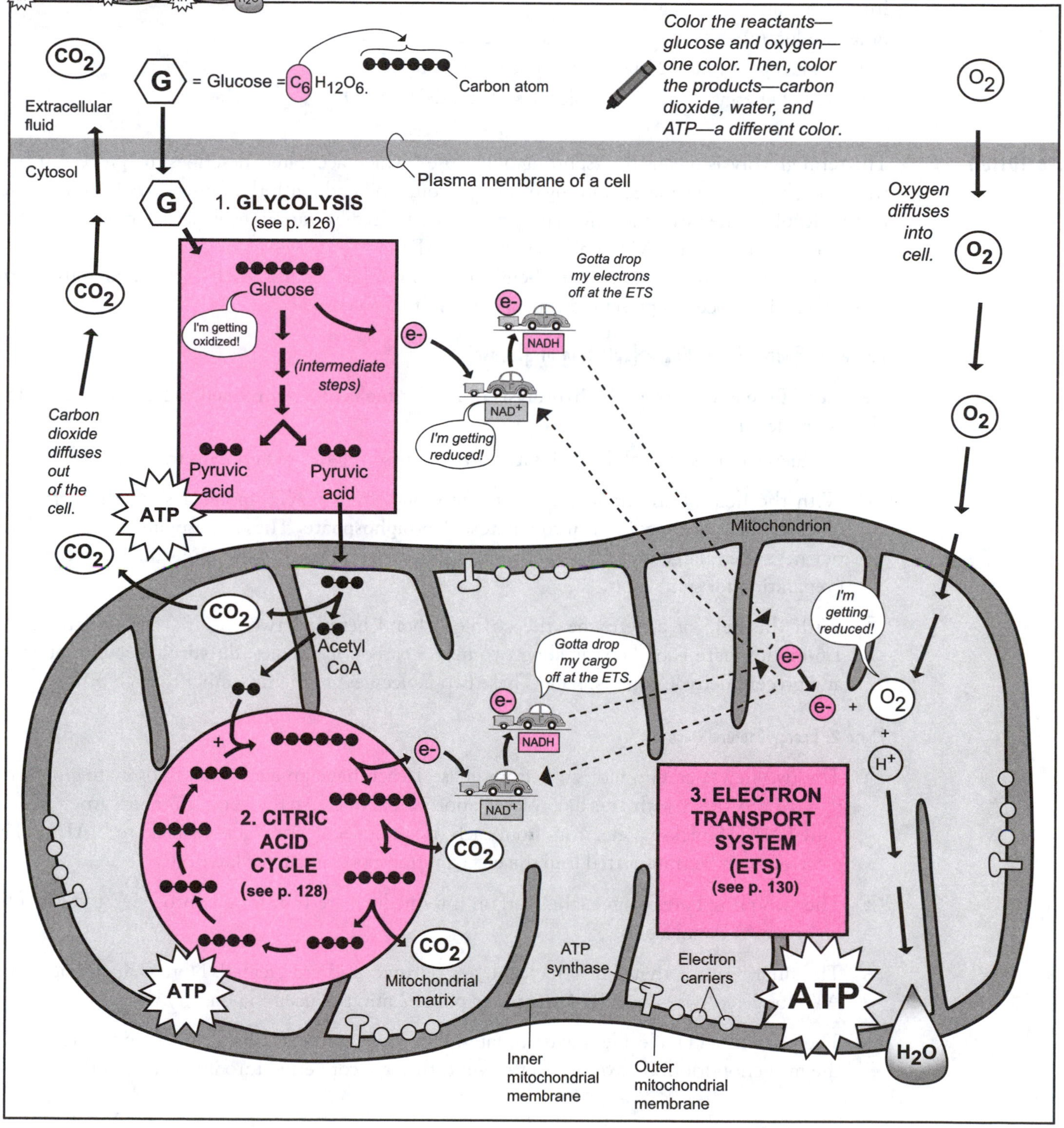

KEY

(e-) = electron

(H) = hydrogen atom

NAD^+ = nicotinamide adenine dinucleotide (coenzyme)

FUN FACT: The carbon atom in each molecule of carbon dioxide gas that you exhale originally came from one of the carbon atoms in glucose.

Overview

Location within cell:	cytosol
Aerobic or anaerobic?	anaerobic
Initial reactant:	glucose
Final product(s):	2 pyruvic acid molecules
Side products:	2 NADH (reduced coenzymes)
Net yield of energy:	2 ATP molecules (4 created; 2 used)

Description

The term **glycolysis** means the "splitting of glucose," which accurately describes the process. It begins with one glucose molecule containing 6 carbon atoms and ends with the formation of two new 3-carbon molecules called **pyruvic acid**. This process is catalyzed by various enzymes in the **cytosol** of cells and yields a net gain of 2 ATP molecules and 2 NADH.

Instead of presenting the various chemical reactions that occur in each of the numerous intermediate steps, the process is presented conceptually in two phases:

Phase 1: Energy Input Phase (splitting of glucose)

Key idea: Free energy from the hydrolysis of 2 ATP molecules is invested to help split the original glucose molecule.

- Glucose enters the cell by **facilitated diffusion** (see p. 46) and remains in the **cytosol.**

① With the help of an enzyme, 2 phosphate groups from 2 ATP molecules are transferred onto glucose, thereby changing it into **fructose-1,6-biphosphate**. These phosphate molecules have a net negative charge and repel each other. This repulsion makes the molecule more unstable in preparation for splitting it.

② With the help of an enzyme, the covalent bond between two specific carbons in fructose-1,6-biphosphate is broken, yielding two new 3-carbon molecules: **dihydroxyacetone phosphate** and **glyceraldehyde 3-phosphate**. These two molecules are highly similar to each other.

Phase 2: Energy Capture Stage

③ The two new 3-carbon molecules from phase 1 each have an additional phosphate group transferred onto them with the help of yet another enzyme. At the same time, a carrier molecule called **NAD^+** picks up electrons from each of the molecules and gets reduced to **NADH**. These electrons can be transported into the electron transport system for later use.

④ The two phosphates from each 3-carbon intermediate are transferred onto ADP to form a total of 4 new ATP molecules.

⑤ The final result is that 2 new molecules of **pyruvic acid** are created. Note that the net gain in ATP for glycolysis is 2 ATP (**4 created** in phase 2 minus **2 used** in phase 1).

⑥ The pyruvic acid molecules have two fates. If oxygen is present, they enter the citric acid cycle in the mitochondrion. If oxygen is not present, they are converted into lactic acid in the cytosol.

Analogy

NAD^+ stands for **N**icotinamide **A**denine **D**inucleotide. It is derived from the vitamin niacin and functions as a *carrier molecule* that transports electrons to the electron transport system (ETS). When NAD^+ is reduced (*gains electrons*), it becomes **NADH**. The **NAD^+** molecule is like a **car pulling a trailer with no cargo**, and **NADH** is like a **car filled with its "electron" cargo**.

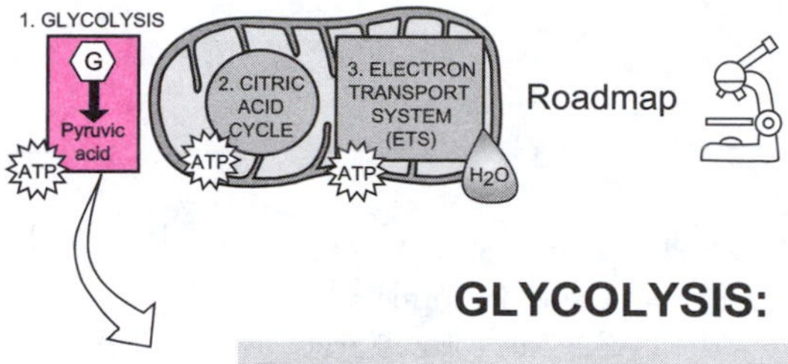

GLYCOLYSIS: CONCEPTUAL OVERVIEW

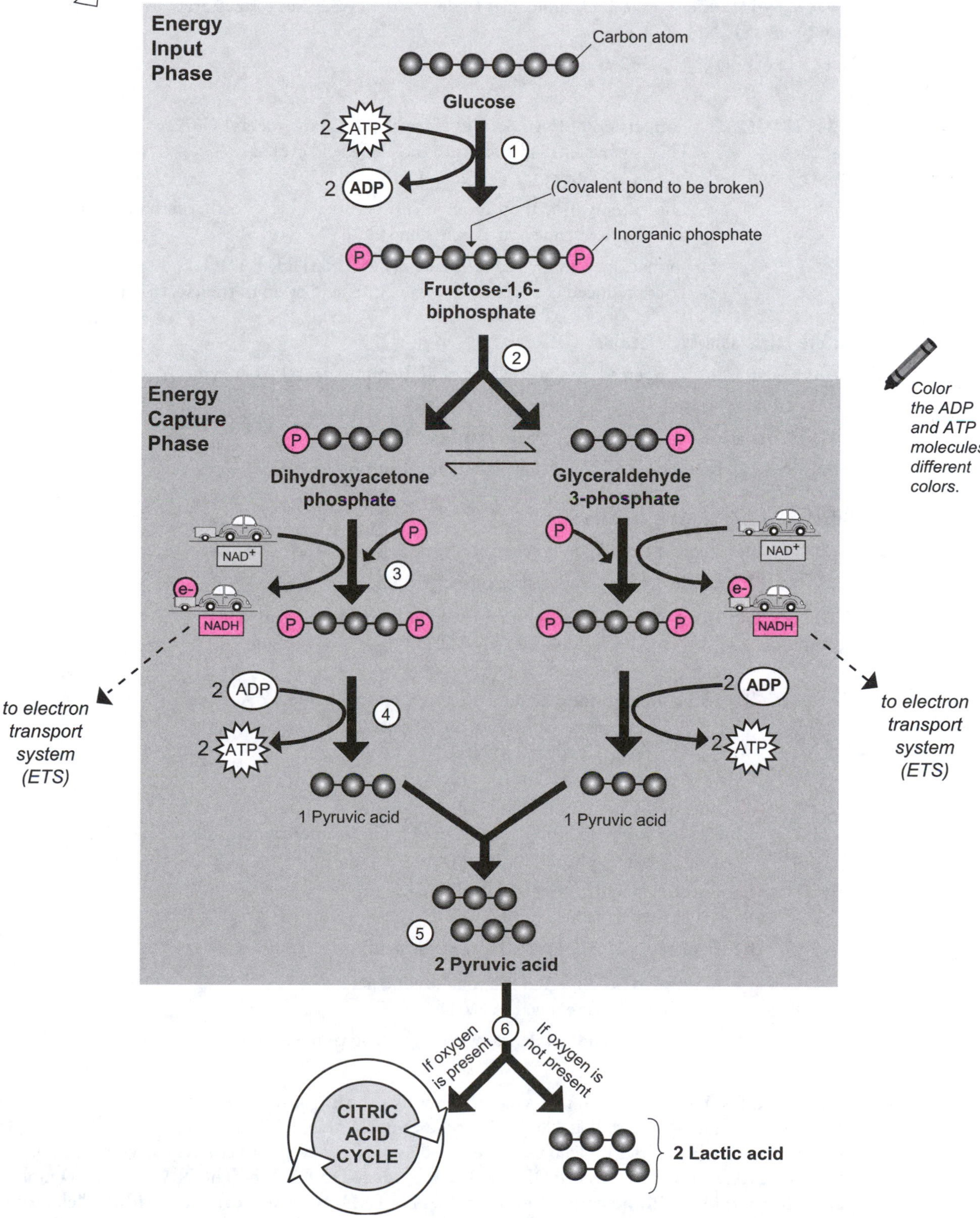

Color the ADP and ATP molecules different colors.

Overview

Location within cell:	mitochondrion
Aerobic or anaerobic?	aerobic (*indirectly*)
Initial reactant:	citric acid
Final product:	oxaloacetic acid (which condenses with acetyl CoA to form citric acid)
Side products: (*for both pyruvic acids from glycolysis*)	NADH, $FADH_2$ (10 reduced coenzymes in total) 6 CO_2 molecules (4 during the cycle; 2 in formation of acetyl CoA)
Net yield of energy:	2 ATP molecules

Description

3 Key events:

1. Conversion of **Pyruvic acid** (*from glycolysis*) into **acetyl CoA.** This is what links glycolysis to the citric acid cycle.
2. Formation of $\mathbf{CO_2}$ as a side product. The carbon in carbon dioxide can be traced back to the carbon in glucose. Think about that the next time you exhale some CO_2!
3. Formation of many **reduced coenzymes (NADH, $\mathbf{FADH_2}$).** These reduced coenzymes link the citric acid cycle to the electron transport system.

Before citric acid cycle begins:

Pyruvic acid is first converted into **acetic acid**, and then into **acetyl Coenzyme A** (CoA).

- $\mathbf{CO_2}$ is released in the process.
- **CoA** is used to convert **acetic acid** into **acetyl CoA.**
- $\mathbf{NAD^+}$ is reduced to **NADH** by picking up electrons.

Citric Acid Cycle: Step by Step:

Note: *Each of the steps in the citric acid cycle requires the help of a different enzyme.*

1. **Citric acid** is converted into **isocitric acid.**
2. **Isocitric** acid is converted into **alpha-ketoglutaric acid.**
   - NAD^+ is reduced to NADH.
   - CO_2 is released in the process.
3. **Alpha-ketoglutaric acid** is converted into **Succinyl CoA.**
   - CoA enters this step to create Succinyl CoA.
   - NAD^+ is reduced to NADH.
   - CO_2 is released in the process.
4. **Succinyl CoA** is converted into **succinic acid.**
   - Phosphate group is transferred from the energy molecule GTP to ADP to form ATP.
   - CoA is released in the process.
5. **Succinic acid** is converted into **fumaric acid.**
   - FAD is reduced to $FADH_2$.
6. **Fumaric acid** is converted into **malic acid.**
7. **Malic acid** is converted into **oxaloacetic acid.**
   - NAD^+ is reduced to NADH.
8. **Oxaloacetic acid** condenses with **acetyl coenzyme A** (CoA) to form **citric acid.**

Analogy

$\mathbf{NAD^+}$ and **FAD** are both small molecules that act like **shuttle buses.** $\mathbf{NAD^+}$ is called Nicotinamide Adenine Dinucleotide, and **FAD** is called Flavin Adenine Dinucleotide. FAD is derived from vitamin B_2 (riboflavin). Both are classified as carrier molecules, represented as different types of vehicles carrying a cargo. By picking up electrons, $\mathbf{NAD^+}$ is reduced to **NADH** and **FAD** is reduced to $\mathbf{FADH_2}$. The $\mathbf{NAD^+}$ or **FAD** molecule is like a **car pulling a trailer with no cargo**, and the **NADH** or $\mathbf{FADH_2}$ is like a **car filled with its "electron" cargo.**

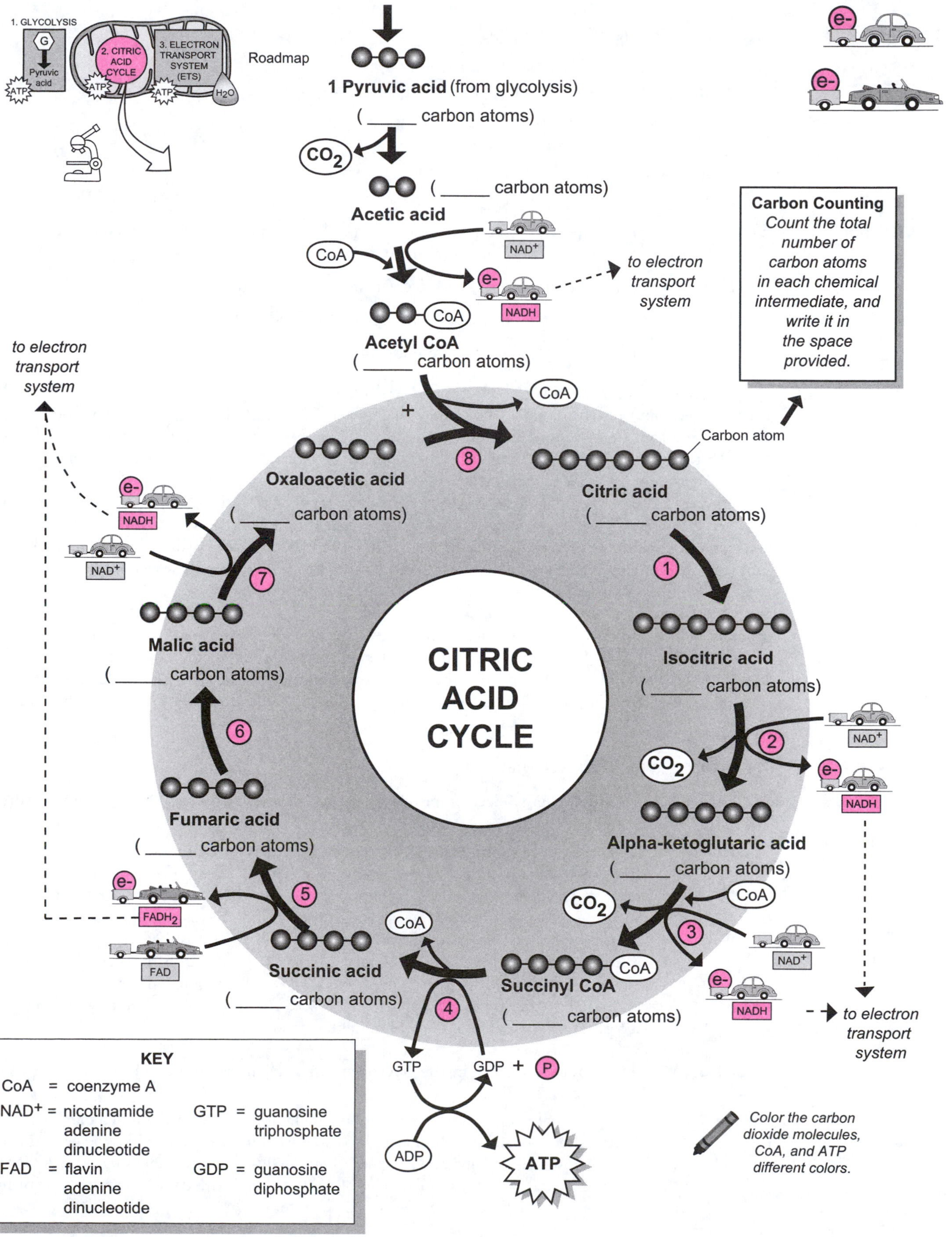
1. GLYCOLYSIS
G
Pyruvic acid
ATP
2. CITRIC ACID CYCLE
ATP
3. ELECTRON TRANSPORT SYSTEM (ETS)
ATP
H2O
Roadmap
1 Pyruvic acid (from glycolysis)
(_____ carbon atoms)
CO2
(_____ carbon atoms)
Acetic acid
NAD+
CoA
e-
NADH
to electron transport system
CoA
Acetyl CoA
(_____ carbon atoms)
Carbon Counting
Count the total number of carbon atoms in each chemical intermediate, and write it in the space provided.
e-
e-
to electron transport system
CoA
+
8
Carbon atom
Oxaloacetic acid
(_____ carbon atoms)
Citric acid
(_____ carbon atoms)
e-
NADH
NAD+
7
1
Malic acid
(_____ carbon atoms)
CITRIC ACID CYCLE
Isocitric acid
(_____ carbon atoms)
6
NAD+
2
CO2
e-
NADH
Fumaric acid
(_____ carbon atoms)
Alpha-ketoglutaric acid
(_____ carbon atoms)
e-
FADH2
5
CO2
CoA
CoA
3
NAD+
FAD
Succinic acid
(_____ carbon atoms)
CoA
Succinyl CoA
(_____ carbon atoms)
e-
NADH
to electron transport system
4
GTP
GDP + P
ADP
ATP
Color the carbon dioxide molecules, CoA, and ATP different colors.
KEY
CoA = coenzyme A
NAD+ = nicotinamide adenine dinucleotide
FAD = flavin adenine dinucleotide
GTP = guanosine triphosphate
GDP = guanosine diphosphate

Overview

Location within cell:	mitochondrion
Aerobic or anaerobic?	aerobic (O_2 used **directly**)
Side products:	12 NADH and $FADH_2$ (total of 12 reduced coenzymes)
Final electron acceptor:	O_2 (oxygen)
Final product:	H_2O (water)
Net yield of energy:	32 ATP molecules

Description

ETS, the **Electron Transport System**, is the last step in cellular respiration, yet it produces far more ATP than either of the first two steps. ETS accomplishes three major things:

1. It uses the energy from electrons (e-) to create a hydrogen ion (*or proton*) gradient.
2. It converts this potential energy into kinetic energy to produce lots of ATP.
3. It regenerates NAD^+ to allow glycolysis and the citric acid cycle to continue.

The reduced coenzymes (**NADH** and **$FADH_2$**) directly link the citric acid cycle to the ETS. They are carrying electrons from the original glucose molecule that started the whole process. These high-energy electrons are delivered to membrane proteins and other electron carriers in the inner mitochondrial membrane, where the ETS process occurs. The electrons are shuttled through a series of electron carriers, where they go from a high-energy state to a low-energy state.

Oxygen serves as the final electron acceptor. As oxygen bonds with the electron, it also bonds to hydrogen ions (H^+) to create a water molecule in the matrix. Where do the hydrogen ions come from? They are in the solution of the matrix. Every solution in the body contains an abundance of hydrogen ions and hydroxyl ions.

Using the energy from the shuttled electrons the electron carriers pump hydrogen ions from the mitochondrial matrix into the inter-membrane space. As a result, a gradient of hydrogen ions is created, which serves as a source of potential energy. The primary way for the hydrogen ions to move quickly out of the inner membrane space is to flow through a pore in a membrane protein called an **ATP synthase**. As the hydrogen ions diffuse down their gradient by flowing through this pore, the potential energy is converted into kinetic energy. This kinetic energy is used to covalently bond a phosphate group (P) to an ADP molecule to produce an ATP molecule. Note that both the phosphate group and the ADP molecule are present already in the solution of the matrix.

2 Key Concepts and Their Analogies

- **Concept #1: The high-energy electron provides energy to the electron carriers to create a hydrogen ion gradient.**
- **Analogy #1:** The **electron** being transported between the proteins in the inner mitochondrial membrane is like a **hot baked potato**. The hot potato has **thermal energy** (*heat*) just like the electron has **energy**. The **electron carriers in the inner mitochondrial membrane** are like a **row of people** standing next to each other. Imagine that the first person tosses the baked potato to the next in line, who then tosses it to the next, and so on. The result is that each person's hands absorb a little bit of the heat from the potato. Similarly, the electron transfers some energy to the electron carriers, which then is used to transport a hydrogen ion (*or proton*) from the mitochondrial matrix to the inner membrane space. The net result is that a hydrogen ion gradient is formed across the inner mitochondrial membrane.

- **Concept #2: The potential energy in the hydrogen ion gradient is converted into kinetic energy and used to create ATP.**
- **Analogy #2:** The **potential energy** in the hydrogen ion gradient is like **water behind a dam**. The **dam** is the **inner mitochondrial membrane**, and the **water behind the dam** is like the **gradient of hydrogen ions** in the inter-membrane space. The **floodgate** in the dam is like the **ATP synthase**, and the kinetic energy from the **flow of water** through the dam is like the **flow of hydrogen ions** through the pore in the ATP synthase protein.

Carbohydrate Metabolism: Cellular Respiration—Electron Transport System (ETS)

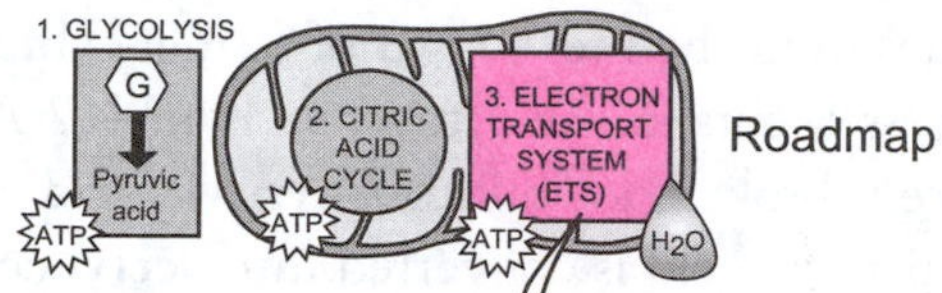

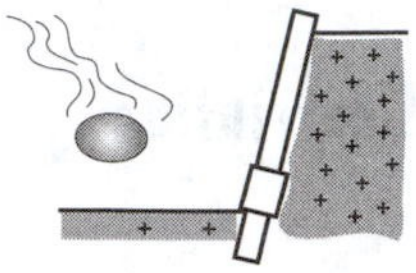

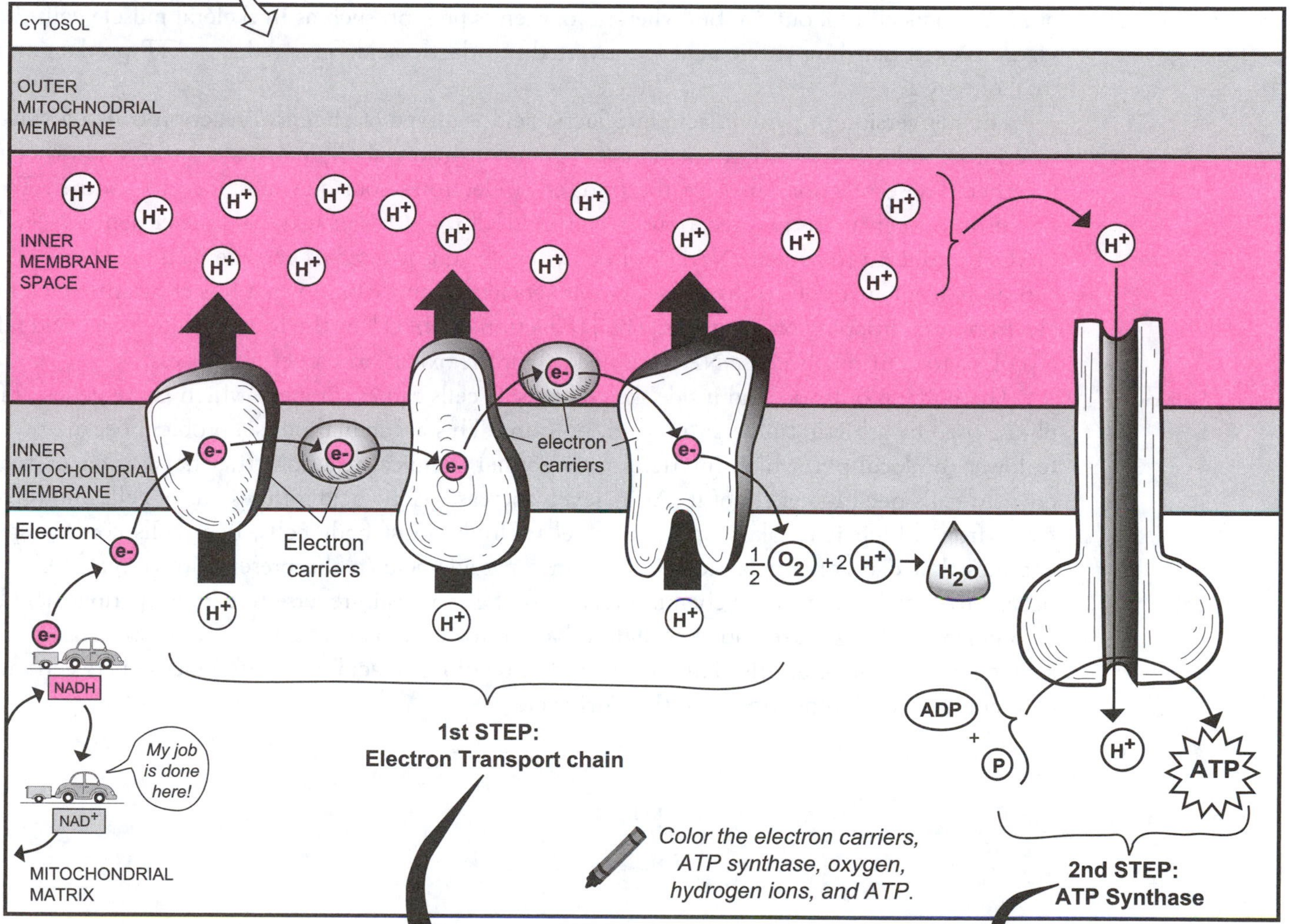

Electron Transport chain: HOT POTATO ANALOGY

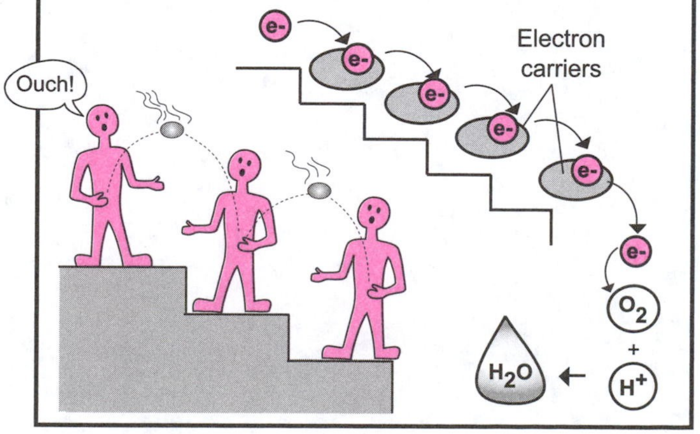

ATP Synthase: DAM ANALOGY

The hydrogen ion gradient is like water behind a dam in the sense that it has potential energy to do work.

Hydrogen ions moving through the ATP synthase protein are like a stream of water moving through the dam's floodgate. This same kinetic energy that drives a turbine and generates hydroelectric power is used by the cell to bond a phosphate group to ADP to form ATP.

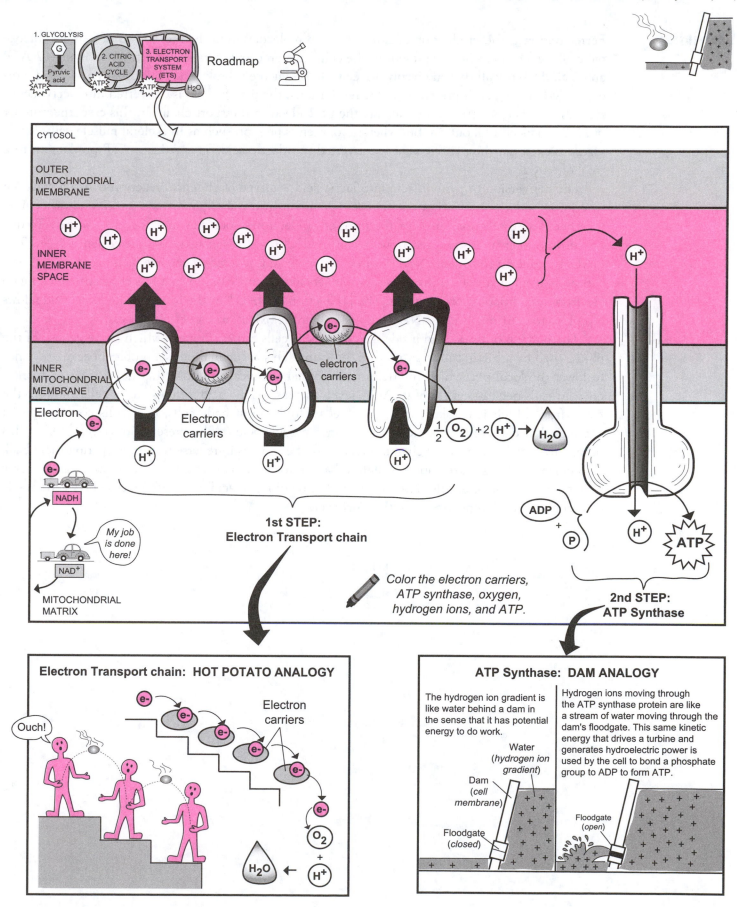

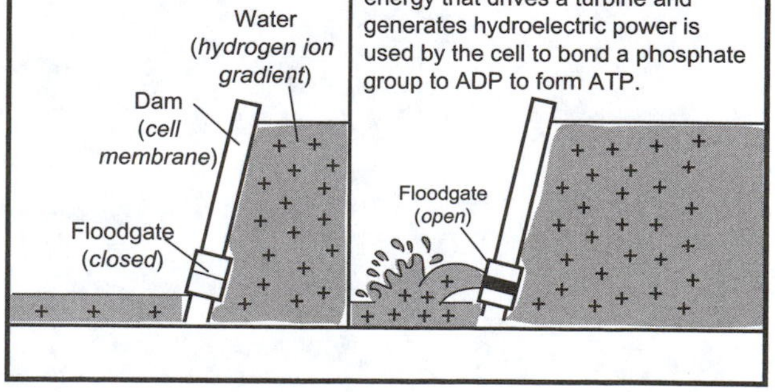

Description

Fermentation is the production of two **lactic acid** molecules from the breakdown of a single **glucose** molecule when no oxygen is present in the cell. This process occurs in the cytosol and yields 2 ATP and 2 NADH. Recall that in glycolysis (see p. 404), glucose is broken down into 2 **pyruvic acid** molecules. When oxygen is present (aerobic respiration), the pyruvic acid is converted into acetyl coenzyme A, enters the citric acid cycle, and then the electron transport chain. In this case, the number of ATP produced is about 36. But when no oxygen is present, such as in skeletal muscle cells during vigorous exercise, the pyruvic acid is converted into lactic acid. This reduces ATP production to a total of only 2.

The conversion of pyruvic acid into lactic acid is a type of chemical reaction called a *reduction* reaction. Oxidation-reduction reactions are common in metabolic pathways and are always paired together. These reactions involve the transfer of electrons from one substance to another. To distinguish between these two reactions, recall OIL RIG = *Oxidation Is Loss, Reduction is Gain*. The NADH is *oxidized* to become NAD^+ while pyruvic acid is *reduced* to become lactic acid. NAD^+ is a carrier molecule with electrons as its cargo. It acts like a shuttle bus in that it is constantly picking up electrons and dropping off electrons. The production of NAD^+ in the formation of lactic acid allows glycolysis to continue because NAD^+ is needed for the oxidation-reduction process.

The buildup of lactic acid inside skeletal muscle cells causes soreness, which is the genesis of the phrase used by athletic trainers: "no pain, no gain." This accumulation is a problem because it tends to lower the local pH, which interferes with normal chemical reactions that have to occur. Consequently, cells need to get rid of it. As its levels increase, lactic acid diffuses out of cells and into the blood from which it is taken up by liver cells. Unlike most body cells, liver cells contain special enzymes that can convert lactic acid back into pyruvic acid in the presence of oxygen. After this occurs, liver cells can run the chemical reactions of glycolysis in reverse to convert pyruvic acid back into glucose. The glucose can then diffuse back into the blood where it can be used as an energy source for other muscle cells. This chemical cycling of lactic acid back into glucose between muscle cells and liver cells is referred to as the **Cori cycle**.

1. GLYCOLYSIS
G
Pyruvic acid
ATP
2. CITRIC ACID CYCLE
ATP
3. ELECTRON TRANSPORT SYSTEM (ETS)
ATP
H_2O
Road map
FERMENTATION
Cytosol
Carbon atom
Glucose
Glycolysis
2 ATP
e-
2 NADH
2 Pyruvic acid
e-
e-
2 NADH
2 NAD^+
2 Lactic acid
Simple diffusion
Blood
Liver cells
KEY
e- = electron
NAD^+ = nicotinamide adenine dinucleotide (*oxidized*)
NADH = nicotinamide adenine dinucleotide (*reduced*)

CORI CYCLE

MUSCLE cell
Blood
LIVER cell
Glucose
Pyruvic acid
Lactic acid
Glucose
Pyruvic acid
Lactic acid

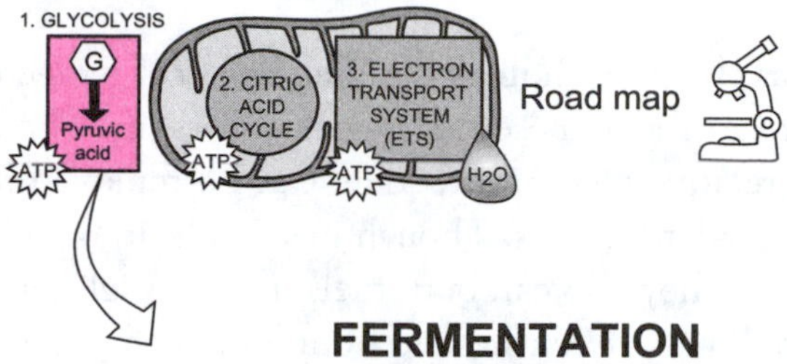

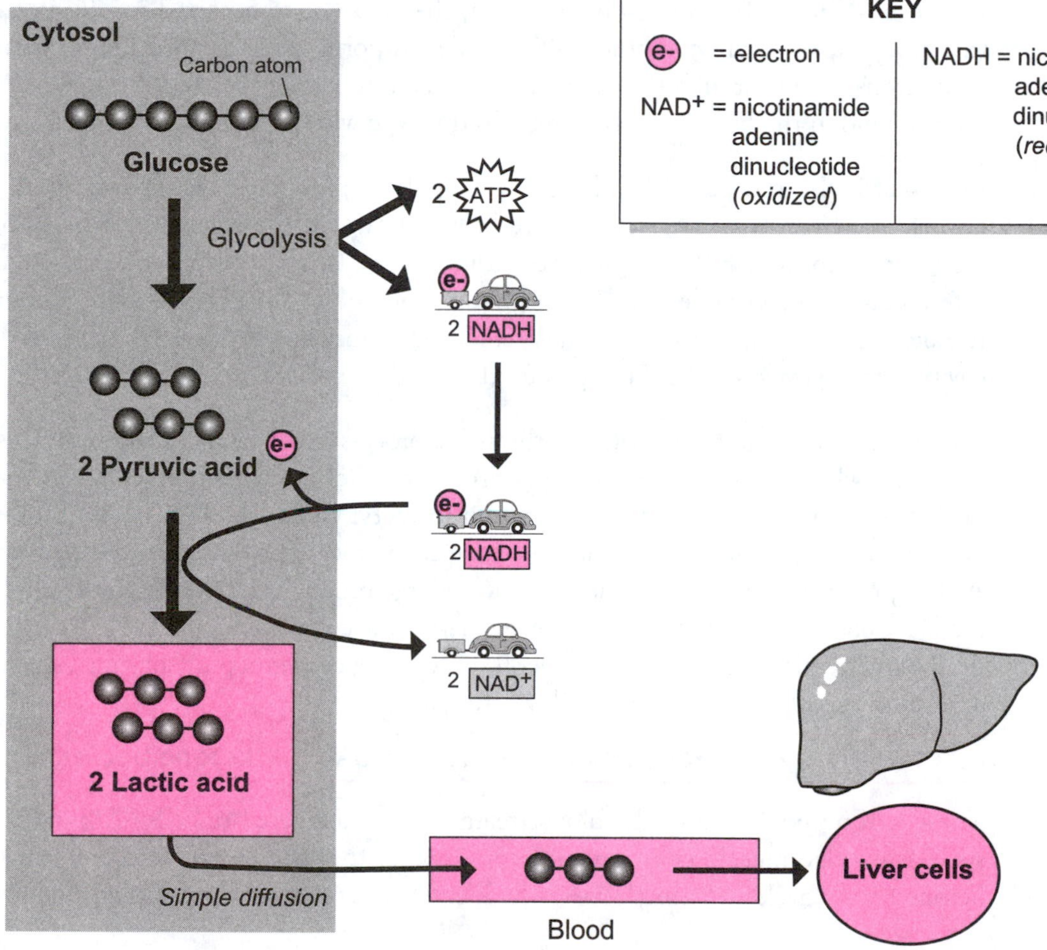

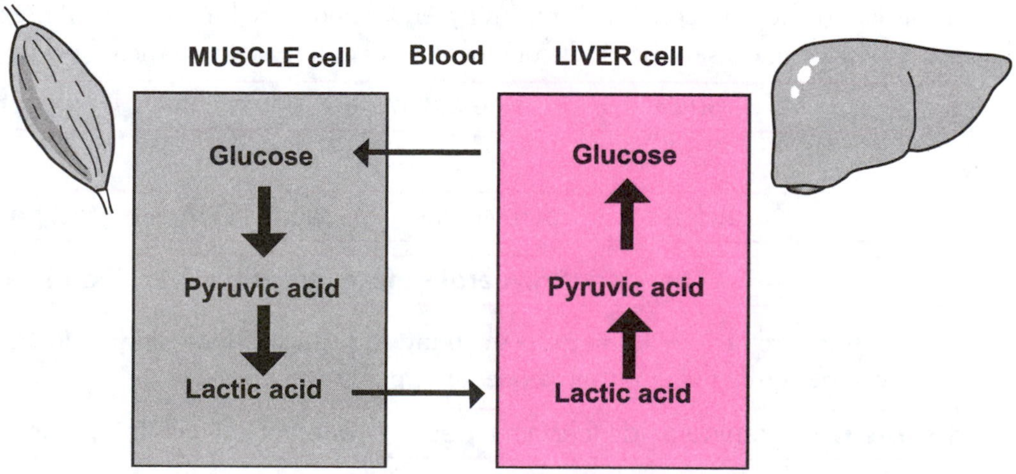

Description

Lipids in the body take many different forms, such as the **phospholipids** and **cholesterol** molecules found in cell membranes, and the **triglycerides** stored in the adipose tissue under the skin. Though the total amount of stored lipid molecules remains relatively constant, some can be removed and either oxidized to be used as fuel or redeposited in other adipose cells. Though many cells in the body prefer to burn a carbohydrate-based fuel such as glucose, they also can burn fuels such as triglycerides or amino acids when necessary. Alternatively, new lipids may be synthesized from other compounds.

Because lipids do not dissolve in water, they are not transported readily in the blood. To solve this problem, lipids are combined with proteins to form spheres called **lipoproteins**, which are more easily transported through the blood and delivered to body cells. Triglycerides are the most common form of lipid in the body, so we will refer to this type when discussing how lipids are metabolized.

Lipolysis

Lipolysis is the process of breaking down lipids to produce ATP (*lipo* = lipid; *-lysis* = breaking down). Via hydrolysis, triglycerides are broken down in the small intestine into their component parts—**glycerol** and **fatty acids**. Once inside the cells, glycerol can be converted into a glycolysis intermediate—**glyceraldehyde-3-phosphate**—and then into **pyruvic acid**. Pyruvic acid is oxidized through the citric acid cycle, producing reduced coenzymes. As these coenzymes are oxidized by the electron transport system, ATP is produced.

Beta Oxidation

Fatty acids follow a different pathway through a process called **beta oxidation**. In this series of chemical reactions, which occurs in the mitochondria, fatty acids are broken down into two-carbon fragments. Each of these fragments then is converted into **acetyl CoA**, which is oxidized through the citric acid cycle, producing reduced coenzymes. As these coenzymes are oxidized by the electron transport system, ATP is produced. In total, this accounts for a large production of ATP. For example, some fatty acids produce about four times more ATP as compared to the complete catabolism of a single glucose molecule.

In liver cells, fatty acids are catabolized to produce a group of substances called **ketone bodies**. This process, called **ketogenesis**, follows this pathway:

Fatty acids → acetyl CoA → **ketone bodies**

These ketone bodies enter the bloodstream and are delivered to cells. To produce ATP, some cells, such as cardiac muscle cells, prefer ketone bodies to glucose. They accomplish this by converting ketone bodies back into acetyl CoA, which enters the citric acid cycle, then electron transport system.

Lipogenesis

The process of synthesizing lipids is called **lipogenesis** (*lipo* = lipid; *-genesis* = generation, birth). This occurs in liver cells and adipose cells, which can convert carbohydrate, proteins, and fats into triglycerides. For example, drinking a lot of soda pop regularly gives the body more simple sugars than it needs, so they are converted into triglycerides and stored in fat cells. An excess of simple sugars such as glucose leads to the production of glycerol or fatty acids in the following pathways:

glucose → glyceraldehyde-3-phosphate → **glycerol**

. . . *or* . . .

glucose → pyruvic acid → acetyl CoA → **fatty acids**

The excess **glycerol** + **fatty acids** → **triglycerides**

Proteins are broken down in the small intestine into various amino acids. Certain types of amino acids can be converted into triglycerides. The pathway for proteins is:

protein → amino acids (*certain types*) → acetyl CoA → fatty acids → **triglycerides**

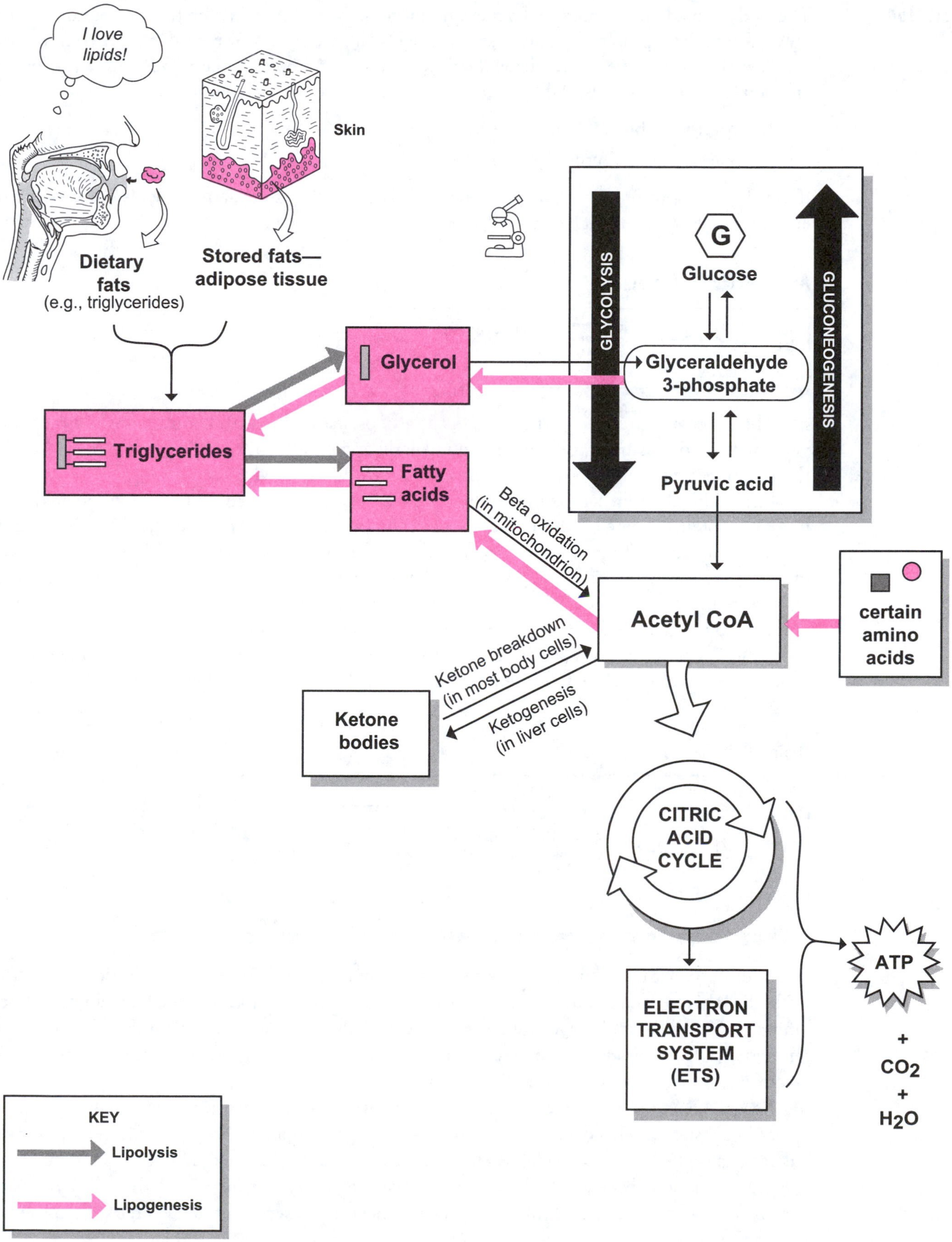
I love lipids!
Skin
Dietary fats (e.g., triglycerides)
Stored fats—adipose tissue
Glycerol
Triglycerides
Fatty acids
GLYCOLYSIS
G
Glucose
Glyceraldehyde 3-phosphate
Pyruvic acid
GLUCONEOGENESIS
Beta oxidation (in mitochondrion)
Acetyl CoA
certain amino acids
Ketone breakdown (in most body cells)
Ketogenesis (in liver cells)
Ketone bodies
CITRIC ACID CYCLE
ELECTRON TRANSPORT SYSTEM (ETS)
ATP
+
CO_2
+
H_2O
KEY
Lipolysis
Lipogenesis

Description

The body is made up primarily of water and protein. A healthy adult body may contain as much as 18% protein. The protein in a chicken sandwich will be broken down gradually into amino acids in the digestive tract. The amino acids then are actively transported (see p. 52) into the bloodstream and delivered to body cells. All body cells need amino acids for two purposes:

1. protein synthesis—to make new proteins within the cells, or
2. to use as a fuel source to make ATP.

If you ingest more protein than your body needs daily, the excess amino acids are converted into either glucose (G) or triglycerides ( ).

Processes

Amino Acids for Protein Synthesis

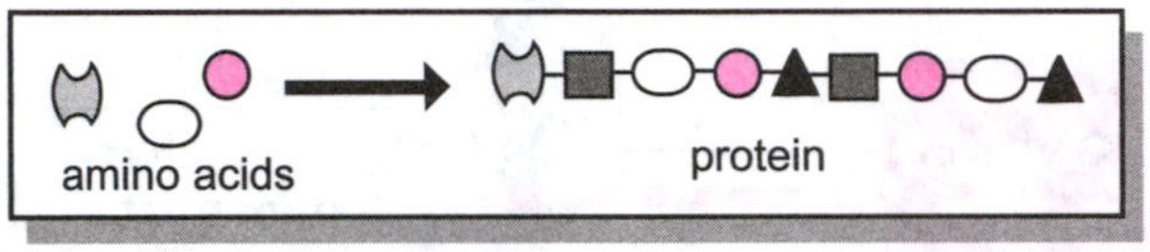

The process of protein synthesis occurs at a ribosome (see p. 40). During human growth and development, proteins have to be produced regularly and rapidly. In the adult, protein synthesis serves the purpose of replacing worn-out proteins and repairing damaged tissues.

Just as the different letters in the alphabet are used to form words, different amino acids are used to make proteins. There are 20 different types of amino acids. The body can synthesize some of them, but most must come from the diet. Some cells, such as hepatocytes (*liver cells*), are more active than others at protein synthesis. Sometimes a cell has to convert one amino acid to be able to create a specific protein. This is achieved by a process called transamination, in which an amino group from an amino acid is transferred to a keto acid with the help of the enzyme transaminase. The result is that a keto acid is converted into an amino acid that can be used in protein synthesis. Some examples of other tissues that are active in protein synthesis are cells in the brain, skeletal muscle, and heart.

Amino Acids as a Fuel Source

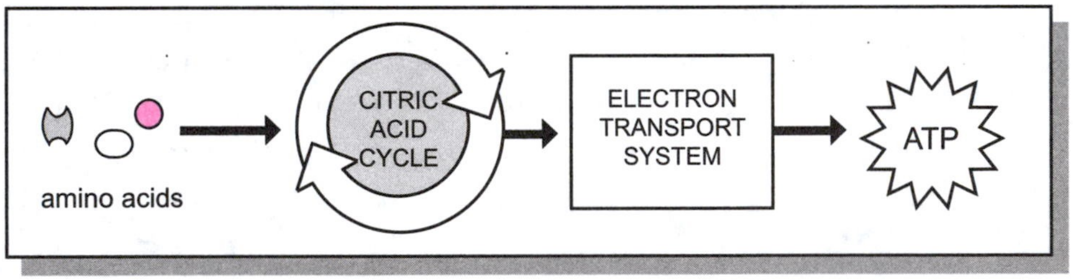

Alternatively, the body cells can use amino acids to produce ATP by entering the citric acid cycle. For this to occur, amino acids must go through a deamination process, whereby they lose their amino (NH_2) group and a hydrogen atom with the help of the enzyme deaminase. The result is that ammonia (NH_3) is produced along with a keto acid. The keto acid can enter the citric acid cycle and electron transport system to produce ATP for the cell.

Because ammonia is toxic to cells, it is converted further into a waste product called urea, a harmless substance. This occurs in hepatocytes when ammonia and carbon dioxide enter a biochemical pathway called the urea cycle, where urea is the primary byproduct. The liver is the major site for these deaminations because the hepatocytes have the proper enzymes to do the job. Urea normally travels through the blood and into the kidneys, where it is filtered out and excreted from the body as part of the urine.

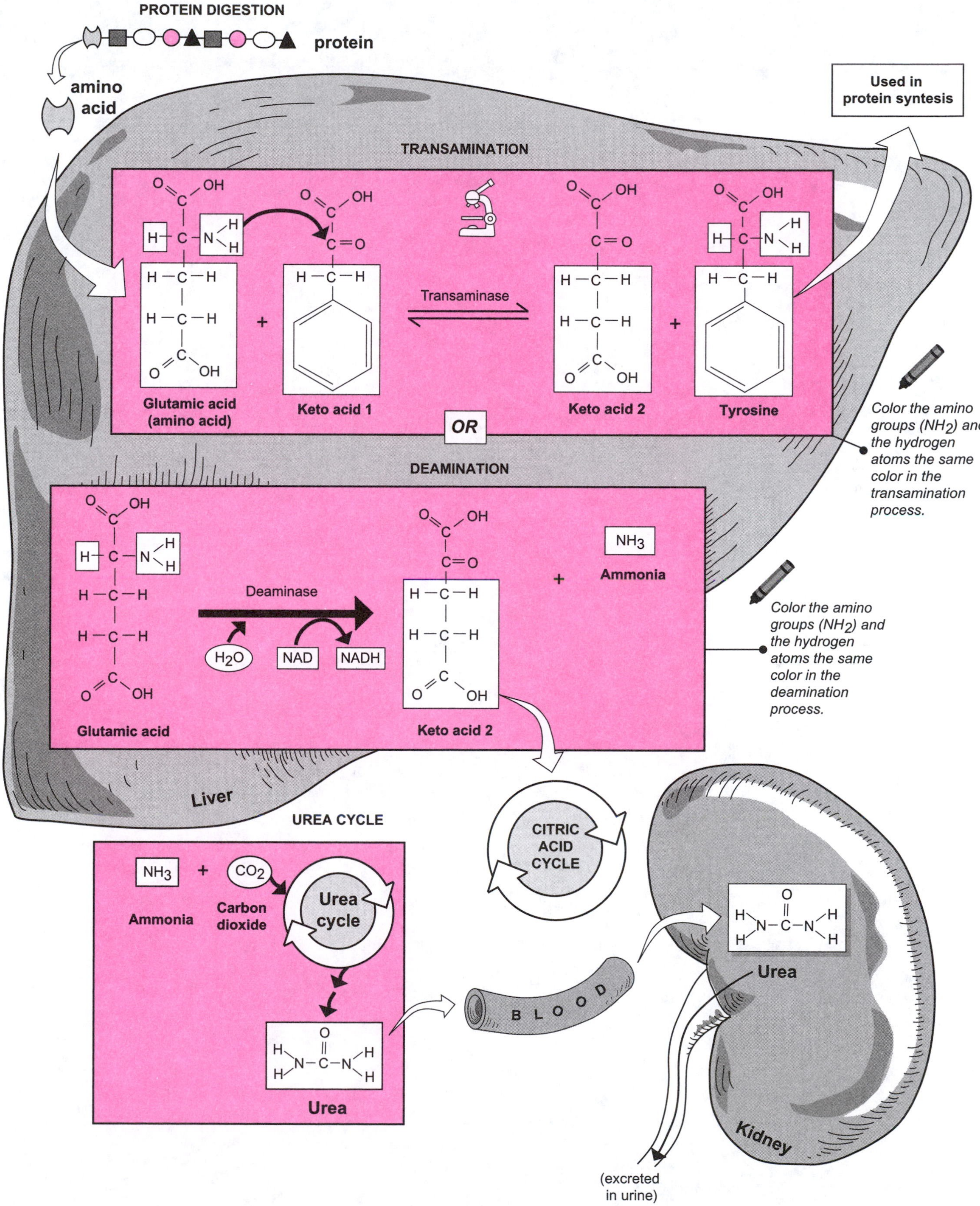

Color the amino groups (NH_2) and the hydrogen atoms the same color in the transamination process.

Color the amino groups (NH_2) and the hydrogen atoms the same color in the deamination process.

Notes

Urinary System

Description

The urinary system is composed of the **kidneys**, **ureters**, **urinary bladder**, and **urethra**. The kidneys are located on either side of the upper lumbar region of the vertebral column. They lie between the dorsal body wall and the **parietal peritoneum** in a retroperitoneal position. They are tightly covered by the parietal peritoneum on their anterior surface and further protected by layers of fatty tissue. Major abdominal organs such as the small and large intestines are all positioned anterior to the kidneys.

The **renal arteries** branch off the **abdominal aorta** and supply oxygenated blood to the kidneys, while the **renal veins** drain blood from the kidneys and empty into the **inferior vena cava**. The blood that enters the kidney via the renal artery contains various waste products that must be removed from the blood. The functional units within the kidney are microscopic structures called **nephrons**. They filter waste products and other substances from the blood and transfer them to a separate tubular system. The fluid that enters the tubular system is called the filtrate. The filtrate is processed in a manner that transports nutrients back to the bloodstream and retains waste products in the tubular system.

When the processing is completed, the liquid inside the tube is called urine. In reality, urine is nothing more than processed blood plasma. The urine then moves out of the nephrons and into the pelvis of the kidneys. Slender, muscular tubes called **ureters** transport urine from the kidney to the urinary bladder. The muscular **urinary bladder** can expand to hold as much as 600–800 ml. of urine. Finally, urine is removed from the body by the process of **micturition** (*urination*), which uses forceful muscular contractions in the wall of the urinary bladder to transport urine into the urethra and out of the body.

Key to Illustration

Major Organs/Structures

1. Kidney
2. Ureter
3. Urinary bladder

Blood Vessels (B)

B1. Inferior vena cava
B2. Abdominal aorta
B3. Renal a.
B4. Renal v.
B5. Common iliac a.
B6. Internal iliac a.
B7. External iliac a.
B8. Gonadal v. *(testicular v. in male; ovarian v. in female)*
B9. Gonadal a. *(testicular a. in male; ovarian a. in female)*

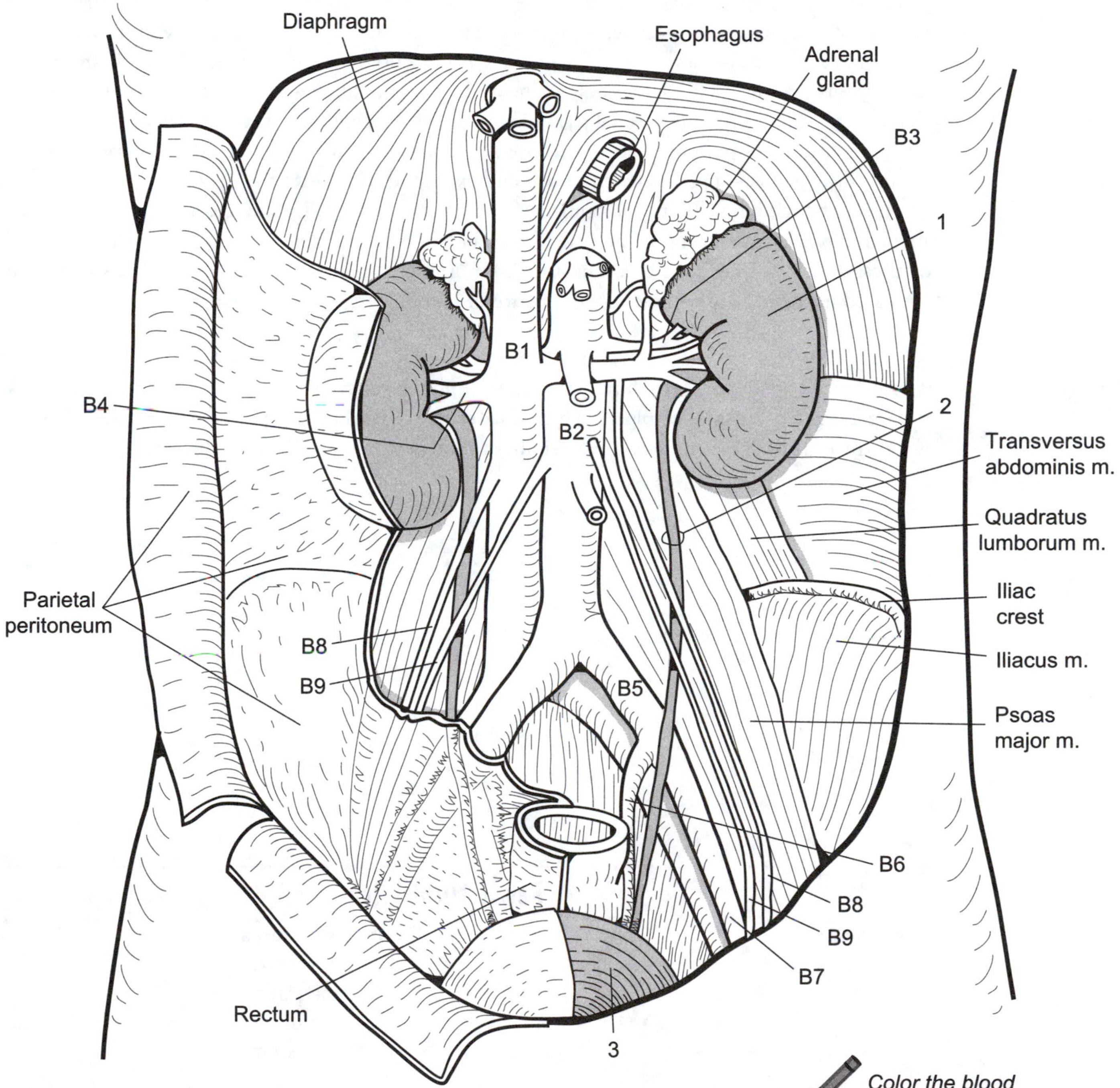

Abdominal cavity
anterior organs and structures removed

Color the blood vessels feeding and draining the kidneys.

Description

The kidneys are covered by a fibrous tissue called the **renal capsule**. The **renal hilus**, a cleft or indentation on the medial surface of the kidneys, is a handy landmark for locating the **ureters**, blood vessels (*renal a.*, *renal v.*), and nerves that serve this organ. Each kidney is divided into three regional areas: cortex, medulla, and pelvis. The **cortex** is the outermost region, like the bark around a tree, and appears granular. The darker colored **medulla** is the middle region, containing numerous funnel-shaped structural units called **renal pyramids,** each with a **papilla** at its tapered end.

Separating the pyramids are projections of cortical tissue called **renal columns**. The central region is the **pelvis**—a pouch that narrows and extends directly into the **ureter**.

Near the medulla, the pelvis branches off into structures called **calyces** (sing., *calyx*). Each cup-shaped **minor calyx** surrounds each papillus of a renal pyramid to receive urine from it. When several of these minor calyces join together, they form a larger chamber called a **major calyx**. As urine is collected in the pelvis, it moves down the **ureter** and into the urinary bladder.

Analogy

The **calyces** (sing., *calyx*) are functionally like a **plumbing system** of smaller-diameter pipes leading to larger ones. Each **smaller pipe** is a **minor calyx**, and each **larger pipe** is a **major calyx**. The flow of water through the pipes follows a path similar to the flow of urine through the calyces.

Key to Illustration

Regional areas

1. Cortex
2. Medulla
3. Pelvis

Other structures

4. Renal papillus
5. Renal pyramids
6. Renal capsule
7. Renal column
8. Minor calyx
9. Major calyx
10. Adipose tissue in renal sinus
11. Renal hilus
12. Ureter

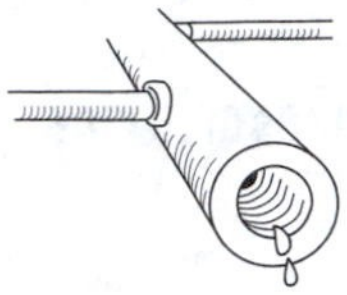

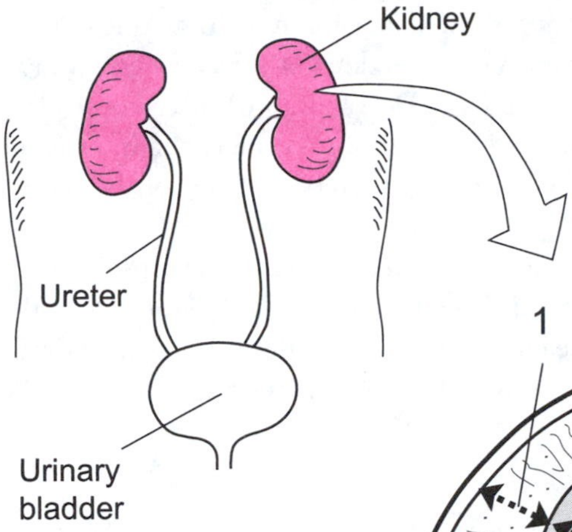

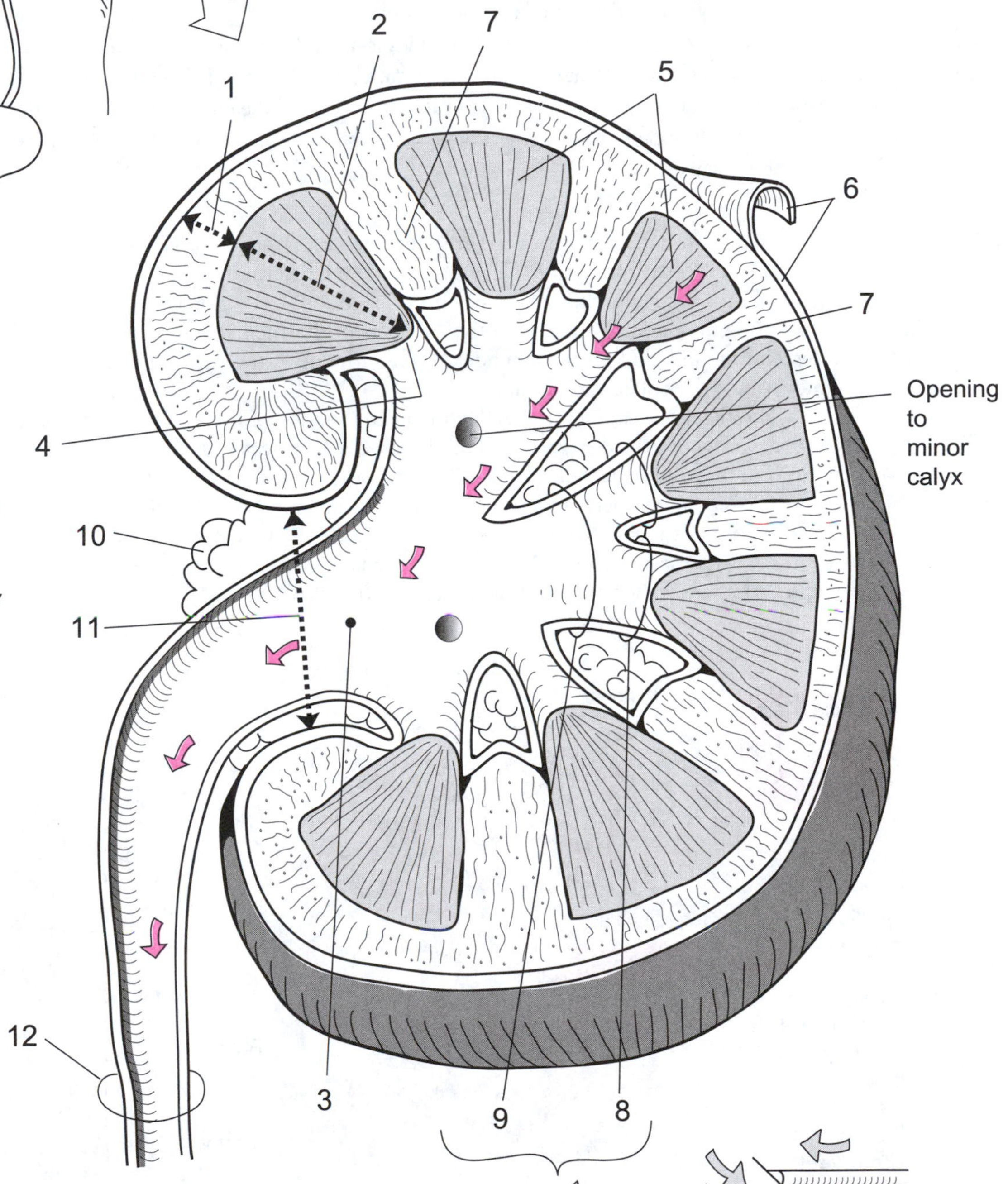

Note:
The colored arrows indicate the pathway that urine follows through the kidney and down the ureter.

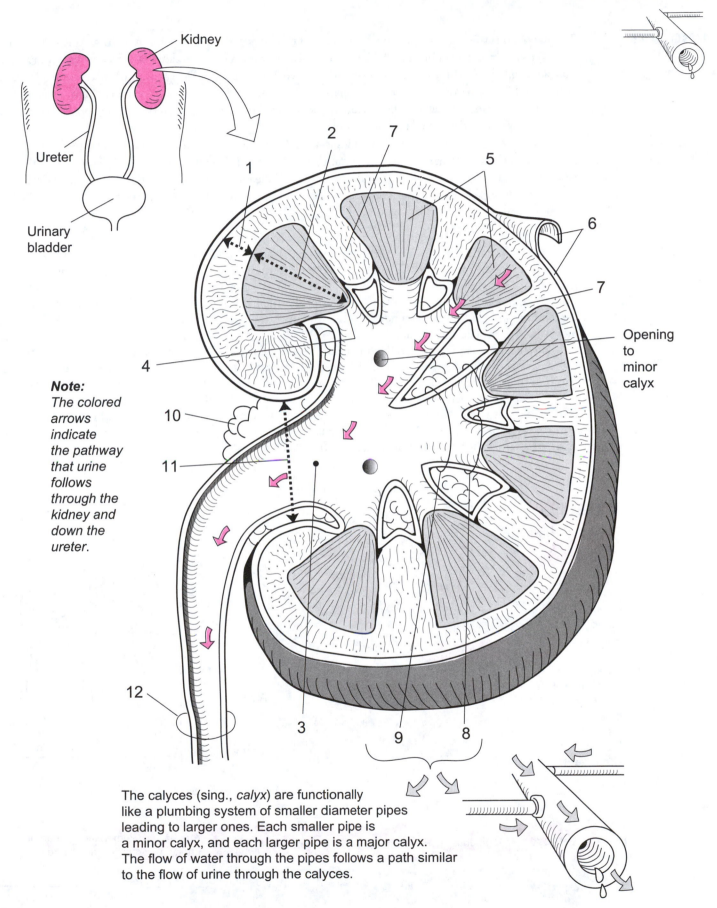

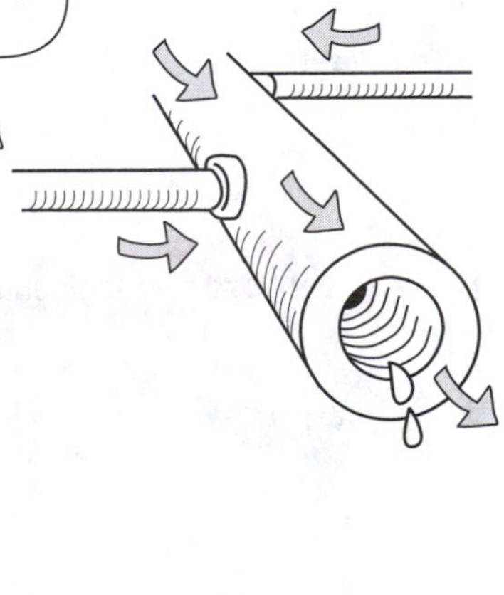

The calyces (sing., *calyx*) are functionally like a plumbing system of smaller diameter pipes leading to larger ones. Each smaller pipe is a minor calyx, and each larger pipe is a major calyx. The flow of water through the pipes follows a path similar to the flow of urine through the calyces.

Description

The **nephron** is the microscopic, functional unit of the kidney. Each kidney has more than one million of these units per kidney. Each nephron is divided into six distinct parts: (1) *glomerulus*, (2) *glomerular capsule*, (3) *proximal convoluted tubule* (PCT), (4) *nephron loop*, (5) *distal convoluted tubule* (DCT), and (6) *collecting tubule*. As blood enters the kidney via the **renal artery**, it branches into smaller vessels until it becomes the **interlobular artery** in the renal cortex. The **afferent arteriole** branches off the interlobular artery and delivers blood to a special coiled ball of capillaries called the **glomerulus**.

Completely surrounding the glomerulus is a cup-like structure called the **glomerular capsule**. Together, the glomerulus and the glomerular capsule form a structure called the **renal corpuscle**. Connected to the capsule is a long tubular system. The glomerulus is highly permeable primarily because of pores in its walls, called **fenestrae**. Wrapped around the glomerular capillaries are cells called **podocytes**—modified simple squamous epithelial cells.

Functions

Each nephron can perform only three **functions**: (1) *filtration*, (2) *reabsorption*, and (3) *secretion*. **Filtration** occurs only in the glomerulus. The blood in the glomerulus is under high pressure, so the plasma is passively filtered into the glomerular capsule. The solution inside the glomerular capsule is referred to as **filtrate**. It flows from the glomerular capsule into the first part of the tubular system, called the PCT. Next it flows down the descending limb and up the ascending limb of the nephron loop. Then the filtrate enters the DCT, and finally moves into the collecting tubule.

The filtrate must be processed while in the tubules of the nephron because it contains both nutrients and waste products mixed together. This is achieved by *reabsorption* and *secretion*. Through **reabsorption**, the nephron returns nutrients such as sodium ions and water to the bloodstream. In **secretion**, unwanted substances in excess amounts, such as hydrogen ions, potassium ions, and antibiotics, are transported out of the blood and into the renal tubular system. After this processing job is completed, the resulting fluid is finally referred to as urine. The urine then follows this pathway: *minor calyx* → *major calyx* → *ureter* → *urinary bladder* → *urethra*. After passing through the urethra, the urine exits the body.

Blood flow through the kidney

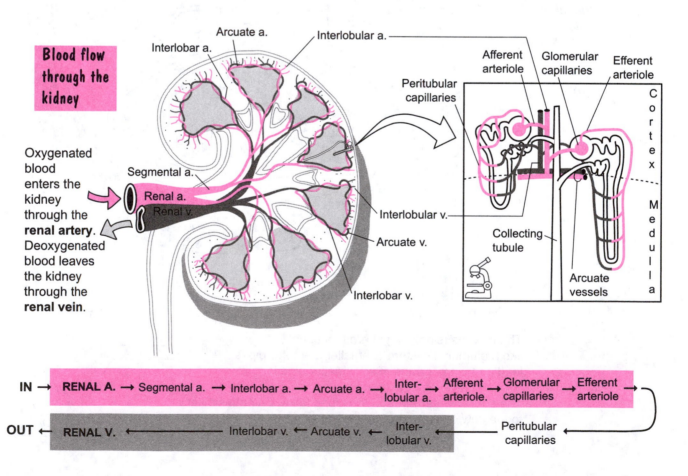

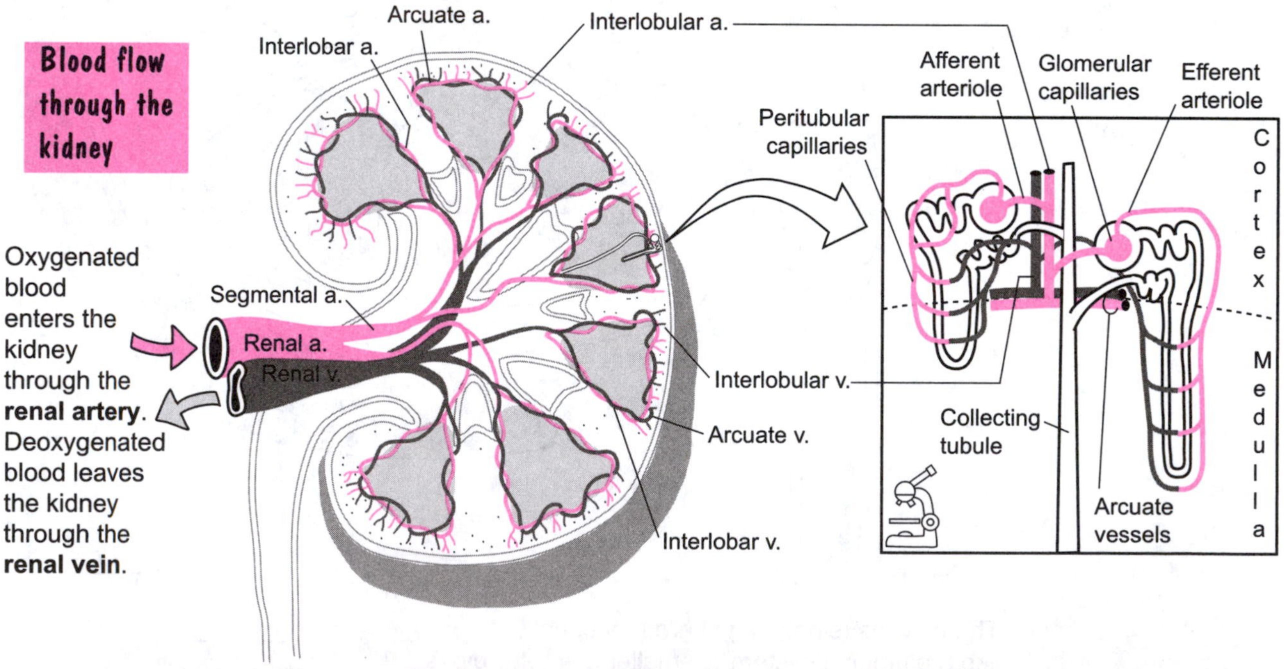

IN → **RENAL A.** → Segmental a. → Interlobar a. → Arcuate a. → Interlobular a. → Afferent arteriole. → Glomerular capillaries → Efferent arteriole → Peritubular capillaries

OUT ← **RENAL V.** ← Interlobar v. ← Arcuate v. ← Interlobular v. ← Peritubular capillaries

Podocyte
Fenestrae
Filtrate
Renal corpuscle
Glomerulus
Glomerular capsule
(Blood OUT)
Efferent arteriole
Inter-lobular a.
Efferent arteriole
Afferent arteriole
(to PCT)
Afferent arteriole
(Blood IN)
Filtrate
Cross Section of Renal Corpuscle
Proximal Convoluted Tubule (PCT)
Distal Convoluted Tubule (DCT)
Renal capsule
Renal corpuscle
Renal cortex
Collecting tubule
Descending limb
Nephron loop
Ascending limb
Renal medulla
Nephron structure
Renal capsule
Major calyx
Ureter
(to urinary bladder)
Kidney, Sagittal Section
Wedge of cortex and medulla
Papilla
Urine
Minor calyx
Note: Black arrows show the direction the filtrate flows through the nephron and the direction urine flows through the kidney.
Nephron: Functional Overview
Glomerulus
Afferent arteriole
Glomeular capsule
Renal tubule
Urine
Blood
Efferent arteriole
Peritubular capillaries
F
R
S
KEY
F = Filtration
R = Reabsorption
S = Secretion

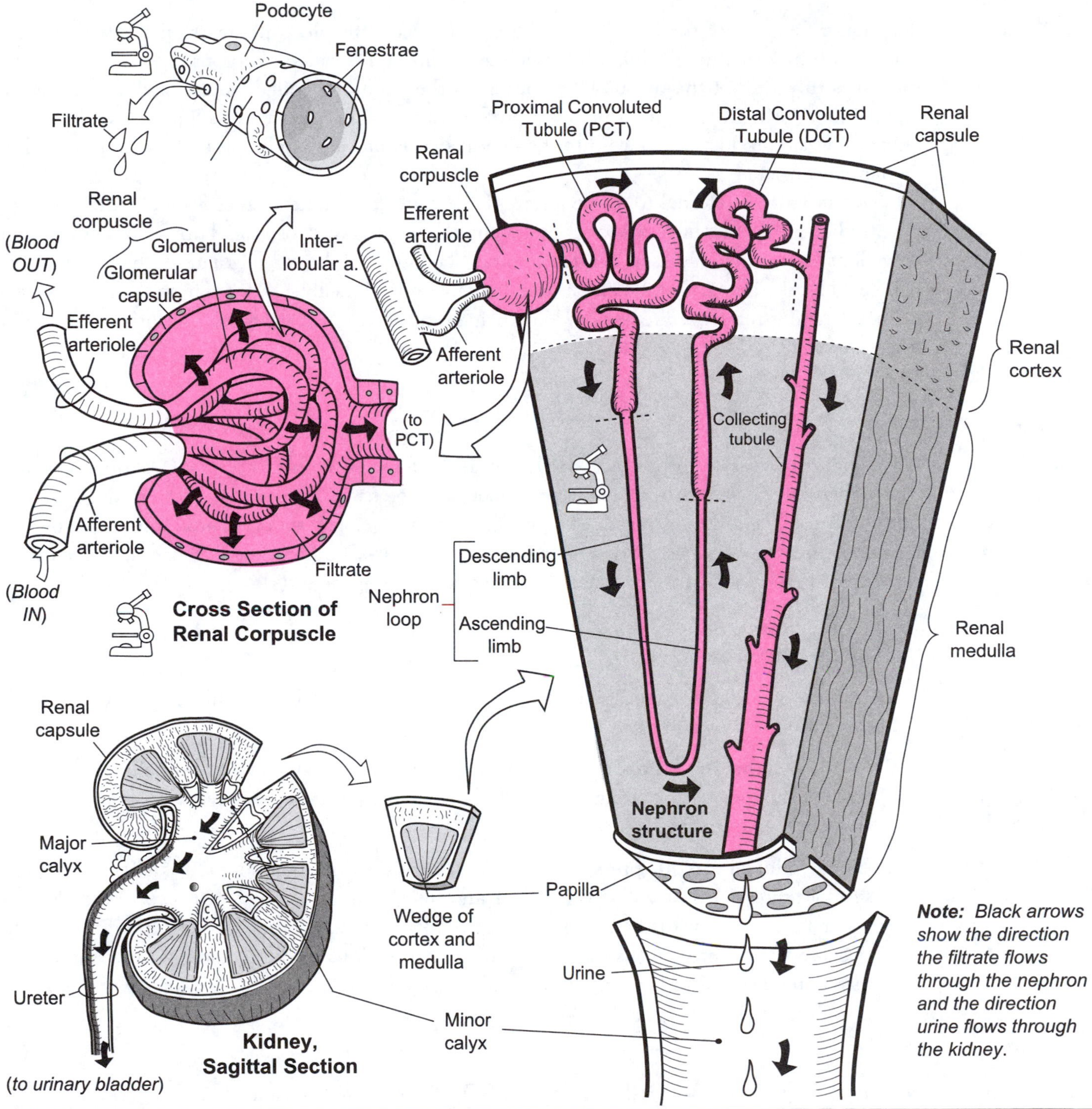

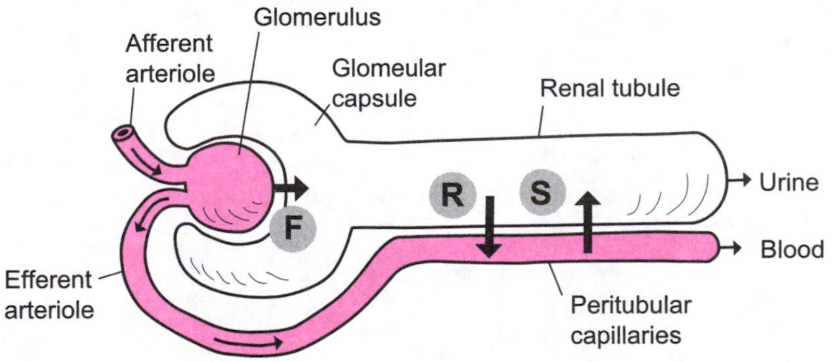

Description

The major function of the kidneys is to filter and process the blood plasma to produce urine. This occurs within the kidney's millions of microscopic filter units called nephrons. The first step in this process is **filtration** of the blood, which occurs in the nephron's coiled ball of capillaries called the glomerulus. During this process, the blood pressure in the glomerulus forces a liquid mixture of nutrients and waste products to move into the cup-like structure surrounding the glomerulus, called the glomerular capsule.

This mixture is referred to as the filtrate. It contains substances such as water, amino acids, glucose, sodium, chloride, urea, uric acid, and many others. Working together, both kidneys filter about 48 gallons (182 liters) of plasma per day. In total, about 99% of this large volume with all its nutrients will be reabsorbed back into the blood, and the remaining 1% will be released from the body as urine. To better understand the process of glomerular filtration, we will examine the forces and counter forces involved.

Force

Glomerular hydrostatic pressure (GHP): This is the force of the blood pressure against the walls of the glomerulus. Filtration is completely dependent on pressure. If the blood pressure drops too low, filtration stops. The blood pressure in the glomerulus averages about 50 mm Hg and cannot be measured directly, so it must be determined mathematically. Think of the GHP like the force exerted by a sumo wrestler as he thrusts himself forward against his opponent(s).

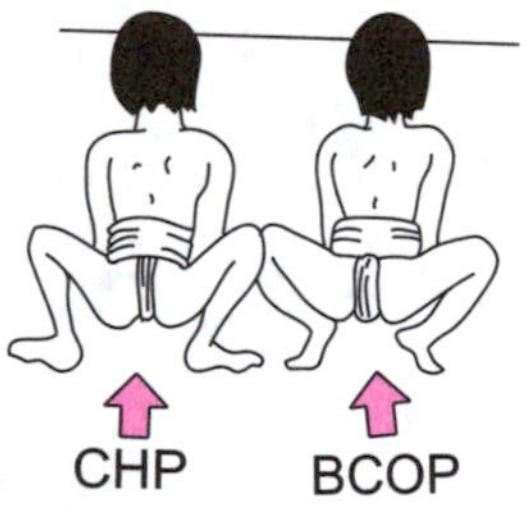

Counter-Forces

Two forces work against the GHP: the CHP and the BCOP. Each is like a boy sumo wrestler pushing against the adult sumo wrestler.

1. **Capsular hydrostatic pressure (CHP)**: Like the force of water in a water balloon, this is the force of the filtrate fluid against the wall of the glomerular capsule. It averages about 15 mm Hg and works against filtration.
2. **Blood colloidal osmotic pressure (BCOP)**: This is the only force that is not a hydrostatic pressure of fluid against a wall. Instead, it is an osmotic pressure, which refers to the tendency of water to move toward the solution with the higher concentration of nonpenetrating solutes. Because plasma proteins like albumin remain in the blood in the glomerulus, the water in the filtrate tends to move back into the glomerulus. This osmotic pressure is a force working against filtration and it averages about 25 mm Hg.

Net Filtration Pressure Formula

Net filtration pressure = GHP − (CHP + BCOP) or "forces = sum of all counter-forces"

= 50 − (15 + 25)
= 50 − (40)
= **10 mm Hg**

In other words, a net positive pressure is maintained so the process of filtration can continue. If this were not the case, renal failure could result.

Nephron Function 1: Filtration

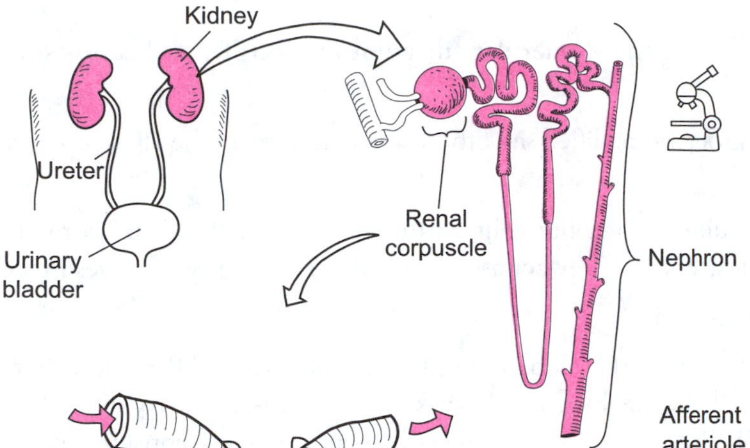

KEY

GHP = glomerular hydrostatic pressure
BCOP = blood colloidal osmotic pressure
CHP = capsular hydrostatic pressure
PCT = proximal convoluted tubule

Note: *To review the process of filtration,* *see p. 50.*

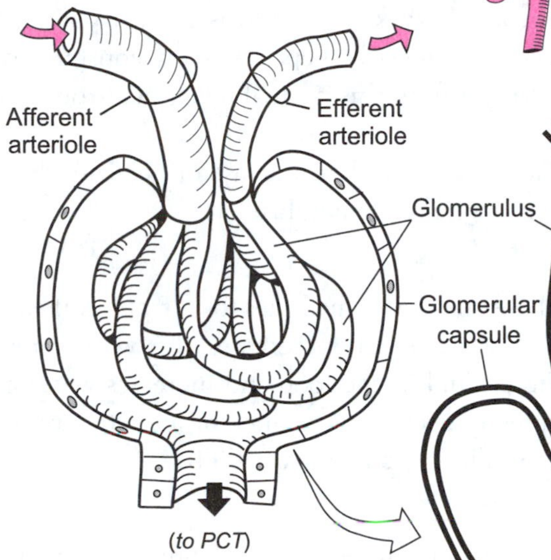

Cross-section of Renal Corpuscle

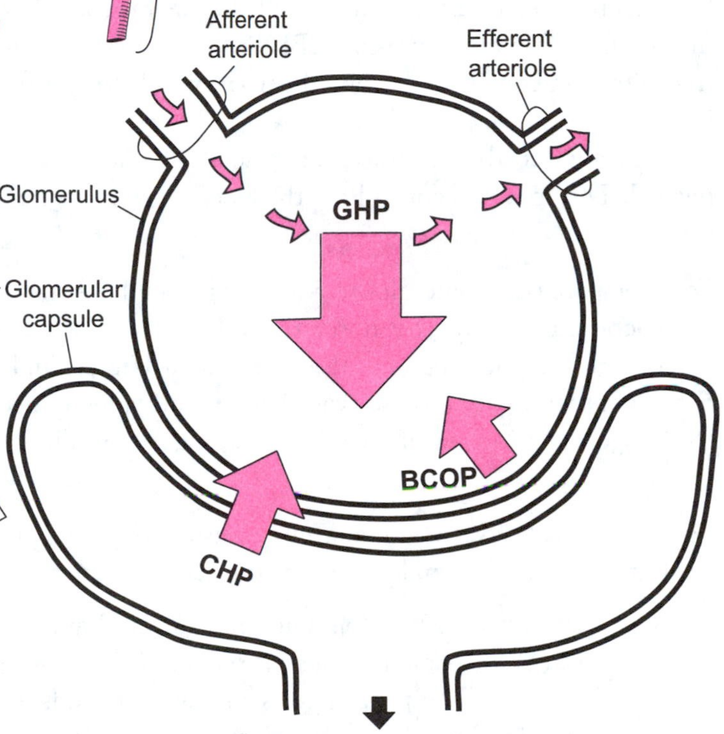

Schematic of Forces and Counter-Forces

Color the glomerulus red to represent the blood that runs through it. Color the space inside the glomerular capsule yellow to represent the filtrate inside it.

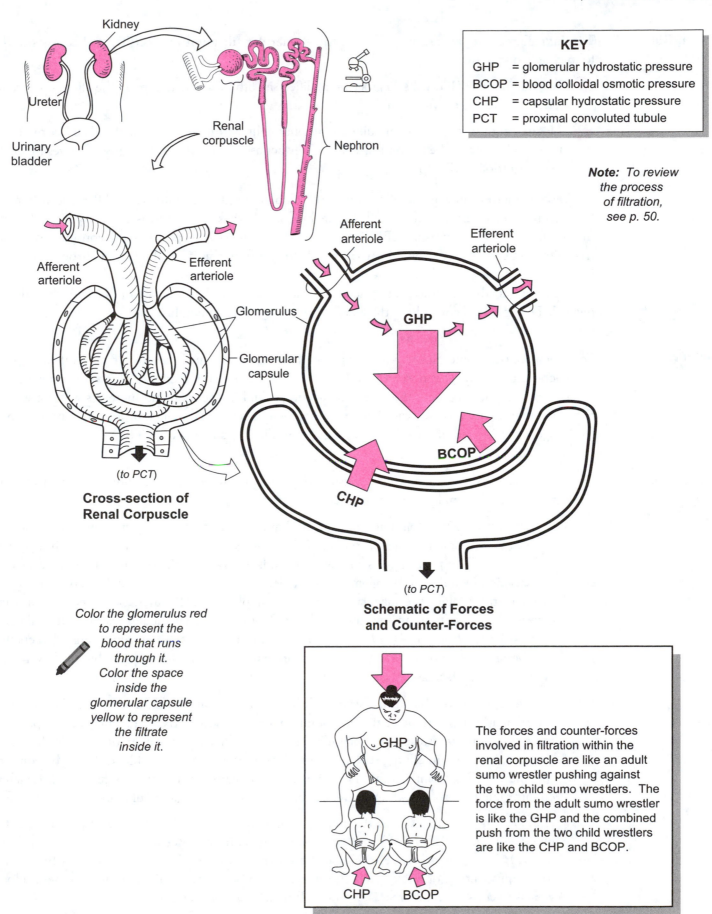

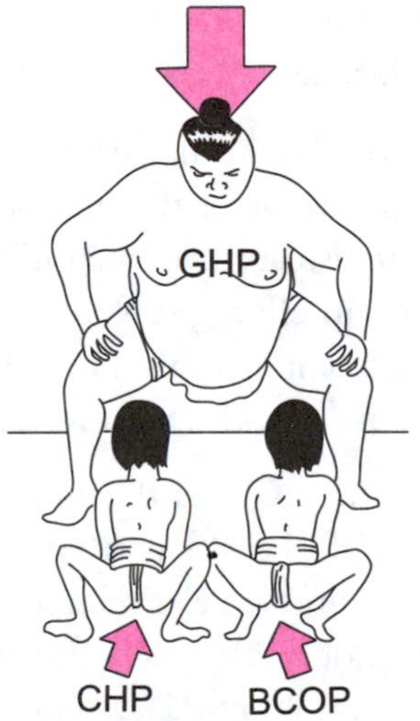

The forces and counter-forces involved in filtration within the renal corpuscle are like an adult sumo wrestler pushing against the two child sumo wrestlers. The force from the adult sumo wrestler is like the GHP and the combined push from the two child wrestlers are like the CHP and BCOP.

Description

The **juxtaglomerular apparatus** is vital to the **glomerular filtration rate (GFR)** and consists of the following two structures:

1. **Juxtaglomerular (JG) cells**: a group of modified smooth muscle cells around the afferent arteriole. They secrete the enzyme renin.

2. **Macula densa:** a group of modified epithelial cells within the wall of the distal convoluted tubule adjacent to the afferent arteriole. They act as *chemoreceptors* that sense changes in solute concentration of the filtrate.

Each glomerulus in every nephron filters the blood to produce filtrate. The GFR is the volume of filtrate produced every minute by all the nephrons in both kidneys and is equal to about 125 ml filtrate/min. The blood pressure (BP) in each glomerulus must be kept relatively constant to ensure that filtration continues. If the GFR rises or falls too far from this normal level, **nephron** function is impaired.

As shown in the illustration, a general scheme is used to either increase or decrease the GFR as needed. This involves controlling the blood flow through the glomerulus.

Analogy

Let's compare the glomerulus to a loop of garden hose that has some holes for filtration. If we expand the incoming end of the hose, this allows more water in and increases the water pressure. If we squeeze the outgoing end, this causes a backup of fluid in the loop and increases water pressure. Similarly, the glomerulus uses vasodilation and vasoconstriction of its afferent and efferent arterioles to change the BP in the glomerulus. By doing so, this directly changes the GFR.

Mechanisms

To control the GFR, three major mechanisms are used: (1) autoregulation, (2) neural regulation, and (3) hormonal regulation. Let's summarize each, in turn.

- *Autoregulation* is the dominant control mechanism at rest. It is subdivided into two mechanisms: (a) *smooth muscle mechanism*, and (b) *tubular mechanism*. The first uses stretching of the wall of the afferent arteriole as a stimulus to induce vasoconstriction of itself. This leads to a decrease in blood flow to the glomerulus and decreases GFR back to normal levels. The second mechanism involves the **macula densa**. For example, when BP increases, the GFR also increases. This results in an increase in filtrate flow in the renal tubules. Because this also increases the level of solute (*such as* Na^+ *and* Cl^-) in the filtrate, the macula densa detects this change and responds by constricting the afferent arteriole. This reduces blood flow to the glomerulus and decreases the GFR back to normal levels.

- *Neural regulation* occurs during periods of physical activity and stressful situations. Because it is the sympathetic division of the autonomic nervous system (ANS) that helps us respond to these situations, it also controls the GFR. Neurons are linked to the afferent arteriole through a reflex pathway. In an effort to deliver more blood to vital organs, less blood should be going to the kidneys. To accomplish this, nerve impulses stimulate the smooth muscle around the afferent arteriole to constrict which decreases the blood flow to the glomerulus and decreases GFR.

- *Hormonal regulation* involves the hormone angiotensin II and is triggered by a decrease in blood volume or BP This is part of the normal renin-angiotensin aldosterone system (see p. 442). Juxtaglomerular cells release the enzyme renin into the bloodstream, which leads to the production of angiotensin II, a powerful vasoconstrictor. By causing peripheral blood vessels to constrict, it increases systemic BP, which increases GFR.

Regulation of the Glomerular Filtration Rate

Juxtaglomerular apparatus

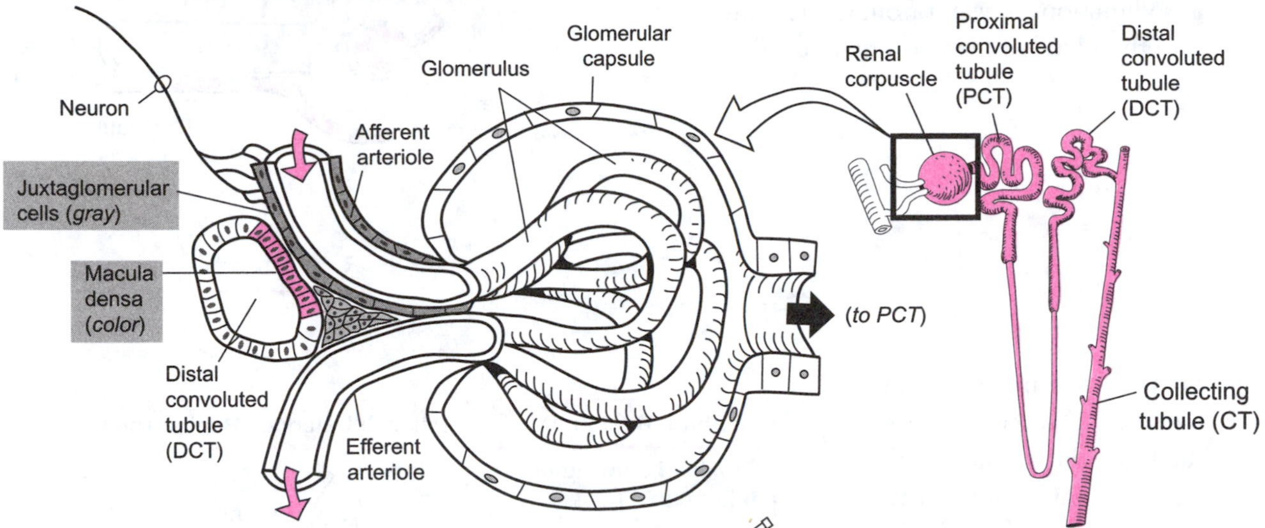

Cross-section of renal corpuscle

Nephron

General Scheme for Controlling the GFR

By increasing or decreasing the BP in the glomerulus

There are two ways to change the BP in the glomerulus	There are two ways to change the water pressure in the garden hose
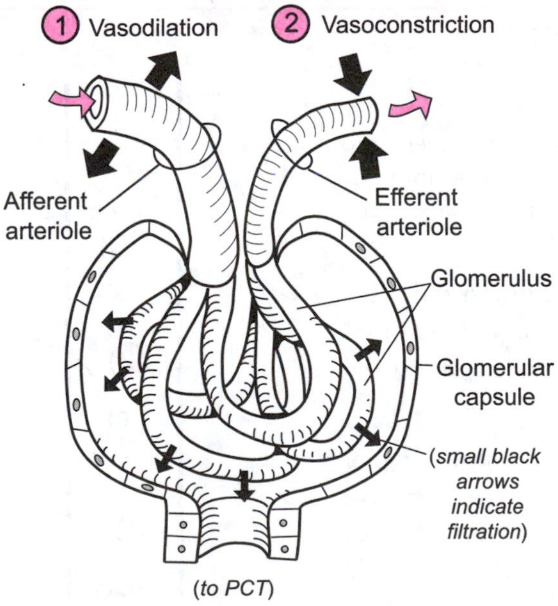	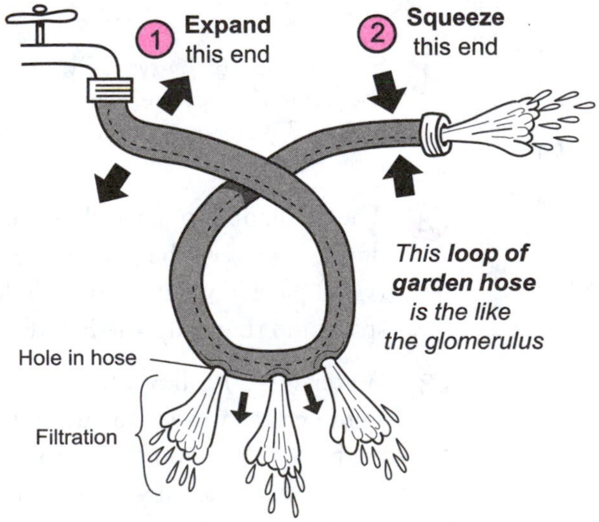

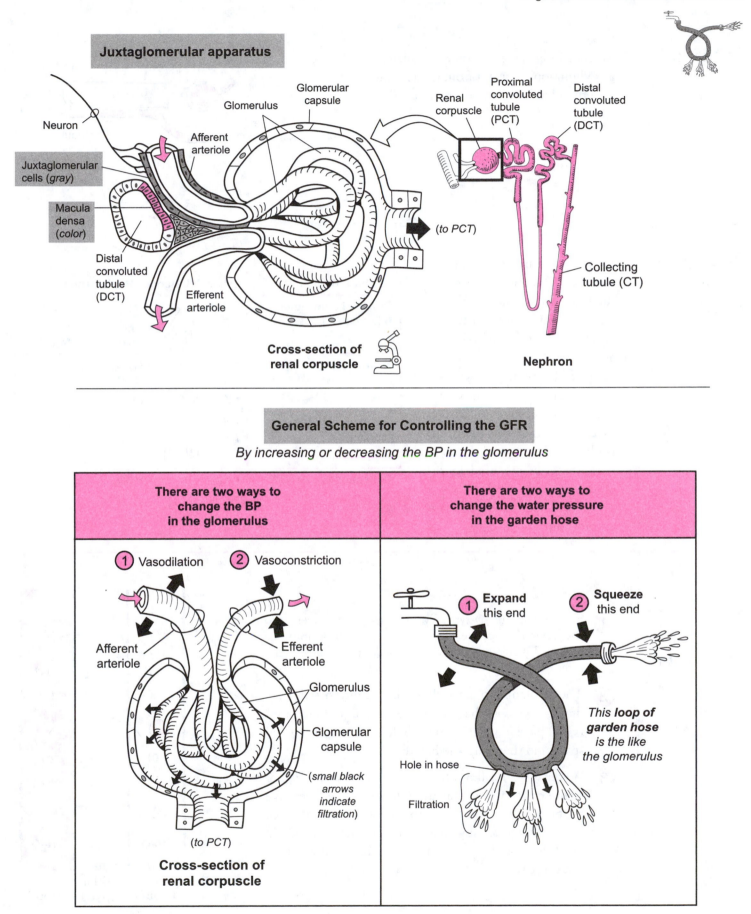

Cross-section of renal corpuscle

Description

Reabsorption is the transport of nutrients such as sodium ions or water from the renal tubule into the blood of the peritubular capillaries.

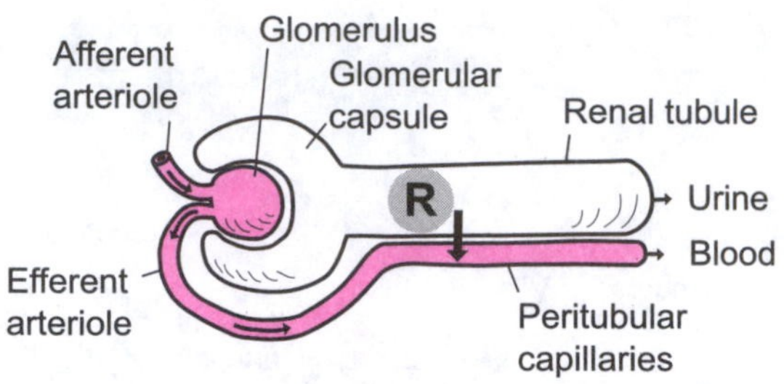

KEY
R = reabsorption

Reabsorption by Nephron Region

Some of the major substances that are reabsorbed include ions such as sodium (Na^+), chloride (Cl^-), potassium(K^+),calcium (Ca^{+2}), magnesium (Mg^{+2}), and bicarbonate (HCO_3^-), along with water, glucose, and amino acids. The table shows what substances are reabsorbed in each part of the nephron. Both active and passive transport processes are used in reabsorption (see p. 44). Some substances, such as glucose, are cotransported.

Nephron Region	Major Substances Reabsorbed
Proximal convoluted tubule (PCT)	• Na^+ • Cl^- • K^+ H_2O • (G) • Amino acids • HCO_3^- • Ca^{+2} • Mg^{+2}
Nephron loop	• Na^+ • Cl^- • K^+ H_2O • HCO_3^- • Ca^{+2} • Mg^{+2}
Distal convoluted tubule (DCT)	• Na^+ • Cl^- H_2O • Ca^{+2}
Collecting tubule (CT)	• Na^+ H_2O • HCO_3^-

General Scheme for Reabsorption

1. Sodium ions (Na^+) typically are transported actively into the **interstitial space**. These ions then diffuse into the blood. This creates a gradient of positive (+) charge in the interstitial space.

2. The gradient of *positive* charge in the interstitial space draws *negatively* charged anions such as chloride ions (Cl^-) passively out of the **tubule**, into the interstitial space and then into the **blood**.

3. Water (H_2O) follows the salt. Because of the higher solute concentration in the interstitial space relative to the tubule, water leaves the tubule via osmosis (see p. 48). Therefore, by creating this gradient of sodium and chloride ions, the nephron has really forced osmosis to occur.

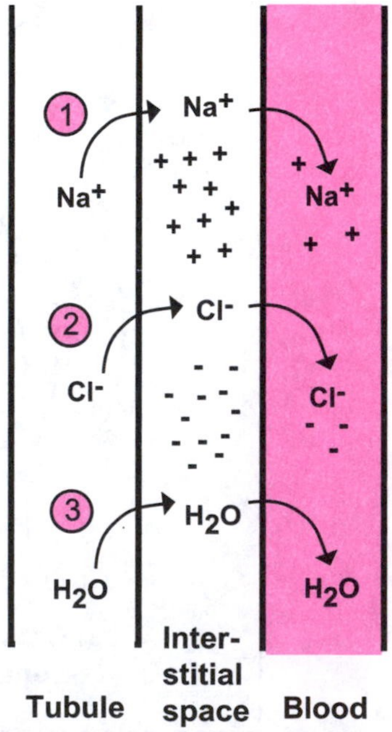

Nephron Function 2: Tubular Reabsorption

Reabsorption in the Nephron

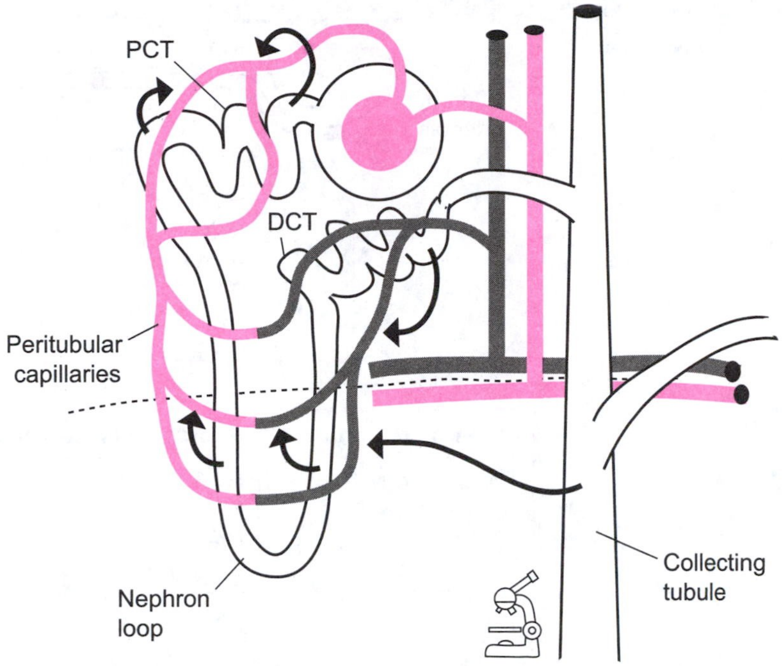

Amount of Filtrate Reabsorbed by Nephron Region

NEPHRON REGION	AMOUNT REABSORBED (out of 125 ml filtrate)
Proximal convoluted tubule (PCT)	100 ml
Nephron loop	7 ml
Distal convoluted tubule (DCT)	12 ml
Collecting tubule (CT)	5 ml

TOTAL REABSORBED: 124 ml

This is based on a rate of filtrate being formed at 125 ml/min by both kidneys. Therefore, 124 of the 125 ml is reabsorbed into the blood. About 80% of this total occurs in the PCT.

Functional Analogy

The reabsorption process in the nephron is like workers sorting different products on a conveyor belt. The substances filtered from the blood in the glomerulus contain a mixture of nutrients (ex: water, glucose, amino acids) and waste products (ex: urea, uric acid). The waste products remain in the tubules (PCT, nephron loop, DCT, and collecting tubule) while the nutrients must be reabsorbed back into the blood. The workers are like the membrane proteins in the tubules that transport specific substances into the blood. The tubule conveyor represents the filtrate moving through the renal tubules and the blood conveyor represents the blood in the peritubular capillaries that surround the tubules. In the end, the waste products are collected in the urine, which is stored in the urinary bladder and excreted from the body while the nutrients travel through the blood to finally be used by body cells.

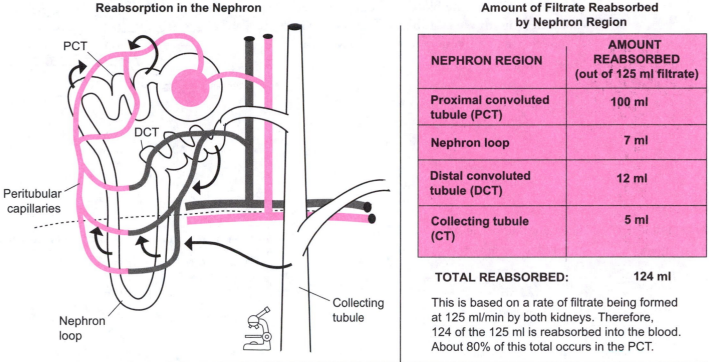

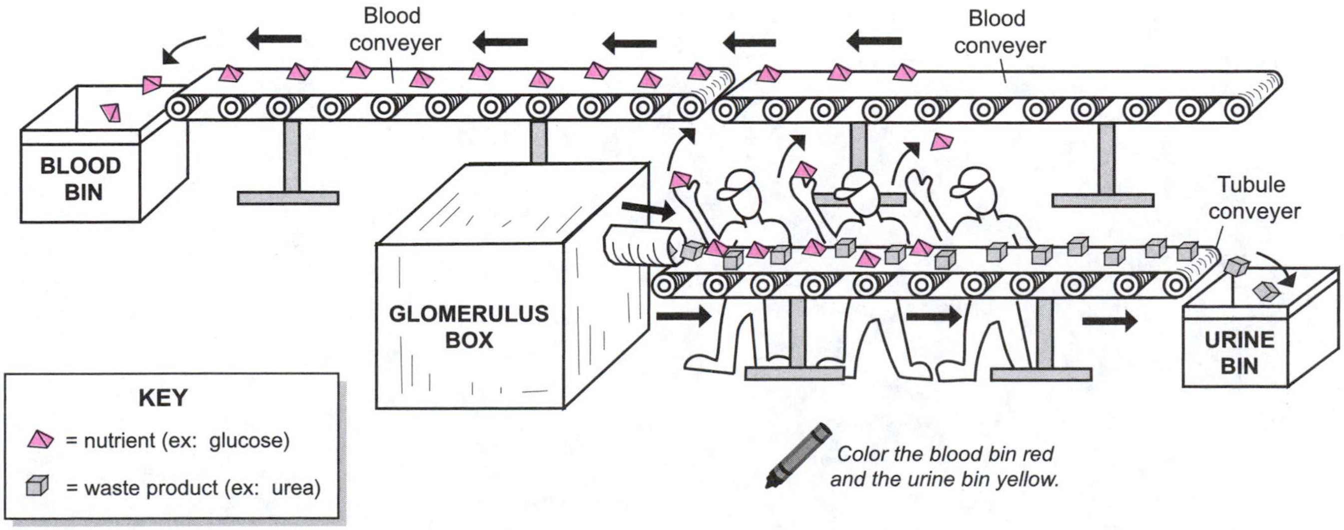

Color the blood bin red and the urine bin yellow.

Nephron Function 3: Tubular Secretion

Description

Secretion is the transport of substances such as hydrogen ions from the blood of the **peritubular capillaries** into the **renal tubules**. The purpose is to remove these substances from the body in the urine.

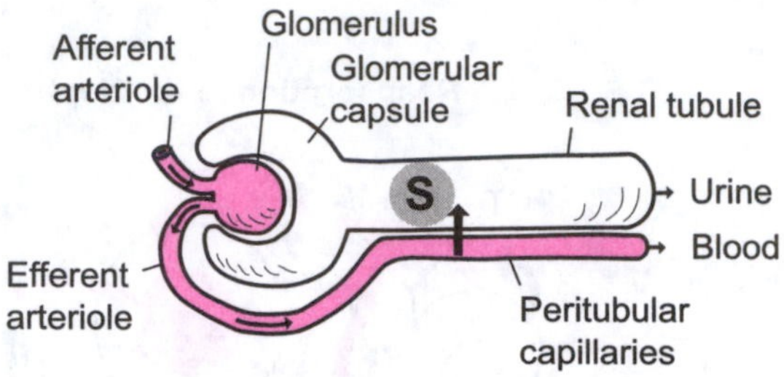

KEY
S = secretion

Secretion by Nephron Region

The table shows the substances that are secreted in each part of the nephron.

KEY
H^+ = hydrogen ions
K^+ = potassium ions
NH_4^+ = ammonium ions

Nephron region	Substance(s) secreted
Proximal convoluted tubule (PCT)	H^+ NH_4^+ Urea Creatinine
Nephron loop	Urea
Distal convoluted tubule (DCT)	(nothing)
Collecting tubule (CT)	H^+ K^+

Outcomes of Secretion

Tubular secretion has three important outcomes:

- to control blood pH: achieved by the secretion of hydrogen ions;
- to rid the body of nitrogen wastes; substances such as urea, creatinine, and ammonium ions are all products of metabolic processes that must be released in the urine; and
- to rid the body of unwanted drugs; drugs such as penicillin also are secreted and released in the urine.

Nephron Function 3: Tubular Secretion

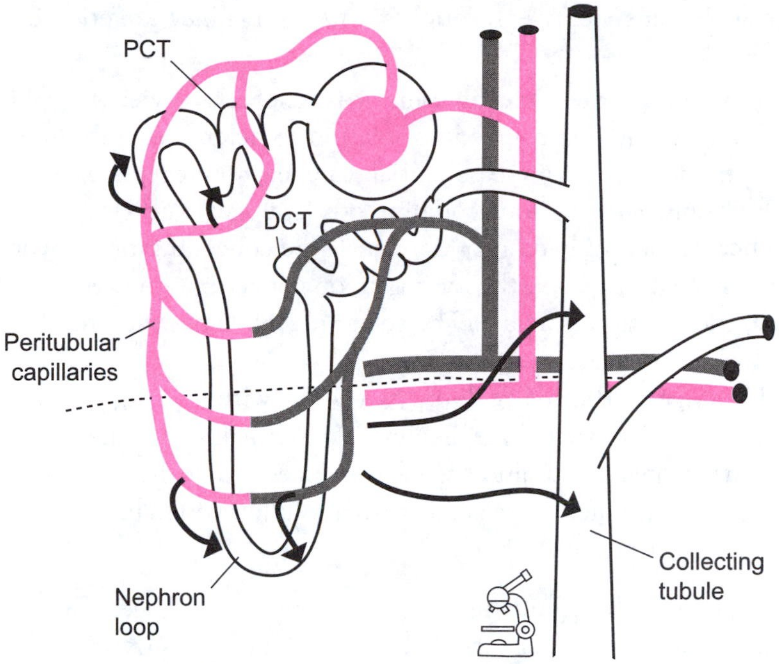

Functional Analogy

The secretion process in the nephron is like workers moving products from one conveyor belt to another. When certain substances such as potassium ions and hydrogen ions reach high levels in the blood, they are transported into the renal tubules and eventually released from the body in the urine. The workers are like the membrane proteins in the blood capillaries that transport substances into the renal tubules. The tubule conveyor represents the filtrate moving through the renal tubules, and the blood conveyor represents the blood in the peritubular capillaries that surround the tubules.

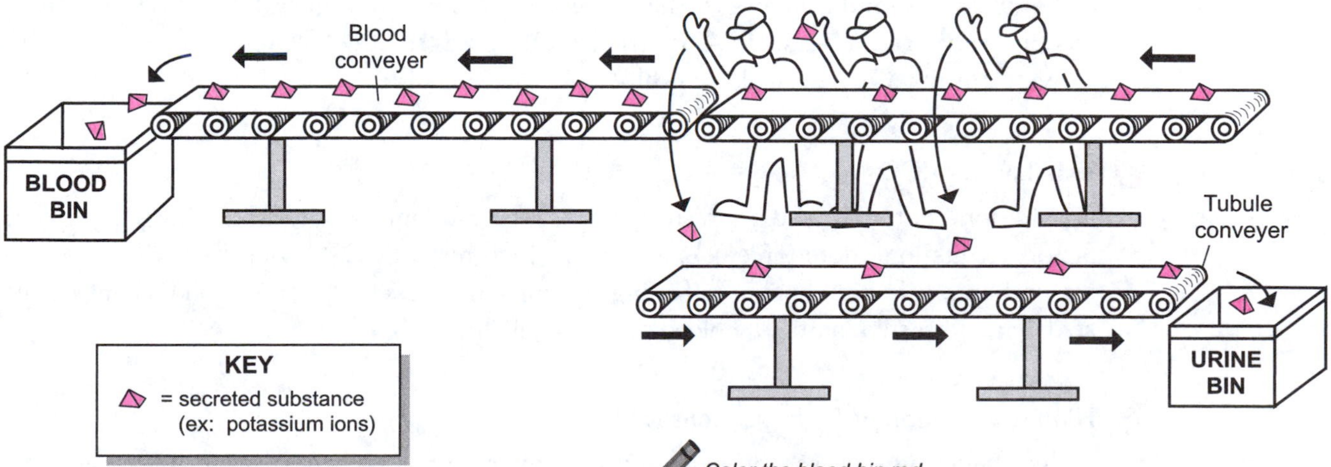

Color the blood bin red and the urine bin yellow.

Description

This module covers the three general mechanisms used to regulate acid-base balance in the body. These are buffer systems, exhalation of CO_2, and kidney excretion of hydrogen ions (H^+).

pH Scale

Before explaining these mechanisms, it is helpful to review the **pH scale**. It directly measures the amount of free hydrogen ions in a solution to determine if the solution is either acidic or alkaline (basic). Because this is an exponential scale, each unit represents a 10x change in the hydrogen ion concentration. All solutions contain both hydrogen ions (H^+) and hydroxide ions (OH^-). When the concentration of hydrogen ions equals the concentration of hydroxide ions (**$[H^+] = [OH^-]$**), a solution is neutral. When the hydrogen ion concentration is greater than the hydroxide ion concentration, the solution is acidic. The solution is alkaline when the hydrogen ion concentration is less than the hydroxide ion concentration.

The range of the pH scale is from 0–14, with 7 being the **neutral point**. Values below 7 are acidic, and values above 7 are alkaline. Weaker acids and bases have pH values closer to the neutral point, and stronger acids and bases have values nearer the ends of the scale. For example, a solution with a pH of 2 is much more acidic than a solution with a pH of 5.

Major Mechanisms

It is vital that body fluids remain in a normal pH range. The **normal blood pH** is **7.35–7.45**. Enzymes and other proteins in the blood are highly sensitive to pH changes and can lose their function if the blood becomes either too acidic or too alkaline. Byproducts from normal metabolism make the blood more acidic. The body uses different mechanisms to raise the pH back to normal. Here is a summary of the three major mechanisms:

1. **Buffer systems**

   A *buffer* is a chemical substance that resists changes in pH. It's like a vice that clamps down on the pH value. The body has several different buffering systems that temporarily bind free H^+, thus making the pH more alkaline. One such system is called the **carbonic acid-bicarbonate buffering system**. When carbon dioxide (CO_2) diffuses into the blood it reacts with water (H_2O) in the plasma to produce carbonic acid (H_2CO_3). This unstable acid then breaks down into hydrogen ions (H^+) and bicarbonate ions (HCO_3^-). When CO_2 levels are high, as during exercise, more carbonic acid is produced and the blood becomes more acidic. But when CO_2 levels are low, the reaction can run in reverse. In short, the HCO_3^- binds the free H^+ to produce carbonic acid and convert it into CO_2. This makes the blood pH more alkaline.

2. **Exhalation of CO_2**

   Respiration centers in the brain stem control the rate and depth of breathing. Increasing these factors results in a more forceful exhalation, which removes CO_2 from the body. As explained in the **carbonic acid-bicarbonate buffering system** above, less CO_2 in the blood means less carbonic acid being formed, making the blood pH more alkaline.

3. **Kidney excretion of hydrogen ions (H^+)**

   Unlike buffering systems, the kidneys actually *remove* the excess H^+ from the body and excrete them in the urine. This makes the urine *more* acidic and the blood *less* acidic. For the details of this process, see p. 436.

Introduction to Acid-Base Balance

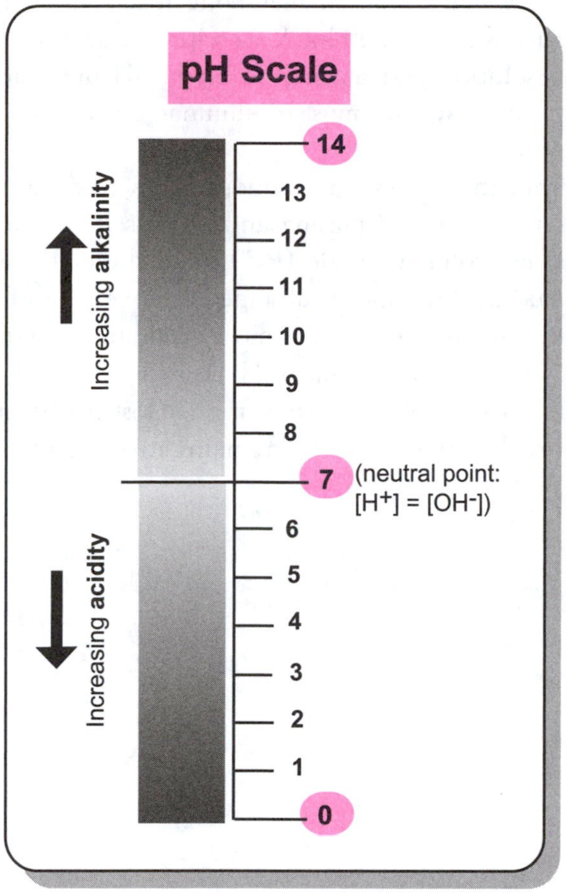

Goal: Maintain normal blood pH:

7.35 – 7.45

S

Change in normal blood pH

Color the numbers indicating the normal blood pH.

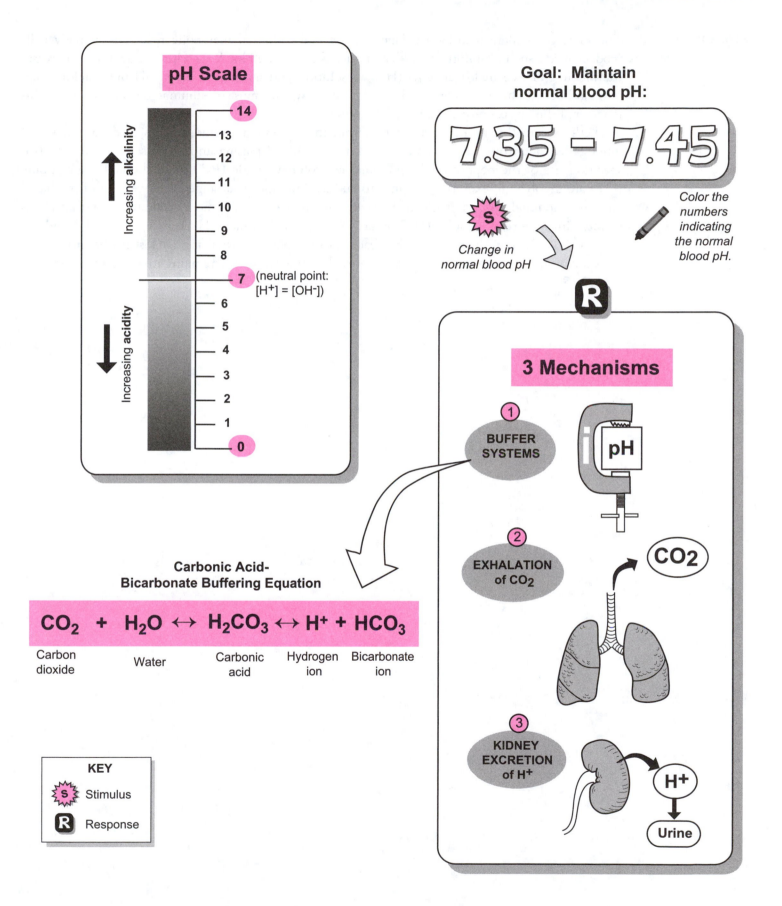

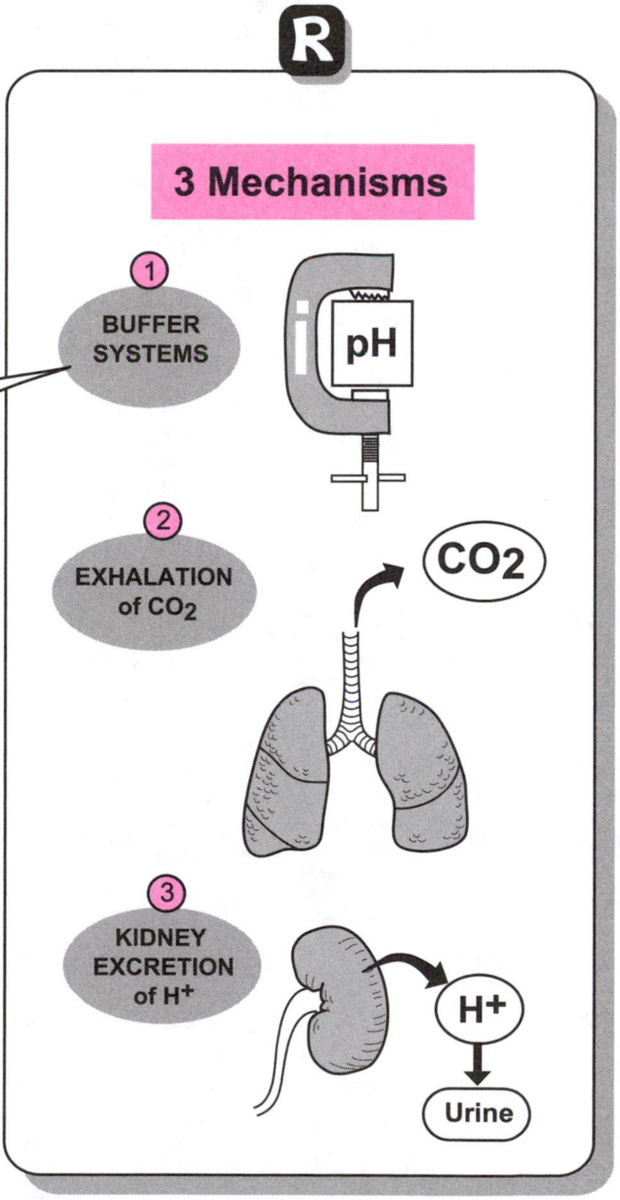

Carbonic Acid-Bicarbonate Buffering Equation

$$CO_2 + H_2O \leftrightarrow H_2CO_3 \leftrightarrow H^+ + HCO_3^-$$

CO_2	H_2O	H_2CO_3	H^+	HCO_3
Carbon dioxide	Water	Carbonic acid	Hydrogen ion	Bicarbonate ion

KEY

S Stimulus

R Response

Description

The kidneys regulate acid-base balance in the body. Normal metabolic reactions in body cells produce acids, such as sulfuric acid, that are deposited in the blood. As with any acid, this increases the amount of free hydrogen ions (H^+) in solution and makes the blood pH more **acidic** than normal. Because these acids are produced day after day, they must be eliminated from the body. This is accomplished by excreting the H^+ in the urine.

Let's explain this idea a bit further. Within the kidneys are millions of microscopic functional units called **nephrons**, which constantly filter the blood plasma and process it into urine. Two specific parts of the nephron—the **proximal convoluted tubule (PCT)** and the **collecting tubule (CT)** are involved in excreting H^+ into the urine. Any time a substance is moved out of the blood and into the renal tubules, it is called *secretion*. Consequently, it is more accurate to say that H^+ is secreted into the **renal tubules** (PCT and CT). At the same time as H^+ is being secreted, bicarbonate ions (HCO_3^-) are being reabsorbed back into the blood so they are not lost in the urine. This constant removal of metabolic acids ensures that the blood pH remains in its normal range of 7.35–7.45.

Renal Regulation of Acid–Base Balance

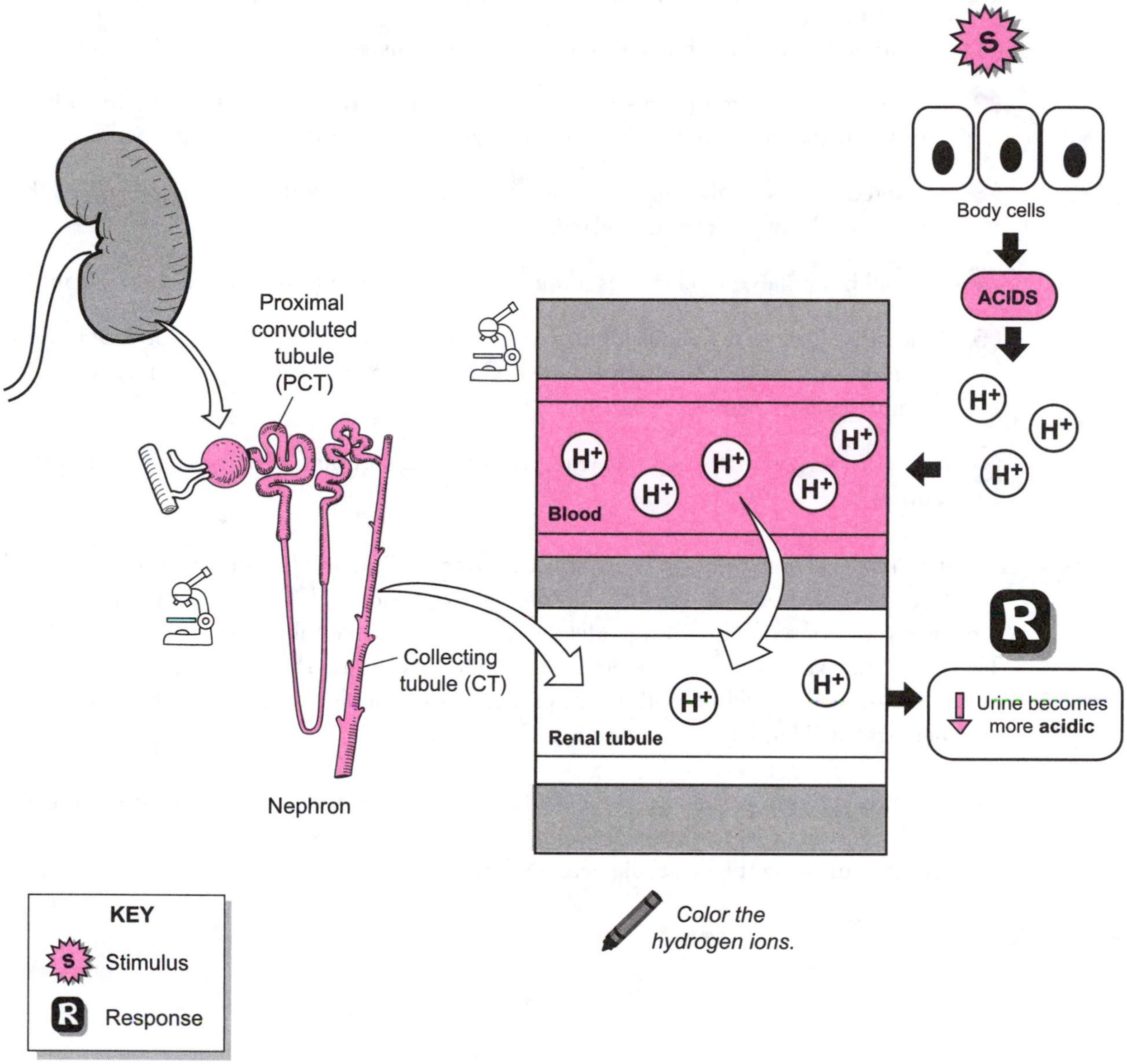

Color the hydrogen ions.

Description

Antidiuretic hormone (ADH) helps conserve water. Imagine a hot day where you are working outdoors, sweating, and getting dehydrated. This triggers the following series of events:

1. Sweating increases the **blood osmotic pressure.** This serves as a stimulus (S).
2. This increase in osmotic pressure is detected by **osmoreceptors** within the **hypothalamus** that constantly monitor the osmolarity ("saltiness") of the blood.
3. Osmoreceptors stimulate groups of neurons within the **hypothalamus** to release **ADH** from the **posterior lobe** of the **pituitary gland**.
4. Like all hormones, ADH travels through the bloodstream to its various target organs.*
5. One of its targets, the **renal tubules** in the **kidney**, contains receptors for ADH. Most of these receptors are in the **collecting tubules (CT)**. When ADH binds to its receptor, it makes the membrane more permeable to water.
6. The final response (R) is an increase in water **reabsorption**, which leads to a decrease in **urine output**.

Now imagine that you drink lots of water after working outdoors to get rehydrated. What happens? The excess water decreases the blood osmotic pressure. This inhibits osmoreceptors, which, in turn, stops secretion of ADH. If no ADH is released, the renal tubules remain impermeable to water. Therefore, the excess water is retained by the renal tubules and the result is an increase in urine output. In short, while ADH helps the body conserve water, it also controls whether the urine is diluted or concentrated. That's one important little hormone!

* Other targets that have receptors for ADH are sweat glands and arterioles. In sweat glands, ADH decreases perspiration to conserve water. In arterioles, it causes the smooth muscle in the wall to constrict, narrowing the vessel diameter and increasing blood pressure.

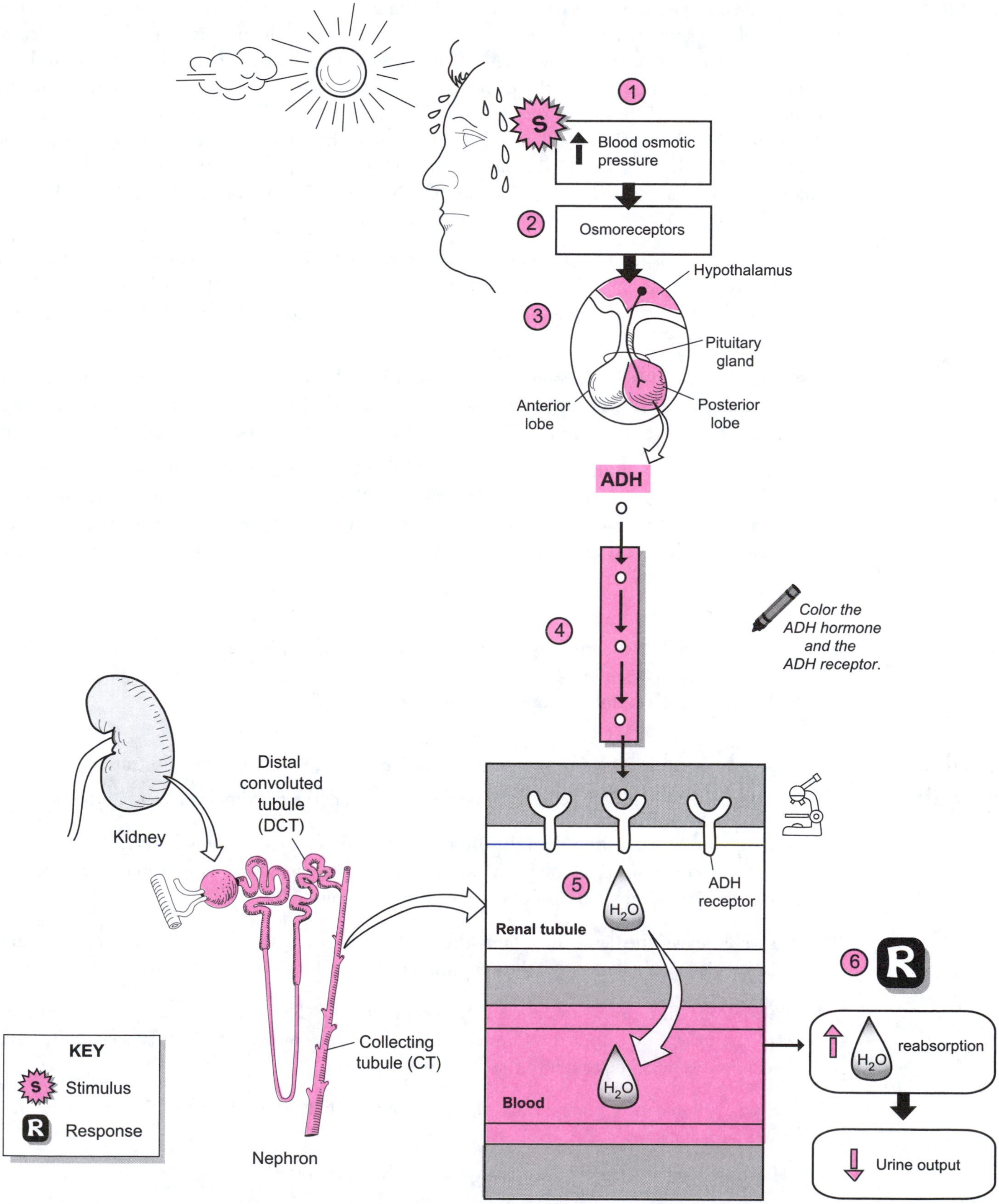
1
S
Blood osmotic pressure
2
Osmoreceptors
Hypothalamus
3
Pituitary gland
Anterior lobe
Posterior lobe
ADH
4
Color the ADH hormone and the ADH receptor.
Kidney
Distal convoluted tubule (DCT)
Collecting tubule (CT)
Nephron
5
H_2O
ADH receptor
Renal tubule
Blood
H_2O
6
R
H_2O reabsorption
Urine output
KEY
S Stimulus
R Response

Description

This module covers the **countercurrent multiplier** mechanism in the **nephron loop** of the **nephrons**. First, though, let's summarize a few key things about the nephron loop. It is composed of a **descending limb (DL)** and an **ascending limb (AL)**. Filtrate from the glomerular capsule flows down the DL and up the AL, before entering the **distal convoluted tubule (DCT)**, followed by the **collecting tubule (CT)**. The DL has a thinner membrane that is permeable to water but not salt, such as sodium ions (**Na^+**) and chloride ions (**Cl^-**). In contrast, the membrane of the AL becomes thicker (as indicated by a heavier line in the illustration). This thicker portion is impermeable to water and contains protein pumps used to actively transport Na^+ and Cl^- out of the AL and into the **extracellular fluid (ECF)**.

- Name? Why is it called the **countercurrent multiplier?** The term *countercurrent* refers to the flow of filtrate in opposite directions in the nephron loop—down the DL and up the AL. The term *multiplier* refers to the fact that it increases (or multiplies) the salt concentration in the renal medulla.
- **Osmolarity**? This is a measure of the amount of dissolved particles in 1 liter of solution. In living systems, smaller units called **millisosmoles (mOsm)** are used. For our purposes, this is a measure of the saltiness of the solution. The higher the number, the saltier the solution, and vice versa.
- Purpose? Notice from the illustration that the ECF deep in the medulla is four times saltier than the ECF near the cortex (1,200 mOsm vs. 300 mOsm). The general purpose of the countercurrent multiplier mechanism is to maintain a constant gradient of salt deep within the renal medulla. Why? This allows the collecting tubule (CT) to reabsorb water via osmosis, thereby concentrating the urine.

Analogy

The salt in your ECF is the same as that in table salt—sodium chloride (NaCl). In solution, sodium chloride ionizes into Na^+ and Cl^-.

A salt shaker will remind you of this connection to everyday life.

Mechanism: Step by Step

For the step-by-step process of this cyclical mechanism, refer to #1–#5 in the illustration:

1. Salty filtrate (300 mOsm) enters the nephron loop from the **proximal convoluted tubule** (PCT).
2. Because the ECF is saltier than the filtrate, water leaves the descending limb via osmosis. Then, it is reabsorbed into the blood. The saltier the ECF becomes, the more water is removed from the descending limb. This loss of water makes the filtrate saltier.
3. As water leaves the descending limb, the filtrate becomes saltier until it reaches a peak of about 1,200 mOsm at the bottom of the nephron loop.
4. As the salty filtrate moves up the ascending limb, salt (Na^+ and Cl^-) is actively transported into the ECF. The saltier the filtrate, the more salt is pumped out. As this occurs, the filtrate becomes less salty, moving into the DCT at 100 mOsm.
5. As salt is constantly pumped out of the ascending limb, the saltier the ECF becomes in the renal medulla.

   This ensures that the medulla is always saltier than the cortex.

Cycle repeats. Go back to step #1.

Countercurrent Multiplier in the Nephron Loop

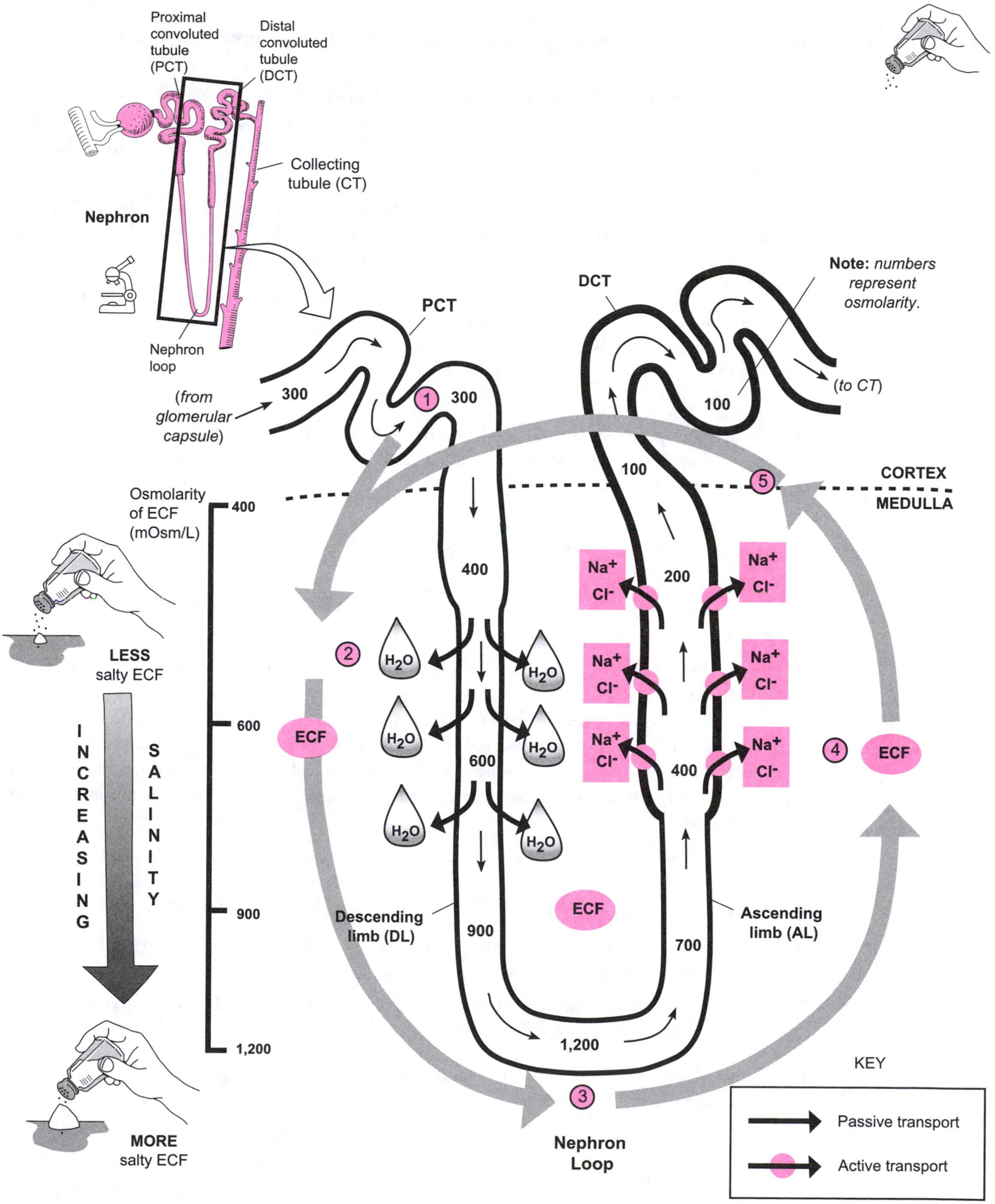
Proximal convoluted tubule (PCT)
Distal convoluted tubule (DCT)
Collecting tubule (CT)
Nephron
Nephron loop
PCT
DCT
Note: numbers represent osmolarity.
(from glomerular capsule)
300
1
300
100
(to CT)
100
5
CORTEX
MEDULLA
Osmolarity of ECF (mOsm/L)
400
600
900
1,200
LESS salty ECF
INCREASING SALINITY
MORE salty ECF
400
200
Na+ Cl-
2
H2O
ECF
600
400
4
ECF
Descending limb (DL)
900
ECF
700
Ascending limb (AL)
1,200
3
Nephron Loop
KEY
Passive transport
Active transport

Description

This module covers the **countercurrent exchanger** mechanism in the **nephron loop** of the nephrons. First, let's go over a few fundamentals:

- Name? Why is it called the countercurrent exchanger? The term *countercurrent* refers to the flow of blood in opposite directions in the **vasa recta**—down one side and up the other. The term *exchanger* refers to the fact that water is exchanged for salt along the length of the vasa recta.
- **Osmolarity**? This is a measure of the amount of dissolved particles in 1 liter of solution. In living systems, smaller units called **millisosmoles (mOsm)** are used. For our purposes, this is a measure of the saltiness of the solution. The higher the number, the saltier the solution, and vice versa.
- Purpose? The general purpose of the countercurrent exchanger mechanism is to maintain the gradient of salt deep within the renal medulla that was established by the countercurrent multiplier mechanism. Without the countercurrent exchanger, the salt would be carried away by the blood in the vasa recta and the gradient would be lost.

Analogy

The salt in your extracellular fluid (ECF) is the same as that in table salt—sodium chloride. In solution, it ionizes into sodium ions (Na^+) and chloride ions (Cl^-). A salt shaker will remind you of this connection.

Mechanism

The **countercurrent multiplier mechanism** (see p. 438) created a salt gradient in the extracellular fluid (**ECF**) of the renal medulla. Here are the steps:

1. The gradient allowed water to be reabsorbed from the **collecting tubule (CT)** by osmosis, thereby concentrating the urine.
2. **Urea** is one of the waste products formed in the nephron tubules. Although most of the urea remains in the tubules, some of it is recycled by diffusing out of the permeable lower end of the CT and into the permeable descending limb of the nephron loop. Because the thickened portions of the nephron (as indicated by a heavier line) in the ascending limb and DCT are impermeable to urea, it can't diffuse out of that portion of the tubule. The net result of this recycling is that urea adds substantially to the total solute concentration in the ECF, which increases the ECF's osmolarity.
3. The blood capillary called the **vasa recta** is adjacent to the **nephron loop** and follows the same looping pattern. As mentioned previously, if the salt were to diffuse into the blood in the vasa recta, the salt gradient could be lost. The countercurrent exchanger stops this problem.
4. As blood flows downward in the vasa recta, salt (Na^+ and Cl^-) diffuses into the blood while water leaves the blood by osmosis.
5. In contrast, as blood flows upward in the vasa recta, water in the ECF moves into the blood by osmosis while salt diffuses out of the blood.

In short, the blood carries away more water than salt. The salt is recycled in the vasa recta, so most of it remains in the ECF, which stabilizes the salt gradient.

Countercurrent Exchanger in the Nephron Loop

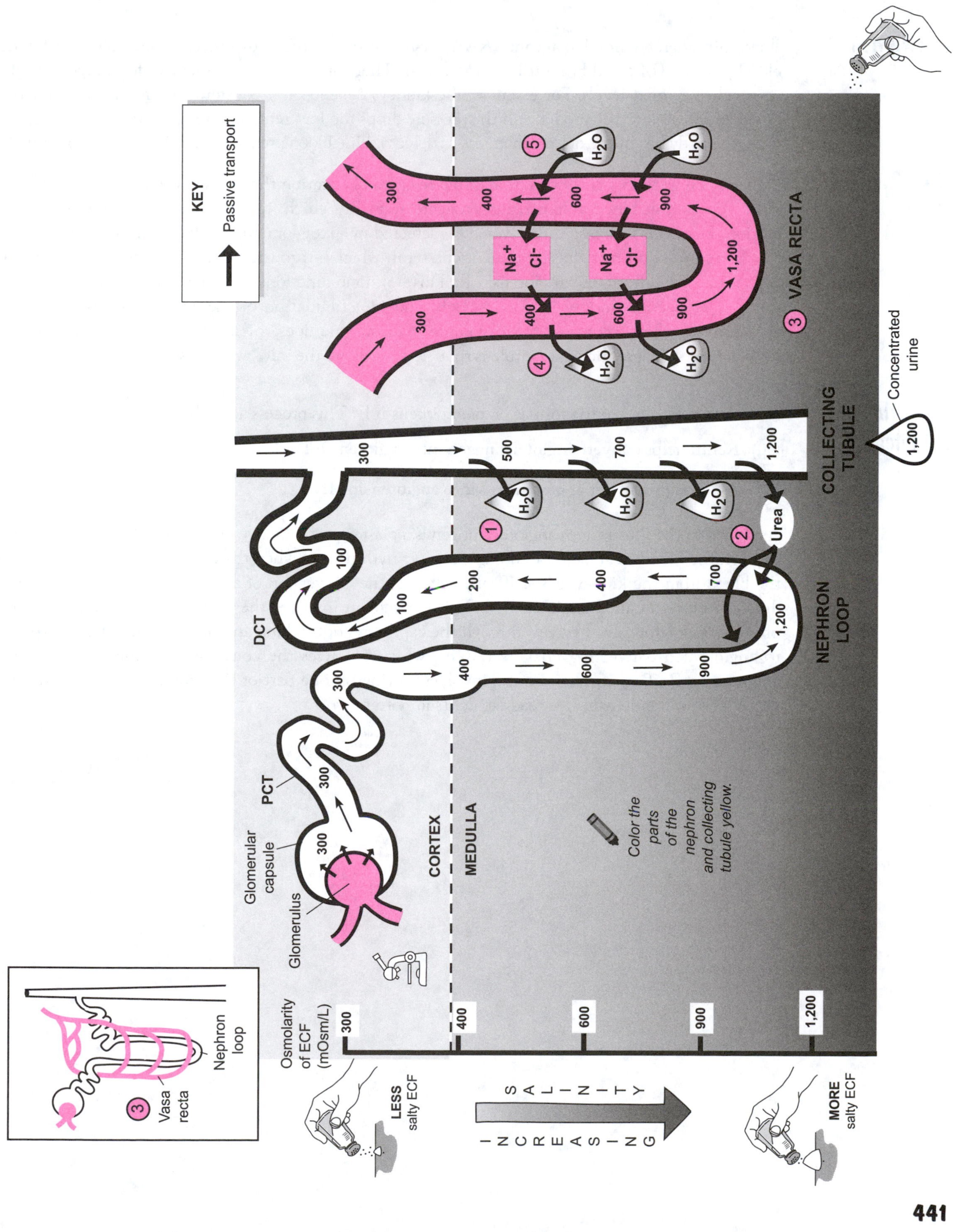

3
Vasa recta
Nephron loop
KEY
Passive transport
Osmolarity of ECF (mOsm/L)
300
400
600
900
1,200
LESS salty ECF
INCREASING SALINITY
MORE salty ECF
Glomerulus
Glomerular capsule
PCT
DCT
CORTEX
MEDULLA
Color the parts of the nephron and collecting tubule yellow.
H_2O
Urea
Na^+ Cl^-
NEPHRON LOOP
COLLECTING TUBULE
3 VASA RECTA
Concentrated urine

Description

The **renin-angiotensin-aldosterone (RAA)** system is one of the body's mechanisms to detect falling blood pressure (BP) and bring it back to normal. There are serious consequences for allowing BP to drop below normal levels. For example, the kidneys depend on a constant, normal BP to filter the blood and remove waste products. If the pressure drops too low, renal failure could result. Not surprisingly, the body has numerous mechanisms to detect falling blood pressure and bring it back to normal.

RAA System: Overview

Let's say you were in a car accident and suffered a bad laceration that caused you to lose some blood. This reduction in blood volume causes a decrease in BP, which, in turn, sets a cascade of events in motion. First, this decrease in BP would be detected by juxtaglomerular cells within the kidney, and they would respond by secreting the enzyme renin which—through a chain of events—leads to the activation of the hormone angiotensin II. This hormone functions to increase BP in two important ways: (1) vasoconstriction of blood vessels, and (2) stimulation of the adrenal cortex to secrete **aldosterone**, which increases blood volume. Aldosterone does this by increasing reabsorption of sodium ions. Because the general rule is that "water follows the salt," water also is reabsorbed.

Formation of Angiotensin II: Detail

How does renin lead to the formation of angiotensin II? This process actually involves two steps:

① **Renin** helps convert **angiotensinogen** into **angiotensin I**

② **ACE** helps convert **angiotensin I** into **angiotensin II**.

As shown in the illustration, angiotensinogen is a plasma protein made by the liver. It normally travels through the bloodstream without doing much of anything. But when the enzyme renin is released into the blood from the kidney, it specifically catalyzes the conversion of angiotensinogen into a slightly different protein called angiotensin I. This conversion involves the cutting off of one part of the protein. As angiotensin I travels through the capillaries of the lung and other tissues, it gets exposed to **angiotensin-converting enzyme (ACE)**, which catalyzes the conversion of angiotensin I into **angiotensin II**. This conversion also involves cutting off one part of the protein and completes the process. Now angiotensin II is ready to do its important work.

RAA System: Overview

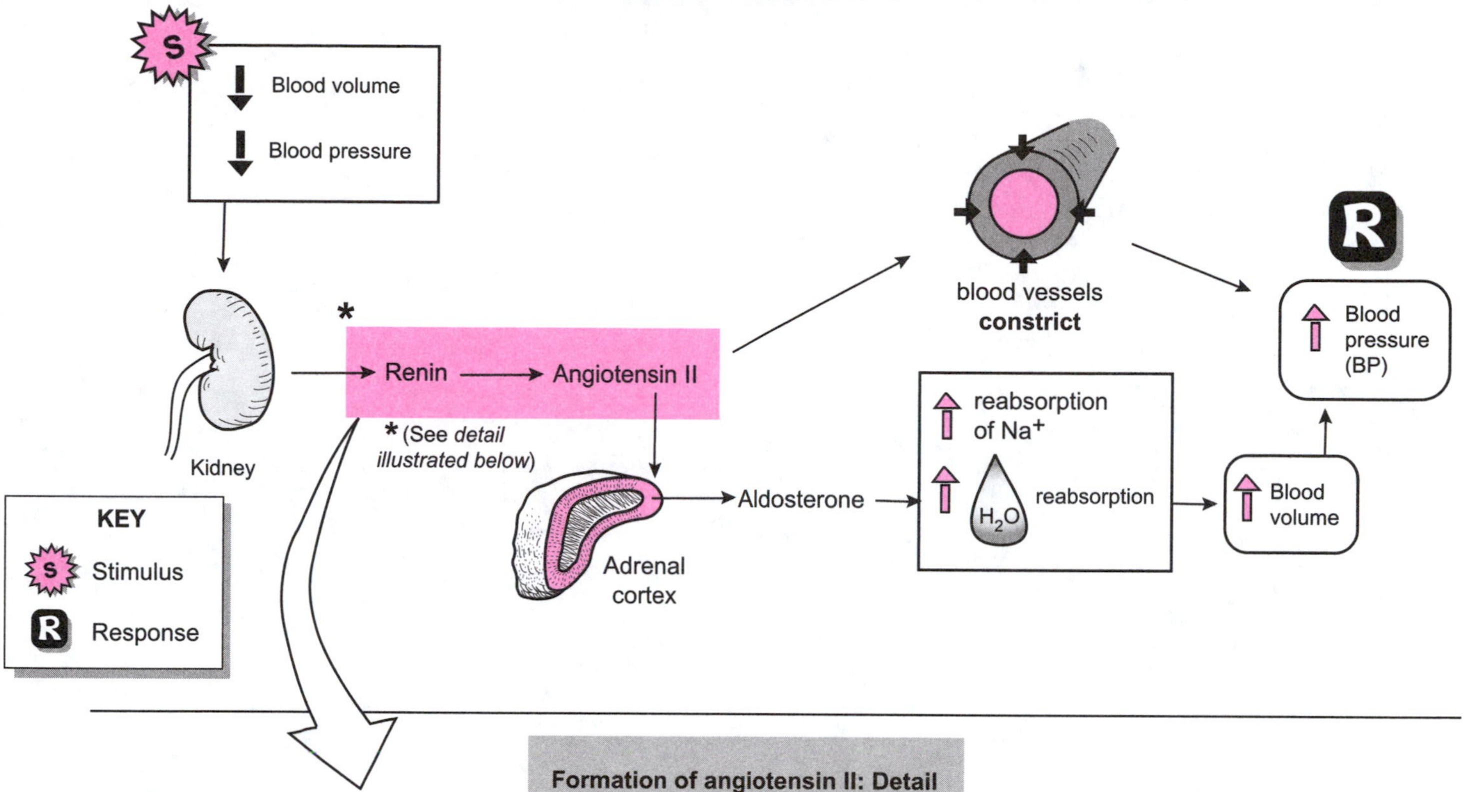

Formation of angiotensin II: Detail

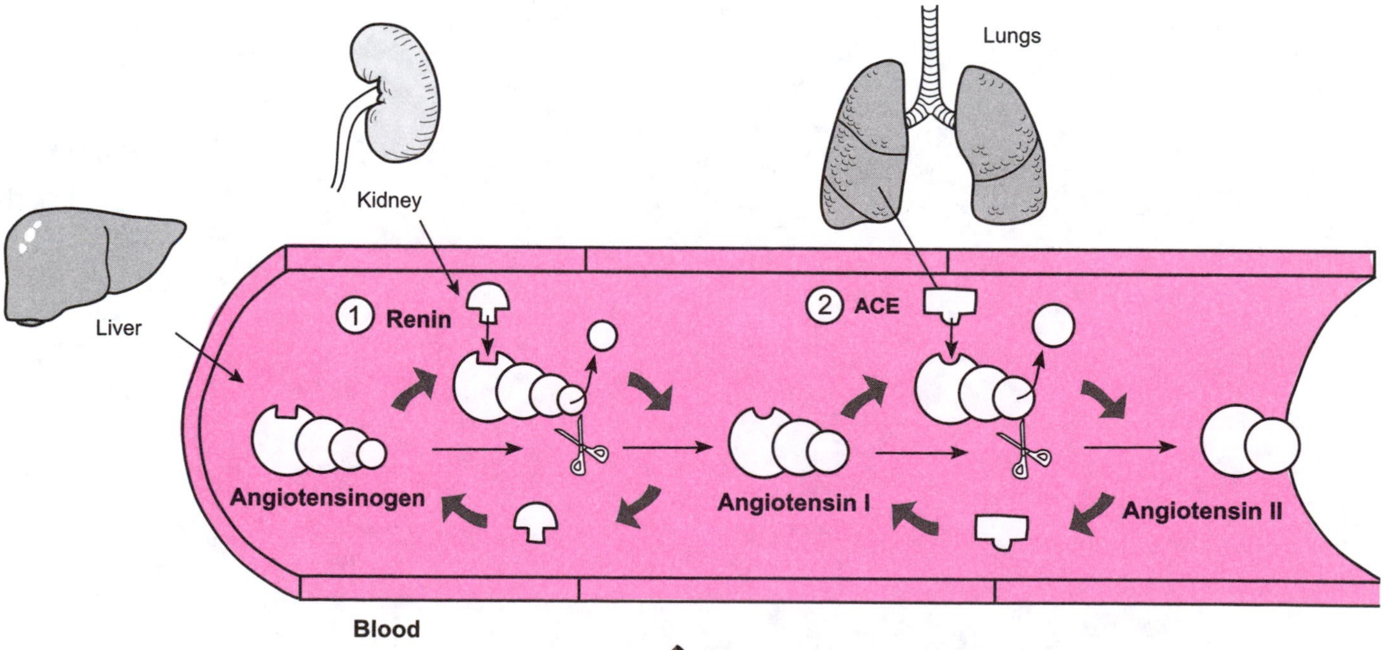

Color the renin, ACE, angiotensinogen, angiotensin I, and angiotensin II different colors.

Notes

Reproductive Systems

Description

The male and female reproductive systems are introduced here through means of comparison. Both reproductive systems are derived from the same general tissues and are composed of the following three general structures:

- **Gonads**: produce the gametes (*sex cells*)
- **Ducts**: transport the gametes to the site of fertilization or outside the body (the "plumbing of the reproductive system")
- **Accessory glands**: add liquid secretions to the reproductive tract to act as a lubricant or a medium in which the gametes are transported

For the external genitalia, similar structures between the sexes are:

- **Penis and clitoris**: both contain erectile tissues
- **Scrotal sac and labial folds**

Comparison

The table below gives a simplified, not comprehensive, structural comparison between male and female reproductive systems.

Structure/Product	Male	Female
Gonads *(glands that produce gametes)*	Testes	Ovaries
Gametes *(sex cells)*	Spermatozoa *(sperm cells)*	Ova *(egg cells)*
Ducts *(transport gametes)*	Epididymis, ductus deferens, and urethra	Uterine tube
Accessory glands *(add liquid secretions to the reproductive tract)*	Seminal vesicles, prostate gland, bulbourethral glands	Greater vestibular glands

Both sexes have a pair of gonads: in the male, two testes within the scrotal sac and, in the female, two ovaries in the pelvic cavity. The testes produce **sperm cells**, while the ovaries produce ova. The duct system in the male is longer and more complex than in the female. It connects the testes to the external urethral orifice, where sperm cells are released from the body. The female has only one main duct—the uterine tube—which connects the ovaries to the uterus. If fertilization occurs, the embryo implants in the **endometrium**—the innermost lining of the uterus.

The three accessory glands in the male produce a collective secretion called the **seminal fluid**. A pair of **seminal vesicles** secrete most of this, the **prostate gland** adds a milky secretion, and the **bulbourethral glands** contribute the smallest amount. In the female, the **greater vestibular glands** secrete a mucus that lines the inside of the **vagina**.

Both sexes use many of the same hormones to regulate reproductive processes. Both use the anterior pituitary hormones—**follicle-stimulating hormone (FSH)** and **luteinizing hormone (LH)** to regulate gonadal activity. The details are explained in other modules. In the testes, hormone-producing cells called interstitial cells secrete the "masculinizing" hormone **testosterone**. During puberty, it stimulates the development of all the secondary sexual characteristics in the male, such as deepening of the voice, enlargement of the genitals, and distribution of body hair. Similarly, the ovaries produce the "feminizing" hormone **estrogen**, which is responsible for the secondary sexual characteristics in the female, such as enlargement of the breasts, widening of the pelvis, and development of body hair. Although it is easy to focus on gender differences, the similarities are numerous.

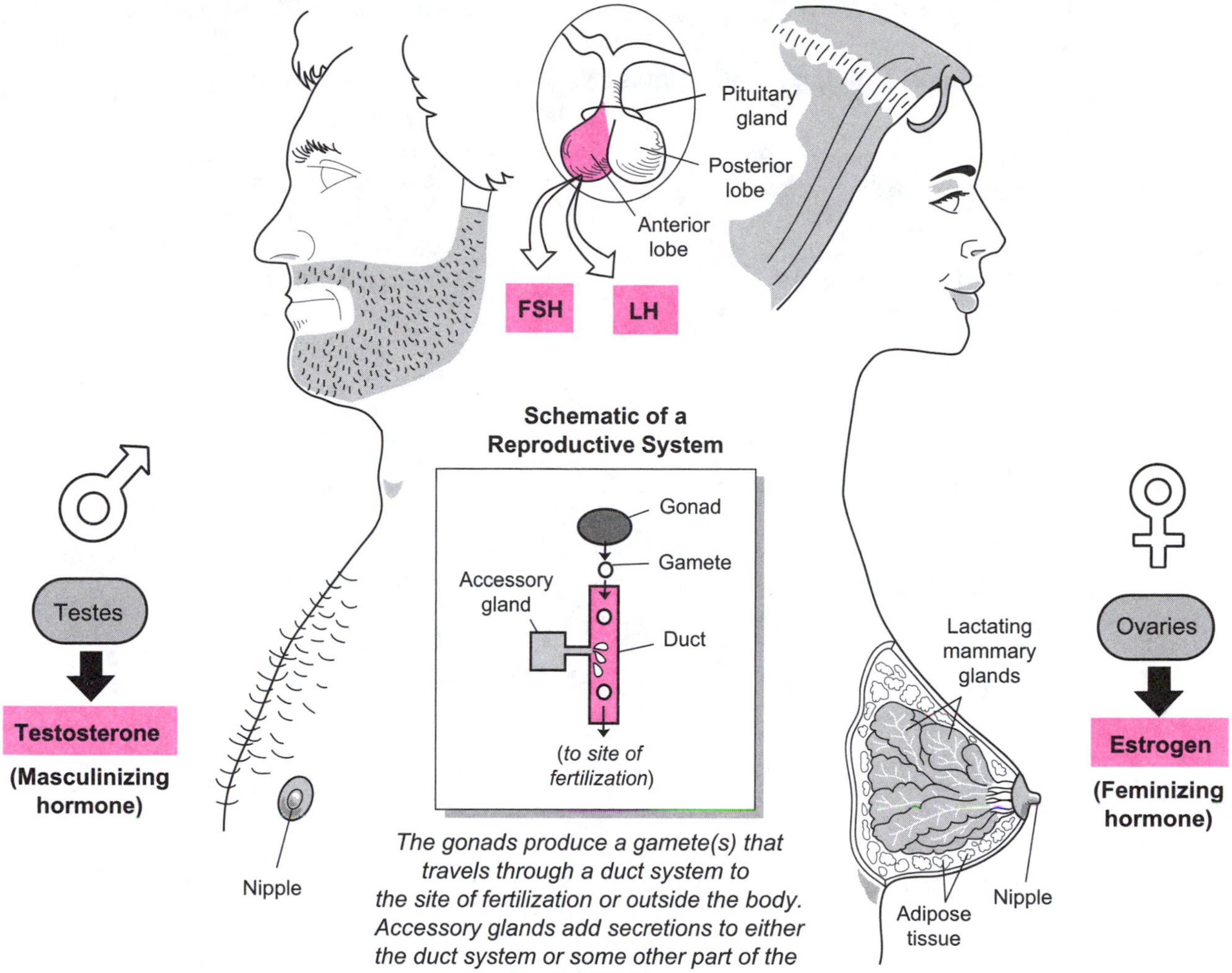

The gonads produce a gamete(s) that travels through a duct system to the site of fertilization or outside the body. Accessory glands add secretions to either the duct system or some other part of the reproductive tract.

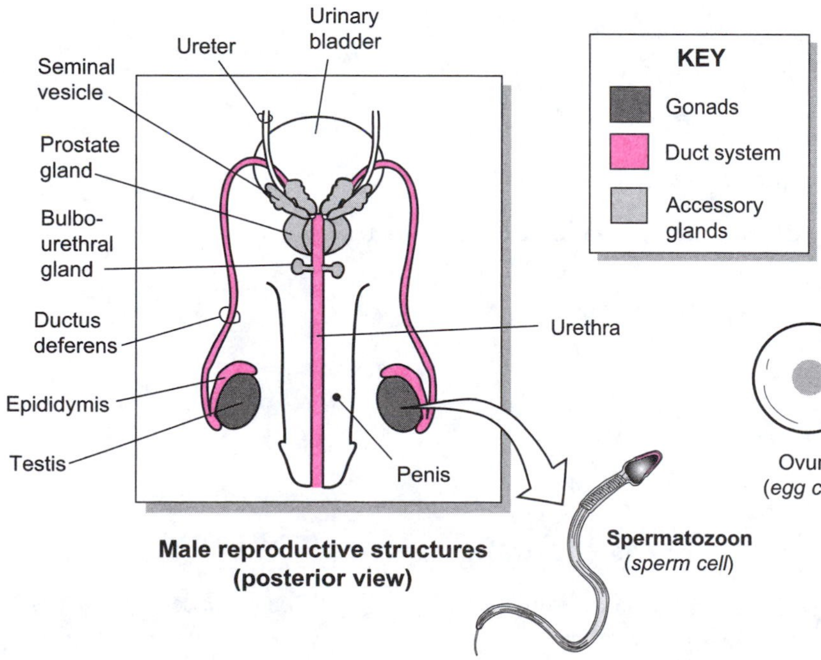

Male reproductive structures (posterior view)

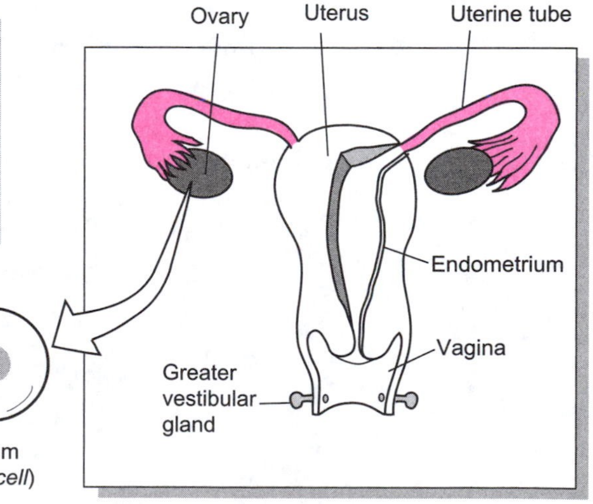

Female reproductive structures (posterior view)

Detailed Pathway

1. Lumen of seminiferous tubule
   ⬇
2. Rete testis
   ⬇
3. Efferent ductules
   ⬇
4. Epididymis
   ⬇
5. Ductus (*vas*) deferens
   ⬇
6. Ampulla of ductus deferens
   ⬇
7. Ejaculatory duct
   ⬇
8. Prostatic urethra
   ⬇
9. Membranous urethra
   ⬇
10. Penile urethra
    ⬇
11. External urethral orifice

Simplified Pathway

Seminiferous tubule
   ⬇
Rete Testis
   ⬇
Efferent ductules
   ⬇
Epididymis
   ⬇
Ductus deferens
   ⬇
Ejaculatory duct
   ⬇
Urethra

Study Tip

Use this mnemonic for the simplified pathway shown above:

"*S*ome *R*eally *E*lderly *E*lephants *D*on't *E*ven *U*rinate!"

Key to Illustration

A. Spermatagonium   C. Secondary spermatocyte   E. Spermatozoon (*sperm cell*)
B. Primary spermatocyte D. Spermatid

Ureters
5
Urinary bladder
Note: Black arrows indicate the pathway spermatozoa follow through the male duct system.
6
Pubic symphysis
8
10
Seminal vesicle
7
Corpus cavernosum
9
Skin
5
Corpus spongiosum
Anus
Prostate gland
Bulbourethral gland
Prepuce
4
Glans
Spermatic cord
Testicular n.
Testicular a.
Testicular v.
3
Sagittal View
11
5
Lobule
2
1
Tail
E
D
C
B
A
Testis detailed view
Neck (midpiece)
Head
Acrosome
Spermatogenesis in the wall of the seminiferous tubule
Spermatozoon (sperm cell)

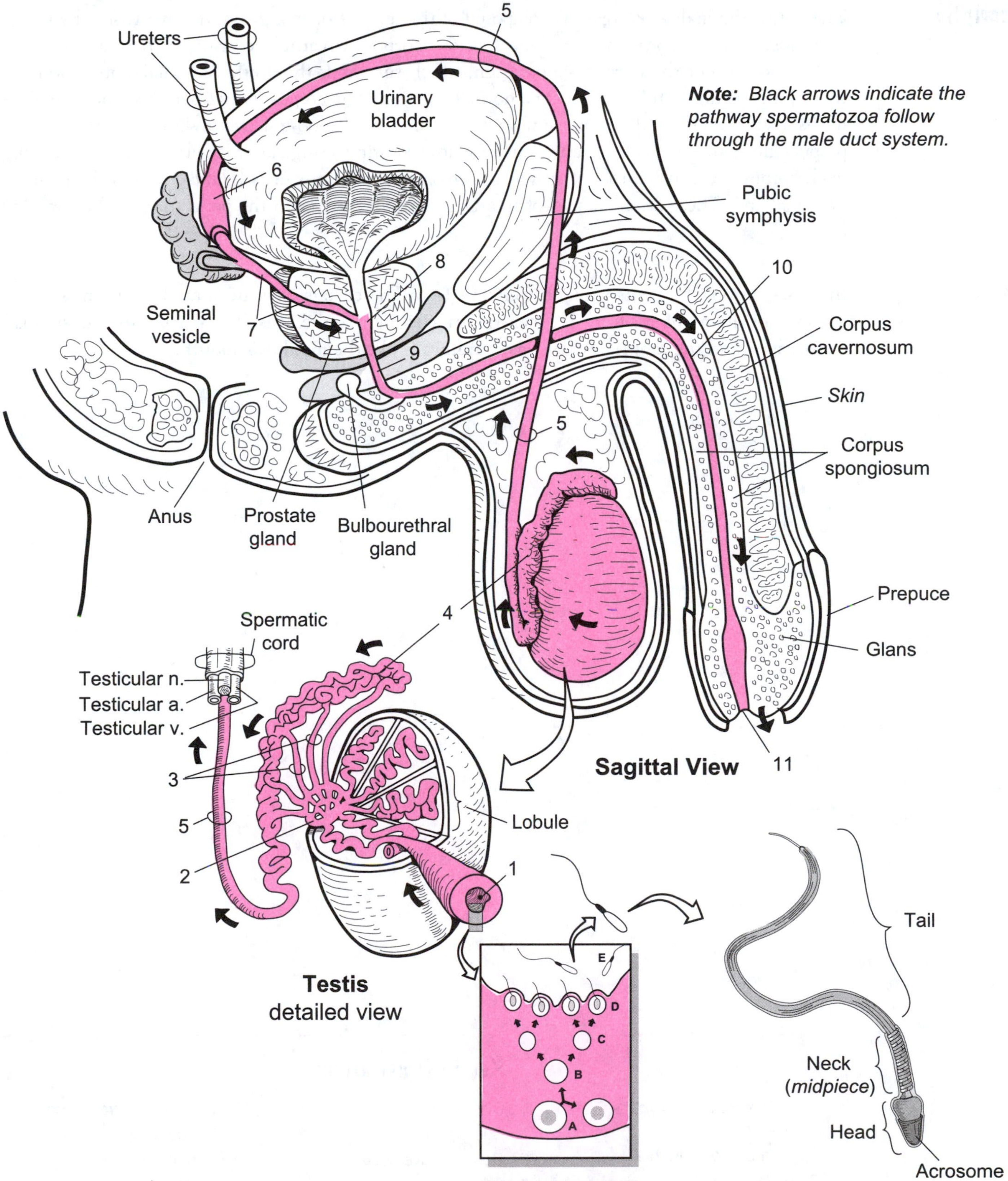

Description

The penis, the male sex organ, is composed of three tubes of spongy connective tissue: two corpora cavernosa and one corpus spongiosum. The two **corpora cavernosa** constitute the bulk of the penis and rest on top of the **corpus spongiosum**, which surrounds the urethra. Sexual stimulation causes these tubes to fill with blood during a normal erection. The corpus spongiosum becomes the **glans** at its distal end and the **bulb** of the penis at its proximal end. A loose sleeve of skin called the foreskin or prepuce normally covers the glans. This is removed during a surgical procedure called a circumcision. Each corpus cavernosum (*plural, corpora cavernosa*) becomes the **crus** (*plural crura)* of the penis at its proximal end. These crura are part of an attachment for the root of the penis to the pubic arch in the pelvis.

Analogy

In cross-section, the penis is like a monkey's face. The **corpus cavernosa** are like the **mask around the eyes** of the monkey's face. The **central arteries** are like the **eyes**. The **corpus spongiosum** is like the **area around the mouth** and the **male urethra** is like the **monkey's mouth**.

Key to Illustration

1. Bulbourethral *(Cowper's glands)* glands
2. Membranous urethra
3. Bulb of penis
4. Crus of penis
5. Subcutaneous dorsal v.
6. Deep dorsal a.
7. Deep fascia
8. Corpora cavernosa
9. Central a.
10. Corpus spongiosum
11. Glans
12. External urethral orifice
13. Tunica albuginea
14. Superficial fascia
15. Urethra

Penis—Visualizing the Cross-Section

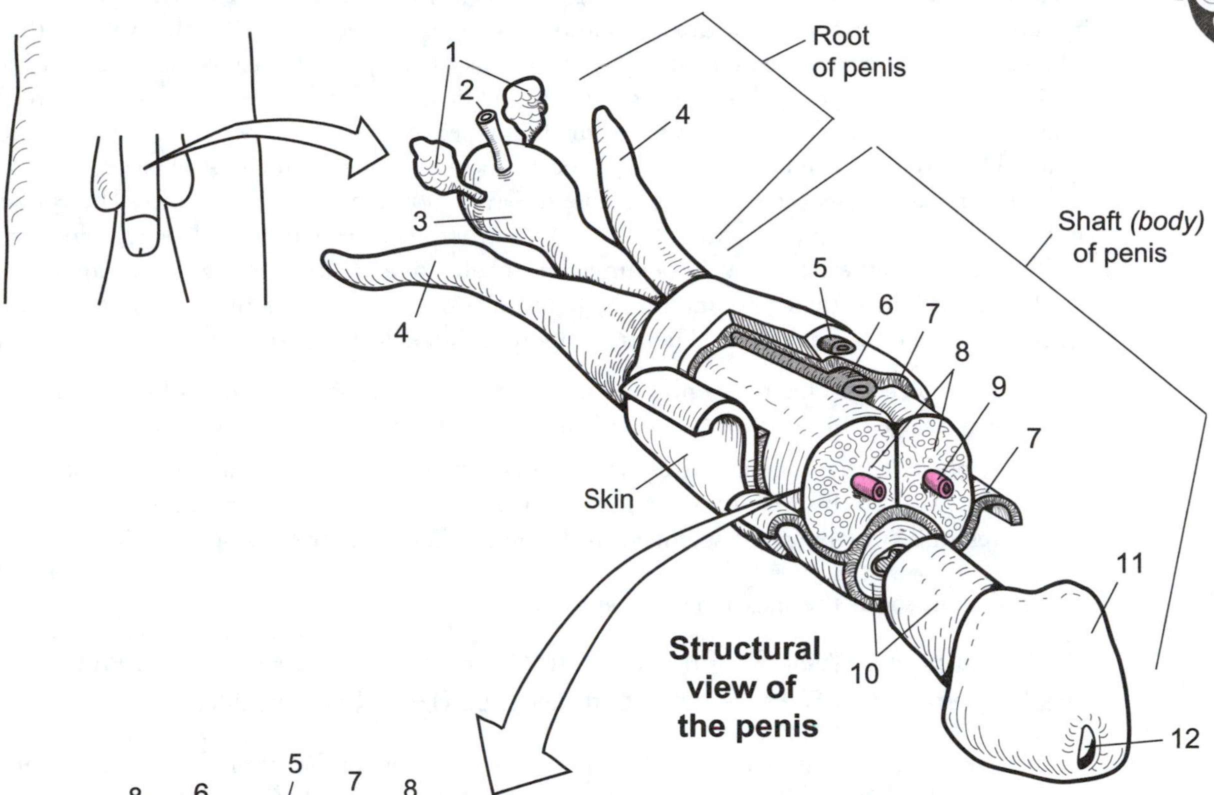

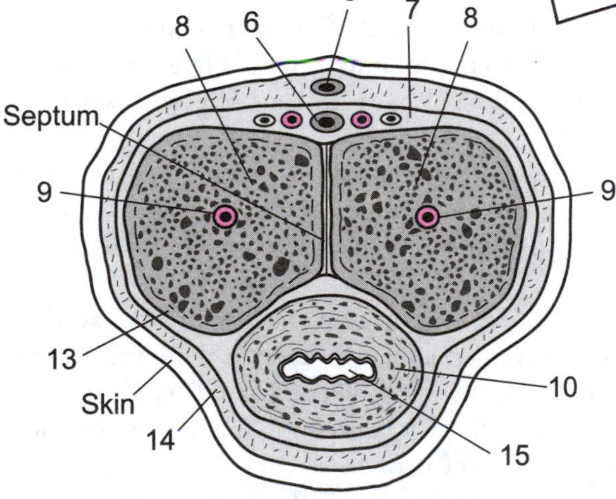

The penis in cross-section looks like...

...a monkey's face.

1. ______________________________
2. ______________________________
3. ______________________________
4. ______________________________
5. ______________________________
6. ______________________________
7. ______________________________
8. ______________________________
9. ______________________________
10. ______________________________
11. ______________________________
12. ______________________________
13. ______________________________
14. ______________________________
15. ______________________________

Description

Spermatogenesis is the process of developing sperm cells in the testes of the male. It begins for the first time in puberty and continues throughout adulthood. The process occurs within the walls of the seminiferous tubules in the testes and takes about 2 to 2.5 months to complete. It begins near the outer portion of the seminiferous tubules, proceeds toward the center, and finally ends when the sperm cells are released into the lumen of the seminiferous tubules. One of the hormones that regulates spermatogenesis is testosterone. It is secreted by **interstitial cells** (*cells of Leydig*), located between the seminiferous tubules.

Sustentacular cells (*Sertoli cells*) compose the walls of the seminiferous tubules. They physically support and nourish the developing sperm cells (DSCs), control the movement of the DSCs, remove wastes, produce the fluid that fills the lumen of the seminiferous tubules, and are involved in other functions. Each sustentacular cell is joined tightly to another by tight junctions that are like rivets that anchor cell membranes together (see p. 62). This tight seal helps maintain the **blood testis barrier**, which has two specific functions.

1. It protects developing sperm cells from potentially harmful substances by preventing proteins and other large molecules in the blood from coming in direct contact with developing sperm cells.
2. It prevents the immune system from being exposed to sperm cell antigens not found on any other body cells. If this were to occur, the immune system would view sperm cells mistakenly as foreign and respond by making anti-sperm antibodies in an effort to destroy the sperm cells.

Stages

This flowchart shows the **stages in spermatogenesis**:

Spermatogonium (SG) → Primary spermatocyte (PS) → Secondary spermatocyte (SS) → Early spermatids (EST) → Late spermatids (LST) → Sperm cells (S)

The **spermatogonium** is a **stem cell** found in the outer wall of the seminiferous tubules within the testes. It undergoes **mitosis** (cell division) to produce two new copies of itself. One of these new cells remains in the outer wall as a new spermatogonium, and the other differentiates into a slightly larger cell called a **primary spermatocyte.** This cell contains the total number or **diploid (2n)** number of chromosomes found in all human body cells—46.

The primary spermatocyte enters a special process called **meiosis**, in which a cell undergoes two cell divisions to produce four final cells, each containing half the number of chromosomes as the original cell. This half-set of chromosomes is called the **haploid (n) number**. The only purpose of meiosis is to produce **gametes** (sperm, ova).

Because the total chromosome number is reduced by half, meiosis is called "reduction division." The primary spermatocyte divides to produce two **secondary spermatocytes**. Each of these cells contains a set of 23 replicated chromosomes. Each secondary spermatocyte then undergoes a second cell division to produce a two new **spermatids**. In total, four spermatids are produced, each containing the haploid (n) number, or 23 chromosomes. Last, the spermatids develop into spermatozoa (sperm cells).

Spermiogenesis is the last part of the spermatogenesis process, in which a spermatid is transformed into a sperm cell. Highlights of this metamorphosis are:

1. **Acrosome formation:** A sac covers the head of the sperm cell and contains the digestive enzymes used to penetrate the outer portion of the ova in the fertilization process.
2. **Mitochondrial reproduction:** The mitochondria replicate themselves and cluster together to form a tightly coiled spiral around the neck of the sperm cell. They will be used to supply large amounts of ATP to propel the sperm cell by moving its flagellum.
3. **Flagellum formation:** The centrioles form the microtubules that make up the flagellum.

The **spermatozoon** or **sperm cell** has three main parts:

1. **Head**: has a nucleus that contains the haploid (n) number of chromosomes (23).
2. **Midpiece**: contains many mitochondria that produce large amounts of ATP to fuel the whipping of the tail.
3. **Tail** (*flagellum*): contains microtubules and is used to propel the sperm cell.

Testis
Seminiferous tubule
SPERMATOGENESIS
2n
Spermatogognium (SG)
2n
2n
Primary spermatocyte (PS)
n
n
Secondary spermatocytes (SS)
n
n
n
n
Early Spermatids (EST)
n
Late Spermatids (LST)
n
Spermatozoa (S) (sperm cells)
Color the 2n cells one color and the n cells another.
Interstitial cells (Cells of Leydig)
Sustentacular cell
Sustentacular cell nucleus
Interstitial space
SG
PS
SS
SS
EST
EST
LST
LST
Tight junction
S
Wall of seminiferous tubule
Lumen of seminiferous tubule
SPERMIOGENESIS
Acrosomal vesicle
Nucleus
Mitochondrion
Centriole
Plasma membrane
Microtubule
Mitochondria
Plasma membrane
Acrosome
Head
Acrosome
Midpiece
Tail
Spermatozoon (sperm cell)

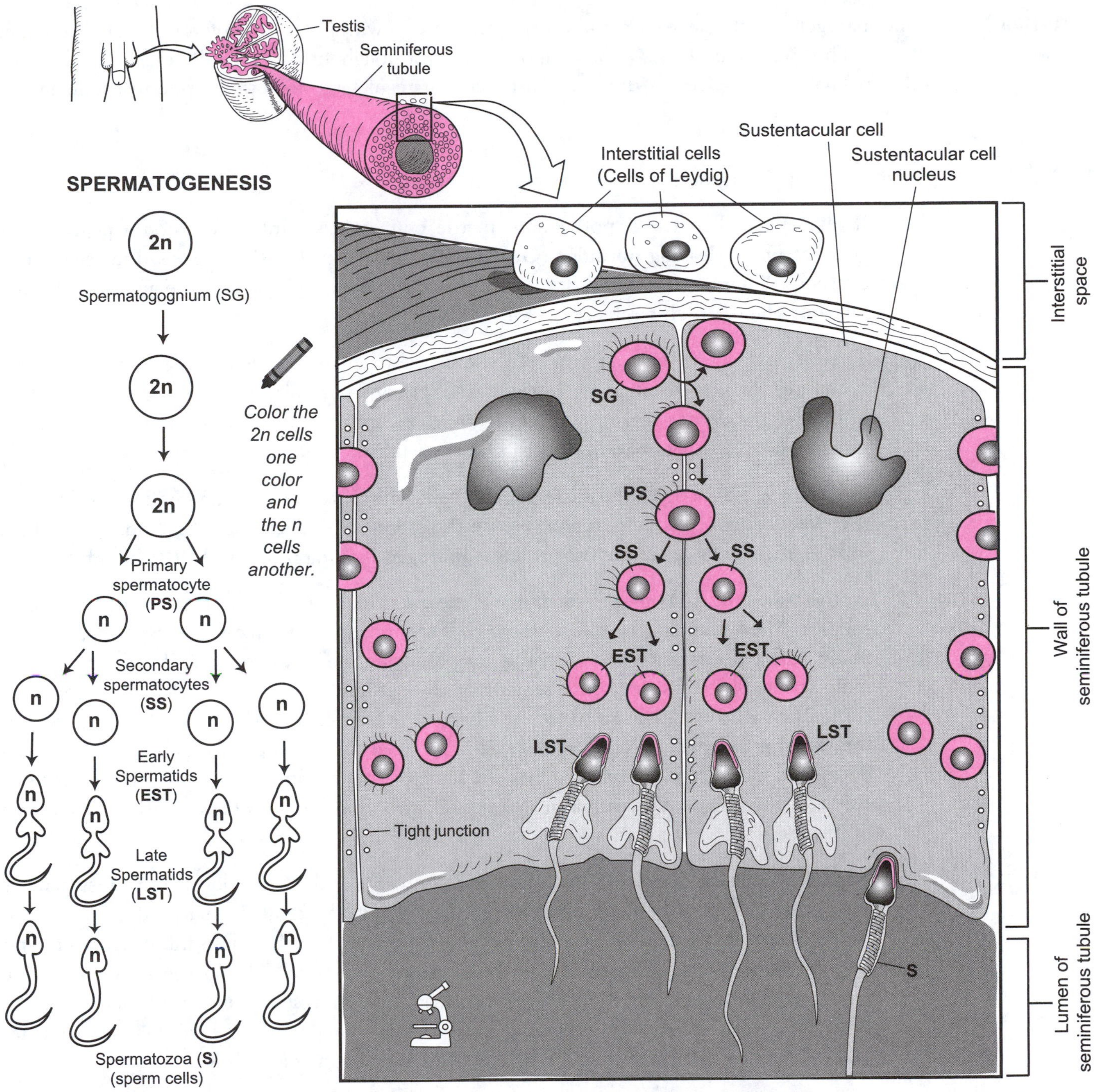

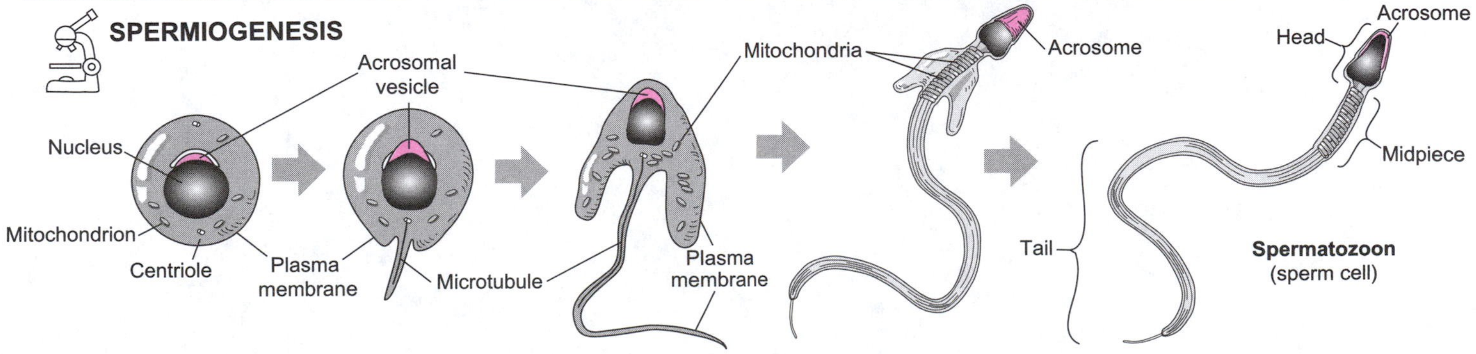

Description

Spermatogenesis, the development of sperm cells, occurs within the seminiferous tubules in the male testes. This process begins at puberty and continues throughout adult life. The group of hormones that regulates this process is the same basic set that regulates the ovarian cycle in the female (see p. 466).

Stages

A step-by-step description of this process is given below:

1. It all begins when the hypothalamus in the brain releases into the blood a **peptide** called **gonadotropin-releasing hormone (GnRH)**. In adult males, GnRH is released in regular bursts every 60–90 minutes. It travels through the blood to target cells within the anterior lobe of the pituitary gland, which has receptors for GnRH.

2. The binding of GnRH induces a response in these cells which is to make two **gonadotropin** hormones: **follicle-stimulating hormone (FSH)** and **luteinizing hormone (LH)**. Like any hormone, these chemical messengers are released into the blood stream and travel to their respective targets that contain receptors for these hormones.

3. In the male, **FSH** targets the **sustentacular cells** within the walls of the seminiferous tubules in the testes. These cells support and nourish the spermatogenetic cells. The binding of FSH also induces these cells to secrete two proteins: **androgen-binding protein (ABP)** and **inhibin.**

4. At the same time, **LH** binds to receptors on the interstitial cells, located between the seminiferous tubules. After binding, LH induces these cells to produce the hormone **testosterone**, which has many different targets. It is responsible for producing the secondary sex characteristics in the male (deepening of voice, enlargement of the genitals, growth of body hair, among other things). In this case, the testosterone binds to the protein **ABP** to form a complex. Once formed, this complex induces the spermatogenic cells to develop into sperm cells. The newly developed sperm cells are released into the lumen of the seminiferous tubule, where they travel through the male duct system until they are released from the body through the external orifice of the penis.

5. Negative feedback is used to shut down the pituitary gland. As the levels of **testosterone** rise, they inhibit the hypothalamus and the pituitary gland. The result is that production of GnRH ceases, which, in turn, leads to no production of either FSH or LH. This inhibition is enhanced by production of the hormone **inhibin** by the sustentacular cells. It has the same effect on the hypothalamus and pituitary as did testosterone. Once the testosterone and inhibin levels fall, inhibition is lost and the cycle begins again. In the adult male, this cycle results in a steady level of testosterone production that fluctuates within a normal range.

Hormonal Regulation of Sperm Cell Production

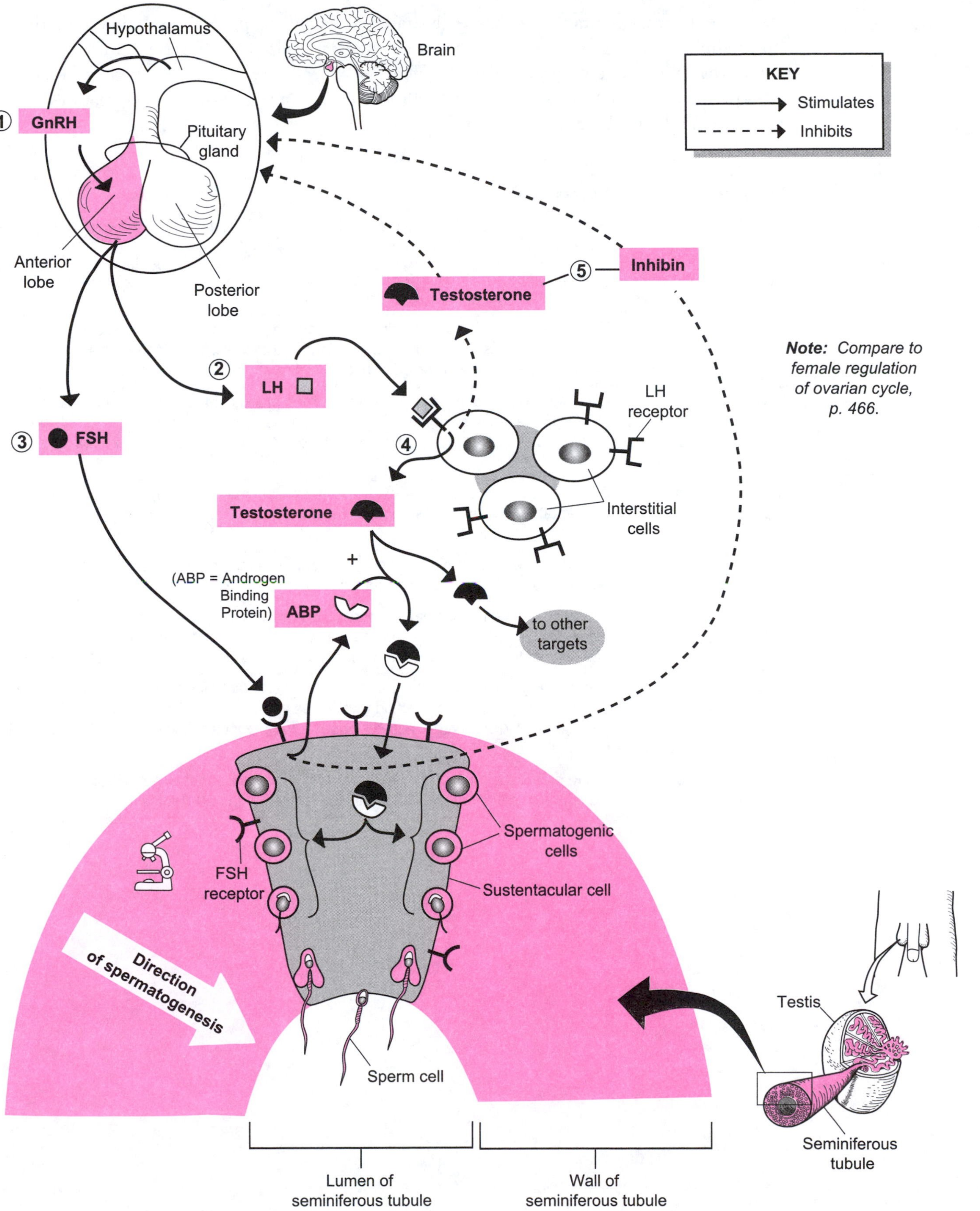
Hypothalamus
Brain
KEY
Stimulates
Inhibits
① GnRH
Pituitary gland
Anterior lobe
Posterior lobe
Testosterone
⑤ Inhibin
Note: Compare to female regulation of ovarian cycle, p. 466.
② LH
③ FSH
④
LH receptor
Testosterone
Interstitial cells
+
(ABP = Androgen Binding Protein) ABP
to other targets
Spermatogenic cells
FSH receptor
Sustentacular cell
Direction of spermatogenesis
Testis
Sperm cell
Seminiferous tubule
Lumen of seminiferous tubule
Wall of seminiferous tubule

Description

The female **external genitalia** (*vulva*) are illustrated on the facing page. Please understand that this is an idealized rendering and does not account for the many variations of normal. In the developing embryo, the external genitalia of males and females appear very similar. Then they differentiate over a period of about 8 weeks. The table will point out *homologous structures*—features that are structurally similar between male and female genitalia.

Structure	Description
1. Mons pubis	A relatively large mound of skin and fatty tissue located anterior to the pubic symphysis. It's covered with pubic hair in adult females
2. Prepuce of clitoris	A hood-like structural extension of the labia minora that covers the glans of the clitoris. This is homologous to the male prepuce (*foreskin*).
3. Glans of clitoris	The tip of the clitoris that contains many sensory nerve endings for sexual pleasure in the female. This is homologous to the glans of the penis. The clitoris is a small, erectile body that engorges with blood during sexual excitation.
4. Urethral orifice (*opening*)	Opening from the urethra located between the glands of the clitoris and the vaginal opening. The urethra is a narrow tube that connects the urinary bladder to outside of the body. Urine collects in the urinary bladder, passes through the urethra, and is expelled from the body
5. Labia minora ("smaller lips")	Smaller, hairless folds located inside the larger labia majora that may have increased pigmentation due to the abundance of melanocytes. They are homologous to the ventral shaft of the penis.
6. Vestibule	The space between the labia minora that contains the urethral opening, the vaginal orifice, and openings to the greater vestibular glands.
7. Vaginal orifice (*opening*)	The opening into the vagina. The vagina is a thick, muscular tube that connects the uterus to outside the body. It acts as the organ to receive the penis during sexual intercourse. It also functions as the birth canal and passageway for menstruation.
8. Openings for the greater vestibular glands	The openings that lead to the pair of greater vestibular glands. During sexual arousal, these glands produce a secretion that serves as a vaginal lubricant. Secretion increases during sexual intercourse. These glands are homologous to the bulbourethral glands in males.
9. Labia majora ("larger lips")	Thick, protruding folds of fatty skin that are homologous to the male scrotum. Outer margins are covered with coarse pubic hair in the adult female.

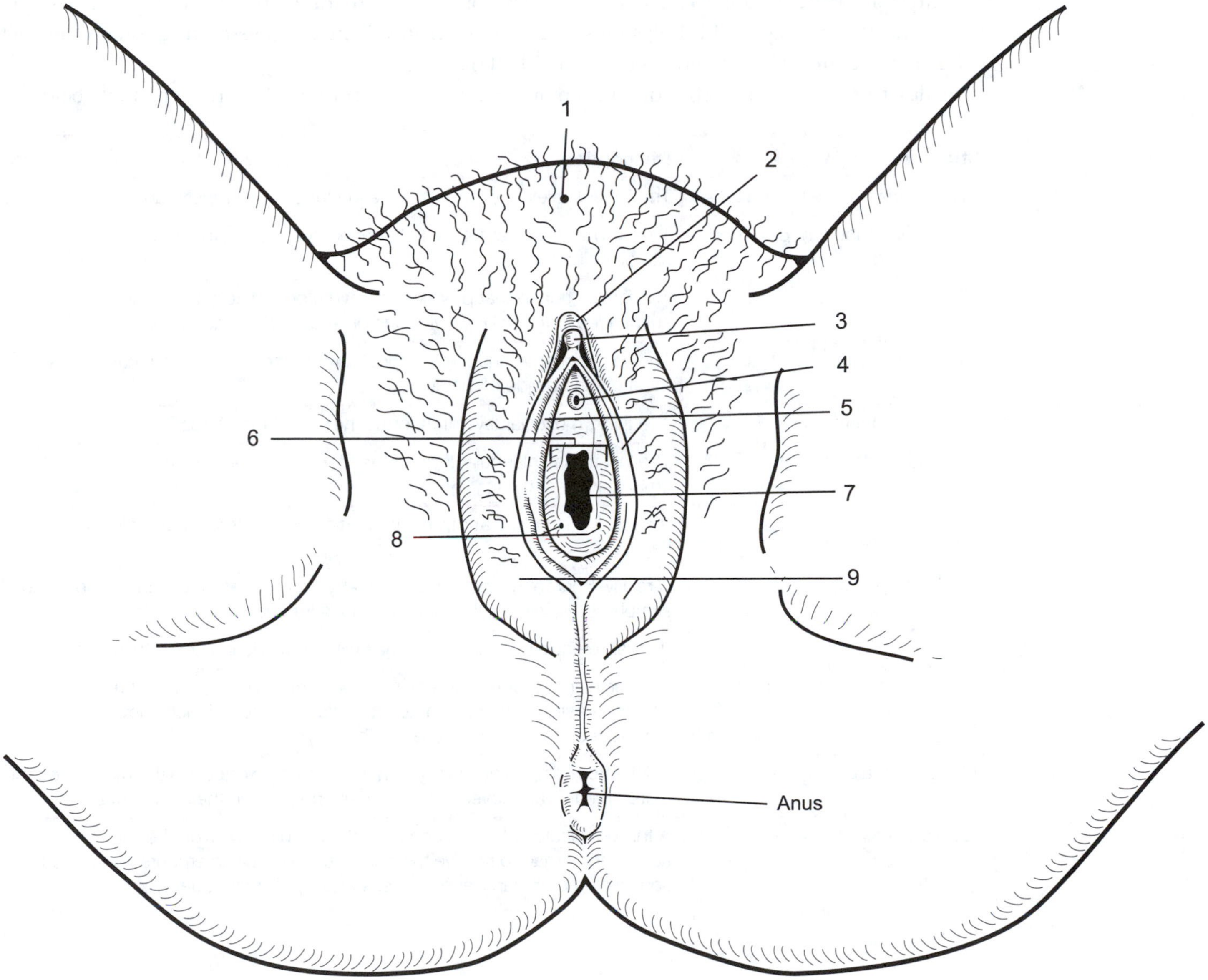

Color the major structures different colors.

1. ______________________
2. ______________________
3. ______________________
4. ______________________
5. ______________________
6. ______________________
7. ______________________
8. ______________________
9. ______________________

Description

The midsagittal view of the female reproductive system allows you to see its relationship to other organs/structures in the body. For example, notice that the uterus is superior to the urinary bladder and anterior to the rectum. Moreover, it allows you to see structures in their full length like the clitoris and the vagina. It also helps illustrate how structures connect one structure to another such as the vagina connecting the uterus to outside the body.

A description of each of the structures numbered in the illustration is given in the table below.

Structure	Description
1. Fimbriae of the uterine tube	The finger-like extensions of the uterine tube nearest the ovary
2. Uterine tube (*fallopian tube*, *oviduct*)	The hollow, muscular tubes that connect the ovaries to the uterus; the site of fertilization
3. Ovary	The female gonad that produces an ovum or egg cell; two ovaries are suspended in the pelvic cavity, one on each side of the uterus
4. Perimetrium	The serous membrane that extends from the peritoneal lining that covers most of the outside of the uterus
5. Myometrium	The thick, muscular layer that forms the wall of the uterus
6. Endometrium	The innermost, glandular layer of the uterus; site where the developing embryo implants
7. Cervix	A neck-like structure at the inferior portion of the uterus that projects into the vagina
8. Mons pubis	A relatively large mound of skin and fatty tissue located anterior to the pubic symphysis; covered with pubic hair in adult females.
9. Clitoris	A small, erectile body that engorges with blood during sexual excitation.
10. Labia minora ("smaller lips")	Smaller, hairless folds located inside the larger labia majora that may have increased pigmentation due to the abundance of melanocytes; homologous to the ventral shaft of the penis.
11. Labia majora ("larger lips")	Thick, protruding folds of fatty skin that are homologous to the male scrotum. Outer margins are covered with coarse pubic hair in the adult female.
12. Vagina	A thick muscular tube that connects the uterus to outside the body; acts as the organ to receive the penis during sexual intercourse and also functions as the birth canal and the passageway for menstruation.

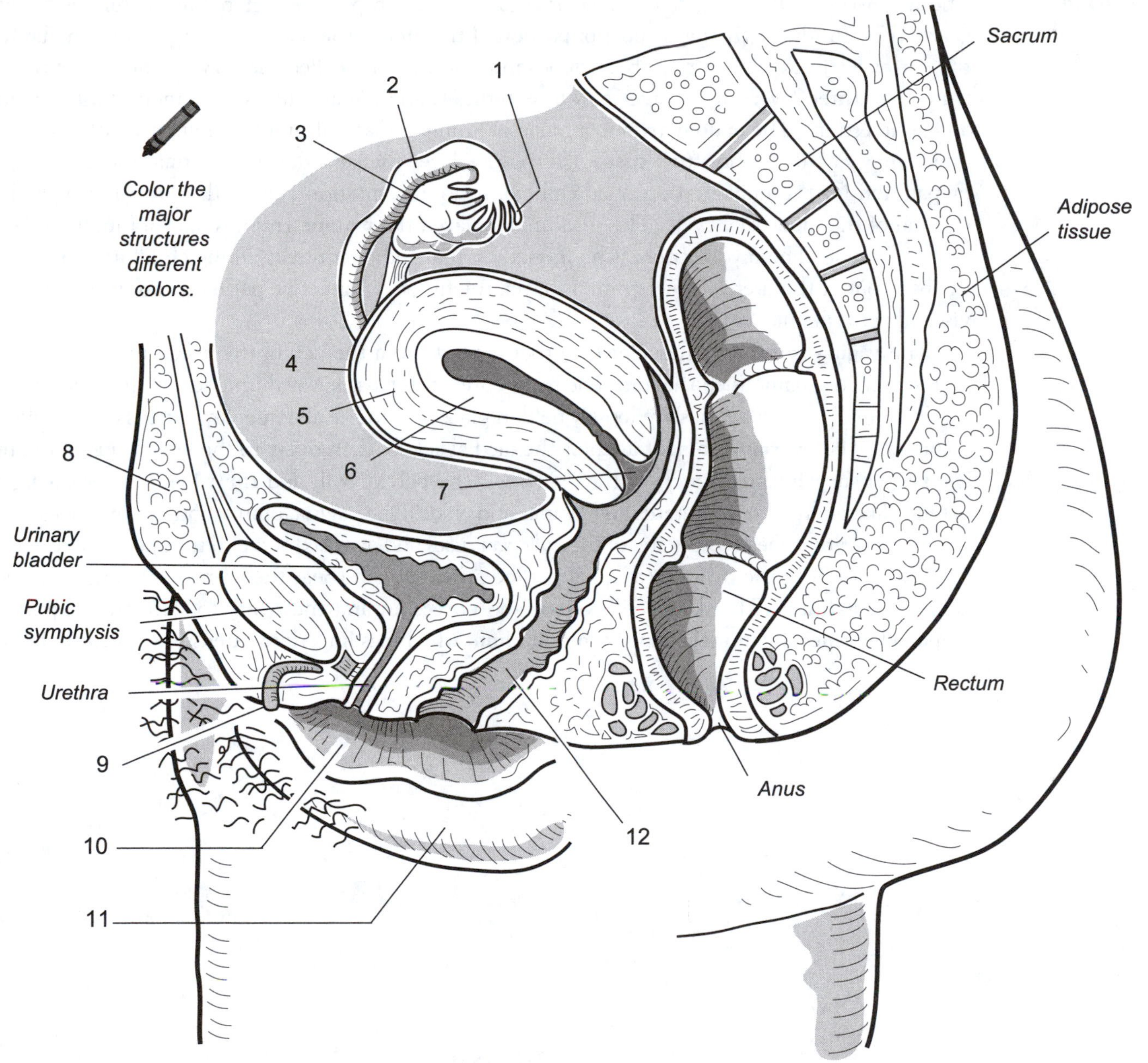

Female Reproductive System
midsagittal view

1. ______________________
2. ______________________
3. ______________________
4. ______________________
5. ______________________
6. ______________________
7. ______________________
8. ______________________
9. ______________________
10. ______________________
11. ______________________
12. ______________________

Description

The **uterus** (*womb*) is a hollow, muscular organ divided into three regional areas: fundus, body, and cervix. The **fundus** is the most superior portion of the uterus, the main portion is called the **body**, and the narrowed, neck-like portion that extends into the vagina is called the **cervix**. The wall of the uterus is made of three layers, from innermost to outermost, and are as follows: endometrium, myometrium, and perimetrium. The **endometrium**, or mucosal lining, is made of simple columnar epithelium and an underlying vascular connective tissue. This entire layer thickens during a normal menstrual cycle in preparation for the implantation of an embryo. If no implantation occurs, this layer is sloughed off at the end of the menstrual cycle. The thick middle layer, the **myometrium**, is made of multiple layers of smooth muscle. The myometrium is hormonally stimulated to contract during childbirth to help move the baby out of the uterus. The outermost layer, the fibrous connective **perimetrium,** is the same as the visceral peritoneum.

The **vagina** is a thick, muscular tube that extends from the cervix to outside the body. It is lined with stratified squamous epithelium. The penis enters this passageway during sexual intercourse.

The small, lumpy **ovaries** are loosely held in place within the abdomen by various connective tissue ligaments. The **ovarian ligament** anchors the medial side of each ovary to the lateral side of the uterus, the **suspensory ligament** anchors the ovaries to the pelvic wall, and the wide, flat, **broad ligament** extends like a tarp over the uterus and ovaries and cradles the vagina, uterus, and uterine tubes.

The **uterine tubes** (consisting of the *Fallopian tubes* and *oviducts*) serve to transport a female gamete (*egg cell; ova*) from the ovary to the uterus. The proximal end (near the uterus) is a long, narrow tube called the **isthmus**. The distal end widens and curves around the ovary to form the **ampulla**. The ampulla becomes the funnel-shaped **infundibulum,** which has fingerlike extensions called **fimbriae**.

Key to Illustration

Uterus
1. Fundus of uterus
2. Body of uterus
3. Cervix
4. Lumen of uterus
5. Endometrium
6. Myometrium
7. Perimetrium
8. Internal os
9. Cervical canal
10. External os

Vagina (V)

Blood Vessels (B)
B1. Ovarian artery
B2. Ovarian vein

Ligaments (L)
L1. Round ligament
L2. Suspensory ligament
L3. Ovarian ligament
L4. Broad ligament

Ovary (O)

Uterine Tubes (U)
U1. Ampulla
U2. Isthmus
U3. Infundibulum
U4. Fimbriae

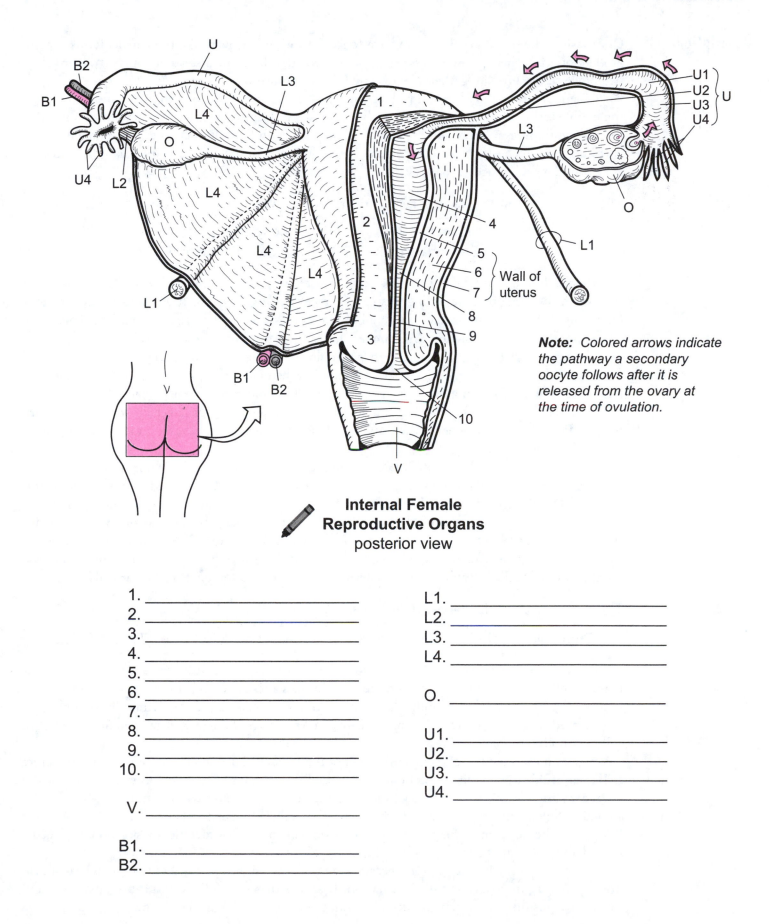

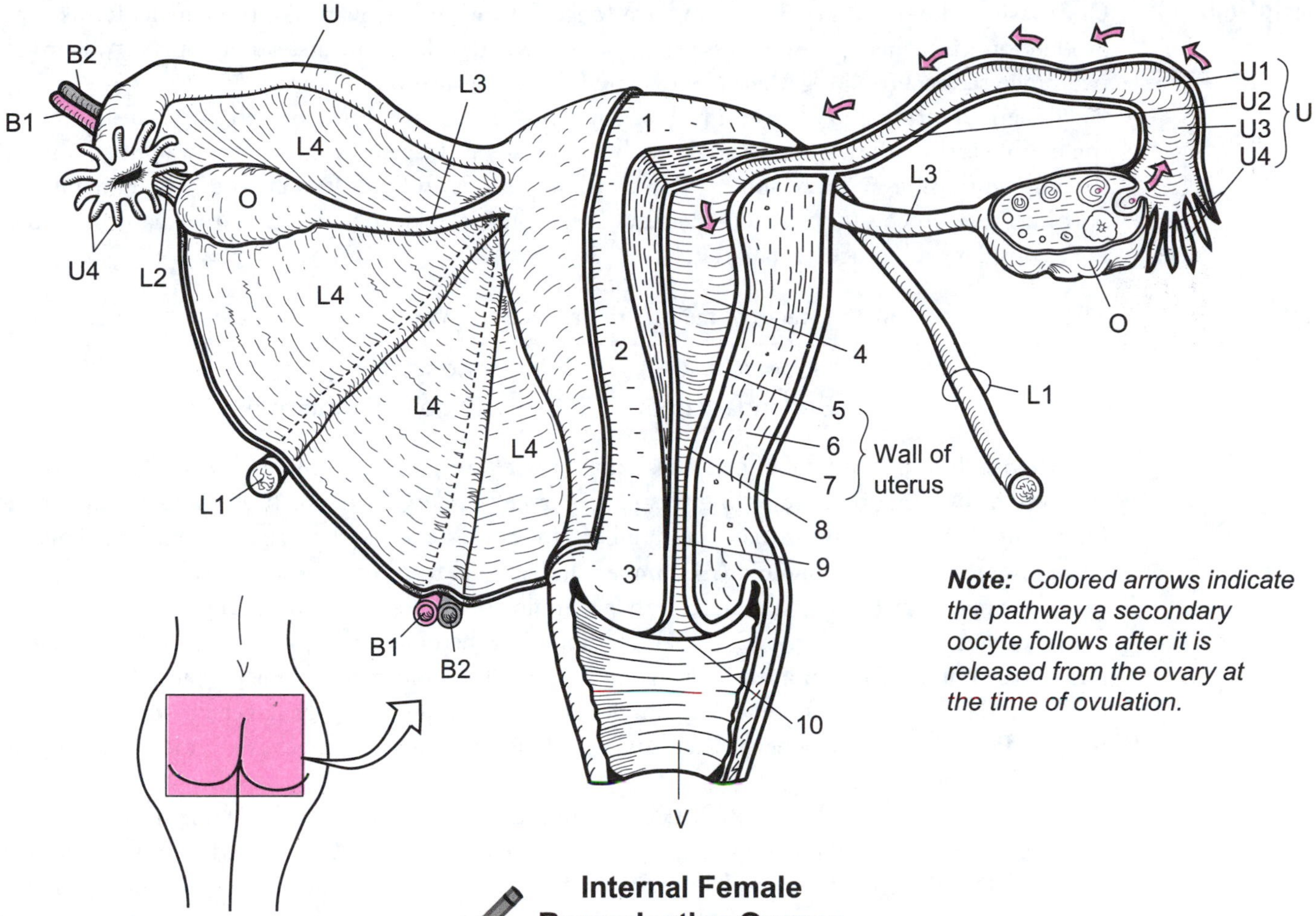

Note: *Colored arrows indicate the pathway a secondary oocyte follows after it is released from the ovary at the time of ovulation.*

Internal Female Reproductive Organs
posterior view

1. ______________________
2. ______________________
3. ______________________
4. ______________________
5. ______________________
6. ______________________
7. ______________________
8. ______________________
9. ______________________
10. ______________________

V. ______________________

B1. ______________________
B2. ______________________

L1. ______________________
L2. ______________________
L3. ______________________
L4. ______________________

O. ______________________

U1. ______________________
U2. ______________________
U3. ______________________
U4. ______________________

Description

Oogenesis is the process of developing **ova** (*egg cells*) within the ovaries of the female. Unlike sperm production in the male, this process begins during fetal development instead of at puberty. It involves two processes occurring together: **meiosis** and **follicle maturation**.

Imagine putting a marble in a balloon and then filling that balloon with water. In this analogy, the marble is the cell that has to go through meiosis to produce an ovum and the water balloon is like the follicle that has to grow and develop by filling itself with fluid. It is important to remember that both of these processes occur together. In other words, meiosis occurs inside the maturing follicles. Let's consider each process separately.

Meiosis

This flowchart shows the stages of **meiosis** to produce an **ovum** within a follicle:

Oogonium (OG) → Primary oocyte (PO) →
Secondary oocyte (SO) → Zygote (if fertilized)

Meiosis is a special type of cell division used solely to produce **gametes** (*sperm, ova*). It is called "reduction division" because during this process a cell undergoes two cell divisions to produce four final cells that each contain half the number of chromosomes as the original cell. This half-set of chromosomes, called the **haploid (n) number**, and is equal to 23 chromosomes in humans.

The **oogonium** (OG) is a stem cell that contains the full set of chromosomes called the **diploid (2n) number**, or 46 ($2 \times n$ or 2×23). It is found in the outer region of the ovaries and undergoes mitosis (cell division) during fetal development to produce millions of cloned copies of itself. Most of these are degenerated, but some are stimulated to grow into a slightly larger cell called a **primary oocyte (PO)**. Though they begin their first cell division of meiosis (meiosis I) in fetal development, they do not complete it.

At birth, about 400,000 to 4,000,000 oogonia (sing., *oogonium*) and POs remain in both ovaries. By puberty, about 40,000 remain and the meiosis process is now strictly regulated by hormones. At this time, one PO is triggered to complete its first cell division of meiosis (meiosis I) every month. This cell division results in the formation of two unequally sized cells—a large, viable **secondary oocyte** (SO) and a small, nonfunctional cell called a **polar body**. The **SO** contains 23 replicated chromosomes. It will undergo its second cell division (meiosis II) only if it is fertilized by a sperm cell. If this occurs, the result is two new cells: a fertilized egg called a **zygote** and a small, nonfunctional **polar body**.

Follicle Maturation

This flowchart shows the stages in **follicle maturation**:

Primordial follicle → Primary follicle → Secondary follicle → Tertiary follicle

Follicle maturation is like a water balloon filling with water until it bursts. The primitive **primordial follicles** are located in the outer region of the ovaries. They consist of a primary oocyte surrounded by a single layer of follicular cells. During the fetal stage, the primordial follicles are transformed into **primary follicles** by gradually thickening the follicle wall with multiple layers of cells around the primary oocyte.

As maturation continues, cells in the wall of the follicle, called granulosa cells, begin to secrete **follicular fluid**, which fills a potential space called the **antrum**. This marks the formation of the **secondary follicles** during adulthood. Like the expansion of a water balloon filling with fluid, the secondary follicle increases in size to become a **tertiary follicle** (*Graafian follicle; mature follicle; vesicular follicle*). At the same time, the primary oocyte has undergone its first meiotic division and produced a secondary oocyte and first polar body.

The tertiary follicle soon ruptures because of a hormonal surge and releases the **secondary oocyte** from the ovary in an event called **ovulation**. Last, the damaged follicle forms a temporary gland called a **corpus luteum**, which produces the hormones progesterone and estrogen.

OOGENESIS:
the process of producing ova

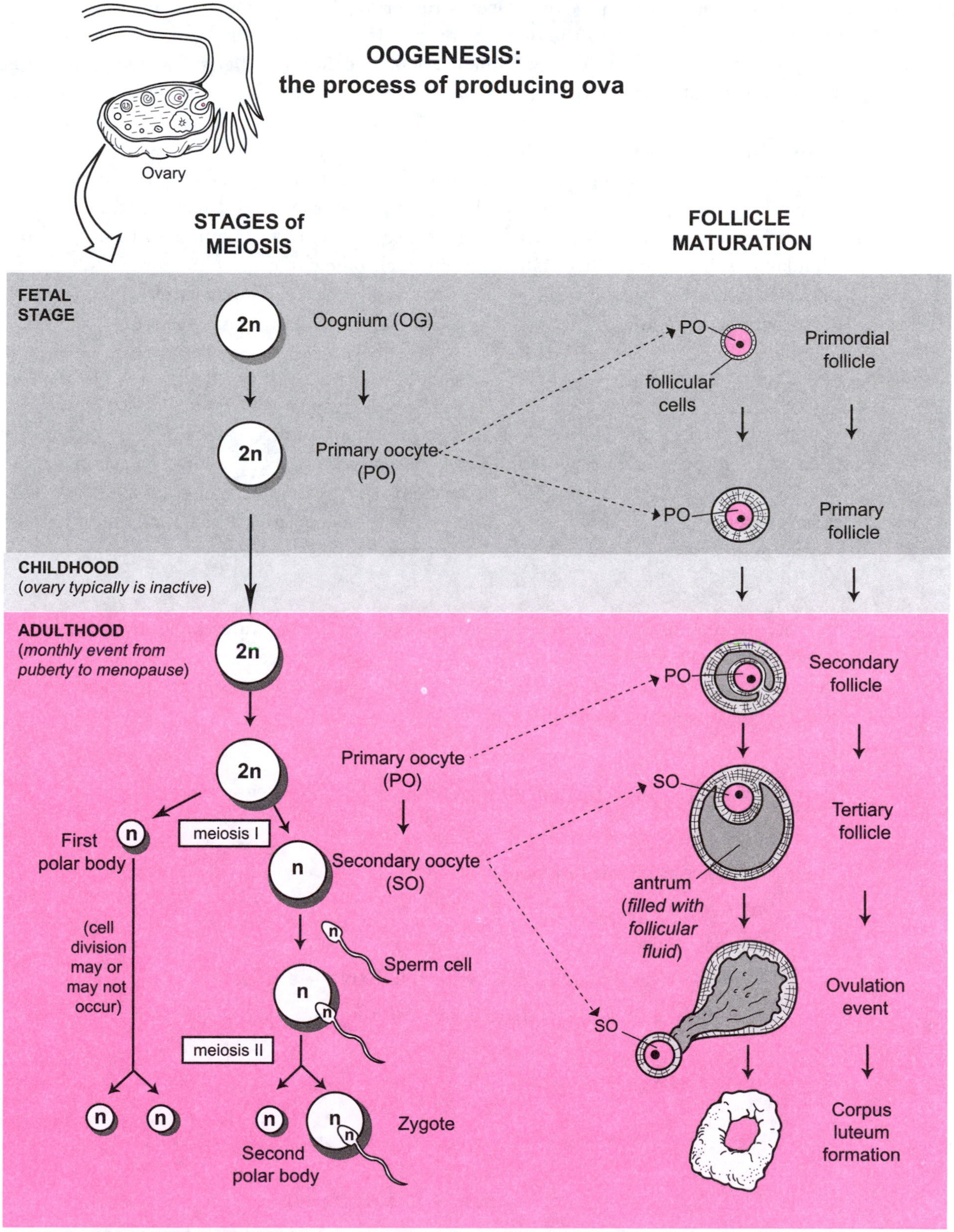

Color the 2n cells one color and the n cells another.

Description

The female has two lumpy oval-shaped structures called **ovaries**, each measuring about 5 cm. in length. Blood is brought to the ovary by an **ovarian artery** and is drained by an **ovarian vein**. The surface of the ovary is covered by a simple cuboidal epithelial layer called the **germinal epithelium**. The two regional areas are the outer **cortex** and the inner **medulla**. Production of **gametes** (*sex cells*) occurs in the **cortex**.

Ideally, the ovaries in a sexually mature female alternate to produce one ovum each month. The specific process for developing an **ovum** (*egg cell*) is called **oogenesis**. This is a part of the ovarian cycle, which refers to all the processes that occur in the ovary during this monthly event.

The ova are produced in special chambers called **follicles**. The process begins when a hormone stimulates an immature follicle called a **primordial follicle** to begin to mature. The wall of the follicle thickens and cells within it begin to produce a fluid called **follicular fluid**, which fills a space called the **antrum**. As the follicle expands, it goes through the following progression:

Primordial follicle → Primary follicle → Secondary follicle → Tertiary follicle

Once a **tertiary follicle** has been formed, a hormonal surge causes it to rupture and an ovum is released in an event called **ovulation**. The specific name for this ovum is a **secondary oocyte**. The broken tissue within the tertiary follicle thickens and becomes a temporary endocrine gland called a **corpus luteum**, which produces a mixture of **estrogens** and **progestins**. Gradually the corpus luteum shrinks and degrades into a small mass of scar tissue called the **corpus albicans**.

Analogy

The **maturation of a follicle** in the ovary is compared to a **water balloon filling with water**. As follicular fluid accumulates inside the antrum, it increases pressure just like water inside the water balloon. When the pressure becomes too great, the follicle ruptures at the **moment of ovulation** just like the **bursting of the water balloon**.

Location

Ovaries are located along the lateral edges of the pelvic cavity.

Function

The ovaries have two basic functions:

1. Produce **gametes** (sex cells)
2. Manufacture and release **hormones**
   - Follicles (primary, secondary, and tertiary) → **estrogen**
   - Corpus luteum → **estrogen, progesterone**

Key to Illustration

1. Primary follicle
2. Secondary follicle
3. Tertiary follicle
4. Ovulation
5. Early corpus luteum (*still forming*)
6. Corpus luteum
7. Corpus albicans

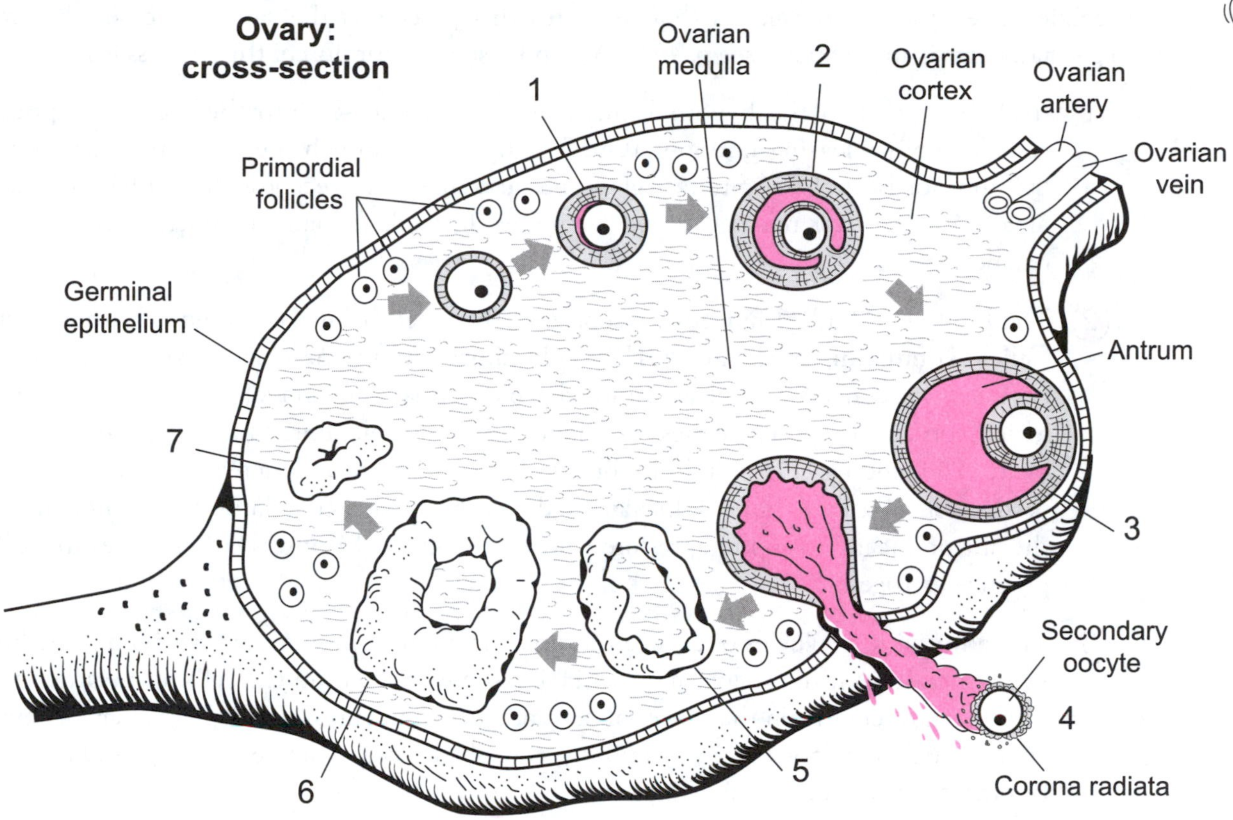

The maturation process for a follicle is like...

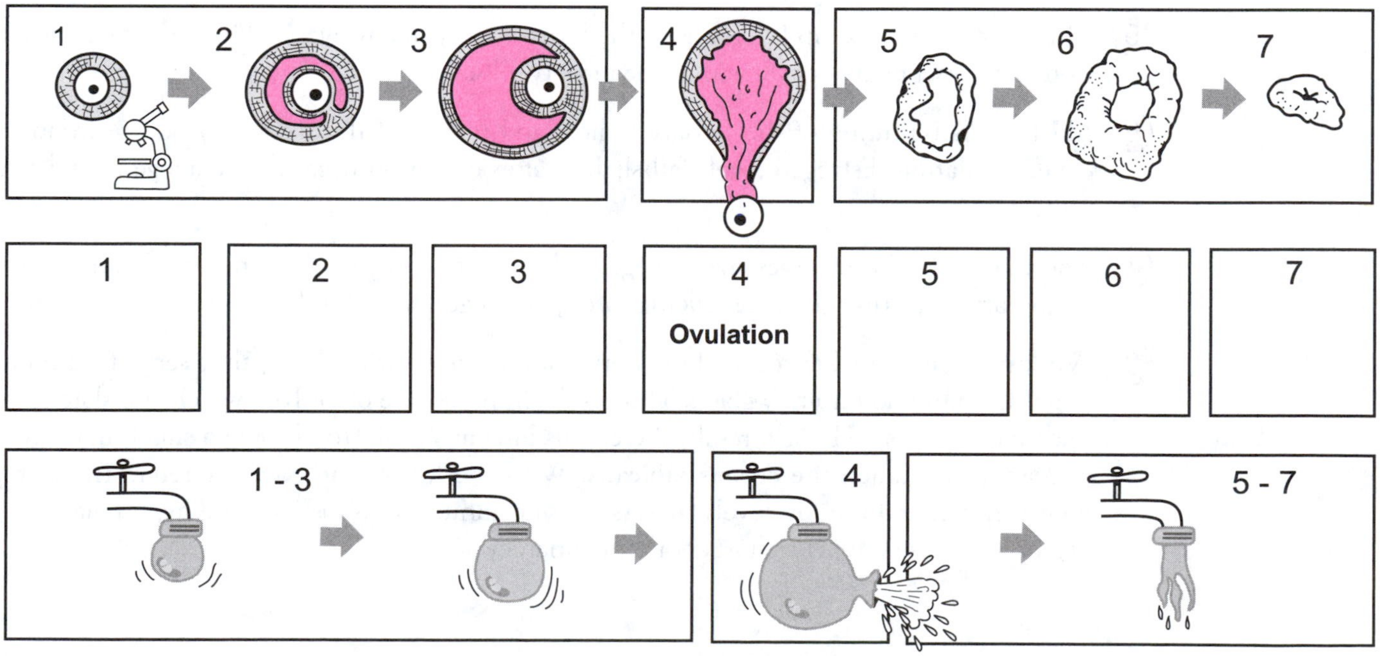

...a water balloon filling up with water and then bursting

Description

The **ovarian cycle** refers the regular, monthly events that occur within the ovary of a sexually active female. The group of hormones that regulates this process is the same basic set that regulates **spermatogenesis** in the male (see p. 454). A step-by-step description of this process is given below:

1. It all begins when the hypothalamus in the brain releases into the blood a **peptide** called **gonadotropin-releasing hormone (GnRH)**. It travels through the blood to target cells within the anterior lobe of the pituitary gland because they have receptors for GnRH. Unlike in the male, GnRH is not released in regular pulses. Instead, the amount released fluctuates in the female.

2. The binding of GnRH induces a response in these cells to make two **gonadotropin** hormones: **follicle-stimulating hormone (FSH)** and **luteinizing hormone (LH)**. Like any hormone, these chemical messengers are released into the bloodstream and travel to their respective targets, which contain receptors for these hormones. In the female, FSH targets the follicles within the ovary. The binding of FSH induces these follicles to mature and increase in size as a result of the production of a fluid called follicular fluid. As the fluid accumulates within a chamber called the antrum, it increases fluid pressure within the follicle. Meanwhile, LH causes the follicles to produce **estrogen**.

3. These initial lower levels of estrogen inhibit the anterior pituitary from *releasing* any hormones while *stimulating* it to produce mostly LH and some FSH within the cells of the anterior lobe. It is similar to stepping on the gas pedal of a car placed in neutral. Though you may be revving the engine, the car is not moving. Similarly, though the cells may be actively producing gonadotropins, they are not releasing any yet.

4. As the **tertiary** (*Graafian, vesicular*) **follicle** is established, it produces more estrogen. These higher levels have the opposite effect of the initial lower levels: they *stimulate* the anterior lobe to release its accumulated LH (and some FSH) in one big surge.

5. This 1- to 2-day spike in LH levels in the blood triggers many events: First, it stimulates formation of the secondary oocyte within the tertiary follicle.

6. **LH** triggers rupturing of the tertiary follicle and release of the secondary oocyte in an event called **ovulation**. Estrogen levels fall slightly after ovulation because of damage to the tertiary follicle.

7. The LH surge also changes the damaged follicle into a **corpus luteum**, which functions as a temporary endocrine gland, producing **estrogen**, **progesterone**, and **inhibin**.

8. Working together, when these three hormones reach a critical level, they serve to inhibit the hypothalamus and pituitary gland. This results in no release of GnRH, which translates into no release of FSH or LH. As a result, the corpus luteum shrinks to become a small, inactive mass of scar tissue, called the **corpus albicans**. With the loss of the corpus luteum, the estrogen, progesterone, and inhibin levels fall. As a result, inhibition is lost and the hypothalamus again begins to release GnRH to start another ovarian cycle.

Hormonal Regulation of the Ovarian Cycle

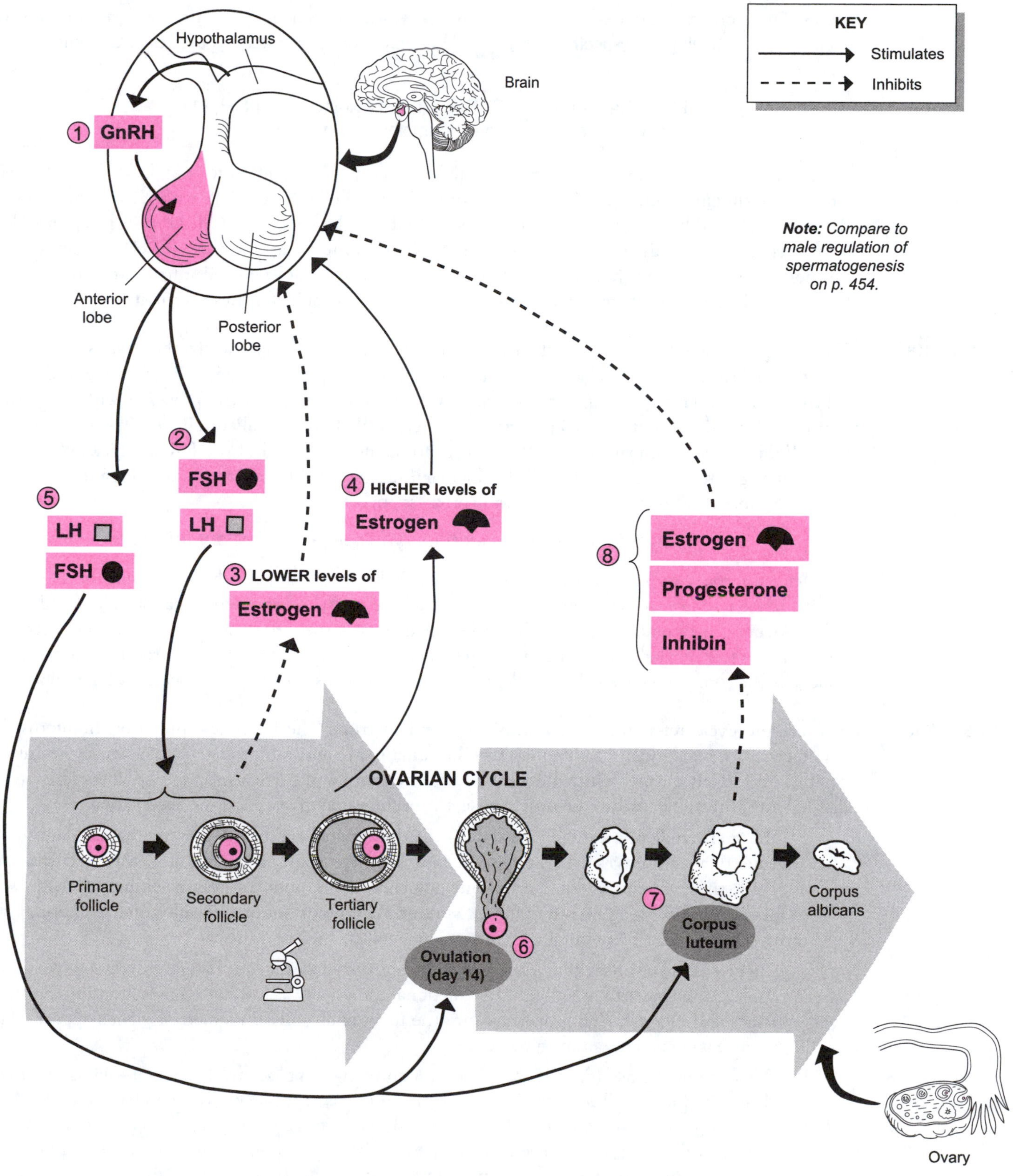

Note: *Compare to male regulation of spermatogenesis on p. 454.*

Description

In a sexually mature female, a predictable series of changes occurs every month within the ovaries and the uterus. These processes are referred to separately as the **ovarian cycle** and the **uterine cycle** (or menstrual cycle) and are regulated by specific hormones. These processes are linked by a series of cascading events. Here is an overview:

Hypothalamus → GnRH pituitary gland → FSH, LH → ovary → estrogen, progesterone → uterus

It all begins when the hypothalamus in the brain releases a **peptide** into the blood, called **Gonadotropin-releasing hormone (GnRH)**. It travels through the blood to target cells within the anterior lobe of the pituitary gland because they have receptors for GnRH. The binding of GnRH to its receptor induces a response in these cells to make the hormones **follicle-stimulating hormone (FSH)** and **leuteinizing hormone (LH)**. Like any hormone, these chemical messengers are released into the bloodstream and travel to their respective targets. Target sites always contain receptors for these hormones.

Ovarian Cycle

The **ovarian cycle** refers to all the changes that occur within the ovary. Ideally, the ovaries in a sexually mature female alternate to produce one ovum each month. The process for developing an **ovum** (*egg cell*) is called **oogenesis**. This is a part of the ovarian cycle, which refers to all the processes within the ovary during this monthly event. The ova are produced in special chambers called **follicles**. The process begins when **FSH** stimulates a **primary follicle** to begin to mature. The wall of the follicle thickens, and cells within it begin to produce a fluid called **follicular fluid**, which fills a space called the **antrum**. As the follicle expands, it goes through the following progression:

Primary follicle → Secondary follicle → Tertiary follicle

Once a **tertiary** (*Graafian, vesicular; mature*) **follicle** has been formed, a hormonal surge in LH, lasting about 1 to 2 days causes it to rupture, and an ovum is released in an event called **ovulation**. The specific name for this ovum is a **secondary oocyte**. The broken tissue within the tertiary follicle thickens and becomes a temporary endocrine gland called a **corpus luteum**, which produces a mixture of **estrogens** and **progestins**. Gradually the corpus luteum shrinks and degrades into a small mass of scar tissue called the **corpus albicans**.

Uterine Cycle

The **uterine cycle** refers to changes within the **endometrium**, the innermost lining of the uterus. The endometrium is subdivided into two layers—the **functional zone**, which lines the lumen of the uterine cavity, and the **basal zone**, which lies beneath it. The entire cycle averages about 28 days. This cycle is divided into three phases: the **menstrual phase**, the **proliferative phase**, and the **secretory phase.**

1. The **menstrual phase** (day 1 to day 7). During this phase, blood flow to the endometrium is reduced, which starves the tissues in the functional zone of their oxygen and other nutrients. As a result, these tissues deteriorate. When they break away from the uterine lining, it causes some blood loss. The sloughing off of these tissues and the associated blood, called **menstruation**, lasts 1 to 7 days. During menstruation the basal zone remains intact.
2. The **proliferative phase** (day 7 to day 13). Estrogens induce all the changes throughout this phase. The major event is to reestablish the functional zone that was lost during menstruation. Epithelial tissues and blood vessels grow back into the functional zone. **Endometrial glands** appear, which produce a mucus containing glycogen.
3. The **secretory phase** (days 15–28). This last and longest phase covers about 14 days. The tissues within the functional zone thicken and increase in vascularization, while endometrial glands enlarge and produce more mucus. The combination of estrogen and progesterone working together are responsible for these changes. The purpose of this thickening of the endometrium is to prepare for possible implantation of a human embryo into the functional layer of the endometrium. The glycogen in the mucus from the endometrial glands will serve to nourish the developing embryo. If no implantation occurs, both the ovarian and uterine cycles begin once again with release of GnRH from the pituitary gland.

Ovaries: Regulation of the Ovarian and Uterine Cycles

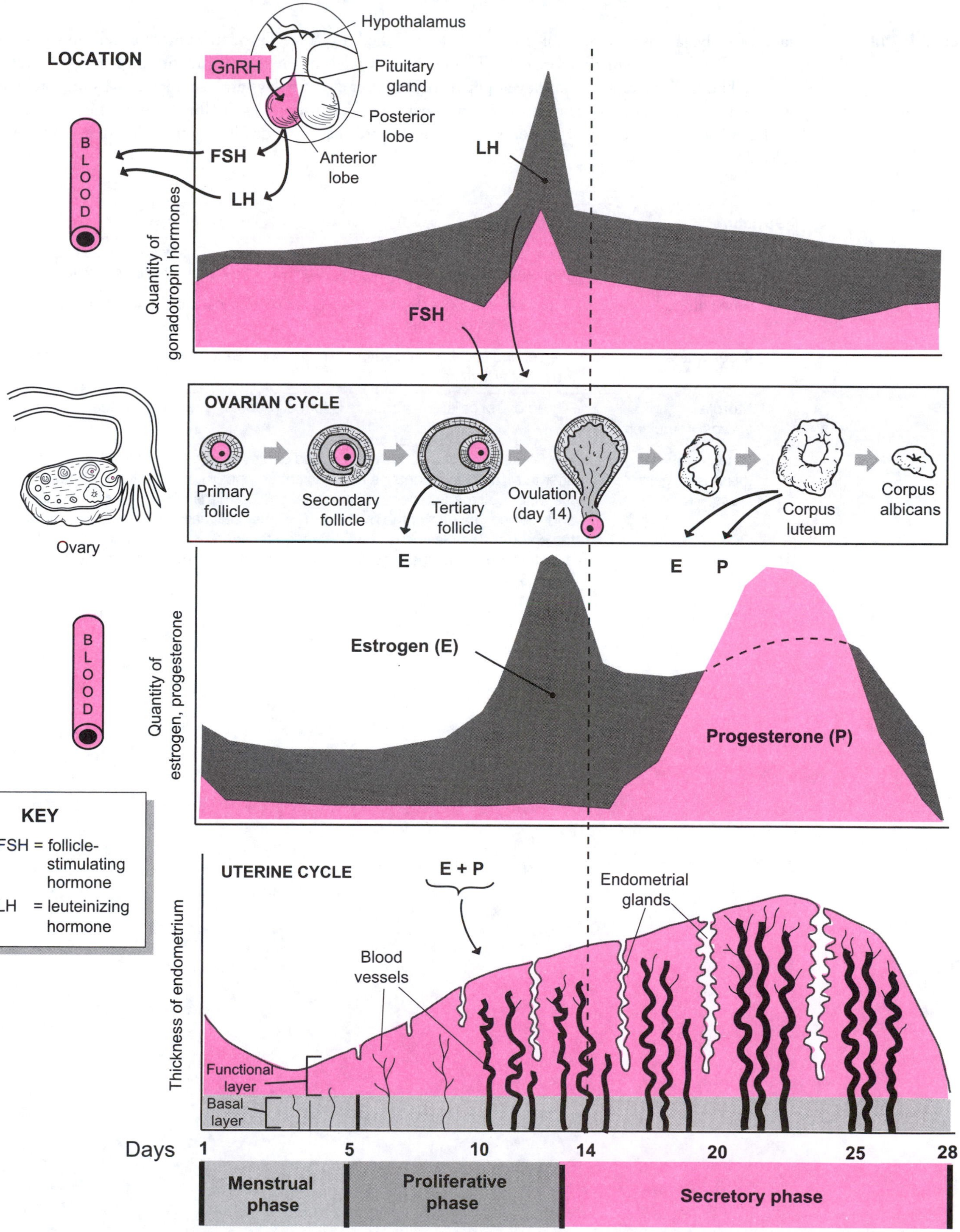

KEY

FSH = follicle-stimulating hormone

LH = leuteinizing hormone

Description

Each of us began as a single cell—a fertilized egg called a zygote. Through numerous cell divisions, this single cell gave rise to many more cells. This eventually leads to the establishment of an embryo and then a fetus. The first stage of human development occurs over about 6 days and is referred to as **preembryonic development**. It begins at conception when a sperm cell fertilizes an egg cell and ends when a blastocyst implants itself into the mother's uterus. Each stage of preembryonic development is described in the table below.

Structure	Description
1. **Zygote** (*zygotos* = union)	The result of a single sperm cell fertilizing an oocyte. This occurs in the uterine tube near the ovary and marks the beginning of a genetically unique organism
2. **2-cell stage**	The zygote's first cell division resulting in two identical cells. This occurs about one and a half days after fertilization.
3. **4-cell stage**	The second cell division creating a total of four identical cells. This occurs about 2 days after fertilization.
4. **Morula** (*moros* = mulberry)	A solid ball of cells; this structure is formed about 3 days after fertilization.
5. **Blastocyst** (*blastos* = germ, *kystis* = bag)	A hollow sphere of cells with a fluid-filled central cavity. It first forms about 4 days after fertilization and contains two different groups of cells: the **trophoblast** and the **inner cell mass**. The flat trophoblast cells form the outer shell of the blastocyst and will become part of the placenta. The inner cell mass will eventually develop into the embryo. The blastocyst travels through the uterine cavity and implants into the endometrium—the innermost lining of the uterus. This implantation event begins about 6 days after fertilization.

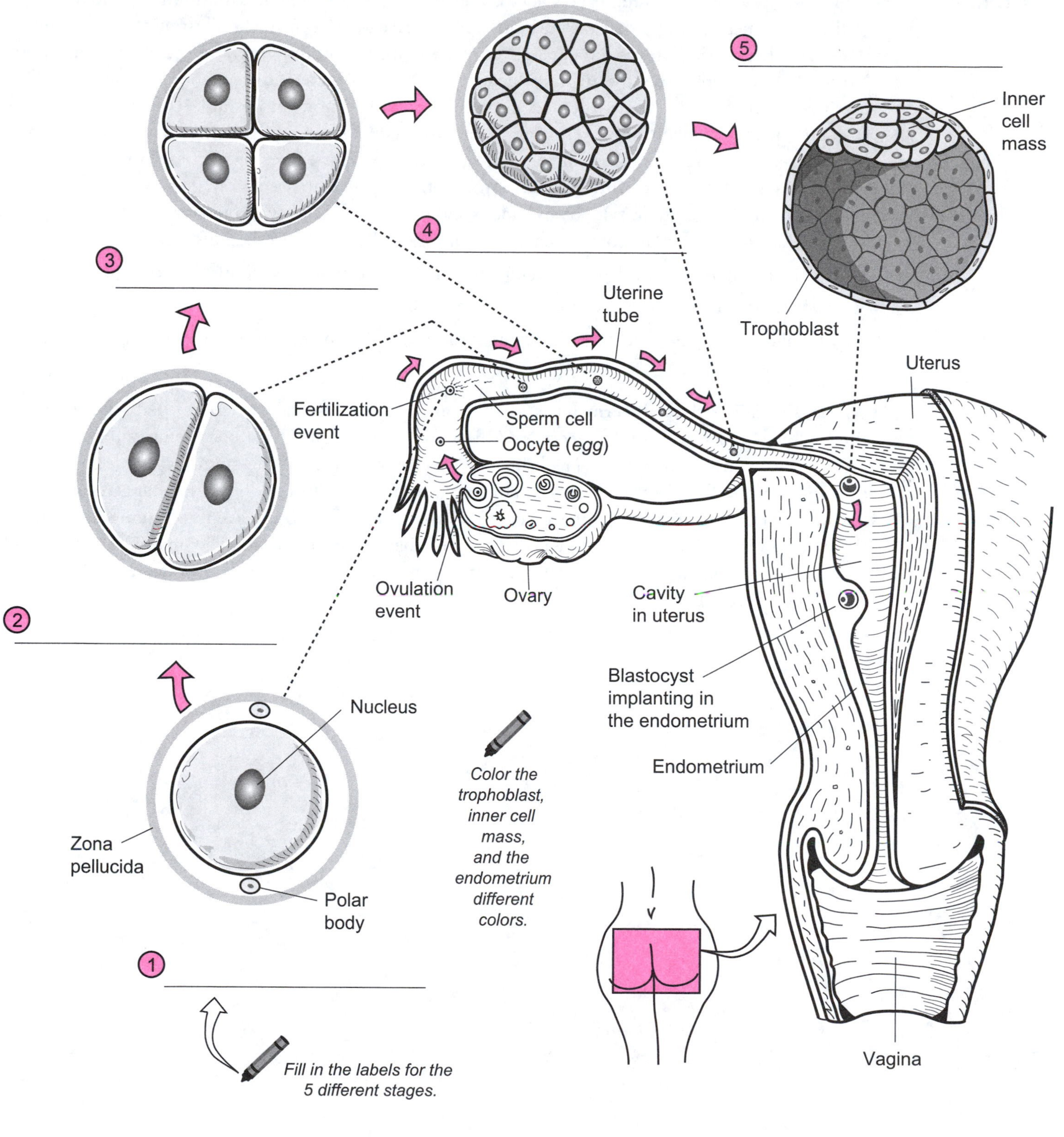

Color the trophoblast, inner cell mass, and the endometrium different colors.

Fill in the labels for the 5 different stages.

Description

The illustration on the facing page shows a woman 9 months pregnant with a full-term fetus. Many of the anatomical changes that occur during pregnancy can be seen in the illustration. For example, the normally fist-sized **uterus** has expanded to an enormous size filling most of the abdominal cavity. Notice how it compresses the **urinary bladder** resulting in increased frequency of urination. This uterine expansion also pushes the abdominal organs and diaphragm upward. This, in turn, causes pressure on the stomach often resulting in heartburn. Another result of the swollen abdomen is lordosis—an increased curvature of the lumbar vertebrae. This causes backaches. Hormonal stimulation causes enlargement of the breasts. Around the nipple, the **areola** often increases in size and darkens in color. The **pubic symphysis** widens and increases flexibility in preparation for birth.

Here are some other anatomical changes that cannot be viewed in the illustration:

- **Heart enlarges**—this leads to an increase in cardiac output to support the growing fetus
- **Weight gain**—a woman can gain as much as 30 lbs.
- **Vagina changes**—increased elasticity and vascularity in preparation for serving as the birth canal

Let's examine the **full-term fetus**. It floats in the **amniotic sac** which is filled with **amniotic fluid**. This liquid serves as a protective cushion and also regulates fetal body temperature. When a pregnant woman says her "*water breaks*" it refers to the rupturing of the amniotic sac, followed by a flow of amniotic fluid out of the body. This often occurs prior to childbirth. Exchange of nutrients and wastes between mother and fetus occurs at a structure called the **placenta**. Linking the fetus to the placenta is a cable-like structure called the **umbilical cord**. It contains the umbilical arteries and vein. Following birth, the placenta detaches from the uterus and is expelled from the body. This is called the **afterbirth**.

Pregnancy

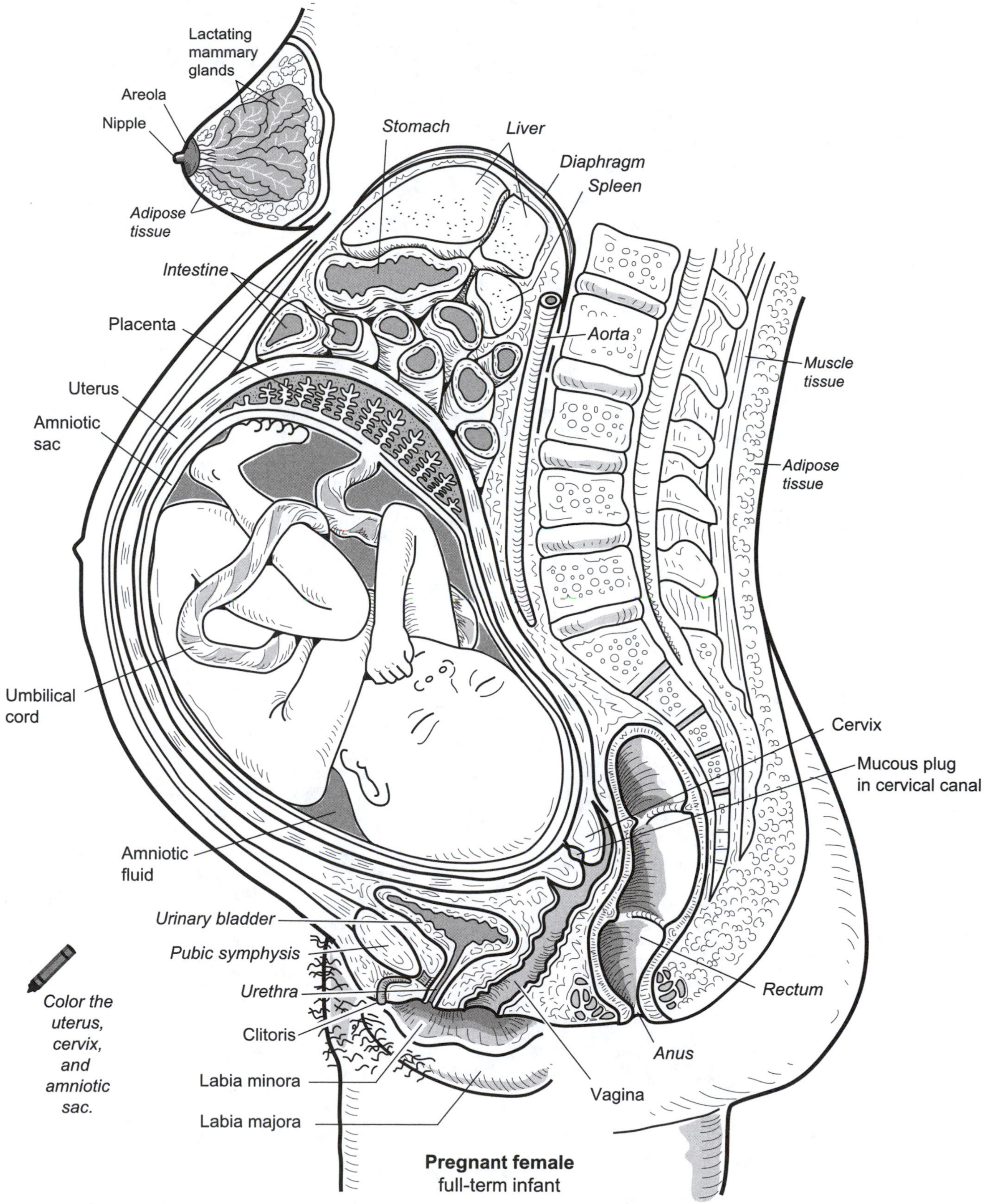

Color the uterus, cervix, and amniotic sac.

Pregnant female
full-term infant

Notes

Glossary of Prefixes and Suffixes

Element	Definition and Example
a-	absent, deficient or without: atrophy
ab-	off, away from: abduct
abdomin-	abdomen
-able	capable of: viable
ac-	toward, to: actin
acou-	hear, acoustic
ad-	denoting to, toward: adduct
af-	movement toward a central point: afferent artery
alba-	pale or white: linea alba
-alg	pain: neuralgia
ambi-	both: ambidextrous
angi-	pertaining to vessel: angiogram
ante-	before: antebrachium
anti-	against: anticoagulant
aqua-	water: aqueous
archi-	to be first: archeteron
arthri-	joint: arthritis
-asis	condition or state of: homeostasis
aud-	pertaining to ear: auditory
auto-	self: autolysis
bi-	two: biceps
bio-	life: biology
blast-	generative or germ bud: osteoblast
brachi-	arm: brachialis
brachy-	short: brachydont
brady-	slow: bradycardia
bucc-	cheek: buccal cavity
cac-	bad, ill: cachexia
calci-	stone: calcification
capit-	head: capitate
carcin-	cancer: carcinogenic
cardi-	heart: cardiac
caud-	tail: cauda equina
cata-	lower, under, against: catabolism
-coel	swelling, and enlarged space or cavity: blastocoele
cephal-	head: cephalis
cerebro-	brain: cerebrospinal fluid
chol-	bile: cholic
chondr-	cartilage: chondrocyte
chrom-	color: chromosome
-cid(e)	destroy: germicide
circum-	around: circumduct
-cis	cut, kill: excision
co-	together: copulation
coel-	hollow cavity: coelom
con-	with, together: congenital
contra-	against, opposite: contraception
corn-	denoting hardness: cornified
corp-	body: corpus
crypt-	hidden: cryptorchism
cyan-	blue color: cyanosis
cysti-	sac or bladder: cystoscope
cyto-	cell: cytology
de-	down, from: descent
derm-	skin: dermatology
di-	two: diarthrotic
dipl-	double: diploid
dis-	apart, away from: disarticulate
duct-	lead, conduct: ductus deferens
dur-	hard: dura mater
-dynia	pain
dys-	bad, difficult, painful: dysentery
e-	out, from: eccrine
ecto-	outside, outer, external: ectoderm
-ectomy	surgical removal: tonsillectomy
ede-	swelling: edema
-emia	pertaining to a condition of the blood: lipemia
end-	within: endoderm
entero-	intestine: enteritis
epi-	upon, in addition: epidermis
erythro-	red: erythrocyte
ex-	out of: excise
exo-	outside: exocrine
extra-	outside of, beyond, in addition: extracellular
fasci-	band: fascia
febr-	fever: febrile
-ferent	bear, carry: efferent arteriole
fiss-	split: fissure
for-	opening: foramen
-form	shape: fusiform
gastro-	stomach: gastrointestinal
-gen	an agent that produces or originates: pathogen
-genic	produced from, producing: carcinogenic
gloss-	tongue: glossopharyngeal
glyco-	sugar: glycogen
-gram	a record, recording: myogram
gran-	grain, particle: agranulocyte
-graph	instrument for recording: electrocardiograph
grav-	heavy: gravid
gyn-	female sex: gynocology
hema(o)-	blood: hematology
haplo-	simple or single: haploid
hemi-	half: hemisphere
hepat-	liver: hepatic portal
hetero-	other, different: heterosexual
histo-	tissue: histology
holo-	whole, entire: holocrine
homo-	same, alike: homologous
hydro-	water: hydrophilic
hyper-	beyond, above, excessive: hypertension
hypo-	under, below: hypoglycemia
-ia	state or condition: hypoglycemia
-iatrics	medical specialties: pediatrics
idio-	self, separate, distinct: idiopathic
ilio-	ilium: iliosacral
infra-	beneath: infraspinatus

Element	Definition and Example
inter-	among, between: intercellular
intra-	inside, within: intracellular
-ion	process: acromion
iso-	equal, like: isotonic
-ism	condition or state: rheumatism
-itis	inflammation: meningitis
labi-	lip: labium majora
lacri-	tears: lacrimal
later-	side: lateral
leuc-	white: leucocyte
lip-	fat: lipid
-logy	science of: cytology
-lysis	solution, dissolve: hemolysis
macro-	large, great: macrophage
mal-	bad, abnormal, disorder: malignant
medi-	middle: medial
mega-	great, large: megakaryocyte
meso-	middle or moderate: mesoderm
meta-	after, beyond: metatarsal
micro-	small: microscope
mito-	thread: mitosis
mono-	alone, one, single: monocyte
mons-	mountain: mons pubis
morph-	form, shape: morphology
multi-	many, much: multicellular
myo-	muscle: myofibril
narc-	numbness, stupor: narcotic
neo-	new, young: neonatal
necro-	corpse, dead: necrosis
nephro-	kidney: nephron
neuro-	nerve: neurolemma
noto-	back: notochord
ob-	against, toward, in front of: obturator
oc-	against: occlusion
-oid	resembling, likeness: sigmoid
oligo-	few, small: oligodendrocyte
-oma	tumor: lymphoma
oo-	egg: oocyte
or-	mouth: oral
orchi-	testicles: cryptorchidism
-ory	pertaining to: sensory
osteo-	bone: osteocyte
-ose	full of: adipose
oto	ear: otolith
ovo-	egg: ovum

Element	Definition and Example
para-	give birth to, bear: parturition
para-	near, beyond, beside: paranasal
path-	disease, that which undergoes sickness: pathology
-pathy	abnormality, disease: neuropathy
ped-	children: pediatrician
pen-	need, lack: penicillin
-penia	deficiency: thrombocytopenia
per-	through: percutaneous
peri-	near, around: pericardium
phag-	to eat: phagocyte
-phil	have an affinity for: neutrophil
phlebo-	vein: phlebitis
-phobe	abnormal fear, dread: hydrophobe
-plasty	reconstruction of: rhinoplasty
platy-	flat, side: platysma
-plegia	stroke, paralysis: paraplegia
-pnea	to breathe: apnea
pneumato-	breathing: pneumonia
pod-	foot: podiatry
-poieis	formation of: hematopoiesis
poly-	many, much: polyploid
post-	after, behind: postnatal
pre-	before in time or place: prenatal
prim-	first: primitive
pro-	before in time or place: prosect
proct-	anus: proctology
pseudo-	false: pseudostratified
psycho-	mental: psychology
pyo-	pus: pyoculture
quad-	fourfold: quadriceps femoris
re-	back, again: repolarization
rect-	straight: rectus abdominis
reno-	kidney: renal
rete-	network: retina
retro-	backward: retroperitoneal
rhin-	nose: rhinitis
-rrhage	excessive flow: hemorrhage
-rrhea	flow, or discharge: diarrhea
sanguin-	blood: sanguiferous
sarc-	flesh: sarcoplasm
-scope	instrument for examination of a part: stethoscope
-sect	cut: dissect
semi-	half: semilunar
serrate-	saw-edged: serratus anterior
-sis	state or condition: dialysis
steno-	narrow: stenohaline

Element	Definition and Example
-stomy	surgical opening: tracheotomy
sub-	under, beneath, below: subcutaneous
super-	above, beyond, upper: superficial
supra-	above, over: suprarenal
syn (sym)	together, joined, with: synapse
tachy-	swift, rapid: tachometer
tele-	far: telencephalon
tens-	stretch: tensor fascia latae
tetra-	four: tetrad
therm-	heat: thermogram
thorac-	chest: thoracic cavity
thrombo-	lump, clot: thrombocyte
-tomy	cut: appendectomy
tox-	poison: toxic
tract-	draw, drag: traction
trans-	across, over: transfuse
tri-	three: trigone
trich-	hair: trichology
-trophy	a state relating to nutrition: hypertrophy
-tropic	turning toward, changing: gonadotropic
ultra-	beyond, excess: ultrasonic
uni-	one: unicellular
uro-	urine, urinary organs or tract: uroscope
-uria	urine: polyuria
vas-	vessel: vasoconstriction
vermi-	worm: vermiform
viscer-	organ: visceral
vit-	life: vitamin
zoo-	animal: zoology
zygo-	union, join: zygote

Glossary

A

abdomen (AB-doh-men) The region between the diaphragm and the pelvis.

absorption (ab-ZORP-shun) The passage of gases, liquids, or solutes through a membrane.

acetylcholine (ACh) (ass-ee-til-KOH-leen) A neurotransmitter secreted by various neurons into synapses; may have an excitatory effect (neuromuscular junction of skeletal muscles) or an inhibitory effect (neuromuscular junction of cardiac muscle tissue).

acetylcholinesterase (AChase) (ass-ee-til-koe-lin-ESS-ter-ase) An enzyme that rapidly inactivates the acetylcholine bound to postsynaptic receptors.

acidic (ah-SID-ik) Describes a solution in which the pH is less than 7; having a relatively high concentration of hydrogen ions.

acromial (ah-KROH-mee-al) The outer end of the scapula; site where the clavicle is attached.

acrosome (AK-roh-sohm) A cap-like structure on the head of a sperm cell; produces enzymes for penetrating eggs.

actin (AK-tin) A thin, protein filament found in skeletal muscle cells; protein component of microfilaments.

action potential An electrical impulse that advances along a plasma membrane of a neuron or muscle cell.

active transport A carrier-mediated process in which cellular energy is used to move molecules against their concentration gradient through a plasma membrane.

adenosine diphosphate (ah-DEN-oh-seen dye-FAHS-fate) An energy molecule composed of an adenine base, a ribose sugar, and two phosphate groups.

adenosine triphosphatase (ATPase) (ah-DEN-oh-seen try-FAHS-fate-ays) (AY-tee-pays) An enzyme that catalyzes the hydrolysis of ATP to form ADP, a phosphate group, and a net release of free energy.

adenosine triphosphate (ATP) (ah-DEN-oh-seen try-FAHS-fate) The energy currency of a cell; composed of an adenine base, a ribose sugar, and three phosphate groups.

adenylate cyclase (ah-DEN-il-ayt SYE-klayz) An enzyme that catalyzes the conversion of ATP to cyclic AMP (cAMP).

adipocyte (AD-i-poh-syte) A fat cell.

adrenergic neuron (ad-ren-ER-jik NOO-ron) A neuron that secretes norepinephrine (*noradrenaline*) or epinephrine (*adrenaline*) as its neurotransmitter.

aerobic Any process that requires oxygen.

afferent (AF-fer-ent) Toward; opposite of efferent.

aldosterone (al-DAH-stair-ohn) A mineralocorticoid; a hormone produced by the adrenal cortex that acts on the kidneys to increase sodium levels in the blood and increase excretion of potassium in urine.

alkaline (AL-kah-lin) Refers to a base, a pH greater than 7; having a relatively low concentration of hydrogen ions.

alveolus/alveoli (al-VEE-oh-luss/al-VEE-oh-lye) Delicate air sacs in the lungs where gas exchange occurs.

amino acid (ah-MEE-no ASS-id) The structural unit of a protein.

amnion (AM-nee-on) One of the extraembryonic membranes; develops around the embryo/fetus forming the amniotic cavity.

amniotic fluid (am-nee-OT-ik fluid) The fluid that surrounds and cushions the developing embryo and fetus.

amniotic sac (am-nee-OT-ik sak) Fluid-filled chamber in which the embryo floats during development.

amylase (AM-eh-layz) An enzyme that breaks down polysaccharides; produced by salivary glands and pancreas.

anabolism (ah-NAB-oh-liz-em) Synthesis reactions that require energy to join small molecules together to form more complex molecules; for example, amino acids bonding together to make proteins.

anaerobic (an-air-OH-bik) Any process that does not require oxygen.

anaphase (AN-nah-fayz) Stage of mitosis when the chromatid pairs separate and the daughter chromosomes move toward the opposite ends of the cell.

antebrachial (an-tee-BRAY-kee-al) Pertaining to the forearm.

antecubital (an-tee-KYOO-bi-tal) Pertaining to the anterior side of the elbow.

anterior (an-TEER-ee-or) Toward the front or ventral; opposite of posterior.

antibody Proteins produced by plasma cells in response to specific antigens; antibodies combine with antigens to render a pathogen harmless; also known as immunoglobulin (Ig).

antigen (ANN-tih-jen) Means "*anti*body *gen*erating substances"; macromolecules that induce the immune system to make antibodies.

antrum (AN-trum) Central chamber.

anus (AY-nus) External opening at the end of the rectum.

aorta (ay-OR-tah) Largest artery in the body; carries oxygenated blood from left ventricle and into the systemic circuit.

apex (AY-peks) Pointed tip of a structure.

apical (AY-pik-al) Located near or relating to the apex or pointed structure; opposite of basal.

apnea (APP-nee-ah) Temporary cessation of breathing at the end of a normal expiration.

aponeurosis (ap-oh-nyoo-ROH-sis) Broad, flat collagenous sheets that may serve as anchor points for skeletal muscle.

appendix (ah-PEND-diks) Small organ connected to the cecum of the large intestine.

aqueous humor (AY-kwee-us HYOO-mor) A fluid that fills the anterior chamber of the eye.

arachnoid (ah-RAK-noyd) Middle meninges that surround the CSF and protect the brain and spinal cord.

arbor vitae (AR-bor VYE-tay) Central area of white matter in the cerebellum.

arteries (AR-ter-eez) Blood vessels that carry blood away from the heart.

arteriole (ar-TEER-ee-ohl) Microscopic blood vessel that connects small arteries to capillaries.

articulation (ar-tik-yoo-LAY-shun) Joint; point of contact between bones.

atom (AT-om) The smallest particle of a chemical element that displays the properties of that element.

atrioventricular (AV) (ay-tree-oh-ven-TRIK-yoo-lar) **bundle** Group of specialized cardiac muscle cells that extend from the AV node to the Purkinje fibers; coordinates contraction of the heart muscle.

atrioventricular (ay-tree-oh-ven-TRIK-yoo-lar) **(AV) node** Secondary pacemaker of the heart; a small mass of specialized cardiac muscle cells that plays a role in the electrical conduction system within the heart.

atrioventricular (AV) (ay-tree-oh-ven-TRIK-yoo-lar) **valves** Pair of heart valves located between the atria and the ventricles; tricuspid and bicuspid; prevent backflow of blood into atria during ventricular contraction.

atrium (AY-tree-um) An upper chamber of the heart; receives blood from the pulmonary or systemic circuit.

auditory (AW-di-toh-ree) Pertaining to the sense of hearing.

auditory tube (AW-di-toh-ree tube) Passageway that connects the nasopharynx with the middle ear.

auricle (AW-ri-kul) Curved, flexible upper portion of the ear; also, expandable flap-like structure of an atrium in the heart.

autonomic (aw-toh-NAHM-ik) **nervous system (ANS)** Part of the nervous system that governs itself without our conscious knowledge; divided into sympathetic and parasympathetic divisions; controls organs through reflex pathways that link to smooth muscle, cardiac muscle, and glands.

axilla (AK-sil-ah) The armpit.

axon hillock (AK-son HILL-ok) Funnel-shaped portion of neural cell body from which the axon extends.

axons (AK-sonz) Long, single process of neurons that conduct nerve impulses toward the synaptic knobs.

B

basal (BAY-sal) Located at or near the base of a structure; opposite of apical.

basement membrane (BAYSE-ment MEM-brayne) A layer of protein fibers that connects the epithelium to the underlying connective tissue.

basic *See alkaline.*

basophil (BAY-so-fil) White blood cell; releases histamine into damaged tissue.

bicuspid valve (bye-KUSS-pid valve) The left atrio-ventricular (A-V) valve in the heart; also known as the mitral valve.

bicuspids (bye-KUSS-pids) Teeth used for crushing, mashing, and grinding; premolars.

bile A secretion of the liver that is stored in the gallbladder and released into the duodenum; physically breaks down fat into smaller droplets.

bipolar neuron (bye-POH-lar NOO-ron) Nerve cell with two distinct processes; one dendrite, one axon.

blastocyst (BLASS-toh-sist) Early stage in embryonic development; a hollow ball of cells consisting of an inner cell mass and an outer cell mass.

blastomere (BLASS-toh-meer) The first cleavage division that produces a pre-embryo consisting of two identical cells.

bolus (BOH-luss) A ball or mass of chewed food that passes from the mouth to the stomach.

Bowman's capsule *See glomerular capsule.*

Boyle's law A physics law stating that the volume of a gas is inversely proportional to its pressure.

brachial (BRAY-kee-al) Pertaining to the upper limb between shoulder and elbow.

brain stem (brayn stem) Part of the brain that contains important processing centers; consists of the medulla oblongata, pons, and midbrain.

bronchiole (BRONG-kee-ohl) A small, tube-like branch of a bronchus; lacks cartilaginous supports, but wall contains smooth muscle.

bronchus (BRONG-kuss) A branch of the bronchial tree between the trachea and the bronchioles.

buccal (BUK-al) Pertaining to the cheek.

bulbourethral glands (BUL-boh-yoo-REE-thral glands) Small, mucus glands located at base of the penis; secretions lubricate the urethra.

bundle branches Two bands of specialized cardiac muscle cells that transmit impulses from the AV bundle to the Purkinje fibers.

bundle of HIS *See atrioventricular (AV) bundle.*

C

calcification (kal-sih-fih-KAY-shun) Process of hardening a tissue with deposits of calcium salts.

calcitonin (CT) (kal-sih-TOE-nin) A hormone produced by the thyroid gland that serves to decrease calcium levels in the blood.

canaliculus/canaliculi (kan-ah-LIK-yoo-luss/kan-ah-LIK-yoo-lye) Microscopic channels between cells, found in compact bone and liver; in compact bone, canaliculi allow diffusion of nutrients and wastes; in liver, bile canaliculi transport bile to bile ducts.

canine tooth (KAY-nyne tooth) *See cuspids.*

capillary (KAP-i-lair-ee) Smallest, simplest blood vessel; microscopic; connects arterioles and venules.

carbohydrate (kar-boh-HYE-drayt) Organic compound containing carbon, oxygen, and hydrogen; sugars, starches, and cellulose.

cardiac (KAR-dee-ak) **cycle** A complete heartbeat or pumping cycle consisting of systole and diastole.

cardiac (KAR-dee-ak) **output** The volume of blood pumped by one heart ventricle each minute.

carotid artery (kah-ROT-id AR-ter-ee) The large artery of the neck that provides a major blood supply to the brain.

carpals (KAR-puls) Wrist bones.

carpus/carpal (KAR-pus/KAR-pul) The wrist.

catabolism (kah-TAB-oh-liz-em) Chemical reactions that break down complex organic compounds into simpler ones with a net release of energy; for example, proteins broken down into amino acids.

catecholamines (kat-eh-KOLE-ah-meenz) A class of neurotransmitters that are synthesized from the amino acid tyrosine; include norepinephrine, epinephrine, and dopamine; may play a role in sleep, motor function, regulating mood, and pleasure recognition.

caudal (KAW-dal) The tail.

CCK *See cholecystokinin.*

cecum (SEE-kum) The pouch located at the beginning of the large intestine.

cell (sell) The basic unit of life.

cellular respiration Metabolic process by which carbohydrates, fatty acids, and amino acids are broken down to produce ATP.

cementum (see-MEN-tum) Bone-like material covering the root of the tooth.

centriole (SEN-tree-ohl) Tiny, cylindrical organelle of a cell; involved with spindle formation during mitosis.

centromere (SEN-troh-meer) The region where two chromatids are connected together during the early stages of cell division.

centrosome (SEN-troh-sohm) The region of cytoplasm that coordinates the activities of centrioles.

cerebellum (sair-eh-BELL-um) Second largest part of the brain; coordinates and refines learned movement patterns.

cerebrospinal (SAIR-eh-broh-SPY-nal) **fluid (CSF)** Liquid found in the central nervous system by filtering and processing blood plasma; circulates in the ventricles, central canal, and subarachnoid space of the brain and spinal cord.

cerebrum (SAIR-eh-brum) Largest region of the brain; origin of conscious thoughts and all intellectual functions; controls sensory and motor integration.

cervical (SER-vih-kal) Pertaining to the neck.

cervix (SER-viks) Neck-like structure at the inferior portion of the uterus; projects into vagina.

chief cells (CHEEF sells) Cells in the gastric mucosa that primarily secrete pepsinogen, which later is converted into the protein digesting enzyme called pepsin.

cholecystokinin (CCK) (koh-lee-sis-toh-KYE-nin) A hormone produced by the duodenal mucosa that stimulates contraction of the gallbladder and secretion of pancreatic juice rich in digestive enzymes.

cholinergic neuron (koh-leh-NER-jik NOO-ron) Neuron that secretes acetylcholine as their neurotransmitters.

chondrocyte (KON-droh-syte) A cartilage cell.

chorion (KOH-ree-on) A membrane consisting of the mesoderm and trophoblast; develops into a membrane of the placenta.

choroid (KOH-royd) Middle, vascular layer of the eye.

chromatid (KROH-mah-tid) Either of two daughter strands of chromosomes that are joined by a single centromere.

chromatin (KROH-mah-tin) Chromosomal material that is loosely coiled, forming a tangle of fine filaments with a grainy appearance.

chromosomes (KROH-meh-sohms) Tightly compacted structures that contain coiled DNA wrapped around histone proteins; normal human body cells contain 46 chromosomes; term means "colored body."

chyme (kyme) A soupy, viscous mixture of ingested substances and gastric juices leaving the stomach.

cilia (SIL-ee-ah) Long folds of plasma membrane that contain microtubules.

ciliary body (SIL-ee-air-ee body) A muscular structure that surrounds the perimeter of the lens of the eye and attaches to it through the suspensory ligaments.

citric acid cycle (SIT-rik ASS-id SYE-kul) An aerobic chemical cycle that begins with the formation of citric acid and results in the formation of oxaloacetic acid; in the process, ATP is produced and carbon dioxide is released; occurs in the mitochondrion.

clitoris (KLIT-oh-ris) Small female organ composed of erectile tissue located behind the junction of the labia majora.

coccygeal (KOKS-ih-jee-al) Relating to or near the coccyx.

coccyx (KOKS-siks) The tailbone; most inferior portion of the vertebral column.

cochlea (KOHK-lee-ah) An inner ear structure that resembles a snail shell; contains nerve endings that are essential to hearing.

collagen (KAHL-ah-jen) The most common type of protein fiber found in connective tissues, serves to strengthen tissues.

colon (KOH-lon) The large intestine.

common bile duct (KOM-mon byle dukt) Formed by the union of the cystic duct and the hepatic duct; carries bile from the liver and gallbladder to the duodenum.

compact bone (KOM-pak bone) Dense bone; contains osteons.

compound (KOM-pound) A substance formed by the union of two or more elements.

connective tissue (koh-NEK-tiv TISH-yoo) One of the four major tissue types; serve to give structural support to other tissues and organs in the body; most contain cells, protein fibers, and ground substance.

cornea (KOHR-nee-ah) The transparent, anterior region of the sclera.

corona radiata (koh-ROHN-ah ray-dee-AY-tah) Follicular cells that surround the oocyte.

coronoid (KOHR-oh-noyd) Pertaining to certain processes of the bone; shaped like a crow's beak.

corpus/corpora (KOHR-pus/KOHR-pohr-ah) Body.

corpus albicans (KOHR-pus AL-bi-kans) Pale scar tissue in the ovaries that replaces the nonfunctional corpus luteum.

corpus callosum (KOHR-pus kah-LOH-sum) Area of the brain that links the right and left cerebral hemispheres.

corpus cavernosum (KOHR-pus kav-er-NO-sum) Two columns of erectile tissue that extend along the length of the penis.

corpus luteum (KOHR-pus LOO-tee-um) An ovarian structure transformed from a ruptured follicle; secretes estrogen and progesterone.

corpus spongiosum (KOHR-pus spun-jee-OH-sum) Erectile body that surrounds the urethra.

cortex (KOHR-teks) Outer part of an organ; adrenal cortex.

cortisol (KOHR-tih-sawl) A glucocorticoid; a hormone produced by the adrenal cortex that helps regulate responses to stress; also known as hydrocortisone.

Cowper's glands (KOW-perz glands) *See bulbourethral glands*.

coxa/coxae (KOKS-ah/KOKS-ee) Pelvic bone or hip bone.

crest (as in bone marking) Slightly raised narrow ridge on a bone; site for muscle attachments.

crural (KROOR-al) Refers to the anterior portion of the leg (below the knee).

cubital (KYOO-bi-tal) Pertaining to the elbow.

cuspids (KUS-pids) Sharp, pointed teeth; canines.

cyclic adenosine monophosphate (SIK-lik ah-DEN-oh-seen mon-oh-FOS-fate) **(cAMP)** A second messenger formed from ATP; composed of an adenine base, a ribose sugar, and one phosphate group; used for signal transduction.

cystic duct (SIS-tik dukt) A tube that leads from the gallbladder toward the liver; unites with the common hepatic duct to form the common bile duct.

cytokinesis (SYE-toe-kih-nee-siss) Physical division of the cytoplasm during cell division.

cytoplasm (SYE-toh-plaz-em) Gel-like material between the nucleus and the plasma membrane; includes cytosol and organelles.

cytoskeleton (sye-toh-SKEL-eh-ton) An integrated network of microtubules and microfilaments in the cytoplasm that gives shape and provides support to the cell and its organelles.

cytosol (SYE-toh-sawl) Fluid portion of the cytoplasm.

D

decidua basalis (dih-SID-yoo-ah bah-SAY-lis) Area of the endometrium that develops into the maternal part of the placenta.

deep away from the surface; opposite of superficial.

deltoid (DEL-toyd) Triangular shape.

dendrite (DEN-dryte) A long cellular extension of a neuron that responds directly to stimuli.

dentin (DEN-tin) The mineralized matrix found in teeth.

dermis (DER-mis) Layer of connective tissue that lies beneath the epidermis.

diaphragm (DYE-ah-fram) Dome-shaped muscle that separates the thoracic and abdominal cavities; a major muscle involved with respiration.

diaphysis (dye-AF-i-sis) Tubular shaft of a long bone.

diastole (dye-ASS-toh-lee) Relaxation of both atria and both ventricles in the heart; opposite of systole.

diffusion (dih-FYOO-shun) The net movement of substances from an area of high concentration to an area of low concentration.

disaccharide (dye-SAK-ah-ryde) A double-unit sugar formed by bonding a pair of monosaccharides together; *ex:* lactose, sucrose.

distal (DIS-tall) Refers to the region or reference away from an attached base; opposite of proximal.

distal convoluted tubule (DCT) (DIS-tal KON-voh-loo-ted TOOB-yool) The part of the nephron distal to the ascending limb of the nephron loop; site for active secretion and selective reabsorption of ions and other substances.

dorsal (DOR-sal) Pertaining to the back; posterior; opposite of ventral.

ductus deferens (DUK-tus DEF-er-ens) A smooth muscular tube that propels sperm from the epididymis to the ejaculatory duct; also known as vas deferens.

duodenum (doo-oh-DEE-num) First and shortest segment of the small intestine—about 10 in. long; receives chyme from the stomach and digestive secretions from the pancreas.

dura mater (DOO-rah MAH-ter) Outermost layer of the meninges.

E

efferent (EF-fer-ent) Away from; opposite of afferent.

ejaculatory duct (ee-JAK-yoo-lah-toh-ree dukt) Short passageway that allows sperm to enter the urethra.

electrocardiogram (ECG or EKG) (ee-lek-troh-KAR-dee-oh-gram) A graphic record of the heart's electrical activity or conduction of impulses; used to evaluate the heart's action potential.

electroencephalogram (EEG) (ee-lek-troh-en-SEF-ah-loe-gram) A graphic record of brain electrical potentials; used to evaluate nerve tissue function and diagnose specific disorders (such as epilepsy).

electrolytes (ee-LEK-troh-lytes) Inorganic substances that break up in solution to form ions and conduct electricity; include acids, bases, and salts.

electrons (ee-LEK-trons) Negatively charged subatomic particles in atoms; located in energy shells outside the nucleus of the atom.

electron transport system (ETS) (ee-LEK-tron TRANS-port SIS-tem) Transfer of electrons along a series of membrane-bound electron carrier molecules in the mitochondria; energy from this process is used to synthesize ATP; aerobic process that produces water.

embolus (EM-boe-luss) A clot that dislodges and circulates through the bloodstream.

embryo (EM-bree-oh) A stage of human development beginning at fertilization and ending at the start of the 8th developmental week.

emulsification (ee-mul-sih-fih-KAY-shun) The process by which bile breaks up fats.

enamel (ee-NAM-el) Hardest manufactured substance in the body; covers the crown of the tooth.

endocardium (en-doh-KAR-dee-um) Simple squamous epithelial inner layer of the heart.

endocytosis (en-doh-sye-TOH-sis) An active transport mechanism that allows extracellular substances to enter the cell by phagocytosis, pinocytosis, or receptor-mediated endocytosis.

endometrium (en-doh-MEE-tree-um) Innermost, glandular layer of the uterine wall.

endomysium (en-doh-MISH-ee-um) Inner layer of connective tissue that surrounds each skeletal muscle fiber.

endoneurium (en-doh-NOO-ree-um) A layer of connective tissue that surrounds individual axons.

endoplasmic reticulum (en-doh-PLAS-mik re-TIK-yoo-lum) The network of intracellular membranes that synthesizes and manufactures membrane-bound proteins.

endosteum (en-DOS-tee-um) A membrane that lines the marrow cavity inside a bone.

enzyme (EN-zyme) A biological catalyst.

eosinophil (ee-oh-SIN-oh-fil) A phagocytic white blood cell; numbers increase during allergic reaction.

epicardium (ep-i-KAR-dee-um) Outer covering of the heart; also called the visceral pericardium.

epidermis (ep-i-DER-mis) Outermost layer of the skin.

epididymis (ep-i-DID-i-miss) A long, coiled and twisted tubule that lies along the posterior border of the testis; stores sperm and facilitates their maturation.

epiglottis (ep-i-GLOT-iss) Flap of elastic cartilage that folds back over the larynx during swallowing.

epimysium (ep-i-MISH-ee-um) Connective tissue layer that surrounds the entire skeletal muscle.

epineurium (ep-i-NOO-ree-um) Outermost fibrous connective tissue sheath that surrounds a peripheral nerve.

epiphysis (eh-PIF-i-sis) Expanded end of a long bone.

epithelial tissue (ep-i-THEE-lee-al TISH-yoo) One of the four major tissue types; serves to cover exposed body surfaces and lines internal cavities and passageways.

erythrocyte (e-RITH-roh-syte) Red blood cell.

esophagus (eh-SOF-ah-gus) Hollow, muscular tube that transports food and liquid from the pharynx to the stomach.

estrogens (ES-troh-jens) Steroid hormone produced by the ovaries; dominant sex hormone in females.

exocytosis (eks-oh-sye-TOH-sis) An active transport mechanism that involves movement of vesicle-bound substances out of the cell; vesicles fuse with the plasma membrane, and contents are released outside the cell.

extracelluar fluid (ECF) Liquid located outside body cells such as plasma, lymph, and interstitial fluid.

F

facet (FASS-et) Flat surface of a bone that forms a joint with another bone.

facilitated diffusion A passive transport process wherein a substance binds to a carrier molecule in the plasma membrane and is moved from an area of higher concentration to an area of lower concentration.

Fallopian tubes (fal-LOH-pee-an toobs) *See uterine tubes.*

fascia (FAY-sha) A connective tissue sheath consisting of fibrous tissue and fat; unites skin to underlying tissue.

fascicle (FASS-i-kul) A single bundle.

feces (FEE-seez) The waste material eliminated from the large intestine; residue of digestion; also known as *stool.*

femoral (FEM-or-al) Pertaining to the thigh.

fetus (FEE-tus) The name given to the unborn young from the eighth week of pregnancy to birth.

fibroblasts (FYE-broh-blasts) The most abundant fixed cells in connective tissue proper.

fibular (FIB-yoo-lar) Pertaining to the fibula.

fight-or-flight-response A group of reactions that ready the body to engage in maximum muscular exertion needed to deal with a perceived threat.

filtrate Fluid produced by the process of filtration, such as that made by the glomeruli in the kidneys.

filtration A process that uses hydrostatic pressure to move water and solutes through a membrane.

fimbria/fimbriae (FIM-bree-ah/FIM-bree-ee) Fingerlike projections, such as those found on the ends of the uterine tubes nearer the ovaries.

first messenger Typically, a nonsteroid hormone that binds to its plasma membrane receptor, thereby inducing a response in the target cell.

fissure (FISH-ur) Deep groove.

follicle (FOL-lih-kul)**-stimulating hormone (FSH)** A hormone that stimulates structures within the ovaries and primary follicles to grow toward maturity; stimulates follicle cells to synthesize and secrete estrogen in the female and stimulates development of the seminiferous tubules in the male.

fontanel (FON-tah-nel) Fibrous area between the cranial bones; "soft spot".

foramen/foramina (foh-RAY-men/foh-RAM-i-nah) A hole or passageway in bone for blood vessels and/or nerves to pass through.

forearm (FORE-arm) The part of the arm between the elbow and the wrist.

fossa (FOSS-ah) Depression.

frenulum (FREN-yoo-lum) Small bridle; thin, flat fold of mucous membrane between two structures, such as the lingual frenulum that anchors the body of the tongue to the floor of the oral cavity.

frontal (FRUN-tal) Toward the anterior.

frontal plane (FRUN-tal playne) Runs parallel to the long axis of the body; divides the body into anterior and posterior sections.

G

G cells (jee sells) Cells found in the stomach; secrete the hormone gastrin.

gallbladder Pear-shaped organ attached to the liver that stores and concentrates bile.

gametes (GAM-eets) Mature male or female reproductive cells—sperm or secondary oocyte.

gastric (GAS-trik) Pertaining to the stomach.

gastric juice (GAS-trik joos) A mixture of secretions that contain water, mucus, hydrochloric acid, and the enzyme pepsin; secreted by exocrine gastric glands.

gastrin (GAS-trin) The hormone that stimulates the secretion of both parietal and chief cells in the stomach.

genitalia (jen-i-TAYL-yah) The reproductive organs.

glomerular (gloh-MARE-yoo-lar) **capsule** Cup-shaped mouth of the nephron; surrounds the glomerulus; receives the glomerular filtrate; also known as Bowman's capsule.

glomerulus/glomeruli (gloh-MARE-yoo-lus/gloh-MARE-yoo-lye) Cluster of capillaries tucked into the glomerular capsule; forms the glomerular filtrate.

glucagon (GLOO-kah-gon) A hormone produced by alpha cells in the pancreas; increases blood glucose levels.

glucose (GLOO-kohs) A monosaccharide; principal source of energy for the cells.

glucocorticoids (gloo-koe-KORE-tih-koyds) A group of hormones produced by the adrenal cortex that help regulate normal glucose levels and increase resistance to stress; *ex:* cortisol.

gluconeogenesis (gloo-koh-nee-oh-JEN-eh-sis) The synthesis of glucose from lipid or protein.

gluteal (GLOO-tee-al) Pertaining to the buttocks.

glycogen (GLYE-koh-jen) A polysaccharide consisting of a long chain of glucose units; found mainly in liver and skeletal muscle cells.

glycolysis (glye-KOL-ih-sis) An anaerobic process that breaks down a glucose molecule into two pyruvic acid molecules; ATP is produced; occurs in the cytoplasm.

Golgi complex (GOL-jee KOM-pleks) Membranous cellular organelle that stores, alters, and packages secretory products such as proteins; gives rise to lysosomes and secretory vesicles; also known as Golgi apparatus or Golgi body.

G protein Membrane protein that reacts with guanosine triphosphate (GTP); activates the membrane-bound enzyme adenylate cyclase.

Graafian follicle (GRAH-fee-en FOL-li-kul) *See tertiary follicle.*

gray matter (gray MAT-er) Regions in the brain and spinal cord containing nerve cell bodies, glial cells, and unmyelinated axons.

growth hormone Hormone produced by the anterior pituitary; promotes bodily growth indirectly by stimulating the liver to produce certain growth factors; also known as somatotropin.

gyrus/gyri (JYE-russ/JYE-rye) Elevated ridge.

H

hard palate (hard PAL-let) Bony roof of the mouth.

haustrum/haustra (HAWS-trum/HAWS-trah) Series of pouches in the wall of the colon; permits distention and elongation.

Haversian system (hah-VER-shun system) *See osteon.*

head Rounded process at the end of a long bone.

hemoglobin (hee-mo-GLOH-bin) A protein containing iron in red blood cells; functions to transport oxygen and some carbon dioxide in the blood.

hepatic duct (heh-PAT-ik dukt) Collects bile from all of the bile ducts of the liver lobes; unite to form common bile duct.

hepatic portal vein (heh-PAT-ik POR-tall vane) Delivers blood to the liver.

hepatic triads (heh-PAT-ik TRYE-ads) Portal areas located at each of the six corners of the hepatic lobule; each triad consists of the following three structures: (1) branch of hepatic portal vein, (2) branch of hepatic artery proper, and (3) branch of the bile duct.

hepatocytes (heh-PAT-oh-sytes) Liver cells.

hepatopancreatic sphincter (heh-PAT-oh-pan-kree-AT-ik SFINGK-ter) Muscular ring that surrounds the lumen of the common bile duct and the duodenal ampulla; seals off the passageway and prevents bile from entering the small intestine; also known as sphincter of Oddi.

hilus (HYE-luss) A depression on an organ that serves as an entrance site for blood vessels, lymphatic vessels, and/or nerves; *ex:* renal hilus.

homeostasis (hoh-mee-oh-STAY-sis) The process of maintaining the relatively stable state in the body's internal environment despite regular changes in the external environment.

hormone (HOR-mohn) Chemical messenger; secreted by an endocrine gland and transported through the blood.

hydroxyapatite (hye-DROK-see-ap-ah-tyte) The chief structural component of bone; formed by the interaction of calcium phosphate and calcium hydroxide.

hypertonic (hye-per-TON-ik) **solution** Solution that contains a higher concentration of nonpenetrating solutes than the adjacent cell; causes cell to shrink.

hypodermis (hye-poh-DER-mis) Layer of loose connective tissue that separates the skin from underlying tissue and organs; also known as subcutaneous layer.

hypothalamus (hye-poh-THAL-ah-mus) Part of the diencephalon in the brain; contains control centers regulating autonomic functions, emotions, and hormone production.

hypotonic (hye-poh-TON-ik) **solution** Solution that contains a lower concentration of nonpenetrating solutes than the adjacent cell; causes cell to swell.

I

ileum (IL-ee-um) Final 12-foot segment of the small intestine.

ilium (IL-ee-um) Largest coxal bone.

incisors (in-SYE-zors) Blade-shaped teeth found at the front of the mouth; used for clipping or cutting.

incus (IN-kuss) Middle ear bone; also called the anvil.

inferior (in-FEER-ee-or) Below; opposite of superior.

inferior vena cava (inferior VEE-nah KAY-vah) One of the great veins; carries blood to the right atrium from the trunk and lower extremities.

inguinal (ING-gwih-nal) Pertaining to the groin.

inhibin (in-HIB-in) A hormone secreted by the testes and ovaries that inhibits the release of follicle-stimulating hormone (FSH) from the anterior pituitary.

inner ear Region of the ear that contains the receptors of equilibrium and hearing.

inorganic (in-or-GAN-ik) Pertaining to compounds that lack hydrocarbons (carbon atoms bonded to hydrogen atoms); *ex:* water, salts.

insulin A hormone produced by the pancreas; promotes the movement of glucose, amino acids, and fatty acids out of the blood and into body cells.

intercalated discs (in-TER-kah-lay-ted disks) Specialized regions that form connections between cardiac muscle cells.

interdigitate (in-ter-DIJ-ih-tayt) To interlock.

intermediate filaments (in-ter-MEE-dee-it FIL-ah-ments) Part of the cell's cytoskeleton; provides strength, stabilizes the position of the organelles, and transports material within cells.

interphase (IN-ter-fayz) Longest stage in the cell life cycle when the cell prepares for division; DNA is replicated in this stage.

interstitial (in-ter-STISH-al) **fluid** Liquid that fills the spaces between cells.

intervertebral disc (in-ter-VER-tee-bral disk) Pad of fibrocartilage that separates and cushions the vertebrae.

intra-alveolar (in-trah-al-VEE-oh-lar) **pressure** Air pressure inside the alveoli of the lungs.

intracellular fluid (ICF) Liquid located within the cells.

ion (EYE-on) A charged particle; *ex:* Na^+.

iris (EYE-riss) Part of the eye that contains blood vessels, pigment cells, and smooth muscle.

ischium/ischia (IS-kee-um/IS-kee-ah) One of the three bones that fuse to create a coxal bone in the hip.

isotonic (eye-soh-TON-ik) **solution** Solution that contains the same concentration of nonpenetrating solutes as the adjacent cell; causes cells to neither shrink nor swell.

J

jejunum (jeh-JOO-num) 8–foot middle segment of the small intestine between the duodenum and the ileum.

K

kidney (KID-nee) Major organ of the urinary system; filters waste products from the blood.

Krebs cycle *See citric acid cycle.*

L

labia (LAY-bee-ah) "Lips"; skin folds on the sides of the opening to the vagina.

lacrimal gland (LAK-rih-mal gland) Almond-shaped tear gland.

lacteal (LAK-tee-al) Lymphatic capillary within a villus of the small intestine.

lactic acid A waste product of glucose metabolism as a result of insufficient oxygen supply; produced in muscles after an intense workout.

lacuna (lah-KOO-nah) A small space where bone or cartilage cells are found.

lamella/lamellae (lah-MEL-ah/lah-MEL-ee) Layer of calcified bone matrix.

lamina propria (LAY-min-ah PRO-pree-ah) The loose connective tissue component of a mucous membrane.

laryngopharynx (lair-ring-go-FAIR-inks) Portion of the pharynx lying between the hyoid bone and the entrance to the esophagus.

larynx (LAIR-inks) Cartilaginous structure that surrounds and protects the glottis; voice box.

lateral (LAT-er-al) Away from the midline of the body; opposite of medial.

lens A part of the eye that lies posterior to the cornea; focuses the visual image on the retinal photoreceptors.

leukocyte (LOO-koh-syte) White blood cell.

ligament (LIG-ah-ment) A type of connective tissue that connects bone to bone.

line A long, narrow strip or mark; in the skeletal system, similar to a crest.

lipids (LIP-idz) Organic molecules containing carbon, hydrogen, and oxygen, such as in fats and oils.

lobule (LOB-yool) A small lobe.

lipase (LYE-payse) A pancreatic enzyme that breaks down lipids.

loop of Henle (loop of HEN-lee) *See nephron loop.*

lumbar (LUM-bar) Lower back.

lumen (LOO-men) Hollow area within a tube.

lungs The pair of respiratory organs that exchanges oxygen and carbon dioxide in the blood.

luteinizing (loo-tee-in-EYE-zing) **hormone (LH)** A hormone produced by the anterior pituitary; in females, acts on ovaries to stimulate ovulation and progesterone secretion from corpus luteum; in males, acts on testes to stimulate the synthesis and secretion of testosterone.

lymph (limf) Fluid transported by the lymphatic system; similar to plasma but contains lower concentration of proteins.

lymph node (limf node) Small oval organ that filters and purifies lymph before it reaches the venous system.

lymphocyte (LIM-foh-syte) Primary cell of the lymphatic system; a type of white blood cell.

lysosome (LYE-so-sohm) Cell organelle that contains digestive enzymes.

M

macromolecules (mak-roh-MOL-eh-kyools) Large, complex molecules made from simpler molecules; *ex:* proteins, nucleic acids.

malleus (MAL-ee-us) One of three ear bones; also called the hammer.

marrow (MAIR-roh) Soft vascular tissue that fills the cavities of most bones.

matrix (MAY-triks) Extracellular substance of a connective tissue.

meatus (mee-AY-tus) Tubelike opening or channel.

medial (MEE-dee-al) Toward the midline longitudinal axis of the body; opposite of lateral.

median plane (MEE-dee-an plane) A section that passes along the midline and divides the body into right and left halves.

mediastinum (mee-dee-as-STYE-num) Region of the thoracic cavity between the lungs.

medulla (meh-DUL-ah) Inner portion of an organ; *ex:* adrenal medulla.

medulla oblongata (meh-DUL-ah ob-long-GAH-tah) Most inferior portion of the brain stem; site of vital control centers for respiratory and cardiovascular systems.

medullary cavity (MED-oo-lair-ee cavity) Hollow space inside the shaft of the bone; marrow cavity.

meiosis (my-OH-sis) Type of cell division that results in the formation of sperm or ova that contain half the number of chromosomes as the parent cell; occurs in the testes and ovaries.

melanin (MEL-ah-nin) Brown skin pigment that absorbs ultraviolet radiation.

melanocytes (MEL-ah-noh-sytes) A fixed cell that synthesizes and stores melanin.

melatonin (mel-ah-TOE-nin) A hormone produced by the pineal gland; involved in regulating the body's sleep–wake cycles.

meninx/meninges (ME-ninks/meh-NIN-jeez) One of the three membranes surrounding the spinal cord or brain.

menstrual phase (MEN-stroo-all fayz) The first phase of the uterine cycle, characterized by menstrual bleeding as a result of shedding the uterine lining called the endometrium.

metabolism (meh-TAB-oh-liz-em) Sum of all chemical activities in the body.

metacarpals (met-ah-KAR-puls) The bones in the palm of each hand.

metaphase (MET-ah-fayz) Stage of mitosis; evident when chromatids line up at the center of the cell.

microfilaments (my-KROH-FIL-ah-ments) Slender protein strands that make up the framework of the cytoskeleton; also called actin.

microtubules (my-kroh-TOOB-yools) Hollow protein tubes found in all body cells; framework of the cytoskeleton.

microvillus/microvilli (my-kroh-VIL-luss/my-kroh-VIL-eye) Small, finger-shaped projections of the plasma membrane of an epithelial cell.

midbrain Portion of the brain stem; provides pathways between brain stem and cerebrum.

middle ear Small cavity between the external ear and inner ear that houses the three small ear bones.

mineraloccorticoids (min-er-al-oh-KOR-tih-koyds) A group of hormones produced by the adrenal cortex that regulate mineral homeostasis, such as sodium and potassium levels in the blood; *ex:* aldosterone.

mitochondrion/mitochondria (my-toh-KON-dree-ohn/my-toh-KON-dree-ah) Cellular organelle that is the site of aerobic cellular respiration; produces ATP for the cell.

mitosis (my-TOH-sis) Process of cell division that results in the formation of two daughter cells that contain the same type and number of chromosomes as the parent cell; divided into 4 stages: prophase, metaphase, anaphase, and telophase.

molars (MOHL-larz) Teeth with flattened crowns with prominent ridges; used for grinding and crushing.

molecules (MOL-eh-kyools) Compound consisting of two or more atoms held together by chemical bonds.

monocyte (MON-oh-syte) A phagocytic white blood cell.

monosaccharide (mon-oh-SAK-ah-ryde) A simple, single-unit sugar such as glucose or fructose.

morula (MOR-yoo-lah) A solid ball of cells formed from mitotic divisions of the blastomeres.

motility (moh-TIL-i-tee) Contraction of smooth muscle in the wall of the digestive tract for the purpose of mixing and propelling digested substances; includes the processes of peristalsis and segmentation.

mucosa (myoo-KOH-sah) Mucous membrane.

mucous (adj.) (MYOO-kuss) Describes lubricating secretions along the digestive, respiratory, urinary, and reproductive tracts.

mucus (n.) (MYOO-kuss) A thick, gel-like substance secreted by glands in the digestive, respiratory, urinary, and reproductive tracts.

multipolar neuron (mul-tee-PO-lar NOO-ron) A neuron that has several dendrites and a single axon.

myelin (MY-eh-lin) A membranous insulation consisting of many layers of glial cell membrane around the axon of a nerve cell; improves the speed of impulse conduction.

myelination (my-eh-lih-NAY-shun) The process of forming a myelin sheath.

myocardium (my-oh-KAR-dee-um) The cardiac muscle in the wall of the heart.

myofibril (my-oh-FYE-brill) Slender fibers found in skeletal muscle and extending the length of the cell; composed of thick and thin filaments.

myometrium (my-oh-MEE-tree-um) The middle muscular layer of the uterine wall.

myosin (MY-oh-syn) Thick protein filament found in skeletal muscle cells.

N

nasal (NAY-zal) Pertaining to the nose.

nasopharynx (nay-zoh-FAIR-inks) The superior portion of the larynx.

neck a slender region, usually connected to the head of a bone.

nephron (NEF-ron) The basic functional unit of the kidney.

nephron loop (NEF-ron loop) A subdivision of the nephron; located between the proximal convoluted tubule and the distal convoluted tubule; also known as the loop of Henle.

nerves Long, cable-like structures that extend throughout the body; bundles of peripheral neurons held together by several layers of connective tissues.

neurilemma (noo-rih-LEM-mah) The cytoplasmic covering around the axon provided by the Schwann cells; found only in the peripheral nervous system.

neuroglia (noo-ROG-lee-ah) A class of cell types in neural tissue that provides nutrients and a supporting framework to neural tissue; does not transmit impulses.

neurolemmocyte (noo-ROH-lem-moh-syte) A neuroglial cell that wraps itself around peripheral axons to form a protective covering.

neurons (NOO-rons) Conducting cells of the nervous system; each consists of a cell body, one or more dendrites, and a single axon.

neurotransmitters (noo-roh-trans-MIH-terz) Chemical messengers released by neurons for the purpose of stimulating or inhibiting target cells.

neutron (NOO-tron) Subatomic particle in the nucleus of an atom; no electrical charge.

neutrophil (NOO-troh-fil) Phagocytic white blood cell; produced in bone marrow.

Nissl Bodies (NISS-ul Bodies) Ribosomal clusters found in neuron cell bodies.

node of Ranvier (node of rahn-vee-AY) Gaps of exposed axon not covered by a myelin sheath; appear at regular intervals in some myelinated axons.

norepinephrine (nor-ep-ih-NEF-rin) **(NE)** A neurotransmitter and a hormone; involved in regulating autonomic nervous system activity; also known as noradrenaline.

nuclease (NOO-klee-ayz) An enzyme that catalyzes the hydrolysis of nucleic acids.

nucleic acid (noo-KLAY-ik ASS-id) Organic molecule found in the nucleus; made up of functional units called nucleotides; e.g., RNA, DNA.

nucleus Prominent organelle of the cell; contains DNA molecules.

O

oblique (oh-BLEEK) Diagonal.

olecranon (oh-LEK-rah-nohn) The point of the elbow.

olfaction (ohl-FAK-shun) The sense of smell.

oocyte (OH-oh-syte) Immature stage of the female gamete.

oogenesis (oh-oh-JEN-eh-sis) Production of female gametes or ova; occurs in the ovaries.

optic chiasma (OP-tik kye-AS-mah) Area near the diencephalon where the optic nerves cross over each other to eventually connect to opposite sides of the brain.

optic nerve (OP-tik nerve) Carries visual information from the retina of the eyes to the brain.

oral (OR-al) Pertaining to the mouth.

oral cavity (OR-al KAV-i-tee) Space within the mouth; includes lips, cheeks, tongue and its muscles, and hard and soft palates.

orbital (OR-bi-tal) Pertaining to the eye.

organ (OR-gan) A group of tissues that perform a specific function.

organ system (OR-gan SISS-tem) Varying number and kinds of organs arranged in a way to perform complex functions; most complex organizational unit of the body; *ex:* respiratory system.

organelle (or-gah-NELL) Small internal structures in cells; divided into membranous and nonmembranous types; *ex:* ribosome.

organic (or-GAN-ik) Pertaining to compounds that contain hydrocarbons (carbon atoms bonded to hydrogen atoms); *ex:* carbohydrates, proteins.

organism (OR-gan-ism) An individual living thing.

oropharynx (oh-roh-FAIR-inks) Portion of the pharynx that extends between the soft palate and the base of the tongue at the level of the hyoid bone.

osmosis (os-MOH-sis) Movement of water across a semipermeable membrane toward the solution containing the higher solute concentration.

osteoblasts (OS-tee-oh-blasts) Cells that lay down the specialized matrix of bone; responsible for the production of new bone.

osteoclasts (OS-tee-oh-clasts) Cells that decompose the matrix of bone.

osteocytes (OS-tee-oh-sytes) Mature bone cells.

osteon (OS-tee-ohn) The basic functional unit of mature compact bone.

oval window (OH-val WIN-doh) Membranous structure in the inner ear; separates the middle and inner ear; stapes connects to it.

ovary (OH-vah-ree) Female gonad; produces ova.

ovulation (ov-yoo-LAY-shun) Release of an egg cell called a secondary oocyte from a tertiary follicle in the ovary.

ovum/ova (OH-vum/OH-vah) Female sex cell; egg cell.

oxidation Loss of one or more electrons from an atom or molecule.

P

palpate (PAL-payt) To examine by feeling part of the body.

pancreas (PAN-kree-ass) Slender, elongated organ that lies in the abdominopelvic cavity; has the dual function of manufacturing digestive enzymes and making hormones such as insulin.

pancreatic juice A mixture of secretions that contain mostly water, salts, sodium bicarbonate, and various digestive enzymes; secreted by the exocrine acinar cells of the pancreas.

papilla (pah-PILL-ah) Nipple-shaped mounds.

parathyroid hormone (PTH) Hormone secreted by the parathyroid glands that is the primary regulator of calcium homeostasis; increases calcium levels in the blood.

parietal (pah-RYE-i-tal) Relating to or forming the wall of an organ or cavity.

parietal cells (pah-RYE-i-tal sells) Cells located in the stomach; secrete hydrochloric acid.

pectoral (PEK-toh-ral) Pertaining to the chest; between sternal and axillary regions.

pedicle (PED-i-kul) Bony structure that attaches the vertebral arch to the body of a vertebra; means "foot".

pelvic (PEL-vik) Relating to the pelvis.

pepsin (pep-SYN) Enzyme made by chief cells in the stomach; functions to digest proteins.

pepsinogen (pep-SYN-oh-jen) An inactive enzyme secreted by chief cells in the stomach; converted to the active enzyme pepsin during gastric digestion.

perimysium (pair-i-MISH-ee-um) A connective tissue layer that divides the skeletal muscle into a series of compartments or bundles called fascicles.

perineal (pair-i-NEE-al) The region between the scrotum and the anus in males and between the posterior vulva junction and the anus in females.

perineurium (pair-i-NOO-ree-um) Connective tissue sheath that surrounds a bundle of nerve fibers within a nerve and holds them together.

periosteum (pair-ee-OS-tee-um) A fibrous layer that covers the bone.

peristalsis (pair-ih-STAWL-siss) Wave-like ripple of muscular contractions along the wall layer of a hollow organ, such as the small intestine.

peroxisomes (per-AHK-si-sohms) Cell organelles that absorb and neutralize toxins.

pH Means "**p**otential **H**ydrogen scale"; a measure of the concentration of hydrogen ions in a solution; scale ranges in value from 0–14 where 7 is the neutral point; values below 7 are acidic and those above 7 are alkaline.

phagocytosis (fag-oh-sye-TOH-siss) Type of endocytosis; process in which microorganisms or other large particles are engulfed by the plasma membrane and enter the cell.

phalanx/phalanges (FAH-lanks/fah-LAN-jeez) The finger bones and toe bones.

pharynx (FAIR-inks) Passageway that connects the nose, mouth, and throat; commonly called the throat.

physiology (fiz-ee-AHL-oh-jee) Science that examines the function of the living organism and its parts.

pia mater (PEE-ah MAH-ter) The highly vascular, innermost layer of the meninges.

pinocytosis (pin-oh-sye-TOH-sis) Type of endocytosis; process in which a plasma membrane invaginates to capture some extracellular fluid and brings it into the cytoplasm in a membranous vesicle.

pituitary (pih-TOO-i-tair-ee) **gland** Small endocrine gland located near the base of the brain that releases many different hormones; consists of two separate glands: anterior pituitary and posterior pituitary; referred to as the "master gland" because many of its hormones control other endocrine glands in the body.

placenta (plah-SEN-tah) The specialized organ within the uterus that supports embryonic and fetal development; also called the afterbirth.

plasma Fluid portion of the blood.

plica/plicae (PLYE-kah/PLYE-kee) Transverse folds of the intestinal lining.

polysaccharide (pol-ee-SAK-ah-ryde) An organic macromolucule; a complex sugar formed by bonding many monosaccharides together in a long chain, such as starch or cellulose.

pons (ponz) Part of the brain stem; connects the brain stem to the cerebellum; contains relay centers and is involved with somatic and visceral motor control.

positive feedback A physiological mechanism that tends to amplify or reinforce the change in internal environment.

posterior (pos-TEER-ee-or) Back; behind; opposite of anterior.

postganglionic neuron (post-gang-glee-ON-ik NOO-ron) The second neuron in the motor output portion of an autonomic reflex pathway; connects a ganglion to smooth muscle or cardiac muscle, or a gland.

preganglionic neuron (pree-gang-glee-ON-ik NOO-ron) The first neuron in the motor output portion of an autonomic reflex pathway; connects the central nervous system to a ganglion.

premolars (pree-MOHL-larz) *See bicuspids.*

prepuce (PREE-pus) The fold of skin that surrounds the glans penis in males and clitoris in females; foreskin.

primary active transport Refers to the transport of a specific substance across the plasma membrane, against its concentration gradient; typically involves a membrane protein, referred to as a "pump," that uses ATP.

progesterone (pro-JES-ter-ohn) Female hormone produced by the ovaries; prepares the endometrium for possible implantation of the embryo and stimulates milk secretion in mammary glands.

prolactin (pro-LAHK-tin) **(PRL)** Hormone secreted by the anterior pituitary; stimulates milk secretion in mammary glands.

proliferative phase (PROH-lif-eh-rah-tiv fayz) Second phase of the uterine cycle; involves reestablishing a portion of the endometrium.

prophase (PRO-fayz) Initial stage of mitosis; evident when the chromosomes become visible.

prostate gland (PROSS-tayt) Small, muscular, rounded organ; produces a weak acidic secretion that contributes to about 30% of the volume of semen.

protease (PRO-tee-ayz) Enzyme that catalyzes the breakdown of proteins into intermediate compounds.

protein (PRO-teen) Organic macromolecule with a long, complex polymer structure and a variety of functions; structural units are amino acids.

proton (PRO-ton) Positively charged subatomic particle; located in the nucleus of an atom.

proximal (PROK-sih-mal) Refers to the region or reference toward an attached base; opposite of distal.

proximal convoluted tubule (PROK-sih-mal kon-voh-LOO-ted TOOB-yool) Part of the nephron in the kidney; located between the glomerular capsule and the nephron loop; primary site of reabsorption of nutrients from the filtrate.

pubic (PYOO-bik) Relating to the region of the pubis.

pubis (PYOO-biss) The articulation between the two coxal bones.

pudendum (pyoo-DEN-dum) Region enclosing the female external genitalia; usually called the vulva.

pulp cavity (pulp KAV-i-tee) A spongy, highly vascular cavity in the tooth.

pupil (PYOO-pill) Central opening in the iris of the eye that allows light to enter the eye.

Purkinje (pur-KIN-jee) **fibers** Specialized cardiac muscle cells that extend out to the lateral walls of the ventricles and papillary muscles; conduct impulses from the AV bundle, stimulating heart contraction.

pyloric sphincter (pye-LOR-ik SFINGK-ter) A muscular structure that regulates the release of chyme from the stomach to the duodenum.

R

ramus (RAY-mus) A branch, such as the vertical section of bone in the mandible that connects it to the skull.

reabsorption Second step in urine formation; movement of substances out of the renal tubules and into the blood.

rectum (REK-tum) Last segment of the large intestine; end of the digestive tract.

rectus (REK-tus) Straight.

reduction Gain of one or more electrons by an atom or molecule.

renal (REE-nal) Pertaining to the kidney.

renin (REH-nin) An enzyme produced by the nephrons in the kidneys; converts angiotensinogen to angiotensin I in the plasma.

rete testis (REE-tee TES-teez) A maze of passageways formed from the seminiferous tubules and located within the mediastinum of the testis.

retina (RET-i-nah) The inner layer of the eye; contains the photoreceptor cells of the eye.

ribosomes (RYE-boh-sohms) Cellular organelles that are the sites of protein synthesis; composed of two subunits.

round window (round WIN-doh) A thin, membranous partition that separates the cochlear chambers from the air spaces of the middle ear.

rugae (ROO-gee) Longitudinal folds that line the inner wall of the stomach.

S

sacral (SAY-kral) Pertaining to the sacrum.

saggital (SAJ-i-tal) Runs along with the long axis of the body.

sagittal plane (SAJ-i-tal plane) Extends from anterior to posterior and divides the body into right and left sections.

sarcolemma (sar-koh-LEM-ah) Plasma membrane of a skeletal muscle cell.

sarcomere (SAR-koh-meer) Contractile unit of a muscle cell.

scapula (SKAP-yoo-lah) Shoulder blade.

Schwann cell (shwon sell) *See neurolemmocyte*.

sclera (SKLAIR-ah) The dense fibrous, white outer covering of the eye.

scrotum (SKRO-tum) Pouchlike sac, divided into two chambers; contains the testes.

sebaceous gland (seh-BAY-shus gland) Holocrine gland that secretes a waxy, oily secretion into the hair follicles.

secondary active transport Typically involves co-transport of substances across the plasma membrane; does not use ATP; relies on the concentration gradients established by primary active transport.

secondary oocyte (SEK-on-dair-ee OH-oh-syte) Ovum released at ovulation.

second messenger An intracellular molecule, such as cyclic AMP, produced in response to a first messenger (*ex:* hormone) binding to its plasma membrane receptor; triggers a chemical chain reaction within the target cell, causing cellular changes.

secretin (seh-KREE-tin) A digestive hormone produced by duodenal mucosa; stimulates secretion of pancreatic juice rich in bicarbonate ions.

secretion Production and release of a chemical substance from a cell; *ex:* neurotransmitters from a neuron.

secretory phase (SEEK-reh-toh-ree fayz) The last and longest phase of the uterine cycle; characterized by thickening of the endometrium in preparation for possible implantation of a human embryo.

segmentation Mixing movement; occurs when digestive reflexes cause a forward-and-backward movement within a single region of the digestive tract.

semen (SEE-men) Fluid released from the penis during ejaculation; contains sperm and seminal fluid produced by various glands.

semicircular canals (sem-i-SIR-kyoo-lar kah-NALS) Three bony rings located in the inner ear; contain fluid-filled ducts that, in turn, contain receptors that function to maintain dynamic equilibrium.

semilunar valve (sem-i-LOO-nar valv) Pair of heart valves located at the exit point of each ventricle; each valve has three pouch-like flaps; consists of the pulmonary and aortic valves.

seminiferous tubules (seh-mih-NIF-er-us TOOB-yools) A tightly coiled structure that is located in each lobule of the testis; site where spermatozoa develops.

septum/septae (SEP-tum/SEP-tee) A terminal partition that divides an organ.

serosa (seh-ROH-sah) A serous membrane that covers most of the digestive tract.

simple diffusion *See diffusion.*

sinoatrial (sye-NOH-AY-tree-al) **(SA) node** Primary pacemaker of the heart; specialized mass of cardiac muscle cells located in the wall of the right atrium that play a role in electrical conduction system within the heart.

sinus (SYE-nus) A cavity or space in a tissue; a large dilated vein.

sodium-potassium pump Protein pump that uses active transport; located in the plasma membrane of most cells; exchanges three sodium ions moved out of the cell for two potassium ions brought into the cell; also known as the sodium-potassium ATPase.

soft palate (soft PAL-let) The portion of the roof of the mouth that lies posterior to the hard palate.

solute Dissolved substance(s) in a solution; *ex:* salt.

solution (suh-LOO-shun) Mixture in which a solute is dissolved in a solvent; *ex:* salt water.

solvent Liquid portion of a solution; dissolves the solute; *ex:* water.

soma (SO-mah) Cell body.

sperm *See spermatozoon.*

spermatic cord (sper-MAT-ik cord) A layer of fascia, tough connective tissue and muscle surrounding the ductus deferens, and the blood vessels and nerves of the testes.

spermatogenesis (sper-mah-toh-JEN-eh-sis) Production of sperm cells; occurs in the testes.

spermatozoon/spermatozoa (sper-mah-tah-ZOH-ohn/sper-mah-tah-ZOH-ah) Sperm cell; male gamete or sex cell.

spindle fiber (SPIN-dul FYE-ber) Network of tubules in the cell that extends between the centriole pairs.

spine slender, pointed process on a bone; site for muscle attachment.

spongy bone (SPUN-jee bone) Type of bone that has a porous network of bony plates; also known as cancellous bone.

stapes (STAY-peez) The inner ear bone; also called the stirrup.

stem cells Non-specialized cells that have the ability to divide continually to form new types of more specialized cells.

steroids A large class of lipids characterized by a carbon ring chemical structure; act as components for cells and regulate body function; e.g.: cholesterol.

stimulus Any agent that induces a response; causes a change in an excitable tissue.

striation (STRYE-ay-shun) A striped or banded appearance such as that found in skeletal muscle cells and cardiac muscle cells.

submucosa (sub-myoo-KOH-sah) A layer of loose connective tissue in the wall of the digestive tract; large blood vessels, lymphatic vessels, and nerve fibers are found here.

sulcus (SUL-kus) A shallow depression.

superficial (soo-per-FISH-al) On the surface; opposite of deep.

superior (soo-PEER-ee-or) Above; opposite of inferior.

surface tension The force of attraction between water molecules as a result of hydrogen bonding; creates a thin surface film on water.

surfactant (sur-FAK-tant) Chemical mixture that contains lipids and proteins and coats the inner surface of each alveolus in the lungs; reduces surface tension.

suture (SOO-chur) The boundary between the skull bones; immovable joint.

synapse (SIN-aps) Specialized junction where a neuron communicates with another cell, such as another neuron or a muscle cell.

synovial fluid (sih-NO-vee-all FLOO-id) Thick and colorless lubricating fluid secreted by synovial membranes.

synovial joint (sih-NO-vee-all joynt) Freely movable joint; most numerous and anatomically most complex joint in the body.

synovial membrane (sih-NO-vee-all MEM-brayne) Connective tissue membrane that lines the spaces between bones and joints; secretes synovial fluid.

systole (SISS-toh-lee) Contraction of both atria and both ventricles in the heart; opposite of diastole.

T

tarsals (TAR-sals) Ankle bones.

tarsus (TAR-sus) Ankle.

telophase (TEL-oh-fayz) Stage of mitosis; nuclear membrane forms, nuclei enlarge, and chromosomes gradually uncoil and disappear.

temporal (TEM-poh-ral) Pertaining to the temples on the sides of the head, above the zygomatic arch.

tendon (TEN-don) A cable-like, fibrous connective tissue that connects skeletal muscle to bone.

tertiary follicle (TER-shee-air-ee FOL-lih-kul) Mature ovum in ovary; also known as Graafian follicle or vesicular follicle.

testis/testes (TES-tiss/TES-teez) The male gonad that produces sperm.

testosterone Principal male sex hormone; produced by interstitial cells in the testes; responsible for growth and maintenance of male sexual characteristics and for sperm production.

thalamus (THAL-ah-mus) Located in diencephalon of the brain; final relay point for ascending sensory information that will be sent to the cerebrum.

thoracic (tho-RASS-ik) Pertaining to the chest.

thrombocyte (THROM-boh-syte) A blood cell that has a role in clotting: also known as a platelet.

thrombus (THROM-bus) Stationary blood clot.

thyroid gland (THY-royd gland) Endocrine gland located near the trachea in the neck; produces the hormones T_3, T_4 and calcitonin.

thyroid stimulating hormone (TSH) Hormone secreted by anterior pituitary; stimulates thyroid gland to produce the hormones T_3 and T_4.

thyroxin (T_4) (thy-ROK-syn) Hormone produced by the thyroid gland; influences metabolic rate, growth, and development.

tissue (TISH-yoo) Group of similar cells that perform a common function.

tonsil (TAHN-sil) A large nodule containing masses of lymphoid tissue: located in the walls of the pharynx.

trachea (TRAY-kee-ah) Long tube between the larynx and the primary bronchi that serves as air passageway; windpipe.

transverse plane (TRANS-vers plane) A division that lies at right angles to the long axis of the body; divides the body into superior and inferior sections.

tricuspid valve (try-KUS-pid valv) The right atrioventricular (A-V) valve in the heart located between the right atrium and right ventricle.

triiodothyronine (T_3) (try-eye-oh-doh-THY-roe-neen) Hormone produced by the thyroid gland; regulates metabolic rate, growth, and development.

trochanter (troh-KAN-ter) Large bump on a bone; site for muscle attachment.

trochlea (TROHK-lee-ah) A bone process that resembles a pulley.

tubercle (TOO-ber-kal) Small bump on a bone;.

tuberosity (too-bah-ROS-i-tee) Bump on a bone; smaller than a trochanter.

tympanic membrane (tim-PAN-ik membrane) The eardrum.

U

umbilical cord (um-BIL-i-kul cord) The vascular structure that connects the fetus to the placenta.

unipolar neuron (YOO-nee-POH-lar NOO-ron) A neuron in which the dendrite and axonal processes are continuous, and the cell body lies off to the side.

ureters (YOOR-eh-ters) Pair of long, slender, muscular tubes that transport urine from the kidneys to the urinary bladder.

urethra (yoo-REE-thrah) A passageway that transports urine from the neck of the urinary bladder to the exterior; in males, also acts as a passageway for spermatozoa to the exterior.

uterine tubes (YOO-ter-in tubes) Hollow muscular tubes that transport a secondary oocyte to the uterus; also known as oviducts or Fallopian tubes.

uterus (YOO-ter-us) Hollow muscular organ that provides mechanical protection, nutritional support, and waste removal for a developing embryo.

V

vagina (vah-JYE-nah) Muscular passageway in the female that connects the uterus with the exterior genitalia.

vas deferens (vas DEF-er-enz) *See ductus deferens.*

vascular (vas-KYOO-ler) Pertaining to the blood vessels.

veins (vanes) Vessels that collect blood from tissues and organs and return it to the heart.

ventral (VEN-trul) Anterior or belly side in humans; opposite of dorsal.

ventricle (VEN-tri-kul) Cavity or chamber.

venule (VEN-yool) Microscopic blood vessel that connects capillaries to small veins.

villus (VIL-us) Finger-like extension of mucous membrane of small intestine.

vitreous humor (VIT-ree-us humor) A gel-like mass that fills the posterior chamber of the eye; helps maintain shape of the eye and give support to the retina.

W

white matter (whyte MAT-ter) Nerve fibers that are covered with the myelin sheath.

Y

yolk sac (yohk sak) The first of the extraembryonic membranes to appear; an important site for blood cell production.

Z

zona pellucida (ZOH-nah pel-LOO-sih-dah) A region formed between the innermost follicular cells and the developing oocyte.

zygote (ZYE-goht) Fertilized egg cell.

Index